MODERN ELECTROSTATICS

Proceedings of the International Conference, 1988, Beijing, China

Edited by

LI RUINIAN

Sponsored by

Electrostatics Committee, Chinese Physical Society

Applied Physics Department, Beijing Institute of Technology

INTERNATIONAL ACADEMIC PUBLISHERS
A Pergamon-CNPIEC Joint Venture

PERGAMON PRESS

OXFORD · NEW YORK · BEIJING · FRANKFURT
SÃO PAULO · SYDNEY · TOKYO · TORONTO

DISTRIBUTORS:

U.K.	Pergamon Press plc, Headington Hill Hall, Oxford OX3 0BW, England
U.S.A.	Pergamon Press, Inc., Maxwell House, Fairview Park, Elmsford, New York 10523, U.S.A.
PEOPLE'S REPUBLIC OF CHINA	International Academic Publishers, Xizhimenwai Dajie, Beijing Exhibition Centre, Beijing 100044, People's Republic of China
FEDERAL REPUBLIC OF GERMANY	Pergamon Press GmbH, Hammerweg 6, D-6242 Kronberg, Federal Republic of Germany
BRAZIL	Pergamon Editora Ltda, Rua Eça de Queiros, 346, CEP 04011, Paraiso, São Paulo, Brazil
AUSTRALIA	Pergamon Press Australia Pty Ltd., P.O. Box 544, Potts Point, N.S.W. 2011, Australia
JAPAN	Pergamon Press, 5th Floor, Matsuoka Central Building, 1-7-1 Nishishinjuku, Shinjuku-ku, Tokyo 160, Japan
CANADA	Pergamon Press Canada Ltd., Suite No. 271, 253 College Street, Toronto, Ontario, Canada M5T 1R5

Copyright © 1989 International Academic Publishers
(A Pergamon-CNPIEC Joint Venture)

All Rights Reserved. No part of this publication may be reproduced, stored in a retrieval system or transmitted in any form or by any means: electronic, electrostatic, magnetic tape, mechanical, photocopying, recording or otherwise, without permission in writing from the publishers.

First edition 1989

Library of Congress Cataloging in Publication Data

International Conference on Modern Electrostatics (1988: Peking, China)
Modern electrostatics: proceedings of the International Conference, 1988, Beijing, China/edited by Li Ruinian; sponsored by Electrostatics Committee, Chinese Physical Society, Applied Physics Department, Beijing Institute of Technology.
p. cm.
"A Pergamon-CNPIEC joint venture."
1. Electrostatics—Congresses.   I. Li Jui-nien.
II. Chung-kuo wu li hsüeh hui.
III. Pei-ching kung yeh hsüeh yüan.
IV. Title.
QC570.I585   1988   537'.2—dc20   89-15208

British Library Cataloguing in Publication Data
Modern electrostatics.
1. Electrostatics
I. Li Ruinian
537'.2
ISBN 0-08-037029-2

The book has been photographically reproduced from the best available copy. The papers were not refereed but were reviewed for their technical contents. Editing was restricted to matters of format, general organization and retyping.

The editors assume no responsibility for the accuracy, completeness or usefulness of the information disclosed in this volume. Unauthorized use might infringe on privately owned patents of publication rights. Please contact the individual authors for permission to reprint or otherwise use information from their papers.

Printed in the People's Republic of China

PREFACE

Though electrostatic field theory was established more than one century ago, modern electrostatics develops from modern physics, mechanics, chemistry, material science, electrical and electronic engineering, biology and other related modern technology and techniques. Based upon modern electrostatics, varied ingenious techniques were developed, which made possible the use of electric field in manipulating various substances, such as multi-phase fluids, granules, particulates, chemical and biological colloidal particles, macro-molecules or even ordinary molecules so as to reach some technological goals, which are otherwise difficult to get to by conventional methods. Modern electrostatics finds its ample scope in the fields of environmental engineering, information techniques, biological engineering, precision chemical engineering, electrical and electronic techniques and industrial safety engineering. It has made significant contributions to energy-saving, the protection of the environment, the raising of efficiency in information communication, the increase of agricultural yields, the improvement of industrial techniques, the invention of new materials and in ensuring the reliability of micro-electronic products and safety in chemical production.

Theoretically modern electrostatics is involved in the electrodynamics of gases, liquids, solids and multi-phase fluids; in the electronic and molecular dynamics of amorphous solids and interfaces of substances; and biological microscopic dynamics. The theory is very difficult to develop. For instance, the gaseous discharge in the atmosphere is commonly used in electrostatic techniques, but its electrodynamic process is highly non-linear. Phenomena observed in liquids and solids under high intensity of electric field is also extremely non-linear. The classical tribo-electric phenomena are hardly predictable. Though the phenomena and processes can be explained in various ways, any accurate and reliable predictions are usually difficult to make.

However, through numerous experiments, many techniques controlling these processes(e.g. tribo-electrification of toners, the signs of flashover in ESP, effective electrostatic treatment for seeds etc.) have been developed and used in practice. The essence of modern electrostatics is practically empirical and experimental. It is an artistic achievement made with the help of modern science and technology. In addition, to industrialize these techniques, a great number of problems in electrical and electronic engineering, precision machinery, material used etc. have to be first solved. Consequently, modern electrostatics involves numerous elaborate know-hows.

In the past half a century or more, scientists and engineers of all over the world engaged themselves in broad developments in electrostatic precipitation, electrophotography, various kinds of electrostatic processes, consummate measuring techniques, multi-functioned power sources and methods of identifying electrostatic hazards and nuisances and the control measures. Electrostatic techniques have since become indispensable in modern social life. In recent years, electrostatic biological techniques have aroused great interests of many researchers. Besides, in order to improve the efficacy and invent new functional materials, theoretical attempts to explore the mechanism of the various processes are made through elaborate researches on electro-hydrodynamics (including electrodynamics of multi-phased substances, e.g. electric field with particulate load, ionic current and gas flow, electrophoresis and dielectrophoresis, electrostriction and electro-induced convection, demusification etc.), particulate dynamics (toner development dynamics, electric curtains, electro-pneumatic processes in powder coating and spraying) and microscopic electrodynamic analysis in amorphous and composite materials, such as electronic processes in layered photoconductors and electrets.

In the last ten or more years, Chinese scientists and engineers carried out a number of researches though most of them were still in initial stages. Applications of modern electrostatics to environmental protection, information techniques, biological engineering and industrial safety engineering are financially supported by the Chinese government and some successes in these fields have been achieved.

More than 250 scientists and engineers from 14 countries and regions will participate in the International Conference on Modern Electrostatics (ICMES). With 18 invited lectures and about 105 contributed papers, the conference will cover the developments in almost all important areas in applied electrostatics. Very lively and beneficial discussions are being expected.

On behalf of ICMES, I am deeply grateful to all the participants, especially those oversea specialists for their valuable contributions to the conference.

Li Ruinian
Professor of Physics of
Beijing Institute of
Technology

Chairman of ICMES

July, 1988

ORGANIZING COMMITTEE

Chairman: Li Ruinian
Co-Chairman: Senichi Masuda

Members: Akasaki Masanori (Japan)
Bailey Adrain (UK)
Cheng Ruguang (China)
Ding Jing (China)
Felici Noel (France)
Hendricks Charles (USA)
Hoshino Yasushi (Japan)
Hughes John (UK)
Inculet Ion (Canada)
Jin Xiangfeng (China)
Löffler Friedrich (FRC)
Min Guanming (China)
Taillet Joseph (France)
Wu Yi (China)
Yan Xingzhong (China)
Yokoyama Masaaki (Japan)

Secretary: Bao Chongguang
Yan Keping

AUTHOR INDEX

Adachi, T.	24, 28, 513	Gao, T.S.	140
Akazaki, M.	24, 28, 97, 513	Gao, X.L.	79, 83
Asai, H.	507	Ge, L.	140
		Ge, Z.H.	301
Bai, X.Y.	144, 161, 166	Gu, X.Y.	59
	176, 215, 219	Guan, X.S.	236
Benyamina, M.	404	Guo, C.J.	289
Berbeco, G.R.	379	Guo, R.S.	71
Beuthe, T.G.	281	Guo, S.C.	236
Bi, Z.J.	395		
Borzeix, J.	400	Hao, G.H.	212
		Hao, J.M.	75
Cai, R.F.	221	Hao, W.Y.	310
Cai, S.G.	477	Hara, M.	97
Chang, J.S.	28, 281, 513	Hayashi, N.	281
Chao, H.X.	75	Hayashi, N.	297
Chen, G.R.	289	Hays, D.A.	327
Chen, Jianlin	350, 435	He, F.	495
Chen, Jingliang	492	He, K.B.	75
Chen, J.X.	312	Hendricks, C.D.	335
Chen, J.Y.	310	Hirata, T.	49
Chen, L.R.	255	Hornady, R.S.	335
Chen, M.H.	489	Hoshino, Y.	275
Chen, R.Y.	45	Hu, D.	207
Chen, T.M.	176	Hu, J.M.	57
Chen, X.G.	41	Hu, M.Y.	79, 83
Chen, Z.L.	166	Hu, Z.G.	79, 83
Chi, Y.P.	465	Huang, H.F.	41
Chu, F.Y.	281	Huang, J.S.	472
Csorvassy, I.	203	Huang, M.X.	215
Cucu, D.	377	Huang, Q.S.	83
Cui, M.	186	Huang, S.Z.	148, 155
		Huang, X.M.	118
Dai, M.Q.	221	Humeau, P.	400
Dai, N.	59		
Darby, K.	3	Isahaya, F.	101
Dascalescu, L.	158, 203	Ishida, K.	507
Deng, X.H.	271	Iuga, AL.	158, 203
Dong, C.G.	395, 415		
Dong, Z.H.	155	Ji, X.G.	339
Dou, W.G.	212	Jiang, S.X.	489
		Jiang, Y.B.	108
Elholm, P.	112	Jin, X.F.	251
Fang, Z.H.	59	Kamachi, S.	431
Felici, N.J.	134	Kazimierz, C.	137
Feng, Z.L.	322	Ke, L.	454
Fu, M.Z.	176	Keila, M.	281
Fujii, M.	404	Kitamura, N.	391
Fujimura, K.	24	Kiss, E.	507, 517
Fujino, M.	247	Kobayashi, S.	339
Fujishima, H.	14	Koda, K.	268
Fushimi, I.	391	Komai, Y.	339
		Kutsuwada, N.	258

Law, S.E.	129	Pan, G.		411
Lee, P.L.	357	Porle, K.		9
Li, A.B.	180	Qian, C.		443
Li, F.H.	122	Qiu, J.B.		251
Li, G.X.	354, 451	Qu, J.B.		407
Li, H.N.	451			
Li, H.Y.	271	Rea, M		20
Li, K.J.	115	Ren, D.F.	215, 224
Li, L.	79	Ren, D.Y.		251
Li, R.N.	350	Ren, X.L.		443
Li, W.M.	180			
Li, X.L.	144, 161, 166	Seya, K.		49
Li, X.R.	495	Shen, M.X.		489
Li, X.S.	122	Shi, C.Y.		441
Li, Y.M.	308	Shi, X.M.		174
Liang, W.J.	411	Shiwa, S.		275
Liang, Y.Z.	174	Song, G.C.		395
Lin, T.	301	Song, Y.P.		183
Liu, B.	481	Sumiyoshitani, S.		97
Liu, D.J.	212	Sun, C.M.		186
Liu, J.T.	481	Sun, D.W.		105
Liu, J.Y.	385	Sun, K.P.		457
Liu, L.M.	57, 105	Sun, Q.W.		236
Liu, S.H.	502	Sun, S.Z.		477
Liu, S.Z.	492	Suzuki, T.		431
Liu, W.B.	492			
Liu, X.Z.	492	Tabata, Y.	427, 431
Liu, Y.C.	219, 224	Tachibana, N.		14
Liu, Y.N.	293, 343	Takahashi, T.	93, 285
Liu, Y.S.	45	Takahashi, Toru		245
Liu, Y.Y.	304, 465	Takahashi, Y.		339
Liu, Z.C.	502	Tan, F.G.	310, 415
Loffler, F.	37	Tan, T.Y.		53
Lu, C.Z.	495, 498	Tanabe, K.		404
Lu, M.F.	407	Tanaka, T.		275
Luo, H.C.	360, 439	Tao, T.		411
		Touchard, G.	364, 400, 404
Ma, A.C.	144, 161			
Ma, J.R.	161, 166, 170	Ueda, M.		391
Makin, B.	207			
Man, S.L.	215, 219, 224	Wan, A.M.		308
Mao, J.Y.	41			
Mao, S.Q.	477	Wang, C.Y.	304, 465
Marcano, L.	400	Wang, Jing		87
Masuda, S.	93, 507, 517	Wang, Juan		176
Matsumoto, N.	247	Wang, Jun		66
Matsuo, Y.	427	Wang, J.Z.		219
Matsuuchi, M.	297	Wang, L.Q.	360, 439
Meng, X.C.	144	Wang, Q.	316, 319
Min, S.K.	255	Wang, Q.Z.	144, 161, 166
Morar, R.	158, 203	Wang, R.H.		108
Munteanu, I,	158	Wang, R.Y.	57, 105, 196,
Murata, Y.	371		196, 316, 319
		Wang, S.L.		395
Nakamura, Y.	258	Wang, S.T.		193
Nakane, T.	49	Wang, W.L.		468
Neamtu, V.	158, 203	Wang, W.X.		108
Ni, Z.Q.	347	Wang, X.D.		322
Ning, W.	215	Wang, X.L.	57, 105
Nomoto, Y.	28, 513	Wang, X.W.		151
Nossent, J.P.	400	Wang, Y.		221
		Wang, Y.C.		183
Oda, T.	93, 285, 331	Wang, Y.M.		221
Oglesby Jr., S.	23	Wang, Y.Q.		251
Ohashi, A.	404	Wang, Y.Y.		265
Ohkubo, T.	28, 513	Watanabe, S.		404
Ohta, M.	275	Wei, D.T.		308
Omodani, M.	275	Wei, D.H.		502

Wu, G.R.	443	Zhang, F.		212
Wu, X.Z.	492	Zhang, F.Z.		255
Wu, Y.	57, 105	Zhang, G.C.		255
Wu, Z.H.	364, 400, 404	Zhang, G.G.		193
		Zhang, G.M.		271
Xia, B.	105	Zhang, G.Q.		32, 63
Xia, H.	385	Zhang, G.S.		423
Xian, F.S.	239	Zhang, H.D.		66, 71
Xiao, F.C.	32	Zhang, H.K.		193
Xie, B.X.	227	Zhang, H.Y.		475
Xie, L.	183	Zhang, H.Z.		461
Xiong, J.G.	360	Zhang, J.F.		140
Xu, L.H.	308	Zhang, J.M.		180
Xu, S.Z.	122, 148, 155	Zhang, L.		53
Xu, Y.	176	Zhang, L.Z.		502
		Zhang, P.		301
Yan, J.	79	Zhang, Q.		229
Yan, J.X.	148	Zhang, S.		489
Yan, L.	144, 161, 166	Zhang, W.M.		112
Yan, S.Z.	215, 219, 244	Zhang, Y.W.		293
Yan, X.Z.	87	Zhang, Z.J.		322
Yanagida, M.	331	Zhang, Z.W.		122
Yang, B.T.	343	Zhao, H.Z.		118
Yang, J.	395	Zhao, L.Q.		368
Yang, T.Q.	174	Zhao, L.Z.	435, 447,	472
Yang, Y.Q.	368	Zheng, J.Q.		239
Ye, J.M.	140, 196	Zhong, D.L.		447
Yin, J.H.	183	Zhong, D.Q.		105
Yokoi, Y.	297	Zhong, G.R.		271
Yokoyama, K.	247	Zhong, X.L.		360
Yokoyama, M.	247	Zhou, F.		364
Yu, Y.F.	451	Zhou, H.L.		140
Yu, Z.G.	105	Zhou, S.Q.		251
		Zhou, S.Y.		232
Zeng, H.H	118	Zhou, Z.H.		477
Zhai, B.X.	502	Zhou, Z.Y.		360
Zhang, B.	170	Zhu, H.S.		186
Zhang, D.G.	45	Zou, D.D.		308

CONTENTS

I. Electrostatic Precipitation

Electrostatic Precipitators in Modern Power Plant 3
K. Darby

Pulsed Energization of Electrostatic Precipitators—A Review of Worldwide
Experience 9
K. Porle

Applications of Electrostatic Precipitator with the Intermittent
Energization 14
N. Tachibana and H. Fujishima

Advanced Electrostatic Applications to Flue Gas Cleaning 20
M. Rea

Electron Irradiation of Flue Gas by Pulsed Corona for Sulfur and Nitrogen
Oxide Reduction 23
S. Oglesby Jr. and E.B. Dismukes

Discussion on the Correction Formula for Deutsch Formula Related to
Collection Efficiency of an Electrostatic Precipitator 24
K. Fujimura, T. Adachi and M. Akazaki

Similarity Phenomena of Electrohydrodynamic Flow Field Inside ESP 28
T. Adachi, T. Ohkubo, Y. Nomoto, J.S. Chang and M. Akazaki

Study on Collection Theory of Electrostatic Precipitators 32
F.C. Xiao and G.Q. Zhang

Electrostatic Enhancement of Particle Collection in Fiber Filters 37
F. Löffler

A Research on the Mechanism of Dust Collection in Electrostatic Lentoid
Fields 41
X.G. Chen, H.F. Huang and J.Y. Mao

A Study of Numerical Simulation for Gas Flow Distribution in the Inlet
to Roof Mounted Electrostatic Precipitator 45
Y.S. Liu, R.Y. Chen and D.G. Zhang

Experiment on Electrostatic Precipitator with Application of Ultrasonic
Agglomeration 49
T. Nakane, T. Hirata and K. Seya

Preliminary Study on the Collection of Fine Particles by Charged Drops 53
T.Y. Tan and L. Zhang

Pulse Energised Electrostatic Precipitator with Transverse Louvers—A
Dust Collection Technique for Fluidized-Bed Boiler 57
L.M. Liu, R.Y. Wang, Y. Wu, X.L. Wang, J.M. Hu et al.

Field Analysis in Electric Precipitators 59
Z.H. Fang, X.Y. Gu and N. Dai

Research on the Distribution of Electric Field in Electrostatic
Precipitator (ESP) 63
Z.F. Wu and G.Q. Zhang

A Study on Two-Dimensional Electric Field Strength and Current Density
in Electrostatic Precipitator 66
H.D. Zhang and J. Wang

Study on Electrode Geometries and Configurations of Wire-Plate ESP
by Electric Field Strength Near the Profiled Collecting Electrodes 71
R.S. Guo and H.D. Zhang

Simulating Calculation for Electric Field Strength Distributions of
Electrostatic Precipitators 75
J.M. Hao, K.B. He and H.X. Chao

Experimental Study of Optimum Wire Spacings at Different Plate Spacings
of EP 79
X.L. Gao, Z.G. Hu, M.Y. Hu, J. Yan and L. Li

Discharge Performances Study of Corona Wires in Electrostatic
Precipitators 83
M.Y. Hu, X.L. Gao, Z.G. Hu and Q.S. Huang

An Attempt to Use Specific Collecting Volume as a Modular Number for
ESP Performance Comparison 87
X.Z. Yan and J. Wang

TSDC Measurements of Coal Fly Ashes 93
T. Oda, T. Takahashi and S. Masuda

Advanced Analysing Model of Charge Carrier Motion Around a Charged
Spherical Object 97
S. Sumiyoshitani, M. Hara and M. Akazaki

Specific Surface Charge Density of Non-Spherical Particle and Charge
Decrease by Partial Self-Discharge in High Electric Field Intensity 101
F. Isahaya

Super-High Pulse Voltage Electrostatic Precipitator 105
R.Y. Wang, L.M. Liu, Y. Wu, D.W. Sun, Z.G. Yu, D.Q. Zhong, B. Xia and
X.L. Wang

The Correct Diagnosis and Optimum Disposal of Flashover Signals in
Electrostatic Precipitators 108
W.X. Wang, Y.B. Jiang and R.H. Wang

Pulse Energization on ESP for the End of Sinter Band 112
W.M. Zhang and P. Elholm

Using Spark Discharge for Ash Removal—Breaking off Agglomerates By
Sparks 115
K.J. Li

The Influence of Electrode Coating on Corona Discharge in Electrostatic
Precipitators 118
X.M. Huang, H.Z. Zhao and H.H. Zeng

Relation Between Spacing of Prickle Electrodes and Corona Current 122
X.S. Li, S.Z. Xu, Z.W. Zhang and F.H. Li

Progress in Dust Collection by Electrostatic Precipitation in View of
Gaseous Pollutants Removal 126
G. Mayer-Schwinning

II. Electrostatic Biological Effects

Applications of Electrostatic Technology in Agriculture 129
S.E. Law

Are Air Ions Biologically Significant? 134
N.J. Felici

The Effects of the Electric Field on Plant Growth 137
C. Kazimierz and B. Piotr

Effects of High-Voltage Electrostatic Field on Growth in Plants 140
J.M. Ye, T.S. Gao, L. Ge, H.L. Zhou and J.F. Zhang

Experimental Study on Promotion of Plant Growth By Electrostatic Field 144
L. Yan, X.Y. Bai, X.L. Li, A.C. Ma, X.C. Meng and Q.Z. Wang

Effect of Static Electricity on Initial Growth of Panax Ginseng 148
S.Z. Huang, J.X. Yan and S.Z. Xu

The Influence of HV Electrostatic Field on the Root System of Plants
and Absorbing Nutritive Elements 151
R.Y. Wang and X.W. Wang

Biological Potential in Plant and External Electrostatic Field 155
S.Z. Xu, S.Z. Huang and Z.H. Dong

Separation and Biostimulation of Soybeans Using High-Intensity Electric
Fields 158
R. Morar, Al. Iuga, L. Dascalescu, V. Neamtu and I. Munteanu

Physiological and Biochemical Experiments in Electrostatic Treated
Seeds 161
X.Y. Bai, A.C. Ma, J.R. Ma, X.L. Li, L. Yan and Q.Z. Wang

Study on Biological Effect of Seeds in Electrostatic Treatment 166
X.Y. Bai, L. Yan, Z.L. Chen, X.L. Li, J.R. Ma and Q.Z. Wang

Study on the Total Activity of Amylase of Sprouting Maize Seeds Affected
By Treatment of Electrostatic Field 170
J.R. Ma and B. Zhang

A Preliminary Study of the Biological After-Effect on Crop Seeds Treated
By Uniform Electrostatic Field 174
Y.Z. Liang, X.M. Shi and T.Q. Yang

A High-Pressure Shock Sterilizing Technique 176
X.Y. Bai, T.M. Chen, J. Wang, M.Z. Fu and Y. Xu

Observation of Changes of Some Biological Indices in Patients with
Occupational Leukopenia Before and After the Negative Air Ion Therapy 180
W.M. Li, A.B. Li and J.M. Zhang

Effect of Negative Ions on Spontaneous and Cyclophosphomidum-Induced
Micronucleus Frequency in Mouse Bone Marrow Polychromatophilic
Erythroblasts 183
L. Xie, J.H. Yin, Y.C. Wang and Y.P. Song

Treatment for Tissue Injury By Use of Micro-Porous Polymeric Electret
Film 186
C.M. Sun, M. Cui and H.S. Zhu

Electrically Stimulated Bone Growth by Electret Film 193
G.G. Zhang, S.T. Wang and H.K. Zhang

Acute Experimental Results of Effects of High-Voltage Electrostatic
Field on Animal Organisms 196
J.M. Ye and R.Y. Wang

III. Novel Electrostatic Techniques and the Prospects of Industrialization

Corona-Electrostatic Separation Processes for Recovery of Conducting
and Insulating Materials From Industrial Wastes 203
L. Dascalescu, R. Morara, Al. Iuga, V. Neamtu and I. Csorvassy

Electrostatic Particle Levitation on a Quadrupole System 207
B. Makin, and D. Hu

Experimental Study on a New Type Electrostatic Separating Device 212
W.G. Dou, G.H. Hao, F. Zhang and B.J. Liu

Study on Dielectrophoretic Oil Purification Technique 215
X.Y. Bai, M.X. Huang, D.F. Ren, W. Ning, S.Z. Yan and S.L. Man

Electrostatic Separation of Fine-Graded Talcum Powder Minerals 219
J.Z. Wang, X.Y. Bai, Y.C. Liu, S.Z. Yan and S.L. Man

Industrial Pilot Test of Electrostatic Purification Technology for
Dry-distillated Shale Gas 221
Y. Wang, S.Z. Wang, R.F. Cai, Y.M. Wang and M.Q. Dai

Electrostatic Separation of Fibre and Metal From Rubber Powder 224
Y.C. Liu, S.Z. Yan, D.F. Ren and S.L. Man

Using Electrostatic Bonding on Semiconductor Devices Technology 227
B.X. Xie

Technology Development and Mechanism Research of Electrostatic Spinning 229
Q. Zhang

The Basic Principle of Eletrostatic Powder Coating 232
S.Y. Zhou

Studies on Artificial Curing of White Spirit By Means of Ultrahigh
Voltage (UHV) Electrostatic Field 236
X.S. Guan, Q.W. Sun and S.C. Guo

Effect of Electrostatic Field on Liquid Surface Tension 239
F.S. Xian and J.Q. Zheng

IV. Fundamentals of Electrophotography

Color Copy Machine 245
T. Takahashi

Novel Organic Photoreceptor Using Organopolysilanes 247
K. Yokoyama, M. Yokoyama, M. Fujino and N. Matsumoto

Transport Properties of Copper Phthalocyanine 251
X.F. Jin, S.Q. Zhou, Y.Q. Wang, J.B. Qiu and D.Y. Ren

The Residual Potential of Xerographic Materials As-Se 255
S.K. Min, G.C. Zhang, L.R. Chen and F.Z. Zhang

A Measuring Method of Electrophotographic Toner Charge 258
N. Kutsuwada and Y. Nakamura

The Influence of As^+ Ion Implantation on the Optical and
Electrophotographic Properties of Amorphous Se Photoreceptor 265
Y.Y. Wang

Advanced Technology of Electrophotography in Japan 268
K. Kōda

The Surface and Interface of α-Se Film Studied By XPS 271
H.Y. Li, X.H. Deng, G.M. Zhang and G.R. Zhong

Ion Flow Control Technology and Its Applications 275
Y. Hoshino, M. Ohta, T. Tanaka, M. Omodani and S. Shiwa

V. Measurement Techniques in Applied Electrostatics and Electrostatic Source

Development of High Temperature Farady Cups for In-Situ Measurements
of Powder Surface Charge Under Thermal Plasma-Powder Flow 281
J.S. Chang, T.G. Beuthe, N. Hayashi, M. Keila and F.Y. Chu

Resistivity Measurements of Dielectric Films in Ion-Rich Atmosphere 285
T. Oda and T. Takahashi

Electrostatic Measurements by Means of Microwaves 289
G.R. Chen and C.J. Guo

Improving the Resolution of Surface Charge Density Measurement 293
Y.W. Zhang and Y.N. Liu

Field Calculation in Ring-Type Electrode Capacitance Transducer for
Particle Fraction Measurements 297
N. Hayashi, M. Matsuuchi and Y. Yokoi

The Principle of Optical Phase Compensation—A New Method for Measuring
Electrostatic Field By Means of Pockels Device 301
P. Zhang, Z.H. Ge and T. Lin

The Study of Powder Electrostatic Accumulation Test Device 304
Y.Y. Liu and C.Y. Wang

On Resistivity Measurement of Conductive Sponge 308
L.H. Xu, D.T. Wei, D.D. Zou, Y.M. Li and A.M. Wan

A New Method for Measuring Charge Density in an Oil-Piping System 310
W.Y. Hao, J.Y. Chen and F.G. Tan

Study on Decay Characteristics for Static Electricity on Both Conductive
and Anti-Static Materials 312
J.X. Chen

The Realization of Superhigh Voltage Pulse with Multiplying Circuit 316
Q. Wang and R.Y. Wang

Research on the High Voltage Multiplying Circuit with Arbitrary
Capacitance 319
Q. Wang and R.Y. Wang

High Voltage Modulating Pulse Electrostatic Power and Its Application
in Electrostatic Demulsification 322
Z.L. Feng, X.D. Wang and X.J. Zhang

VI. Fundamental Research of Static Electrification and Discharge

Contact Electrification Between Metals and Polymers: Effect of Surface
Oxidation 327
Dan A. Hays

Observation of Local Discharge Phenomena on Charged Surface of Dielectric
Films 331
T. Oda and M. Yanagida

Experimental Search for Free Quarks in Matter 335
C.D. Hendricks and R.S. Hornady

Separating Discharge in Film/Roller System 339
X.G. Ji, Y. Komai, Y. Takahashi and S. Kobayashi

Charge Transport Processes at Corona-Charged Polymer Film Surface 343
B.T. Yang and Y.N. Liu

Calculation of the Onset Voltage of Streamer From a Grounded
Hemispherically Capped Cylinder to Negative Well-Charged Dielectric
Plate 347
Z.Q. Ni

Standing Ionization Wave in the Positive Column of Glow Discharge of
Oxygen 350
J.L. Chen and R.N. Li

The Effects of Small Isolated Conductor on Propagating Brush Discharge 354
G.X. Li

On the Problem of Charge Distribution on a Conductor and the Calculation
of Geometric Factor of an Emitting Cathode 357
P.L. Lee

Research and Simulated Test on Space Electrostatic Field 360
H.C. Luo, X.L. Zhong, J.G. Xiong, Z.Y. Zhou and L.Q. Wang

A Method of Numerical Analysis and Calculation of Streaming Current
By a Hopping Model 364
F. Zhou, Z.H. Wu and G. Touchard

Research on Electric Field in Storage Tank with Numerical Method 368
Y.Q. Yang and L.Q. Zhao

Electronic States and Contact Charging of Polymers 371
Y. Murata

VII. Electrostatic Pollution in Electronic Industries

Electrostatic Pollution and Control Technology in Super Clean Rooms 377
D. Cucu

Fundamentals of Static Control for the Electronics Industry 379
G.R. Berbeco

Reliablity Study on MOS Electrostatic Protection from Charge on Human
Body 385
H. Xia and J.Y. Liu

VIII. Electrostatic Hazards in Petro-Chemical Industries and the Prevention Measures

Relaxation Time of Electrified Petroleum Surface Potential in Storage
Tanks with Different Capacity 391
N. Kitamura, I. Fushimi and M. Ueda

Techniques of Anti-Electrostatic Hazard of the Petrochemical Industry
in China and Outlook 395
S.L. Wang, C.G. Dong, J. Yang, G.C. Song and Z.J. Bi

Static Electrification Due to Liquefied Natural Gas Flows 400
G. Touchard, P. Humeau, L. Marcano, Z.H. Wu, J. Borzeix, J.P. Nossent
and S. Watanabe

Relationship Between Additive Concentration and Streaming Current 404
S. Watanabe, M. Fujii, K. Tanabe, A. Ohashi, Z.H. Wu, M. Benyamina and
G. Touchard

A Study on the Safe Potential of Charged Hydrocarbon Oils 407
J.B. Qu and M.F. Lu

Safety Surface Potential for Hydrocarbon Products 411
C.Y. Shi, W.J. Liang, G. Pan and T. Tao

Effect of Material Surface Character on Electrification of Fuel Oil 415
F.G. Tan and C.G. Dong

IX. Electrostatic Hazards and the Prevention Measures

Electrostatic Hazards of Explosive Articles and the Control Measures 423
G.S. Zhang

Applications of Electrically Conductive Fibers to Electrostatic Safety 427
Y. Tabata and Y. Matsuo

Control of Incendiary Discharges Occurring from Electrostatic Eliminator 431
Y. Tabata, T. Suzuki and S. Kamachi

A Study on Incendivity of Discharge from Charged Insulating Plate with
Different Electric Polarities 435
J.L. Chen and L.Z. Zhao

Simulated Electrostatic Test for the Fuel Tank of Q-6 Pursuit 439
H.C. Luo and L.Q. Wang

The Research on Electrostatic Spark Sensitivity in Suspending Black
Powder Dust 443
C. Qian, G.R. Wu and X.L. Ren

The Problem Concerning How to Analyse and Identify the Electrostatic
Explosion Accident 447
D.L. Zhong and L.Z. Zhao

Electrostatic Shielding Effect of Box Wagon for Ammunition 451
Y.F. Yu, H.N. Li and C.G. Li

Static Electicity of Human Body 454
L. Ke

Evaluation of the Safety Property on Tankers During the Drainage of
Ballast Water 457
K.P. Sun

The Relation Between Aircraft Precipitation Static Electricity and Flight
Environment 461
H.Z. Zhang

Experimental Study of the Electrostatic Accumulation of Explosive Powders 465
C.Y. Wang, Y.P. Chi and Y.Y. Liu

Dangers from Static Electricity in Black Powder Production and Its
Countermeasures 468
W.L. Wang

The Safety Electrostatic Potential of Non-Incendiary Discharges from
Charged Insulator 472
J.S. Huang and L.Z. Zhao

The Principle and Application of α Ion Source Static Eliminator 475
H.Y. Zhang

Po-210 Static Eliminator and its Industry Application 477
S.Y. Cai, Z.H. Zhou, S.Q. Mao and S.Z. Sun

Study on the Critical Voltage and Electrostatic Elimination in Release
of Hydrogen 481
J.T. Liu and B. Liu

X. Electrostatic Materials

The Research, Development and Application of Electrically Conductive
Fibre 489
M.X. Shen, M.H. Chen, S. Zhang and S.X. Jiang

The Studies of the Properties of PVC Conducting Plastic and Its
Application 492
X.Z. Wu, J.L. Chen, X.Z. Liu and W.B. Liu

Use of Piezo-Sensitive Material to Make Anti-Static Detonators 495
X.R. Li, F. He and C.Z. Lu

Research on Antistatic Pressure-Sensitive Materials 498
C.Z. Lu

A Study on Using Radiative Effect to Improve the Electrostatic Properties
of Polymer Surface 502
S.H. Liu, Z.C. Liu, B.X. Zhai, G.H. Wei and L.Z Zhang

XI. Ozone Techniques

Quick Disinfection of Handpiece in Dental Use 507
S. Masuda, E. Kiss, K. Ishida and H. Asai

Experimental Study of Improvement in Ozone Yield in a Parallel Plate
Ozonizer with a Rotating Plate Electrode 513
Y. Nomoto, T. Ohkubo, T. Adachi, J.S. Chang and M. Akazaki

Investigation of Discharge Current of Surface Discharge Type Ozoniser 517
E. Kiss and S. Masuda

I. Electrostatic Precipitation

ELECTROSTATIC PRECIPITATORS IN MODERN POWER PLANT

K. Darby
Lodge-Cottrell Division
Dresser UK Limited
Birmingham B3 1QQ, England

Abstract

This Paper describes a total energy management system for precipitator control to minimise dust emission and also the power consumed by the precipitator. The system includes a number of operating features of which the most important are the automatic voltage control and the provision for pulse modulation. Information is given on precipitator performance on widely varying ash resistivities.

Introduction

The trend of pollution legislation for power stations is to reduce the dust emission permitted requiring greater efficiency of the dust arresting equipment.

The preferred dust collecting device continues to be the electrostatic precipitator and the effect of the demand for higher efficiency is to substantially increase the power consumption and precipitator size. There is great emphasis not only on meeting the legislative requirements during carefully controlled proving tests, but also on a continuous basis. In some cases when the permitted dust emission limit is exceeded the boiler is either shut down or the load reduced until the permitted limit is reached.

A further factor in some cases is that coal may be supplied from any one of a number of sources, according to cost and availability. For example, in Europe coal could be of local supply but could also originate from the United States, South America, South Africa or Australia. The result is a range of fly ash with very different precipitation characteristics [1]. Coal from Australia may contain 0.5% sulphur compared to 1.5% to 2% for indigenous UK coals. The net effect is that the precipitator must be designed for the worst ash characteristics. As the electrical characteristics of the ash vary considerably, the precipitator energisation control system must be able to maintain maximum dust removal efficiency for all conditions. With the advent of acid rain control requiring FGD systems to be fitted also, there is a financial incentive to use the low sulphur coals, thus reducing the degree of desulphurisation required.

The problem of the different types of ash and their effect on the precipitator is discussed in some detail below. The differing operating conditions require sophisticated controls with facilities greatly in excess of those associated with the simple automatic voltage controllers used in the past. The result has been the development of a total energy management system, based on ten years experience of micro-processor based automatic voltage controllers, and five years experience with continually up-dated total energy management systems.

The Total Energy Management System

A schematic arrangement of the Total Energy Management System is shown in Fig.1. This shows a three field precipitator, each field with its own bus section controller. These are connected to the local supervisor unit on the precipitator and there is a similar remote supervisor in the boiler control room. In practice the supervisor units can monitor upwards of 100 bus section controllers. The system is described in greater detail in an earlier Paper [2].

Bus Section Controllers

The basic building block on which the total energy management system is designed is the local micro-processor at the bus section controller unit. The functions of this control are shown in Fig. 1. The advent of mass-produced relatively low priced micro-processor units presented the precipitator designer with an opportunity to control each bus section in isolation and also to facilitate inter-communication between the local bus section control unit and the local supervisor controller which will be described later.

The bus section controller which incorporates the automatic voltage control must be capable of operating not only as an integral part of the total system but also as a "stand alone" unit in the event of a failure in any other part of the Total Energy Management System. Thus in this situation each of the bus section controllers would continue to operate the TR sets such as to give the maximum efficiency for the precipitator.

The automatic voltage control is probably the most important part of the system since its only function is to maximise dust removal efficiency. Various forms of automatic control have been developed over the last half century but fast response controllers first started to appear in the 1950's with the development of new generations of electronic devices. The principle on which the AVC operates is described in detail below.

Local Supervisor Unit

The local supervisor unit is a "state of the art" computer packaged for an industrial environment. To facilitate data graphics and data display the local supervisor unit has a high resolution colour monitor and an optional printer allows for hard copies to be made for data and graphic displays if required. Logging of information can be made on to disk storage; the unit is fitted with a twin disk drive.

Linkage of the local supervisor unit into the total precipitator control system is shown in Fig.1. A single local supervisor unit can control all of the bus section control unit on a large boiler installation.

Remote Supervisor Unit

The local supervisor unit can in turn be linked to an optional remote supervisor unit which has a similar colour monitor to that of the local supervisor unit but has only a limited keyboard. This keyboard is sufficient to allow an operator to call up the various screens of data displayed, graphic displays and alarms but will not permit the re-setting of any of the bus section control parameters; these can only be set at the local supervisor unit or the bus section controller units. This eliminates interference with the control system by unauthorised personnel.

The principal features which can be incorporated into this system, include:

1) Automatic Voltage Control Units for individual bus section efficiency maximisation.

2) Optimum performance combined with minimised energy input including pulse modulation.

3) Full display of operating parameters with option for including DC output wave forms.

4) Bus section related rapping control.

5) Bus section related hopper temperature and dust level monitoring.

6) Minimum gas temperature monitoring to prevent TR energisation below the dewpoint of the gas.

7) Integrated communication of the bus section controller unit with a local supervisor unit for overall data gathering.

8) Optional communication link with the central computer for the boiler for overall plant monitoring.

Items 1 and 2 are the basic controls for dust emission and power consumption.

Items 3 to 8 listed above play important parts in ensuring dust emission is maintained as required. For example,[4] rapping; the cycle of operation of the rapping system is related to the dust concentration at the inlet to the field. Excessive rapping can cause serious increases in dust emission; too little rapping allows build up of dust on the electrode system which also increases emission [3]. In the event that a field upstream becomes inoperative, then the rapping cycle of downstream fields is altered accordingly.

Also if a high dust level is signalled, there is a danger that dust will short-out the electrode system. The control system will alert the Operator to the fault condition and can also switch off the field until the condition is rectified.

Other features are concerned with data gathering covering the important factors in the precipitator operation. Such information can be used to identify faults when these exist and provide proof, if needed, of the continuous healthy operation of the plant. The opacity meter in particular is often used as a continuous monitor of dust emission. In many plants in America the stack plume opacity is a feature of the precipitator guarantee, together with dust emission. Due to variations in particle sizing of the dust for different coals, the opacity meter reading is not an absolute measure of dust emission. Calibration for the coals fired makes it a useful device for the continuous monitoring of emission.

Electrical Characteristics of the Ash

In order to understand the requirements of automatic voltage control, the effect of the different fly ash electrical characteristics must be considered.

The corona discharge characteristics of the precipitator vary considerably with the ash from different coals. This in turn alters the efficiency of dust removal such that the size of precipitator needed to maintain a given efficiency can vary by as much as 3:1.

Figure 2 shows the discharge electrode voltage/corona current characteristics of fly ash for a range of fuels which could be fired over a period of time at a power station in the UK..

Curve 1 is for a low sulphur (0.45%) low sodium bituminous coal mined in Scotland. This gives a low corona current corresponding to the peak value of the electrode voltage; the precipitator operation at this point becomes very erratic and any further increase in voltage reduces the efficiency due to heavy arcing.

In contrast Curve No. 2 is an Australian coal which exhibits reverse corona in an extreme form. The voltage rises only to 32 kV and levels off, the current rising to any level which it is possible to feed into the precipitator without flashover. Although it would appear that Curves 1 and 2 correspond to very different ash, in practice both have highly resistive characteristics and in both cases the sulphur is of the order of 0.45% with low sodium. It is interesting to note that when the flue gas for the boiler firing the coal for No. 1 Curve is conditioned by the addition of sulphur trioxide [1], then the characteristic is shown in Curve No. 5. Flashover is more controllable and the current-voltage characteristics are similar to those obtained in Curves 3 and 4 which relate to fly ash with good precipitation characteristics. In the case of Curve No. 2 when water was sprayed into the combustion chamber to increase dewpoint without changing the gas temperature then this changed its characteristics to those shown in Curve No. 6, giving much more favourable precipitation.

The comparison between the ash corresponding to Curves 1 and 2 is interesting, as both are caused by a highly resistive condition of the ash. Using conditioning agents (1), water or sulphur trioxide, results in both having much improved precipitation characteristics with effective migration velocities of the order of 10 cm/sec.

These Curves represent a wide range of operating conditions which the control system would have to deal with on an operating plant where fuels from different sources are fired.

The automatic voltage control, which is part of the total energy management system, must operate the precipitator supply rectifier to give the electrical input to the precipitator corresponding to the maximum dust removal efficiency at all times.

Total Energy Management System (Practice)

The way in which the control system operates is now described. The following factors are directly related to dust emission and power consumption and can be incorporated in the system.

The AVC is the most important part of the TEMS as it is the controller of dust emission. It has been demonstrated that, regardless of the dust characteristics, if the discharge electrode voltage is set to the maximum mean discharge value, then the precipitator will give the best efficiency possible under the conditions of dust within the precipitator.

Automatic Voltage Control System

In Fig. 3 is shown a graph giving the power input plotted against corona current, mean discharge electrode voltage and dust emission. The Curves shown were derived from data provided by the Central Electricity Generating Board from tests carried out to test the validity of this control theory. The advantage of this technique is that by controlling at the maximum mean discharge electrode voltage, the highest dust removal efficiency possible is maintained at all times.

The mean discharge electrode voltage here is defined as the average value of the discharge electrode voltage over a significantly long period, long enough to include a degree of flashover if this occurs. When the mean voltage decreases with increased power input, this is due to the increasing effect of flashover. At the maximum mean voltage point there is usually some flashover. This is beneficial in the operation of the precipitator as it has been shown[4] that such flashes have a similar effect to the rapping mechanism, but much more localised. Dust is dislodged in the proximity of the arcs without significantly increasing the loss of dust due to re-entrainment.

The Lodge-Cottrell automatic voltage control unit is incorporated in the bus section controller. There is one such unit for each TR set and its function is to select the peak voltage for the different dust condition illustrated in Fig.2 and hence the highest precipitator efficiency.

Alternatively where no flashover takes place (Fig 2, Curve No. 2), it will seek the point where the discharge electrode voltage no longer rises significantly, as applied voltage increases.

The automatic voltage control unit operates through a signal representative of the discharge electrode voltage taken through a voltage divider. The control cycle starts by initiating an increase in primary voltage to the TR set (Fig 3). The control system senses whether this resulted in an increase in discharge electrode voltage; if so a further increase takes place followed by successive cycles until the peak condition is reached where the voltage curve flattens off and begins to fall. On the negative side of the Curve any further increase in the primary voltage will result in a decrease in discharge electrode voltage which is detected by the control system circuitry. In this situation the control system backs off roughly double the amount that is advanced in the sensing cycle. This action, often referred to as the 'hill-climbing' technique, will maintain the conditions of the precipitator very close to the peak value shown in the graphs. If conditions change, eg fuel, operation etc., so that a different characteristic curve appears, then it will immediately seek the new peak.

The system takes into account variation in the intensity and duration of the flashovers by using the principle of measuring the mean discharge voltage. It is not dependent on the accuracy of the voltage measured since it works by difference.

Pulse Energisation and Pulse Modulation

Pulse energisation and pulse modulation are two systems which have been developed with the objective of reducing the effect of high resistivity dust on precipitator performance thus enabling the precipitator to operate at a higher level of efficiency on the worst dusts. These developments of the electrical circuitry are possible due to the development of solid state switching devices such as the thyristor used to control the primary input to the TR set. These devices can be included in the overall control programme of the local supervisor unit.

Pulse Energisation

With highly resistive dusts where electric charge accumulates within the deposited dust layer and the gas within the layer becomes ionised, a stream of positive ions is produced which passes from the dust layer to the discharge electrode in the opposite direction to the flow of negative ions produced by the discharge electrodes. This effect, known as reverse ionisation, not only reduces precipitator efficiency substantially, but also increases the total current, and hence the power consumption, of the precipitator very considerably.

With pulse energisation a separate supply is provided which can provide pulses of energisation with an amplitude in the range of up to 50 kilovolts above the base DC volts, duration 50 to 200 micro-seconds and a repetition rate of 0 to 250 pulses per second. This is illustrated in Fig 4(A). When such pulses are applied the conventional TR set voltage is reduced until the reverse ionisation threshold is reached. The short duration high voltage pulses ensure improved dust charging. This, together with the reduction of reverse ionisation, improves dust collection efficiency and power consumption is reduced also.

The pulse energisation system requires a separate power supply, the cost of which is comparable to the conventional TR set and, therefore, the use of pulse energisation would roughly double the cost of the electrical energisation of the precipitator and is not insignificant. Experience using this technique has produced extremely impressive improvements on dry process cement plant[5]. Efficiency was increased by pulsing from 92 to 99.7%, when pulsing only two fields of a three field unit. In the case of power stations with highly resistive dust the results have been less dramatic. Typically emissions of 400 mg/Nm³ have been reduced by a factor of 2 to 2.5 times which can still be significant so far as the precipitator performance is concerned. For example this could be equivalent to increasing the efficiency of dust removal from 96 to 98%. This would need a precipitator size increase of 20% to achieve the same result.

Pulse Modulation

Pulse modulation involves the interruption of the power supply as illustrated in Fig. 4 by switching off, for pre-determined numbers of cycles of the power thyristors feeding the TR set. The diagram shows the effect of switching off 1 - 4 complete cycles. Since the precipitator electrode system is a condenser, the voltage does not collapse. There is a reduction in the rate of voltage decay as the corona discharge decreases. At corona threshold only insulator leakage would occur and the field would be static, ie non-discharging. The nett effect of this system is that for highly resistive dusts, power consumption is reduced to a fraction of that for full energisation. The automatic voltage control system can operate normally to give the highest mean value of electrode voltage. Peak and hence charging voltage is maintained and there will be flashover.

Dust emissions of the level of 400 to 500 mg/Nm³ are reduced by up to 2 - 3 times. The result is similar to pulse energisation but the control circuit producing the charge is part of the micro-processor of which the AVC is part also. The cost of incorporating pulse modulation is consequently negligible.

Fig. 5 shows the effect of pulse modulation on a power station plant firing a 0.45% sulphur coal with low sodium content. The graph shows a typical curve of efficiency plotted against specific power input, that is corona power/Am³ of gas. This is a typical curve for fly ash with favourable precipitation characteristics originating from medium sulphur bituminous coal. It is interesting to see the effect of pulse modulation on a highly resistive ash. With no interruption of the supply, the emission of the precipitator which is roughly 25 mg/Nm³ meets the guarantee for the plant, but with a high power consumption due to reverse ionisation. As the number of cycles switched off increases, the power consumed actually falls below that of the characteristic curve for favourable dust and there is a slight improvement in the emission.

This plant was designed for, and was giving, the very high efficiencies shown, that is in excess of 99.8%, with correspondingly low emissions. The normally anticipated halving of emission is not obtained. This is probably due to other factors which affect dust emission, such as re-entrainment due to rapping, slight sneakage of gas past the field, giving a base emission not influenced by the electric field[3].

On Fig 5 is shown also Curve D which corresponds to the effect of pulse modulation with a precipitator operating on a favourable dust from medium sulphur bituminous coal. This plant is designed for a lower efficiency. Since there is no reverse ionisation with this dust the effect of reducing the firing cycles is to reduce power and efficiency of precipitation in line with the standard Curve (A). Efficiency is reduced from 98.8 to 97.2% for a corona power reduction of 5:1.

Consider now the application of pulse modulation to a precipitator designed for 99.5% efficiency on a low sulphur low sodium ash coal which is then required to operate on a high sulphur coal with favourable precipitation characteristics. In the case of the high resistivity ash the Curve would follow that shown on Fig. 5, point (a) to (b) to (c), with increasing interruption of the supply. Such a precipitator would need to be designed with an emv in the range of 4 to 5 cm/second.

When such a plant is fed with dust resulting from the firing of a higher sulphur coal giving a migration velocity of about 10 cm/second, then the anticipated efficiency under these conditions with full energisation would exceed 99.9% by a substantial amount but power consumption could double (dotted projection of power curve). However, the application of pulse modulation would move down the Curve such that at the 99.5% efficiency needed, there would be substantial reduction in power consumed in the order of 50%. Thus it will be seen that for either high or low resistive dust pulse modulation can reduce the power consumed by the precipitator very substantially and still maintain the required precipitator efficiency of dust removal. Substantial power savings can be obtained by the alternative method of reducing the voltage to the TR sets, but the problem with this method is that flashover is entirely eliminated at the reduced voltages. It has been shown[4] that flashover is beneficial in keeping the plant operationally clean. For example, Laboratory tests showed that flashover produced vibration in the plates, local to the arc, of a similar order to the rapping system.

```
                        TABLE 1

Boiler:      600 MW

Number of Transformer Rectifier Sets: 72
   36 - 55 kV  600 mA
   36 - 55 kV  900 mA

Total installed capacity:   422 kVA - AC

Anticipated Consumption:    (80%) 3380 kVA - AC
```

A third alternative for the favourable fuel condition would be to de-energise fields successively still maintaining the same precipitation efficiency. This would have the advantage that flashover would occur in a normal fashion on other fields.

The TEMS system can use any of these systems to control power consumption and dust emission but Pulse Modulation is the preferred system.

The significance of power saving by whatever method is illustrated by Table 1 which shows the installed power for a 600 MW boiler designed for an efficiency of 99%. The usage of power is estimated at 0.4 MW so clearly reducing this by 50% represents a considerable economy, particularly as in the case of the highly resistive dust there is no loss in efficiency.

In each of the methods given above in contrast to the AVC system of selecting for the high voltage which is known to coincide with the best precipitator efficiency, measures of the changes caused by pulse modulation, pulse energisation or other means of power reduction can only be really assessed by the use of a calibrated opacity meter in the outlet flue. The control system can either aim to keep the opacity below a certain level or have a specific target - in the United States, commonly 10% to 20%. The problem in using the opacity meter signal is that, in contrast to the voltage of the electrodes where a change is immediately effective in altering dust removal efficiency, the opacity meter which is placed some distance downstream of the precipitator is subject to a delay. The gas passing through the field in which the changes are made can take up to 15 seconds to reach the opacity meter. This must be taken into account when the programme is drawn up for the total energy control system. In contrast to the change of conditions catered for by the automatic control system the type of change envisaged for detection by the opacity meter would result from changes of coal of a substantial nature. Such changes could be expected to take place over a matter of fractions of an hour, or even hours, and the effect would persist, probably for several days. In this situation the response time of the system is not important.

While the AVC units will be within the bus section section control and operate independently, other controls such as pulse modulation would operate according to the programming of the local supervisor. This is necessary as such systems controlled by the opacity meter signal, need a different approach incorporating an integrated, systematic control for the bus sections.

References:

(1) Busby, HGTB and Darby, K
'Efficiency of Precipitators as Affected by the Properties and Combustion of Coal'
Institute of Fuel Journal - May 1963

(2) Russell-Jones, A
'A Total Energy Management System'
Third International Conference on Electrostatic Precipitation, Abano, Italy - October 1987

(3) Darby, K
'Criteria for Designing Electrostatic Precipitators'
Second Symposium on Transfer and Utilisation of Particulate Control Technology
Denver, USA - July 1979

(4) Darby, K
'The Use of Pilot Precipitators for Process and Design Development
Second CSIRO Conference, Australia- August 1983

(5) Darby, K
'Pulse Energisation - An Alternative to Conditioning for Highly Resistive Dusts'
Second International Conference on Electrostatic Precipitation, Kyoto, Japan
November 1984

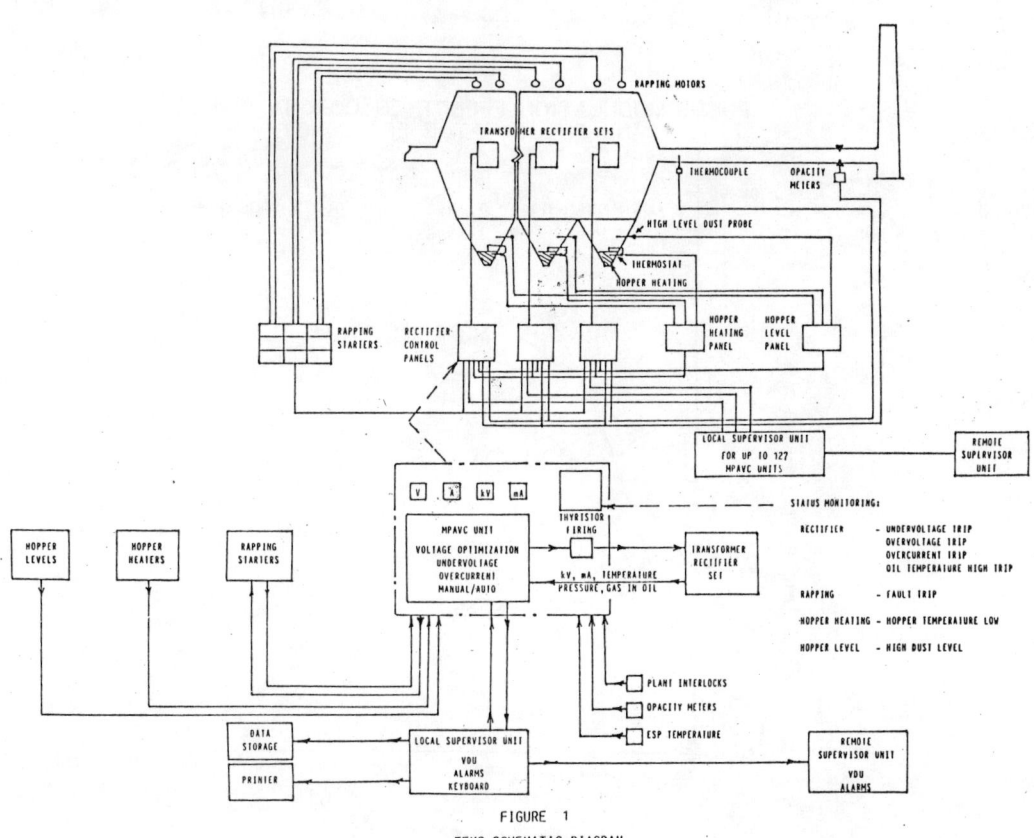

FIGURE 1
TEMS SCHEMATIC DIAGRAM

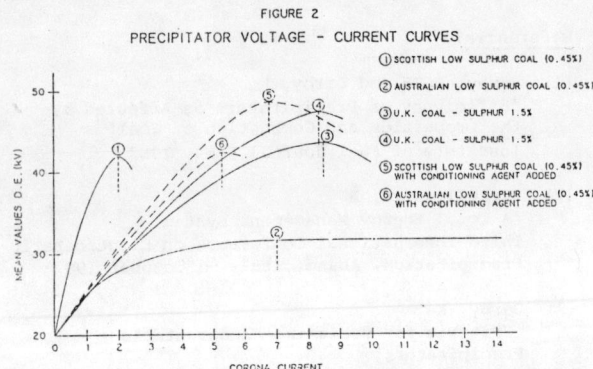

FIGURE 2
PRECIPITATOR VOLTAGE - CURRENT CURVES

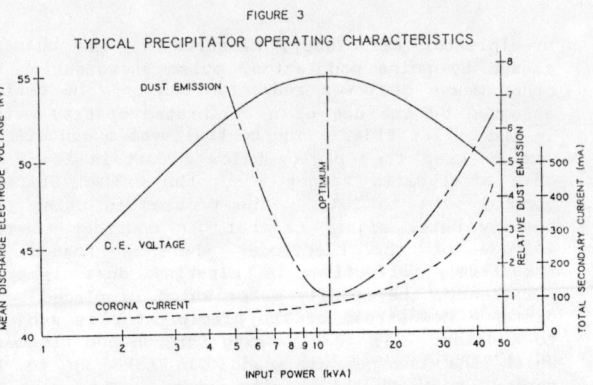

FIGURE 3
TYPICAL PRECIPITATOR OPERATING CHARACTERISTICS

FIGURE 4

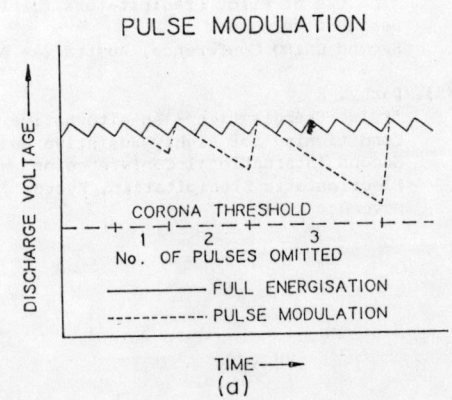

PULSE MODULATION
(a)

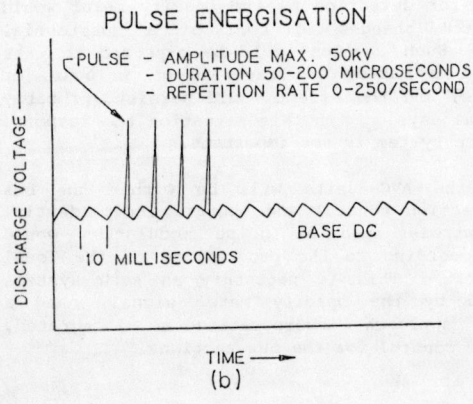

PULSE ENERGISATION
(b)

FIGURE 5

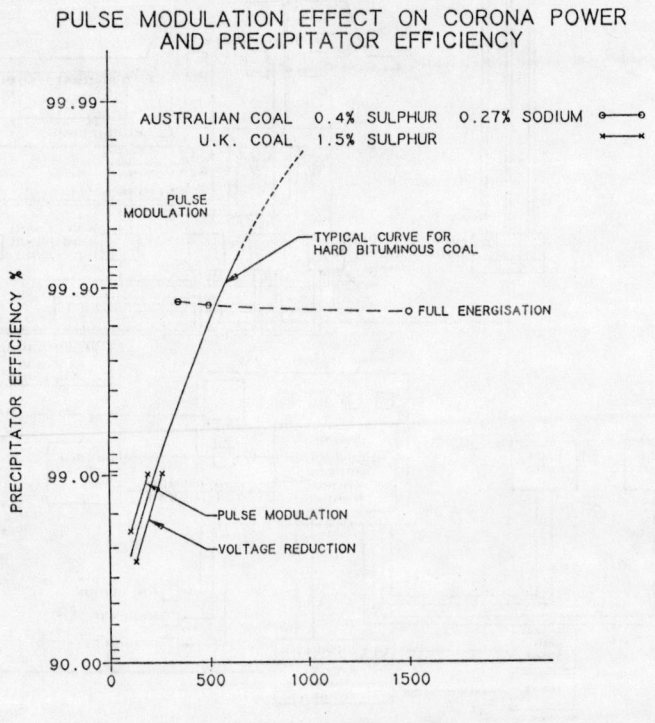

PULSE MODULATION EFFECT ON CORONA POWER AND PRECIPITATOR EFFICIENCY

-8-

Proceedings of International Conference on Modern Electrostatics

PULSED ENERGIZATION OF ELECTROSTATIC PRECIPITATORS -- A REVIEW OF WORLDWIDE EXPERIENCE

Kjell Porle, Product Manager
Particulate Control Technology
Fläkt Industriella Processer AB, Växjö, Sweden

Abstract

With environmental protection legislation becoming increasingly stringent, Electrostatic Precipitators (ESPs) have to be sized larger and larger to meet lower emission requirements. Pulsed energization is now offering substantial improvements in performance. Experience from a large number of plants has shown that power consumtion of transformer/rectifiers for the precipitator's high-voltage supply can be reduced 70 to 90 percent for many applications, without increasing outlet emissions. For medium- and high-resistivity applications, the original size of the ESP can usually be reduced by 30 to 40 percent. The paper also compares pulsing with some other methods -- NH_3 and SO_3 conditioning -- for reducing ESP size.

Introduction

Today, the electrostatic percipitator (ESP) is well established throughout the world as an efficient means of air pollution control. The first commercial unit started operation in 1907. Since then, field experience, in combination with research and development, has helped create a much more efficient and reliable product. Coal-fired power stations are among the major users of ESPs, accounting for about two-thirds of all ESP installations. R&D work on ESPs has been extensive during the past 10 to 15 years. New concepts, including precharging, three-electrode systems, wider spacing, new energization methods, gas conditioning and improved electrode designs have been investigated as well as commercialized to a great extent. Reasons for the intensified R&D work are:

1. The evident shift from high-sulphur coals to low-sulphur grades since the energy crises. As a result, larger ESP sizes are required.

2. New and more stringent emission regulations worldwide.

3. Increasing competition from fabric filters as the main dust collector.

This paper will describe new energization methods that have been used usccessfully on fullscale precipitators. Remarkable power savings up to 90 percent have been achieved in several installations, and outlet emissions have been reduced by as much as 85 percent, compared to conventional energization methods.

Pulsed energization - theory

Until recently the aim of ESP energization was to supply high DC voltage with unspecified ripple. The ESP was charged via a single-phase transformer and fullwave rectifier (T/R) without further smoothing of the DC. For low resistivity fly ashes the highest collecting efficiency is obtained when the corona current is maximum. In a few cases three-phase transformes have been used. The voltage wave form has less ripple than for the single-phase ones and the intention was to increase the corona current as well as the electric field. Contrary to this, the newly developed pulsed energization methods superimpose voltage pulses in specific forms on the DC, resulting in a pulsed corona current. Experience has shown that this concept improves the performance considerably especially for high resistivity fly ashes.

The concept of pulsed energization was pioneered more than 30 years ago, but was never commercially viable due to hardware limitations. However, today's high-voltage components and micro-processors have made pulsing economically attractive. Several types of pulsed systems are now in commercial operation in various parts of the world. Most pulsed systems are designed for conventional ESPs. The main benefit of pulsed energization was long thought to be improved collecting efficiency, particularly in cases where high-resistivity fly ash should be precipitated from the flue gas. Coals with low sulphur and sodium content normally generate diffficult high resistivity fly ash. When upgrading or retrofitting existing ESPs to comply with new emission regulations, the options have been conditioning or enlargement of the ESP. Conditioning of the flue gas means primarily a way of reducing the resistivity. This can be done for example by increasing the moisture content or by injection of NH_3 or SO_3 in the flue gas.

Pulsed energization now offers a number of specific benefits in addition to improved performance. These include : power savings; installation with minimal ESP downtime; minimum supervision or maintenance; low investment since only the T/R set is changed and the internals of the ESP do not have to be modified. Various designs of internal components have been used, some with excellent results and some with moderate improvements.

For good ESP performance the corona current must be well distributed across the entire collecting electrode area. Experience shows a certain minimum current is needed to ensure at least some flow to the most quiescent areas of the collecting electrodes connected to one transformer/rectifier. Each fly ash layer on the collecting plates exerts a resistance to the flow of electrical current. The electrical field strength in the dust layer increases with the current. Local sparking occurs in the dust layer even at comparatively small currents, if the resistivity is too high. This is called back corona and results in reduced collection efficiency. Back corona disappears if current in the ESP, or the resistivity by means of gas conditioning is reduced. However, a reduced current results in a bad current distribution and the efficiency of the ESP goes down.

Fig. 1 and 2 show examples of one reasonable and one very bad current distribution along the collecting plate. A plate type discharge electrode (fig. 2) shows high local current densities and large areas with zero current for about the same average current density as in fig. 1. The electrode in fig. 2 is therefore much more sensitive to high resistivity dust and back corona conditions can easily develop. In ref. (1) is shown that the distribution improves with higher current densities. However, the improvements is much more pronounced for the wire in fig. 1. Poor current distribution for low current densities or the occurence of severe back-corona for high current densities can be reduced by pulsed operation. As it takes some time -- in the order of some seconds for a high-resistivity fly ash -- for the charges in the ash layer to disappear, the current density, where back corona starts, is equal to the time-average current density as a first approximation. Thus, high short peak currents at low frequencies from the discharge electrode have little effect on back corona. By inreasing the voltage, good current distribution is obtained at regular intervals while the average current density is still low.

The quantitative improvement obtained by pulsed operation cannot be evaluated theoretically. Extensive testing is therefore required. Experience has shown that peak voltages can be increased during a short pulse, compared to conventional energization without causing spark overs. Peak voltages and current distribution will improve in direct relation to the reduction in pulse duration. The result is higher collection efficiencies. The average current is reduced, implying power savings.

Commercial system

Commercial T/R sets providing pulsed power supply have been on the market since 1980. Currently, the main suppliers are companies based in Japan, U.S., Sweden, Germany and Denmark. Most are ESP manufacturers. Since the approach to pulsing varies between the suppliers, for example the high voltage wave forms, it is difficult to make direct comparisons. To date, very few, if any, comparative tests have been made to evaluate the cost and performance benefits of the sets offered by different manufacturers. However, fig. 3 illustrates a key parameter -- the pulse width. There is a significant difference in the enhancement factor between pulses of microsecond and millisecond width. w_k is a measure of the precipitability of the fly ash and is defined by:

$$\eta = 1 - e^{-(w_k \times A/Q)^k};$$

normally $k = 0.5$, where η is the particulate collection efficiency of the ESP, A is the collecting area (m^2) and Q is the gas flow (m^3/sec.). For high resistivity fly ash, w_k might be 12 to 15 cm/s, while for low resistivity it might be 60 to 70 cm/s. The required collecting area for a given efficiency is inversely proportional to w_k. Enhancement factor of 2 indicates a reduction of EP-size to half of the original. Data in fig. 3 is taken from results presented by Mitsubishi, Japan (1); Research-Cottrell, U.S. (2); F.L. Smidth Denmark (3); and from Fläkt, Sweden (4).

The equipment used for microsecond pulses, also called full pulse, is fairly expensive -- several times the price of a conventional T/R set. Different methods of forming these pulses are described in references 2, 3 and 4. All involve storage of the energy to be transferred to the ESP in a capacitor in the T/R set. The Research-Cottrell unit uses a spark-gap to switch the pulsed energy to the ESP. This creates a maintenance requirement, and energy not being used in the ESP is disposed of externally. The F.L. Smidth design uses an energy-saving concept in which unutilized pulsed energy in the ESP is stored in the T/R set and reused. In both these systems separate T/R sets for the DC supply have to be incorporated. Fläkt uses a third variant in which the DC voltage and the pulses are formed in a single unit (see later description).

The number of microsecond pulse generators in operation worldwide was close to two hundred in May 1988 and the technology is now well established. The dominating applications so far are within the cement industry and power stations (coal-fired boilers). The number of units installed for retrofit -- upgrading of old ESPs -- accounts for about half of the total.

An interesting application is the dedusting of the flue gas after preheater cement kilns. The conventional approach has been to use a conditioning tower (water injection) in front of an ESP. This reduces resisitivity and the gas temperature is $150^\circ C$. The lack of water in many places and maintenance problems with the towers call for a better solution. A hot ESP ($350^\circ C$) without conditioning and equipped with pulsed energization is in many cases an attractive solution.

The lower enhancement factors for millisecond pulses, see fig. 3, are justified by the less-expensive equipment used. A kind of varying secondary voltage is obtained by pulsing the input power to the T/R set. The concept is often refered to as "Intermittent Energization (IE)". Conventional thyristor-controlled T/R sets can be easily modified just by adding a microprocessor control system. Most new sets are already equipped with suitable microprocessors. A retrofit is often justified by the energy savings alone. Fläkt has extensive experience with these type of control systems, with more than 2.800 units currently in operation for a wide range of applications worldwide.

Various models of IE have been investigated by C. Landham et al (7) and the report mentions that the various wave forms give different improvements. Also the type of discharge electrode was important.

Fläkt experience

Fig. 4 and 5 show the microsecond pulse unit working according to the Fläkt Multipulse concept (MP). A capacitor after the high-voltage rectifier is charged. The energy is transferred via inductance to the ESP through tryristors. The energy swings back and forth until an essential part has been utilized in the ESP. Internal energy losses are low.

The pulse unit is designed to supply a DC base voltage equal to the corona onset voltage. No corona current is generated in between the pulses. Each burst -- an oscillation -- consists of closely spaced multiple pulses. Experience has shown that a low burst frequency with several pulses per burst gives better ESP performance than evenly spaced single pulses. Fig. 6 and 7 illustrate the millisecond pulse method, using Fläkt's Semipulse concept (SP). The example shows a SP mode where only each third half wave is allowed to pass. This is called a charging ratio (CR) of 1:3. Best performance has often been obtained with CRs from 1:7 to 1:11. By maintaining the same amplitude on each half wave in fig. 7 as in fig. 6, the average current decreases to one third of the original. The voltage waveform in fig. 7 has a more pronounced ripple (pulsing) than in fig. 6. Typical results from comparison tests with different energization concepts are shown in fig. 8. A 3-field ESP, which is installed after a coal-fired boiler in Australia, normally operates with back corona, due to very high fly ash resistivity, above 10^{13} ohm cm. This means that performance deteriorates for currents above 75 microamp/m^2 for conventional energization. When switching to SP in two fields, minimum emissions declined to 50 percent combined with reduced current (power saving). A further reduction in emissions to 28 percent was obtained when two MP units where used instead of SP. During even more severe conditions reductions to below 20 percent were repeatedly recorded. The table (below) gives examples of results at other coal-fired stations.

Plant	Country	Boiler load MW	Concept	Emission Reduction %	Power saving %
A	Denmark	150	MP	66	70
A	Denmark	150	SP	30	80
B	Denmark	630	SP	30	90
C	India	200	MP	85	70
D	Japan	600	SP	55	90
E	Australia	660	SP	50	50

The total power saving for plant B is between 800 and 1000 kW.

Conditioning

By injecting, for example, SO_3 or NH_3 in the flue gas before the ESP, the fly ash properties can usually be improved, resulting in higher collecting efficiencies. Particle agglomeration is enhanced and resistivity is lowered. Normal injection rates are from 15 to 30 ppm. Special equipemt for evaporation (NH_3), oxidation (SO_3), air mixing, etc, has to be installed beside the ESP. The installation costs are not negligible and maintenance and operational costs are fairly high. Precautionary measures differ from country to country, since chemicals have to be handled. However, for some application where the ESP has to be substantially upgraded, conditioning may still provide better results than pulsing and can therefore be justified. It is not possible to save power, however, since the conditioning equipment requires power and the ESP consumes more energy when such conditions prevail. For new plants, pulsing is the only realistic alternative.

NH_3 injection is fairly easy to operate as it can be switched on and off without any problems. Some power stations use it just on an intermittent basis. SO_3 injection must be on a continous basis as the risk for corrosion is high during start up and shut down of the equipment. Commercial reliable systems of both kinds are available today.

Conclusion

Operating experience during the past few years has shown that pulsed energization is an economical means to improve ESP performance and to cut energy consumption. Pulsing permits the sizing of smaller ESPs for high-resistivity fly ash while offering substantial power savings. Compared with other methods for improving the performance of existing ESPs, such as NH_3 and SO_3 conditioning, pulsing is a more reliable and economical alternative to reduce emissions sharply and to cut energy consumption.

Reference list

1. Porle K. On Back Corona in Precipitators and Suppressing it Using Different Energization Methods Third International Conference on ESPs, Albano, Italy, October 1987.

2. Feldman, P.L., A.A., P.J., and Thanh, L.C. Present Status of Research-Cottrell Pulse Energization Technology. EPRI Conference on Electrostatic Precipitator Technology for Coal-Fired Power Plants. Nashville, Tenn., July 1982.

3. Lausen, P. and Lind, H. Application of Pulse Energization on Fly Ash Precipitators. Tiz-Fachberichte, Vol. 106, No 12, 1982.

4. Porle K. and Bradburn K. Full-Scale Experience with Pulsed Energization of Electrostatic Precipitators. Fifth Symposium on the Transfer and Utilization of Particulate Control Technology, Kansas City, Missouri, August 1984.

5. Matsumoto Y., Sugiura S., Ando T., Teramura N. Development of Mitsubishi Intermittent Energization System (MIE) for ESPs for Coal-Fired Boilers. Paper released by Mitsubishi Heavy Industries, Ltd, Japan.

6. Mauritzson C., Porle K. and Kirsten M. Experience with Pulsed Energization of Precipitators for a Wide Range of Processes and Operating Conditions. Third International Conference on ESPs, Albano, Italy, October 1987.

7. Landham, E.C. Jr., Du Bard J.L., Piulle, W. Pilot-Scale Evaluation of ESP Intermittent Energization. The 6th Symposium on the Transfer and Utilization of Particulate Control Technology, New Orleans, Louisiana, U.S.A., February 25-28 1986.

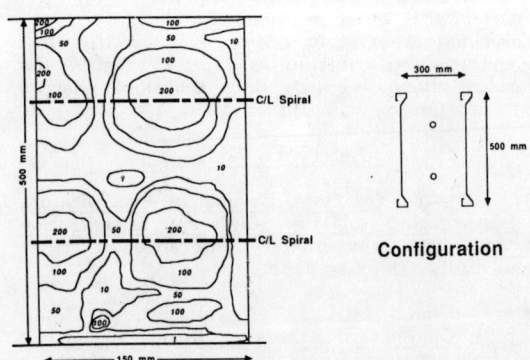

Fig. 1. Current distribution from spiral discharge electrodes. i=280 μA/m2 average.

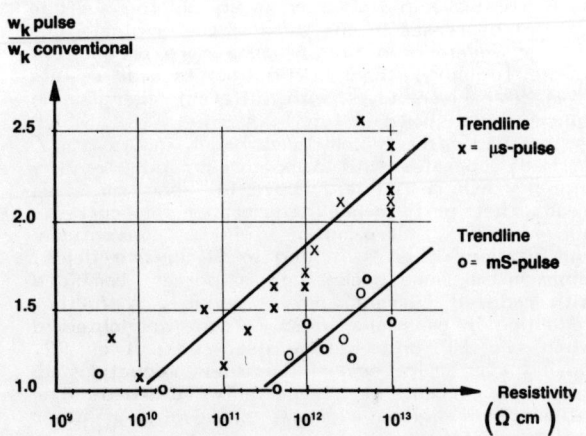

Fig. 3. Enhancement factor.

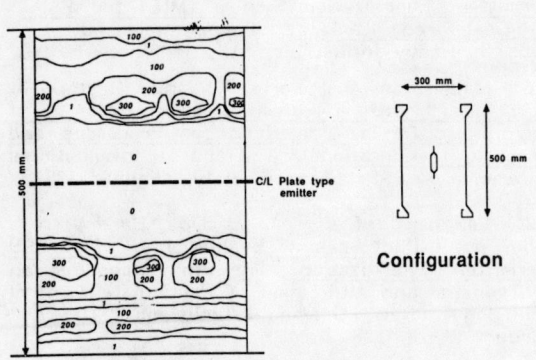

Fig. 2. Current distribution from plate-type discharge electrodes. i=226 μA/m2 average.

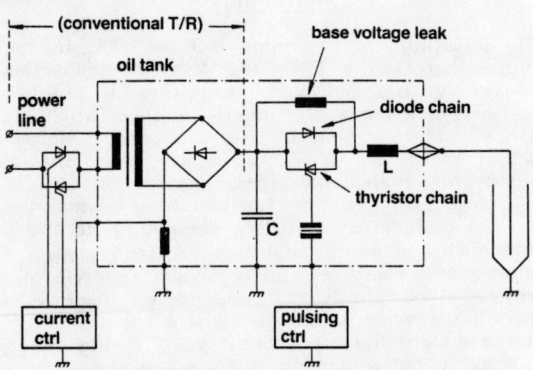

Fig. 4. Principal circuit diagram for Multipulse System.

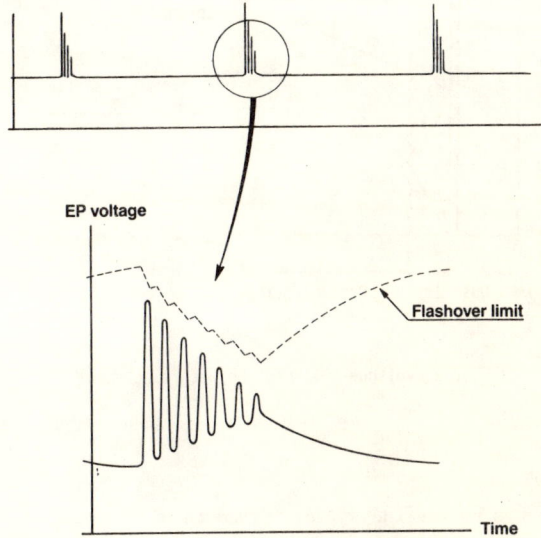

Fig. 5. Multipulse high-voltage waveform.

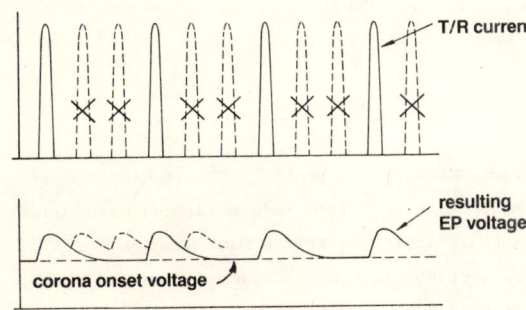

Fig. 7. Semipulse T/R operation.

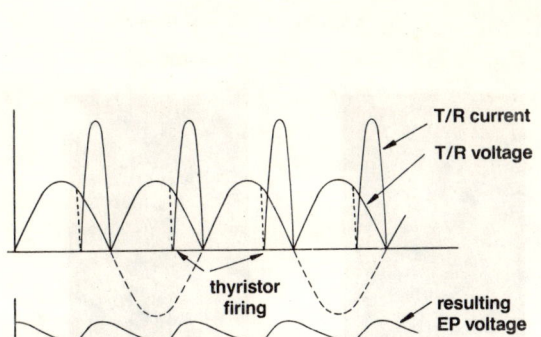

Fig. 6. Conventional full wave T/R operation.

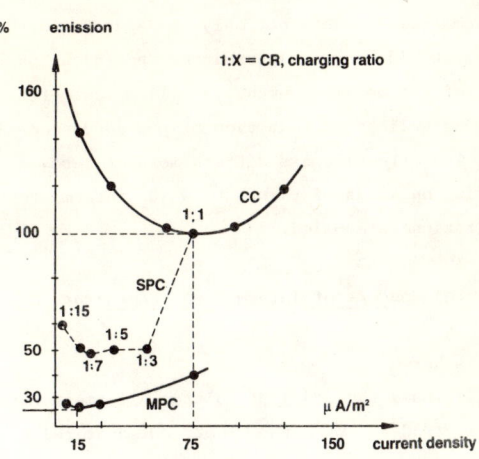

Fig. 8. Performance comparison between conventional (CC), semipulse (SP) and multipulse (MP) energization.

APPLICATIONS OF ELECTROSTATIC PRECIPITATOR WITH THE INTERMITTENT ENERGIZATION

N.Tachibana and H.Fujishima
Mitsubishi Heavy Ind.,Ltd. Kobe Shipyard & Machinery Works
1-1-1, Wadasaki-cho, Hyogo-ku, Kobe, 652 JAPAN

Abstract

Intermittent Enegization is now well known technology to mitigate the collection performance degradation caused by back discharge and to save remarkable energy consumption of electrostatic precipitators.

This paper decribes and outline of the intermittent energization principle and its effect based on the operating results from various kinds of commercial application.

Introduction

Electrostatic preciptator (ESP) is widely used for collecting dust from industrial and waste gas. A major difficulty in ESP is back discharge on the collecting electrode surface due to the high resistivity dust. To prevent or suppress back discharge phenomenon, several means, such as operating gas temperature control (Ex. hot side ESP for coal fired boiler), gas conditioning and energization control, etc. are taken.

Intermittent energization, one of the most economical means, was commercially applied and succeeded in 1979 [1], and has been applied for many commercial plants not only in Japan but also over the world since. Intermittent energization has an effect on improvement in collection efficiency as well as on reduction of power consumption. The effectiveness differs more or less depending on kinds of plants to which intermittent energization is applied.

General Concept of Intermittent Energization

Back Discharge

Fig.1 shows the model of voltage allotment between electrodes. When the electrical dust resistivity is extremely high, the voltage of the dust layer on the collecting electrode (Vd) increases and insulation breakdown occurs within the layer. This phenomenon is called back discharge and its onset condition is generally explained by the following formula.

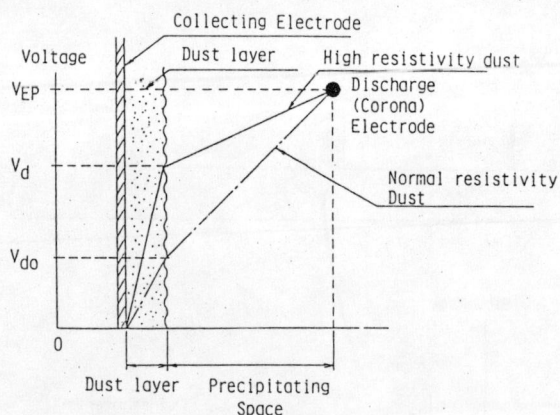

Fig. 1 Voltage allotment between electrodes

$$E_{ds} \lesssim j(E_{ds}) \cdot \rho_d(E_{ds}) \qquad (1)$$

Where E_{ds} = breakdown field strength of the dust layer, $j(E_{ds})$ and $\rho_d(E_{ds})$ = current density and dust layer resistivity at the breakdown of the layer. [4]

Fig.2 (a) shows thr flying paths of the ordinary resistivity dust and (b) shows that of the high resistivity dust (occurring back discharge). Once back discharge occurs, particles met with positive ions at the vicinity of the collecting electrode, lose their their negative charge and are pushed back towards the discharge electrode so that ESP collection performance degrades.

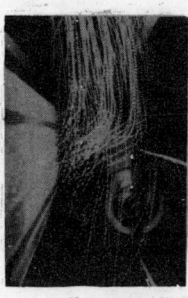 

Fig.2(a) Flying path with ordinary resistivity dust layer on the collecting electrode

Fig.2(b) Flying path with high resistivity dust layer on the collecting electrode

Principle of the Intermittent Energization

Fig.3 shows a typical V/I characteristics of an industrial precipitator with low (1) and high (2) resistivity dust. Once back discharge occurs, promptly the current increases (indicated with the solid line 2).
When the voltage is managed to rise rapidly, there is an over-shoot of voltage beyond the highest value of solid line 2, as indicated with the dodded line 3, and in the next moment, the voltage drops back to the solid line 2 and current goes up.

This phenomenon can be explained in terms of the electrical relaxation time of the dust layer. (2)
Modelling the dust layer by the equivalent circuit consisting of a resistor R and a capacitor C as shown in Fig.4 (a), the voltage allotment of the layer V_d is expressed by the following Eq. (2), when a constant corona current ($j=I_0$) is supplied to the dust layer from time $t = 0$ (reffer to Fig.4 (b)) :

$$V_d = R \cdot I_0 \cdot (1-e^{-t/t_0}) \quad (2)$$

where t_0 is the time constant ($t_0 = C R$) and is calculated by Eq. (3).

$$t_0 = 8.85 \cdot \varepsilon_s \cdot \rho_d \times 10^{-14} \quad (sec) \quad (3)$$

where ε_s = specific dielectric constant of the dust layer, ρ_d = specific resistivity of the dust layer (ohm-cm). For instance, $t_0 = 0.27$ sec when $\varepsilon_s = 3$ and $\rho_d = 10^{12}$ ohm-cm. The starting time of back discharge from the initiation of current

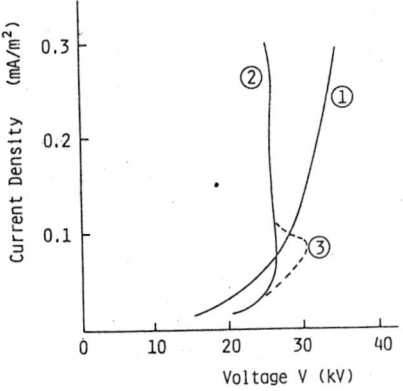

① Electrodes without Back Discharge
② Electrode with Back Discharge
③ Voltage Over-Shoot at rapid voltage rise

Fig. 3 V-I Characteristics of Precipitation Field

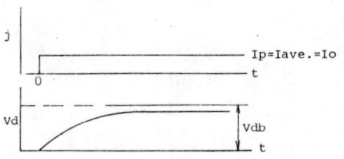

(b) Step response for conventional energization

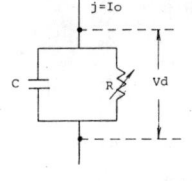

(a) Equivalent circuit of a dust layer

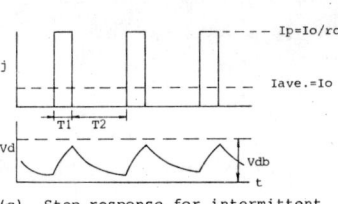

(c) Step response for intermittent energization

(V_{db}; insulation breakdown voltage of a layer)

Fig. 4 Equivalent circuit representing the electrical behaviour of a dust layer on an collecting electrode

supply can be expressed [1] from Eq. (1) and (2)

$$t_1 = t_0 \ln [1/\{1 - E_{ds}/(\rho_d j_0)\}] \quad (4)$$

For example, $t_1 = 0.109$ sec, when $t_0 = 0.27$ sec, $E_{ds} = 10kV/cm$, $j_0 = 0.3$ mA/m^2, $\varepsilon_s = 3$ and $\rho_d = 10^{12}$ ohm-cm. The t_1 is long enough to seitch thyristors on and off by a simple circuit [2].

The principle of the intermittent energization is therefore to control power supply deenergizing prior to the voltage V_d reaches the critical level V_{db} and reenergizing when V_d comes back to the initial low level, with the duty ratio $r_c = T_1/(T_1 + T_2)$ where T_1 is on and T_2 is off-time. refer to Fig. 4 (c)).

Applications and Operating Results

General View

Since the first commercial application of intermittent energization in 1979, more than 130 sets of infermittnet energization control units have been put into operation for new ESPs and existing ESPs.

Intermittent energization, in any applications, in any applications, gives remarkable power saving. And for high resistivity dusts, it gives improvement of collecting performance.
The effect can be classified into following two types.
Type I : Back discharge occurs for high restivity dusts ($10^{11} \sim 10^{13}$ $\Omega \cdot cm$) from iron ore sinter plants and low sulfur coal fired boilers.

ESP outlet dust burden decrease by 30 up to 50 percent less than that of conventional energization, and power can be saved by 50 up to 95 percents. Intermittent energization is effective for both collection performance and power saving.

Type II : No back discharge occurs for medium resitivity dusts (less than 10^{11} $\Omega\cdot$cm) from midium and high sulfur coal fired boilers, oil fired boilers, kraft pulp recovery boilers and cement plants etc. Collection efficiency keeps nearly same or slightly decreases as compared with that of conventional energization, however power can be saved almost 50 percents.

For examples of two types, Fig.5 (a) Fig.5 (b) hows typical V/I characteristics, and the improvement of collection performance in relation to duty ratio respectively. The improvement of collection efficiency is evaluated by the enhancement factor $H^{(4)}$ which is defined as the ratio of the modified migration velocity W_k for intermittent energization to that for conventional energization. (K = 0.5 in this paper.) Moreover, intermittent energization can save power more than that of decreasing current in conventional energization, as shown in Fig.6.

A1 High resistivity coal-fired Boiler (Type I)
A2 Medium resistivity coal-fired Boiler (Type II)
B   Oil-fired Boiler (Type II)
C   Sinter plant (Type I)

Coal Fired Boilers

Precipitability of fly ash depends generally on the dust resitivity due to sulfur contents of coal and alkali metal contents of ly ash. Japan with few coal resources is necessarily dependent on imports. Most of imported coals with low sulfur have high resistivity and significantly impairs ESP performance. More than 50 sets of ESPs have equipped with intermittent energization control units which have been functioned for fly ash from more than 80 kinds of coals.

Typical 4 operating results are shown in Fig.7, 8 and Tab.1, which are selected from a series of tests for a 33 MW boiler firing 10 kinds of coals. (3 of them classified into Type I and the rest into Type II.) In the case of coal B and C, elonging to Type I, with heavy back discharge, remarkable ESP performance enhancement is achieved by intermittent energization. Especially for coal B, approximately 1.5 of enhancement factor is achieved at the duty ratio 1/15 and its power consumption is reduced to only 5 percent of that as compared with conventional energization.
On the other hand, in the cases of coal A and D, belonging to type II , do not improve thier collection performance, however power consumption can be decreased a lot.

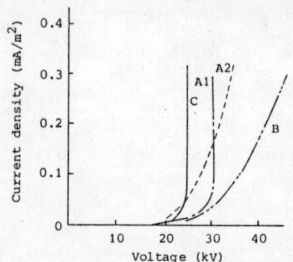
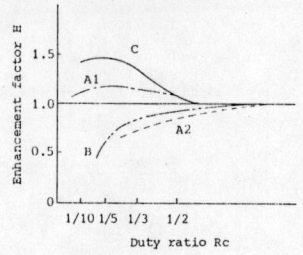

Fig. 5(a) V/I Characteristics

Fig. 5(b) Performance improvement in relation to duty ratio

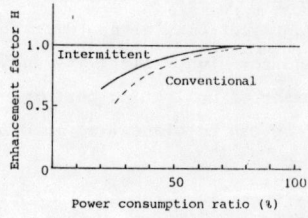

Fig. 6 Comparison of energy saving for typical data of Type II

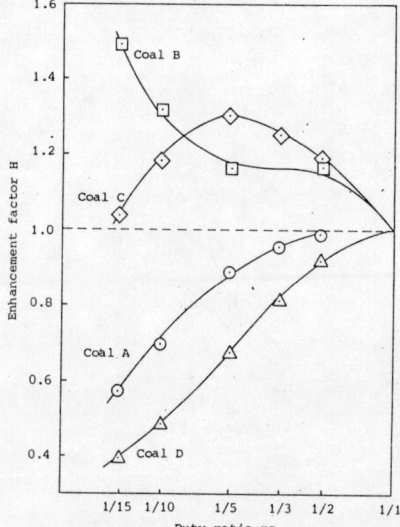

Fig. 7 Operating results of ESP performance improvement for a coal fired boiler relation to duty ratio

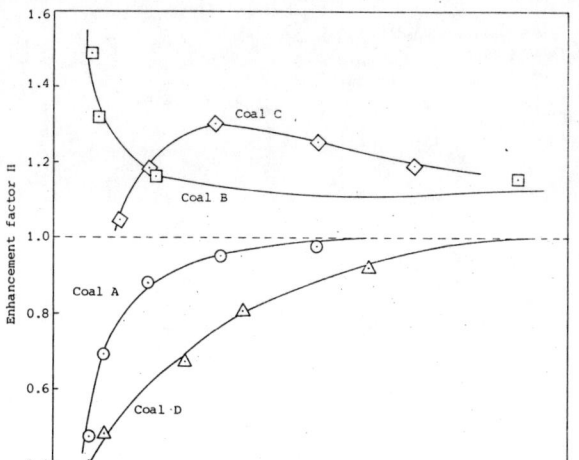

Fig. 8 Operating results of ESP performance improvement for a coal fired boiler in relation to power consumption ratio

Tab.1 ESP operating results for a coal fird boiler

Test No.		Test 1	Test 5	Test 6	Test 8
Coal		Coal A	Coal B	Coal C	Coal D
Product		North America	Austraria	Austraria	South Africa
Heat value	kcal/kg	6880	6470	6480	6590
Ash	%	11.42	17.2	8.1	13.4
Sulfur	%	0.55	0.51	0.32	1.09
Ash Composition					
SiO_2	%	59.0	78.6	58.6	44.6
Al_2O_3	%	18.9	13.5	25.6	26.5
Fe_2O_3	%	3.06	0.89	1.46	5.45
CaO	%	6.67	0.19	0.12	7.43
MgO	%	1.81	0.09	0.16	2.21
Na_2O	%	0.67	0.05	0.22	0.28
K_2O	%	1.14	0.40	0.08	0.51
SO_3	%	0.22	0.29	< 0.01	0.81
Wk value for conventional	cm/s	60.77	1.46	24.01	71.78
Wk value for intermittent	cm/s	59.36	2.17	31.30	66.08
Wk enhancement H		0.98	1.49	1.30	0.92
Duty ration		1/2	1/15	1/5	1/2
Type		II	I	I	II
Power consumption ratio (intermittent/conventional)		0.36	0.05	0.22	0.43

Oil Fired Boilers

Dust from oil fired boiler has comparably low resistivity. Intermittent energization gives only power saving effect as shown in Fig.9.

Kraft Pulp Recovery Boilers

Intermittent energization applied for recovery boiler gives only power saving effect due to its medium resistivity. Fig.10 shows that collection efficiency does not increase, but power can be saved more than that of conventional energization. Furthermore, intermittent energization is effective in dislodging the sticky and bulky dust by electrode's rapping.

Iron Ore Sinter Plants

Intermittent energization applied for iron ore sinter plants gives, withoutexception, both improvement of collection efficiency and power saving. Operating results from 7 retrofit plants

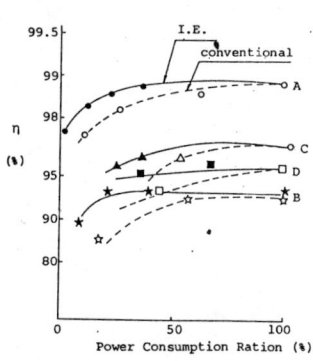

Fig. 9 Operating results of ESP for oil fired boilers

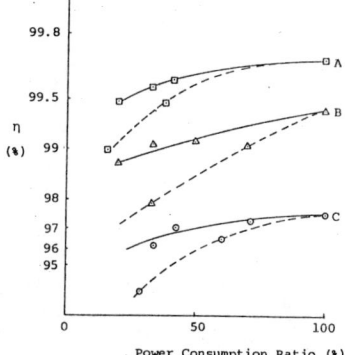

Fig. 10 Operating results of ESP for kraft pulp recovery boilers

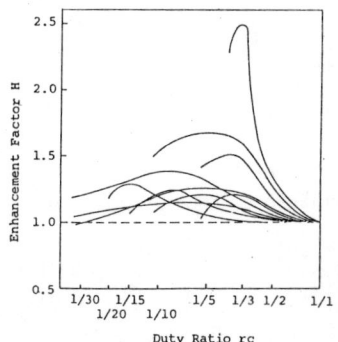

Fig. 11 Operating results of ESP for iron ore sinter plants

are plotted in Fig.11. In a certain plant, tests are done twice in summer and in winter respectively. It is interesting that the duty ratio for the maximum collection performance changes depending on the ambient atmosphere. As for power comsumption saving, the most remarkable reduction from 296 kW to 33 kW has been succeeded, and that makes retrofit cost of intermittant energization less than ESP operating cost per year.

Optimum Adjustment Technique

In order to optimize the collection performance of ESP, several methods related to intermittent energization have been developed, for example, as follows :

Full or Half Cycle Intermittent Energization

Intermittent energization was originally developed as that with full cycle intermittent waves, and later that with half cycle intermittent waves has been avilable. The comparison between full and half cycle waves is shown in Fig.12. When the dust resistirity is very high, half cycle is often more effective than full cycle, because, in the case of full cycle, Vd (voltage allotment of a dust layer) takes higher value with the charge of two continuous waves and thus back discharge easily occurs as compared with half cycle. Fig.13 shows one of the most effective operating results of half cycle waves obtained at the same plant as shown in Fig.7,8 and Tab.1 firing mixed coal (30 % of Coal B and 70 % of Coal C).

Vp·Vm Adjustment

Dust migration velocity W is roughly expressed by the following formula.[1] :

$$W \propto V_p \cdot V_m \quad (5)$$

where Vp = peak voltage of ESP, Vm = mean voltage of ESP. (Refer to Fig.14 that shows the relation between Wk and Vp·Vm obtained at the same plant as shown in Fig.7,8 and Tab.1 firing Coal A.) Therefore, the optimization of ESP performance is achieved when the product of Vp·Vm is maximized by the adjustment of parameters of intermittent energization such as, for example, the duty ratio.

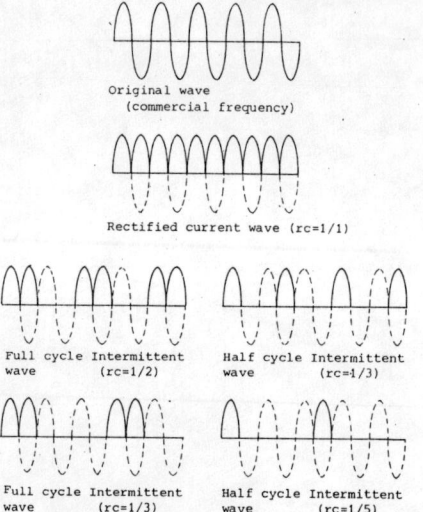

Fig. 12 Primary current wave of full cycle and half cycle intermittent energization

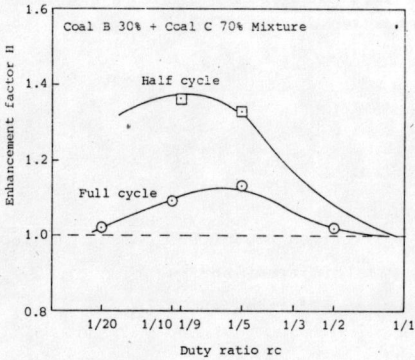

Fig. 13 An operating result of ESP performance improvement by half cycle intermittent energization

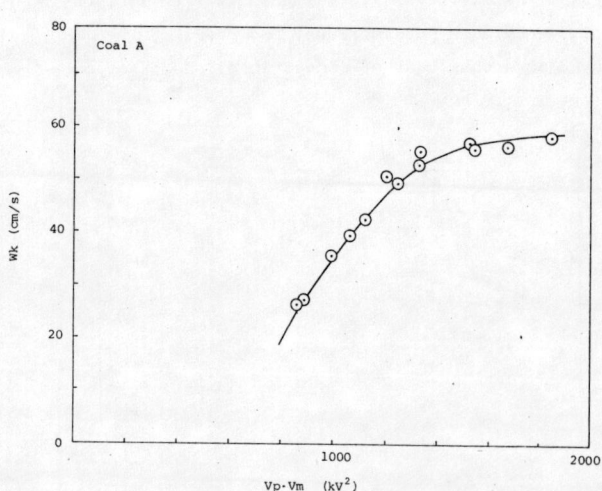

Fig. 14 Relation between Wk and Vp·Vm

ESP Emission Control System

Recently, it is often required to control the emission systematically among the waste gas treating equipment such as DeNOx, DeSOx, and ESP. Under such a situation, to keep and optimum and/or certain constant value of ESP outlet burden is requested more and more.

One of the solution is the control of duty ratio of intermittent energization, using micro computer, with feedback signal of opacity meter installed at ESP outlet duct. The flow and control model is shown in Fig.15 and an operating results is shown in Fig.16. In this case, after starting control, ESP emission is controlled to be kept within a fixed range by changing duty ratio from 1/61 to 1/3 in accordance with boiler operation conditions, such as, transience of the boiler load and soot blow operations. Duty ratio control has an advantage in saving power consumption as compared with current control of conventional energization, as shown in Fig.6.

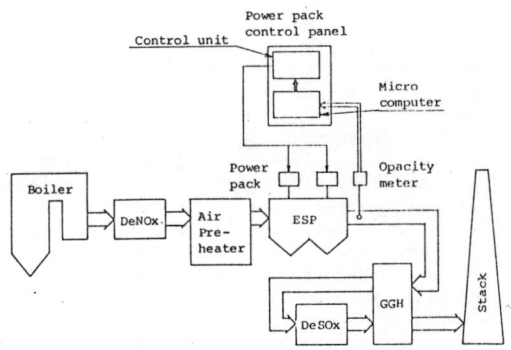

Fig. 15 Flow and control model of ESP emission control system

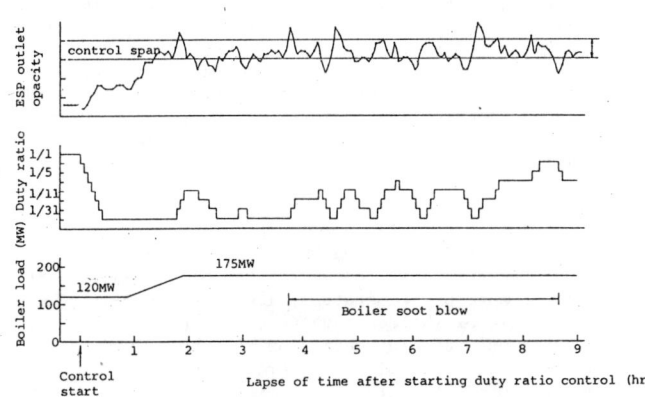

Fig. 16 Operating result of duty ratio control

Conclusion

1. Intermittent energization is one of the most economical means to prevent or supress back discharge in ESP.
2. More than 130 sets of control unit with intermittent energization are pplied in commercial plants since 1979.
3. The effect of Intermittent energization is saving power consumption and, for high resistivity dusts, increase of collection efficiency.
4. Half cycle intermittent energization is often more effective in ESP performance than full cycle even at the same duty ratio.
5. The effect of performance enhancement differs depending on dust properties for each application.
   In iron ore sinter plants, the maximum improvement of enhancement factor is 2.5, and in low sulfur coal fired boilers, 1.5.
   In medium dust resitivity coal fired boilers, kraft pulp recovery boilers, oil fired boilers and cement plants, collection performance does not improve, however does gain more than 50 percent of power saving under duty ratio 1/2.
6. ESP emission control system has been succeeded, which controls the duty ratio of ESP intermittent energization to keep an optimum and/or certain constant value of ESP outlet dust burden in accordance with boiler operation conditions such as change of boiler load, soot blowering etc,.

References

(1) T. Ando, N. Tachibana, Y. Matsumoto., A new energization method for electrostatic precipitators Mitsubishi intermittent energization., Proc. of the 4th symposium on the transfer and utilization of particulate control technology. Houston Tx. USA, Oct. 1982.

(2) H.J. White., Industrial electrostatic Precipitation. Addison - Wasley publishing Co, 1963., pp.325-326

(3) T. Ando, US patent 4,410,849., Oct 18, 1983

(4) Paul L. Feldman and Helmut I. Milde, Pulsed energization for enhanced electrostatic precipitation in high-resistivity application, symposium on transfer and utilization of particulate control technology, July 1978.

ADVANCED ELECTROSTATIC APPLICATIONS TO FLUE GAS CLEANING

Prof. Massimo Rea,
Department of Electrical Engineering,
University of Padova, Italy

Abstract.

During the last ten years the electrostatic precipitator performances have achieved a significant improvement as consequence of improved electrical energization: i.e. by applying voltage waveforms composed of short pulses superimposed to a continuous value.

The use of pulse voltages in electrodic structures similar to those used for ESP leaded to recognize the possibility to use the pulse corona discharge as mean for the removal of several gaseous pollutants such as NO_x, SO_2, mercury vapours, etc.

This communication aims to shortly inform about the state of the art of those advanced electrostatic technologies and to suggest further investigations and interesting applications of electrostatics to flue gas cleaning.

The improvement of ESP performances

Since several years it has been recognized that back corona which develops during the collection of high resistivity dust is the main responsible of the reduced performances of precipitators in several field of application such as in the cement industry and in power stations burning of low sulphur coals. Following such an understanding, the idea developed to apply more sophisticated electric energization in order to inhibit the onset of back corona.

In the second half of seventies prof. Senichi Masuda of the University of Tokyo worked out a number of technical solutions for controlling back corona, all based on the ideal separation of the two basic mechanisms occurring into the electrostatic precipitator: the electrical charging of particles and their transport onto the plate.

The production of space charges, needed for the charging of particles is better achieved through an impulse corona, i.e. through the corona produced by a voltage pulse; on the other side the drift of particles to the plate is better achieved with a continuous voltage of level low enough not to produce a strong electric field into the dust layer which forms on the plate surface.

Already in 1956 prof. Harry White suggested to use pulsing voltages for the energization of ESP, but at that time the technology was not able to produce economically and reliably enough suitable pulse voltage generators.

Following the investigations of prof. Masuda, at the beginning of eighties, manufacturers developed industrial equipments able to supply industrial electrostatic precipitators with a voltage waveform composed by a number of pulses superimposed to a continuous value.

From the conceptual point of view, the electric circuit used to produce such voltage waveforms is an oscillating circuit formed by a storage capacitor, a resonant inductor and the capacitance of the precipitator; a suitable switching device closes the circuit for a single period of current, thus producing on the capacitance of the precipitator a voltage pulse as described in figure 1. The pulse generator is finally coupled though a condenser to a DC voltage generator.

Technically two alternative are available: the first uses a low voltage switch device and a step-up transformer, the second uses a high voltage switch avoiding the transformer (see fig. 2a and 2b).

In Table I the general characteristics of a commercial pulser for ESP are reported as an example.

Table I: general characteristics of a pulser for ESP

DC base voltage	40 to 50 kV
Crest value of the voltage pulse	50 to 60 kV
Frequency of the voltage pulses	10 to 200 Hz
Duration of a single pulse	50 to 250 s

Using such pulsing voltage waveforms, several experiments have been carried out first on pilot plants and finally on industrial units. The results were always positive even if not so much emphasized probably because of the reluctance of manufacturers of precipitator to change the consolidated design procedures.

The most complete and thorough tests on the application of pulsers to an industrial process are probably those carried out by ENEL (the Electric Power Authority in Italy), with the scientific consultancy of the Department of Electrical Engineering of the University of Padova, at the Marghera Power Station and recently presented during the Third International Conference on Electrostatic Precipitation which was held in Abano (Italy) between october 25-29, 1987 [1]. They can be summarized as in Table II.

It is evident that the improvement of collecting performances strongly depends on the characteristics of coal and particularly on its resistivity. When the dust resistivity is over 10^{12} ohm cm the effective migration velocity increases by a factor between 2 and 2.5 while the electric power decreases by a factor 3 to 5 following the otimization of emitting electrodes with conventional energization. In case the resistivity is below 10^{12} ohm cm the results show that it is possible to reach the same removal efficiency with two third of precipitator length and a power consumption which decreases by a factor between 2 and 5.

Even if a fair good knowledge has been reached by manufacturers how to control such a sophisticated voltage waveform generators; i.e. how to adjust the DC voltage level the crest value of the pulses and their repetition rate for obtaining the best precipitator performances, still a further development is needed and expected. Also further investigations are supposed in order to otimize the electrodic structure to this new form of electric energization. However, provided that the major cost of the electric power supply is easily recovered by the lower energy consumption, a large use of such devices is expected in the next future.

The pulse plasma induced chemical processes (PPCP)

Some decades ago a number of scientific investigations were carried out in order to understand the chemical reactions induced by the irradiation of a flue gas with an Electron-Beam. The results showed

Table II: Summary of results obtained by ENEL [1].

dust resistivity $\times 10^{11}$ ohm cm	10 - 39	25 - 57	3 - 7
input dust conc. gr/Nm^3	13.6	11.7	6
outlet dust conc. with conv. energ. mgr/Nm^3	155	500	23
outlet dust conc. with pulse energ. mgr/Nm^3	23	46	20$^{(1)}$
power consumption with conv. energ.$^{(2)}$ kW	40 - 26	51 - 44	37 - 14
power consumption with pulse energ. kW	8	10	7(1)

(1) results obtained energizing two fields instead of three.
(2) first figure refers to standard pin emitting electrodes, second one refers to increased diameter round emitting electrodes.

that the high energy electrons produce a number of active radicals and ions such as HO, HO_2, O^+ ecc. which induce a series of chemical reactions leading to the oxidation of NO and SO_2 and the formation of nitric and sulphuric acids. In case of ammonia injection, ammonia sulphate and nitrate are formed as solid particles which can be quite easily removed and used as fertilizers [2], [3].

The corona discharge is also a source of electrons, the energy of which depends on the electric field at the moment of the onset of the discharge; using very fast rising voltage waveforms the energy of the electrons produced by the corona discharge are in the range 10 to 15 eV. These energy levels are high enough for inducing the formation of the mentioned radicals thus suggesting corona as a suitable source of electrons for the removing of gaseous pollutants from the industrial effluents.

Actually, experiments carried out with a pulse corona reactor supplied with 1000 Nm^3/h of flue gas from a coal burning boiler suggest the technical feasibility of the process while economical projections suggest its competitiveness in comparison with the traditional processes for removing the SO_2 only [4].

The results of the experiments, carried out by ENEL at the Marghera Power Station with the scientific consultancy of the Department of Electrical Engineering of the University of Padova, are synthetically reported in fig. 3 where the NO_x removal efficiency is plotted, for different initial concentrations, versus the energization of the gas by the corona discharge.

Recently prof. Senichi Masuda reported the results of similar experiments carried out on an electrostatic precipitator installed downstream of a refuse incinerator in Tokyo [5]; besides similar results when $DeNO_x$ and $DeSO_x$ are concerned, a very effective removal of mercury vapour has been remarked, suggesting a wider application of the process.

References

[1] G.Dinelli, F.Mattachini, V.Bogani, A.Baldacci, R.Tarli
Industrial Demonstration of Pulse Energization on Electrostatic Precipitators.
Third Int. Conf. on Electrostatic Precipitation, Abano (Padova), oct. 1977.

[2] N.W.Frank, K.Kawamura, G.A.Miller
Design Notes on Testing Conducted During the Period June 1985 - September 1986 on the Process Demonstration Unit at Indianapolis, Indiana.
IAEA-TECDOC-428, Report of a Consultants Meeting held in Karlsruhe, oct. 1986

[3] P-Fuchs, B.Roth, U.Schwing
Removal of NO_x and SO_2 by the Electron Beam Process.
IAEA-TECDOC-428, Report of a Consultants Meeting held in Karlsruhe, oct. 1986

[4] L.Civitano, G.Dinelli, M.Rea
Removal of NO_x and SO_2 from Combustion Gases by Means of Corona Energization.
Third Int. Conf. on Electrostatic Precipitation, Abano (Padova), oct. 1977.

[5] S.Masuda, Y.Wu, T.Urabe, Y.Ono
Pulse Induced Plasma Chemical Process for $DeNO_x$ and Mercury Vapour Control of Combustion Gases.
Third Int. Conf. on Electrostatic Precipitation, Abano (Padova), oct. 1977.

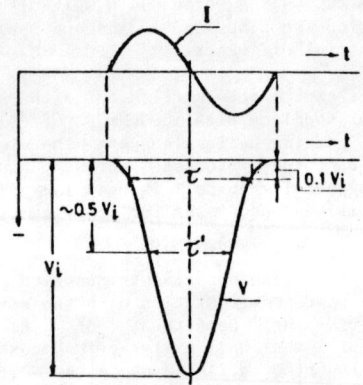

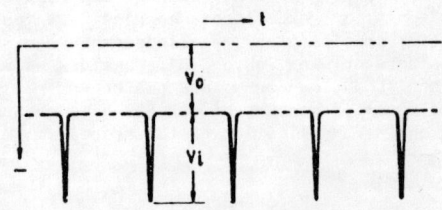

Fig. 1: Typical current and voltage waveshapes produced by a pulse generator based on an oscillating circuit concept.

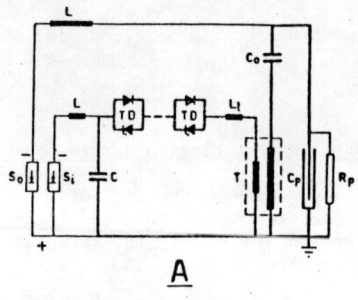

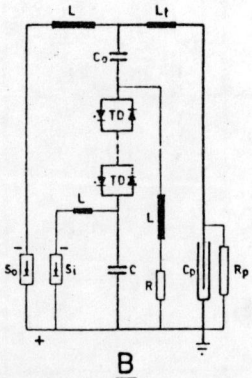

Fig. 2: Schemes of pulse voltage generators
a) using a voltage step-up transformer
b) using a high voltage thyristor switch

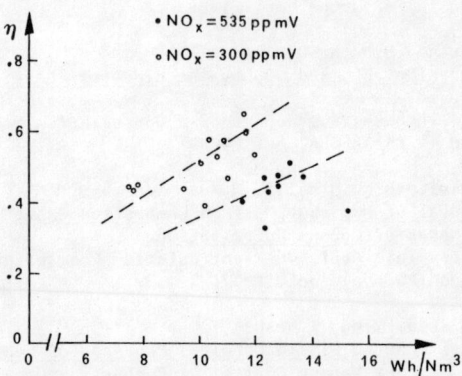

Fig. 3: NO_x removal efficiency as function of energization. Concentration of SO_2: initial = 450 ppmv, final 60 ppmv. Ammonia injected = 10000 ppmv

ELECTRON IRRADIATION OF FLUE GAS BY PULSED CORONA FOR SULFUR AND NITROGEN OXIDE REDUCTION

Sabert Oglesby, Jr. and Edward B. Dismukes
Southern Research Institute
P.O. Box 55305, Birmingham, Alabama 35255-5305 U.S.A.

ABSTRACT

An alternative method to use of scrubbers and selective catalytic reduction for removal of gaseous pollutants is electron irradiation of flue gases from combustion processes. Two methods of electron irradiation that have been studied both theoretically and in experimental facilities include electron beam and pulsed corona systems. Electron beam irradiation involves electrons with high initial energies - of the order of 10^5 to 10^6 electron volts. Pulsed corona systems, on the other hand, produce electrons with energies in the range of 2 to 20 electron volts. Operationally, the two processes differ in that in E-beam systems each electron has enough energy to produce large numbers of reactive species for pollutant removal. Pulsed corona systems, on the other hand, require that each electron be re-energized after impact to an energy level sufficient to generate active species in sufficient numbers.

Because of the high energy levels in E-beam systems, it has been estimated that about half of the electrons produce ions and half produce radicals and excited molecules. Ion generation requires electron energies of the order of 15 to 20 eV whereas radicals produced by dissociation of molecules requires energies in the range of 2 to 5 eV. Since pulse corona systems operate at lower electron energies, most of the active species are produced by dissociation of molecules.

Efficiencies of electron irradiation systems are reported as G values defined as the number of pollutant molecules removed per 100 electron volts. Thus for E-beam systems, the maximum theoretical G value would be around 10 assuming one reactive specie is required for each molecule of pollutant removal. Because of the lower energy required for pulsed corona systems to produce an active specie, the maximum G value required for pollutant removal would be around 20.

Reaction with OH radicals is assumed to be the only significant mechanism for SO_2 reduction. Recent studies of reactions in the atmosphere indicate that the conversion of SO_2 to sulfuric acid involves a process for regeneration of OH radicals, and analysis of the data would appear to indicate that similar regeneration of OH radicals may occur in flue gas systems. If so, it would not be necessary to generate an OH radical for each molecule of SO_2 and hence the theoretical G value limit for both E-beam and pulsed corona systems could be higher. Offsetting this higher G value is the effect of competing reactions that remove available OH radicals from the process.

Most of the research on E-beam systems has been concerned with nitrogen oxide removal. The chemistry for conversion of nitric oxide (NO) to nitric acid has not been clearly established. Evidently, however, the chemistry involving NO is far more complex than that involving SO_2; that is, more than one active specie participates in NO removal, whereas only the OH radical is important for SO_2.

Pulsed corona systems reported in the literature utilize spark gap pulsers of the type developed by Masuda. Although these pulsers are convenient for laboratory studies, they are too inefficient as means for operating high voltage, short duration pulse supplies for commercial sulfur and nitrogen oxide removal systems.

Most of the experimental studies of sulfur and nitrogen oxide removal systems utilizing pulsed corona systems indicate G values of less than 10 for sulfur oxide and less than 5 for nitrogen oxide.

Southern Research Institute has built an experimental pulser based on spark gap principle for studies of pulsed corona system of electron irradiation. Work with this facility is very preliminary at this time. Initial data indicate that optimization of the pulse characteristics to achieve maximum efficiency is necessary as is measurement of the energy actually delivered to the reactor. Work is continuing in an effort to define conditions necessary to achieve maximum reduction of pollutant level at minimum energy input.

Discussion on the Correction Formula for Deutsch Formula Related to Collection Efficiency of an Electrostatic Precipitator

K.Fujimura,[1] T.Adachi[2] and M.Akazaki[3]

1) Dept. of Product Mech. Eng., Faculty of Eng., Yamaguchi University,
   2557, Tokiwadai, Ube, Japan 755
2) Dept. of Elec. Eng., Faculty of Eng., Oita University,
   700, Dannoharu, Oita, Japan 870-11
3) Graduate school of Engineering Science, Kyushyu University,
   6-1, Kasuga-Koen, Kasuga, Japan 816

Abstract

Correction formulas of Deutsch formula for calculating the collection efficiency of an industrial ESP have some constants, and these constants change depending on the distribution factor of dust particle size. The relation between distribution factor and value of constant, however, has not been clarified.

In order to know the value of constant suitable for arbitrary distribution factor of dust particle size, the effect of distribution factor on the constant was calculated. The suitable value of constant for given dust can be obtained easily by using the diagrams derived from this calculation.

Introduction

Deutsch formula for calculating the collection efficiency of an Electrostatic precipitator (ESP) were induced for dust composed of same size particles. Then, some corrections were attempted in order to apply this formula to the plate type industrial ESP collecting the dust of wide range particle size distribution.

Matts formula is generally used as the correction formula for Deutsch formula, and the authors also proposed other type one in the previous paper.

Each formula has one constant respectively, which changes depending on the distribution factor of dust particle size. In the case of using these correction formula, abundant experience or know-how are required to determine the value of constants, because relations between these constants and distribution factor were not made clear numerically.

In order to know the value of constants suitable for arbitrary distribution factor of dust particle size, the effect of distribution factor on constants of correction formulas was calculated.

Correction formula for Deutsch formula

Deutsch formula[1] which is original equation for the collection efficiency of ESP is

$$\eta = 1 - \exp(-wf), \quad (1)$$

where η is collection efficiency, w is migration velocity of dust particle(m/s) and f is specific collection area(SCA) (s/m).

In this paper, two correction formulas for Deutsch formula are studied as follows
(1) Matts formula[2]

$$\eta = 1 - \exp[-(w_M f)^{K_M}] \quad (2)$$

or

$$P = 1 - \eta = \exp[-(w_M f)^{K_M}], \quad (3)$$

where P is penetration of dust, w_M is apparent migration velocity(AMV) of dust(m/s) and K_M is constant connected with distribution characteristics of dust particle size.

(2) New correction formula

$$\eta = 1 - \exp[-(w_F f)^{(f_0/f)^{K_F}}] \quad (4)$$

or

$$P = 1 - \eta = \exp[-(w_F f)^{(f_0/f)^{K_F}}]. \quad (5)$$

This formula was proposed by authors in the previous paper[3]. Where, w_F is AMV of dust(m/s), K_F is constant connected with distribution characteristic of dust particle size, and f_0 is constant which has same dimension with f. Eq.(4) agrees with measured value of collection efficiency for fly ash, where the value of K_F is 0.166 and f_0 is 30s/m. On the Matts formula the value of w_M and K_M must be changed in response to the change of SCA f, even if the same dust is collected. On the other hand, values of w_F and K_F of the new correction formula need not be changed according to the change of value of f on the same dust.

Calculation of particle size distribution

It is assumed that dust particle size distribution is logarithmic normal distribution. Then, distribution characteristics is expressed by straight line A-A as showen in Fig.1, which horizontal axis is logarithmic scale of particle diameter D_P and vertical axis is probability scale of residual rate R.

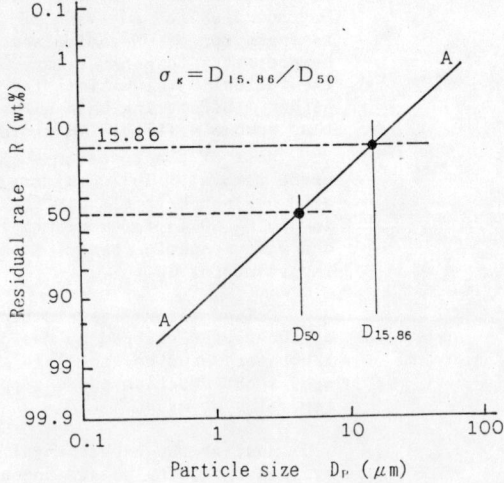

Fig.1. Representation of logarithmic nomal distribution.

Position of straight line A-A is determined by geometric standard deviation σ_K and particle diameter of 50 weight percent(wt%) residual rate D_{50}. Then, distribution characteristic of dust particle size is described by one pare of value of σ_K and D_{50}.

When the values of σ_κ and D_{50} are given, residual rate of arbitrary particle size D_P is calculated as follows

$$R \leq 50 \text{ wt\%}; \quad R=\phi(y)\times 100 \text{ \%}, \quad (6)$$
$$R \geq 50 \text{ wt\%}; \quad R=[1-\phi(y)]\times 100 \text{ \%}, \quad (7)$$

where, $\phi(y)$ is normal distribution function on y, and approximated as follows

$$\phi(y) \doteq 1-\frac{1}{2}(1+d_1y+d_2y^2+d_3y^3\cdots+d_6y^6)^{-16}, \quad (8)$$

where, $y \geq 0$ and

$d_1 = 0.04986\ 73470$ $d_4 = 0.00003\ 80036$
$d_2 = 0.02114\ 10061$ $d_5 = 0.00004\ 88906$
$d_3 = 0.00327\ 76263$ $d_6 = 0.00000\ 53830$,

$$y = \frac{\log(D_P/D_{50})}{\log \sigma_\kappa}. \quad (9)$$

Particle size distribution on actual dust

Values of σ_κ and D_{50} on several actual dust were calculated from the data obtained in literature, and plotted in Fig.2. In this figure, inside of square frame A-B-C-D is the practical range of σ_κ and D_{50}, and if the subject of study is limited to fly ash, the necessary range is reduced to inside of frame E-F-G-H.

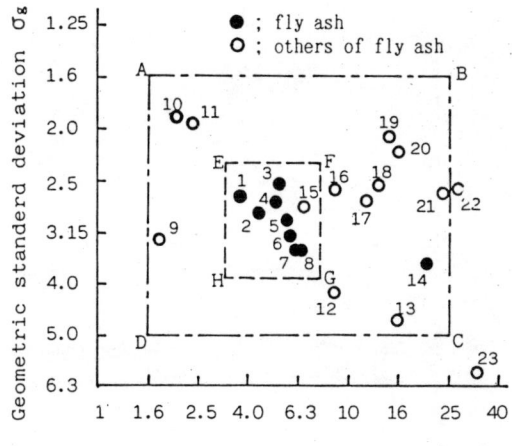

Fig.2. Value of σ_κ and D_{50} for typical dust.

Individual particle size calculation method

Calculating procedure

Collection efficiency corresponding to σ_κ and D_{50} of a given dust is calculated by the procedure which called "Individual particle size calculation method" as follows -
(1) Distribution range of dust particle size is divided into a large number of band.
(2) Distribution rate of individual particle size is calculated by Eq.(4), which shows wt% of dust particles in each band. Fig.3 is a calculated exampl for distribution rate,
(3) Partial collection efficiency of each band are calculated according to given SCA by assuming all particles in same band have same diameter. In this calculation, Deutsch formula of Eq.(1) is used.
(4) The products of distribution rate of individual particle size and partial collection efficiency are calculated for each band.
(5) Sum of these products is collection efficiency for given dust on the given SCA.
(6) f~η characteristic is obtained by repeat of same procedure chaining the value of SCA.

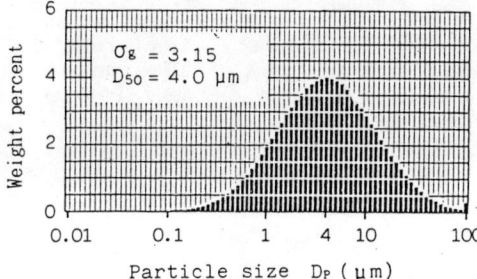

Fig.3. Calculated example for individual particle size distribution rate.

Dividing of distribution range of dust particle size

Distribution range of dust particle size is divided into 81 pieces of band according to preferred number[4],[5] as Table 1. Where, No.1 band includes the all particles under 0.0106 μm of diameter and No.80 band includes the all particles above 95 μm.

Table 1. Dividing of distribution range of dust particle size.

Band No.	Border diameter (μm)	Mean diameter (μm)
0		0.0100
1	0.0106	0.0112
2	0.0118	0.0125
3	0.0132	0.0140
	0.0150	
:	:	:
19		0.090
20	0.095	0.100
21	0.106	0.112
22	0.118	0.125
	0.132	
:	:	:
78		80.0
79	85.0	90.0
80	95.0	100.0

Migration velocity of particle

Migration velocity w in the Eq.(1) is calculated as[6]

$$w=[E_0E_PD_P/4\pi\mu\ (1+A\frac{2\lambda}{D_P}), \quad (10)$$

where E_0 is charging field, E_P is precipitating field, D_P is diameter of dust particle and μ is viscosity of gas. Inside of ( ) is the Cunningham correction factor, and D_P and λ are constants concerned to gas. Then let

$$k=E_0E_P/4\pi\mu, \quad (11)$$

Eq.(10) may be written as follows

$$w=kD_P\ (1+A\frac{2\lambda}{D_P}), \quad (12)$$

where units of variables are as follows: E_0 and E_P; kV/cm, μ;P, w;m/s, k;1/s, D_P;μm, λ;μm and A;non dimension.

k is coefficient of particle migration velocity. In this paper, the value of k is put at 0.05 1/s, which is equivalent value to standard operating range of ESP, and A is put at 0.86 and λ is 0.1 μm.

Results of calculation

Relations between Matts formula's constant and distribution characteristics of dust particle size

According to given values of σ_κ and D_{50}, f~η curve was calculated by individual particle size calculation method, and Matts formula's $f \sim \eta$ curve was calculated under the condition that this curve passes through the points A_1 and A_2 on the former $f \sim \eta$ curves as shown in Fig.2.

In this case, constant K_M and AMV w_M are calculated by the following equations, by using SCA f_1, f_2 and penetration P_1, P_2 on points A_1, A_2,

$$K_M = \ln(\ln P_1/\ln P_2)/\ln(f_1/f_2), \quad (13)$$

$$w_M = (-\ln P_1)^{1/K_M}/f_1 \quad (14)$$

or

$$w_M = (-\ln P_2)^{1/K_M}/f_2. \quad (14)'$$

These two $f \sim \eta$ curves do not coincide strictly speaking, and have the shift of A, B and C as shown in Fig.4. However, if f_1 is 40s/m and f_2 is 120s/m, these shifts become very small, and two curves coincide practically. Accordingly, the calculated values of K_M and w_M when f_1 is 40s/m and f_2 is 120s/m are taken to be the values of constant and AMV corresponding to given σ_g and D_{50}.

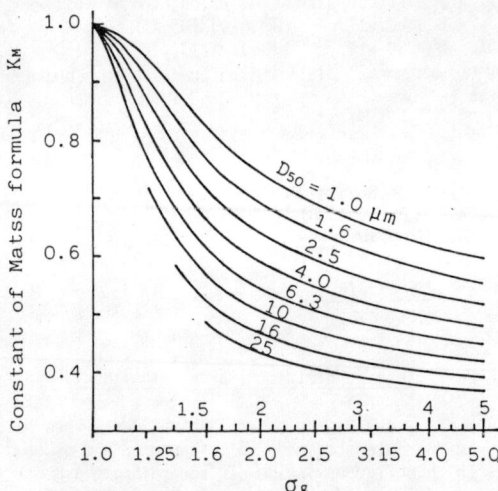

Fig.5. Diagram for determining K_M from σ_g and D_{50}.

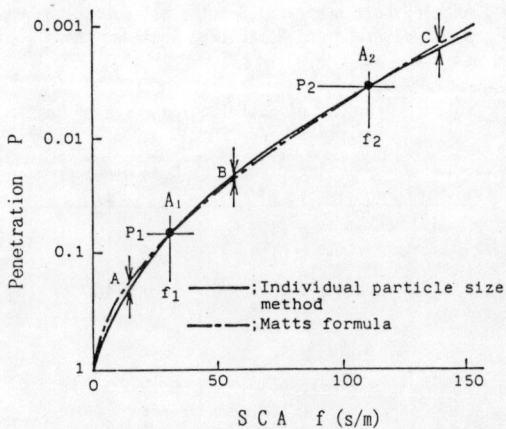

Fig.4. General characteristic of Matts formula.

Table 2 is results of calculation for K_M changing the values of σ_g and D_{50}. From this table, following items was clarifided.

(1) The value of Matts formula's constant K_M changes according to the values of σ_g and D_{50} which is the characteristics of dust particle size distribution.

(2) When σ_g equals 1, the value of K_M becomes 1 independently of values of D_{50}.

(3) Square flame in Table 2 is equivalent to the flame E-F-G-H in Fig.2 as the range of σ_g and D_{50} for fly ash. The values of K_M in the flame agree well with experimentally admitted value of 0.5 for fly ash on the large SCA range[7].

Fig.5 and Fig.6 are diagrams for determining the value of K_M and w_M. From these diagram, the values of K_M and w_M corresponding to the values of σ_g and D_{50} are decided easily.

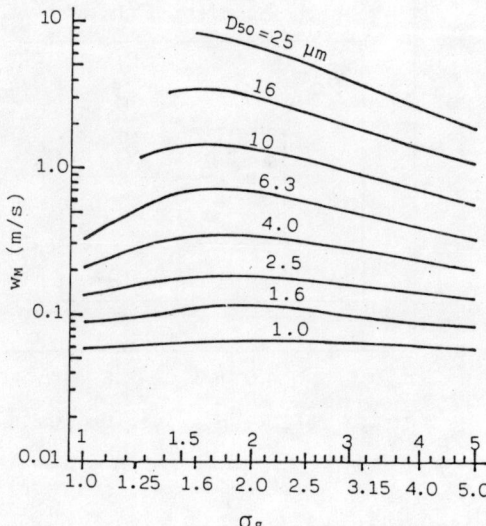

Fig.6. Diagram for determining w_M from σ_g and D_{50}.

Relations between new correction formula's constant and distribution characteristics of dust particle size

The constant K_F for New correction formula is calculated by same method as the case of Matts formula. However, $f \sim \eta$ curves of new correction formula agree with the $f \sim \eta$ curves of individual particle size calculation method only in the range of large SCA, therefore, f_1 is taken as 100s/m and f_2 is 120s/m.

The results of calculation are shown in Table 3. From this table, following items were clarified.

(1) The value of new correction formula's constant K_F changes according to the values of σ_g and D_{50}.

(2) When σ_g equals 1, the value of K_F becomes 0 independently of values of D_{50}.

(3) Square flame in Table 3 is equivalent to the flame E-F-G-H in Fig.2 as the range of σ_g and D_{50} for fly ash. The values of K_F in the flame agree with 0.166 for fly ash which

Table 2. Results of calculation for K_M changing the values of σ_g and D_{50}.

σ_g	D_{50} (μm)					
	1.6	2.5	4.0	6.3	10	16
1.0	1.000	1.000	1.000	1.000	---	---
1.25	0.905	0.867	0.821	0.772	0.718	---
1.6	0.772	0.720	0.665	0.615	0.568	0.523
2.0	0.692	0.641	0.591	0.546	0.505	0.467
2.5	0.639	0.591	0.546	0.506	0.470	0.435
3.15	0.600	0.557	0.515	0.479	0.445	0.415
4.0	0.572	0.532	0.494	0.460	0.329	0.401
5.0	0.553	0.516	0.480	0.448	0.419	0.392

confirmed by authors experimentally[3]. Fig.7 and Fig.8 are diagrams for determining the value of K_F and w_F from σ_κ and D_{50}.

Table 3. Results of calculation for K_F changing the values of σ_κ and D_{50}.

σ_κ	D_{50} (μm)					
	1.6	2.5	4.0	6.3	10	16
1.0	0.000	0.000	0.000	0.000		
1.25	0.0374	0.0460	0.0550	0.0632	0.0711	
1.6	0.0859	0.0953	0.103	0.110	0.115	0.119
2.0	0.117	0.125	0.131	0.136	0.140	0.142
2.5	0.140	0.147	0.152	0.156	0.159	0.161
3.15	0.158	0.164	0.169	0.172	0.175	0.176
4.0	0.172	0.178	0.183	0.186	0.188	0.189
5.0	0.184	0.189	0.194	0.197	0.198	0.200

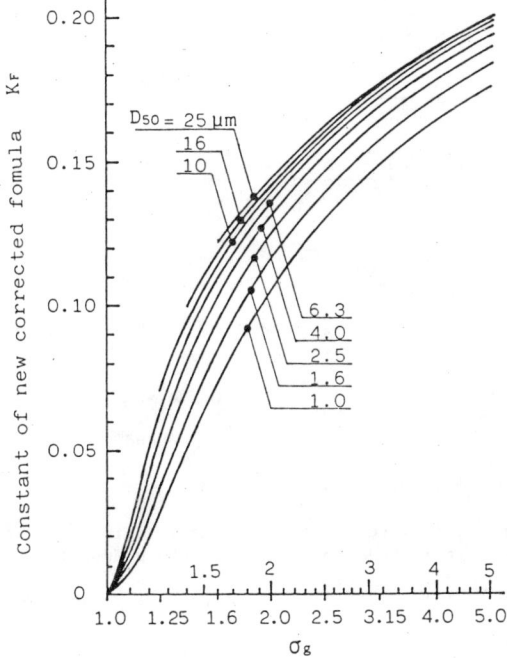

Fig. 7. Diagram for determining K_F from σ_κ and D_{50}.

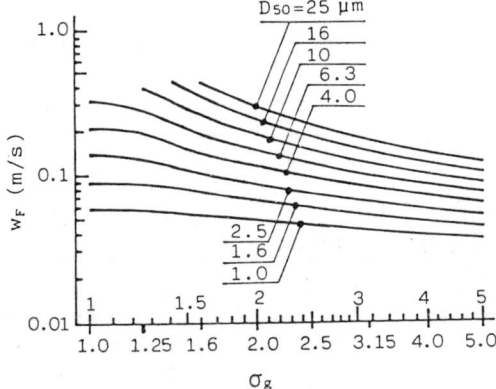

Fig. 8. Diagram for determining w_F from σ_κ and D_{50}.

w_M and w_F determined from Fig.6 and Fig.8 are equivalent to the values which calculated as the coefficient of particle migration velocity k given by Eq.(12) is 0.05 1/s. Therefore, values of w_M and w_F change when value of k changes from 0.05. However, the values of K_M and K_F do not change practically by the change of k. K_F and w_F calculated by above mentioned procedure can be used on any value of SCA, however, K_M and w_M must be used only in the range of large SCA (about 80s/m or more).

Conclusion

Relations between characteristics of dust particle size distribution and values of constants of Matts formula and new correction formula were studied. In this calculation, individual particle size calculation method was used assuming that dust particle size distribution is logarithmic normal distribution. Conclusion of study are as follows.
(1)The values of correction formula's constants K_M and K_F change following to change of dust particle size distribution.
(2)Calculated values of K_M and K_F for fly ash agree with performance data.
(3)The values of K_M and w_M concerning given values of σ_κ and D_{50} are known from Fig.5 and Fig.6, and K_F, w_F are known from Fig.7, Fig.8.
(4)K_F and w_F calculated by above mentioned procedure can be used on any value of SCA, but, K_M and w_M must be used only on the range of large SCA.

Acknowledgements

The autheres are pleased to acknowledge the continuous encouragement they have recived from Prof. Dr. M.hara of Kyusyu University and Prof. Dr.Y.Nomoto of Oita University.

Nomenclature

- D_P ; Diameter of dust particle (μm)
- D_{50}; Particle diameter of 50wt% residual rate (μm)
- E_O, E_P ; Charging field and precipitating field (kV/cm)
- f_O ; Constant in the new correction formula (s/m)
- K_M, K_F ; Constant in the Matts formula and new correction formula (non dimension)
- k ; Coefficient of particle migration velocity (1/s)
- P ; Penetration of dust (non dimension)
- w ; Migration velocity of dust particle (m/s)
- w_M, w_F ; Apparent migration velocity of dust in the Matts formula and new correction formula (m/s)
- η ; Collection efficiency.(non dimension or %)
- μ ; Viscosity of gas (P)
- σ_κ ; Geometric standard deviation (non dimension)
- $\phi(y)$; Normal distribution function on y (non dimension)

References

[1] W.Deutsch: Ann. der physik.,68(1922) 335
[2] S.Matts: Proc. of CSIRO Conf. on Electrostatic precipitator, 3-1(1978)
[3] K.Fujimura and T.Adachi: Proc. of the Inst. of Electrostatics Japan, 6,5(1982) 312
[4] Japanese Industrial Standards Committee:Japanese Industrial Standard (Jis),Z 8601,Japanese Standards Association, Tokyo (1954)
[5] K.Fujimura: Journal of Japan Society for Design & Drafting, 20,125 (1985) 1
[6] H.J.White:Industrial Electrostatic Precipitation, p.155, Addison-Wesley, Massachusetts (1963)
[7] The Institute of Electrostatics Japan: Handbook of Electrostatics, p.507, Ohm sha, Tokyo (1981)

SIMILARITY PHENOMENA OF ELECTROHYDORODYNAMIC FLOW FIELD INSIDE ESP

T. Adachi[1], T. Ohkubo[1], Y. Nomoto[1], J.S. Chang[2], and M. Akazaki[3]

1) Dept. of Elect., Eng. Faculty of Eng., Oita University, 700, Dannoharu, Oita, Japan 870-11
2) Dept. of Eng. Phys., Faculty of Eng., McMaster University, Hamilton, Ontario, Canada L8S 4M1
3) Graduate School of Engineering Science, Kyushu University, 6-1, Kasuga-Koen, Kasuga, Japan 816

ABSTRACT

The flow electro-hydrodinamic field inside an Electrostatic Precipitator (ESP) is usually effected by ionic wind produced by corona discharge. In this paper, the effects of corona discharge on the flow field are studied numerically in a wire-duct type ESP with 20 cm plate spacing, where the flow field inside ESP is numerically analyzed by using Navier-Stokes equation. Time-dependent vorticity equation is calculated by alternating direction implicit method (ADI) while Poisson's type equation for stream function by finite element method. The boundary conditions of the flow near plate electrode are based on the engineering standpoint of view. The stream lines inside ESP and the vortex occurring near collecting plate electrode are visualized by Schlieren method with two concave mirrors.

The results show that the patterns of non-dimensional flow field inside ESP agree well with experimental observations for the same EHD number. When applied voltage and the velocity of flow field are changed for the same EHD number, the similarity phenomena of flow patterns are observed.

1. INTRODUCTION

In an Electrostatic Precipitators (ESP), the behavior of dust particles is related closely not only to the electric field profiles but also to the gas flow profiles. In corona discharging field, unipolar ions drifting from the ionization region near discharge electrode to collecting electrode, and collide with neutral molecules in air flow and give it's momentum energy to neutral molecules. Thus, corona discharge generates a secondary flow, which is called ionic wind [1-4]. Since the gas flow profile effects the behavior of fine particles, especially submicron particles, it is considered that flow field in ESP plays an important role in the collection mechanism of submicron particles.

In this paper, electrohydrodynamic (EHD) flow field in a wire-duct type ESP with 20 cm plate spacing is investigated by using numerical analysis of Navier-Stokes equation for incompressible viscous fluid. The equation is analyzed by using vorticity and stream function based on the assumption that the external force acting on the fluid is only Coulomb force density. Coulomb force density is calculated by the electric field analysis [5] with Deutsch assumptions [6]. Time-dependent vorticity equation is calculated by Alternating Direction Implicit (ADI) method while Poisson's type equation for stream function by finite element method [7].

The stream lines inside ESP and the vortex near collecting electrode are visualized by Schlieren method with two concave mirror. The calculated patterns of flow field inside ESP are agree well with experimental results for the same electrohydrodynamic (EHD) number. When applied voltage and the velocity of gas flow inside ESP are changed for the same EHD number less than 2.5, the similarity phenomena of flow patterns are observed.

2. ANALYSIS METHOD OF FLOW FIELD

Navier-Stokes equation and continuity equation for incompressible viscous flow can be expressed by following set of nondimensional equations [2,3].

$$\frac{\partial \omega^*}{\partial t^*} = -\frac{\partial \Psi^*}{\partial y^*}\frac{\partial \omega^*}{\partial x^*} + \frac{\partial \Psi^*}{\partial x^*}\frac{\partial \omega^*}{\partial y^*}$$
$$+ \frac{1}{R_{ehd}}\left(\frac{\partial^2 \omega^*}{\partial x^{*2}} + \frac{\partial^2 \omega^*}{\partial y^{*2}}\right)$$
$$- \frac{\partial \rho^*}{\partial x^*}\frac{\partial \phi^*}{\partial y^*} + \frac{\partial \rho^*}{\partial y^*}\frac{\partial \phi^*}{\partial x^*} \quad (1)$$

$$\left(\frac{\partial^2 \Psi^*}{\partial x^{*2}} + \frac{\partial^2 \Psi^*}{\partial y^{*2}}\right) = -\omega^* \quad (2)$$

Nondimensional quantities in these equation are defined as follows

$$x^*=x/D,\ y^*=y/D,\ t^*=tU_{ehd}/D,\ \omega^*=D\omega/U_{ehd}$$
$$\rho^*=\rho/\rho_{io},\ \phi^*=\phi/V,\ \Psi^*=\Psi/(U_{ehd}D) \quad (3)$$
$$\rho_{io}=J_pD/\mu V,\ U_{ehd}=\sqrt{J_pD/\mu \rho_g}$$

where D is wire-to-plate spacing, ω is vorticity, Ψ is stream function, ρ is charge density, ϕ is potential, ρ_g is density of air, μ is mobility of ions, ν is kinetic viscosity, R_{ehd} is EHD Reynolds number, U_{ehd} is characteristic ionic wind velocity and ρ_{io} is characteristic charge density, and superscript * indicates the nondimensional quantity. Since the peak corona current density on the plate electrode J_p is measurable, the characteristic quantities in the above equations can be estimated from the experimental values.

During the analysis of the flow field inside wire-duct type ESP, the following assumptions are used: (A) Only Coulomb force density ρE is taken into account as external force in a Navier-Stokes equation (B) The flow field inside the ESP is two-

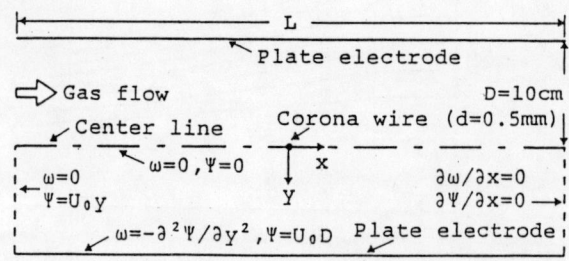

Fig.1 Calculation region and boundary conditions

dimensional steady flow. (C) The effect of gas flow on the electric field distribution is negligible. (D) The flow field inside ESP is not effected by the existence of small corona wire.

Vorticity in Eq.(1) is numerically analyzed by ADI method, which is stable for relatively large ratio of time and space increment. Equation (2) is calculated by finite element method for Poisson's type equation with the estimated vorticity profiles calculated from Eq.(1).

The calculation model of wire-to-duct type ESP with 2D =20cm and boundary conditions are shown in Fig.1. Since the configuration of ESP is symmetric with respect to the center line, the flow field of lower half region is analyzed by dividing it to 161x 21 mesh points. The boundary condition of vorticity on the plate electrode is based on the following expression.

$$\omega_{i,M} = \frac{-\Psi_{i,M-2}+2\Psi_{i,M-1}-\Psi_{i,M}}{2h^2} \quad (4)$$

Where M denotes the mesh point number on the plate electrode. This is not non-slip condition on the plate and corresponds to analyze the flow field from outside of the boundary layer, which is several mm for U_0=0.8 m/s. Under initial condition of uniform inlet gas velocity throughout ESP space, Eq.(1) and (2) are alternately advanced in time until the variation rate of vorticity in time becomes smaller than the specified value.

3. EXPERIMENTAL APPARATUS

Schematic diagram of experimental apparatus is shown in Fig.2. The experimental apparatus consists of sirocco fan, buffer section, straightener section and wire-duct type ESP. The straightener section is made by soda straws of 6mm diam. and nets with 18 mesh. The plate spacing 2D of the ESP is 20 cm used in industrial ESP and the collecting plate electrodes 100x60 cm. One corona wire of 0.5 mm diameter is placed at the center between two plate electrodes.

In case of negative discharge, corona tufts exist along corona wire and corresponding to it, emitting points of air flow are formed along corona wire. So, it is considered that the flow field is three dimensionally disturbed by negative corona. On the other hand, corona field and flow field for positive corona are nearly two dimensional. In order to study the basic characteristics of flow field, positive dc high voltage is applied to corona wire of 0.5 mm diam. placed at the center of ESP.

The stream lines of flow field inside ESP are visualized by Schlieren system which consists of spark light source and two concave mirrors of 30 cm. Five nickel-chromium wires of 0.15 diam. and 20 cm length are placed at 30 cm upstream from corona wire. In order to make density gradient of air flow, the wires are heated by the current of 0.7 A/wire for visualization of the stream lines inside ESP space. On the other hand, for visualization of vortex near plate electrode, heating element of 5 mm diam. is placed flush with the electrode under corona wire.

4. RESULTS AND DISCUSSIONS

4.1 EHD number inside ESP

The electric field inside ESP is calculated by the electric field analysis with Deutsch assumptions [5]. In the analysis, Peek's equation for corona onset field strength and mobility of $1.4\times10^{-4} m^2/(V.s)$ for positive ion are used. The calculated corona characteristics, current density distribution and electric field distribution on the plate electrode are in good agreement with the measured values [5]. Therefore, the distributions of Coulomb force density used in flow field analysis are calculated by this method.

In this paper, EHD number ($N_{ehd}=U_{ehd}/U_o$) is calculated by using U_{ehd} defined in Eq.(3). This number N_{ehd} indicates the relative intensity of ionic wind compared to inlet gas velocity and gives similarity of flow field in a non-dimensional Navier-Stokes equation. U_{ehd} in Eq.(3) is obtained as a function of applied voltage by the electric field analysis. The dependence of EHD number on applied voltage for various inlet flow velocities is shown Fig.3.

4.2 Numerical results of the flow fields

The ratio of time increment and space increment, r=1.0 is used in the calculation program of ADI method. The relative difference of stream function for it compared to that for r=0.5 is less than 10^{-3} at each mesh points.

The calculated stream lines are shown in Fig.4. In the calculation, the length of ESP is set to 8D, while the stream lines in the region of 5D are shown

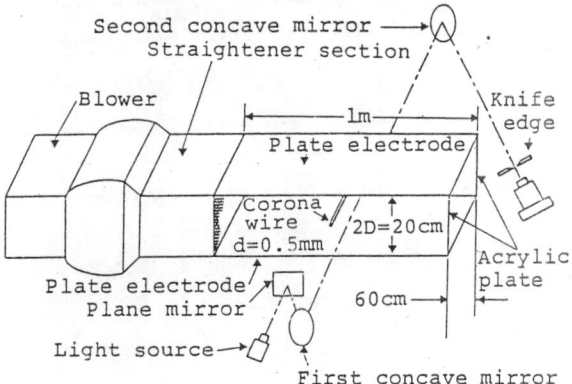

Fig.2 Schematic diagram of experimental apparatus

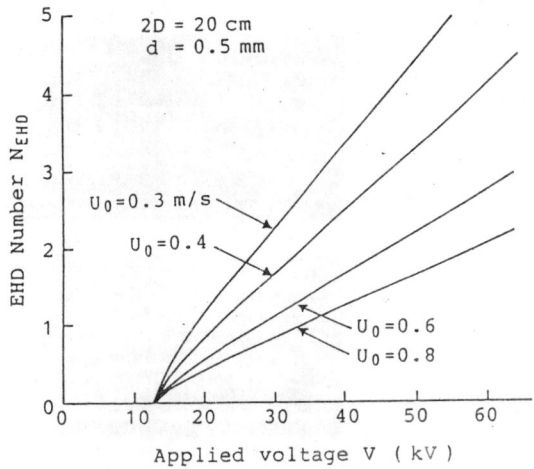

Fig.3 Dependence of EHD number on applied voltage with inlet flow velocity as avariable.

in this figure. The relative difference between the calculated stream functions for 161x21 mesh points and 51x11 mesh points is less than 2×10^{-3}. The stream lines are slightly drawn toward corona wire and returned toward plate electrode for N_{ehd} less than 2.0. In case of N_{ehd} greater than 2.0, a vortex is generated near plate electrode under corona wire due to large effect of ionic wind.

The gas velocity is accelerated near corona wire and is decelerated near plate electrode. For $N_{ehd}=1.27$ shown in Fig.4 (a), maximum velocity is 1.2m/s ($1.5U_o$) in the vicinity of corona wire and minimum velocity is 0.54m/s ($0.68U_o$) near plate electrode under corona wire. For N_{ehd} greater than 2, gas flow velocity under corona wire is in the reverse direction of inlet gas velocity.

It is found from the results of numerical analysis that the configurations of non-dimensional stream function are nearly similar for the same N_{ehd}.

4.3 Experimental results

Fig.5 shows Schlieren photographs of heated air from heating wires for positive corona. The measured corona onset voltage of this ESP is 13.1kV. The flow field for positive corona is observed as stationary laminar flow. The effect of heated air on gas flow is negligible small since in case of no applied voltage, it is observed to flow parallel to plate electrode for U_o greater than 0.2m/s. Therefore, it is considered that the streak lines of heated air give stream lines inside ESP.

The stream lines, in the same manner as obtained in the flow analysis, are drawn toward corona wire and returned toward plate electrode. The configurations of the visualized stream lines show similarity phenomena for the same N_{ehd}.

From Fig.5, it is observed that the pattern of EHD flow inside ESP depends on EHD number N_{ehd} and if N_{ehd} is same, the pattern is qualitatively

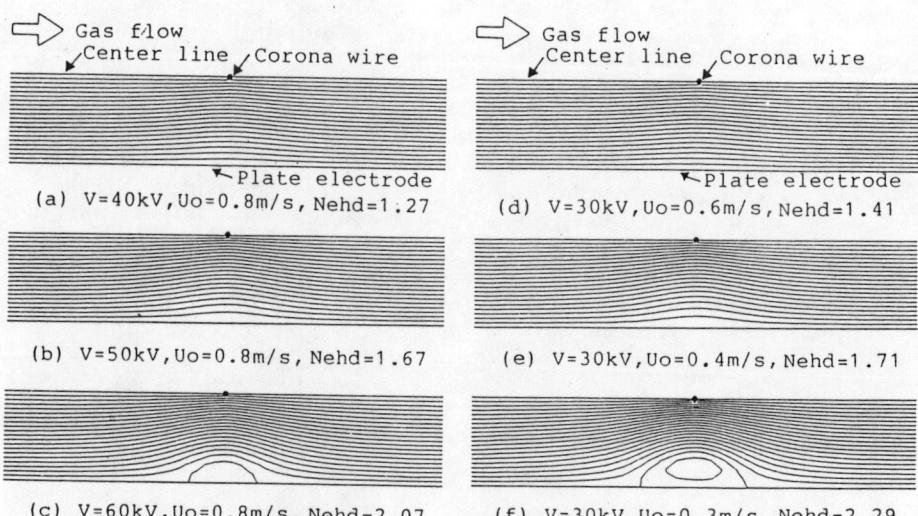

Fig.4 This figure shows the calculated stream lines. The patterns of stream lines depend on EHD number.

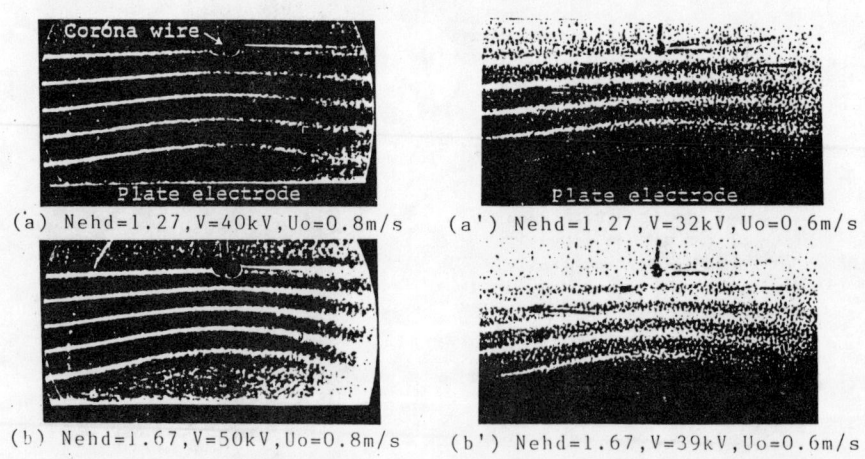

Fig.5 Schlieren photographs of heated air from upstream heating wires for positive corona. The white lines in the photographs show stream lines in ESP.

similar. Regardless of many assumptions used in the flow analysis, the calculated stream lines are in good agreement with the experimental results except for neighbourhood of plate electrode under corona wire.

The pattern of stream lines effected by ionic wind due to corona discharge is shown in Fig.6. Stream line number L1,L2 and etc are the distance from plate electrode to stream lines under corona wire.

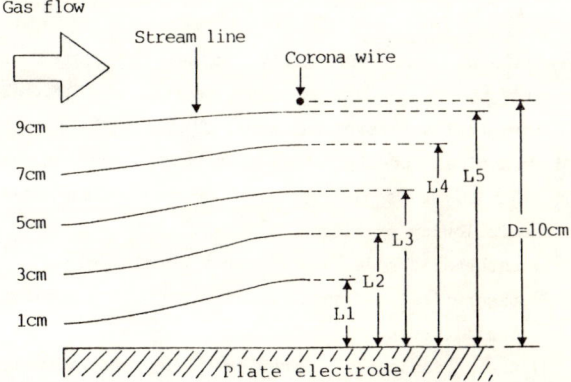

Fig.6 Pattern of stream line inside ESP

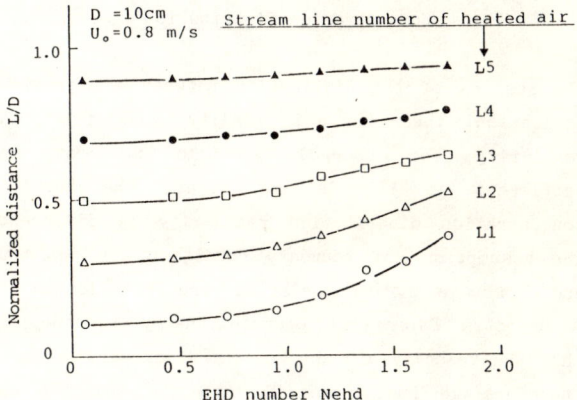

Fig.7 L/D - Nehd characteristic

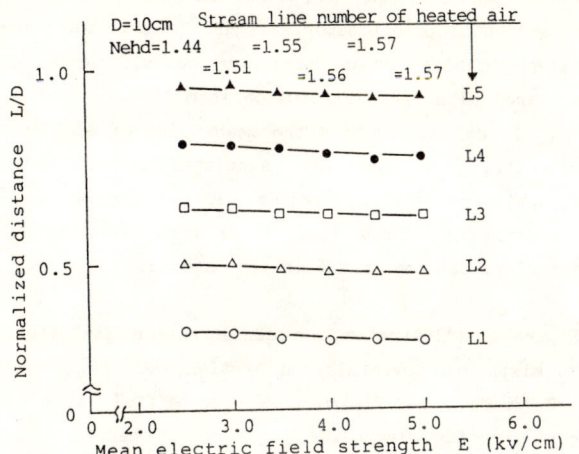

Fig.8 L/D - E characteristic

As shown in Fig.7, normalized distances L/D near plate electrode (example ; L1 and L2) are increase with increasing N_{ehd} of EHD number. It is considered that a vortex is formed under the corona wire and this vortex become large with increasing N_{ehd}. Fig.8 show the characteristics of normalized distance L/D and mean electric field strength E (=V/D) for the constant value of N_{ehd}=1.44. Similarity phenomena of stream lines for the constant value of N_{ehd} is judged from the result that L/D in Fig.8 has a constant value when E and Uo are changed.

5. CONCLUSIONS

The effects of corona discharge on the flow field are numerically analyzed by using the Navier-Stokes equations. The calculated results are compared with the stream lines of flow field inside ESP and the vortex near collecting plate electrode observed by Schlieren method. The results are summarized as follows.
(1) The stream lines are drawn toward corona wire and returned toward plate electrode for small N_{ehd}. In case of large N_{ehd}, a vortex is generated near plate electrode due to large effect of ionic wind.
(2) The patterns of the flow field inside ESP depend on EHD number. This patterns are nearly similar for the same EHD number when mean electrin field strength E and gas velocity Uo are changed.
(3) The analysis method in this paper has merits since it can simulate the electric flow fields inside ESP without experimental data.

ACKNOWLEDGEMENTS

The authors are pleased to acknowledge the continuous encouragement they have received for useful discussions with Dr. M.Hara of Kyusyu University and Dr. T.Yamamoto of Research Triangle Institute in USA. The authors also wish to express thanks to Mr. S.Akamine and Mr. T.Enokida of Oita University for technical help with the experimental programme.

REFERENCES

[1] T.Adachi, S.Masuda, K.Akutsu : "Velocity Distribution of Negative Ionic Wind in the Corona Discharging Space with Needle to Plate Electrode, " J. Inst. Elect. Eng.,Japan, 97-A, (1977) 37
[2] A.Yabe, Y.Mori and K.Hijikata: "EHD Study of the Corona Wind between Wire and Plate Electrodes, " AIAA Journal, 16, 4 (1987) 340
[3] T.Yamamoto and H.R.Velkoff :"Electrohydrodynamics in an electrostatic precipitator," J.Fluid Mech., 108 (1981) 1
[4] T.Yamamoto : "Some Aspects of Efficiency Theory for Electrostatic Precipitators," 2nd Inter. Conf. on Electrostatic Precipitation, (1984) 8A-3
[5] T.Ohkubo, Y.Nomoto, T.Adachi and K.J.McLean : "Electric Field in a Wire-Duct Type Electrostatic Precipitator", Journal of Electrostatics, 18 (1986) 289
[6] M.P.Sarma and W.Janischewskyj : "Analysis of Corona Losses on DC Transmission Lines : I - Unipolar Lines," IEEE Trans. Power Appar. Syst., PAS-88 (1969) 718
[7] T.Ohkubo, T.Murakami and T.Adachi : "Analytic and Experimental Study of Flow Field for Wire-Duct Type Electrostatic Precipitator," J. Inst. Elect. Eng., Japan, 106-A (1986) 377

STUDY ON COLLECTION THEORY OF ELECTROSTATIC PRECIPITATORS[*]

Xiao Fuchun
Department of Mining Engineering
Jiangxi Institute of Metallurgy
Ganzhou, Jiangxi, China

Zhang Guoquan
Department of Mining Engineering
Northeast University of Technology
Shenyang, Liaoning, China

Abstract

A mathematic model of particle turbulent dispersion has been set up and boundary condition has been correctly determined based on the research of gas velocity and particle concentration fields in ESP in this paper. Particle effective migration velocity and a new collection efficiency formula have been obtained. The effect of electric wind is ingeniously introduced to the turbulent mixing coefficients. Thereby, a general electrostatic collection model including the multiple effect of electrostatic force, gravity, turbulent mixing and electric wind has been developed.

Introduction

Electrostatic precipitator, as a kind of high efficiency, saving energy, low cost in operation and strong applicable purifying device has been widely used in many industrial fields. But, the theory for guiding the design of electrostatic precipitator is slowly developed. The established Deutsch model [1] for the collection process, which is still widely used to size and assess the performance of ESP, make the crude assumption that there is complete mixing in planes transverse to the main flow direction (no axial diffusion). The difference between the computed result by Deutsch formula and the collection efficiency in practice is very large. The effective migration velocity inversely calculated from practical efficiency is only 1/2-1/10 times the theoretical value. In order to overcome the inadequacy of Deutsch theory, many authors had put forward different collection mathematic models. The theories of Cooperman [2] and Leonard [3] are representative. Even though some assumptions of Deutsch theory have been given up, the assumption of boundary condition and the values of turbulent diffusion coefficient are not rational. Based on small and short laboratory precipitator the efficiency value calculated by Leonard theory is larger than Deutsch result. We consider that a theoretical model fitted in Laboratory and practical precipitators has not been set up. In order to solve this problem we must mainly start on the following problems:

1) mathematic model and boundary condition,
2) theoretical formula of turbulent mixing coefficients,
3) the analogous problem in experimental technology.

Gas Velocity and Particle Concentration Distribution

The reasons of the distinction between all previous collection theories and reality are that the regularity for charged particle movement and settlement in ESP is not clear, the particle concentration distribution regularity is different, the assumptions of boundary condition are not true and some parameters selected are not fitted or incorrect. Especially Deutsch theory hypothesized that the velocity and the concentration were uniform. Therefore we must research into the velocity and the concentration fields in ESP.

There are 9 rows, 45 holes in the top of the experimental precipitator. Figure 1 shows the tested values in sixth cross section by hot-wire anemometry compared with the distribution form $[V=V_{max}(y/b)^{1/m}]$. V_{max} is calculated from the mean value of each point velocity, then inversely calculated by the formula $V_{mean}=[m/(m+1)] \cdot V_{max}$, where m is 8 and 10 respectively. From Fig. 1 we can know that the turbulent tensity in ESP is very strong.

Figure 2 illustrates the concentration distribution in sixth and seventh cross section where $V_{mean}=1m/s$, $w=0.236m/s$, $i=0.21mA/m$. We can find that the concentration increases from the channel center

---

[*] Projects supported by the Science Fund of the Chinese Academy of Sciences

to both sides. From Figs. 1 and 2, the assumption of Destschian velocity and concentration distribution is false.

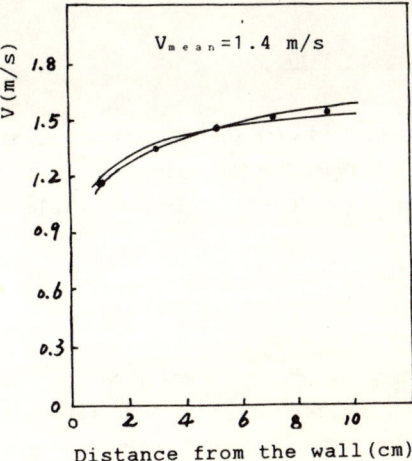

Fig. 1 Comparison of the velocity

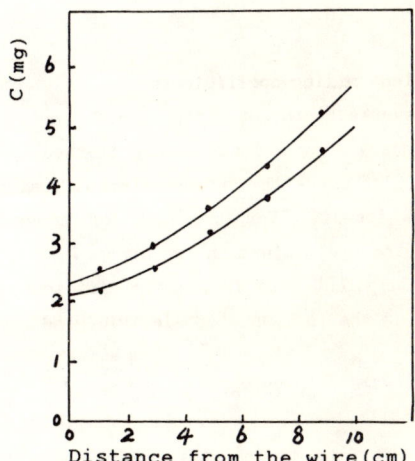

Fig. 2 Concentration distribution of particles

Study of Electrostatic Collection Theory

1. Transport phenomena

In order to correctly describe particle transport in ESP, two kinds of basic transport phenomena, diffusion is the process by which the marker concentration is changed when the velocity distribution over any cross section is uniform. There is a spreading of markers in the longitudinal direction in a channel or a pipe when the velocity distribution over the cross section is nonuniform. This longitudinal dispersion may be defined as the spreading of marked fluid particles by the combined action of a nonuniform velocity distribution and diffusion.

2. Mathematic and physical model

Although charged particle transport is a very complicated process in ESP, there are the following kinds of fields in total: fluid field, diffusion field and force field (electrostatic force and gravity). The charged particle is settled by the combined action of these fields.

Taking a differential volume element in ESP, according to mass conservation law, the number of total particles entering it equals that leaving from it, then

$$Dx\frac{\partial^2 c}{\partial x^2} + Dy\frac{\partial^2 c}{\partial y^2} + Dz\frac{\partial^2 c}{\partial z^2} - v\frac{\partial c}{\partial x} - w\frac{\partial c}{\partial y} - u\frac{\partial c}{\partial z} = 0 \quad (1)$$

If all particles transport in the channel are considered, substituting the mean velocity V for v in eq. (1), the diffusion term due to this varied velocity will change into dispersion term. Then eq. (1) can be written as

$$Ex\frac{\partial^2 c}{\partial x^2} + Ey\frac{\partial^2 c}{\partial y^2} + Ez\frac{\partial^2 c}{\partial z^2} - V\frac{\partial c}{\partial x} - w\frac{\partial c}{\partial y} - u\frac{\partial c}{\partial z} = 0 \quad (2)$$

Neglecting the effect of gravity, then eq. (2) changes into a two-dimensional problem:

$$V\frac{\partial c}{\partial x} - Ex\frac{\partial c}{\partial y} + w\frac{\partial^2 c}{\partial x^2} - Ey\frac{\partial^2 c}{\partial y^2} = 0 \quad (3)$$

Eq (3) is a two-dimensional dispersion equation that charged particle follows in electrostatic field where E_x, E_y corresponds to longitudinal, transverse turbulent mixing coefficient respectively.

3. Boundary conditions

(1) X direction (Gas flow direction)
Assuming uniform concentration at X=0 yields

$$X = 0, \quad C = C_0 \quad (4)$$

(2) Y direction (settling direction)
In the plane composed of the corona wires, particle net flux penetrating the plane is

$$Y = 0, \quad Ey\frac{\partial c}{\partial y} - wc = 0 \quad (5)$$

It is known from above experiment that particle concentration near wall region is higher than that close to corona wire region. The particle in high concentration region due to turbulent mixing must transport to the center. It is believed that at some time, some cross section and per unit length, the net particle flux, M, to the wall equals the theoretical flux, wc, minus the diffusion flux, $E_y(\frac{\partial c}{\partial y})$, that is

$$Y = b, \quad E_y\frac{\partial c}{\partial y} + M = wc \qquad (6)$$

The net particle flux, M, can be theoretically given as follows:

$$M = -\frac{\partial c}{\partial x}Vb \qquad (7)$$

Substituting eq. (7) to eq (6), then dividing by concentration c yields

$$w = \frac{E_y}{c}\frac{\partial c}{\partial y} - \frac{1}{c}\frac{\partial c}{\partial x}Vb \qquad (8)$$

The net particle migration velocity (effective migration velocity) in the y direction due to both the electric force and the turbulent diffusion, can be written [3]

$$W_e = w - \frac{1}{c}\frac{\partial c}{\partial y} \qquad (9)$$

Comparing eq. (8) with eq. (9) yields

$$W_e = -\frac{1}{c}\frac{\partial c}{\partial x}Vb \qquad (10)$$

Eq. (10) is the formula of particle effective migration velocity with new boundary condition. The concept of particle effective migration velocity is clear and it plays an important role in our new theory.

4. Solution of the mathematic model and efficiency formula

(1) Solution of the mathematic model
Eq. (3) is solved by the standard technique of separation of the variables. Substituting the boundary condition, the solution is

$$C(x,y)=B[\frac{2}{P_{ey}}\cos(\theta\frac{y}{b})+\sin(\theta\frac{y}{b})]e^{-(F\frac{wx}{Vb}+\frac{P_{ey}y}{2b})} \qquad (11)$$

where:

$$F=\frac{V}{w}\frac{P_{ey}}{2}\{[1+\frac{w}{V}\frac{P_{ey}}{P_{ex}}(1+(\frac{2\theta}{P_{ey}})^2)]^{\frac{1}{2}} - 1\} \qquad (12)$$

$$tg\theta=(4F\frac{\theta}{P_{ey}})/[(\frac{2\theta}{P_{ey}})^2 - 2F+1] \qquad (13)$$

$$P_{ex} = \frac{Vb}{E_x} \qquad P_{ey} = \frac{wb}{E_y} \qquad (14)$$

(2) Collection efficiency formula
From eq. (11) the collection efficiency formula can be written as

$$\eta = 1 - \frac{\bar{C}(L)}{C_o} = 1 - e^{-F\frac{wL}{Vb}} \qquad (15)$$

Comparing eq. (15) with Deutsch formula, an additional factor F is found in the exponent. It is believed that particle effective migration velocity from eq. (15) is

$$W_e = F w \qquad (16)$$

Substituting eq. (11) to eq. (10) also yields

$$W_e = F w \qquad (17)$$

These two equations are perfectly the same. This shows that the boundary condition and eq. (10) are perfectly correct. Deutsch formula is only a special form (F=1). It is nothing surprising that it can not exactly calculate the collection efficiency in ESP.

5. Turbulent mixing coefficients

(1) Transverse turbulent mixing coefficient
The secondary flow and turbulence induced by electric wind can also cause the influence on diffusion besides gas flow itself. For the sake of convenience we consider the coefficient in two steps:

a. Turbulent diffusion coefficient without considering the influence of electric wind

$$E_y = \frac{1}{4}C_f Vb/S_{ct} \qquad (18)$$

where S_{ct} is the turbulent Schmidt number. Mustafa [4] had gained the relation of S_{ct} and Reynolds' number, Re, in a pipe through much experimental research, as follows:

$$S_{ct}=0.74 Re^{0.04} \qquad (19)$$

For the precipitators in practice the frictional resistance coefficient, C_f, should be replaced by the local resistance coefficient, ζ, then eq. (19) changes into

$$E_y = 0.3378\zeta Vb/Re^{0.04} \qquad (20)$$

b. Influence of the electric wind on the fluid flow
Yamamoto et al. [5] had introduced electrohydrodynamic (Reynolds' number R_{EHD}) in studying the effect of the electric wind on fluid flow. The ratio of R_{EHD} and R_e is N_{EHD}

$$N_{EHD} = R_{EHD}/R_e \quad (21)$$

Leonard et al. [6] had further studied the effect of the electric wind on fluid flow. The specified N_{EHD} had been achieved as follows:

$$N_{EHD} = i/(\ell\rho kV^2) \quad (22)$$

where: i — corona discharge current in total
 ℓ — corona wires length in total
 k — ionic mobility

Because the effect of the electric wind can intensify the turbulent mixing of gas flow in electrostatic field, considering the total Reynolds' number in the channel will yield

$$R_{et} = R_e + R_{EHD} = (1 + N_{EHD}) R_e \quad (23)$$

According to the definition of R_e number we can think that R_e number increases by $(1+N_{EHD})$ times. The mean velocity also increases by $(1+N_{EHD})$ times. Considering the effect of electric wind, E_y expression is

$$E_y = 0.3378(1+N_{EHD})\zeta Vb/[(1+N_{EHD})R_e]^{0.04} \quad (24)$$

(2) Longitudinal dispersion coefficient

Dispersion is mainly caused by nonuniform velocity. The dispersion in ESP is different from common fluid dispersion. Charged particle transport coefficient produced due to nonuniform velocity is

$$E_{x1} = (bV^2)/(2w) \quad (25)$$

For anisotropic turbulence, longitudinal turbulent diffusion coefficient is commonly larger than transverse diffusion coefficient. Let E_{x2} equal $2E_y$, charged particle longitudinal dispersion coefficient is

$$E_x = E_{x1} + E_{x2} = (bV^2)/(2w) + 2E_y \quad (26)$$

For the sake of consistency, we call E_x the longitudinal turbulent mixing coefficient, where $E_{x2} \ll E_{x1}$.

Comparison between New Theory, Old Theory and Tested Result

New theory has been perfectly set up by determining the two turbulent mixing coefficient. F value in new efficiency formula, under the given voltage, is mainly determined by the local resistance coefficient, ζ. For illustrating problem, the different ζ values are selected to compare with other typical theories. The model precipitator geometry parameters are, the height h=0.3m, the length L=0.6m, the semi-channel width b=0.1m. The voltage across the two is, $V_o=4\times10^4$V. The value of f, D in Cooperman theory is 0.5, 1m²/s respectively. ζ value in new theory is selected as 0.2, 0.6, 1.0 respectively.

Fig. 3 shows the computed fractional efficiency for V=2m/s. It is clear from this Figure that the larger the ζ value is, the smaller the efficiency is and when $\zeta=0.2$ the efficiency curve of new theory is almost close to that of Deutsch theory. It can be expected that when $\zeta<0.2$, η_{new} is larger than η_D. Therefore, ζ is mainly analogous parameter in experimental technology (N_{EHD} is also an analogous parameter). Thus, new theory is a general theory model. It can be used not only in small and smooth model precipitators but also in practical precipitators.

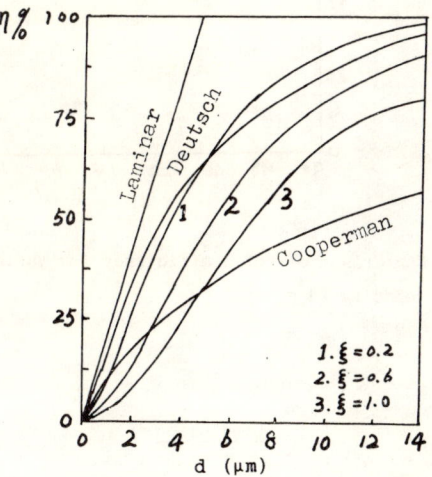

Fig. 3 Schematic of theory comparison (V=2.0m/s)

The local resistance coefficient, , tested in the model precipitator is 0.453. Figures 4 and 5 show the tested efficiency values compared with the computed efficiency values with new theory while =0.453. It can be seen from the two Figures that the computed efficiency values are very close to the tested results.

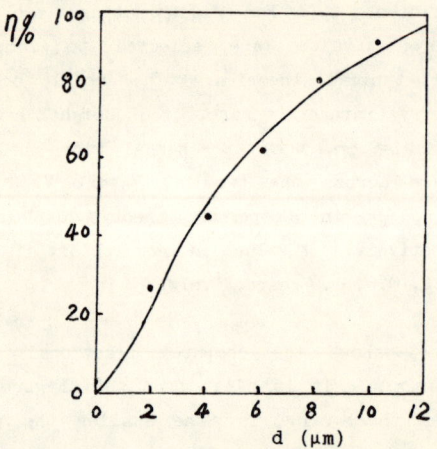

Fig. 4 Comparison for fractional efficiency (V=1.2m/s)

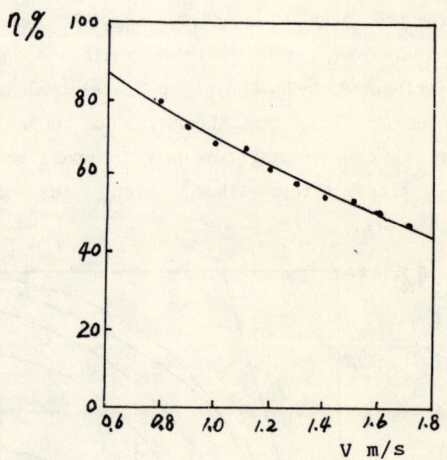

Fig. 5 Comparison of total efficiency for various velocity (d = 5 μm)

Conclusion

1. Gas velocity and particle concentration fields in ESP are nonuniform. Charged particle settlement satisfies eq. (3).
2. The collection efficiency formula for charged particle is eq. (15). The effective migration velocity formula is eq. (10) or eq. (16).
3. Transverse turbulent mixing is caused by gas flow itself and the electric wind. Its coefficient formula is eq. (24). Longitudinal turbulent mixing coefficient formula is eq. (26).
4. New theory is a general electrostatic collection theory model including the comprehensive effect of electrostatic force, gravity, turbulent mixing and electric wind.

References

1. H.J. White, Industrial Electrostatic Precipitators, Addison Wesley (1963).
2. P. Cooperman, A New Theory of Precipitator Efficiency, Atmos. Environ., Vol. 5, pp541-551 (1971).
3. G.L. Leonard, M. Mitchner and S.A. Self, Particle Transport in Electrostatic Precipitators, Atmos. Environ., Vol. 14, pp1289-1299 (1980).
4. El-Rifai Mustafa, Dispersion and Transport in Turbulent Pipe Flow, D. Ph., The University of Oklahama, Ph. D., Eng. Chem., pp124-149 (1965).
5. T. Yamamoto and H.R. Velkoff, Electrohydrodynamics in an Electrostatic Precipitator, J. Fluid Mech., Vol. 108, pp1-18 (1981).
6. G.L. Leonard, M. Mitchner and S.A. Self, Experimental Study of the Electrohydrodynamic Flow in Electrostatic Precipitators, J. Fluid Mech., Vol. 127, pp123-140 (1983).

ELECTROSTATIC ENHANCEMENT OF PARTICLE COLLECTION IN FIBER FILTERS

Prof.Dr.-Ing.F.Löffler

Universität Karlsruhe, Institut für Mechanische Verfahrenstechnik und Mechanik

Abstract

Deep-bed electret filters are used to separate small solid or liquid particles from gas flows. Compared with conventional deep-bed filters of similar pressure loss, they operate with substantially higher collection efficiencies. During the course of filtration, however, the electrical charges of the deposited particles neutralize the electret fibers, temporarily decreasing the collection efficiency. In this paper, experimentally determined fractional separation functions of different types of electret filters are presented for the particle size range from 10 nm to 10 μm at a filter face velocity of 10 cm/s. The results of long-term filtration experiments show that a complex time-dependent behavior exists for different filter materials.

Introduction

Electret filters are a special type of deep-bed filter consisting of fibres which are permanently electrically charged. The collection features of these filters are characterised, on the one side, by high initial collection efficiencies, but on the other, by a complex dynamic behaviour. Depending upon the charge and aggregate condition of the aerosol to be filtered and upon its particle size range, the dynamic behaviour of the collection efficiency and of the pressure loss varies significantly.

In the first part of this paper experimentally determined collection efficiencies as a function of the filter loading are presented. The second part deals with the theoretical evaluation of the collection efficiencies of fibrous electret filters and compares these with the experimental results. Two electret filter type have been investigated:

(1) An electrostatically spun filter, comprising of a 30 g/m² layer of electret fibres (D_f = 9 μm) covered on both sides by coarse uncharged fibres |1|.

(2) A split-fibre filter (in two qualities, 100 and 200 g/m²; D_f = 25 μm) |2|.

The initial collection performance of these filters is excellent as we can see from fig. 1 where the collection efficiencies of two different electret filter types in comparison to a conventional glass fibre filter are plotted |3, 4|. The decisive contribution of electrostatic interaction can clearly be detected especially in the transition regime between non-electrical diffusion and inertia mechanisms, i.e. in the particle size range between 0.05 μm and 5 μm. The conclusion to be drawn is that when collecting uncharged or charged particles the separation efficiency

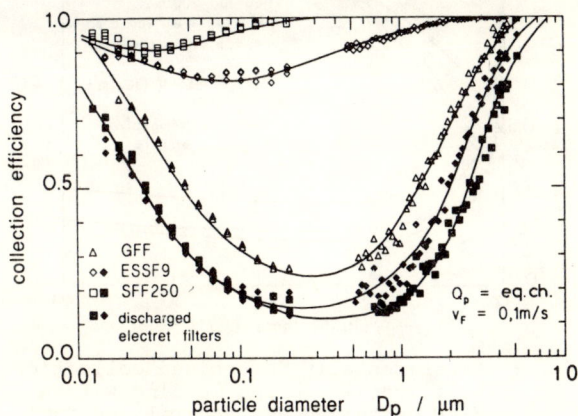

Fig. 1 Fractional separation curves of two electret filters and a glass-fibre filter possessing nearly identical pressure losses (20 Pa at v_F = 10 cm/s)

of an electrostatic enhanced fibre filter, is significantly higher than that which can be acchieved with conventional filter media.

Long-term Behaviour

With progressive filtration time, however, the filters portray different collection efficiency trends |5, 6|. In the following, the results of long-term filtration experiments are given for an aerosol which was, on the one hand in the so-called "equilibrium charge state" (neutralized) and on the other hand in an almost uncharged state |5|. This test aerosol consisted of a NaCl condensation aerosol with a nearly log-normal size distribution (mean particle diameter 0.1 μm, geometric standard deviation 1.4).

When loading with the neutralized aerosol, both filter types behaved in a similar manner (fig. 2). The high initial collection efficiencies decreased at first sharply, and then more gradually to very low values, corresponding to the collection efficiency of an uncharged filter. A further improvement of collection performance is only apparent after extensive continuous loading. Contrary to this, when filtering uncharged aerosols, the initial collection efficiencies were lower due to the absence of Coulomb forces. A slight efficiency drop occurred at the beginning which can be explained by the extremely low residual electrical charge of the aerosol, or by a screening-effect of the deposited particles. With progressive filtration time, the collection improves; for the spun filter very rapidly, and for the split-fibre filter only very gradually. This difference is caused by the fact that the split-fibre filter consists of substantially coarser fibres, the captured particles therefore not significantly modifying their geometry. The finer spun fibres act together with the deposited particles as collection elements, which are also "mechanically" very efficient (mainly interception and diffusion collection).

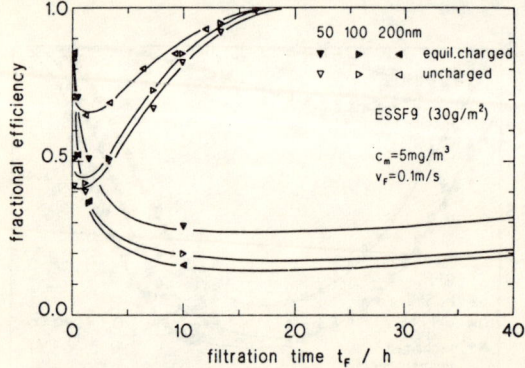

Fig. 2 Experimentally determined collection efficiencies during loading with a fine uncharged and neutralized NaCl aerosol

  a) Electrostatically spun filter (ESSF9)

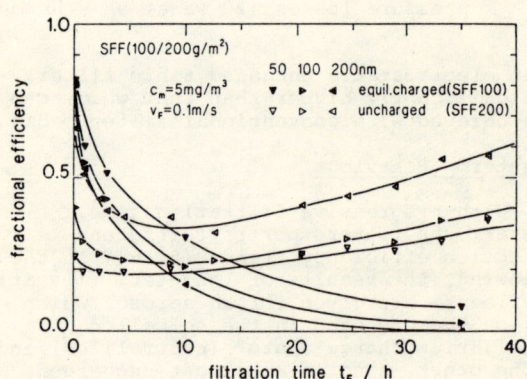

  b) Split-fibre filter (SFF 100/200)

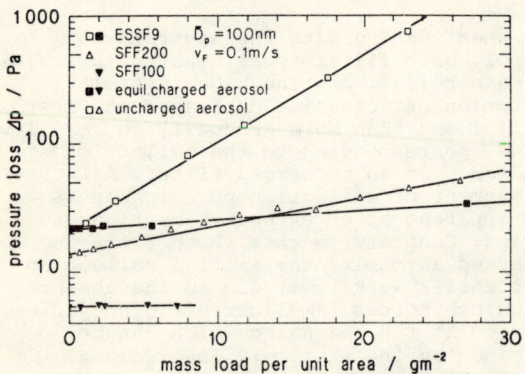

Fig. 3 Experimentally determined pressure drop values during loading the electret filters with fine NaCl Aerosols in different charge states

This can be clearly followed by the pressure drop trends. Fig. 3 portrays these trends for all four instances, plotted as a function of the dust mass collected per unit area. The increase of pressure drop across the filter remains low or even unmeasureable for the filtration of neutralized aerosols. Concerning uncharged aerosols, however, it increases noticably for the split-fibre filter and dramatically for the spun filter, so that clogging can even occur.

Theoretical Model

Using the simulation model presented by Baumgartner und Löffler |5| the collection efficiencies of the electret filters were also calculated with the experimental parameters as input data. In fig. 4a, the collection efficiency trends of single fibres are shown,

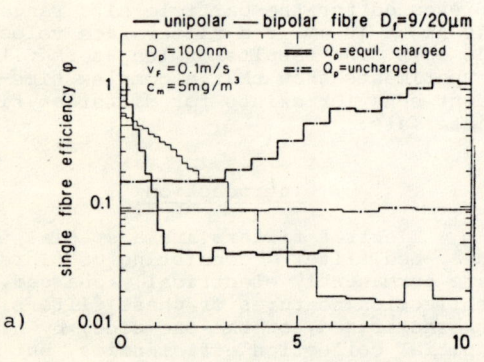

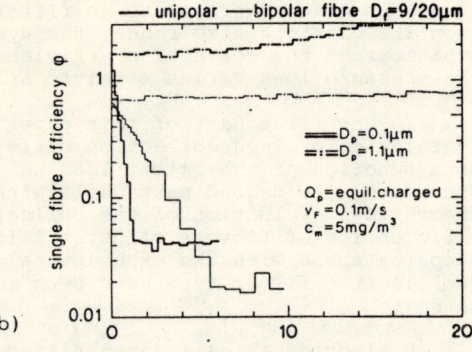

Fig. 4 Calculated collection efficiency trends of single fibres, when loading with:
  a) neutralized and uncharged 0.1 μm particles (left)
  b) neutralized 0.1 and 1.1 μm particles (right)

when loading with neutralized and uncharged aerosols respectively. The curves, when loading with neutralized aerosols of different particle size (0.1 and 1.1 μm), are illustrated in fig. 4b. As can be seen, the calculated trends of the single fibre correspond to those of the filters determined experimentally. Whilst the neutralized aerosol causes a steep decrease of collection efficiency, an initial slight, and following steep increase results when filtering the uncharged aerosol. Experimental results of the loading with particles in the size range of 1 µm have been presented in |5|, the calculated trends (fig. 4b) can be seen to correspond, at least in the case of the spun filter.

In a further development, a simple numeric process, based upon the load-dependent collection of a single fibre has been devised, allowing the efficiency of a complete fibre layer to be calculated. Here, the layer is sub-divided into a number of thin so-called "unit layers" and the decrease of particle concentration from unit layer to unit layer is incrementally calculated. The ratio of particle concentration, emerging from the bottom unit layer to that being offered at the top represents the degree of overall penetration P (1-P being the overall collection efficiency). Each of the calculations are conducted over finite time intervals, so that the collection efficiency trend is best portrayed in the form of a step-function.

The measured and calculated collection efficiencies of the spun and split-fibre filters are compared in figs. 5 and 6.

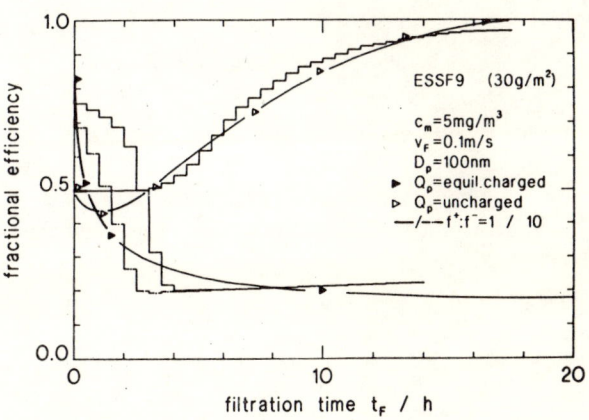

Fig. 5 Calculated and measured collection efficiency trends of the elektrostatically spun filter (ESSF9)

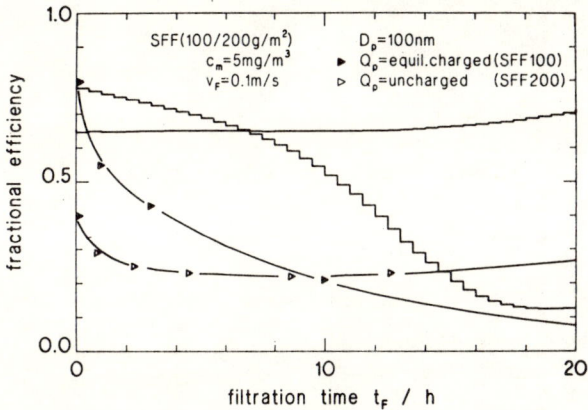

Fig. 6 Calculated and measured collection efficiency trends of the split-fibre filters (SFF 100/200)

As a result of the simplifications within the simulation calculation and also the presumption of an <u>ideal</u> filter structure, the curves cannot be expected to coincide. In the case of loading with a neutralized aerosol, typical deviations result: Firstly, the calculated curves drop substantially slower than those of the experiments and then abruptly fall to nominal values. The experiments, however, show that this collection minimum is more gradually approached. This oviously results from inhomogenities within the layer, the denser regions are penetrated to a lesser degree by the aerosol and therefore loose their charge more slowly (flat curve declination). The opposite occurs in the thinner sections: these fibres are rapidly neutralized, and therefore quickly loose their high collection efficiency (initially steep curve drop).

In the case of the uncharged aerosol, the model calculation of the load dependence of the spun filter is unexpectedly exact. With the exception of a short minimum just at the beginning, which cannot be simulated due to the model assumption (particle electrical charge = 0), the calculated values correspond very well to those measured. This is even more surprising, as with a high fibre loading, the two-dimensional simulation generally overemphasizes the effect of the interception collection.

The large difference between the calculated and measured values when loading the split-fibre filter with uncharged aerosol is, at the present, unexplainable. Although the trends, with their slight increase for longer filtration durations, correspond; the level of the calculated collection efficiency, is however, much too high. A possible reason for this, is that the filter sample used for this experiment was unrepresentative.

Although the calculated and measured collection efficiencies do not yet completely coincide, the model allows a good description of the general trends. In order to achieve even better agreement, however, the non-ideal fibrous layer structure must be taken into consideration.

Acknowledgement

These investigations were financially assisted by the Deutsche Forschungsgemeinschaft (Projekt No. Lo 142/10).

References

|1| Schmidt, K., Melliand Text.Ber. 6 (1980) 495-497

|2| Van Turnhout, J.; Van Bochove, C.; Van Veldhuizen, G.J.; Staub Reinh.Luft 36 (1976) 1, 36-39

|3| Baumgartner, Hp.; Löffler, F.; Proc.International Conf. on Electrostatic Precip., Abano Italy, Oct. 1987

|4| Baumgartner, Hp.; VDI-Fortschr.Ber. Reihe 3, Nr. 146, 1987

|5| Baumgartner, Hp.; Löffler, F.; Proc.4th
World Filtr.Congr., Ostend (1986)
2.11-2.22.

|6| Blackford, D.B.; Bostock, G.J.;
Brown, R.C.; Loxley, R.; Wake, D.;
4th World Filtr.Congr., Ostend (1986)
7.27-7.33.

A RESEARCH ON THE MECHANISM OF DUST COLLECTION IN ELECTROSTATIC LENTOID FIELDS

Chen Xuegou   Huang Huifen   Mao Jinyuan

Wuhan University of Technolgoy

14 Rose Road, Wuhan, Hubei, China

Abstract

This paper introduces a new type of electrostatic lentoid fields (ESLFs). Once the charged particulates come to ESLFs, they will soon be sucked into dust collecting chambers, from which they are incapable of escaping, but impact with each other, agglomerate into masses, and descend into dust buckets. As it is simple in construction, no rapping device is required. Wide adaptability for dust specific resistivity, low once-investment and maintenance cost as well as low electric energy consumption, and a high efficiency of dust collection may be obtained.

Introduction

Since the first electrostatic precipitator, numerous contributions on the theoretical investigations and technical improvements such as pre-charging, wide-spacing ducts and pulse-energization etc. have been carried out.

This paper firstly introduces the basic characteristics of dust collection in electrostatic lentoid fields (ESLFs), and finally, a brief evaluation to the experimental results with a pilot-scale model as well as a brief prospects to this new technology are given.

Basic Characteristics of Electrostatic Lens Fields

Electrostatic lens fields are the kind of electric fields with axis-symmetry. Fig. 1 shows a kind of such fields excited by parallelly interleaving a thin metal plate with a round aperture into a two-plate air capacitor [1]. Three electrodes are charged to potentials U_1, U_2 and U_0 respectively. U_0 is the common end of positive polarity and grounded, whereas $|U_1| > |U_2|$.

By adjusting the voltages on these electrodes and the plate separating distances d_1 and d_2, the ellipsoidal potential distribution could be varied to satisfy the requirement $E_2 > E_1$, so that the ellipsoidal surfaces oversail towards the weak field side, as shown in Fig. 2. The real curves are a system of hyperbolic electric field lines [1,2].

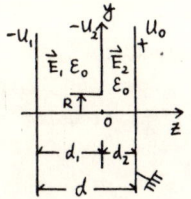
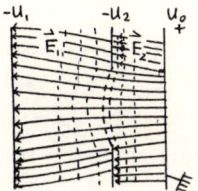

Fig. 1 Round aperture interleaving into plate capacitor to excite electrostatic lens fields.

Fig. 2 Electrostatic lens field in the neighborhood of round aperture, $E_2 > E_1$.

If in the field space, there is not any singularity point (i.e. no point charge, surface charges and bipolar layer), the spatial potential can be analytically expanded in power series. As the fields are of axis-symmetry, there are only even power terms of r, thus in the neighborhood of round aperture, the spatial potential (Scherzer's formula) is expressed by:

$$\psi(z,r) = \sum_{k=0}^{\infty} (-1)^k \frac{1}{(k!)^2} \left(\frac{r}{2}\right)^{2k} V^{(2k)}(z)$$

$$= V(z) - \frac{1}{4}V''(z)r^2 + \frac{1}{64}V^{(4)}(z)r^4 + \ldots \quad (1)$$

$$E(z) = -\frac{\partial \psi}{\partial z} = -V'(z) + \frac{1}{4}V^{(3)}(z)r^2 + \ldots \quad (2)$$

$$E(r) = -\frac{\partial \psi}{\partial r} = \frac{1}{2}V''(z)r - \frac{1}{16}V^{(4)}(z)r^3 + \ldots \quad (3)$$

Here $V(z)$ refers to the potential function of point $(z,0)$ on Z-axis.

Taking the first-order approximation in the neighborhood of axis-symmetry, the forces of electric field on electron can be given as:

$$F(z) = -eE(z) = eV'(z) \qquad (4)$$

$$F(r) = -eE(r) = -\frac{e}{2}V''(z)r \qquad (5)$$

It can be seen that the axial force $F(z)$ is proportional to $V'(z)$, the radial force $F(r)$ is proportional to r and $V''(z)$; when $V''(z) > 0$, then $F(r) < 0$, the electron is attracted towards the axis and intersects at some point on Z-axis; when $V''(z) < 0$, then $F(r) > 0$, the electron moves away from Z-axis.

When the electron with initial velocity V_z moves by an angle ψ towards Z-axis as shown in Fig. 3, the focus f, as similar to the lens focusing in optics, can be proved [3]:

$$f = \frac{r}{\psi} = \frac{2m_e V_z^2}{e(E_2 - E_1)} \qquad (6)$$

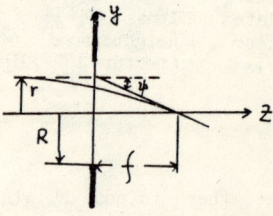

Fig. 3 Visualization of electron focusing in ESLFs.

Here E_1 and E_2 stand for the field strength in both left- and right-hand sides as shown in Fig. 2; m_e stands for the mass of an electron and e for the charge quantity of an electron.

From equation (6), it is easily seen that the less the kinetic energy of an electron, the larger the $(E_2 - E_1)$, and that the smaller the focus f, the easier the focusing of the electron.

Similarly, for the negative charged particulate with mass m and charge quantity $q = ne$ (n: any positive integer), the following equation can be derived:

$$f = \frac{r}{\psi} = \frac{2mV_z^2}{q(E_2 - E_1)} \qquad (7)$$

From equation (7), when negative charged particulate is under the lens field action, it moves towards the Z-axis similarly. But charged particulates intersect at different points on Z-axis, for the mass m and charge quantity q and initial velocity $V(z)$ are not the same, so the motion of charged particulates towards Z-axis is called axis-approaching effect.

In conclusion, there are three basic characteristics in electrostatic lens fields: firstly, they are symmetry to the axes of round apertures; secondly, in the neighborhood of round aperture, the equi-potential surfaces are of ellipsoidal, those perpendicular to these surfaces are system of hyperboloids, in which lie the field lines; and thirdly, the lens fields have the focusing effect on electrons and axis-approaching effect on the charged dust particulates.

Mechanism of Dust Collection in Electrostatic Lentoid Fields (ESLFs)

ESLFs were developed on the basis of electrostatic lens fields together with the consideration of the requirements in dust collection. They are composed of three kinds of electrodes: i.e. the main corona electrodes, the lens electrodes and the dust collecting electrodes. One row of main corona electrodes and two rows of lens electrodes constitute the wide-spacing duct; one row of dust collecting electrodes and two rows of lens electrodes constitute the dust collecting chambers. All electrodes are covered with a certain surface charges and emit different coronas. Corona current fields superposing on the electrostatic lens fields thus gives the electrostatic lentoid fields (ESLFs). Solving for ESLFs, only Poisson's equation can be used, and with the aid of computers, a solution of digital approximation of the fields may be obtained [2]. This paper emphasizes the discussing of the mechanism of dust collection in ESLFs only. Other paper will discuss the effect of gas fluid dynamics on dust collections.

ESLFs possess the actions as in electrostatic lens fields. The charged particulates (with a certain velocity V_z) once come into these fields, they will move counter the hyperbolic field lines in directions, pass through round apertures, and go into the dust collecting chambers. Due to the strong axis-approaching effect of ESLFs, a large amount of charged particulates forms a flow of dust particulate groups, just like that there are so strong attractive forces at the lens openings, sucking all the particulate groups into the chambers. From equations (4) and (5), the axial force $F(z)$ and radial force $F(r)$ that act on the charged particulates are in proportion to $V'(z)$, $V''(z)$ and r. If the applied voltages are increased, all the values of E_1, E_2, q, $V(z)$, $V'(z)$, $V''(z)$, $F(z)$ and $F(r)$ will correspondingly be increased. Thus, the amount of dust particulates sucked into the chambers will also be increased. From equation (7): the larger the values of $(E_1 - E_2)$ and q, the smaller the particulate kinetic energy, the stronger the axis-approaching effect to particulates, and the larger the amount of charged particulates sucked into the chambers. However, a small part of particulates will still be deposited on the outside surfaces of lens electrodes. As the thickness of dust layers increase to a certain extent, they will be collapsed into larger masses and descended by weight into dust buckets with negligible secondary flying of dust.

Since lens electrodes at proper potential will discharge proper amount of negative ions towards the positive dust collecting electrodes. If any positive ions, positive charged particulates or neutral particulates come along the directions of dust collecting fields lines to the lens openings, they will confront with the flow of negative ions and of large amount of negative charged particulate groups that are sucked by ESLFs into the lens openings. All these particles will be neutralized and then charged into negative, and move back into the interior of dust collecting chambers. In a word, once large amount of negative charged particulates are sucked by ESLFs into dust collecting chambers, they can escape no more from the lens openings, such an effect is called electric field sealing at lens openings.

In dust collecting chambers, there are strong electric fields between dust collecting electrodes and lens electrodes. Under their actions the charged particulates will be absorbed on the dust collecting electrodes. As the absorbed dust layer will decrease the fields strength, if layer thickness is increased to a certain extent, the dust collecting fields can collect only very little of dust. But if using special design of dust electrodes, and applying a certain voltage, corona shells will be excited over the positive dust collecting electrodes. Under the action of electric fields, the negative charged particulates can touch the positive electrodes no more, but will come into collision with positive ions, thus producing different particles: negative charged particulates, positive ions, neutral particulates, and positive charged particulates. All these particles proceed on the procedure of impaction, agglomeration, combination into masses, increase in weight, decrease in charge quantity and kinetic energy. Once the mass weight exceeds over the electric field forces and gas viscous resistances, it will descend to lower potential level. During the process of descending, the masses will still come into collision with other floating particles, until they fall below the bottom-rim of electric fields and then drop directly into dust buckets. To this stage, the procedure of dust collection may be said completed.

The mechanism of dust collection of ESLFs may be condensed in a few words: the charged particulates are sucked into dust collecting chambers by ESLFs; the electric field sealing effect appears at the lens openings; proper positive corona shells cover the dust collecting electrodes; the charged particulates can only take the way of impaction, agglomeration, combination and descension successively under the action of dust collecting eletric fields; and the masses dropped into dust buckets at last.

Based on the theory and trial tests as mentioned above, a pilot-scale model of ESLFs was developed through large amount of dust collecting experiments on several kinds of dust particulates with different specific resistivities, and different sizes. The dust collecting efficiency has been reached from 97% to 99% and can still be adjusted higher; it bas been proved that dusts of both high and low specific resistivity are collected with higher efficiency; no rapping device is required; less steel is used; technic equipment is simple; low investment and less maintenance cost are needed; and less electric energy is consumed. It is capable to develop to aerosol collection. All these prove a broad application and good prospect of ESLFs.

References

1. Ying Genyu: Electrinic Optics, pp 10-21, 1984.
2. G.A. Nagy et al.: Introduction to the theory of space-charge optics, 1985.
3. I.S. Grant, W.R. Phillips: Electromagnetism, pp 108-150, 1975.
4. Feng Cizhang: Static Electromagnetic Fields, 1984.
5. Xie Guangrun: High-Voltage Electrostatic Fields, 1987.
6. Chen Jidan, Liu Ziyu: Physics in Dielectrics, 1982.

A STUDY OF NUMERICAL SIMULATION FOR GAS FLOW DISTRIBUTION IN THE INLET TO ROOF MOUNTED ELECTROSTATIC PRECIPITATOR

Liu Yushi, Chen Renyi, Zhang Diguang

General Research Institute of Building & Construction
Ministry of Metallurgical Industry, China

Abstract

Installing the electrostatic precipitator on the roof of workshop is one of the effective method to clean the fugitive gases. Uneven gas flow distribution may make the collection efficiency decrease largely. Modeling experiment using equipment to find even distribution needs comparatively more expense. In this paper, a mathematical model is established by solving Navier-Stokes equation to replace, to some degree, the model using equipment.

Introduction

Among various techniques of particulate control for the fugitive gas in a workshop, it is a cost-effective method to install an ESP on the roof. The harmful gas gathered under the roof can be exhausted by buoyancy of relatively hot gas itself without using a fan. Hence, a considerale amount of electricity can be saved. However, flow distribution in the inlet is not uniform due to complex flow pattern within the workshop as well as the obstruction of water handling device in ESP inlet. This nonuniform distribution results in a significant decrease of collection efficiency. Therefore, a flow model study is needed for designing flow control device of each site-specific manpower. If we can establish a mathematical model through numerical simulation, a computer may be used to obtain required data with less cost and manpower. Taking Navier-Stokes equation as the governing equation and setting various boundary conditions, velocity and pressure distribution data can be calculated. From the calculated results we can determine proper configuration of the ESP inlet and its diffusing device to achieve uniform flow and low pressure drop.

Usually the roof condition of an existing workshop is so complicated that the fugitive gas flows out through a number of exits. In this case, roof ESP of unit type is required for each exit. For a new workshop with integral roof, one large ESP can be provided, if necessary. Here is a discussion on one unit of roof ESP. Fig. 1 shows a typical inlet configuration. Nomenclatures used are as follows:

u, v: velocity along x, y direction
ψ : stream function
ω : vorticity
ν, μ: kenematic, dynamic viscosity
ν_t: turbulence kinematic viscosity
ν_{eff}: effective kinematic viscosity
k : turbulence energy
l : Prandtl mixing length
U : resultant velocity, $U=(u^2+v^2)^{\frac{1}{2}}$
ρ : density
p : pressure
U : reference velocity
l_0: reference length
Re: Reynolds number
$\bar{u}, \bar{v}$: time averaged velocity along x, y direction
u', v': pulsating velocity along x, y direction
ε : iterative error limit
ψ_{diff}: iterative error of stream function
ω_{diff}: iterative error of vorticity
P : iterative error of pressure
σ : mean square deviation
Q : flow rate

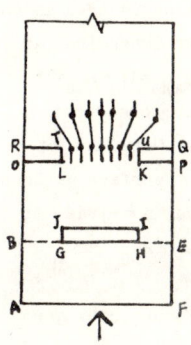

Fig. 1. Typical inlet configuration

Fundamental Mathematical Model

The Navier-Stokes equation of instantaneous turbulent flow field is

$$\frac{\partial}{\partial x_j}(\tilde{u}_i \tilde{u}_j) = -\frac{1}{\rho}\frac{\partial \tilde{P}}{\partial x_j} + \nu \frac{\partial}{\partial x_j}\left(\frac{\partial \tilde{u}_i}{\partial x_j}\right) \quad (1)$$

From equation (1), we can obtain

$$\frac{\partial \psi}{\partial y}\cdot\frac{\partial \omega}{\partial x} - \frac{\partial \psi}{\partial y}\cdot\frac{\partial \omega}{\partial y} = \frac{\partial^2}{\partial x^2}(\nu_{eff}\omega) + \frac{\partial^2}{\partial y^2}(\nu_{eff}\omega) - 4\frac{\partial^2\psi}{\partial x \partial y}\cdot\frac{\partial \nu_{eff}}{\partial x \partial y} + 2\frac{\partial^2\psi}{\partial y^2}\cdot\frac{\partial \nu_{eff}}{\partial x^2} + 2\frac{\partial^2\psi}{\partial x^2}\cdot\frac{\partial^2 \nu_{eff}}{\partial y^2} \quad (2)$$

and $\quad \nu_{eff} = \frac{1}{R_e} + C_\nu K^{\frac{1}{2}} l \quad (3)$

Here, C_ν is a constant. In addition,

$$\nabla^2 \psi = -\omega \quad (4)$$

$$u = \frac{\partial \psi}{\partial y} \quad (5)$$

$$v = -\frac{\partial \psi}{\partial x} \quad (6)$$

Equations (2)-(6) form a group of governing equations. Near the walls, 'l' in equation (3) increases with distance from the wall according to a linear distribution, while at distance farther from the wall, 'l' remains constant. From the relation

$$2K = 3U^2 \quad (7)$$

where $U=(u^2+v^2)^{\frac{1}{2}}$, we can obtain distribution of K.

Transform the above governing equations into proximate difference equations by integration.

Area for calculation is AFQR as shown in Fig. 1. Boundary conditions are determined as follows:
(a) on side walls, $\psi_{AOLT}=1$, $\psi_{FPKU}=0$;
(b) on the central obstruction, take average values of stream functions at certain interior points near front surface as boundary values;
(c) in the inlet area, $\psi = -\int v dx + 1$;
(d) at the outlet, $\psi_p = 2\psi_{Q_1} - \psi_{R_1}$.

Boundary conditions of vorticity are:
(a) on the solid wall,
$$\omega_p = -\left[\frac{3(\psi_E - \psi_p)}{(\Delta n)^2} + \frac{\omega_E}{2}\right];$$

(b) in the inlet, =0;
(c) at the outlet,
$$\omega_p = \frac{1}{3}(\omega_{Q_1} - \omega_{R_1})$$

(d) on the protruding corner, use average value of vorticities at two adjacent points as boundary values.

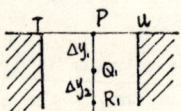

Fig. 2 Mesh points near the outlet

Results of Numerical Simulation and Model Experiment

Conditions for calculation to be considered convergent are as follows:

$$\psi_{diff} = |\psi^{(n)} - \psi^{(n-1)}|_{max} \quad (8)$$

$$\psi_{diff} = |\omega^{(n)} - \omega^{(n-1)}|_{max} \quad (9)$$

$$(\psi_{diff} \text{ and } \omega_{diff}) \leq \varepsilon \quad (10)$$

Use the mathematical model established to make model experiments and observe flow conditions for different inlet velocity distributions.

Fig. 3 indicates the calculated velocity distribution at outlet TU and the measured inlet distribution to be in fair approximation.

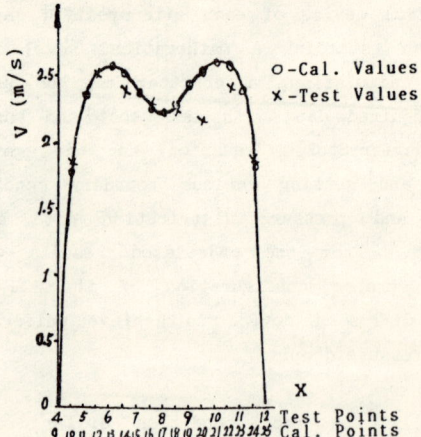

Fig. 3 Velocity distribution at the outlet

Measures Taken to Achieve Uniform Flow in Inlet

Based on the calculated data (as shown in Fig. 3) a scheme for uniform gas distribution can be determined. In order to minimize flow resistance, a set of diffusing plates are used to achieve uniform distribution. Using continuity equations of integral form in each flow channel, we obtain simultaneous equations as follows:

$$\begin{aligned} A \cdot V_1 &= x_1 \cdot V \\ &\cdots\cdots \\ A \cdot V_n &= x_n \cdot V \\ \sum_{i=1}^{n} x_i &= 600 \end{aligned} \qquad (11)$$

Solving the above equations we have

$$x_i = \frac{V_i}{\sum_{i=1}^{n} V_i} \cdot 600 \qquad (12)$$

Use mean square deviation to indicate degree of uniformity,

$$\sigma = \frac{1}{n} \sum_{i=1}^{n} \left(\frac{V_i - V_0}{V_0} \right)^2 \qquad (13)$$

$$V_0 = \frac{1}{n} \sum_{i=1}^{n} V_i \qquad (14)$$

When $n=8$, adjust the inclination of diffusing plates according to the above-mentioned scheme and take measurement. The value of σ decreases from 0.7610 without diffusing plates to 0.4087 with diffusing plates.

Increasing the number n without increasing too much pressure drop can further lower the value of σ. No further discussion will be made here on this subject.

Pressure Drop

For such problem of gas flow, the authors had made some calculations by both methods of laminar flow and turbulent flow, and the results meet quite well. Here we discuss only on the method of laminar flow. From dimensionless laminar N.S. equation, pressure Poisson equation is given by

$$\nabla^2 P = -\zeta \qquad (15)$$

and

$$\zeta = 2\left[\left(\frac{\partial u}{\partial x} \right)^2 + \frac{\partial u}{\partial y} \cdot \frac{\partial v}{\partial x} \right] \qquad (16)$$

Since the velocity distribution has been obtained by calculation, the problem of pressure field is summed up into solving Poisson equation.

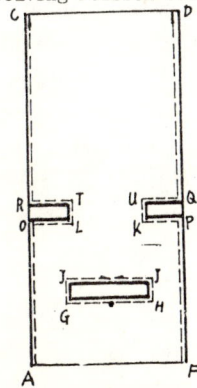

Fig. 4. Area for pressure calculation

Boundary conditions of pressure are:
(a) on solid walls, $\frac{\partial p}{\partial n} = 0$;
(b) in the inlet section AF (see Fig. 4), $P_0 = 1$;
(c) at the outlet section CD, considering N.S. equation in y direction, we have

$$P_{i,N} = P_{i,N-1} - \frac{1}{\beta}\left[(UV)_{i+1,N-1} - (UV)_{i,N-1} \right] - \left[V_{i,N}^2 - V_{i,N-1}^2 \right] \qquad (17)$$

(d) assume that there is a stream line along the solid walls. Use Bernoulli equation for this stream line as an additional condition,

$$P_2 = P_1 + \frac{1}{2}(V_1^2 - V_2^2) - f_r \cdot \Delta \tau \qquad (18)$$

Here, $\Delta \tau$ is the length of one differential grid along the walls.

Use iteration. When iteration meets the following equation, we take the iteration as convergent.

$$P_{diff} = \frac{|P^{(N)} - P^{(N-1)}|_{max}}{|P^{(N)}|_{max}} \qquad (19)$$

and

$$P_{diff} \leq \varepsilon \qquad (20)$$

Let P_c be average pressure at the outlet CD, then pressure drop across the inlet and outlet is

$$\Delta P = P - P_c \qquad (21)$$

Varying the vertical position of central obstruction changes the pressure drop, values of which can be obtained by calculation.

Conclusions

It has been proved that $\psi - \omega$ method can be used effectively to solve Navier-Stokes equation for the problem of flow through a chamber, as discussed in this paper. In engineering practise, boundary conditions are usually complicated. It is a practical idea to establish a simple turbulence model, which makes calculation easy to converge. As to pressure drop, we have discussed on numerical simulation for the purpose of engineering application. Gas distribution in the inlet of roof mounted ESP modified through numerical simulation was verified by model experiment to have significant improvement. It is seen that computer-aided numerical simulation for problems of such a subject is feasible.

EXPERIMENT ON ELECTROSTATIC PRICIPITATOR WITH APPLICATION OF ULTRASONIC AGGLOMERATION

Tomoo Nakane, Takashi Hirata and Koichiro Seya

Department of Electrical Engineering, College of Industrial Technology, Nihon University, 1-2-1 Izumicho, Narashino, Chiba, 275 Japan

ABSTRACT

This paper examines an effect of application of ultrasonic agglomeration on the electrostatic precipitator. As a result, the collection efficiency of sub-micron particles was increased with using 22 kHz ultrasonic waves. Diluted sub-micron particles were the only object of this experiment.

INTRODUCTION

An electrostatic precipitator (ESP) usually has a low collection efficiency for sub-micron particles.[1] It is considered that a growing sub-micron particles to enough size by ultrasonic agglomeration[1] is one of the methods to operate ESP to obtaining a good collection efficiency. Hueter and Bolt were theoriticaly investigated[2] on the sonic agglomeration for the particles suspended in air. Inside the ESP, 22 kHz high-intensity ultrasonic sound field was produced and high collection efficiency was obtained.

EXPERIMENTAL APPARATUS

The outline of experimental equipment was shown in figure 1. The size of pipe type ESP was 6 cm diameter and 44 cm long. the discharging wire, 0.12 mm diameter, was located at the center of ESP. Between wire (−) and the pipe (+ earthed), a voltage of 0−7 kV was applied. Figure 2 shows the volt-ampere characteristics of the ESP. The high-intensity ultrasonic field was produced by the vibrator with 80 mm diameter stepped circular vibrating plate,[3] which was located at 6 cm apart from out let side of ESP. The frequency of ultrasonic sound 22 kHz. Figure 3 shows the distribution of sound pressure level. Figure 4 shows the relation between input power of ultrasonic vibrator and sound pressure level. From figure 3 and 4, it can be seen that the sound field was high enough at inside the ESP. The particles used in this experiment were made by burning the insense sticks, the average particle size was measured about 0.4 microns. The particles were mixed with clean air. The flow rate in the ESP was kept at 0.75 m/s. The number of particle was measured by the particle counter at the out let of ESP.

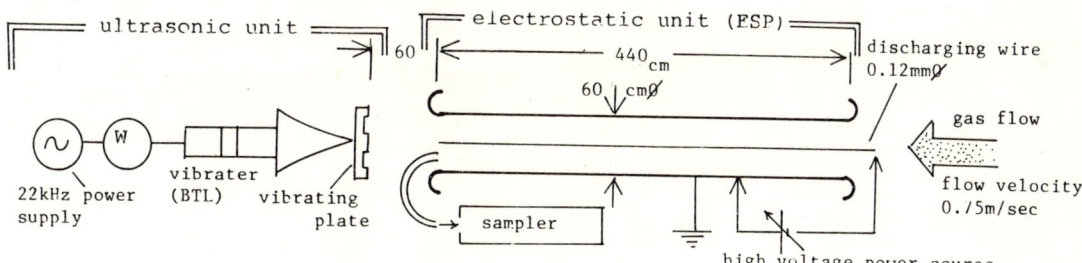

Fig.1 Schematic diagram of experimental apparatus.

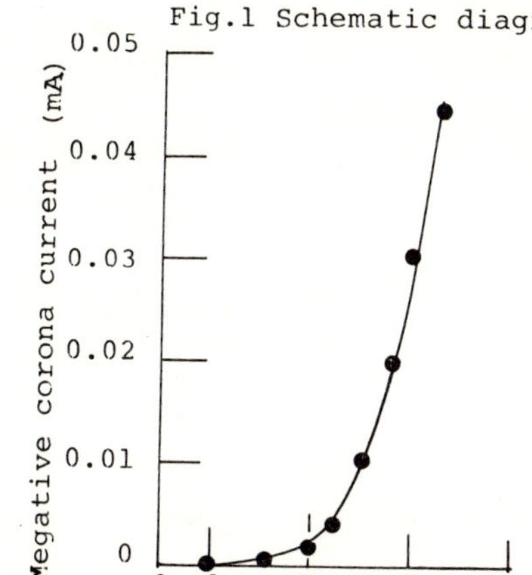

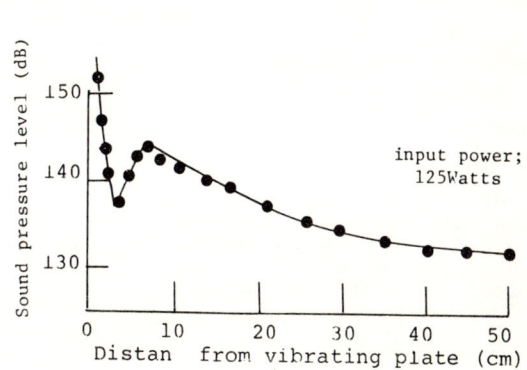

Fig.2 Volt-ampere characteristic of the ESP.

Fig.3 Sound pressure distribution.

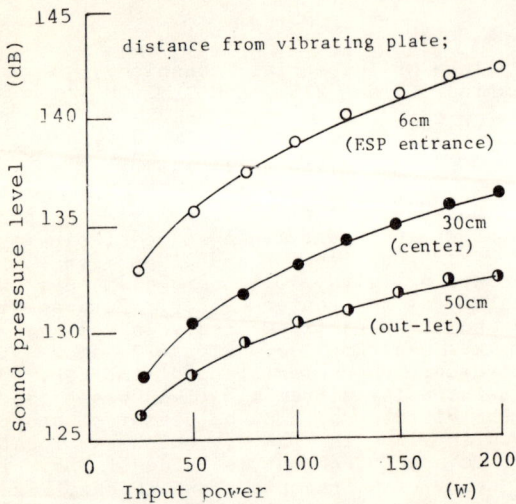

Fig.4 Relation between input power of ultrasonic vibrator and sound pressure level.

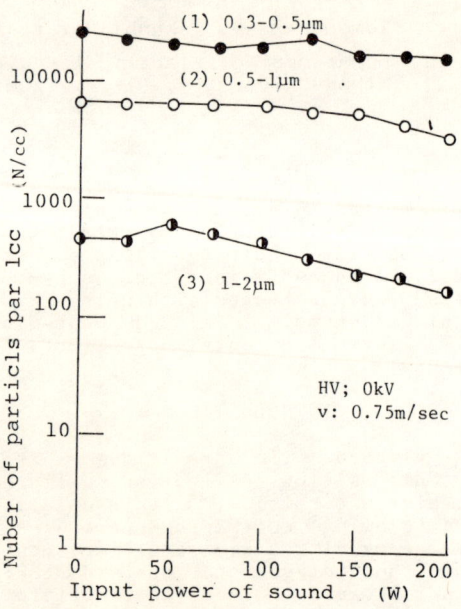

Fig.5 The shift of the number of particles in flow out gas when the applied voltge of 0kV.

EXPERIMENTAL RESULTS

The experimental results were shown in figures 5-9. The verticle axes show the number of particles per 1cc draw off gas. The transverse axes show input power of ultrasonic vibrator. In these figures, ● marks were obtained in case of particle size of 0.3-0.5 microns, ○ marks were 0.5-1.0 microns. In figures 5-7, ◐ marks was measured in case of particle size of 1-2 microns. Figure 5 was acquired without applied voltage, figures 6-9 were obtained with applied voltage of 1,2,3,4 kV respectively.

Figure 5 was obtained without applied voltage. When without application of sound waves, the number of particles which came out from ESP were 32500 N/cc. The particle size of 0.3-0.5 microns was 25000 N/cc, 0.5-1.0 was 7100 N/cc, 1.0-2.0 microns was 400 N/cc. Then ultrasonic was irradiated, the number of particles were decreased on each particle sizes. For instance, line (1) in this figure, the number of particles were decreased to 20000 N/cc with application of ultrasonic sound waves of 200 W input power to the vibrator. In case of line (2), the number of particle were decreased to 400 N/cc, in case of line (3) was decreased to 200 N/cc. It can be noted that as the sound waves were irradiated (input power of 200 W) the number of particles was 25% decreased (the number was 24200 N/cc).

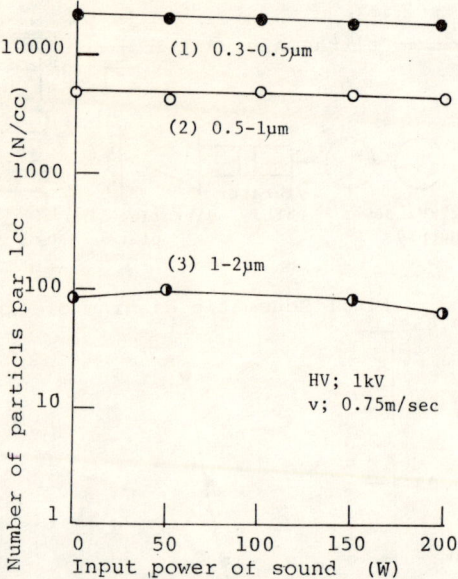

Fig.6 The shift of the number of particles in flow out gas when the applied voltage of 1kV.

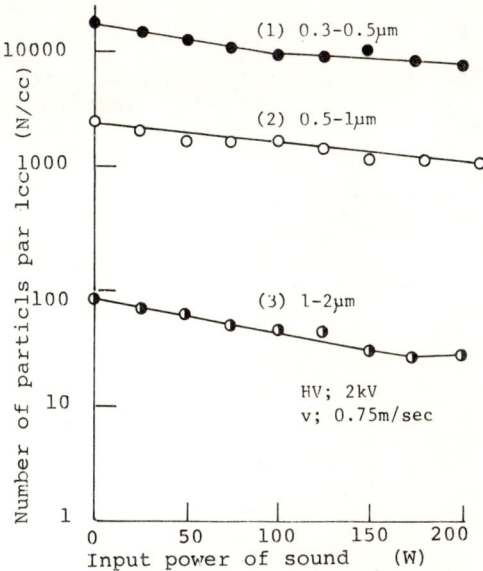

Fig.7 The shift of the number of particles in flow out gas when the applied voltage of 2kV.

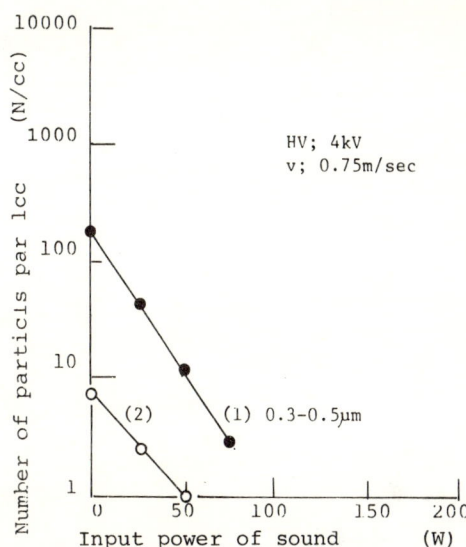

Fig.9 The shift of the number of particles in flow out gas when the applied voltage of 4kV.

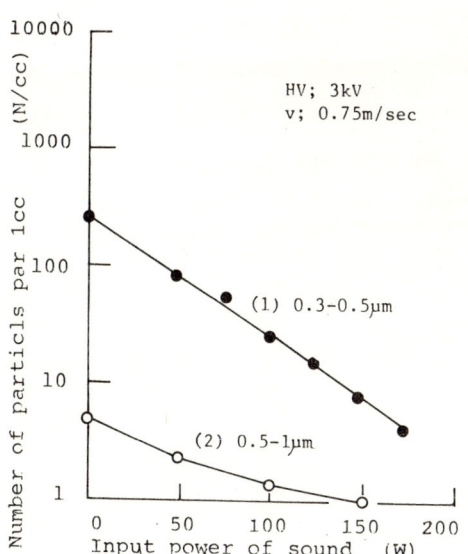

Fig.8 The shift of the number of particles in flow out gas when the applied voltage of 3kV.

Then figure 6 and 7 were obtained with applied voltage of 1, 2 kV. It can be seen that the number of particles were decreased by application of ultrasonic sound waves, they show a similar tendency to figure 5. As an applied voltage was increased to 3, 4 kV, figure 8 and 9 were acquired. An ultrasonic was extremely effected in case of applied voltage of 3 and 4 kV.

In figure 9, the numbers of particles were decreased to 159 N/cc (0.3-0.5 microns), 8 N/cc (0.5-1.0 microns) with applied voltage of 4 kV, and further application of ultrasonic (input power of 75 W), the number of particles was decreased to 3 N/cc (0.5-1.0 microns).

Then investigate about the results of figure 9, the collection efficiency was calculated by the following equation,

$$\eta = \frac{Ic - Pd}{Ic} \times 100 \ (\%) \qquad (1)$$

where η is the collection efficiency, Ic is the aerosol concentration of initial melting and Pd is aerosol concentration of operating ESP. Figure 10 was obtained from the results of figure 9 with calculation of equation (1). The vertical axes of the figure shows the collection efficiency and the transverse axes shows input power of the ultrasonic vibrator. From this figure, the higher input power of the ultrasonic vibrator, the higher collection efficiency was obtained. For

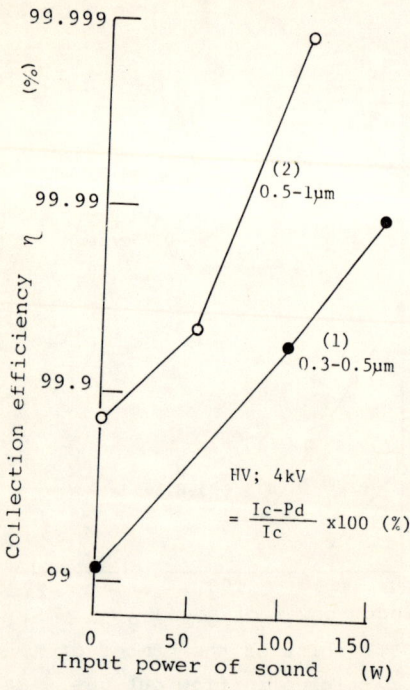

Fig.10 The collection efficiency of ESP with varying sound intensity.

instance, in case of the particle size of 0.3-0.5 microns, the collection efficiency was 99.14 % without sound waves. As ultrasonic (input 100 W) was added, the collection efficiency was increased to 99.98 %, so that it can be noted that the application of ultrasonic was very effective to collect sub-micron particles.

CONCLUSION

The application of ultrasonic to the ESP was very effective to increase the collection efficiency of sub-micron particles. In this experiment, the collection efficiency was much higher than the results which had been reported [4] and some obscure passages points were occurred.

(1) It is known that an ultrasonic agglomeration was occurred at a sound pressure of more than 150 dB, however in this experiment, it was caused at 130-140 dB (except the neighbourhood of ultrasonic unit).

(2) In general, ultrasonic agglomeration is very effective in case of thick aerosol gas. Even in thin aerosol gas, its effect was high in this experiment.

(3) In the previous report [5], the application of ultrasonic agglomeration to the ESP was effective in case of lower applied voltage, however in this experiment, it was effective in case of higher applied voltage.

For further examination, the effect of sound waves except for agglomeration, for instance a shock wave effect or hydrodynamic effect of sound, should investigate.

REFERENCES

1) "Electrostatic Handbook" Ohm Sha 542-543 (in Japanese)
2) Hueter and Bolt "Sonics" John Willy & Sons Inc. 211-214
3) Otsuka and Seya 1986 Proc of 12th ICA L2-3
4) Nanane and Seya 1986 Proc Int of Electrostatics Jpn Vol.10 No.5 331-337
5) Nakane, Otsuka and Seya 1984 Proc Second ICEP 734-738

PRELIMINARY STUDY ON THE COLLECTION OF FINE PARTICLES BY CHARGED DROPS

Tan Tianyou   Zhang Lian
Associate Professor   Lecturer

Wuhan Metallurgical College of Construction
Wuhan, Hubei, China

Abstract

This report describes one of the most efficient means to collect fine particles. The theoretical analysis of spray charging as well as particle removal efficiency have been presented with the primary experiment, which lays the foundation for further study.

Introduction

Various dust collectors have been used and improved to control air pollution, but a large amount of fine particles which are the most harmful to human health are still being emitted into the atmosphere due to either the inefficiency or expensive cost. However, the collection efficiency is largely increased by means of spray charging in the collecting space. With a large amount of charging drops the effect of electrostatic force could be enhanced and collection efficiency could be increased. The research study in this paper consists of two parts as following:

(1) Study on mechanism of spray charging and the collection efficiency.

(2) Based on the measurement of drops charge/mass ratio with different charging methods, the optimized spray charging is proposed.

Theoretical analysis on collection efficiency

a. Mathematical model of collection efficiency

All those particles approching the drop inside the swept area with radius, S, will impinge on the drop as indicated in Fig.1. For a noncharged drop the swept radius, S, is less than the radius of drop, R, even more than R due to the electrostatic forces between the charged drop and charged (or noncharged) particles. The single drop collection efficiency, E, is then in proportion to S^2. If the particles are uniformly distributed in the gas stream, collection efficiency, E, may be defined as: [2],[3]

$$E = \frac{\pi S^2 \sigma}{\pi R^2} \qquad (1)$$

where σ —attachment coefficient. If it assumed all particles will stick to the drop upon striking it, $\sigma = 1$;

S—swept radius of drop;
R—physical radius of drop.

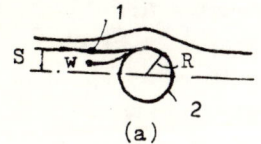

(a)

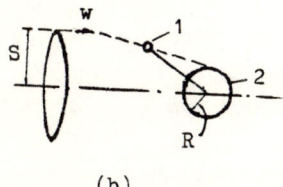

(b)

Fig.1 Trajectory of particle collected by drop

(a) noncharged drop;   (b) charged drop;
1 particle;   2 drop.

The charged spray scrubber collection efficiency then can be expressed by the ratio of the total swept area of all charged drops to the area of the cross section of srubber as follows:

$$\eta = \frac{n \pi S^3}{A} \sigma \alpha \qquad (2)$$

where n — drops number per m^2 cross section;
A — cross section area of the scrubber;
α — overlap coefficient of the drops swept area, usually 1.

Assume the count concentration of particles in scrubber is N per m^3 of gas, and the thickness of section of scrubber associated with a drop is 2S, then

$$n = N\, 2S\, A \quad (3)$$

Thus equation (2) becomes

$$\eta = N2SA\pi S^2 \sigma\alpha / A = 2S^3 N\sigma\alpha\pi \quad (4)$$

Equation (4) shows that scrubber efficiency depends on the swept radius, S, to a great degree.

b. **Swept area of chrged drops**

The particle trajectory is given by the momentum balance:

$$F - P_D = m \frac{dV}{dt} \quad (5)$$

$$P_D = 3\pi\mu d_p V/C' \quad (6)$$

where F — external force on the particle (electrical force etc.);
P_D — resistance force;
m — mass of particle;
v — velocity of particle;
t — time;
μ — the dynamic viscosity of gas;
d_p — particle diameter;
C' — cunningham slip correction factor.

The particles accelerating process is neglected because the particle is very small. When the external force equals resistance, the particle velocity, v, can be obtained.

$$v = \frac{F\, C'}{3\pi\mu d_p} \quad (7)$$

Assume that in the scrubber gas and drops have the same velocity in the same direction. The particle carried by the gas moves a distance, S', during the time it stay in the scrubbber.

$$S' = vt = \frac{FC't}{3\pi\mu d_p} \quad (8)$$

The swept radius, S, is then:

$$S = S' + 1/2\, d_w \quad (9)$$

where d_w — drop diameter.

When the drop is not large, S' can be used approximately as S:

$$S = S' = \frac{F\, C'\, t}{3\pi\mu d_p} \quad (9')$$

c. **Electrostatic forces**

The particles near the charged drop are affected by electrostatic forces. There are two cases according to if the particle is charged or not.

1. **Charged particles**

The Coulombic force between charged drops and charged particles is:

$$F = \frac{Q\, q}{4\pi\varepsilon_0 r^2} \quad (10)$$

where Q — drop charge;
q — particle charge;
ε_0 — permittivity of free space;
r — distance between drop and particle.

As the particle gradually moves closer to the drop r varies. Assume r approximately equals half of swept radius, S, thus equation (10) becomes:

$$F = \frac{Q\, q}{\pi\varepsilon_0 S^2} \quad (11)$$

Substitue equation (11) in equation (9'),

$$S = \left(\frac{Q\, q\, C'\, t}{3\pi^2 \varepsilon_0 \mu\, d_p}\right)^{1/3} \quad (12)$$

And removal efficiency is from equation (4):

$$\eta = 2/3 \frac{Q q\, C' t}{\pi\varepsilon_0 \mu\, d_p} N \sigma\alpha \quad (13)$$

2. **Uncharged particles**

When particles are not charged, a charged drop's image force is: [4],[5]

$$F = \left(\frac{\varepsilon_p - \varepsilon_w}{\varepsilon_p + 2\varepsilon_w}\right)\left(\frac{d_p^3\, Q^2}{16\pi\varepsilon_0 r^5}\right) \quad (14)$$

where ε_p — dielectric constant of the particle;
ε_w — dielectric constant of the drop.

If r considered as S/2, from equation (14) and equation (9') S may be expressed by:

$$S = \left[\left(\frac{\varepsilon_p - \varepsilon_w}{\varepsilon_p + 2\varepsilon_w}\right)\left(\frac{2 d_p^2 Q^2 C' t}{3\pi^2 \mu\varepsilon_0}\right)\right]^{1/6} \quad (15)$$

From equation (4) remaovel efficiency becomes:

$$\eta = \left(\frac{\varepsilon_p - \varepsilon_w}{\varepsilon_p + 2\varepsilon_w}\right)^{1/2} (8/3)^{1/2} \left(\frac{C'\, t}{\mu\varepsilon_0}\right)^{1/2} d_p Q N\sigma\alpha \quad (16)$$

Apparatus

The schematic diagram of the apparatus is shown in Fig.2 with a plexiglass tube of 60mm in diameter and 1000mm in length, dust feeder, blower, precharging system, vaccum pump, and Farady cage. the spray water drops from the injector could be inductirely charged with the ring elecrode 3, and be charged directly. Due to the gravitation these drops pass through the tube and then into the Farady cage, the charge/mass ratio could be measured.

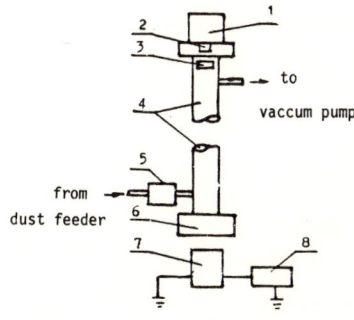

Fig.2 Experimental apparatus
1 — water box;
2 — injector;
3 — induction ring electrode;
4 — plexiglass tube;
5 — dust precharging device;
6 — hopper;   7 — Farady cage;
8 — electrometer.

Experiment Results

a. <u>Direct charging</u>

When negative D. C. high voltage is applied directly to the injector, the measured charge/mass ratio is linearly dependent on the voltage.

When the external voltage is less than 10kV as shown in Fig.3. The trjectory of the falling drop becomes curved when voltage is greater than 14kV, a fine stream from the injector is formed when the voltage is greater than 16kV.

b. <u>Induction charging</u>

Negative high voltage is applied to the induction ring electrod, above which the injector is located. Experiments show that as drops pass through the

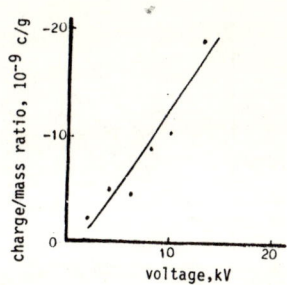

Fig.3 Relationship between charge/mass ratio and voltage for direct charging

ring, they becomes positively charged as shown in Fig.4. When the applied voltage is greater than 16kV, the drops become fine and move towards the ring.

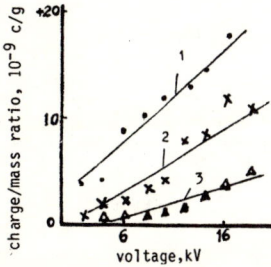

Fig.4 Relationship between charge/mass ratio and voltage for inductioon charging

1 — long injecting needle with 35mm ring diameter;

2 — long injecting needle with 25mm ring diameter;

3 — short injenting needle with 25mm ring diameter.

By induction charging in which the water is grounded, a higher charge/mass ratio can be obtained even if the voltage is the voltage is not very high. The charge polarity of water drops is opposite to the applied voltage as shown in Fig.5. Based on the obtained charge/mass ratio results, it could be pointed out the induction charging method could be efficiently used for spray charging.

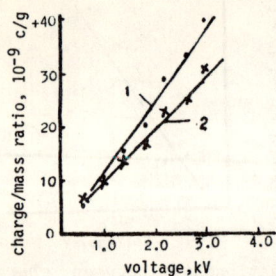

Fig.5 Relationship between charge/mamss ratio and voltage for induction charging with water grounded

1 — long injecting needle with 25mm ring diameter;
2 — short injecting needle 25mm ring diameter

REFERENCES

[1] 莺义夫: Handbook of Electrostatics, 1972

[2] C.N. Davis: Air Filtration

[3] Shui - Chow etal: Spray Charging and Trapping Scrubber for Fugitive Particle Emission Control, P. B. 82-115304, 1981

[4] H. F. Kraemer et al: Collection of Aerosol Particle in Presence of Electrostatic Fields, Industrial and Engineering Chemistry, vol. 47 No. 12, 1955

[5] Tan Tianyou, Liang Fengzhen: Industrial Ventilation and Dust Control, China Building Industry Press, 1984

PULSE ENERGISED ELECTROSTATIC PRECIPITATOR WITH TRANSVERSE LOUVERS— A DUST COLLECTION TECHNIQUE FOR FLUIDIZED-BED BOILER

Liu linmao, Wang Rongyi, Wu Yan, Wang Xilu, Hu Jingming
Ma Furong, Sun Dawei and Yu Zhigang
(Northeast Normal University, Changchun, China)

ABSTRACT

A solution to collecting hight-concentration and high resistivity dust by changing relectodes configuration of the precipitator and the pattern of energization is dealt with in this paper. Experiments showed that the collection efficiency can be increased by adding transverse louvers to the electrostatic precipitator with conventional electrode configuration and that the size of the precipitator can be reduced.

Fluidized-bed boilers can burn low grade coal effectively. This is a matter of significance in coal economy. But its impact on air environment hinders it from being widely utilized due to high dust concentration. On the basis of conventional horizontal type ESP, we integrate pulse energization and transverse louvers for purposes of overcoming the back-corona phenomena in high-resistivity dust collection and increasing collection efficiency as well as the gas velocity, resulting in reduction of ESP dimension and hence the capital cost. Moreover the hybrid technique is adaptable to retrofit existing ESP.

LABORATORY EXPERIMENTS

We installed a facility (Fig.1) in the laboratory. The star wires of cross section 4mm×5mm are taken as discharge electrodes. The pulsed high voltage power supply which can be adjusted from 0 to 180KV [1] was used for energization. The louvers used in the experiment include three types as shown in Fig.1b, 1c and 1b. The louver open area in type (b) and the gas flow rate are both adjustable. The dust used in the experiment is 500# cement. Collection efficiency is measured by weighing method.

RELATION BETWEEN LOUVER OPEN AREA AND PRESSURE DROP

Transverse louvers added to ordinary plate type ESP could increase the collecting area without increasing the

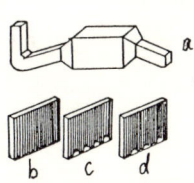

Fig.1 The experiment facility and louvers

dimension of ESP, but ESP pressure drop will be increased more or less. We investigated experimently the relation between pressure drop and open area of louvers. Results is shown in Fig.2. Pressure drop is negligible when % open area of diffusion plates is smaller than that of louvers. The transverse louvers would play an important role in pressure drop when %open area of diffusiion plates is larger than that of louvers.

Fig. 2 Relation of pressure drop against % open area of louvers
A - % open area of louvers
pp - Louver pressure drop, %

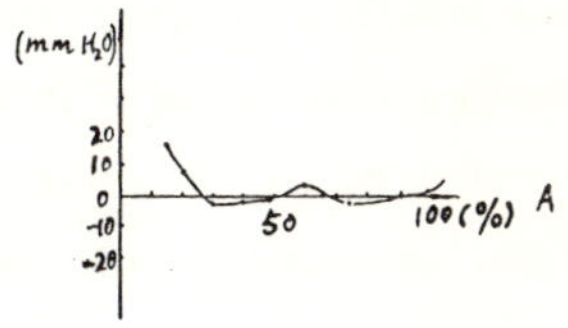

Fig.2

RELATION BETWEEN COLLECTION EFFICIENCY AND OPEN AREA OF LOUVERS

We investigatied general relation between cllection efficiency and open area of the louvers. The average applied voltage is 60KV, and average corone current is 0.8mA. The results are shown in Fig.3 and Fig.4. It is seen that a higher collection efficiency occurs in the region where % open area of louvers larger than that of diffusion plates. This range of higher efficiency relates to %open area of the diffusion plates.

-57-

Fig.3 Relation between collection efficiency and open area of louvers
η - Collection efficiency
A - Open area of louvers
1. % open area of diffusion plates, 17%

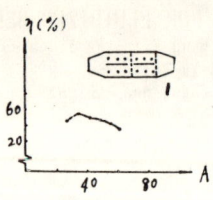

Fig.3

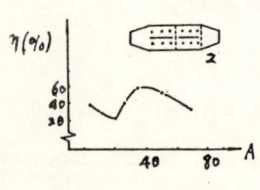

Fig.4

Fig. 4 Relation between collection efficiency and % open of louvers
η - Collection efficiency
OR- Open area of louvers
2. % open area of diffusion plates, 32%

INFLUENCE OF ELECTRODE CONSTRUCTION ON COLLECTION EFFICIENCY

The arrangement of electrodes in an ordinary horizontal type ESP is that the negative and positive electrodes are in parallel with the direction of gas flow. On such basis the louvers are placed perpendicular to the direction of gas flow. We compared the performance of several different electrode structures by experiments. The average voltage applied is 60KV. The results obtained are listed in Table 1. It is evident that by adding transverse louvers the collection efficiency is increased.

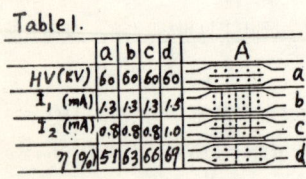

Table 1. Influence of electrodes arrangement on collection efficiency
A - Electrodes arrangement
HV- High voltage
I_1- Current without dust
I_2- Current with dust
η - Collection efficiency

RESULTS IN PRACTICAL APPLICATIONS

ESPs with pulsed energisation and transvere louvers have been succesefully used in dust precipitation of fluidized-bed boilers and become available with a series of commercial products. Take one for a 4t/hr fluidized-bed boiler as an example, The gas flow is 10000-13000Nm³/hr, dust concentration is 40-70g/m³, gas temperature is 130-190°C and ash resistivity is 10^{11}-10^{13} Ω·cm. The poximate chemical analyses of the coal burned are, moisture content 11.2%, Ash-52.06%, volatile matter 20.77% and heating value 2412 kcal/kg. Two stages of dust collecors including a cyclone as the first stage, and pulse energized ESP (Fig. 1a) whith transverse louvers (Fig.1c) as the second are adopted. Other features of the ESP are, hammer type rapping system, discharge wire spacing 600mm, cross section of elestric field 4m², gas residence time 7 sec, average current density 50μA/m², power consumption 1 KW, ESP emission 80~100mg/Nm³ and collection efficiency 99.8~99.9%. Since commencement of operation more than one year ago, the installation has performed very well.

CONCLUSIONS

Applications of pulse energised ESP with transverse louvers both in laboratory and in industrial sites has exhibited outstanding advantages. Its efficiency increased evidently without increasing gas pressure drop, as compared with ordinany hovizontal ESP. Pulse energization enables it suitable for collection of high concentration and high resistivity dust. Its application to fluidized-bed boilers achieves good results with dust emission lower than the regulation 200mg/m³. Widening the electrode spacing can reduce power consumption, since power consumption in ESP is proportional to the capacitance between discharge and collection electrodes.

In a word, super-high voltage plused ESP equipped with transverse louvers is a seccesseful modification of ordinary plate-type ESP.

REFERENCES

[1] Wu Yan, Wang Rongyi and Liu Linmao, Miniature pulsed voltage source and its applications. 日本国静电学会讲演论文集。(1986.10)

FIELD ANALYSIS IN ELECTRIC PRECIPITATORS

Zhenghu Fang, Xiangyu Gu, Ning Dai

Zhejiang University

Hangzhou, China

Abstract

As the corona discharge and the charging and collection of dust particles are closely related to the electric field distribution in precipitators, a detailed analysis of the later will surely improve precipitator's design and collection efficiency. This paper is devoted to analyse the corona field equations, the existence and uniqueness of their solutions. A numerical method of solution and proper choice of boundary conditions are also recommended. Two dimensional case studies involving the use of FEM-FDM and the charge simulation method are presented.

General Description

It is well known that the corona discharge, the charging and collection of dust particles are closely related to the electric field distribution between the electrodes in precipitator. The studies of the related quantities such as the electric potential φ, electric field intensity E, space charge density ρ and the current density j are of primary importance in improving their design, optimizing the structure and performance. But due to the actual operating conditions are too complicated theoretical studies have not been widely carried out. This paper is devoted to the theoretical study of corona field equations including the existence and uniqueness of their solutions, the choice of appropriate boundary conditions, etc. The distributions of φ, E and ρ for a 2-dimensional field under no load condition are also presented. Some practical problems such as the arrangement of electrodes, wide spacing problems and the shape of corona electrode are also discussed.

The Analysis of Electric Field

Mathematical Model

Consider the wire-plate arrangement as illustrated in Fig.1. The geometrical shape of the electrodes does not affect the field

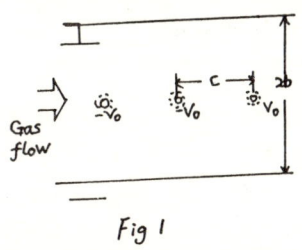

Fig 1

equations but boundary conditions. Between the electrodes there are negative ions, electrons and charged particles with their densities ρ_i, ρ_e and ρ_p respectively. When dust particles are larger than 0.5 μm, field charging is dominative, the diffusion current can be disregarded, and ρ_e due to the small value is also negligible. Let ρ be the equivalent charge density and K the equivalent mobility, then we have:

$$\nabla \cdot \vec{E} = \rho/\epsilon_0 \quad \text{or} \quad \nabla^2 \varphi = -\rho/\epsilon_0 \quad (2-1)$$

$$\nabla \cdot \vec{j} = (K_i \rho_i + K_p \rho_p)\vec{E} = K\rho\vec{E} \quad (2-2)$$

$$\nabla \cdot (K\rho\vec{E}) = 0 \quad (2-3)$$

At no load, $\rho_p = 0$ and K_i is nearly constant, the above equation can be written as:

$$\nabla(\nabla^2 \varphi) \cdot \nabla \varphi + (\nabla^2 \varphi)^2 = 0 \quad (2-3')$$

Under loaded conditions, ρ and K depend on flow distributions and are functions of space coordinates. The equations will take the form of

$$\nabla \cdot (K\rho\vec{E}) = \nabla(K\rho) \cdot \vec{E} + K\rho \nabla \cdot \vec{E} = 0 \quad (2-4)$$

The boundary conditions for φ and ρ are:

$\varphi|_s$ = const.   s is the surface areas of the electrodes

$\rho|_{s2} = \rho_{so}$ s2 is the surface areas of corona electrode

$$(2-5)$$

Existence and Uniqueness of Solution

This problem has so far not been treated. If we put equation (2-4) in the form of

$$\nabla(K\rho) \cdot \nabla \varphi = (K\rho)^2/(\epsilon_0 K)$$

and let $K\rho = u$, considering the component of $\nabla \varphi (E_x, E_y, E_z)$ being functions of u and the coordinates, we have

$$\frac{\partial u}{\partial x}E_x + \frac{\partial u}{\partial y}E_y + \frac{\partial u}{\partial z}E_z = -\frac{u^2}{\epsilon_o K} = h(u,\rho) \quad (2-6)$$

which is a quasi-linear partial differential equation, and by virtual of Cauchy method, we can prove that under given boundary conditions, it has a unique solution. The same conclusion can be drawn from the fixed-point theory in functional analysis.[1]

Analysis of Solution Methods

Due to complex electrode configuration, only numerical methods are feasible [2,3,4]. Since the equation is a nonlinear partial differential equation of third order, simplifications are necessary. Now we take one dimensional problem corresponding a wire-pipe arrangement for analysis. In numerical analysis, directly solving the field equation is scarcely used for its complication, the often used method is to solve the simultaneous equations by iteration. There are three modes of iteration under the same initial assumption of $\rho^{(o)}$ distribution. In the first mode, we solve the Poisson's equation for $\phi^{(k)}$ at first, then substitute the results into the continuity equation to solve $\rho^{(k)}$. If $|\rho^{(k+1)} - \rho^{(k)}|$ not less than the accuracy ε, repeat the iteration once again until $|\rho^{(k+1)} - \rho^{(k)}| < \varepsilon$. According to the numerical analysis theory, this method ensures stable convergance and sufficient accuracy. For second mode, first solve the continuity equation for $\phi^{(k)}$, then substitute into the Poisson's equation to solve $\rho^{(k)}$. Because the iteration process involves the solution of differential equation and differentiation is more sensitive to error accumulation than integration, the improper choice of initial value $\rho^{(o)}$ may cause the solution divergence and instability. The third mode is equivalent to solving two elliptic partial differential equations for $\phi_a^{(k)}$ and $\phi_b^{(k)}$, the stability is no problem, but this requires an appropriate error correction function $\delta(\rho^{(k)}) = f(|\phi_a^{(k)} - \phi_b^{(k)}|)$ to modify the $\rho^{(k+1)}$, which is hard to determine, hence there is no apparent gain in using this mode. As a result mode 1 is most preferable and our experience in carrying the calculations by computer has proved this.

Case Study

Field computation of the star-shaped corona electrode

Assume K=const. at no load, the equations are

$$\nabla^2 \phi = -\rho/\epsilon_o \quad (2-7)$$
$$\nabla \rho \cdot \nabla \phi = \rho^2/\epsilon_o \quad (2-8)$$

with boundary conditions same as (2-5). Considering symmetry, only a quarter of the space between 2 corona electrodes and the dust collection plate is taken into consideration. In some papers so far published had taken the second condition in equation (2-5) as $E|_{s2} = E_o$, E_o is the corona-onset field strength. Here we use $\rho|_{s2} = \rho_{so}$, the reason is that E_o depends too much on the ambient conditions while ρ_{s2} is more stable. By using the perturbution theory, the approximate value of the potential for low corona current can be found, from which the first approximation of i is

$$i = \frac{K}{b^2 \ln(d/r_o)} V(V-V_o)$$

where $d = 4b/\pi$, V_o is the corona-onset voltage, r_o is the equivalent radius of star-shaped electrode. Then ρ_{so} can be found by

$$\rho_{so} = \frac{i}{2\pi r_o KE}$$

Under given boundary conditions, a quadrilateral 8 point isoparametric FEM is used to solve the equation (2-7) and the finite difference method (FDM) for equation (2-8). The recurrent formula for the space charge density is

$$\rho_o = \frac{1}{2}\left[\sqrt{\epsilon_o\left(\frac{E_{xo}}{h_x} + \frac{E_{yo}}{h_y}\right)^2 + 4\epsilon_o\left(\frac{E_{xo}}{h_x}\rho_1 + \frac{E_{yo}}{h_y}\rho_2\right)} - \epsilon_o\left(\frac{E_{xo}}{h_x} + \frac{E_{yo}}{h_y}\right)\right]$$

Note that ρ_o in the above expression is the charge density calculated at the node, h_x and h_y are the step length along the x and y axes respectively, E_{xo} and E_{yo} are the components of E at the points, ρ_1 and ρ_2 are the densities at the two neighbouring points. The field values obtained by iteration method mentioned in mode 1 are shown in Fig.2 (for which b=0.15m, c=0.12m, K=2.2×10⁻⁴ m/s.v). It can be seen that ρ and E near the corona electrode decay rapidly while the variation of ρ, unlike E, is much smoother. Table 1 gives the calculated values of average current density on the collecting plate and those by measurement for comparison.

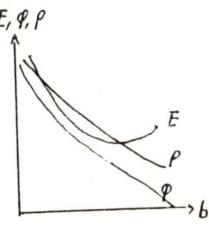

Fig. 2

Table 1

V (kv)	$\rho_{so} \times 10^{-4}$ (c/m^3)	$J_p \times 10^{-3}$ a/m^2 (cal.)	$J_p \times 10^{-3}$ a/m^2 (mea.)
50	0.260	0.467	0.470
55	0.358	0.665	0.670
60	0.480	0.900	0.910
65	0.600	1.240	1.200
69	0.803	1.520	1.460

Analysis for wire-plate electrodes

The analysis is based on the superposition principle. The corona field is resolved into static field and space charge field. Charge simulation method and FDM are used to suit the characters of the respective fields. This method has the advantages of rapid convergence, high accuracy and small storage capacity of computer. Calculation have been carried out for different electrode spacings and the results are list below (table 2)

Table 2, distribution of Jp in different electrode distances

b(m)	c/b	V(kv)	$J_{pa} \times$ (a/m^2)	J_{min}/J_{max}
0.15	0.50	39	2.95×10^{-4}	0.3584
0.15	0.63	39	3.00×10^{-4}	0.2619
0.15	0.80	39	2.957×10^{-4}	0.1999

It has been seen that the current distribution is more uniform on small spacing. The nonuniformity increases with spacing causing the total current to decrease. For a definite spacing b between the corona electrode and the collecting plate there is an optimum corona electrode spacing c_m (often expressed as c_m/b) for a maximum current V-I characteristics for b=300mm. These curves show that optimum c_m/b lies between 0.53--0.68 when J_{pa} ranges from 10^{-4} a/m^2 to 5×10^{-4} a/m^2 (the curve of $J_{pa}=10^{-4}$ is not shown in Fig.3). The figure 0.53--0.68 is quite close to the value of c/b=0.6, chosen by most of the precipitator manufacfacturers abroad. For wire-plate spacing between 200mm and 250mm, which are regarded as wide spacing, c_m/b decreases with the increase of spacing, but at same J_{pa} the calculated field distribution (See Fig.4) and the minimum field intensity for the wide spacing are similar to that of the ordinary design except the former field intensity is lower near the corona electrode and higher near the collecting plate. This type of field distribution is beneficial for collecting dust particles. i.e. to enhance the efficiency.

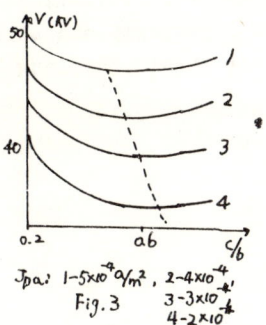

J_{pa}: 1-5×10^{-4} a/m^2; 2-4×10^{-4};
Fig. 3   3-3×10^{-4};
          4-2×10^{-4}

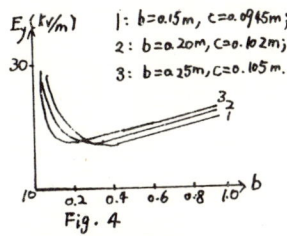

1: b=0.15m, c=0.0915m;
2: b=0.20m, c=0.12m;
3: b=0.25m, c=0.105m.

Fig. 4

Conclusion

1. Corona field equation is discussed including the existence and uniqueness of solution. The algorithm for solving 1-dimensional problems are classified into 3 modes, one of which is recommended to be most

satisfactory.

2. $\varphi|_s$ and $\rho|_{s2}$ are chosen as boundary conditions and a method for determining the boundary charge density $\rho|_{s2}$ is suggested. Two case studies are made, yielding reasonably accurate results.

3. Several field distribution characteristic which be useful to the design and performance analysis of precipitators are discussed.

References:

1. Gu Xiangyu:"The solution of ionic field equation and its application in study of E.P.",Master thesis,Zhejiang University, 1987;
2. S.Oglesby:"Electric Precipitators", P.15--51 ,1978;
3. P.Cooperman:Trans.AIEE,Part I, Vol.79, p.47--50,1960
4. J.R.MaDonald:J.Appl.Phys.,Vol.48,No.6, p.2231--2243,1977;
5. Dai Ning:"The study of ionic field in E.P. and its operation property and structure parameter",Master thesis,Zhejiang University,1988;

RESEARCH ON THE DISTRIBUTION OF ELECTRIC FIELD IN ELECTROSTATIC PRECIPITATOR (ESP)[*]

Wu Zhangfa   Zhang Guoquan

Department of Mining, Northeast University of Technology
Shenyang, Liaoning, China

Abstract

In this paper, the distribution of electric field in wire-plate type ESP was discussed by comparing the theory and measured results. The defects of the past electric field theory were overcome. The new formulas of both the electric field intensity distribution and electric potential distribution were obtained on the basis of the theoretical analyses, and the research in characters of electric field was conducted thoroughly. As a result, it was found that the electric field distribution of wire-plate type ESP is concerned with work voltage, the structure parameter of electrode, particle concentration and space position, and that space charge density has important effects on the characters of electric field.

Introcution

The electric field intensity of ESP has important effects on the charging and collection of particles. Many scholars have made research in the distribution of electric field intensity (or electric field) in wire-plate type ESP generally used in industry. However, because of the non-symmetry electrode structure of this type ESP, it is difficult for researching on the distribution of electric field in it. So far, satisfactory results have not been obtained yet. An analytic method [1] was applied to research in the distribution of electric field under the assumption that space electric density is uniform, and an analytic solution in series form was obtained by P. Cooperman. A numerical method [1] was also applied to that after 'effective mobility' was introduced to the analysis, and a numerical solution was obtained by Leutert. In this paper, a new analytic method was introduced under the assumption that both the ionic electric density and particle electric density are non-uniform, and the new calculation formulas of electric field and important conclusions about the characters of electric field were obtained.

Theoretical Analysis

The two-dimension equation of electric potential distribution is given by

$$\frac{\partial^2 V}{\partial x^2} + \frac{\partial^2 V}{\partial y^2} = -\frac{p}{\varepsilon_0} \qquad (1)$$

Boundary Condition:
$V(0,0)=V_o$ $V(x,b)=0$
$E_x(0,0)=E_y(0,0)=0$ $E_x(c,0)=E_y(c,0)=0$
$E_x(c,y)=E_x(x,b)=0$ $E_y(x,0)=E_x(0,y)=0$

And from current Continuity equation:

$$\nabla_j = \nabla(P \cdot \vec{b} \cdot \vec{E}) = 0$$

We can obtain

$$p^2 = \varepsilon_0 \left(\frac{\partial V}{\partial x} \cdot \frac{\partial p}{\partial x} + \frac{\partial V}{\partial y} \cdot \frac{\partial p}{\partial y} \right) \qquad (2)$$

After dispersing equation and making a repeated calculation and according to the equation $E_y = -\left(\frac{\partial V}{\partial y}\right)$, both the distribution of electric potential and the distribution of electric field intensity were obtained.

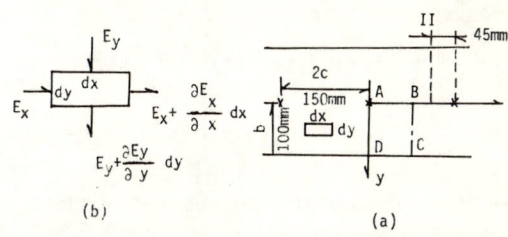

Fig. 1 The mathematical model of two-dimension electric field

---

[*]Projects supported by the Science Fund of the Chinese Academy of Sciences.

The analytical solution

From Gauss theorem, we can obtain

$$\frac{dE}{dy} = -\frac{p_{ek}}{\varepsilon_o} + \frac{i_a}{k\varepsilon_o E} \qquad (3)$$

The solution is:

$$E(y) = E(b) + \frac{p_{ek}}{\varepsilon_o}(y-b) + \frac{i_a}{k \cdot p_{ek}} \cdot$$

$$\cdot \ln\left|\frac{k \cdot p_{ek}(y-b) + k\varepsilon_o E(b)}{k\varepsilon_o E(b)}\right| \qquad (4)$$

The above logarithmic item can be expanded into series with Taylor expansion, i.e.

$$\ln\left|(y-b)kp_{ek} + k\varepsilon_o E(b)\right| = \ln\left|k\varepsilon_o E(b)\right| +$$

$$+ \frac{y-b}{\varepsilon_o E(b)} p_{ek} - \frac{1}{2}\left(\frac{y-b}{\varepsilon_o E(b)}\right)^2 p_{ek}^2 +$$

$$+ 1/3\left(\frac{y-b}{\varepsilon_o E(b)}\right)^3 p_{ek}^3 + \cdots + \frac{(-1)^{n-1}}{n} \cdot$$

$$\cdot \left(\frac{y-b}{\varepsilon_o E(b)}\right)^n p_{ek}^n + \cdots \qquad (5)$$

and

$$q_b = \frac{3n_s}{n_s+2} \pi \cdot \varepsilon_o \cdot d^2 E(y) \qquad (6)$$

$$p_{ek} = q_b \cdot C_n(y) \qquad (7)$$

The distribution of electric field intensity is

$$E(y) = \frac{\sqrt{i_1/4k\varepsilon_o bc} \cdot y}{1 - M \cdot C_p(y) \cdot (v-b)} \qquad (8)$$

The distribution of electric potential is:

$$U_{(y)} = \frac{\sqrt{i_1/4k\varepsilon_o bc}}{M \cdot \bar{C}_p}(y-b) +$$

$$+ \sqrt{i_1/4k\varepsilon_o bc} \cdot (1/M^2 \cdot \bar{C}_p^2 + b/M \cdot \bar{C}_p) \cdot$$

$$\cdot \ln\left|1 + M \cdot \bar{C}_p(b-y)\right| \qquad (9)$$

where, the characteristic parameter of particle:

$$M = \frac{18 n_s}{(n_s+2) \cdot \rho \cdot \bar{d}_g \cdot e^{(1/2) \cdot \ln^2 \delta_g}} \qquad (10)$$

The weight concentration distribution of particle:

$$C_p(y) = C_o \exp[-a/f\,(w/v\,x_o - y)] \qquad (11)$$

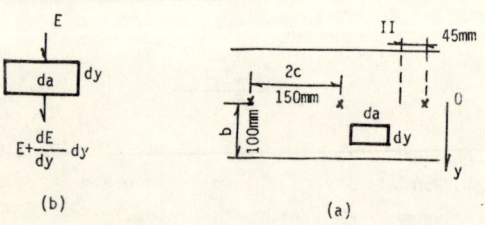

Fig. 2 The mathematical model of one-dimension electric field

The Results of Theory and Experiment*

Both the electric field intensity distribution and electric potential distribution on the representative section II under the condition that $V_o = 25\,kv$ and $C_o = 7.47\,g/m^3$ were showed in Fig. 3.

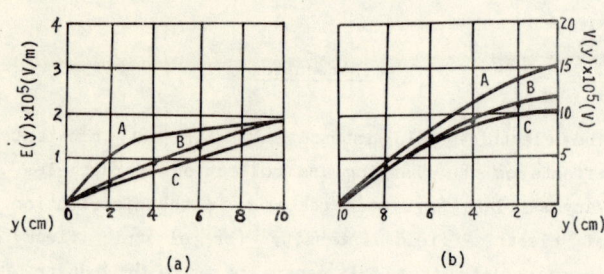

Fig. 3 The electric field distribution on the section II under the condition that $V_o = 25\,kv$ and $C_o = 7.47\,g/m^3$.
A. Leutert's numerical solution
B. measurement
C. one-dimension solution

Conclusions

The following conclusions were obtained on the basis of theoretical analyses and experimental research for the electric field distribution of wire-plate type ESP:

1. The electric field distribution of wire-plate

---

*The figures describing the characters of electric field were omitted, only the related conclusions were given.

type ESP is concerned with work voltage, the structure parameter of electrode, particle concentration and space position.

2. Electric field intensity gradually increases along the direction from the plane consisted of electrode wires to electrode plate except for the region near electrode wire where the electric field intensity is large. It is uniform near electrode plate. The distribution of electric field intensity exists a minimum (the distribution of electric potential exists a turning-point) on the section opposite to electrode wire.

3. Electric potential gradually decreases along the direction from the plane consisted of electrode wires to electrode plate, and it is zero at electrode plate.

4. The increment of space charge density (or particle concentration) results in: (a) The electric field intensity and its gradient decrease near electrode wire but increase near electrode plate. (b) The distance between minimum-point of E(y) [the turning-point of V(y)] and electrode wire decreases. (c) Space electric potential increases.

5. The variety of space charge density (or particle concentration) changes the electric field intensity and its gradient near electrode wire and electrode plate, but it does not change the uniformity of electric field intensity near electrode plate.

Nomenclature

V_o : work voltage
P_{ek} : the charge density of particle
$E(b)$: the electric field intensity at plate
K : ionic mobility
P : the total charge density
E_x : the x component of electric field intensity
E_y : the y component of electric field intensity
i_a : the average current density at plate
i_l : linear current density
q_b : the saturate charge of particle
ε_o : the dielectric constant in vacuum
$\bar{b}$: effective mobility
n_s : the relative dielectric constant of duct
J : current density
d : the diameter of particle
$\bar{d}_g$: the geometric average diameter of particle
σg : geometric standard deviation
$C_n(y)$: the quantity concentration distribution of particle
$C_p(y)$: the weight concentration distribution of particle
C_o : the weight concentration of particle at entrance
$\bar{C}_p$: the average weight concentration of particle in electric field
f : the half of sectional area between the two electrode plates
a : the height of electrode plate
b : the distance between the electrode wire and electrode plate
C : the half of distance between the two electrode wires
ρ : the density of duct
w : the theoretical migration velocity of particle
v : the average velocity of air-stream in electric field.

References

1. Sabert Oglesby, Jr. Grady B. Nichols: Electrostatic Precipitation, the first edition, pp.36-42, the Publishing House of Water Conservancy and Electric Power, 1983, China.
2. Zhang Guoquan: A research on the collection efficiency of ESP, Nonferrous Metals, the English edition, pp.1-7, (2) 1984.

A STUDY ON TWO-DIMENSIONAL ELECTRIC FIELD STRENGTH AND CURRENT DENSITY IN ELECTROSTATIC PRECIPITATOR

Zhang Hongdi   Wang Jun

Beijing Municipal Institute of Labour Protection, Beijing, China

Abstract

The two-dimensional electric field strength (E) and current density (J) are measured and analysed in round-wire to flat-plate and other profiled wire-plate geometries ESP. For spike wires to flat-plate, the E of near-plate space is calculated by a semi-empirical formula $E(\theta)=E(0)\cos^2\theta$, and the J by introducing Warburg's Law. Furthermore, except for project and concave parts, the E and J distributions in the near-space of profiled-plates can be estimated by the distributions of flat-plate with other geometrical aspects the same.

Introduction

In the performance of electrostatic precipitator, the spatial distributions of electric field strength (E) and current density (J) are two important factors which predetermine the collection efficiencies. So far, many experimental methods on detecting E and J have been carried out, including the dropping ball technique used by J.S. Lagarius [1], the stationary-probe method of G.W. Penney and R.E. Matick [2] and plate dividing method conducted by O.J. Tassicker [3]. On the strictly theoretical basis and with the advantage of high projecting speeds and minute deflection distance, P. Cooperman [4] has brought into use the projecting ball method and measured some two-dimensional E and J distributions in a specific ESP. Since P. Cooperman's test, very little study has been concerned with the use of projecting ball method. In our research, by applying high performing voltage and reducing the capacitance of the cable connecting the Faraday Cage to the electrometer, a good agreement has been reached between experimental and theoretical results.

Research System

A schematic drawing of the experimental apparatus is shown in Fig. 1. The probe is operated by firing a steel ball of 4mm in diameter from a projecting device, through the corona zoe, and into a Faraday Cage where the charge picked up is measured by an electrometer.

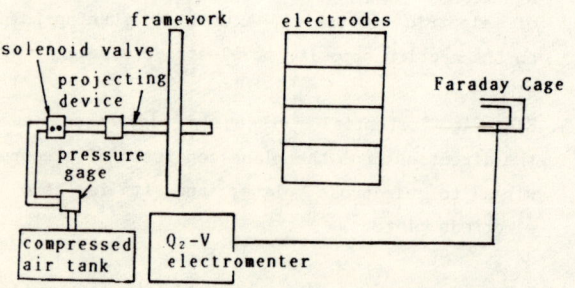

Fig. 1 Schematic diagram of apparatus.

The charge picked up by a steel ball submits to the Lwas of particle-charging in unipolar fields which may be written, in electrostatic units (esu.), as

$$V^{-1} = V_s^{-1}(1 + v/v_o) \quad (1)$$
$$V_s = 3a^2 E/C \quad (2)$$
$$v_o = \pi L J/E \quad (3)$$

where V=voltage in the system
   v=velocity of the ball
   C=capacitance of the system
   L=length of the trajectory within discharge space
   a=radius of the ball

By the linear relationship of V^{-1} and v, v_o and V_s^{-1} can be acquired, then E and J are derived.

Experimental Results and Discussions

Take a line along the middle wire as Z coordinate, a line perpendicular to plate as X, a line through the center of wires as Y, plate spacing as $2S_x$ and wire spacing as $2S_y$. The E of round-wire to flat-plate ESP along Y=0 line can be calculated by differentiating the potential distribution formula, derived by P. Cooperman [5] with well-distributed spatial electric charges.

$$E_x = -U_0 \frac{\pi}{S_x}\sin(\frac{\pi x}{2S_x}) \sum_{m=-\infty}^{\infty} \frac{\cosh[\pi(y-2mS_y)/(2S_x)]}{\{\cosh[\pi(y-2mS_y)/(2S_x)]\}^2 - \{\cos\pi x/(2S_x)\}^2}$$

$$\div \sum_{m=-\infty}^{\infty} \ln\{\frac{\cosh(\pi m S_y/S_x) - \cos[\pi a/(2S_x)]}{\cosh(\pi m S_y/S_x) - \cos[\pi a/(2S_x)]}\}$$

$$+ \frac{JS_y \ln[4S_x/(\pi a)]}{\pi \varepsilon_0 b V_0} X - \frac{1}{4}\frac{JS_y \ln[4S_x/(\pi a)]}{\pi \varepsilon_0 b V_0}(S_x - X) \quad (4)$$

The last term is added in order to take the E of the opposite direction generated by the charges of $X-S_x$ space into consideration.

In Fig. 2, theoretical and experimental results show a good agreement in the range of X=80-140mm, but depart from each other at X=0-80mm.

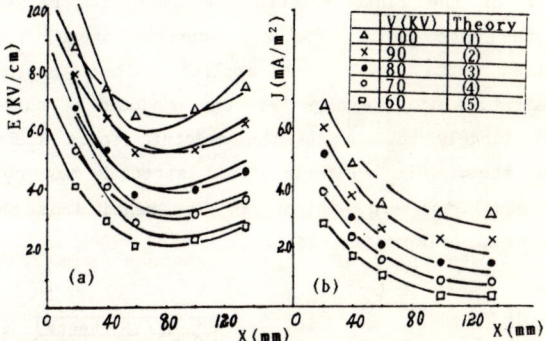

Fig. 2 φ5wire—flat-plate geometry, $S_x=S_y=150$mm
(A) Comparison of measured and calculated E along y=0 line.
(B) Measured J along y=0 line.

Because of the assumption of uniform space charge distributions, Fig.3 shows that the calculated values in near-plate space are not in accordance with the measured results even though the values of J in formula (4) have been taken as the measured values of the corresponding points.

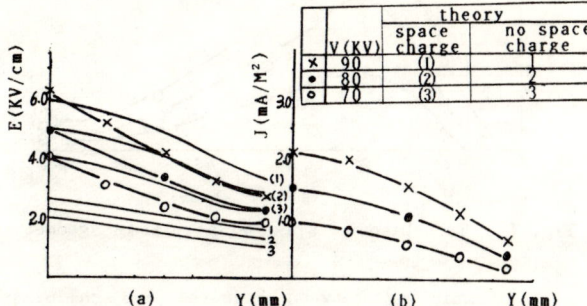

Fig. φ5wire—flat-plate geometry, $S_x=S_y=150$mm
(A) Comparison of measured and calculated E along x=135mm line.
(B) Measured J along x=135mm line.

Table 1 shows E values at point (135mm, 0).

Table 1. Comparson between measured and calculated E at point (135mm, 0), $S_x=S_y=150$mm, φ5wire— flat-plate geometry.

Voltage	Experimental	Calculated	Absolute error	Relative error
(KV)	(KV/cm)	(KV/cm)	(KV/cm)	(%)
60	2.86	2.95	0.09	3.2
70	4.02	3.93	0.09	2.3
80	4.77	5.12	0.35	7.4
90	6.19	6.27	0.08	1.3
100	7.39	7.95	0.56	7.6

Fig. 4 gives the two-dimensional spatial distributions of E and J.

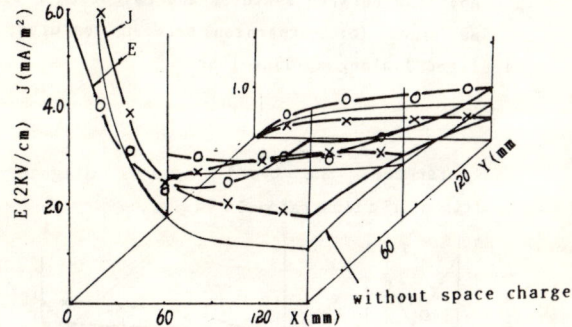

Fig. 4 Spatial distributions of measured E and J and calculated E (without space charge) under V=90KV, $S_x=S_y=150$mm, φ5wire—flat-plate geometry.

After considering the non-two-dimensional features of spike wires, experimental studies are concentrated on the region of near-plate space.

Fig. 5 indicates E and J distributions in the near-plate space of saw-tooth spike to flat-plate geometry, in which the theoretical E curves are calculated by

$$E(\theta) = E(0)\cos^2\theta \quad (5)$$

with E(0) being the electric field strength on the line opposite spike series, E(θ) being that at any given lines parallel to the line of E(0) and θ being the angle between the two planes formed by the spike series and the two lines; the theoretical J curves are calculated from the extension of Warburg's Law [6] that originally fits for the model of a single spike to flat-plate geometry. Warburg's Law can be expressed as follows:

$$J(\theta) = J(0)\cos^5\theta \qquad (6)$$

with $J(0)$ being

$$J(0) = \varepsilon_0 \delta (U-U_o)^2 / L^3 \qquad (7)$$

The relative error between theoretical and experimental values of E and J is within 20%.

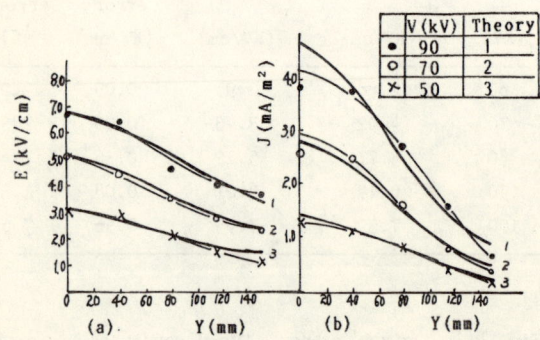

Fig 5 Saw-tooth spike -flate geometry $s_x=s_y=150mm$.
(a) comparsion between measured and calcualed E along x=135mm line. (b) comparsion between measured and calculated J along x=135mm line

As a reference, the two-dimensional diagram of E and J is also illustrated in Fig. 6.

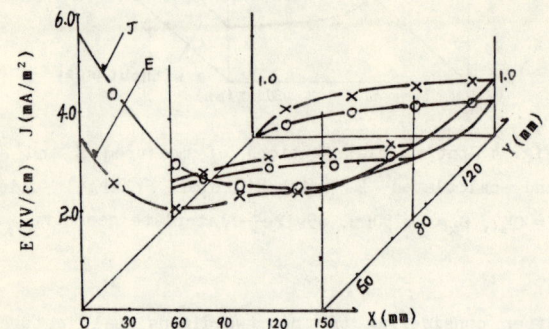

Fig. 6 Spatial distributions of measured E and J under V=70KV, $S_x=S_y=150mm$, saw-tooth spike —flat-plate geometry.

We define the structural parameters of tube-based spike as this: the minimum distance of the adjoining tufts on different tubes being d, the distance of the tufts on opposite sides of the same tube being 1 and the distance of the adjacent tubes being $2S_y$.

Fig. 7 shows the good agreements between theoretical and experimental values of E and J while having a width of $2S_y=480mm$ (corresponding to d=355mm).

However, Fig. 8 indicates that when $2S_y=300mm$ (corresponding to d=155mm), the theoretical and experimental values have a rather big gap, it is clear that the superimposed E of the two adjoining tufts make these differences. Tests show that if keeping d>200mm, good agreements can be achieved.

As for the particular field configuration of the tube-based spike, it is explained as this: different curvatures produce different E values, by the nature of the non-intercrossing characteristics of the electric force lines and by the symmetry of the structure of tube-based spike, it can be deduced that the electric force lines from the tube and plate will accumulate at the area around central parts of the plate between the two tufts series on the same tube. The zero current around the central parts under some applied voltages seems apparent, since the negative ions produced at tufts will largely move along the electric force lines from them. The electric field strength may not be equal to zero, which can be judged from the vector superimposition of E.

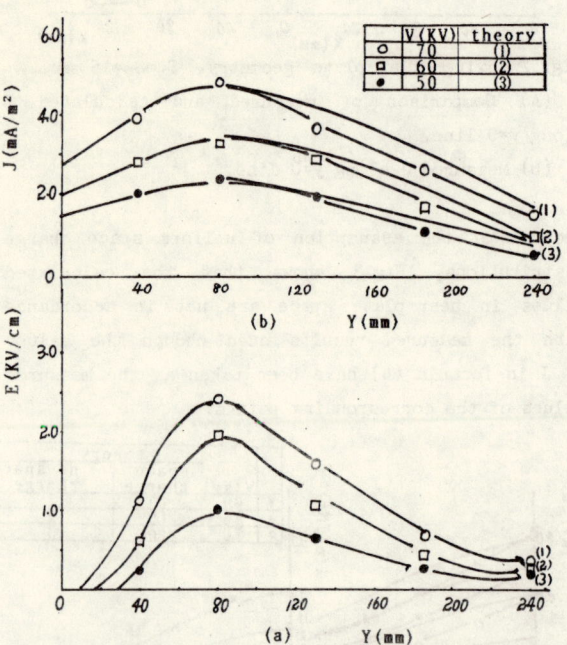

Fig. 7 Tube-based spike—flat-plate geometry, $S_x=150mm$, $S_y=240mm$.
(A) Comparison between measured and calculated E along x=135mm line.
(B) Comparison of measured and calculated J along x=135mm line.

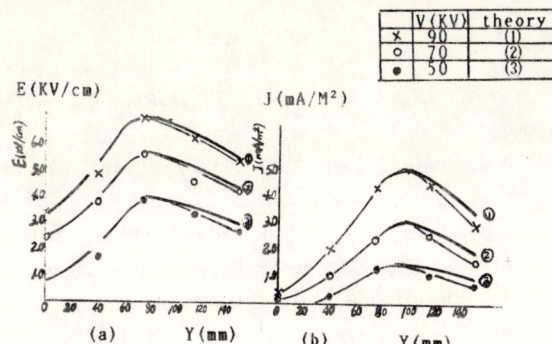

Fig. 8 Tube-based spike—flat plate geometry, $S_x = S_y = 150$ mm.
(A) Comparison between measured and calculated E along x=135mm line.
(B) Comparison between measured and calculated J along x=135mm line.

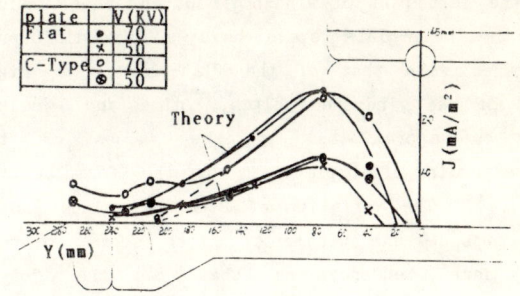

Fig. 10 Comparison of measured J between C-type profiled-plate and flat-plate in near-plate space, $S_x = 150$ mm, $S_y = 240$ mm, tube-based spike as corona wire.

For tube-based spike to Z-type profiled-plate geometry, see Figs. 11 and 12.

For tube-based spide to C- and Z-type profiled-plate geometries, a comparison is given between the profiled-plate and the flat-plate with other conditions unchanged.

From Figs. 9 and 10, apart from some particular points, the E and J distributions in the near-plate space of C-type profiled-plate coincide fairly well with that of the flat-plate and can also be estimated by the formulae (5) and (6). Inside the wind-proof ditch, the E values are very small and the J values tend to be zero. Outside it, the E and J values are larger than those on the corresponding points of the flat-plate, and with the increase of the magnitude of applied voltages, this difference seems more obvious.

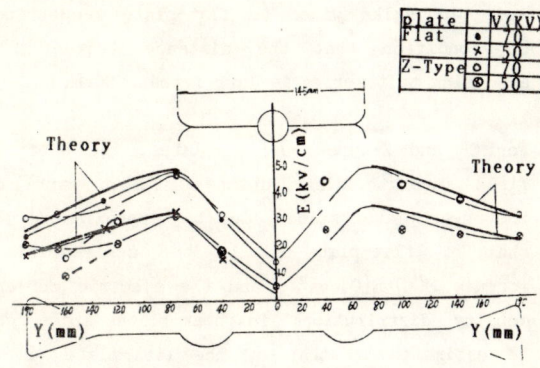

Fig. 11 Comparison of measured E between Z-type profiled-plate and flat-plate in near-plate space, $S_x = 150$ mm, $S_y = 190$ mm, tube-based spike as corona wire.

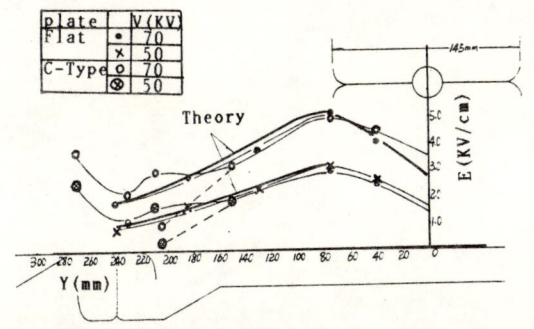

Fig. 9 Comparison of measured E between C-type profiled-plate and flat-plate in near-plate space, $S_x = 150$ mm, $S_y = 240$ mm, tube-based spike as corona wire.

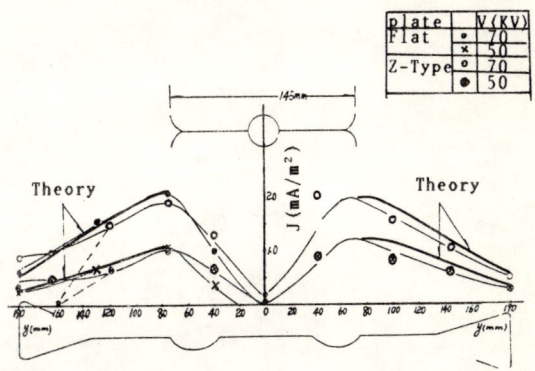

Fig. 12 Comparison of measured J between Z-type profile plate in near-plate space, $S_x = 150$ mm, $S_y = 190$ mm, tube-based spike as corona wire.

On the left part of wind-proff ditch, the E values in the near-plate space have a big difference compared with that of the flat-plate. On right part of the wind-proof ditch, both E and J values can be approximated by the values on the corresponding flat-plate or by the formulae (5) and (6). The variation of E and J values inside and outside the wind-proof ditch seems to have the same tendency as that of the C-type profiled-plate, but not so obvious.

Conclusions

1. The electric field strength distributions follow the semi-empirical formula $E(\theta)=E(0)\cos^2\theta$ and the electric current density distributions follow the extended Warburg's Law in the near-plate space of spike wires to flat-plate geometries, on condition that the distance between the adjoining tuft series is larger than 200mm.

2. For C- and Z-type profiled-plates, the electric field strength distributions in most parts of the near-plate space can be approximated by that of flat-plate or by the semi-empirical formula $E(\theta)=E(0)\cos^2\theta$, and the electric current density distributions in near-plate space can be estimated by that of the flat-plate or by extended Warburg's Law.

References

1. J.S. Iagarius, Trans. A.I.E.E., 78, 427-433, 1959.
2. G.W. Penney and R.E. Matick, Trans. A.I.E.E., Part 1, 79, 91-99, 1960.
3. O.J. Tassicker, Ph.D. Dissertation, Dept. of Elec. Eng., Wallongong University College, Univ. New South Wales, Australia, 1972.
4. P. Cooperman, Trans. A.I.E.E., 75, 64-67, 1956.
5. P. Cooperman, Trans. A.I.E.E., Part 1, 79, 47-50, 1960.
6. P.A. Lawless, K.J. Mclean, L.E. Sparks and G.H. Ramsey, Journal of Electrostatics, 18, 199-217, 1986.

STUDY ON ELECTRODE GEOMETRIES AND CONFIGURATIONS OF WIRE--PLATE ESP BY ELECTRIC FIELD STRENGTH NEAR THE PROFILED COLLECTING ELECTRODES

Guo Ri-sheng   Zhang Hong-di
Beijing Municipal Institute of Labour Protection
Beijing, P.R. China

Abstract

RS--tube based spike discharging electrode, C--shaped and Z--shaped profiled collecting electrodes are commonly used in electrostatic precipitators (ESPs). The experimental study on the configuration with these electrodes has been made by measuring the distributions of the electric field strength (E) near the collecting electrodes, and the results show that there are two large drops of E, one is directly opposite the tube part of the discharging electrode, and another, in the slotes of the collecting electrodes. Finally, a novel configuration is put forward and the results seem satisfactory.

Introduction

The geometries of the electrodes and their configurations in ESPs greatly affect the collection efficiency. In the evaluation on the electrode configurations, two criteria are used, which are the voltage--current characteristics and the spatial distribution of current density or electric field strength, especially that near the collecting electrodes in ESP.

The study on the electrode configurations was started from the end of 1950's. In 1962, H. Isahaya [1] investigated four discharging electrodes by measuring the spatial field strength in an ESP. In 1978, Misaka.T. et al [2] conducted experiments on the field strength near a flat plate for different plate-to-plate spacings and confirmed "wide spacing effect". In 1985, C.E. Åkerlund [3] reported his investigation results for C--shaped and G--shaped profiled plate by a conductive paper technique which can be used to measure the field strength on the plates.

In 1986, the author [4] analysed by experiment the distribution of electric field strength (E) near the grounded flat plate for four corona electrodes (including round wire, saw-tooth, star-shaped and RS--tube based spike). The study showed that the RS--tube based spike with the strongest average E, is the best.

On the basis of our previous study, this paper investigates the effect of RS--tube based spike on the E distribution near the collecting electrodes assumbled with flat plates, C-shaped profiled plates and Z-shaped profiled plates. Finally, a novel set of electrodes is designed and tested.

Experiment Arrangements

Though the E generated by a tube based spike is a three-dimensional field, the E near the collecting electrodes can be approximately considered to be two-dimensional due to the fairly small distance of two adjacent spikes alongside the discharging electrode, and could be measured with a ball-dropping technique [1] [2] [4].

The E distribution near the collecting electrodes has been measured and analysed for different voltages (from oneset to 100 kv), plate-to-plate spacings (from 30 cm to 50 cm), ratio of plate-to-plate spacing to wire-to-wire spacing (from 0.6 to 2.2) and collecting electrodes (flat plate, C-shaped, Z-shaped).

Fig. 1 gives the outlines of the electrodes in the experiments.

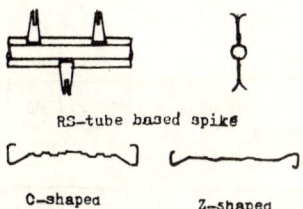

Fig.1 Outlines of the electrodes used in the experiments

Results and Discussions

The Effect of Applied Voltages

Fig.2 illustrates the E distribution near the flate plate for different applied voltages. $\bar{X}$ represents the ratio of the coordinate at testing point, X, to the half of wire-to-wire spacing, C, i.e. $X/C/2$. So, the point at $\bar{X}=0$ is just opposite the tube of the discharging electrode, $\bar{X}=1$, opposite midway between two adjacent the electrodes.

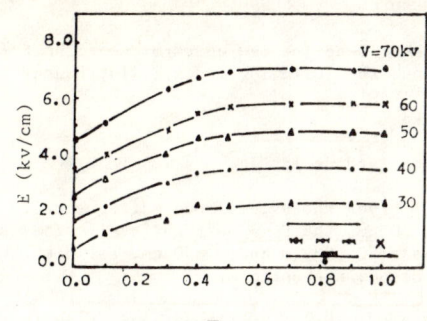

Fig.2 Distribution of E near flat plate for different applied voltages V, plate-to-plate spacing D=30 cm, wire-to-wire spacing C=20 cm

It is seen from Fig.2 that there is a drop in E around $\bar{X}=0$, which can not be filled up by merely increasing the applied voltages. The average E values (Ep) versus applied voltages (V) are shown in Fig.3. When D (plate-to-plate spacing) is constant, Ep has a linear relationship with V and with the increase of D, the straight lines slope down to the V-coordinate. K is the ratio, D/C.

The Effect of Wire-to-wire Spacings

Fig.4 indicates the E distribution for different wire-to-wire spacings. It can been found that there are two drops in the curves, one is at $\bar{X}=0$, the width ($\bar{X}d$) of which increases with the reduction of C and another one is at $\bar{X}=1$ the width of which decreases

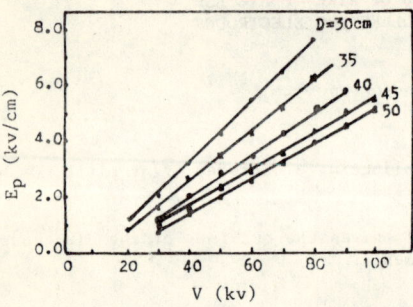

Fig.3 Relation between Ep and V for K=1.5

with the reducing of C, and when C is reduced to less than 23 cm the second drop is vanished, while the former drop can not be eliminated by simply changing wire-to-wire spacings.

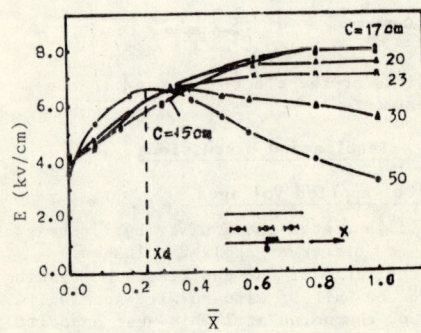

Fig.4 Distribution of electric field strength E near flat plate for different wire-to-wire spacings C, D=30 cm, V=70 kv

To further describe the uniformity of the E distributions, the following index is introduced

$$\beta = S/E_p$$

where, S--standard deviation.

Table 1 gives the values under different C. When C is around 30 cm, the E is most uniformly distributed. In the cases of C<23 cm and C>30 cm, the distributions of E become non-uniform.

From Table 1 and Fig.4 altogether, it is found that the poor uniformity of the E distributions for C<23 cm and for C>30 cm is due to the large drops in the E.

Table 1 Values for different wire-to-wire spacing C

C(cm)	15	17	20	23	30	50
β	0.19	0.19	0.17	0.15	0.15	0.27

Fig.5 shows the relation between Ep and K with different D. When D and V are fixed, if K is less than about 1.5, Ep increases with the increase of K; if K is greater than about 1.5, Ep keeps as a constant. This tendency of Ep can be explained from Fig.4. When K is smaller (corresponding to larger C), the two drops mentioned above all exist and cause Ep weakened. With increase of K, the Ep rises as the result of the reduction of the drop at $\bar{X}=0$. And when K is greater (corresponding to less C), only the drop at X=0 exists

and, it's width and the maximum value of E all increase. Their effects on Ep are contrary and cancel out each other in magnitude, so Ep keeps unchanged.

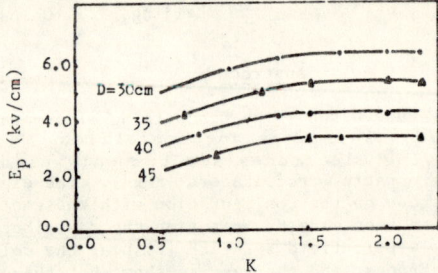

Fig.5 Relation between the average field strength Ep and the ratio K, for V=70 kv

The Effect of Different Plate-to-plate Spacings

When K keeps constant, the E distributions under different D are shown in Fig.6. Under the varied D, the distribution pattern retains unchanged, with the drop at $\bar{X}=0$. Fig.7 indicates the relationship between Ep and D, with the same V and K. The increase of D results in the reduction of Ep the decline tendency of which becomes gentle as D is within the range from 40 cm to 50 cm.

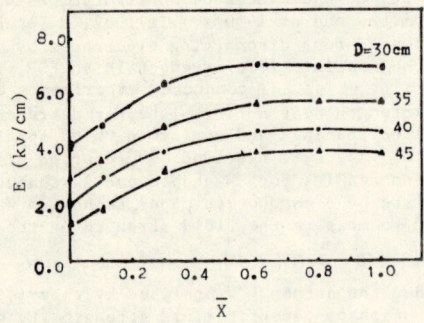

Fig.6 Distribution of the electric field strength E near the flat plate for different plate-to-plate spacings D, V=70 kv, K=1.5

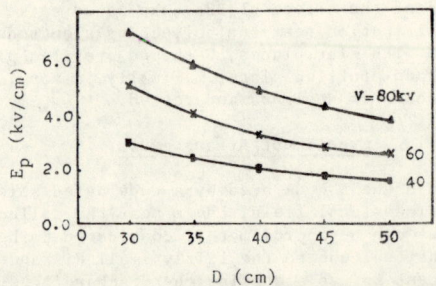

Fig.7 Relation between the average field strength Ep and plate-to-plate spacings D for K=1.5

For Different Profiled Collecting Electrodes

480 mm wide C-shaped, and 385 mm wide Z-shaped profiled collecting electrodes have been used in tests. A discharging electrode serves a collecting

electrode. Fig.8 demonstrates the typical results with D=35 cm and D=30 cm respectively.

From Fig.8 and Fig.9, in addition to the drop of E around the point opposite the tube of the discharging electrode, the E at the protrusion parts and slots of the profiled collecting electrodes also decreases to some degree. The E at the slots is reduced by 40%. These are due to the fact that the wire-to-wire spacings are too large, which are limited by the width of the collecting electrodes, and the electric shield effect is very strong at the slots. In practical application, the protrusions are exposed to the flow for brushing and the slots are shielded electrically, both cases have the disadvantages for dust precipitation.

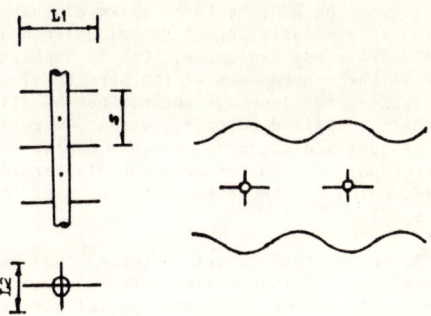

Fig.10 Structures and configuration of the cross-positioned barb discharging electrode and sine-wave shaped collecting electrode

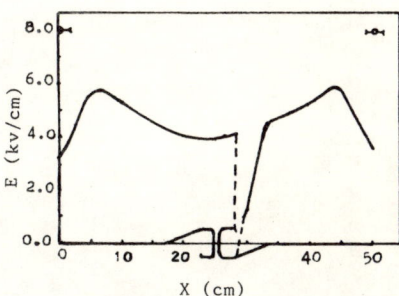

Fig.8 Distribution of the electric field strength E along the C-shaped profiled collecting electrode for V=70 kv, D=35 cm

The structural parameters of the novel discharging electrode, L1, L2, and S, are optimised according to the distributions of field strength and current density near the collecting electrode. The obtained optimum E distribution is shown in Fig.11, which is quite uniform.

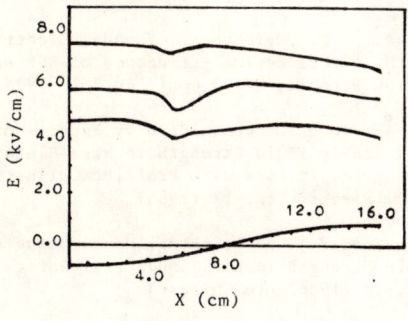

Fig.11 Distribution of field strength E along the sine-wave shaped collecting electrode for the cross-positioned barb discharging electrode

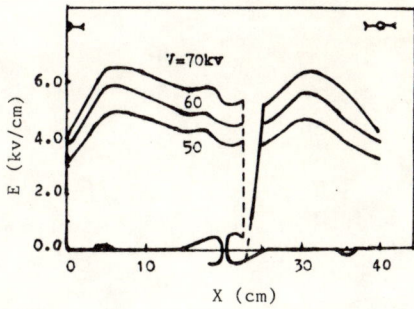

Fig.9 Distribution of the electric field strength E along the Z-shaped profiled collecting electrode, for D=30 cm

Improved Geometries of Discharging Electrode and Collecting Electrode

As mentioned above, there are two large drops of E near profiled collecting eletrodes for RS-tube based spike to C-shaped or Z-shaped electrodes. In order to improve the bad distributed E, a cross-positioned barb discharging electrode and a sine-wave shaped collecting electrode are proposed as shown in Fig.10.

Table 3 gives the comparison of the E distribution between the configurations of cross-positioned barb—sine-wave shaped and RS-tube based spike—flat plate electrodes (D=30 cm, C=30 cm). The average E with the improved configuration is greater than that with the original one by 10%, and the β value is reduced by 50%.

Table 3. Comparison of E distributions between configurations for cross-shaped to sine-wave shaped (I) and RS-tube based spike to flat plate (II) electrodes

V(kv)	Titles	I	II
50	Ep (kv/cm)	4.4	4.2
	β	0.06	0.15
60	Ep (kv/cm)	5.8	5.2
	β	0.07	0.17
70	Ep (kv/cm)	7.1	6.4
	β	0.05	0.17

Conclusions

1. For the RS-tube based spike discharging electrode used presently, the E distribution near collecting electrode has two drops. One is opposite the tube, which is the consequence of the structural demerits of the discharging electrode and can not be eliminated by varying the applied voltages, wire-to-wire spacings and collecting electrodes, another one is opposite the point midway between two adjacent discharging electrodes, which can be moved up by reducing the wire-to-wire spacing.

2. In the configurations for RS-tube based spike to C-shaped or Z-shaped electrode, because of the restrictions of the width and spacial structures of the profiled collecting electrodes, the E values are small at their protrusions and slots, especially in the slots where the E value is only about 40% of the average E.

3. With the novel configuration of cross-positioned barb to sine-wave shaped electrodes, **stronger magnitude and more uniform distribution of electric field strength near the profiled collecting electrode could be obtained.**

References

[1] H. Isahaya, Analysis of corona Field Intensity Distribution by Steel ball Dropping Method in Electrostatic Precipitators, J.I.E.E.J. 2 (1962)

[2] Misaka. T, K. Sugimoto, H. Yamada, Electric Field Strength and Collection Effiecency of ESP Having Wide Collection Pitches, CSIRO Conf. on ESP, 1978

[3] C.E. Åkerlund, Determination by Experimental Method of the Static Field Strength in Wire-Plate Electrostatic Precipitators with Profiled Collecting Electrodes, J. Electrostat. 17 (1985)

[4] Zhang Hong-di, Guo Ri-sheng, Measurement of Electric Field Strength in ESP, Industrial Safety and Dust Control, 12 (1986) (in Chinese)

SIMULATING CALCULATION FOR ELECTRIC FIELD
STRENGTH DISTRIBUTIONS OF ELECTROSTATIC PRECIPITATORS

Hao Jiming   He Kebin   Chao Hongxun
(Department of Environmental Engineering, Tsinghua University)

ABSTRACT

Based on experiments and the calculation for wire--flat plate geometry, this paper presents the concept of equivalent value of diameter and the method of attenuation function to calculate the electric field strength distribution for some new electrode systems frequently used in China. Fairly good agreement is found between calculated results and measured ones in verification calculations with data of four full scale electrostatic precipitators(ESPs) used in China.

INTRODUCTION

The calculation for electric field strength distribution in the duct of ESPs is one of the main parts of its mathematical models. However, there are some calculation methods only for wire--tube and wire--flat plate geometries[1][2].

Recently, various new types of discharge electrodes (such as star-shaped stripe, saw-tooth spike, tube-type spike and rhombus stripe) and collecting eletrodes (such as Z-type, C-type and ripple-type profiled plates) are used extensively. Experimental results have showed that the geometry of collecting electrodes has little effect on electric field distribution near collecting electrodes. The experimental results with flat plates are fit for C-type and Z-type profiled plates[3]. Therefore the calculation for C-type and Z-type profiled plates can approximately be represented by that for flat plates. Orthogonal experimental results have showed that effect of discharge electrodes on the electric field strength distributions is second only to that of applied voltages among various factors[3]. Consequently, more attention should be paid to discharge electrodes during calculating the electric field strength distributions. Simulating calculation models of electric field strength distribution for several discharge electrodes frequently used in China are presented in this paper.

CALCULATION OF ELECTRIC FIELD
STRENGTH DISTRIBUTIONS FOR
STAR-SHAPED STRIPE AND SAW-TOOTH SPIKE

To describe the duct arrangement briefly, it is necessary to define B as discharge electrodes spacing in direction of gas flow, C as collecting electrodes spacing, Y as coordinate position measured toward collecting electrode with the discharge electrode as origin, and X as coordinate position measured parallel to collecting electrode with the discharge electrode as origin.

The experimental curves of field strength distribution near collecting electrodes for several discharge electrodes in common use have been showed in Fig. 1 to 4.[3] These figures demonstrate that the field strength distributions for star-shaped stripe and saw-tooth spike are extremely similar to those for wires. Therefore it is believed that the field strength distributions for star-shaped stripe and saw-tooth spike can be expressed as those for two wires with different diameters, respectively. In other words, the field strength distributions for star-shaped stripe and saw-tooth spike can be calculated directly by the methods for wires but the equivalent values of diameter corresponding to the star-shaped stripe and saw-tooth spike must be properly considered and determined respectively. The equivalent value of diameter of a discharge electrode is defined as the diameter of a wire whose volt-ampere characteristic under the same conditions satisfactorily coincide with that of the discharge electrode.

Fig.5 and Fig.6 present several voltage-current curves obtained by experiments for star-shaped stripe, saw-tooth spike and four wires. By means of the least square method, regressions of voltage-current curves have been carried out for those four wires and power function has been selected as a basic function for regression. The voltage-current curves for wires with diameters between 0.2mm and 1.6mm, and between 2.2mm and 3.5mm can approximately be thought of as two curve families associated with the diameter D. Consequently, the equation for these curve families can be given as follows.

$$I = aV^b \qquad (1)$$

where a and b approximately are linear functions of the diameter D. The equations for two curve families can be obtained from the data in Fig.5 and Fig.6 and showed in Table 1.

Table 1 Equations of curve families

diameter (mm)	equations*
0.2-1.6	$I=(3.983-1.926D) \times 10^{-2} V^{(0.375D+2.212)}$
2.2-3.5	$I=(13.68-2.7D) \times 10^{-2} V^{(0.166D+2.49)}$

* D is the diameters of discharge electrodes in unit mm, I and V are total current and applied voltage of electrode systems, respectively, in mA and 10kv.

In the light of the data for star-shaped stripe and saw-tooth spike in Fig.5 and Fig.6, it is presented by the least square method that the equivalent values of diameter of star-shaped stripe and saw-tooth spike are 2.7mm and 0.8mm, respectively.

On the basis of the numerical method for wire-flat plate geometry[4], calculations of electric field strength distributions for star-shaped stripe--flat plate and saw-tooth spike--flat plate electrode system have been carried out by using the equivalent value of diameter obtained above. Comparisons between calculated results and experimental result along two special planes are showed in Fig.7 to 10. The electric field strengths are measured experimentally with the ball droping technique for star-shaped stripe and the ball projecting method for saw-tooth spike[5]. These figures demonstrate that the calculated values fairly coincide with experimental ones. Consequently, it is feasible to calculate the field strength distribution for some new discharge electrodes by means of the idea of equivalent value of diameter, especially for stripes and relatively simple spikes.

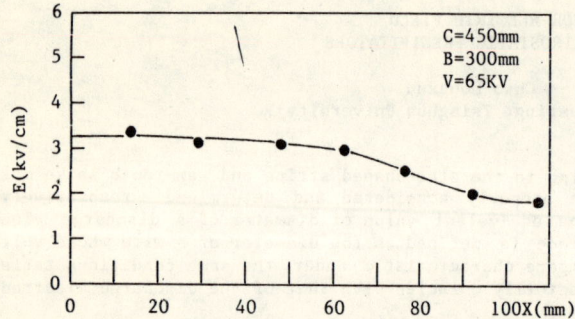

Fig. 1 Field strength distribution for saw-tooth spike --flat plate geometry (source: reference 3)

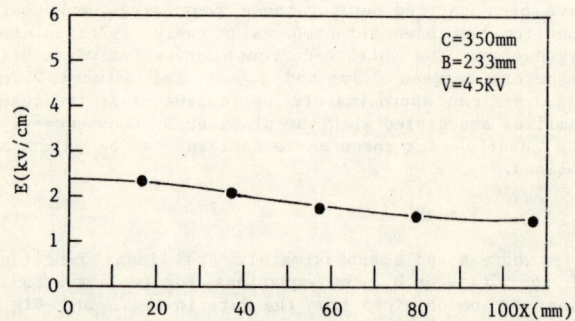

Fig. 2 Field strength distribution for Φ 6.0 wire-flat plate geometry (source: reference 3)

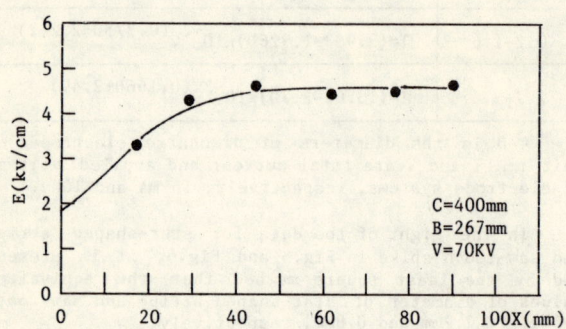

Fig. 3 Field strength distribution for tube-type spike --flat plate geometry (source: reference 3)

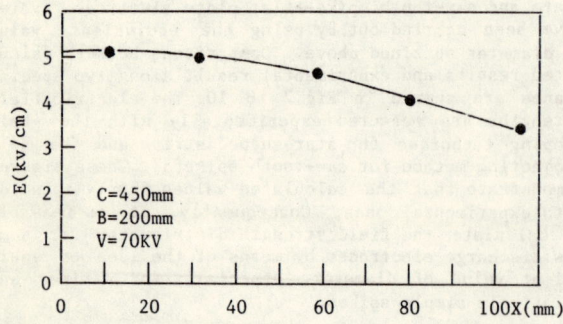

Fig. 4 Field strength distribution for star-shaped stripe--flat plate geometry (source: reference 3)

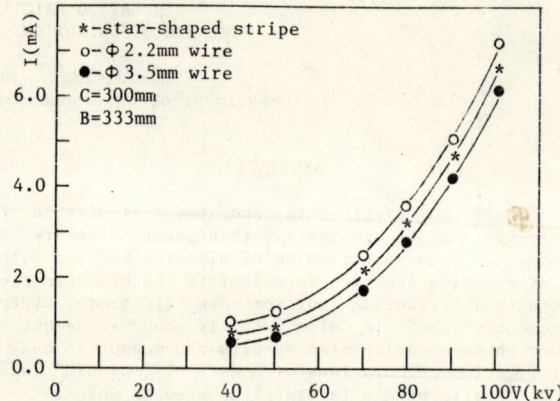

Fig. 5 Voltage-current curves for star-shaped stripe and wires

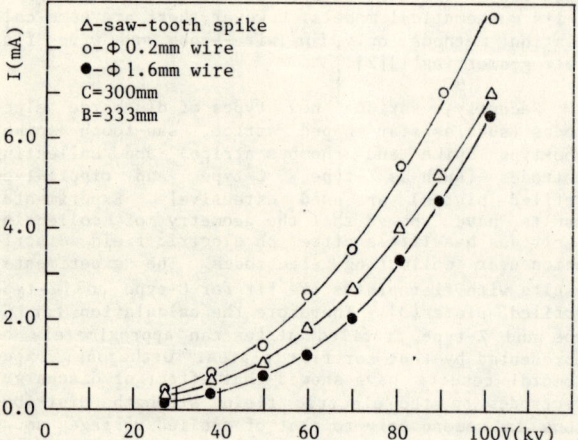

Fig. 6 Voltage-current curves for saw-tooth spike and wires

CALCULATION OF FIELD STRENGTH DISTRIBUTION FOR TUBE-TYPE SPIKE

Fig.2 and Fig.3 show that the field strength distribution near collecting electrode for tube-type spike is quite different from those for wires, and its distinguished feature is that there is a sharp drop for field strength values at the point where X is zero. According to the qualitative analysis for electrolines formed by tube-type spike--flat plates geometry, the idea of attenuation functions can be used, that is, the field strength distributions for tube-type spikes can approximately be calculated by the method for two wires which lie in the position of both tips of tube-type spike and relative attenuation functions. Based on experiments an attenuation function can be defined after considering various factors, including the discharge electrodes spacing, the collecting electrodes spacing, and applied voltages, and so on. In the condition of 300mm in duct spacing, an attenuation function has been defined according to experimental data obtained with the ball droping technique and given as follows

$$W_n = \begin{cases} 1.0 & \text{(when } n=1\text{)} \\ W_{n-1} - 0.365/(M-1) & \text{(when } n=2,3,\ldots,M\text{)} \end{cases} \quad (2)$$

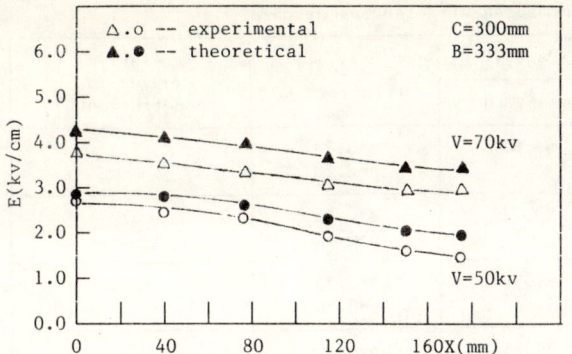

Fig. 7 Field strength distribution for star-shaped stripe--flat plate geometry near collecting electrodes

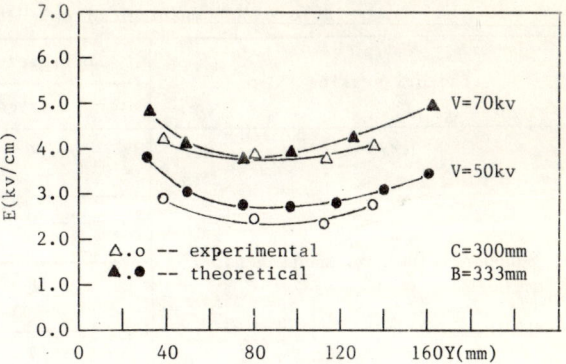

Fig. 9 Field strength distribution for star-shaped stripe--flat plate geometry along X=0

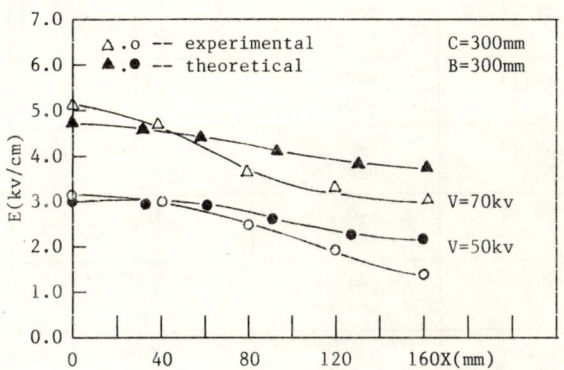

Fig. 8 Field strength distribution for saw-tooth spike --flat plate geometry near collecting electrodes

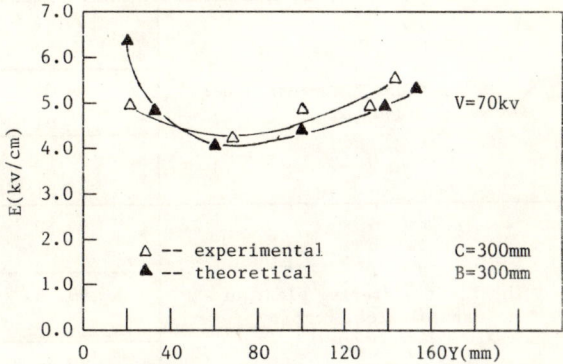

Fig.10 Field strength distribution for saw-tooth spike --flat plate geometry along X=0

$$W_n' = \begin{cases} 1.0 & \text{(when } n=1) \\ 1.0/[(0.75-8.0\times10^{-6}V) \times n] & \text{(when } n=2,3,\ldots,M) \end{cases} \quad (3)$$

Where Wn is the attenuation function in the direction from the tip to centre of tube-type spike (no unit), Wn' is the attenuation function between the tips of two tube-type spikes (no unit), n is the number of calculated points along horizontal line between two tube-type spikes, M is the total number of calculated points along horizontal line between two tube-type spikes and V is applied voltage (V).

A comparison of experimental and calculated values of field strength for tube-type spike--flat plate geometry is shown in Fig.11. Experimental values are obtained with the ball droping technique and calculated values are obtained by the method of attenuation functions. From Fig.11, it is found that calculated values fairly well agree with the experimental ones for four applied voltages. The method of attenuation function presented here can be used for the discharge electrodes which have complicated structures, however, the technique of determining attenuation functions should be improved further.

A VERIFICATION CALCULATION WITH INDUSTRIAL ESPs

Based on the mathematical model of ESPs developed by SRI, U.S.A.[4], the performance of ESPs equipped in four Chinese factories has been estimated by means of

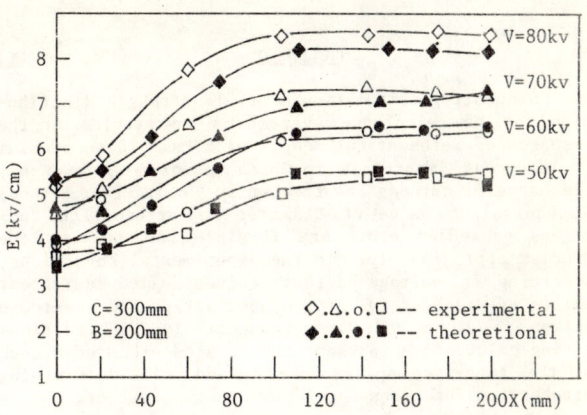

Fig.11 Field strength distribution for tube-type spike --flat plate geometry near collecting electrodes

the method discussed above. In the four factories the discharge electrodes include star-shaped stripes, tube-type spikes, and wires. The calculated and measured efficiencies of ESPs used in these four factories and the relative errors of penetrations are given in Table 2. It is found that the largest relative error is 52 percent. As compared with the results of samilar research in other countries[6][7], it is clear that this calculated result is reliable and testifies that the simulating calculation methods discussed above are feasible.

Table 2 A comparison of calculated and measured efficiencies of ESPs.

factories using ESPs	efficiencies of ESPs (%)		nonideal* factors	relative errors of penetration(%)
	calculated	measured		
Shandong aluminium casting plant	99.64	99.62	$\sigma=0.25$ $s=0.10$	5.3
	99.70	99.72	$\sigma=0.25$ $s=0.1$	6.7
	99.58	99.36	$\sigma=0.25$ $s=0.10$	52.4
	99.77	99.76	$\sigma=0.25$ $s=0.10$	4.3
Taiyuan steel plant	96.60	96.00	$\sigma=0.25$ $s=0.10$	17.6
Sintering plant of Anshan steel company	99.12	99.42	$\sigma=0.25$ $s=0.10$	34.0
	99.37	99.33	$\sigma=0.25$ $s=0.10$	6.3
	98.43	99.00	$\sigma=0.25$ $s=0.10$	36.3
The first sintering plant of capital steel company	99.80	99.89	$\sigma=0.25$ $s=0.10$	45.0
	99.86	99.09	$\sigma=0.25$ $s=0.10$	21.4
	99.80	99.86	$\sigma=0.25$ $s=0.10$	30.0

* where σ is the normalized standard deviation of the velocity distribution and s is the fractional amount of gas sneakage.

CONCLUSION

Simulating calculation of field strength distribution for new electrode systems is an problem in the research of mathematical models of ESPs. This problem is of great importance to China in which various new discharge electrodes are frequently used. The methods of equivalent value of diameter and attenuation functions presented above are feasible for solving the problem approximately. In the experimental conditions, according to voltage-current curves the equivalent values of diameter of star-shaped stripe and saw-tooth spike are 2.7mm and 0.8mm respectively. In the simulating calculation for some complicated electrodes, such as the tube-type spike, the method of determining attenuation functions should be studied further.

REFERENCE

[1] M.J. White: Industrial Electrostatic Precipitation, Addison-Wesley, Reading, Mass, 1963

[2] G. Lentert and B. Bohlen: The Spatial Trend of Electric Field Strength and Space Charge Density in Plate-Type Electrostatic Precipitators, Staub, 1972, 32(7), 27

[3] Guo Risheng: An Experimental Research on Electric Field Strength near Collecting Electrodes of Electrostatic Precipitators, (in Chinese), M.S. dissertation, 1986

[4] M.G. Faulkner and J.L. Dubard: A Mathematical Model of Electrostatic Precipitation (Revision 3): Volume 1, Modeling and Programing, Graut No. R806216010, May, 1982

[5] Wang-Jun: An Experimental Research on Electric Field Strength and Current Density in Two-Dimensional Space of Electrostatic Precipitator with 300mm in Duct Spacing, (in Chinese), M.S. dissertation, 1987

[6] P.B. 279625

[7] C.A. Galler: Electrostatic Precipitator Reference Manual. CS-2809 Project 1402-4, January, 1983

EXPERIMENTAL STUDY of OPTIMUM WIRE SPACINGS at DIFFERENT PLATE SPACINGS of EP

Gao Xiang-lin Hu Zhi-guang Hu Man-yin Yan Jian Li Ling

North China Institute of Electric Power

Abstrct

Corona discharge experiments with 1.28 mm diameter round wires, star-shaped wires, RS wires and sawtooth wires were carried out at different plate spacings. effects of the wire spacing on the average plate current density were investigated. there obtained the empirical formulae relating the optimum wire spacings with various plate spacings.

Introduction

In EP, it is imperative to arrange properly the plate spacings and wire spacings for enhancing the dust collection efficiency. Currently, the electrode alignment commonly applied in industry is to have one RS wire, two sawtooth wires or two star-shaped wires matched with one plate. Here presented are the description of the discharge performance experiments with 1.28mm diameter round wires, star-shaped wires, RS wires and sawtooth wires as discharge wires and conducted by varying the wire spacings at different plate spacings. By analysing the effect of varying the wire spacings on the average plate current density, there conclude the empirical formulae defining the optimum wire spacings at various plate spacings.

Experimental Rig and Measurements

Electric field of one duct was simulated in laboratory scale. The region corresponding to the middle part of the electrode plate of an area of 1000*300mm^2 should be an electric field freed from disturbance. The plate spacing 2b will be varied to 300, 400, 500 and 600mm. At each of these spacings, adjusted at will the spacings 2c of such corona wires as 1.28mm diameter round wires, star-shaped wires, sawtooth wires and RS wires and measured the corona voltage, V-I characteristics and the break-down voltage. The experiments were conducted in air medium and with the passive electrode in the geometry of the flat plate.

Experimental Results and Analysis

The Regularity of $\bar{j}$, The Average Plate Current Density, Varying at Different Wire Spacings 2c

When the plate spacing maintained constant, the average plate current density, $\bar{j}$, varies regularly and simularly at certain voltage with the corona wire spacing 2c, no matter the discharge wires being of round, star-shaped, sawtooth-shaped or barb-shaped. As shown in Fig.1 and Fig.2, the value of $\bar{j}$ comes to be very small, as wire

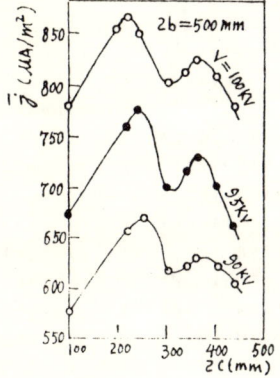

Fig.1. $\bar{j}$ vs. 2c for round wire under different operating voltages

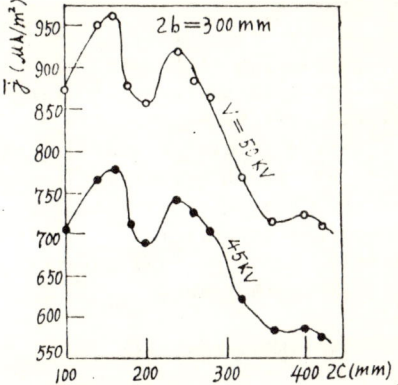

Fig.2. $\bar{j}$ vs. 2c for RS wire under different operating voltages

spacing 2c is very small, and increases to

the maximum, the front peak value, as 2c increases, thereafter, decreases gradually, as 2c increases further, and rises again to a rear peak value, as 2c increases to a certain big value, hence, drops gradually again, as 2c continuous to increase. Here, the rear peak value is smaller than the front one. A third peak value may also occur at certain plate spacing, e.g. 2b=300, 400mm, with RS wire as corona wire. As the operating voltage increases, the front peak value will shift slightly toward the direction where the smaller 2c locates. The above mentioned phenomena should be attributed to the weakening of electric field of corona region due to the fact that the electric field of corona wires may interfere mutually, when their spacing being very small and that the effect of electric charges in space may diminish and deform the corona region so as to decrease the discharge intensity of corona wires. Although the effective length of corona wires may increase in certain condition which will diminish the total corona current at the same time. In EP with electrode of geometry of wire-plate type, the general regularity of corona discharge is as following: The plate current density being the maximum at where just opposite to the corona wire, the minimum, the region between the adjacent wires. that means, there existing a j valley which, being enlarged as the wire spacing 2c coming to be very large. In this case, although each corona wire will discharge in its full extent the total corona current, however, will decrease, e.g. $\bar{j}$ will decreases, resulting a poor utilization of the plate.

In the view of obtaining the maximum total corona current or the largest $\bar{j}$, there exists the optimum wire spacing, with which, $\bar{j}$ will be the highest. Owing to the fact that there are many factors to exert influences on the $\bar{j}$, such as, the electric field of corona wires, the effect of space electric charges, and also, the effective utilization of plate etc., there will appear two or three peak $\bar{j}$ values, the front one, however, will be larger than the rear.

The Optimum Wire Spacings

The wire spacings 2c, of 4 different geometries of corona wires, round, star-shaped, sawtooth-shaped and barb-shaped etc., plotted against 2b, are shown in Table 1 and Fig.3. It is necessary to point out that in this paper the 2c of RS wires is

Table.1

types \ 2c \ 2b	round wire	star-shaped wire	saw-tooth wire	RS wire
300	170	170	160	160
400	210	210	200	180
500	225	225	225	225
600	250	250	250	250

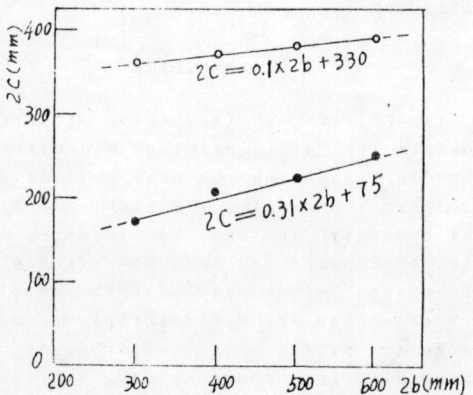

Fig.3. The optimum wire spacings under different plate spacings 2b

the distance between the two closer rows of barbs, and not that between the center lines of supporting pipes of the adjacent wires. As shown in Table.1 and relevant experimental data, the values of 2c, corresponding to the front peak values on the $\bar{j}$~2c curves of 4 different types of corona wires, are basically the same. The corresponding points of the front peak value shift to the right on the curves as the plate spacing increases.

As shown in experiments, when the plate spacing 2b is 300, 400, 500 or 600mm, the corresponding optimum wire spacings, 160~180, 190~220, 220~230, or 240~260mm respectively. By numerical analyzing, the relation between the various corona wire spacing 2c, corresponding to the front peak value, and the plate spacing 2b can be formulated as:

$$2c = 0.31 * 2b + 75 \quad (mm) \quad (1)$$

When the plate spacing 2b is 300, 400, 500, 600mm, the corona wire spacing 2c corresponding to the rear peak value, 360, 370, 380, or 390mm respectively. The relation can be formulated as:

$$2c = 0.1 * 2b + 330 \quad (mm) \quad (2)$$

The wire spacing 2c corresponding to the front peak value on $\bar{J} \sim 2c$ curves is the optimum wire spacing. With this wire spacing, the strongest corona discharge, thereof the maximum total corona current and the average plate current density will be produced. The 2c region corresponding to the front and rear peak, i.e. to that defined by the formulae (1) and (2) is the effective zone in this wire spacing. Within this region, the corona current, hence the $\bar{J}$ will be larger, while beyond this region, with the 2c smaller or larger than those corresponding to the front peak or rear peak values respectively, the corona current will be attenuated dramatically.

Basing upon the formula (1), for the presently prevailing large size, 480mm wide C type plate, and when the plate spacing 2b is 300mm, let suggest to deploy 3 sawtooth wires instead of 2 as usual. For RS wires, let also propose to arrange more densely than usually do. With the closer corona wire space, not only the total corona current and $\bar{J}$ will increase, but also the field strength in average and in vicinity adjacent to the plate, thus enhancing the migration velocity of the charged dust particles. Besides, the dense arrangement will have the plate current density distribute more uniformly, a favourable condition to alleviate or prevent the back corona in dust particles of high resistivity.

The Relation Between Operating Voltage Limits and Plate Spacings

The effective operating voltage means a voltage difference between the break-down voltage Vp and the starting corona voltage Vc. The larger the difference, the more stable the operation of EP.

1. With constant plate spacing, as the wire spacing increases, the break-down voltage Vp rises slightly, while the starting corona voltage Vc drops a little. That means, as the wire spacing increases, the effective operating voltage rises slightly.

2. With constant wire spacing, as the plate spacing increases, the break-down voltage Vp rises remarkably while the starting corona voltage Vc increases a little. That means, as the plate spacing increases, the effective operating voltage rises prominently. Therefore, the operation of EP with wide plate spacings is stable.

Conclusions

1. With the plate spacing 2b remaining constant, as the wire spacing 2c increases, there occur more than two peak values of the average plate current density. The value of $\bar{J}$, within a wide range of 2c, will be large, and will decrease sharply with too small or rather large a spacing of 2c.

2. With plate spacing 300, 400, 500 or 600mm, the optimum wire spacing 2c, corresponding to the front peak $\bar{J}$, will be 160~180, 190~200, 220~230, 240~260mm respectively, no matter the geometry of corona wire being round, star-shaped, saw-tooth-shaped or barb-shaped, will be formulated as:

$$2c = 0.31 * 2b + 75 \quad (mm)$$

The wire spacing 2c, corresponding to the rear peak $\bar{J}$ will be formulated as:

$$2c = 0.1 * 2b + 330 \quad (mm)$$

Then, with the plate spacing 2b=300~600mm, the 2c corresponding to the rear peak $\bar{J}$ will be 360~390mm. The 2c region corresponding to that between the front and rear peak $\bar{J}$, is the effective zone to arrange the corona wire. Within this region, the value of $\bar{J}$ is larger, and, the maximum, corresponding 2c values are called optimum wire spacings. The above conclusion will be beneficial to the design of EP.

3. With constant wire spacing, as the plate spacing increases, the break-down voltage Vp rises dramatically while the starting corona voltage Vc rises slightly. That means, the effective operating voltage increases remarkably, and hence, the operation of EP is stable.

Therefore, the wide plate spacing and dense wire spacing are favourable to improve the migration velocity and hence the dust-collecting efficiency, and is also helpful to prevent and alleviate the back corona in dust particles of high resistivity.

References

[1]. S.Abert Oglesby,Jr.et al.,Electrostatic Precipitation,Marcel Dekker,Inc. 1978.

[2]. Gao Xiang-lin et al.,TWo Demensional Electric Field Characteristics in Electrostatic Precipitators,Proceedings of 1987 Annual Conference on Electrostatics.

DISCHARGE PERFORMANCES STUDY of CORONA WIRES in ELECTROSTATIC PRECIPITATORS

Hu Man-yin Gao Xiang-lin Hu Zhi-guang Huang Qi-shun

North China Institute of Electric Power

Abstract

Effects of the geometry and arrangement of the barbed wires, such as the barb spacing, the barb length, the diameter of supporting pipes, the relative position of barbs to the plate, parallel or perpendicular etc. on the discharge performances were comprehensively studied. In conclusions, three types of barbed wires have been recommended.

Introduction

Corona wires are important component in electrostatic precipitators. Their performances bring direct effect to the dust collecting efficiency. By experiments on a laboratory simulating rig of EP, the discharge performances, such as starting corona voltage, the break-down voltage, the V-I characteristics, and the uniformity of plate current density etc. with different geometry and arrangement of corona wires such as the barb spacing, the barb length, the diameter of supporting pipes, and the relative position of barbs to the plate, parallel or perpendicular etc. were studied. Finally, aiming at promoting the migration velocity of dust particles and thereof the dust collecting efficiency, three types of barbed wires have been recommended.

Experimental Method

The experimental rig used was a simulation of one electric field duct of EP. The lengths and the spacings of the pin-shaped barbs of the experimental corona wires were adjustable. The plate spacing was fixed at 300mm. The supporting pipes of the pin-shaped barbed wires were of two types, ⌀8mm and ⌀25mm. The starting corona voltage, break-down voltage, V-I characteristics and distribution of the plate current density etc. were measured when the geometry or arrangement of the corona wire was varied by adjusting the lengths and spacings of the pin-shaped barbed wire and by improving the structure of sawtooth-shaped and RS wires. Experiments were carried out in air medium.

Experimental Results and Analysis

Experimental Study of The Pin-shaped Barbed Wire

1. The Pin Spacings of Pin-shaped Barbed wire

(1) Break-down Voltage V_p and Starting Corona Voltage V_c

In experiments, the adjacent corona wire spacing 2c adopted is 250mm, the pin length, 15mm, and the supporting pipes, ⌀8 mm. As the pin spacing increases, the V_p rises slightly, and the V_c minutely, thereof, the effective operating voltage, a little. The V_p of pin-shaped barbed wires with pin parallel to the plate is higher than that with pin perpendicular to the plate. Generally, the mentioned difference in V_p is small with small pin spacing, and large with large pin spacing. The V_c with pin parallel to the plate is slightly higher than that with pin perpendicular to the plate.

(2) Discharge Strength

When the imposed voltage V=50kv, the relation curves of average plate current density $\bar{j}$ against the pin spacing were shown in Fig.1. As shown, the smaller the

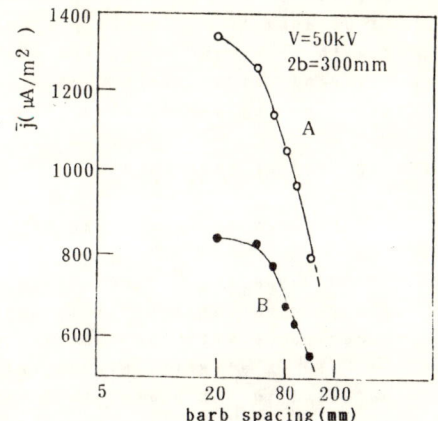

Fig.1. The average plate current density j vs. barb spacings
A: barbs vertical to the plate
B: barbs parallel to the plate

pin spacing, the larger the corona current. The corona current with pin perpendicular to the plate was usually 30~50 percent greater than that with pin parallel to the plate.

(3) Maximum Corona Power

The maximum corona power is defined as the product of Vp and the corresponding corona current Ip. However, in this paper, it is represented by that of the (Vp-10kv) and the corresponding corona current, as shown in Fig.2. From the curve, it is ob-

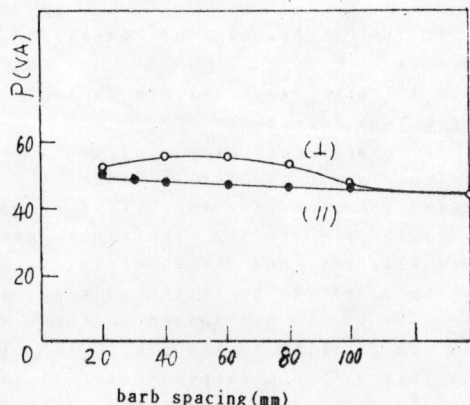

Fig.2. The maximum corona power P vs. barb spacings

served that the maximum corona power decreases as the barb spacing increases, when the barbs parallel to the plate and that the corona power is maximum as the barb spacing, 40~60mm, when the barbs perpendicular to the plate.

Considering all the situations mentioned above, it is recognized that the barb spacing of 40mm or so is the preferable spacing, with which the Vc is low while Vp , high, thereof, the discharge intensity and the maximum corona power, also higher.

2. The Effects of Barb Lengths

(1) The Effects of Barb Lengths on Vc and Vp

When the barb spacing fixed at 60mm and the barb lengths varying, it is shown that the Vc decreases with the increase of barb lengths, and is slightly higher, with the barbs parallel to the plate, than that perpendicular to the plate, the Vp decreases gradually with the increase of barb lengths, and is higher, with the barbs parallel to the plate, than that, perpendicular to the plate, the longer the length of barbs, the greater the mentioned difference of Vp. As the barbs are parallel to the plate, they do little with the effective operating voltage, while they are perpendicular, their increases will result in the decrease of the effective operating voltage. Thus, there shoud be a limit of the barb length.

(2) Effects of Barb Lengths on V-I Characteristics

The V-I characteristics curves of different barb lengths are shown in Fig.3.

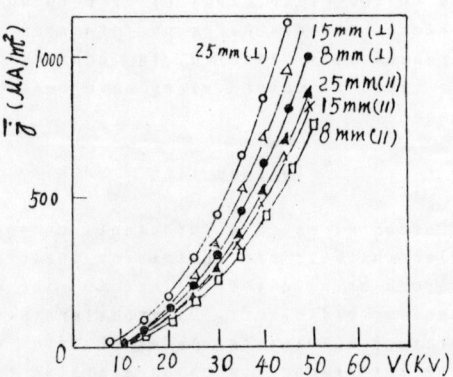

Fig.3. V-I characteristics of different barb height

The curve slope increases with the increase of barb lengths and moreover, the corona current, with barbs vertical to the plate, is much greater than that, parallel to the plate. Therefore, in the electrode arrangement, it is more suitable to have the plate with the barbs vertical to it deployed at the first and second electric field where the dust concentration is large.

Basing upon the above discussion, and taking into account all the effects brought to the starting corona voltage, the break-down voltage and V-I characteristics etc., it is recognized that the barb length of 15~25mm is more appropriate.

3. Effects of Barbed Wire Supporting Pipes

The different diameter of barbed corona wire supporting pipes will play a certain part in the discharge performances. The discharge performance experiments were carried out, with barbs parallel to the plate, with pin-shaped barbed wires of barb length 25mm, barb spacing 100mm and with supporting pipes of diameter of ∅8mm and ∅25mm. When the imposed voltage was 50kv, the corona current produced with the supporting pipe of diameter ∅8mm was 18.6 percent greater than that of diameter ∅25 mm. It is obvious that the slender supporting pipe shows a better discharge performance. In practice, under the condition of providing with sufficient mechanical strength, the more slender the supporting pipe, the better the discharge performance.

4. The Structure of Pin-Shaped Barbed Wires

By investigating the above results of experiments carried out with plate spacing 300mm, it is recognized that with the pin-shaped barbed wires to be 40mm, the pin length, 15∼25mm, when the pin of barbed wires is vertical to the plate, it is suitable for supporting pipe diameter to be ∅6 ∅10 mm, while the pin is parallel to the plate, the pin length should appropriately increase.

Studies on Discharge Performance of RS Wires and Sawtooth Wires

1. Effects of Barb spacings

As shown in Fig.1, under the same operating voltage, the smaller the barb spacing of corona wires, the greater the discharge strength. Thus, to have the barb spacing of RS wires changed now from 100mm to 50mm, the discharge strength will increase by 17 percent, under V=50kv.

2. Improving of RS Wires and Sawtooth Wires

Based upon the above experimental results, the ways to increase the discharge strength of RS wires are the following: (1) reducing the diameter and the width of the supporting pipe, (2) reducing the barb spacing, and (3) increasing the barb length. Of the above means, the second one is most easily to be put into prctice, and also remarkably with improved performances. thus let suggest to have the barb spacing of RS wires changed from 100mm to 50mm.

The ways to increase discharge strength of sawtooth wires are as following: (1) reducing the barb spacing, and (2) increasing the barb length. Let suggest to have barb spacing of sawtooth wires changed from 50mm to 25mm.

Comparison of Discharge Performances of New Types Corona Wires

1. Comparison of Discharge Strength

The discharge strengths of the above suggested three kinds of new type corona wires, the pin-shaped barbed wire, the modified RS wire and the modified sawtooth wire, and the two kinds of original types of corona wires, the original sawtooth wire and the RS wire are shown in Fig.4. The ex-

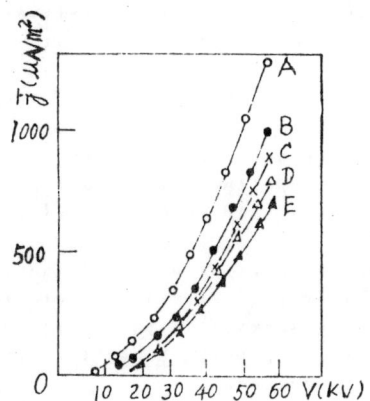

Fig.4. V-J characteristics
A: pin-shaped barbed eires
B: modified sawtooth wires
C: sawtooth wires
D: modified RS wires
E: RS wires

periments were conducted with plate spacing 2b=300mm, and corona wire spacing 2c=480mm for RS wires, 2c=240mm for other types of wire.

When a voltage of 50kv was imposed, the discharge strength of modified RS wire was 19.6 per cent greater than that of the original type, that of modified sawtooth wire, 5.2 per cent greater than that of the original, and that of pin-shaped barbed wire, 41.1 per cent greater than that of the original sawtooth wire.

2. Comparison of Distribution of Plate Current Density

The uniformity of the plate current density is an important criterion in evaluating the performances of corona wires. In this paper, the methods of root-mean-square value and the maximum relative error were adopted to judge the uniformity of plate current density distribution. As results shown, the plate current density distribution uniformity of the pin-shaped barbed wires and that of the modified sawtooth wire are the best, that of modified RS wire is the next, and that of original RS wire, the inferior.

Conclusions

1. In the case of pin-shaped barbed wire, when the pin length is constant, the break-down voltage increases slightly with the increases of pin spacing, and the discharge strength rises with the reduces of pin spacing, and increases slowly, with the pin spacing decreased to a certain value.

2. When the pin spacing of the pin-shaped barbed wire remains constant, the discharge strength rises, the break-down lowers and the effective operating voltage falls with the increases of pin length.

3. The supporting pipes of the barbed wires will bring certain effect to the discharge performances, the more slender the supporting pipes, the better the discharge performances.

4. The discharge strength of the corona wire is greater and the break-down voltage, lower with the barbs vertical to the plate than that respectively, parallel to the plate.

5. Three kinds of corona wire are recommended: (1) pin-shaped barbed wires with the supporting pipe ∅6∼∅10mm, pin length 15mm, pin spacing 40mm; (2) the modified RS wire, to have the barb spacing changed from 100mm to 50mm; (3) the modified sawtooth wire, to have the barb spacing changed from 50mm to 25mm.

6. The comparisons of the above mentioned three kinds of new types of corona wire with the RS wire and the sawtooth wire are the following: (1) In the order of discharge strength, the pin-shaped barbed wire, the modified sawtooth wire, the original sawtooth wire, the modified RS wire, the original RS wire; and (2) in the order of superiority of the plate current density distribution uniformity, the modified sawtooth wire, the pin-shaped barbed wire, the original sawtooth wire, the modified RS wire, the original RS wire.

References

[1]. Gao Xiang-lin et al, Two Dimensional Electrc Field Characteristics in Electrostatic Precipitators, Proceedings of 1987 Annual Conference on Electrostatics.

AN ATTEMPT TO USE SPECIFIC COLLECTING VOLUME AS A MODULAR NUMBER FOR ESP PERFORMANCE COMPARISON

Yan Xingzhong   Wang Jing

Safety & Environmental Protection Research Institute,
Ministry of Metallurgical Industry
Wuhan, China

Abstract

This paper proposes that specific collecting volume (SCV) can be used as modular number for the anlysis and comparisons of ESP performances. It also proposes that ESP's precipitation efficiency should be compared under the conditions of equal SCV, and steel and power comsumption of ESP should be compared under equal efficiency condition. Such modular number and analytic method have been employed in the performance analysis and comparisons of fourteen sets of ESP. The results of analysis and comparisons are more reasonable and pertinent than that of usual methods.

The Problems of Using Specific Collecting Area as a Modular Number

Over a long period of time the specific collecting area (SCA) has been used as a modular number to analyse and to compare the performances of ESP's. This modular number stems from Deutsch formula:

$$P = \exp(-fw) \tag{1}$$

In above formula P is the particle penetration, w is the migration velocity of particle, f is the collecting area per volumetic flow rate of flue gas filtrated, namely SCA.

It also can be believed that w represents the average running velocity of dust particles toward precipitation electrode for particular flue gas, dust particles and technologies of precipitation, and f represents a kind of measure about ESP equipments.

Because the value of P calculated by f and theoretically deduced w differ much from actual value, there are many modifications of Deutsch formula. Whatever the results may be calculated, if the P and f are practical measurement data and Deutsch formula are used to calculated w, such value of w represents not only the statistic average value of particle moving velocity approaching precipitation electrode of that ESP, but also becomes a main empiric ESP's design value which has been accumulated by many ESP's designers.

Along with the development of ESP technology, there are some problems in using SCA as a modular number for ESP performance comparison. For instance as the width between electrodes in wide spacing ESP is enlarged, the SCA will be tremendously decreased if SCA were used as a modular number for comparison of ESP performances, though the precipitatin performances of some wide spacing ESP is not so effective, it's migration velocity w can be easily larger than that of common spacing ESP.

In addition, many kinds of electrodes with different shapes and combinations have been developed, such as housetype precipitation plate, transverse arranging precipitation electrodes etc. Further difficulties would be met with in accounting for such kinds of ESP by SCA. Therefore, it is much important to find another suitable modular number and proper method to analyse and to compare the performance of ESP.

The Meaning of SCV and It's Relation with ESP Performances

The basic reason why SCA is not a suitable modular number to analyse and compare the ESP performances is that the precipitation area of precipitation electrode doesn't interrelate with the space of ESP chamber. Practically, the length, width and height of a ESP chamber effect its precipitation preformances. Lengthening the length or spreading the filtration cross-section area of ESP in definite volumetric flow rate of flue gas can get the same results i.e. lengthening the treatment time. Therefore the ESP filtrating volume per-volumetric flow rate of filtrating flue gas, namely specific collecting volume (SCV) which are connected with all above factors, can be believed to be a basic measure of any ESP equipment.

Though there are some objections against Deutsch formula, the fact can easily be proved theoretically or practically that the dust precipitation along the ESP chamber still follows the law of exponentials. If the modular number SCA is replaced by SCV, Deutsch formula (1) can be trasformed as following:

$$P = \exp[-(V/Q) \times (w/b)] \quad (2)$$

where V,Q represent the effective precipitation volume and the volumetric flow rate of filtrating flue gas repectively V/Q is the SCV, b represents the half width between collecting surface. w/b represents the average migration velocity of unit electrodes spacing. In any set of ESP, w/b can be looked as a constant K, which is the function of dust particle diameter d, resistivity ρ, flue gas humidity ϕ, temperature T and the ability of dust precipitation for the ESP, as following:

$$-\ln P/(V/Q) = w/b = f(d,\rho,T,\phi\cdots) = K \quad (3)$$

When two or more ESPs run nearly in the same condition, that means in the similar d, ρ, ϕ, T, $\cdots$. The difference of value K can denote differenct ability of dust precipitation of ESPs.

The problems in application of formula (3) are that the calculated value p is frequently smaller than the practical observations due to the increase of the accidental influences with enlargement of ESP dimension. That means the value of K must be corrected with different SCV. From a set of values p and V/Q of running ESPs used for filtrating the flue gas of alumina kiln, boiler of power stations and copper reverberator can be worked out following formula

$$K = 2.64(V/Q)^{-0.5} - 0.16 \quad (4)$$

The correlation coefficient for above equation r=0.986. If there is a set of ESP which SCA is V_0/Q_0 and corresponding $-\ln P/(V_0/Q_0)$ is K_0, therefore, the K corresponding any value of V/Q is

$$K = K_0 + 2.64[(V/Q)^{-0.5} - (V_0/Q_0)] \quad (5)$$

The relation between P and V/Q corresponds following equation:

$$-\ln P = 2.64(V/Q)^{+0.5} + V/Q[K_0 - 2.64(V_0/Q_0)^{-0.5}] \quad (6)$$

Due to the penetration changed with variation of V/Q, to compare the ESP precipitation performance have to be under the condition in equal V/Q. If the common SCV for precipitation performance comparison is $(V/Q)_c$, the value of K for comparison will be K_c and it can be calculated by equation:

$$K_c = K_0 + 2.64[(V/Q)^{-0.5}_c - (V_0/Q_0)^{-0.5}] \quad (7)$$

The Relations Between SCV and ESP's Steel and Energy Comsumption

There are several terms such as the net weight and power comsumptions for one set of ESP, the steel and power comsumptions of unit volumetric flow rate of filtrating flue gas, and the steel and power comsumptions of unit effective filtrating volume denote ESP's comsumption performances. But all of such terms can mainly concern ESP's constructions only and they do not concern the connection of comsumption with precipitation performances. The precipitation efficiency of ESP is proportional to the effective filtrating volume, so the reasonable comparisons among ESP's steel and power comsumptions should be taken under equal level of dust precipitation efficency. The SCV in the same dust precipitation efficiency may be obtained from equation (5) and (6). If the penetration among the compared ESPS is assumed to be P_c, the corresponding SCV will be $(V/Q)_k$, it is a root of following equation:

$$[K_0 - 2.64(V_0/Q_0)^{-0.5}] \times (V/Q)_{p_c} + 2.64(V/Q)_{p_c}^{-0.5} + \ln P_c = 0 \quad (8)$$

Due to the net weight M of ESP is proportional to the gross volume of ESP and the gross volume is proportional to effective filtration volume V. So the steel comsumption corresponding $(V/Q)_{p_c}$ will be $C_s[t/(m^3/s)]$, which can also be called as the steel comsumptions under equal efficiency

$$C_s = (V/Q)_{p_c} \times M/V \quad (9)$$

For dry channel type ESP energy comsumption generally consists of power comsumption, the power comsumption of auxiliary electric equipments and the power comsumption due to the resistance of gas flow.

From the observation of running ESPs, the power comsumption by high voltage supply or ionizing power comsumption will be about 50 times as large as that of low voltage supply and the ionizing power comsumption is proportional to the length of corona wire. In the other words it is proportional to the effective filtrating volume. If the ionizing power comsumption is P_i and additonal power comsumption by the auxiliary equipments is 5% of P_i, then electric power comsumption P_e per unit effective filtrating volume will be $P_e/V = P_i(1+0.05)/V$ kw/m^3.

Generally the pressure drop in channal-type ESP is 250-300 Pa. In addition to the pressure drop at inlet and outlet of the equipment, the total pressure drop reaches 400-500 Pa. But this can not repel the possibility of pressure drop occured due to different shapes. Let the pressure drop in ESP be H, then the power comsumption due to the gas flow can be evaluated as $P_g = Q \times H/0.75 \times 1000$ kw. The gas flow power comsumption per effective filtration volume will be P_g/V kw/m^3. Then the summation of power comsumption under equal efficiency is C_p

$$C_p = (V/Q) pc \times (P_e + P_g)/V$$
$$= (V/Q) pc \times [P_i(1+0.05)/V + QH/750V] \quad kw/(m^3/s) \quad (10)$$

A Practice of ESP Performances Comparison

Due to the charactoristics of flue gas and dust particle much effect the running performance of ESP, therefore the performance comparisons focus on the ESPs which run wholly for cleaning the flue gas of open hearth funrnence in the process of blast oxygen. The parameter of 14 ESPs for comparison shown as Table 1.

The SCV for the precipitation performance comparison is arbitary chose to be $10 m^3/(m^3/s)$, then the value K of each can be calculated by equation (5).

In order to ensure variation of penetration by changing V/Q, the value of $-\ln P$ is calculated by equation (6) and $-\ln P$-V/Q curve is formed as Fig.1. For a great majority of ESPs the penetration will be decreased with the increase of V/Q in the range

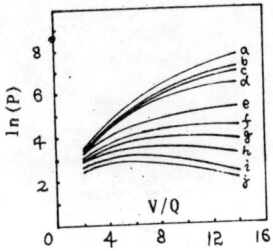

Fig.1 The cure of precipitation performance for 14 ESPs
a - No 1,13; b - No 2;   c - No 3
d - No 4   e - No 5   f - No 6
g - No 7   h - No 8   i - No 12
j - No 9,10,11,14

V/Q=0-15m/(m/s). This means that the efficiency of particle precipitation will gradually increase ($-\ln P$ also increase) with the increase of V/Q, but there are such phenomena that the penetration of several sets of ESPs which are under bad precipitation performances are increased by enlargement of V/Q., after V/Q attending some limit values, Such phenomena can be explained as that the particle precipitation performances for such kinds of ESPs will be decreased due to the accidental factors as the ESP's dismensions are enlarged to some limits. The existance of such phenomena leads to false value K_{10} which will become very large though the observation particle precipitation performance K_0 is very low and it's corresponding specific collecting V_0/Q_0 is very large.

The method to get proper k_{10} is to find out the position of point V_0/Q_0 in the $-\ln P$-V/Q curve. If the position of point V_0/Q_0 sets ahead of the maximum point on the $-\ln P$-V/Q curve, the K_{10} value calculated can be used for performance comparison. If the position of point V_0/Q_0 sets behind the maximum point of that curve, the value K_0 can only be used for performance comparison. The slope of curve $-\ln P$-V/Q at any point will be as:

$$d(-\ln P)/d(V/Q) = 2.64 \times 0.5 (V/Q)^{-.5} + [K_0 - 2.64(V/Q)^{-0.5}] \quad (11)$$

substitute V_0/Q_0 to equation (11)
$$K_0 - 2.64 \times 0.5 (V_0/Q_0) > 0$$

The slope is positive and K_{10} will be treated as precipitation performance comparison value
$$K_0 - 2.64 \times 0.5 (V_-/Q_0)^{-0.5} < 0$$

The slope is negative and K_0 will be treated as comparison value.

Based on such method calues K_{10} and K_0 used for comparison are listed in Table 1. All of such

Table 1 The observation & analysis for the ESPs filtrating the flue gas of open hearth furnace with oxygen blast

No. of equipment		1	2	3	4	5	6	7	8	9	10	11	12	13	14
Capatity of furnace		130	300	300	300	300			500	500	500	500		400	450
Queutify of gas Q m³/n		150,000	322,000	336,960	349,450	343,117	141,672	443,000	350,600	172,962	157,058	474,438	215,600	214,000	535,000
Effective filtrative area m²		38.82	54.53	76.54	88.73	88.77	35	123	60	60	60	100	39.9	70	125
Effective field length m		3.2x4	2.4x4	2.6x4	2.6x4	2.6x4	3.2x4		3.5x3	3.5x3	3.5x3	3.5x4	3.2x4	2.5x3	1.83x5
Efficiency %		97.5-99.	595.86	99.81	99.8	99.37	99.3	95.65	99.5	99.7	99.7	99.6	99.7	99.46	99.4
Net weight of equipment Mt			500	480	614	500			220	220	220	480			
Pressure drop P_a			940.8							539	479.4	313.6			
Average penetration P_0		0.015	0.041	0.0019	0.0011	0.0013	0.007	0.0435	0.005	0.003	0.003	0.004	0.013	0.0154	0.006
Effective collecting volume V_0 m³		496.9	523.5	796	922.8	923.2	448	1722	630	630	630	1400	499.2	525	11436
SCV m³/(m³/s) V_0/Q_0 m³/(m³/s)		11.93	5.85	8.5	9.51	9.69	11.38	13.99	6.74	13.11	14.44	10.62	8.34	8.83	7.7
$-\ln(P_0)$		4.2	3.18	6.27	6.81	6.66	4.96	3.13	5.3	5.81	5.81	5.52	4.34	4.17	5.12
K_0		0.35	0.54	0.74	0.72	0.69	0.44	0.22	9.82	0.44	0.4	0.52	0.52	0.47	0.66
K_{10} (or comparison K_0) (V/Q)$_{PC}$ 0.01 m³/(m³/s)		0.35	0.28	0.67	0.70	0.68	0.49		0.62	0.55	0.54	0.54	0.44	0.41	0.54
M/t t/m³				4	3.76	3.92	6.97		4.47	5.52	5.52	5.34			5.52
M/Q t/(m³/s)			0.96	0.60	0.67	0.54			0.35	0.35	0.35	0.34			
C_s t/(m³/s)			5.59	5.53	6.33	5.75			2.26	4.53	5.04	3.64			
				2.4	2.5	2.1			1.57	1.93	1.93	1.83			
P_i kw			132.9	72.1					125.9	93.5	97.0	156.4			
P_c kw			139.6	79.7					132.2	98.2	101.8	164.2			
P_g kw			112.6							34.5	27.4	55.1			
P kw			251.8							132.7	129.2	219.3			
P/V kw/m³			0.48							0.21	0.21	0.16			
P/Q kw/(m³/s)			1.89							2.76	2.96	1.66			
C_p kw/(m³/s)										0.52	1.13	0.84			

figures construct a precipitation performance comparison historgrams in Fig.2.

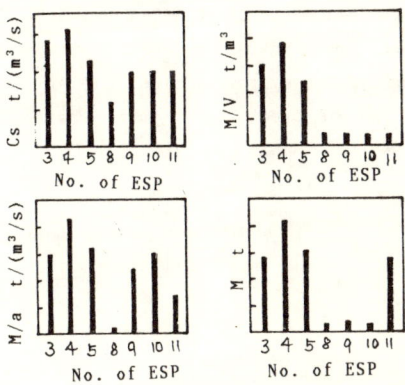

Fig 2 The histograms about steel consumption comparision

less qautities of flue gas, such results of comparison don't really reflect the priorities of ESP's precipitation performances. There are significant differences between the results of precipitation performance comparison of K_0 and K_{10}. During comparison with K_0, the No. 8, ESP wins 1st place, the value K_0 equals to 0.82. But under the basic of equal SCV, the value of K_{10} for No.8 ESP becomes to 0.62, and drop places to 4th. It is clear that the results of comparison in the equal SCV will really express the performance of ESP's precipitation abilities.

The average value of K_{10} is equal to 0.15 for 14 sets of ESP. It is possible to divide the range of K_{10} into three grades, such as $K_{10} > 0.6$, th quality of ESP performance is in high grade. $K_{10} = 0.4-0.6$, in middle range and $K_{10} < 0.4$, in low range.

In order to compare the steel and energy comsumption of ESPs, it is necessary to calculate the steel and energy comsumption per effective filtration volume and SCV which corresponds an arbitrary value of precipitation efficiency.

The arbitrary value of precipitation efficiency is chosed to be $\eta = 99\%$. Then $P_c = 0.01$, $-\ln P = 4.6$. Substitute above figures to formula (9), the (V/Q) can found out and listed in Table 1.

Due to the precipitation efficiency of some ESPs is less then 99%, the value of $(V/Q)_{P_c=0.01}$ can not be calculated. From $(V/Q)_{P_c=0.01}$, steel and power comsumptions per effective filtration volume, the steel and power comsumption under the same efficiency can be calculated. All the comparison data such as the weight of ESP equipment, the power comsumption of equipment, the steel and power comsumption per filtration volume are listed in Table 1, they will be compared with the steel and power comsumption per gas flow rate and the steel and power comsumption under the basis of the same ESP efficiency for each set of ESP, the comparison results have formed steel and power comsumption comparison histograms in Fig.3, Fig,4.

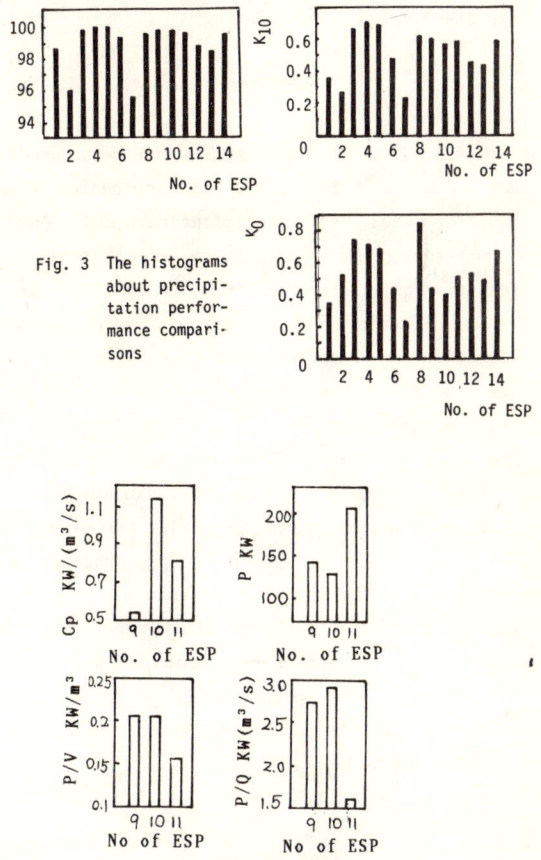

Fig. 3 The histograms about precipitation performance comparisons

Fig 4 The histograms about energy consumption comparison

There are out standing differeacs between the different kinds of comparison methods. For example, the performance of steel comsumption of No.8,9,10 ESP are the same in regard to the mass of equipments or the steel comsumption per unit precipitation volume. But there are significant difference among those 3 sets of ESP when the precipitation effeiciency and the capacity of gas filtrating are considered. The steel comsumption on the basis of equal precipitation efficiency for No.8 ESP is 1.57, which is much less than other two sets of ESP, because it is in best precipitation performance.

It is also possible to divide the steel and energy comsumption on the basis of the same precipitation efficiency into several grades such as the steel comsumption under the same efficiency < 1.6 t/(m^3/s) which can be treated as low steel comsumption, 1.6-2.1 t(m^3/s) as middle range, and > 2.1 t(m^3/s) as high steel comsumption.

Discussion

From above analysis several results can be ensured:
1. Due to the SCV is closely interconnected with the performances of precipitation efficiency, steel and energy comsumption for every shape and combinations of electrodes in ESPs and SCV also is the basic measurement for ESP equipments either in dimention or running capacity, therefore SCV can be dealt to be a modular number for ESP performance comparison.
2. The comparisons of ESP performance need some comparative basis, not only the characteristics of flue gas and airborne particle filtrated by ESP, but also the SCV should be kept similar to compare the precipitation performance, and the precipitation efficiency should also be kept similar to compare the steel and energy comsumptions.
3. The results of performances analysis and comparison of fourteen running ESPs, show that the proposed comparison methods are more reasonable and comparable than usual.

References

1. Yuan Hongrui: Discussion about the production of ESP to correspond the international standard. The description about draft of national ESP standard, 1983
2. Chen Shuyuan: The experience of running ESP filtrating the flue gas of 100 t open hearth furnace. The proceedings of the conference about ESP's power source, electrostatic measurement and precipitation, 1983
3. Wang Juxian: The report about particle dicharge measurement for open hearth furnace in WISCO. The proceedings of the symposium about ESP for filtrating the flue gas of open hearth furnace with blasting oxygen, 1985

TSDC MEASUREMENTS OF COAL FLY ASHES

Tetsuji Oda, Tadashi Takahashi and Senichi Masuda*
Department of Electrical Engineering, the Faculty of Engineering, the University of Tokyo
3-1 Hongo-7chome, Bunkyo-ku, Tokyo 113, Japan

Abstract

The Thermally Stimulated Discharge Currents(TSDCs) of various coal-ash layers charged by the needle corona were studied to understand the charge-storage mechanism, charge-conduction and the electrical properties of coal-fly ashes, related to the ESP characteristics. The real fly ashes sampled from the commercial large thermal power plant, residual ashes burned in a laboratory furnace and fly ashes artificially contaminated by various chemical materials and so on. The pressing force to form the layer is found not so to affect the TSDCs performance. On the other hand, the humidity is very effective to the charge storage in the powder. The contaminant effect to the TSDCs is not so large as expected. The large homo-current at 240 or 270 °C is observed when sulfur is added to the fly ash. When the natural fly ash is heated again at 800 °C, 10 % weight loss, and the increase of the homo-current at 150 °C and the hetero-current at 200 °C is observed. However, the TSDCs peak change by adding the raw coal to the residual ash is not yet clear now.

Introductions

The electrical properties of coal-fly ashes are assumed to be very effective to the separation performance of the electrostatic precipitators (ESP) in a long time range. When the electrical charge is stored in the collected fly-ash layer on the collection electrode, the internal electrostatic field increases to occur the back corona and so on. Some works on the resistivity of those fly-ashes were performed for that reason[1,2]. Some chemical ion effects on the resistivity change by time at high temperature was demonstrated already. As the TSDCs measurement is one of the very effective technique to understand the electrical properties of the solid, especially, the charge storage mechanism of such dielectrics[3], the authors proposed to measure the TSDCs of corona-charged fly ash layers and have developed the high temperature TSDC measurement system for powder by using the personal computer to control the temperature and to reduce the noise by the digital averaging and so on. By using such system, the authors have identified the TSDCs signal of fly ash for the first time[4]. Some different TSDCs for the different coal fly ashes are observed. It was also found that the electrostatic charge stored in the powder is conserved more than a week showing a little bit peak shift to higher temperature.

The pressure strength (to press the powder layer) effect, humidity effect on TSDCs of fly ash powder layers for a long time storage and so on, will be described in this paper. To understand the chemical ion effect on the conduction mechanism of the powder and charge trapping in powder space, some typical chemical asids or alkali such as sulfuric asid, nitric asid, hydrochrolic asid and sodium hydroxide are added to fly ashes. As an extreme example, pure sulfur itself is also mixed to fly ashes. Some comparison between the real fly ash and the residual ash after being heated for 2 hours in the laboratory furnace (which has no residual carbon and so on) is also reported here. Such effect is checked by adding the raw coal to residual ash after perfect burning of that. The TSDCs of coal itself is also examined as the reference. Such test results and comparisons are shown later.

*He is the Preisdent of Fukui Institute of Technology now.

Experimentals

TSDCs Measuring System of Powder

The experimental setup is as same as the former systems[4]. The test cell shown in Fig. 1 is designed to be used as the temperature range of up to 500 °C. As the main insulating materials, the quartz pipes and plates are used as possible for such high temperature. The upper electrode 1 mm above the powder surface (called open TSDC mode or contactless in general) is 24 mm in diameter and the vessel is 34 mm in diameter with 2 mm thickness of the powder. The photograph of that is shown in Fig.2. The temperature control and data processing principle are just as same as the normal TSDCs equipment[5]. The whole measurements are carried out in vacuum conditions from room temperature to about 500 °C.

Samples and Their Preparation

Three different fly ashes of Chinese Coal(A)[0.22%], Australian Coal(A) [0.27%] and South-African Coal(A) are used as fly ash mainly where percentages in[ ] are equivalent SO_3 gas contents of them.

The residual ashes of four different coals, Chinese Coal(B), Australian Coal(B), South African Coal(B) and North American Coal(B) which are heated at 800 °C for more than 2 hours by the laboratory furnace are also examined.

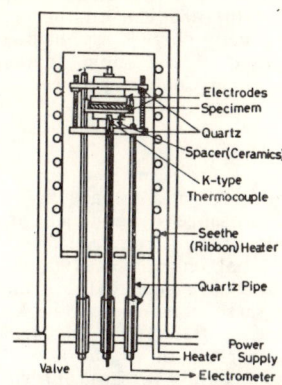

Fig.1 An experimental apparatus of TSDC.

Fig.2 A photograph of the experimental apparatus.

To understand the pressure effect, the sample powder is weighed and filled in the vessel which is laid on a weight meter and is pressed by the plate with a fixed force mechanically.

The fly ashes are contaminated by chemicals, such as
(1) being dipped in sulfuric asid, hydrochloric asid, nitric asid or sodium hydroxide.

The fly ash of 2.8 g is weighed and filled in a vessel (34 mm in diameter) and each is dipped in 0.1 mol liquid to absorbe 1.4 g solution. The vessel with dipped sample is dried for more than one day naturally.
(2) being mixed with 5 weight% pure sulfur sufficiently and filled in the vessel.
(3) being mixed with 2.5 or 5 weight% raw Coal (the exactly a little bit different) and filled in the vessel.

To understand the relation to the magnetic separation process, the ashes are divided into two parts, magnetic component and non-magnetic part by the magnet. Every sample in the vessel is charged by the negative corona-ions for 30 min at 145 °C before measurement of TSDCs.

Experimental Results

TSDCs of Fly Ashes

Typical examples of TSDCs of fly ashes (Chinese and Australian Coals) charged at high temperature are shown in Fig.3. In the case of Australian Coal, a large homo-current is observed at about 140 °C and a hetero-current peak (positive current in this setup) is also seen at about 200 °C. Other current peaks at higher temperature range, such as homo peak at about 240 °C and hetero peak at about 300 °C. The exact explanation for those current peaks, especially at high temperature range (more than 240 °C), are not yet obtained. In the case of Chinese Coal, any apparent TSDC peaks are not yet observed at low temperature range (less than 200 °C) Only peaks at high temperature range can be detected.

Figures 4 (a) and (b) are the comparison between two fly ash powder layers charged at high temperature (180 °C) and charged at the room temperature for Chinese Coal and Australian Coal, respectively.

Pressure Effects

The pressure effects are examined for various conditions. Figures 5 (a), (b) and (c) are those examples indicating the total pressure dependence of TSDCs of fly ashes charged at differnt temperatures. TSDCs of fly ash of Australian Coal charged at 100 °C or at 150 °C shows no siginificant pressure effects on TSDCs in Fig.5 (a) and (b). A large homo-current peak at 250 °C for the sample pressed by 827wg/cm^2 in Fig.5(a). However, as it is not observed in Fig.5(b), it is assumed not to be due to the pressure effect but to be due to the occational instability. Fly ash of Chinese Coal without charging shows no TSDCs signals as in Fig.5(c). Therfore, pressure does not affect the TSDC at all although the charge storage mechanism of the fly ash must be bulk-trap or surface-trap where both are assumed to be very sensitive to the pressure.

The Effects of Humidity on Charge Storage

As the electrical conductivity of the powder is strongly dependent on circumstances, especially, humidity of air and so on, such water vapour effects to the TSDC curves are also examined in Figs.6 and 7. Figure 6 shows the saving conditions of the powder before charging. That is, the line with (1) in Fig.6 is a TSDCs curve of the fly ash which was kept in a standard container before being filled in a charging cell. In the case (2), the pcwder was filled in the charging

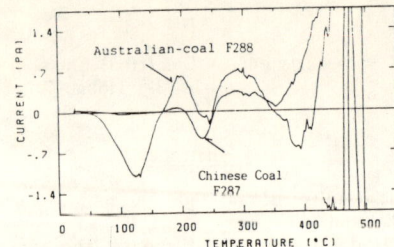
Fig.3 TSDCs of fly ashes (Australian and Chinese coals).

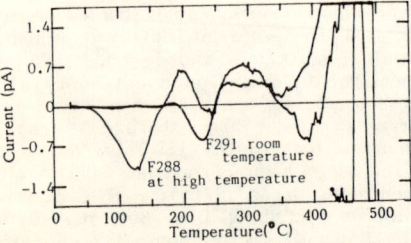
Fig.4(a) TSDCs of ash of Australian coal charged at the room and high (180 °C) temperatures.

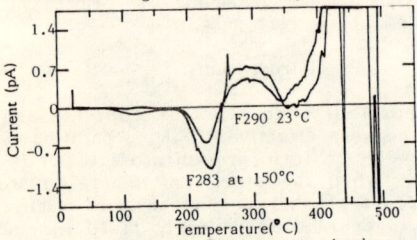
Fig.4(b) TSDCs of ash of Chinese coal charged at the room and high (180 °C) temperatures.

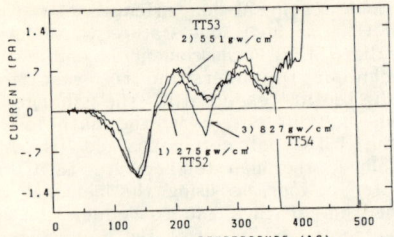
Fig.5(a) Pressure effcts on TSDC (Australian, 150 °C).

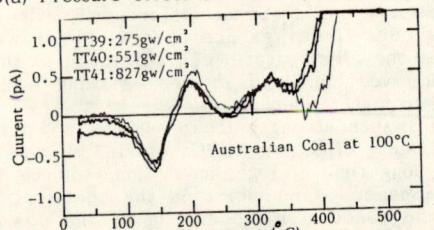
Fig.5(b) Pressure effcts on TSDCs (Australian, 100 °C).

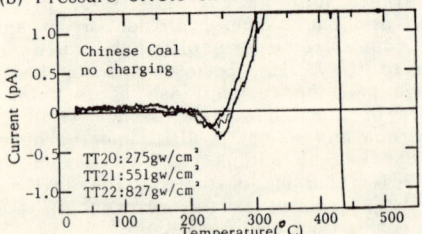
Fig.5(c) Pressure effects on TSDCs (Chinese, non-charged).

cell and stored in room air for 10 days before charging. In the case of (3), the powder was filled in the charging cell and stored in a desiccator for 10 days before charging. Figure 7 is the TSDCs results of ash for storing time effect after the charging. That is, a curve (1) is TSDCs spectra just after the corona-charging and (2) is that of the ash stored for 5 days after the charging in the room condition. A curve (3) is the TSDCs of charged fly ash stored in the desiccator for 5 days after the charging. In both Figures, the powder which was stored in the room air before or after the charging process, may have a little bit large conductivity with water molecular. The PSDC peaks of hetero- and homo-currents at low temperature side, decrease a little in that case. When the fly ash is stored in a desiccator, the charge in an ash layer is mostly conserved as likely as electrets. Drying the sample powder before charging is also found to be very effective to keep the charge with high resistivity.

Residual Ash of Coal Burned in a Laboratory

The TSDCs of artificially obtained ashes of different kinds of coal, that is, burned at 800 °C in the furnace for more than 2 hours where the fly ash is assumed to be heated up to more than 1000 °C for a shorter time, are shown in Fig.8. The same country name do not mean the same coal. The TSDCs of South-African Coal (B) and North American Coal (B) are very similar. Homocurrent peaks at about 180 °C and heterocurrent peaks at about 270 °C are observed. One comparison between the TSDC of fly ash and the residual ash of Australian Coal (B) is shown in Fig. 9. In the case of Chinese Coal, there is large difference between that of fly ash and that of residual ash in Fig.10. The TSDCs of charged fly ash (A) have no apparent homocurrent peak at 145 °C and heterocurrent peak at 200 °C which are very typical in the case (B). When the fly ash is reheated at 800 °C to be burned enough, the TSDCs of that fly ash become very close to that of Chinese Coal (B). That of Australian Coal (B) is in mixing intermedeate properties of Chisese and South-African Coals.

Chemical Additive Effect

The TSDC characteristics of the fly ash dipped in chemical asid or alkali are Fig.11 for the fly ash of Australian Coal (A) and Fig.12 for that of Chinese Coal. Sulfuric asid and hydrochloric asid which are very effective to the electrical conduction of the dielectric powder do not so much affect the TSDCs curves of them from several repeated experimental results as expected, indicating that the negative ion or base do not so affect the TSDCs. The adding effects of nitric asid or sodium hydroxide are assumed to be existing. That is, a small heterocurrent peak by adding effect of nitric asid at 210 °C is observed and a spike-like peak at high temperature range which is assumed to be due to the sodium conduction in bulk, is observed for the ash of Chinese Coal(A). The nitric asid effect may be some chemical transition effect by the asid. The pure sulfur (5 wt%) is added to fly ashes, for sulfur is melted at 145 °C (its melting point is 112 °C) and sulfur covers each particle surface as similar to in ESP. The TSDCs curves are also exhibited in Fig.13 where the homocurrent peak at 150 °C was expected to be affected by the sulfur, but not so apparent indeed. A heterocurrent peak at 200 °C decreases a little bit and a very large homocurrent peak between 240 - 270 °C is identified clearly. It is assumed to be due to the evaporation effect of sulfur. Therefore the origin of a large homocurrent at about 150 °C is not yet clear.

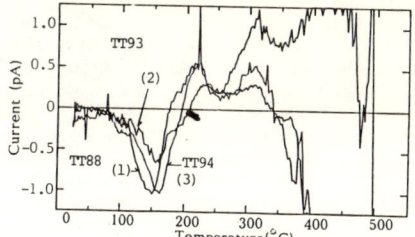
Fig.6 Effects on TSDCs of storing before the charging.

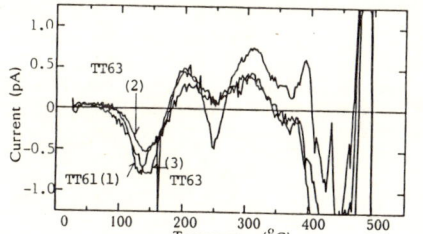
Fig.7 Effects on TSDCs of storing after the charging.

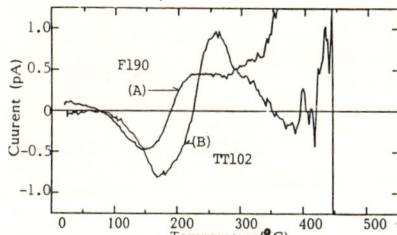
Fig.8 TSDCs of fly-(A) and residual(B) ashes of South-African Coals.

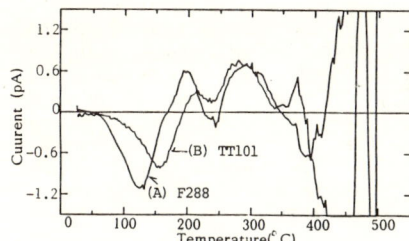
Fig.9 TSDCs of A and B of Australian Coals.

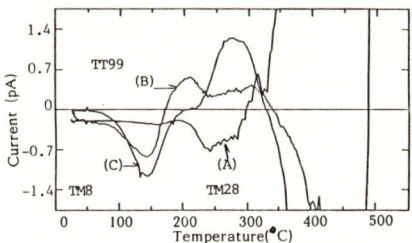
Fig.10 TSDCs of A, B and C(heated A) of Chinese Coals.

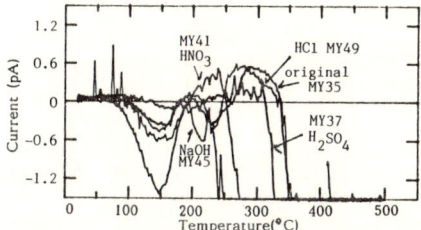
Fig.11 Chemical effects on TSDCs (Australian).

Mixing Effects of Raw Coal

As the ignition loss of the Chinese coal fly ash at 800 °C is very large as 10 % and only the fly ash of Chinese Coal shows current peaks at 150 and 200 °C just after the heating by the furnace, raw coal was added to the fly ash. Those examples are Figs.14 and 15. In the case of Australian Coal, each peak seems to increase opposite as expected. Those phenomena are not explained by the model that addition of the raw coal increase the conductivity and decrease the peak. Some changes of the filling condition of ash may cause such effects. The outgassing from raw coal is not observed as the current which may be misunderstood as the TSDCs.

Related to Magnetic Properties

As one of the very effective beneficiation technologies of coal, Dry type Fluidized High Gradient Magnetic Separation Process is assumed to be very effective to improve the ash content and so on[6]. The relation between the electrical properties (especially, TSDC) and magnetic properties were also studied. The powder of residual ash of coals heated in the small furnace was divided into two parts: one is a magnetic component and another is a non-magnetic component by using the commercial permanent magnet. Two examples are shown in Figs. 16 and 17 where the TSDCs signal of the original ash is larger than that of separated each part, with good agreement of the fact that the resistivity of the mixture ash is not the average of those of each component (magnetic part and non-magnetic part)[7]. There is a tendency that the peaks of the non-magnetic part is a little bit shifted to the higher temperature side, but not so apparent. The enough explanation is not yet obtained.

Conclusions

The TSDCs signals from the ash of coal are identified for different coals (classified by the producing place). The TSDC signal is found to be very sensitive to their electrical properties and is a very effective method to check the electronic states and so on. The signals are affected by the storage conditions, such as humidity and so on. However, the chemical ion effects, pressure effects and magnetic effects on the TSDCs are only observed some traces and exact relationship is not yet understood still now.

References

1) R.E.Bickelhaupt:"Influence of Fly Ash Compositional Factors on Electrical Volume Resistivity"EPA-650/2-74-074 (1974)
2) Ed.by S.Masuda:Proc.2nd Int.Conf.ESP pp228-235,pp283-331 etc.(1985)
3) for example:Ed.by P.Braunlich "Thermally Stimulated Relaxation in Solids" Springer-Verlag (1979) or Ed.by G.Sessler "Electrets" 2nd.Ed. (1986)
4) T.Oda et al:"TSDC Measurements of Fly Ashes from Pulverized Coal Combustion" Proc.2nd.Int.Conf.ESP pp540-547(1985)
5) T.Oda:"Measurement of Thermally Stimulated Discharge Current-Analog and Digital" Proc.Inst.Electrost.Japan pp223-231(1984)
6) T.Oda et al:"Pulverized Coal Beneficiation by a Flidized High Gradient Magnetic Separation Process With Slotted Steel Plates", IEEE Tran.Magn. Mag-23, pp2767-2769 (1987)
7) T.Oda et al:"Electrical Properties of Fly Ash from Pulverized Coal Combustion -related to Its Magnetic Components"Proc.Annual.Meeting of Inst.Electrost.Japan pp9-12(1982) in Japanese.

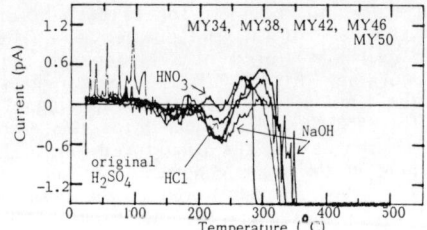
Fig.12 Chemical effects on TSDCs (Chinese).

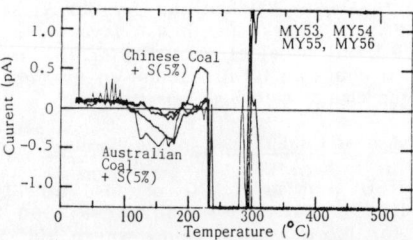
Fig.13 The TSDCs effects of adding sulfur to fly ashes.

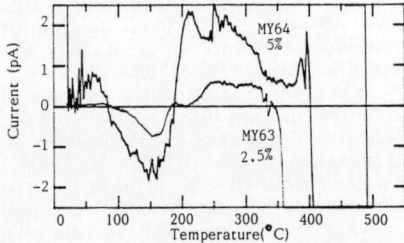
Fig.14 The raw coal adding effects on TSDCs(Australian).

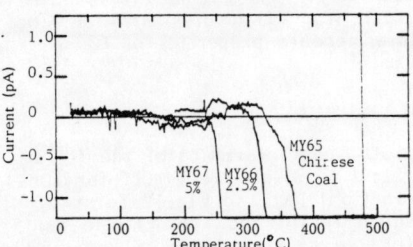
Fig.15 The raw coal adding effects on TSDCs(Chinese).

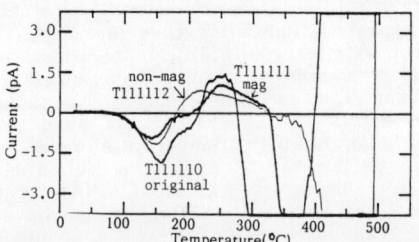
Fig.16 Magnetic effects on TSDCs (Australian).

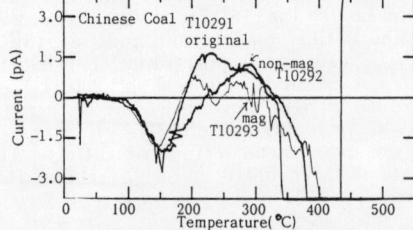
Fig.17 Magnetic effects on TSDCs(Chinese).

ADVANCED ANALYSING MODEL OF CHARGE CARRIER MOTION AROUND A CHARGED SPHERICAL OBJECT

S. Sumiyoshitani

Department of Elec. Eng.
Faculty of Engineering
Fukuoka University
Nanakuma, Johnan-ku
Fukuoka, Japan 814-01

M. Hara

Department of Elec. Eng.
Faculty of Engineering
Kyushu University
Hakozaki, Higashi-ku
Fukuoka, Japan 812

M. Akazaki

Department of Ener. Con.
Graduate School of Eng. Sci.
Kyushu University
Kasugakoen, Kasuga
Fukuoka, Japan 816

Abstract

Studies on the motion of the charge carrier around a charged spherical object have been conducted in detail in several areas, such as particle charging, dust particle collection by the spherical collector, charge build-up on rain drops and so on.

Most of the former studies about these have been done in two-dimensional model that is axisymmetric.

In this paper, the generalized three-dimensional analysing models of the charge carrier motion around a charged spherical object to be utilized in some areas and calculated results are presented.

Introduction

The charging of dust particles in electrostatic precipitators[1], the collection coefficient of dust particles by the collector in charged droplet scrubber [2] and electrofluidized bed[3], the electrification phenomena under DCUHV transmission lines[4], the build-up of charge on rain drops in generation of the thunder cloud[5] and the like, have all been analysed minutely by the similar models. Those models consist of four elements: charge carrier, charged spherical object(sometimes collector), ambient gas flow and electric field. The conditions of these four elements vary respectively and variously combine in each phenomenon.

The orientation of the flow and electric field around the spherical object is important to treat these models. If the orientation is parallel, the two-dimensional model is sufficient, but if the orientation is not, the three-dimensional model is required.

In analysing these models, most of the former studies have been done in two-dimensional models[2], though recently the three-dimensional models have been used only in the case of the perpendicular orientation of the electric field and gas flow without taking account of the motion of the spherical object driven by the electric field[3].

The generalized three-dimensional model that the flow and electric field cross arbitrarily are proposed systematically together with the two-dimensional model.

Outline of the model

The model consists of four factors described above, that is, charge carrier, charged spherical object, ambient gas flow,v_f and electric field,E. The spherical object is movable or not, whose radius and charge are R_c and Q_c respectively. In this model, the spherical object is assumed to be fixed and the relative gas flow,u and the relative motion of the charge carrier to the spherical object are considered.

$$u = v_f - (v_{cx} + v_{ce}) \quad (1)$$

where v_{cx} and v_{ce} are the velocities of the spherical object driven by the gravity and electric field respectively. As the flow pattern of u, potential flow,u_p or Stokes flow,u_s is used generally.

The charge carrier described here are ion or charged aerosol that is not so large to disturb the gas flow and whose radius and charge are R_p (generally $R_p \ll R_c$) and Q_p, though in the case of ion R_p and Q_p are not defined usually. The charge carrier is driven by u and the electrical force,f.

$$f = f_c + f_{ic} + f_{ip} + f_e + f_{icp} \quad (2)$$

where f_c is Coulombic force, f_{ic} charged particle image force, f_{ip} charged collector image force, f_e external electric field force, f_{icp} electric dipole interaction force[2].

The equation of the motion of the charge carrier except for the ion is written by

$$m_p \alpha = -K_s(v - u) + f \quad (3)$$

where m_p, α, K_s and v are mass, acceleration, viscous factor and velocity of the charge carrier. If the charge carrier is inertialess, m_p is assumed to be zero. In the case of the ion, the equation of the motion is

$$v = u + bE' \quad (4)$$

where b is the mobility of the ion in the electric field and E' the sum of the external and Coulombic electric field. Eqs.(3) and (4) are resolved in two or three dimension. And the trajectories of the charge carrier relative to the charged spherical object are obtained. In most cases, the limiting trajectories that divide the collisional and collisionless trajectories to the spherical object are required, because the limiting trajectories are used to decide the cross-sectional area of the charge carrier flowing in the spherical object. And this cross-sectional area is used to know the various physical characteristics, such as the dust collection coefficient to the spherical collector, the current streaming in the spherical object, the charge on the rain drop and so on.

The inflow of the charge carrier to the spherical object is judged from the distance condition of the charge carrier to the spherical object taking into account the interception and Brownian motion of the charge carrier, $r < R_c + R_p + R_b$. R_b is the apparent incremental radius of the charge carrier by Brownian motion. The starting and ending positions of trajectory calculation must be decided carefully according to the required accuracy, because the limiting trajectories vary in accordance with those positions.

Two and three-dimensional models

Figs.1 and 3 are macroscopic model of phenomena. In Fig.1, all directions of gas flow, electric field and the motion of the charged spherical object and the charge carrier are parallel to one another. As the phenomena occur axisymmetrically in this case, it is possible to simulate the phenomena in two dimension as shown in Fig.2. And the components in (r,θ) coordinate of Eqs.(3) and (4) are given in Table 1 which has been used popularly[2].

In Fig.3, two different situations are illustrated for convenience. One is the case that the directions of the gas flow, electric field and the motion of the charged spherical object and the charge carrier are all parallel to the same plane, for example, the surface of this paper. The other is the case that all directions of those are quite arbitrary in three-dimensional space. The model of each case is shown in Figs.4 and 5 respectively. In Fig.4, **u** and **E** are parallel to the x-z plane. In Fig.5, **u** and **E** coss the x axis arbitrarily in three-dimensional space. As the phenomena in these cases do not occur axisymmetrically, the three-dimensional treatment is required. The components in (r,θ,ϕ) coordinate of Eqs.(3) and (4) are given in Table 2. $h(\psi)$, $m(\psi)$ and $n(\psi)$ should be used in the case of Fig.4. $h(\theta,\phi)$, $m(\theta,\phi)$ and $n(\theta,\phi)$ should be used in the case of Fig.5. These expressions of flows and electrical forces are straightforwards obtained by making extention from two to three dimension and rotation of each component around one or two axes in turn.

Results of calculation

Typical calculated results by using the proposed three-dimensional models are shown in Figs.6 and 7. Fig.6 shows the limiting trajectories at four corners and the resulting cross-sectional area being able to flow in the spherical object at the center in the case of ion under conditions of electric field crossing the gas flow with the angle of 135° and the stationary spherical object. Figures at four corners illustrate the same calculated results changing the direction of view. From the similar calculations the existences of stagnation points and minor loop trajectories are fairly interested.

Fig.7 shows the limiting trajectories and the resulting cross-sectional area being able to collide with the spherical collector in the case of charged aerosol under the conditions of the directions of the falling charged spherical collector, gas flow and electric field crossing one another perpendicularly and the spherical object driving by the electric field. In the similar calculations, it appears that the electric field increase the cross-sectional area under some conditions and decrease under the others.

Conclusions

The analysing models of the charge carrier motion around a charged spherical object in three dimension are proposed. And the usefulness of these models are shown by practical calculations.

Acknowledgments

The authors wish to thank Dr. T. Fujimura for his useful discussions and Messrs K. Nakazono and M. Yamabayashi for their technical assistance. This work was supported in part by a Grant-in-Aid for Scientific Research from the Ministry of Education, Science and Culture, Japan.

References

[1] H.J.White: AIEE Trans. 70(1951)1186-1191.
[2] K.A.Nielsen and J.C.Hill: I&EC,Fund. 15(1976)149-163.
[3] M.Shapiro and G.Laufer: I&EC,Fund. 23(1984)164-170.
[4] M.Hara and M.Akazaki: J.Electros. 4(1978)349-365.
[5] F.J.W.Whipple and J.A.Chalmer: Quart.J.Roy.Met.Soc. 70(1944)102-119.

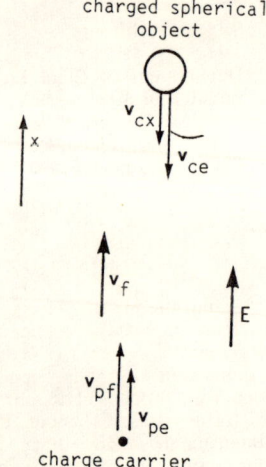

Fig.1. Macroscopic model of phenomena, all directions are parallel to one another.

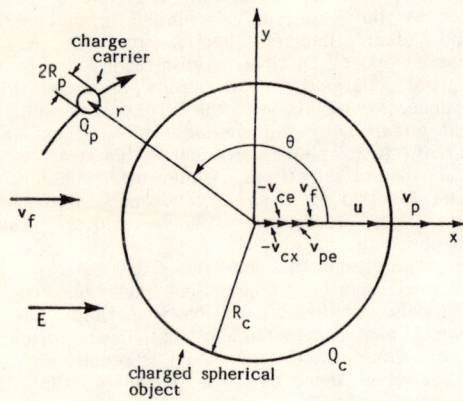

Fig.2. Two-dimensional analysing model.

Table 1. Dimensionless expressions of flows and electrical forces in two dimension.

	r	θ
u_p^*	$\cos\theta(1-1/r^3)$	$-\sin\theta(1+1/2r^3)$
u_s^*	$\cos\theta(1-3/2r+1/2r^3)$	$-\sin\theta(1-3/4r-1/4r^3)$
f_c^*	$1/r^2$	0
f_{ic}^*	$1/r^3 - r/(r^2-1)^2$	0
f_{ip}^*	$-1/r^5$	0
f_e^*	$\cos\theta(1+2\gamma_c/r^3)$	$-\sin\theta(1-\gamma_c/r^3)$
f_{icp}^*	$(1-\gamma_c/r^3-3\cos^2\theta(1+\gamma_c/r^3))/r^4$	$-\sin\theta\cos\theta(2+\gamma_c/r^3)/r^4$

-98-

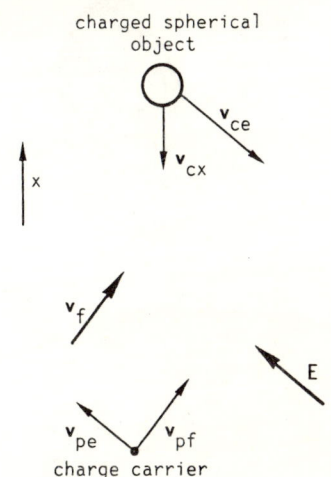

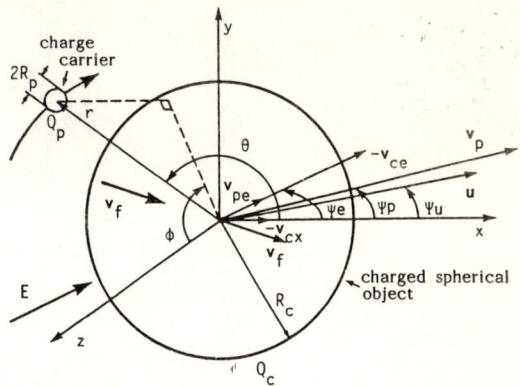

Fig.4. Generalized three-dimensional analysing model-I, both **u** and **E** are parallel to the x-z plane.

Fig.3. Macroscopic model of phenomena, there are two cases that all directions are parallel to the same plane and all directions are arbitrary in three-dimensional space.

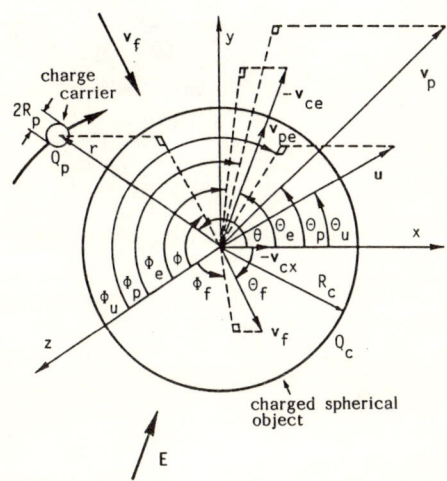

Fig.5. Generalized three-dimensional analysing model-II, both **u** and **E** cross the x axis arbitrarily.

Table 2. Generalized dimensionless expressions of flows and electrical forces in three dimension.

	r	θ	φ
u_p^*	$h(^\Psi_{\Theta,\Phi})(1-1/r^3)$	$-m(^\Psi_{\Theta,\Phi})(1+1/2r^3)$	$n(^\Psi_{\Theta,\Phi})(1+1/2r^3)$
u_s^*	$h(^\Psi_{\Theta,\Phi})(1-3/2r+1/2r^3)$	$-m(^\Psi_{\Theta,\Phi})(1-3/4r-1/4r^3)$	$n(^\Psi_{\Theta,\Phi})(1-3/4r-1/4r^3)$
f_c^*	$1/r^2$	0	0
f_{ic}^*	$1/r^3 - r/(r^2-1)^2$	0	0
f_{ip}^*	$-1/r^5$	0	0
f_e^*	$h(^\Psi_{\Theta,\Phi})(1+2\gamma_c/r^3)$	$-m(^\Psi_{\Theta,\Phi})(1-\gamma_c/r^3)$	$n(^\Psi_{\Theta,\Phi})(1-\gamma_c/r^3)$
f_{icp}^*	$(-3h^2(^\Psi_{\Theta,\Phi})(1+\gamma_c/r^3)+1-\gamma_c/r^3)/r^4$	$-h(^\Psi_{\Theta,\Phi})m(^\Psi_{\Theta,\Phi})(2+\gamma_c/r^3)/r^4$	$h(^\Psi_{\Theta,\Phi})n(^\Psi_{\Theta,\Phi})(2+\gamma_c/r^3)/r^4$

$h(\Psi) = \cos\Psi\cos\theta - \sin\Psi\sin\theta\cos\phi$, $h(\Theta,\Phi) = \cos\Theta\cos\theta + \sin\Theta\sin\Phi\sin\theta\sin\phi + \sin\Theta\cos\Phi\sin\theta\cos\phi$

$m(\Psi) = \cos\Psi\sin\theta + \sin\Psi\cos\theta\cos\phi$, $m(\Theta,\Phi) = \cos\Theta\sin\theta - \sin\Theta\sin\Phi\cos\theta\sin\phi - \sin\Theta\cos\Phi\cos\theta\cos\phi$

$n(\Psi) = \sin\Psi\sin\phi$, $n(\Theta,\Phi) = \sin\Theta\sin\Phi\cos\phi - \sin\Theta\cos\Phi\sin\phi$

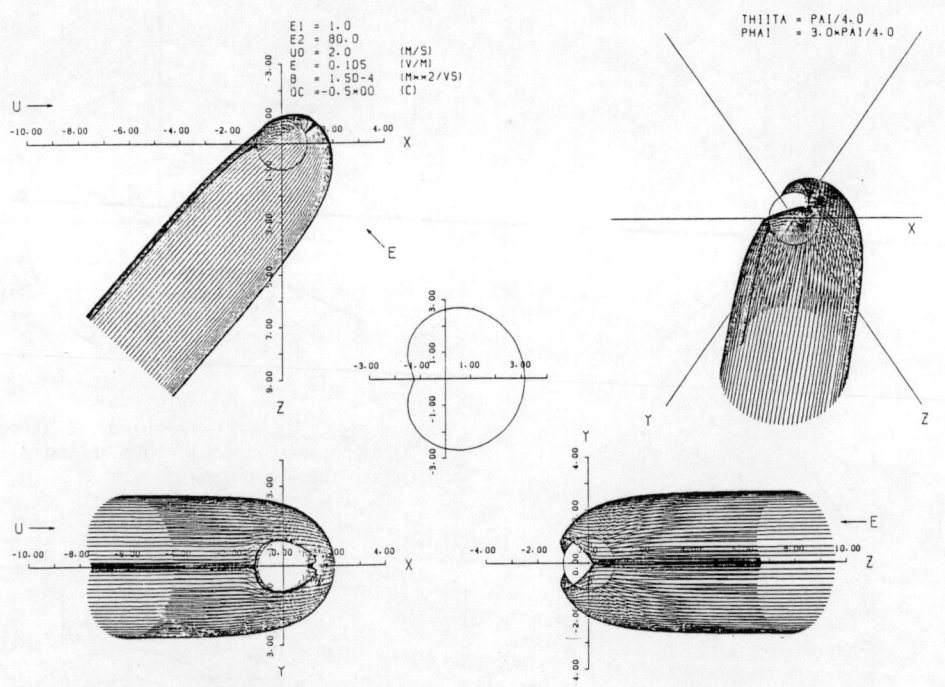

Fig.6. Limiting trajectories and resulting cross-sectional area flowing in the spherical object in the case of the ion.

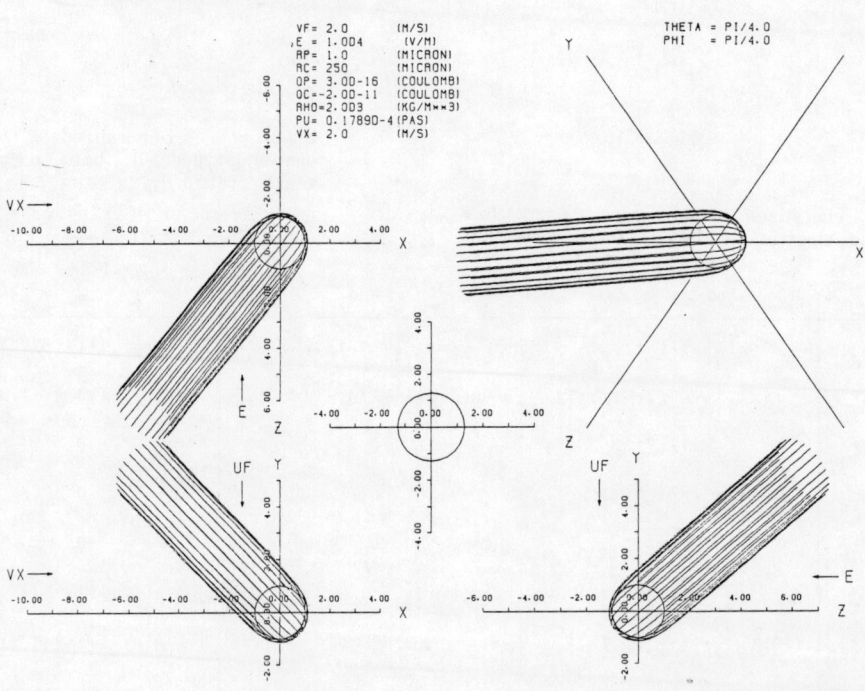

Fig.7. Limiting trajectories and resulting cross-sectional area colliding with the spherical object in the case of the charged aerosol.

Proceedings of International Conference on Modern Electrostatics

SPECIFIC SURFACE CHARGE DENSITY OF NON-SPHERICAL PARTICLE AND CHARGE DECREASE BY PARTIAL SELF-DISCHARGE IN HIGH ELECTRIC FIELD INTENSITY

Fumio Isahaya
Electrostatics & Aerosol Technology Consultant
1-2-10 Himonya, Meguro-Ku, Tokyo 152, Japan

Abstract

The surface charge density Q/S (Q:saturation charge of particle, S:surface area of particle) of non-spherical particle model was measured with the use of a particle model dropping Faraday cage methode and compared with that of a spherical particle model. First the particle model was gravitationally dropped into a corona electric field, then passed through a uniform electrostatic field. The measured Q/S of regularly triangular prism became almost the same as that of the sphere, furthermore the Q/S of column, star-shaped pillar and regular hexahedron were 4.2, 4.5 and 11% respectively lower than that of sphere. The results clearly showed that the charge decrease phenomena by a partial self-discharge in the non-spherical particle model occured critically when the Q/S and the external electric field exceeded to the extent of 6 to 10 $\mu\mu C/mm^2$ and several kV/cm, particularly in negatively charged non-spherical particle model.

Introduction

The geometrical shape of dust particles generated in an industrial process presents mostly a non-spherical one due to their mechanism of generation and chemical composition and they become the shape of their own.

In order to make clear the electrical charging characteristics of non-spherical particle in a single-stage and two-stage type electrostatic precipitator, the effect of the shape of non-sperical particle on the saturation charge of particle in a corona electric field and uniform electrostatic field were measured with a particle model dropping Faraday cage method, particularly in comparison with that of spherical particle[1],[2],[3]. The column, regular hexahedron, regularly triangular prism and star-shaped pillar as a non-spherical particle model were used and made of both a 18Ni-8Cr stainless steel as a conductive particle model and a Boron-Silicate glass as a dielectric particle model.

Particularly, it was interesting to find out how important was the effect of the partial enhancement of electric field intensity at the sharp edge of non-spherical particle to the saturation charge and that the charge decrease by the partial self-discharge caused by the superposition of the external electrostatic field on the surface electric field of the charged particle itself would occur [2],[3].

Experimental Apparatus

As shown in Fig. 1, first the particle model was gravitationally dropped into the corona electric field of coaxial cylinder type electrode, then passed through the uniform electric field of parallel plates type electrode, and finally received by the cup electrode in the Faraday cage which was connected to an electrometer for measuring the electric charge of particle model. When the particle charge of dielectric particle model was measured, the cup electrode was filled with water. The relative humidity of ambient air was kept at approx. 40%.

Fig. 2 illustrates the geometrical shape, size and surface area of particle models used in this experiments. These particle models were made the aspect ratio of approx. 1 in the direction of X, Y, and Z axis.

Experimental Results

Figs. 3, 4 and 5 show the measured saturation charge of particle model against the corona electric field intensity relationship with the parameter of the shape of particle model. Also, Figs. 3, 4 and 5 are for the case of the positively, negatively charged conductive particle model and positively charged dielectric particle model respectively.

Figs. 6, 7, 8, 9 and 10 show the measured saturation charge of particle model against the external uniform electrostatic field intensity relationship with the parameter of the negative saturation pre-charge of particle for the case of sphere, column, regular hexahedron, regularly triangular prism and star-shaped pillar respectively.

The corona electric field intensity of Ei corresponded to the corona current density of i for the coaxial cylinder type electrode as indicated in Figs. 3 to 10 can be given by the use of the following equation:

$$Ei^2 = (2i/K) + (Ecr_0/r)^2, \quad (1)$$

where K is the ion mobility, Ec the electric field intensity of corona discharge electrode at corona onset, r_0 the radius of corona discharge electrode, r the radial distance from the center of corona discharge electrode. That is, the value of i= 6.3, 12.5, 18.8 and 25 μA/cm corresponds to the value of Ei=2.38, 3.35, 4.11. and 4.74 kV/cm respectively.

Discussion and conclusion

(i) Specific surface charge density

From the results shown in Fig.3, when the particle is negatively charged in the corona field intensity of 2 kV/cm, the surface charge density Q/S and the specific surface charge density Qn defined as the ratio of the Q/S of non-spherical particle model to the Q/S of spherical particle model are tabulated in Table 1 for each particle model. The results shown in Table 1 have led to the following conclusions;

(A) The Q/S of regularly triangular prism is almost the same as that of the sphere, the Q/S of column and star-shaped pillar is slightly lower than that of sphere and the Q/S of regular hexahedron is approx. 11% lower than that of sphere. That is, if the difference to the extent of approx. ±10% in the particle charge can be neglected in the practical view point in an industry ESP, furthermore when the aspect ratio in the direction of X, Y, Z axis of particle nearly equals to 1 and the corona electric field is kept at constant, the linear relationship between the saturation charge and the surface area of particle is observed.

(B) While, when the aspect ratio of particle is considerably different from 1.0 such as the ellipsoidal particle of revolution at the major axis.[4] as shown in Table 1, for the case of the ratio of minor axis to major axis of 0.068 and 0.1, the Qn becomes 2.6 and 0.44 respectively. It can be said that when the longitudinal axis of the conductive fibriform or pearl-chain form particle coincides with the direction of corona eleciric field, depending on the ratio of the minor axis to the major axis, the Qn will become a considerably higher or lower value in comparison with that of sphere.

(C) As shown in Figs. 3 and 4, for the case of positively or negatively charged non-spherical particles except for the star-shaped pillar, the linear relationship between the saturation charge and the intensity of corona electric field to the extent of approx. 5 kV/cm are observed. For the case of star-shaped pillar, the saturation charge is critically decreased when the corona electric field is in excess of approx. 2.4 kV/cm in a negative corona and 4 kV/cm in a positive corona. Because the electric field at the sharp edge of particle is enhanced partially and thereat a partial self-discharge occurs.

(D) For the case of Boron-Silicate glass as a non-spherical dielectric particle model as shown in Fig. 5 (positive charging), the linear relationship between the saturation charge and the intensity of corona electric field is also obserbed, however the proportionality to the surface area of particle is disappeared. The reason is not cleared yet.

(2) Partial self-discharge

Figs. 6 to 10 show the effect of external uniform electrostatic field on the spherical and non-spherical particle charge. These measured results have led to the following conclusions;

(A) For the case of sphere, as shown in Fig. 6, no partial self-discharge occurs to the extent of approx. 10 kV/cm in an external electrostatic field intensity E_0 both the positively and negatively charged particle. Hereupon, the electric field intensity E_n in the normal direction at a surface of conductive sphere particle of radius a_0 can be given by the following equation,

$$E_n = 3E_0 \cos\theta \pm Q/a_0^2$$
$$= 3E_0 \cos\theta \pm 3E_i. \quad (2)$$

Therefore, the maximum value of E_n exists at $\theta=0$, thus when E_0 and E_i become 10 kV/cm and 4.74 kV/cm respectively, no partial self-discharge for conductive sphere occurs to the extent of $|E_n|_{max} = 3 \times (10+4.74) \approx 44.7$ kV/cm at the least.

(B) For the case of non-spherical particle as shown in Figs. 7 to 10, the critical value of surface charge density Q/S and external electric field intensity E_0 for a partial self-discharge onset are tabulated in Table 2.

(C) Particularly, as shown in Figs. 9 and 10, it can be seen that the polarity of negatively charged particle is reversed an opposit sign in the range of a high external electrostatic field.

References

[1] Isahaya, F.: Analysis of the Corona Field Intensity Distribution in Electrostatic Precipitators by the Steel Ball Dropping Method, The Electrotechnical Journal of Japan, Vol.8, No.3/4, 1963, p.151~157

[2] Isahaya, F.: Decrease of Particle Charge by Partial Self-Discharge in Corona Electric Field and Effect of Geometrical Shape of Particle, the Proceedings of Annual Meeting of I.E.E.E. of Japan, 1962, No.787

[3] Isahaya, F.: Decrease of Pre-Charged Non-spherical Particle by Partial Self-Discharge in Electrostatic Field and Effect of Polarity, the Proceedings of Annual Meeting of I.E.E.E. of Japan, 1964, No.997

[4] Mizuno, A. and Fukuma, M.: Decrease in Charge of Sharp-Edged Ellipsoidal Particles by Self-Discharge, IEEE Trans. on Industry Application, Vol. IA-21, No.1, January/February 1985

Table 1. Comparison of specific surface charge density Q_n for non-spherical particle models

Shape	$S(mm^2)$	$Q(\mu\mu C)$	$Q/S(\mu\mu C/mm^2)$	Q_n
Sphere	49.00	260	5.31	1.0
Column	29.44	150	5.09	0.958
Regular hexahedron	96.00	453	4.72	0.889
Regularly triangular prism	61.86	330	5.33	1.004
Star-shaped pillar	84.82	430	5.07	0.955

Ellipsoid of revolution at major axis

Model 1 (minor axis 0.47 / major axis 7.01 ≒ 0.068)
    8.14   113.5   13.94   2.63

Model 2 (minor axis 0.37 / major axis 3.76 ≒ 0.1)
    3.45   8.08   2.34   0.44

Model 3 (minor axis 2.08 / major axis 4.51 ≒ 0.5)
    24.90   154.0   6.18   1.16

Table 2. Critical value of surface charge density and external electrostatic field intensity for partial self-discharge onset

Polarity	Column	Regular hexahedron	Regularly triangular prism	Star-shaped pillar
Positively charged	--	--	$Q/S \geq 9.7$ $E_0 \geq 6.0$	$Q/S=5.5$ $E_0 \geq 8$
″	--	--	--	$Q/S=10.5$ $E_0 \geq 2$
Negatively charged	$Q/S=10.2$ $E_0 \geq 8$	$Q/S=6.04$ $E_0 \geq 6$	$Q/S=6.2$ $E_0 \geq 6$	$Q/S=5.66$ $E_0 \geq 2.4$
″	--	$Q/S=11.25$ $E_0 \geq 4$	$Q/S=10$ $E_0 \geq 4$	--

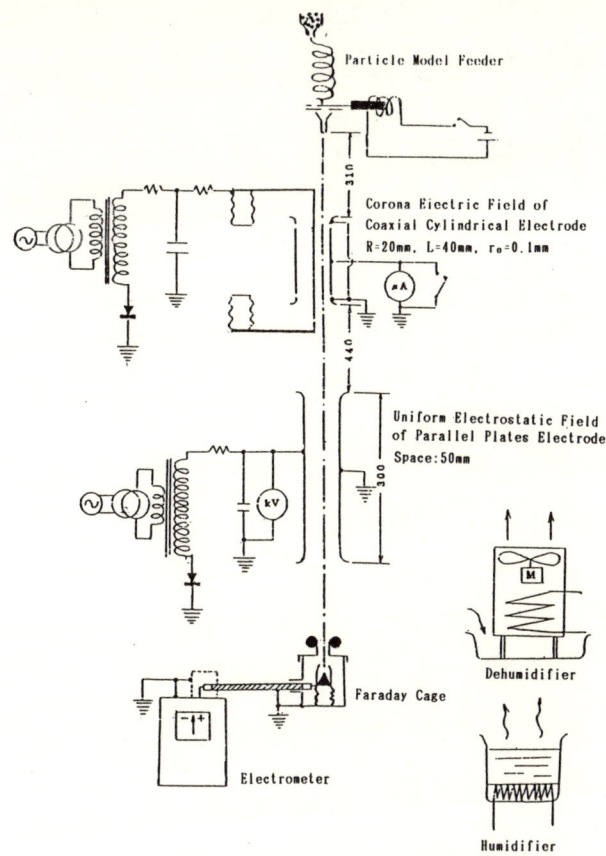

Fig. 1. Experimental apparatus

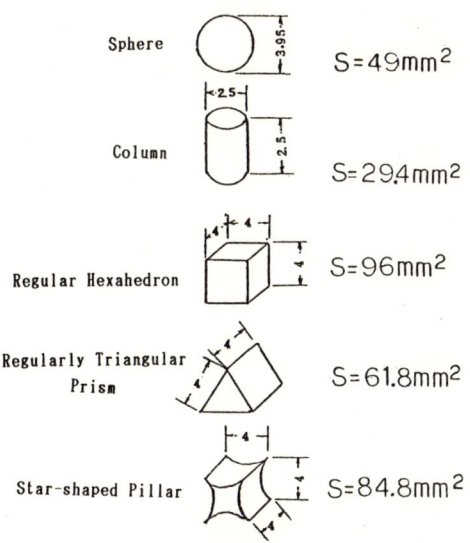

Fig. 2. Geometrical shape, size and surface area of non-spherical particle models in experiment

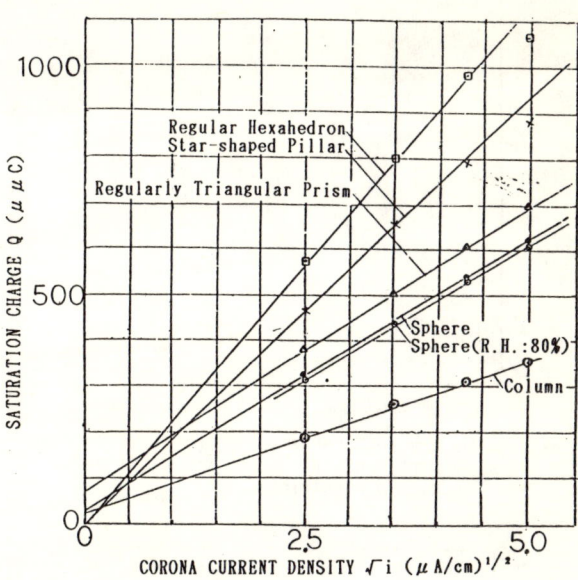

Fig. 3. Saturation charge versus positive corona current density for non-spherical particle models (conductive)

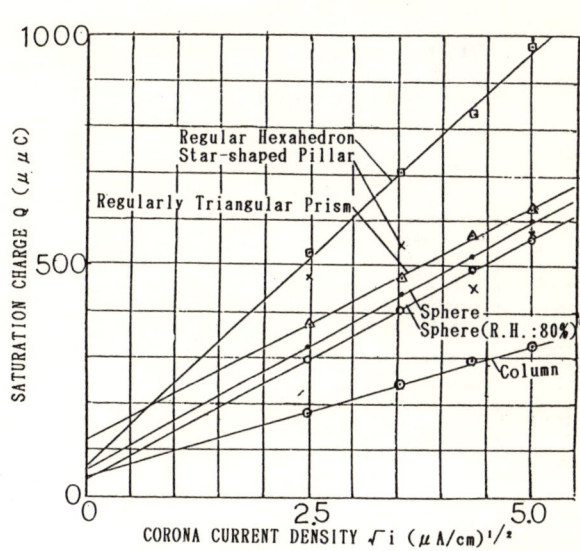

Fig. 4. Saturation charge versus negative corona current density for non-spherical particle models (conductive)

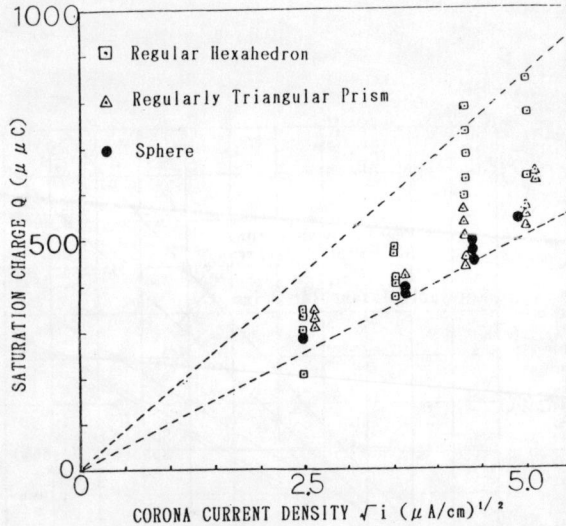

Fig. 5. Saturation charge versus positive corona current density for non-spherical particle models (dielectric)

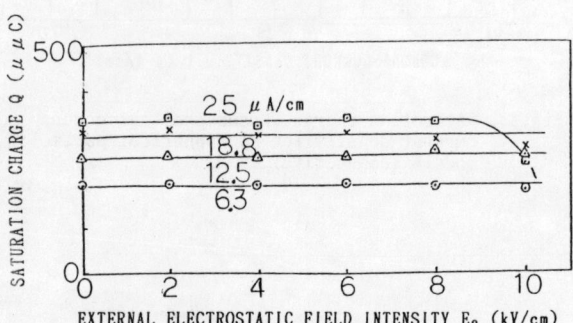

Fig. 7. Effect of external electrostatic field intensity on saturation charge for negatively pre-charged column particle (conductive)

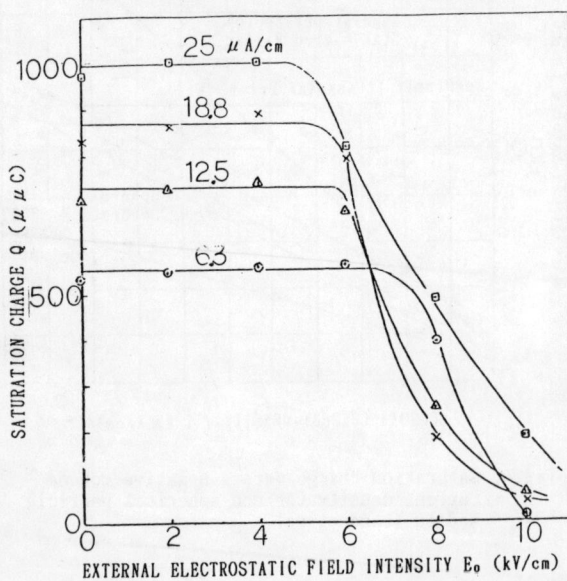

Fig. 8. Effect of external electrostatic field intensity on saturation charge for negatively pre-charged regular hexahedron particle (conductive)

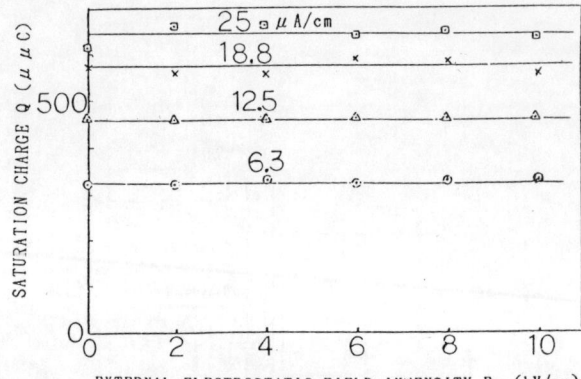

Fig. 6. Effect of external electrostatic field intensity on saturation charge for negatively pre-charged spherical particle (conductive)

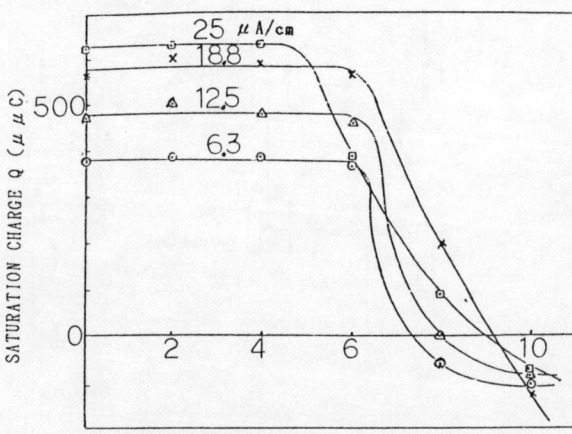

Fig. 9. Effect of external electrostatic field intensity on saturation charge for negatively pre-charged regularly triangular prism particle (conductive)

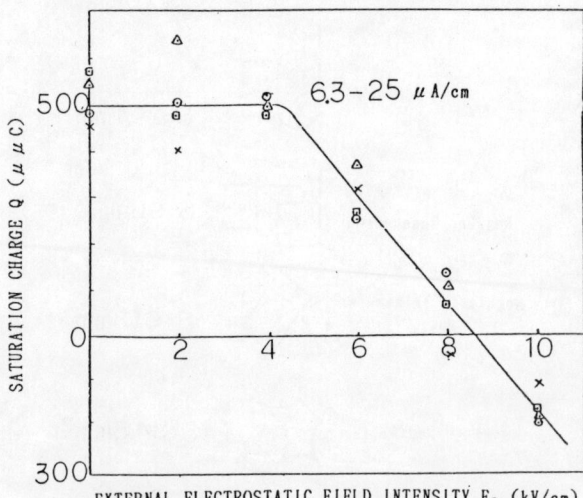

Fig. 10. Effect of external electrostatic field intensity on saturation charge for negatively pre-charged star-shaped pillar (conductive)

SUPER-HIGH PULSE VOLTAGE ELECTROSTATIC PRECIPITATOR

Wang Rongyi Liu Jinmao Wu Yan Sun Dawei
Yu Zhigang Zhong Dianqang Xia Bin Wang Xilu
(Northeast Normal University, Changchun, China)

ABSTRACT

Supper-high voltage puste energization for ESP and its corona discharge characteristics was studied. This technique has been utilized succefully on the sites of cement industry, fluidized-bed boiler and others. Results show that super-high voltage energization combined with wider spacing and pulse energization is appropriate to collecting high resistivity dust. As the requirements for voltage waveform and pulse repeatition frequency could be lowered, the super-high voltage pulsed power supply is cost saving, easy to manufacture and hence suitable for extensive applications.

INTRODUCTION

Both pulse-energization and Super-high voltage electrostatic precipitation with wide spacing are the new techniques developed in recent years, and important research subjects for improving ESP performance. The ESP with wide spacing is more advantageous in structure and performance, compared with conventional ESP. For example, effective migration velocity can be increased [1] [2]. The pulse energized ESP has lower power consumption, higher collection efficiency, and suitable for high resistivity dust. In general, pulse energization is adaptable to various types of ESPs [3] [4]. By combining super-high voltage wide spacing with pulsed energization, the unique super-high voltage pulsed energization electrostatic precipitation technology still retains their advantages.

DESIGN OF SUPER-HIGH VOLTAGE PULSED POWER SUPPLY

The Key to implementation of super-high voltage pulsed precipitation is the design of its power supply. For the present practice, pulsed power supply is obtained by superimposing a pulsed voltage on DC voltage, which is maintained around the corona starting voltage. The DC voltage plays the role of cleaning electric field ofter each pulse without causing gas breakdown. The pulse amplitude is 50~70KV, pulse width 50-200μs and repeation frequency 25~400Hz. Both repeation frequency and pulse width of the pulsed power supply for precipitation are mainly used to adjust the average current density at the collection plate. In case of high resistivity dust, average current density is adjusted to prevent the occurrence of backcorona. As for pulse widths there is no strict requirements. The time constant is $T=9.85\rho E\times10^{-14}$ sec. When dust resistivity is $\rho=10^{11}$ Ω-cm, relative premittivity of dust is $E=2$, then the time constant is $T=1.8$ sec. This is much longer than the pulse width of 10 ms for half wave rectification of commercial frequency (50Hz). To ensure the pulse width lower than 10ms under supper-high pulsed voltage, repeatation frequency is used to regulate the average current density in used to regulate the average current density in designing the power supply for purposes of simplicity, low cost, and being adaptable to industrial applications. Figure 1 is the schematic circuit diagram of super-high voltage pulsed power supply. The secondary of step-up transformer t1 is in series with the primary of step-up transformer T2 to ease handling of insulation, as shown in Fig.1.

Fig.1 Schematic diagram of Super-high
voltage pulsed power supply.

T1, T2,-Transformer
D1, D2, D3,-Sillicon diode
C- Capacitor
R- Measuring resistance
B- Control box

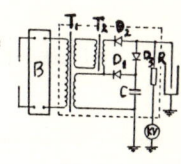

Fig.1

With T1 supplying DC voltage and T2 supplying pulse voltage, the pulsed voltage is superimposed on DC voltage. By adjusting transformation ratios of T1 and T2, various proportions of DC voltage and pulse voltage can be obtained. The frequency, amplitude and pulse width of the pulse voltage output are controlled by the primary of T1. By using two-way thyristor AC voltage regulation, 50Hz super-high pulsed voltage can be obtained. Let the secondary voltage of T1 plus that of T2 higher than the voltage required, then the conduction angle of the thyristor is less than 90 degrees during operation, and the pulsed width is less than 2ms. A narrow pulse voltage with controlable frequency can be obtained by using the controlling circuit given in the paper [5]. The currents flowing through D2 and D3 are small because voltage across the capacitance c serves only for cleaning the electric field. The load current flows mainly through D1. T1, T2, D1, D2, D3 and the measuring resistance R were placed in an insulating oil tank, treated in vacuum, and then sealed tigtly to increase dielectric strength and the life time.

CORONA DISCHARGE CHARACTERISTICS IN AIR

In laboratory, we studied corna discharge characteristics of the pulsed power supply by using the facility shown in Fig.2, and compared it with DC power supply. The results are given in Table 1 and Table 2.

Fig.2 The experimental facility
1- Super-high voltage pulsed power supply
2- Control box
3- Electrometer
4- Oscillograph
5- Corona discharge test device
R1, R2, R3- Measuring resistances

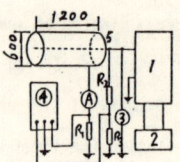

Fig.2

TABLE 1. V-I characteristics of DC power supply

Voltage (KV)	33.5	44	53.5	63	72	87
Current (mA)	0.01	0.03	0.15	0.25	0.40	0.65

Results show that the relation berween peak voltage and peak corona current of the pulsed power supply is the same as V-I characteristics of the DC power supply. By using two-way thyristor to regulate the primary voltage of the

TABLE 2. V-I characteristics of pulsed power supply

average voltage (KV)	41	47	53	57
peak voltage (KV)	66	84	99	110
DC Component (KV)	39	45	52.5	60
peak current (mA)	0.5	0.83	1.04	1.25
DC Component (mA)	0.03	0.058	0.083	0.1
average current (mA)	0.04	0.07	0.15	0.2

pulsed power supply, the conduction angle of the thyrister increased, pulse width became widen, DC component increased, and the ratio of peak current and average current, which ranges generally form 5 to 10, decreased. The wavaforms of discharge current and pulse voltage presented by dual-trace oscillogragh are shown in Fig.3.

Fig.3 Waveforms of voltage and aurrent
1 - Pulsed voltage
2 - Pulsed current
D1- DC voltage
D2 -DC current   Fig.3

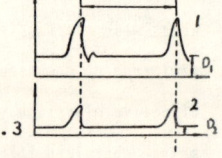

It is noted that corona discharge takes place only at the half wave front of pulse voltage and the lagging of pulse corona current is steeper.

CONCLUSIONS

Electrostatic precipitation with super-high voltage pulse energization has advantages of low power consumption, high collection efficiency, and being appropriate to collect high resistivity dust. In general, it fits for varieties of dust with the same advantages, The key to the question is Super-high voltage pulsed power supply. The ratio of DC component and pulse component should be set appropriately for different resistivity dust. For collecting high resistivity dust, DC voltage should be lower than the corona starting voltage, while for medium resistivity dust, the DC voltage may be higher than that.

Thus, enhanced electric field for cleaning can be obtained behind the peak of voltage pulse. It still has advantages of higher spark voltage and lower power consumption, as compared with the DC power supply. The supper-high voltage pulsed power supply designed has been applied seccussefully to fluidized-bed dust precipitation and cement precipitation with good results. In a word, super-high voltage pulsed precipitation which combines the pulsed energization with super-high voltage wide spacing is a new and very promising technique.

REFERENCES

[1] 伊藤良三シス广ムのスシコ电一式静电集尘装置，产业机械；52年10月

[2] 刘林茂，板式电收尘器中的电场分布，东北师大学报。1988年第三期。

[3] Masuda, S., et al, Application of Boxer-Charge in Pulsed Electrostatic Precipitator, Conf. Rec of IEEE/IAS Ann. Meet. (1980).

[4] Porle, K., Reduced Emission and Energy Consumption with Pulsed Energization of Electrostatic Precipitators, J. Electrostatic, Vol. 16, Nos. 2+3 (1985).

[5] 王强，超高压脉冲在高压倍加电路上的实现， 东北师大研究生毕业论文，1987年。

[6] 刘林茂等，横向百叶窗式脉冲静电除尘器， 在本次会议上发表 1988年10月

THE CORRECT DIAGNOSIS AND OPTIMUM DISPOSAL OF FLASHOVER SIGNALS IN ELECTROSTATIC PRECIPITATORS

Wang Weixue Jiang Yabin
Environmental Protection Research
Institute of Electric Power

Wang Ronghua
Zhejiang Jinhua Electronic
Instruments Factory

Abstract

With the studies of the methods of diagnosing and disposing of the flashover in EP both at home and abroad, the paper clarifies the drawbacks in these methods. To start with the variation regularity of the working voltage and current during the flashover in electric field, the paper concludes a set of effective methods in making use of the distortions of the wave form as identification marks and thereof followed by sampling, numerical calculating and logical comparing with the aid of a flashover without necessarily locking up the main circuit of the silicon controlled voltage regulater. As shown in results of its actual industrial operation, it is obvious that the described method will not only get rid of the drawbacks of the other methods in flashover diagnosing and treatment but also will win maximum extension of the effective power supply time in a electric field frequenty troubled by flashover and the enhancement of working voltage and improvement of the dust collecting efficiency.

1. Issue raised

The flashover phenomena frequently occured during EP operation will play a direct part in the operating values of voltage and current and hence the efficiency of dust collection. Many researches into the methods to distinguish and handle the flashover signals have been made at home and abroad.

In the silicon controlled voltage regulator with analog control imported in 70s, the spotting-sign is the high frequency current signal occurred during electric field flashover, mostly sampled by means of a transistor differential circuit. After sampling, the control system lockons up the main circuit of silicon controlled voltage regulator for 40-60ms for each flashover signal acquired in the main circuit and then allows the voltage to rise again slawly, as shown in Fig. (1)

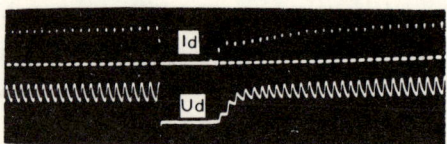

Fig. (1)

Thus, the shortcoming of this device is the notable decrease in secondary voltage and current and impairment of the dust collecting efficiency.

Ever since the micro-computer control technology has been applied to the EP, the experts of the western countries forst suggested a method of varied disposal of flashover according to the strength.

Fig (2) is the photos of the wave form of the variation of working voltage and current, before and after the occurrence of flashover, taken from the equipment imported Sweden Flakt during its operation in China and show the method of diagnosis and disposal.

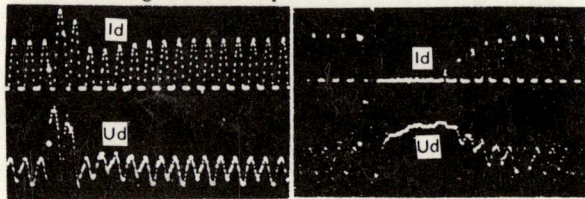

Fig. (2)

Under the influences of the western experts, the method of distinguishing and handling as shown in Fig. (3) is adopted in some micro-computer controlled power supply devices.

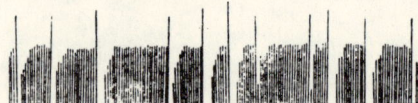

Fig. (3)

The idea of disposing of flashovers variedly according to the magnitude of the signals as shown in Fig. (2), (3) is reasonable. The key point in the method, however, is to make use of the magnitude of the sudden rise of the amplitude of the secondary current of the fundamental wave after flashover as the identification marks and that will involve the following drawbacks:

(1) The secondary voltage corresponding to the suddenly raised wave amplitude of the working fundamental currernt will fall to zero, and will cause a voltage loss in these waves and faillure in dust collecting.

(2) The power of these suddenly raised fundamental current waves will be dissipated in the form of heat loss in the system of power supply.

(3) No matter whether the sudden rise of the current fundamental wave or the sharp fall of the voltage, all will make the equipment suffer from the impulse of over current and over-voltage of high frequency.

(4) The problem of working voltage loss remains basically unsolved in the case of frequent occurrence of flashover as these methods will still have the main ciruit of the controlled silicon locked up to certain degree.

The American Belco. Co. noticed the above mentioned shortcomings (1) - (3), but still meeded to lock up the silicon controlled circuit, as shown in Fig. (4).

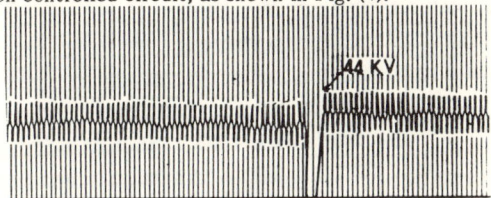

Fig. (4)

Aiming at overcoming the above mentioned shortcomings thoroughly to win the possibly most effective power supply time during the flashover of electric field, to enhance the field's working voltage and thereof to improve the dust collecting efficiency, an investigation of the methods of correct identification and optimum treatment of the flashover signs was carried out.

2. Method to identify the flashover signs in electric field.

By investigating the great amount of the photos of wave forms of electric field's working voltage and current, taken from the actual industrial operation of EP, the authors found out that any occurence of flashover will simultaneously result in the distortions of working voltage and current waves on the flashover instant as shown in Fig. (5).

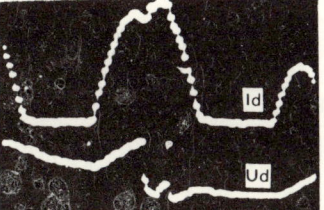

Fig. (5)

Further studies also revealed that all the resulted distortions of the working voltage and current during flashover follow a definite regularity. This regularity can be described by the four variation characteristics of the working voltage and current waves on the flashover instant in EP. With the aid of micro-computer, the values of very half of the voltage and current wave will be respectively sampled and then numerically calculated and logically compared with the corresponding characteristics to distinguish and judge the resulted flashover signals comprehensively. As proven by practices, any flashover signals thus picked out was correct only if the accuracy in sampling reaches the demanded degree and the method of numerical analysis and logical comparison is proper.

Without the method mentioned above, a sudden rise of the fundamental current wave as shown in Fig. (6) may be hard to avoid and that will be a miss judgement of the flashover signs.

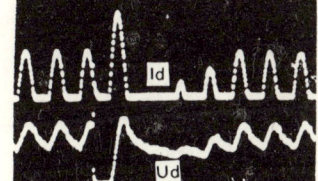

Fig. (6)

As shown in Fig. (7), the signal of working voltage and current in the field should accord with its variation characteristics. Besides, there should not be abrupt rise in current fundamental wave. Only in this case, it is considered to be a correct identification of the flashover signal by the computer. (Note that in rare case, when flashover emerges at the front of a sinusoidal wave, the amplitude of fundamental wave of the current appears to have some increase.)

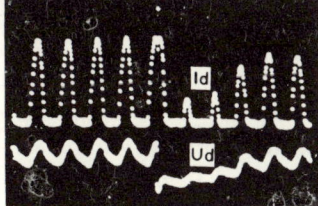

Fig. (7)

3. Methods to deal with the flashover and the results in industrial operation.

From Fig. (7), it is obvious that in disposing of the flashover sign after its occurrence, it is not necessary to lock up the silicon controlled circuit but only, in the next half of the wave, to have the field working voltage lowered by diminishing appropriately the conduction angle and then raised quicklly again. In this way, the voltage can resume its value same as that before flashover in about 40ms.

How can the system have the dielectric insulation strength return to its normal value without the locking up the main circuit of the voltage regulating controlled silicon after flashover? We gave the explanation of the mechanism as follows:

From Fig. (7), it can be found that the instant when the flashover occurs or when the field voltage suddenly falls to zero is the time when the field dielectric insulation strength begins to recover as well. There is time for ionic combination in the half wave in which the flashover occurs. In the closely following half wave there is still time for ionic combination because the conduction angle has been diminished so that conduction of the controlled silicon and the operation of the voltage is delayed for a periods. It is proved from our practical industrial operation that these two periods are enough for the field dielectric strength to return to normal. There is no need to lack up the voltage regulating controlled silicon in the main circuit. Fig. (8) shows the wave-forms of the secondary field voltage and current in both common voltage regulator with its controlled silicon locked up during the flashover and our control system without the controlled silicon locked up.

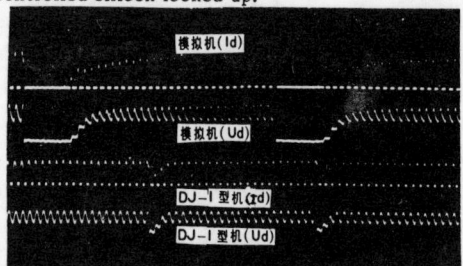

Fig. (8)

Having used this method of identification and treatment, possibly most time of effective power supply can be gained so that the operating voltage of the field is enhanced as shown in Fig. (9). Table 1 gives the actual operating parameters for both the present method and the method by which the controlled silicon of main circuit is locked up for half a period (10ms) and four halves of period (40ms) respectively. Under the experimental conditions regarding Table 4, for each half (10ms) of wave of locking up the voltagae controlled silicon in the main circuit, the voltage output loses at least 1kv. This result agrees with the theoretical calculation.

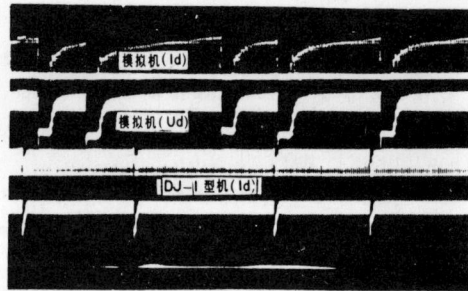

Fig. (9)

In the same electric field of EP in Jian Bi Power Plant where in Jiangsu is conducted a contrast test in operating parameters of the equipment described in this paper and the power supply controlled with a micro-computer imported form Sweden Flakt Co. The comparison of the results shown in Table 2 indicates that adoption of this technique to diagnose and dispose of flashover can yield even higher output of voltage and current than that the other power supply can provide under the same conditions. In order to indicated the effects in enhancement of dust collection efficiency gained by this technique under the conditions of industrial operation, collection efficiency is measured (Fig. 10). A home-made silicon controlled voltage regulator allows an exit dust concentration of $1004mg/m^3$, while the operation method given in this paper provides a dust penetration as low as $561mg/m^3$. The precipitation efficiency increases from 94.65% to 98.38%.

Table 1

	$U_1(V)$	$I_1(A)$	$U_d(KV)$	$I_d(mA)$	SP	SCR
no locking up	330	54	47	270	149	143
	304	49	46	247	152	138
	310	51	46	253	149	139
	316	53	46	263	157	140
	342	61	46	299	154	139
locking up for 1/2 period	324	55	45	264	148	140
	296	48	45	238	155	135
	310	51	45	246	151	136
	282	46	44	228	153	130
	240	38	44	189	153	124
locking up for 2 period	320	53	42	234	152	132
	300	48	42	218	160	127
	284	45	41	202	149	123
	276	45	40	199	151	122
	294	48	41	216	159	127

(a)△   Fig. (10)   (b)△△

Table 2

Name of equipment	Recording time	A-1# Electric Field			
		U1	I1	Ud	Id
Imported from Flakt Co.	9:00	350	160	52	750
	9:05	340	150	51	700
	9:10	340	150	51	700
	9:15	330	140	50	670
	9:20	350	150	51	700
DJ-1 Type*	9:30	365	200	55	1000
	9:35	370	205	56	1050
	9:40	370	205	56	1050
	9:45	370	205	56	1050
	9:50	370	205	56	1050

* The EP power supply equipment controlled by the method described in this Paper.

△ controlled by usual silicon
controlled voltage regulator
spark rate 150/min
conc. at entrance 30.11g/NM^3
conc. at exit 1004mg/NM^3

△△ controlled by
DJ-1 Computer
spark rate 150/min
conc. at entrance
35.107g/NM^3
conc. at exit
561mg/NM^3

PULSE ENERGIZATION ON ESP FOR THE END OF SINTER BAND

Zhang Weiming
Central Research Institute of Building and Construction
Ministry of Metallurgical Industry, Beijing, China

Peter Elholm
F. L. smidth & Co. A/S
77, Vigerslev Alle, DK-2500 Valby, Copenhagen, Denmark

Abstract

High resistivity dust is exhausted in the sintering process at CTISC. Under the conventional dc supply, the performance of the electrostatic precipitator for the end of sinter band showed unhealthy with the emissions higher than the limit. Pulse generators were substituted for the existing dc supplies in 1987, increasing the efficiency from 96% to 99% and reducing the emissions from more than 350mg/Nm³ to less than 150 mg/Nm³. The Wk enhancement factor 2 was obtained. The energy saving was about 70%.

The down-draft sinter band at the Chinese Taiyuan Iron and Steel Company (CTISC) is used for producing self-fluxing sinter. Limestone and lime are added to the raw material to keep up proper alkalinity (the ratio between the contents of Calcium oxide and silicon dioxide) in sinter with the aim of increasing the productivity of blast furnace. The dedusting equipment for the end of sinter band is a two-stage system involving a cyclone and an electrostatic precipitator. The back corona takes place with high dust resistivity during the operation of the precipitator. Under the conventional dc supply, the collecting efficiency is lower and the emissions are higher than the state-prescribed limit.

Considering the schemes to retrofit the dedusting system, it is difficult to enlarge the size or to increase the number of electrical fields of the precipitator as the place at sintering plant is limited and it will cost much money. Finally, it is intended to supply the field of the precipitator with pulse generator instead of the conventional dc supply. In order to avoid blind acting the feasibility of pulse energization was tested on the precipitator for the end of sinter band in July 1986 by the mobile pulse testing unit. The test results show that pulse energization is definitely more effective than the dc supply. Two regular pulse generators were installed on the roof of the precipitator in November 1987. After a week of commissioning, they were put into normal operation. The data of measurements show that the effectiveness of pulse energization is in keeping with the test results.

Dedusting System

Sintering is one of the processes in iron making industry for briquetting of powder materials. It is a process that after the feeding of mixture of powder material, fuel and flux a little of melt will be formed by means of burning the fuel, thus gluing the loosened materials to lumped sinter.

Fig 1 is a scheme diagram of sintering equip-

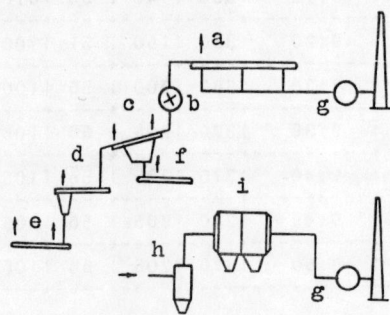

Fig 1 Scheme diagram of sinter band

ment and dedusting system at CTISC. In the down-draft sinter band (a), the air is extracted beneath the layer of materials on the band by an extractor (g). The air passes through the layer of materials and the mixture is sintered with burning of fuel during its conveyance. It is not permissible that big lumped sinter will be directly fed to the blast furnace. Therefore, after the end of sinter band there is a crusher (b) to crush the sinter to suitable pieces. The crushed sinter is then led to a vibrating screen (c) to sieve the powder and the unperfectly burned mixture. After that, the lumped sinter is cooled by a ring cooler (d). Finally, the materials fall down on a rubber belt (e) and continue to the blast furnace. The sifted materials are recovered by chain conveyer (f) as raw material to the band. It can be seen that there are seven dust points in the system involving the dedusting house at the end of sinter band, the front and end of the screen, the chain conveyer, the ring cooler and its hopper, and the feeding station for rubber belt.

The dedusting system consists of two stages of cleaning equipments and air ducts. The first stage is a cyclone (h) with diameter 4300 mm to collect the coarse particles. The second stage is an electrostatic precipitator (i) with two fields of which the cross section is 40m² collecting the fine particles. The duct spacing is 300 mm, the total area of collecting plates 2026m², the original value of gas volume 140000-180000 m³/h, and the gas velocity 1-1.25 m/s. Each field is supplied with a 72 KV, 700mA conventional dc supply.

The existing precipitator was renovated prior to testing the feasibility of pulse energization, for example retrofitted the gas screen for improving the gas distribution, readjusted the distance between the plates and the wires, regulated the rapping system, and repaired all dampers on ducts in order to control the exhausted air of the duct hoods. In addition, the dust samples were taken during the test period. The analysis of dust composition shows that the ratio between the contents of CaO and SiO_2 is 1.7-2.5. The resistivity is higher than 1×10^{13} ohm cm under the ambient temperature 140℃ due to the higher alkalinity.

Pulse Generator

The main circuit of the power supply system used on the precipitator for the end of sinter band is shown in fig 2 (1). The pulse system consists of a

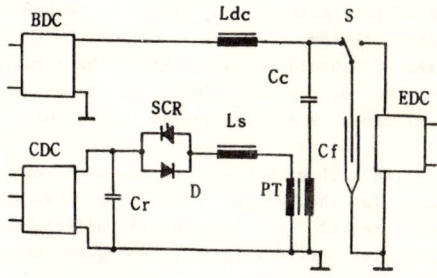

Fig 2. Principle circuit

basic dc supply BDC and a pulse generator. The operating dc voltage on the field of precipitator is supplied by BDC. The pulse generator includes a charger CDC suppling the storage capacitor Cr, a fast switching thyristor SCR and a feed back diode D, a series inductance Ls and a pulse transformer PT. The blocking inductance Ldc is used for preventing the pulse voltage from entering the dc supply. A coupling capacitor Cc will block the dc operatiog voltage from the pulse transformer and couple the pulse voltage to the precipitator represented by the capacitance Cf. The storage capacitor Cr, the series inductance Ls together with the pulse transformer leakage inductance, the coupling capacitor Cc, and the precipitator capacitance Cf form a series oscillatory circuit.

The storage capacitor Cr is charged to a controled dc level by the charger. The switching thyristor SCR is turned on and the precipitator capacitance Cf is charged to maximum pulse voltage level by the first half-period of the oscillatory current. Because of the series oscillation, the energy supplied to Cf and not used for the corona discharge is returned to Cr through the diode D by the current in the second half-period. During this interval, the thyristor switch is turned off and the current in the pulse ciruct is blocked until the next ignition of the thyristor switch. The energy is stored in Cr during the interval between the pulses and is used for the next pulse.

In order to contrast the dedusting effects under different energization, a switch S was set up between the pulse system and the conventional dc supply EDC to switch over the power supply from one to the other during the period of test or the operation of the pulse system.

Effects of Pulse Energization

Table 1 shows two sets of electrical data

Table 1. Electrical data (1 Field/2 field)

operat. mode	conven. dc supply	feasib. test	conven. dc supply	pulse supply
dc voltage (kV)	45/41	29/28	39/37	26/25
aver. curr. (mA)	79/65	34/27	65/65	25/20
pulse peak (kV)	-	43/45	-	40/43
pulse freq. (pps)	-	36/26	-	27/25
pulse width (µs)	-	125/125	-	144/148

corresponding to the period of test and the operation of pulse system. All the figures listed in the table are average values. The actual voltage of corona onset for the existing precipitator is approximately 30 kV. Under the pulse energization, the basic dc voltage is maintained at 25-29 kV slightly below the onset value. The maximum of pulse voltage is ranging from 40 to 45 kV with pulse repetition frequency 25-36 pps. The pulse duration is in the vicinity of its rating without any change in the operation. Fig. 3 shows the wave forms of pre-

The measurements of the precipitator performance are synthesized in table 2. It is shown that the emissions at the outlet of precipitator are always higher than 150 mg/Nm³ for different gas

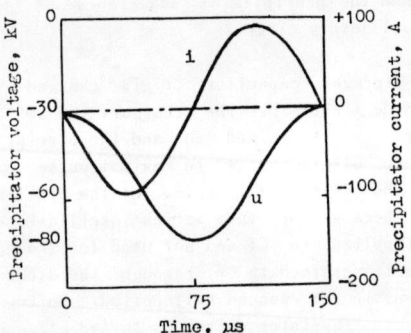

Fig 3. Wave forms of voltage and current cipitator voltage and current.

Table 2. Performance of precipitator

date	gas vol. m³/h	gas temp. ℃	gas veloc. m/s	supply mode#	dust conc. mg/Nm³	Wk fact.	eff. %
1986.7	135000	123	0.94	dc/dc	478	-	95.3
1986.7	135000	123	0.94	ps/ps	72	2.3	99.3
1986.7	158000	105	1.1	dc/dc	370	-	96.0
1986.7	158000	105	1.1	ps/ps	109	1.8	99.0
1986.7	158000	108	1.1	dc/dc	397	-	97.2
1986.7	158000	108	1.1	ps/dc	106	1.3	98.5
1986.7	158000	108	1.1	dc/ps	147	-	98.1
1986.7	190000	95	1.30	dc/dc	258	-	97.0
1986.7	190000	95	1.30	ps/ps	120	1.4	98.5
1987.11	136000	97	0.94	ps/ps	89		98.8
1987.11	164000	98	1.14	dc/dc	787	-	94.6
1987.11	164000	98	1.14	ps/ps	144	2.1	98.5

dc: dc supply, ps: pulse supply, 1 field/2 field

volume, temperature and velocity under the conventional dc supply. But the performance of the precipitator would be upgraded under the same gas condition by pulse energization instead of dc supply.

During the testing period of feasibility, the opening of the damper for the extractor and the dampers on air ducts were adjusted to treat mainly the hot gas from the end of sinter band, screen and chain with total volume 135000 m³/h. The outlet temperature was 123℃. Switching over the switch S, the emissions could be reduced from 478 mg/Nm³ under dc supply to 72 mg/Nm³ with pulse energization. The efficiency increased from 96.3% to 99.3% and Wk enhancement factor 2.3 was obtained. Mixing the cold and the hot air, the gas volume treated was about 158000 m³/h. If the two fields were energized by two pulse generators, the emissions could be reduced from 370 mg/Nm³ to 109 mg/Nm³. The efficiency increased from 96.0 to 99.0%, and Wk enhancement factor was 1.8. If the two fields were supplied by a pulse system and an existing dc supply separately, the performance of the precipitator could be improved too. When the gas volume was continuously increased to 190000 m³/h, the temperature reduced to 95℃, better results still could be obtained under pulse energization, comparing with dc supply.

After the regular pulse generators put into operation, it was already early winter, the gas temperature and humidity were lower than that in test period. Corresponding to the gas volume 136000 m³/h, the emissions was about 90 mg/Nm³, the efficiency 98.8%. If the gas volume was increased to 164000 m³/h, the emissions could be reduced from 787 mg/Nm³ to 144 mg/Nm³. Wk enhancement factor 2.1 was obtained. Thus it can be seen that the operating effects of pulse energization are in keeping with the test results of feasibility of the mobile testing unit.

In addition, the measurement of energy consumption shows that the energy saving is 70%, if the pulse generators are applied to the precipitator under the same gas condition (2).

Conclusion

The performance of the precipitator at CTISC is unhealthy due to the high dust resistivity under dc supply. The dust concentration at the outlet of precipitator is always exceeding the limit.

Substitute pulse energization instead of dc supply, the performance can be improved under the same gas condition without any change of the basic structure of precipitator. In order to satisfy the need of productivity of sinter at CTISC, the gas volume treated should be 160000 m³/h. The emissions can thus be reduced by a factor of three with Wk enhancement factor 2 and the energy saving 70%.

The retrofitting project on the electrostatic precipitator for the end of sinter band has shown its environmental and economical effectiveness during a shorter period with pulse energization. The experience achieved from the project would be benifitial to those existing precipitators needed to upgrade in the industrial areas.

References

(1) S.Olesen, P.Lausen: An Energy Conserving Pulse System for Electrostatic Precipitators. Presented at the conference COAL TECHNOLOGY, EUROPE'81, Cologne, W. Germany, June, 1981.

(2) P. Lausen: Energy Conserving Consideration on Pulse Energization of Electrostatic Precipitators. US-Japan Seminar on Measurement and Control of Particulates Generated from Human Activities. Nov. 1980. Kyoto, Japan.

USING SPARK DISCHARGE FOR ASH REMOVAL
—BREAKING OFF AGGLOMERATES BY SPARKS

Li Kejia
Dalian Electronics Research Institute, China

Abstract

A preliminary study is reported on the dust removal effect by means of spark energization for ESP, which can make a complete ash detachment from electrodes. The main reasons for this may come from the high frequency oscillations of the wires caused by the spark discharge and from the induced shock wave toward the plates. The power is supplied with high spark-rate tracing property.

Introduction

Because of high resistivity and strong adhesiveness of the dust, the "swelling" phenomena of corona wires exist in the performance of ESPs presently used in the fields of metallurgical, power and cement industries. Since it spoils the normal operation of ESP, a great number of studies has been carried out to solve this problem, such as the improvement of rapping methods, the utilization of electric brushes and the conditioning of flue gases, etc. Yet none of them has achieved satisfactory results. In recent years, we found the method of ash removal by spark discharge to be a promising one, and have achieved good results in the ESPs used for sintering furnaces, coal-fired power plants, clinker grinding mills and cement kilns, etc.

1. Influence of accumulation of ash deposit on electrodes upon ESP performance

1.1 Effect of "swollen" electrode wires on corona discharge

Generally, electric field of ESP is formed by a negative tip to positive plate geometry. From distribution of electric field lines we can see, that magnitude of electric field at the place near wire tips is the highest, and it gets lower rapidly in approaching the plate, and is the lowest in the space near the plate. After "swelling" of electrode wire, the barbs of barbed wire and the edges of star wire are covered with ash deposit (agglomerates), thus curvature radius of electrode increases, the electric field lines disperse, the magnitude of electric field becomes lower and in the end corona discharge stops, corona goes out, which causes abnormal operation of ESP. In this situation, even supply voltage is as high as 60KV or more, the current nearly equals zero or only several milliamperes.

1.2 Effect of electrode plates covered with ash deposit on breakdown voltage

During normal operation of ESP, when ash layer on electrode plates has a certain thickness, it should be removed by rapping. But if resistivity of collected ash is high, the charged particles can not release their charges in time. This will not only increase adhesive force of ash layer on electrode, but also greatly reduce breakdown voltage of ESP. It is reported that breakdown voltage of electric field will be reduced by a factor of 50%, as thickness of ash layer with high resistivity on plate increases 1 millimeter. This is because a streamer discharge occurs in electric field.

It is hard for fly ash with high resistivity collected on the plate to release its negative charges to positive electrode plate, and a counter electric field between ash layer and plate is formed, which causes point partial breakdown in ash layer. This is the first step of streamer discharge. Then the breakdown point spreads laterally and forms a discharge region composed of many discharge points. A great number of positive ions produced during discharge migrates to negative electrode, which causes spark discharge at a very low voltage. This moment, ESP works at a state of frequent occurence of sparks, and the spark-rate can reach approximately one thousand times per minute, while operating voltage greatly decreases.

From above analysis we can see that accumulation of ash deposit on electrodes is the important factor affecting ESP performance, which makes it unable to work efficiently for a long time. For example, ESP at a cement clinker grinding mill can keep its normal working condition only for several hours. Even workers clean up electrical fields twice a week, the situation of ESP has not been improved. Can the problem be solved by rapping? Here we must discuss in two respects, i.e. acceleration attenuation of rapping during transference and adhesive force of ash layer on electrode.

2. Reasons for accumulation of ash deposit on electrodes

2.1 Acceleration attenuation of rapping during transference

Acceleration of rapping is transferred to all parts of electrode through anvil, frame and fastening pieces. During transference the rapping acceleration will be attenuated in varying degrees, and this leads to the reduction of rapping efficiency. Some reports indicate, for instance, that acceleration at one point of electrode may several times or more than ten times differ from acceleration at another point of the same electrode.

2.2 Analysis of adhesive force of ash layer on electrode

Adhesive force of ash layer on electrode is composed mainly of three forces: electrostatic attractive force, interfacial force of liquid content in ash layer and frictional force. When the charged fly ash accumulates on electrode, it should release its charges to electrode through resistance of ash layer, which makes electrostatic attractive force decrease to zero. But, in fact, release of charges becomes difficult while resistivity of collected ash increases. A layer of charges opposite to electrode in polarity is always maintained in ash layer. Thus results a great electrostatic attractive force of ash layer for electrode, the value of which enormously depends on resistivity, temperature and humidity of collected ash.

Photo 1 demonstrates the "swollen" wires of ESP performed at a clinker grinding mill, using rapping method for dust removal. The ineffectiveness will exterminate the corona discharge and finally the performance will be stopped to dedust by workhand. Photo 2 shows the state of the electrodes at the same ESP, after utilizing spark discharge for dust removal. In half a year's operation, the results seem excellent.

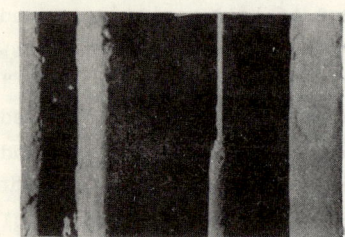

Photo 1. "Swollen" wires of ESP at a cement clinker grinding mill

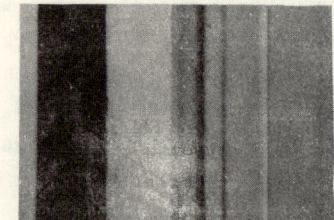

Photo 2. Clean electrodes resulting from spark discharge

3. Physical mechanism of using spark discharge for ash removal

A number of long-term industrial tests has been made on ESPs at the rear of iron ore sintering machine, the coal-fired furnace and the cement clinker grinding mill to prove the ash removing function of spark discharge. The result shows, that spark discharge is able to efficiently remove collected ash (break off agglomerates) from electrodes. How it does? The physical mechanism is discussed below.

3.1 Effect of electrodynamic force

When two parallel conducting wires are electrified, an electrodynamic force occurs on both of them due to interaction of current and magnetic field. Affected by electrodynamic force, two wires will be deformed. If the current is constant, two wires will be in a new equilibrium. But when the current on any one wire of them changes, the electrodynamic force changes too, which makes the wire tremble. The level of trembling depends on the level of current changing.

Normally, during operation of ESP, a certain amount of current passes through each electrode wire, which remains in a state of relative rest among the surrounding wires. But if this wire produces spark discharge, a thousandfold current changing will happen which makes the wire tremble with high acceleration and small shift. Thus the accumulated ash deposit will be removed from the wire.

3.2 Effect of shock wave produced during spark discharge

The spark discharge is a process of one-shot complete ionization of interelectrode gas. During ionization occur physical phenomena of light, sound and heat. The course of producing heat is accomplished in a very short time, which causes the thermal expansion of gas in the region of spark discharge, and the shock wave produced from this makes ash layer fall from electrodes. We have found from on-the-spot inspection, that this shock wave is very effective especially on removing ash accumulated on electrode plates.

3.3 Effect of spark discharge on reducing electrostatic attractive force of ash layer for electrode

Analyzing the adhesive force, we have said, that it is hard for the charged fly ash with high resistivity to release its charges. Thus results an electrostatic attractive force of ash layer for electrode, which makes the rapping a rather difficult job. But during spark discharge, in the space and the ash layer occurs a large number of positive and negative ions, which will be recombined with charges loaded by ash layer. This causes the obvious reduction of electrostatic attractive force. Therefore, the serious problem involving ash deposit with high resistivity has been solved.

In order to most effectively remove the dust from electrodes, the spark-rate must be greatly increased, e.g. the spark-rate must be raised to as high as several hundred to several thousand times per minute to eliminate dust on the electrodes of ESP in a cement clinker grinding mill. This leads to the special design and manufacture of the high voltage power set in which the impact of frequent overload is fully considered and the reliability of the power pack is intensively ensured to be steadily operated at high spark-rate condition.

References

(1) J.R.McDonald and A.H.Dean, "Electrostatic Precipitator Manual", Noyes Data Corporation, New Jersey, U.S.A. 1982
(2) Tan Tian You and Liang Feng Zhen, "Industrial Ventilation and Precipitation" (in Chinese), Arch.Ind. Press, Beijing, China, 1984
(3) Chen Ming Shao, Wu Guang Xing et al., "Basic Theory of ESP and its Application" (in Chinese), Arch. Ind. Press, Beijing, China, 1981
(4) Li Ke Jia and Hou Shou Xian, "Study on Theory of ESP Power Supply" (in Chinese), Chinese Journal Mine Technology, No.2, 1982

THE INFLUENCE OF ELECTRODE COATING ON CORONA DISCHARGE IN ELECTROSTATIC PRECIPITATORS

Huang Xuemin, Zhao Hongzuo, Zeng Hanhou
Xi'an Institute of Metallurgy and Construction Engineering

Abstract

The effects of EP electrodes' anti-corrosion coating on the corona discharge have been investigated in laboratory devices. Experimental results indicate that the resistivity of coated layer on the electrodes is an essential factor effecting the corona discharge. The adverse effects brought by several kind of coatings upon the corona discharge were overcome by adding to the coating certain amount of conductive carbon black.

Introduction

To prevent EP electrodes from metallic corrosion which may likely occur during the long time taked from being manufactured to being eventully installed at field, certain kinds of coatings are painted on their surfaces by the manufactures. For EP electrodes with such surface coatings, their corona currents exceed greatly the designed value under the normal operating conditions, and even rise to such an extent to trip off the power supply under an imposed voltage far below the designed rated value. Thus, such EPs can not be normally operated.

In this paper, the effects of electrodes coated layer on the corona discharge have been investigated, and measures to overcome the adverse effects brought by the anti-corrosion coated layer upon the corona discharge have been found out.

The Possible Mechanism

In the corona region, when the high concentration of ozone may cause oxidation of some kind of organic compounds contained in the coating on discharge wire, there will result some additional kinds of gas molecules. If such molecules with ionization potential different from that of normal air may occur to be in a sufficient amount, the corona starting voltage and the corona current may thus change correspondingly from normal magnitude to some extent. The scratches, pits and other roughness remained on the discharge electrode coated surface may lower the corona starting voltage by a factor ranging from about 0.5 or 0.9. On the other hand, the increase of diameter of the discharge electrode due to the presence of the coated layer may tent to raise it. However, this effect is likely to be insignifcant, as the thickness of the coated layer is usually very small relative to the wire diameter. The coated layer on corona wire may lead to change further the energy given to the electron to emit from the solid surface. This energy is acknowledged to be supplied by the electric field and the photon. Therefore, the change in energy may in some case lead to the generation of electrons sufficient to initiating a avalauche process.

When the coatings subjected to the oxidation in the corona glow, there will present certain kinds of additional gaseous molecules to change the gas composition. If the effective gas ions mobility thus resulted, would increase, the current will in crease, otherwise, will decrease.

The effects of the collecting electrode coating on the V-I characteristics may be mainly determined by the specific resistivity of the coated layer. If the resistivity is low, the voltage drop across the coated layer will also be low, and then the its effects on V-I characteristics will be neglected. If it is high, there will ressult a back corona or a back discharge at the coated layer pores to incease the corona current and to reduce the spark voltage.

Experimental Apparatus

In experiments, there types of electrodes geometry have been adopted:

1. With the z-shaped plate as the collecting electrodes and the twisted-square wires, as the corona electrodes. The duct sizes being 1.2m high, by 0.3m wide. With the wires, spaced by 0.2m, and of a length 1m.

2. With the parallel plates as collecting electrodes and the round wire, as corona wires. The duct width being 0.3m. With the wires, spaced by 0.2m and of a diameter 0.9mm.

3. With the cylinder of a diameter 10.6cm and of a length 74cm as the collecting electrode, and the round wire of a diameter 0.9mm as the corona electrode.

Experimental Results

Experiments have been carried out in utilizing lubricating oil, asphalt, alkyd varnish, phenolic antirusting paint, insulating oil and the specific paint prepared by the Ping Ding Shan Electrostatic Precipitator Plant etc as anti-rusting coatings on EP electrodes with the following results:

a) Effect of coated asphalt layer on the V-I characteristics

Before coating, asphalt was dissolved in gasoline. The solution was then painted on the electrodes with brush. The coated layer was exposed to atmosphere for weeks to evaporate the volatile compounds. Experiments were made to determine the effect of asphalt coating on the electrode V-I characteristics. In the Z-shaped plate and twisted-square wires geometry, the corona starting voltage in the case of asphalt coated wires comes to be about 3KV lower than in the case of clean wires. The discharge current, in the case of coating both the plate and the wires, increases significantly, while the corona starting voltage keeps similar to that in the case of coating the wires only. As imposed voltage exceeding about 45KV, the corona current difference between the case of coated and clean-electrodes becomes evident. At the voltage of 60KV, the current is 2.1mA for clean electrodes, 2.7mA, only wires coated, and 4.2mA, both the wires and plates coated. The spark voltage, 75KV in the case of clean electrodes, will remain essentially the same value with only the wires coated by asphalt, and will reduce to 60KV with both the wires and plates coated.

In the cylinder and round wire geometry, the voltage current curves similar to those of the plate-wire geometry, described above are shown in Fig 1. An interesting feature in the V-I curve for the coated cylinder is the inflection of the curve at larger values of voltage. That means, there is serious back corona phenomena in coated cylinder.

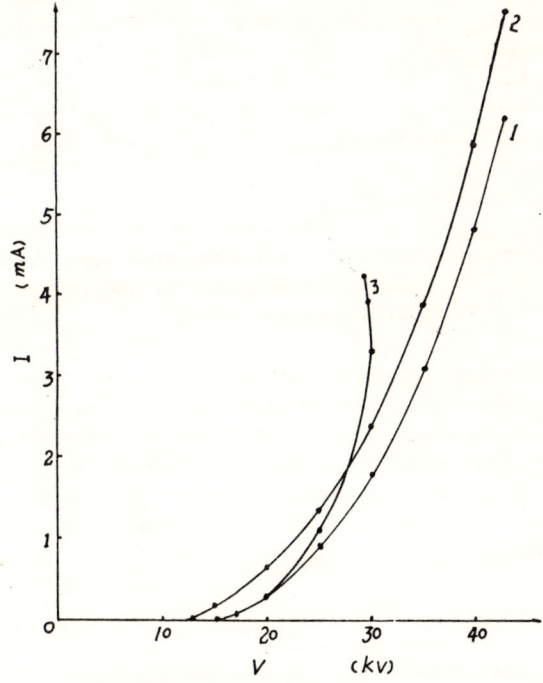

Fig. 1. Variation of V-I curve for cylinder-wire geometry
1. clean electrodes. 2. wire coated. 3. both wire and cylinder coated.

To upgrade the conductivity of the asphalt coating, investigations have been made with the addition of appropriate amount of conductive carbon black. When the specific resistivity of the coating in solidified state comes round to be 10^7 ohm-cm, the adverse effect of the coating is basically overcome. The experimental results in the cylinder-wire geometry are shown in Table 1.

Voltage(KV)	I for clean electrodes(mA)	I for wire coated(mA)	I for cylinder coated (mA)
15.0	0.005	0.012	0.001
25.0	0.28	0.29	0.24
25.0	0.866	0.860	0.870
30.0	1.73	1.72	1.74
35.0	2.97	2.94	3.00
40.0	4.67	4.65	4.70
45.0	7.00	7.00	7.10

Table 1. V-I relationship in cylinder-wire geometry

b) Effect of Ping Ding Shan paint on the corona discharge

The Ping Ding Shan anti-rusting paint is composed of commercial ready-mixed alkyd paint, alkyd varnish, carbon black and white spirit. The specific resistivity of the solidified coated layer is round to be 10^7ohm-cm. When the twisted-square wires coated with the Ping Ding Shan paint, the corona starting voltage reduces slightly while the corona discharge current, comparing with that of clean electrodes, shows no evident changes at higher voltage. When the plates coated with the paint, no obvious change in corona discharge current is observed.

c) Effects of phenolic antirusting paint on corona discharge

The specific resistivity of the solidified layer of this paint is very high. In the experiments, the paint was thinned by gasoline and added by approprite amount of carbon black. The specific resistivity of the paint thus modified will be reduced round to 10^7ohm-cm. The effects of the paint both the commercial and the modified on the corona current were measured. The experimental results, as shown in Table 2, indicate that the phenolic paint has the corona current increased evidently and that the addition of carbon black get the adverse effect of coatings on corona current rid of.

Voltage (KV)	I for clean electrodes (mA)	I for wire coated with the paint (mA)	I for wire coated with the modified paint (mA)
15.0	.006	0.28	0.10
20.0	.305	0.83	0.44
25.0	0.90	1.56	1.02
30.0	1.87	2.70	1.95
35.0	3.20	4.20	3.05
40.0	4.90	6.30	4.85
45.0	7.50	8.90	7.30

Table 2. Effects of phenolic antirusting paint on the V-I characteristics

d) Effects of the other coatings

In the cylinder-wire geometry, when the round corona wire being coated with lubricant oil, little effect on the corona starting voltage and corona current will be detected.

The resistivity of the ready-mixed alkyd paint is round to be 7.8×10^{11} ohm-cm. When the corona wire was coated with this paint, the corona starting voltage decreased by an amount of 2KV, while the corona current increased significantly, comparing respectively with those in the case of clean wire.

In the Z-shaped plate-twisted square wire geometry, the corona starting voltage and the corona current showed no variation as the transformer oil being coated on the twisted-square wire, while the corona current decreased evidently as the same oil being coated on the collecting plates.

e) Effects of asphalt coating on the distribution of electric field

In the flat plate-round wire geometry, the distribution of electric field strength in the vicinity of the plate in the interelectrode region was determined by means of dropping ball technique[6] and calculated numerically by the method of McDonald et al[7].

Fig. 2 presents the values of electric field strength at the imposed voltage of 45KV, both measured and calculated. In which, curve 1 represents the calculated values for the clean electrodes, while curve 2, these for electrodes, both plate and wire coated with asphalt layer. As shown, in the exterior corona glow region the field strength is augmented in the case of both the plate and wire coated and also the case of wire alone coated.

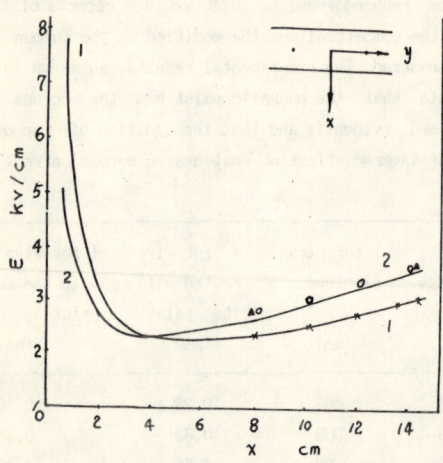

Fig 2. Effects of the electrode coating on field strength distribution
× the measured value with clean electrodes
▲ the measured value with wire alone coated
○ the measured value with both plate and wire coated

In Table 3, the data of the electric potential distribution determined by the method of probe[8], 0.4mm in diameter, at the imposed voltage of 45KV are shown. The data,

distance to the plate (cm)	y = 0.0 cm		
	$I=0.77\sim0.79$mA U for clean wires and plates (KV)	$I=1.19\sim1.24$mA U for wires coated (KV)	$I=1.19\sim1.22$ U for wires and plates coated (KV)
1.0	5.93	6.80	7.75
2.0	9.15	10.8	11.8
3.0	12.7	14.0	16.0

distance to the plate (cm)	y = 10.0 cm		
	$I=0.77\sim0.79$mA U for clean wires and plates (KV)	$I=1.19\sim1.24$mA U for wires coated (KV)	$I=1.19\sim1.22$ U for wires and plates coated (KV)
1.0	4.65	5.40	6.00
2.0	7.90	9.40	9.80
3.0	10.7	12.4	—

Table 3. Effects of electrodes coated layer on the potential distribution in the vicinity of collecting plate

in spite of some errors included, will, however, illustrate qualitatively the fact that due to the presence of a coated layer on corona wires there will be an augmentation of field strength in the vicinity of the plate and that due to the presence of a coated layer on plate in addition to that on the wire, there will be a further increase of potential there.

Discussion

As well known, in spite of the fact that the lubricating oil is very effective in being coated on EP electrodes and also, very cheap and available every where, it is able to antirust only for a short period and so, can't be widely used. The Ping Ding Shan paint will have the corona starting voltage decrease slightly, and the corona current and sparking voltage change little.

The coated layer, of asphalt, phenolic antirusting paint, or alkyd ready-mixed paint, all present a high resistivity. The discharge electrodes with such coated layer will have a less smooth surface than the clean electrodes. The roungness there resulted and there of the possible changes in surface-emitting capability of electrons may be the cause for lowering the corona starting voltage and increasing the corona current.

When negative ions arriving at the coated layer surfaces, the negative electric charges, being prevented from releasing due to high specific resistivity there and thus accumulated at the coated surface pores to the considerable amount, may result in the back corona to increase the corona current and reduce the spark voltage. Therefore, by adding adequate amount of carbon black into the above mentioned antirusting paint to reduce its resistivity will keep at the coated surface the corona current at normal level.

Conclusions

Effects of electrode antirusting coating on corona discharge characteristics have bee investigated experimentally. The factors which play the principal part are the specific resistivity, thickness, surface roughness and porosity of the coated layer, the high specific resistivity problem of certain coatings can be solved by adding into appropriate amount of carbon black. The choice of coating materials may be determined by the antirusting performances and their costs.

Reference

1. J. S. Lagarias, Trans. Am. Inst Electr. Eng. 78, 427(1959).

2. P. Cooperman, Trans. Am. Inst. Electr. Eng. Part 1, 75, 64(1956)

3. J. R. McDonald, W.B.Smith, H.W. Spencer III, and L.E.Sparks, J. Appl. Phys. 48, 2231−2242(1977).

4. O.J.Tassicker, Proc. IEE. 121, 213(1974).

RELATION BETWEEN SPACING OF PRICKLE ELECTRODES AND CORONA CURRENT

Li Xiang-sheng Xu Shao-zeng Zhang Zhi-wei Li Feng-hua

Electrostatic Institute, Northeast Normal University, Changchun, China

Abstract

This paper states the relation between the distribution of prickle electrodes and corona current and the optimized value of the spacing which have been obtained through a great number of experiments and studies. The optimized value could be explained by streamer theory, The fitting of the experiment curve is realized by a simple function, and a semi-empirical formula for the volt-ampere characteristic is given within a certain range, which will provide a theoretical basis for developing of whirlwind precipitator.

Introduction

Engineering practice and theoretical analyses show that the perfect effectiveness of a precipitator working with a single-mechanism will not be achieved. So the techniques and devices combining with multiple mechanism are being developed and used rapidly. For example when the dust, whose diameter is $2\mu m$, is in an electrostatic precipitator with the electric field strength of 3-5kv/cm, or in a whirlwind precipitator with the tangential velocity of 15-18m/s, the electric field force acting directly on the dust is two-orders greater than centrifugal force. Therefore the technique of combining electrostatics with others is developing very rapidly, such as the combinations of electrostatic technique with water spray, whirlwind and filter with water spray, and electrostatic technique with whirlwind etc[1].

Thanks to the combination between the electrostatic technique and whirlwind technique, the dust exerted by the centrifugal force can be cleaned out continuously, and the thickness of dust layer can be controled thinner than the limited thickness of back--corona, meanwhile shaking process can be concealed. This combination has provided a new approach for overcoming the problems of secondary rise and back-corona that can hardly be overcomed by electrostatic precipitator itself. It is advantageous to reduce the volume of electrostatic precipitator, simplify its structure, reduces its cost and save space. It also provides the new feasible technique for improving the existing devices, such as whirlwind water film precipitator, whirlwind precipitator etc, which are widely used in the electric power industry, paper making industry and boiler in china [2].

Prickle corona electrode is more and more used in electrostatic precipitator to force stall back corona, because its specific corona current is stronger and there is a small current density on the collectong electrode, while cleaning out the weak electric dust.

Experimental and Theoretical Investigation

Experiment Conditions and Devices

In order to analyse whirlwind electrostatic precipitator, We should firstly find out the relation between the distribution of prickle electrode and corona current. This is because the precipitation efficiency of the precipitator is raised along with the increase of the emitting power of corona discharge. It is necessary to find out the optimized distribution of prickle corona electrodes. Let the separation between two electrodes(inner and extermal cylindes) and the total area of the prickle array remain unchanged, the applied voltage between the electrodes is taken as a variable parameter, the change of total corona current is measured when the spacing of prickle is adjusted. The experimental setup is shown in Fig.1.

In the figure, the metal cylinder (1) is a collecting electrode which is grounded through an ammeter (5), while the inner cylinder (2) is made of insulating material, serving as a support of prickle electrodes whose two ends

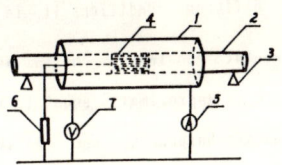

Fig 1. experimental principle

are mounted on insulator (3). The prickle array (4) is fixed on the oiter surface of the internal cylinder and connected to direct current high--voltage power supply(6), via cable, an electrostatic high-voltage meter (7) of type QV-8 is used to measure applied voltage.

The Relation Between the Spacing of the Prickles and Corona Current

During experiment, the separation between the two coaxial cylinder is 110mm, then the applied voltage is adjusted to measure total corona current. If the applied voltage is 40KV, 50KV and 60KV respectively, the experimental curves between the spacing (L) of the prickles and the corona current i, are shown in Fig. 2. Where the curves of the corona current is corresponding

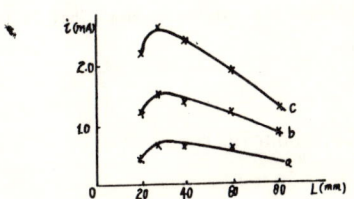

Fig. 2. The relation between the prickle intervals and total corona currents

to the applied voltage of 40KV. similarly curve "b" and curve "c" are measured and drawn under the condition of the applied voltage of 50KV and 60KV respectively. From the three curves, it can be concluded that,

1. When the separation of the two electrodes and the applied voltage are fixed, the total corona current increases along with increasing of the spacing of prickle electrodes. There exists a maximum value when L ranges from 20mm to 40mm, which agrees with reference and the maximum value is on the left side of the point 30mm;

2. The rising gradient of the curve is greater than its falling gradient, therefore, when L<30mm, the value of di/dL is very large;

3. The higher the applied voltage, the more obvious the above mentioned two points;

4. The value of the optimized spacing of the prickle electrode is independant on the applied voltage;

5. The ratio of the peak current value to the current difference between the peak current value and the value corresponding to the space of 20mm, is increased gradually as the applied voltage is reduced.

Theoretical Explaination of the Optimzed Spacing Presented by Prickle Electrodes

This phenomenon could be explained with H.Raether's-treamer theory[4]. Suppose a single prickle under a high negative voltage is perpendicular to a plate electrode. When the distance between them is large, there is an inhomogeneous corona around the end of the prickle. Nagative corona spots appear in the ionization space, stretching out like a "feather-fan". This is called a negative streamer. The "feather-fan" becomes larger as the applied voltage increases.

If there is an array of prickles, we will have an errray of "feather-fan". Each "feather-fan" has eight to ten branches. All the branches repel one another and never intersect. So, as the distribution of prickle becomes dense, the "feather-fan" will weaken each other Therefore there exists a "beat" condition for these feather-fans corresponding the finest value for the corona current.

Fit of the Curve Between Prickle Electrode Spacing and Corona Current

It is very difficult to calculate theoretically the relation of the spacing of prickle electrodes and the corona current. Even with the simpler method of equivalent charge, we must also use computer to calculate, and the solution can only partly fit the curve. Therefore, it is applicable and valuable in engineering design to analyse the experimental solution qualitatively and to fit the experimental curve with simple function, e.g. the Deutsch formula.

In figure 2, curve "c" can be fitted with the following formula, that is

$$i = 1.91 L^2 (2.4)^{-0.71L} \quad (1)$$

where i is the total corona current of the experimental system in Fig.1, L is the spacing of electrodes. The comparison between the calculated solution by formula (1) and the data of the experiment is shown in foltiwing table.

L(cm)	1	2	3	4	6	8
i(mA)cal	1.02	2.20	2.66	2.54	1.71	0.85
i(mA)exp	1.20	2.20	2.63	2.40	1.92	1.30

From the table above, it can be seen that the calculated values can basicaly fit the experimental data, ranging from 2.0cm to 6.0cm which are often used for praciticd cases. By means of formula (1), its other characteristic could be analysed mathematically, for example, calculating its maximum value point.

Let $di/dL = 0$, then we get

$$1.91 \times 2L(2.4)^{-0.71L} - 0.71(2.4)^{-0.71L} 1.91 L^2 = 0$$

$$L_{max} = 2.82 cm$$

Array of Prickles and the Voltage -- current Characteristic of Cylinder System

Since Warburg [5] and Townsed, the corana discharge phenomenona have been widely studied for both point to plate and coaxial cylinder electrodes systems.

The current-voltage charaterictic could be expressed as,

$$i = AV(V - M) \quad (2)$$

where i is corona current, V is applied voltage, A and M are experimental constant. But so far, we have not seen any report on the case of prickle array and cylinder electrodes system. Based on the meased data analysis by computer under different conditions it is proved that if the experimental constant is chosen as $A = 10^{-3}$, and $M = 20 - 30 KV$, which is approximatly equal to the threshold voltage of corona discharge, then eqn. (2) still stands.

If the spacing of the pricktes is $20\sqrt{2}$ mm, the comparison between measured voltage-current characteristic and the calculated curve by eqn.(2) is shown in figure 3.

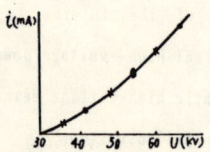

Fig 3. × refers to experimental value, while the solid line is calculated form eqn(2)

From the shape of the curve, we can see that it is in agreement with voltage-current charactersitc of ordinary electrostatic precipitators. It can be thought that the voltage-current charcteristic of prickle array and the cylinder system is situated between the cases of needle--plate and wire-plate electrodes. The curves of voltage-current characterisitic of both needle--plate and wire--plate electrodes are parabolic ones. That is to say, eqn.(2) is suitable.

Conclusions

The spacing of needled prickles of the Soviet--made electrostatic whirlwind precipitator [7] is 20mm. The length of the prickles is equal to that we used. The spacing of needled prickled of some domestic electrostatic precipitator is about 100mm. The spacing of the prickles used by Russians is approximately equal to the optimized value about $20\sqrt{2} \approx 28$mm, but is comparatively small. When the applied voltage is 60KV, its value of corona current is 20 percent smaller than the value for the optimized spacing. When the precipitator works at

50KV, the value is about 23 percent smaller. But, in china, the spacing of prickles is above 100mm, and the value of corona current is about 30 percent of the optimized value. To electrostatic whirlwind precipitation technique, we should not come to a negative conclusion due to the inconspicuous efficiency resulting from simple combination. Because the gradient of the voltage--current characteristic curve is greater in the rising stage, the spacing of the prickles should not be chosed as 20mm, instead, it should be chosen as in 30mm or so. Once the optimized spacing is fixed, no matter how high or low the applied voltage is, the value of the current corresponding to the voltage is always maximum. As the increase of current causes the increase of the electric power of the electrostatic precipitator, the efficiency of the precipitator will be improved finally.

References

[1] Yama ta ko. "chemical factory" Vol, 26 No.11 32-37 (1982)
山田弘《化学工场》 Vol, 26 No.11 32-37 (1982)

[2] Zhang Zhi-zheng "electrical science and technology in Guangding" 3—4 (1984)
张志正《粤电科技》3-4(1984)

[3] B.N. Wusov "The purification of Industrial Gas, Precipitator and Filter"
《Science and Technology Publishing House of heilong-jiang Province》 471 (1984)
B·H·乌索夫《工业气体净化与除尘器过滤器》
黑龙江科学技术出版社 471 (1984)

[4] Yang jun-ji 《Gas Dischatge》 science publishing house 205—211 (1983)
杨津基《气体放电》科学出版社 205-211 (1983)

[5] H·J·White 《Industrial Electrostaic Precipitation》 71 (1963)

[6] A·D·Moore 《Electrostatic and Its Application》 188-191 (1973)

[7] 《Safety in Industrial labour》 No.3 36—37 (1975)
《工业劳动安全》No.3 36-37 (1975)

Progress in Dust Collection by Electrostatic Precipitation
in View of Gaseous Pollutants Removal.

G. Mayer-Schwinning
Gwinnerstr. 27/33,
6000 Frankfurt/M-60
W. Germany

Abstract

In the Federal Republic of Germany the clean gas standards have been considerably tightened in recent years.
As far dust emissions are concerned, coal-fired power stations with a thermal generating capacity of more than 300 MW must not exceed a value of 50 mg/m^3 (s.t.p.) while refuse incineration plants must remain within the limit of 30 mg/m^3 (s.t.p.). Apart from the dust emissions, there are limit values for heavy metals which have to be observed, so that the clean gas requirements to be met in many industrial sectors are < 5 - 20 mg/m^3 (s.t.p.). Moreover, gaseous pollutants such as SO_2, HCl and HF have to be removed from the waste gases; so $CaCO_3$, CaO and $Ca(OH)_2$ are usually added to the waste gas in dry form or in the form of a suspension.

It is shown on the basis of the present state of the art that the stipulations of the pollution control legislation for large boilers can be met by electrostatic precipitators and that fine dust and heavy metals are collected to a sufficiently high degree. The contribution considers recent developments which bring about a fundamental improvement of the collection in the precipitator.

Further, the advantaged use of microcomputers versus analog type controls for high voltage control, is presented. A superordinated, central process computer assumes the functions of control, supervision, and optimization of an inplace precipitator, thereby minimizing clean gas dust content and overall electrical energy consumption.

Noxious gases have to be separated together with the dust to an ever-increasing extent. With dry processes, fluidized beds are arranged upstream of the electrostatic precipitator for the adsorption of, for instance, HCl, HF, and SO_2. Using the CFB technology for a boiler integrated removal of gasous pollutants the precipitator is positioned in the hot gas area downstream of CFB boilers.

II. Electrostatic Biological Effects

APPLICATIONS OF ELECTROSTATIC TECHNOLOGY IN AGRICULTURE

S. Edward Law
Agricultural Engineering Department
University of Georgia
Athens, GA 30602 USA

Abstract

Direct application of forces to small particles *via* interaction with electric fields offers much potential for efficient materials handling in agriculture. Droplet charging and electrostatic deposition methods and devices have been developed to exploit electric force fields for greatly enhancing by up to 7-fold the basic droplet-deposition process in pesticide spraying on crops. Electric deflection fields have been successfully utilized for many separation and sorting functions in agricultural processing, while 3-phase traveling electric fields are being developed for conveyance and separation of agricultural particulates in the size realm of seeds.

Introduction

The production and processing of agricultural materials requires handling of numerous small-sized substances in the forms of individual seed, fertilizer granules, spray droplets, pollen, animal-housing dust, granulated products such as sugars, and milled products such as flours and feeds. Conveyance, alignment, sorting and separating, depositing, evenly distributing, coating, collecting, *etc.* are operations frequently required. These operations are most often accomplished by means of forces applied to the particulates through indirect means such as viscous drag by the gaseous or liquid host medium. Indirect methods such as pneumatic transport are energy-inefficient in that generally 0.1-1 mass units of air must be accelerated and conveyed for each mass unit of material transported [19]. Air-cleaning operations in animal-confinement areas and in process-plant emissions offer other examples of energy-intensive particulate removal in which high air-carrier velocities are needed for particle impaction, for centrifugal acceleration, or in which high pressure drops must be sustained as in mechanical filtration [21]. In certain of these operations, a rational engineering approach would be the direct coupling of materials-management forces to the particles through an applied electromagnetic force field. The relative ease and economical means of establishing intense electric fields of appreciable spatial extent have resulted in a preponderance of technical development in this direction as compared with magnetic fields [18]. The objective of this paper is to present a simplified basis of operation for electric force-field management of particulates, and to review several significant examples of its utilization for agricultural purposes.

Principles of Operation

Electrical forces can be exerted upon either charged or uncharged bodies by various basic means. Dielectrophoresis is widely used for movement and separation of uncharged, polarizable particles immersed in a fluid medium is which the relative difference in the dielectric constants of the fluid and the particle can be maximized for practical purposes [20]. Motion occurs only in highly divergent electric fields of intensity generally exceeding 100 V/cm, such that the equal but opposite polarization charges on either end of a given particulate are acted on by a more intense electric field on one end. The resultant motion is in most cases toward the more intense field region, is independent of field polarity, and may occur equally well under ac fields. While the dielectrophoretic effect depends upon the cube of particle diameter, and thus acts well on course particles, limited usages have occurred for agricultural purposes because of the somewhat small migration velocities attained by particles in the relatively viscous liquid media often encountered. Dielectrophoresis may, however, find expanded applications for micro-harvesting operations in emerging areas of biotechnology in which materials of cellular size and larger must be efficiently collected.

Electrophoresis as a means for imparting force and resultant motion to charged particulates *via* interaction with an applied electric field has, by contrast, found many practical and successful usages in industry and agriculture [9,18]. Such usages include xerography, ink-jet printing, abrasive-paper manufacture, electrostatic precipitation of air pollutants, paint coating, crop spraying, contaminant removal, *etc.* A definitive work on charged-particulate control, emphasizing agricultural pesticide spraying, has been presented to consider the basic phenomena acting [10]. In order to incorporate electric forces for controlling airborne particulate motion, two physical conditions must be met: a.) each particle must become significantly charged (*e.g.*, 10^6 electronic charges on a 50 μm particle); and b.) the charged particle must then be acted upon by an electric field E of appreciable strength in volts per meter. For a particle charged to q_p coulombs, Figure 1 summarizes the primary type electric force fields which can be exploited, singly

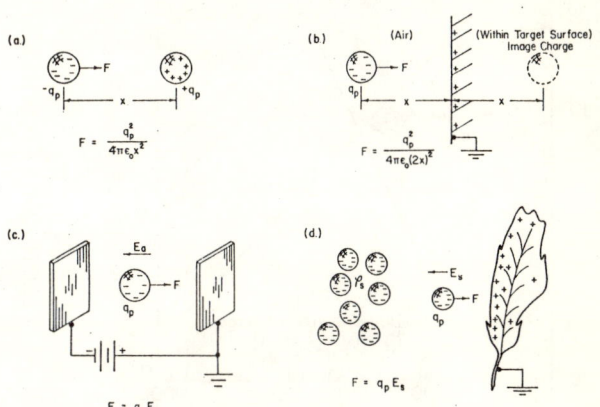

Fig. 1. Electric forces which may be exploited for control of charged particle motion: a.) inverse square; b.) image charge; c.) external field; and d.) space-charge field. Source: (10).

or superimposed, for exerting force in newtons $F = q_p E$ onto charged particulates--i.e., for effecting electrophoresis of a magnitude useful for practical purposes.

Imparting the necessary net electrical charge on particulates is generally accomplished by one of three widely used methods. First, electrostatic induction of charge onto a conductive particle can be achieved if the particle is allowed to break contact with an earthed dispenser (nozzle, spinner, chute, etc.) in the presence of an intense electric field terminating on the earthed dispenser [11]. Generally, material having a resistivity no greater than ca. 10^8 ohm cm will undergo satisfactory induction charging in the short time characterizing particle dispersion from the earthed source.

Second, ionized-field particle charging can successfully impart free charge onto already-dispersed particles of either a conductive or non-conductive nature. Charging equations are well developed for this process in which a high-voltage dc corona discharge from a sharp wire or point ionizer-electrode generates a highly-ionized field of space through which particles are carried for air-ion attachment. Saturation particle charge, which is generally attained in 2-10 ms, depends only upon the particle's dielectric constant K as a physical property and increases with the square of particle diameter [23]. This mode of particle charging is most widespread throughout industry as a practical and reliable method. Figure 2 indicates characteristic half-saturation charge values attained, for example, on a water droplet of spray (dielectric constant $\simeq$ 80). Also shown are charge-to-mass ratio values. While effectively designed induction-charging devices often exceed the charging performance of the ionized-field process, Figure 2 provides a conservative order-of-magnitude estimate of practically attainable particle charges for calculation and comparison purposes.

Also included in Figure 2 is a plot of the ratio of calculated values of the force which can be exerted upon such charged particles by a range of applied electric fields up to 40 kV, as compared with the gravitational force acting on such particles. It is seen that the electric force field becomes dominant for the small size realm under 100-200 μm. For example, the electric force exerted by the 40 kV/m field on a 40 μm diameter particle for control of its motion exceeds the gravitational force by 4-fold. The electric field values used for the Figure 2 calculations are readily achieved by numerous high-voltage dc power supplies routinely utilized in industrial electrostatic processes. Space-charge fields in crop spraying are often of two-fold greater value.

The third commonly encountered particulate-charging method is triboelectrification, often referred to as frictional charging. While much remains to be learned about this process, it is generally thought that the degree of charge exchanged by two dissimilar materials brought into contact and then separated, is related to the differing work functions of the two materials and often expressed in terms of an electrical double layer concept [7]. Although triboelectric charging is successfully relied upon in a number of specialized industrial processes, it can be very unpredictable and highly dependent upon atmospheric humidity. The nuisances and hazards associated with inadvertant triboelectric charging of many common powders are well known.

In summary, it is theoretically possible to efficiently achieve electric force-field control over the motion of many type particulates routinely encountered in agricultural production and processing. Practical means exist for accomplishing this through proper engineering-design choices from several particle-charging techniques and modes of electric-field establishment. The following two sections briefly consider two agriculturally related applications of this electric force-field management of particulates which are under research and development at the University of Georgia.

Electrostatic Crop Spraying

In the production of food crops worldwide, approximately 1.5×10^9 kg of chemical pesticides are presently used to provide protection against insect, disease and weed pests. Exploitation of electric force fields to increase the mass-transfer efficiency in the application of these crop sprays can provide profound improvements relating to reduced economic cost for pesticides, lowered energy consumption as embodied within the formulated chemicals, and lessened environmental concerns regarding the misuse of these highly toxic substances. Many diverse industrial painting and coating operations have long benefited from electric forces. From basic principles, an electrostatic crop-spraying system has been developed at the University of Georgia which specifically satisfies the severe engineering-design requirements of safe outdoor spraying of three-dimensional living plants from a mobile unit [9]. Figure 3 pictures a 14-nozzle row-crop prototype

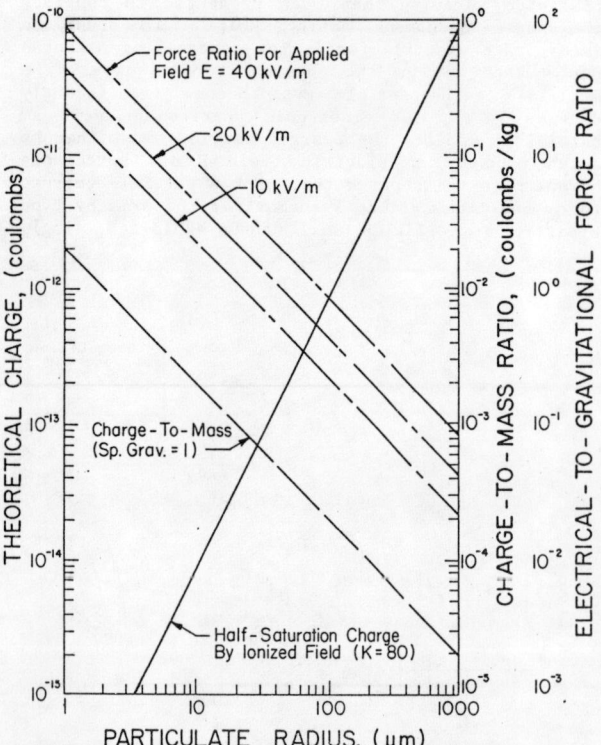

Fig. 2. Particulate charge, charge-to-mass ratio and force ratio values technically achievable on agricultural materials. Source: (10).

Fig. 3. Fourteen-nozzle electrostatic crop sprayer dispensing agricultural pesticides.

earthed target points and edges such as leaf-tips [15]. A transient experimental analyzis of the 600-1200 ms electrostatic crop-spraying event, as recorded in Figure 5, shows the detrimental effect which these target-based gaseous discharges have upon the spray cloud's space-charge field [2]. Induction of a target point causes a partial collapse of this charged-droplet-driving field. Since the severity of the induced corona is polarity dependent, negatively charged spray is confirmed to maintain a 27 percent more intense space-charge field, and to give a 20 percent greater deposition, than did a positively charged spray cloud for coating target objects exibiting earthed points. For ameliorating the target-discharge problem, Cooper and Law [1] report a bipolar spray-charging strategy for space-charge control at the periphery of a 3-dimensional spray target, while Law and Bowen [12] report a dual particle-specie process for enhancement of the space-charge deposition within the electrically shielded confines of such targets (e.g., the canopy regions of living plants).

machine incorporating this system. It utilizes electrostatic-induction charging nozzles [8,13] to electrify conductive pesticide liquids, turbulent-air transport and penetration of the charged spray to regions within the electrically shielded plant canopy, and primarily relies upon space-charge-field deposition of the spray droplets onto the plant surfaces. The laboratory mass-transfer results of Figure 4 confirm the effectiveness of this system's electric force-field management of charged sprays [16]. Also evident in Figure 4 is the inherent limitation in this electrostatic-deposition process caused by corona discharges induced to flow from

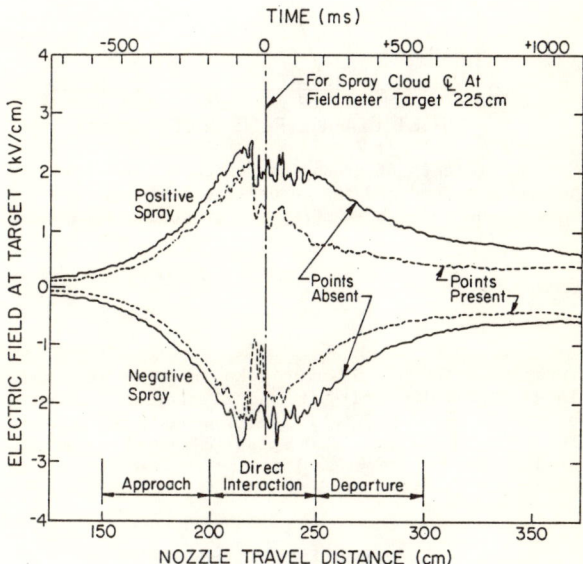

Fig. 5. Effects of earthed point upon the transient behavior of the space-charge field imposed at a target being electrostatically sprayed by a nozzle passing at 5 km/h dispensing 4mC/kg spray. Source: (2).

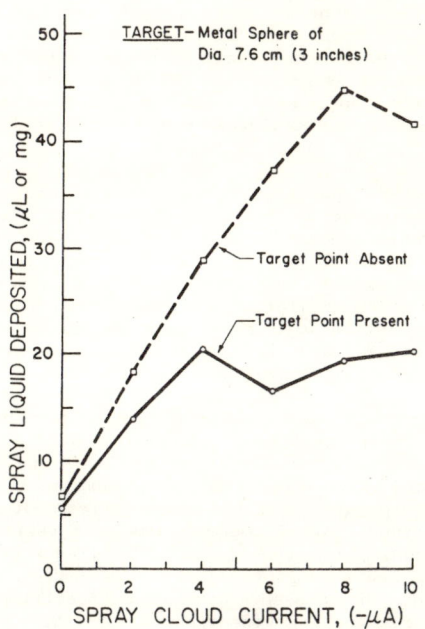

Fig. 4. Electrostatic enhancement of spray deposition and its limitation by induced target corona. Source: (16).

Results for a wide variety of idealized and living-plant targets generally document 2- to 7-fold deposition improvements attributable to the harnessing of electric forces directly upon the spray particles [9]. Developments for agriculture are currently extending this technology beyond row-crop usage into greenhouses, animal housing, orchards, and vineyards.

For greenhouses use of the electrostatic-spraying technology, a recent study has evaluated various electrical resistance and capacitance characteristics of spray targets in order to establish what constitutes adequacy of target grounding [14]. Figure 6 depicts the effects which the resistance of the charge-leakage path to earth has upon both the voltage elevation and the charged-droplet deposition achieved on a target undergoing electrostatic

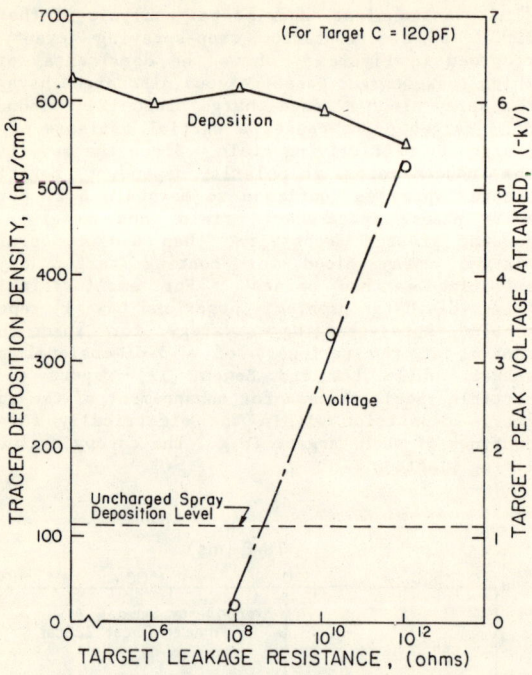

Fig. 6. Effects of grounding resistance upon the peak voltage attained and the deposition achieved on a 120 pF target undergoing a -4 mC/kg electrostatic spraying. Source: (14).

well described in the literature [4,18]. This static-field method and such devices do, however, have a major shortcoming in the limited effective distance over which the particle-controlling field is able to act.

To overcome this deficiency, research currently underway at the University of Georgia is aimed at developing specifically for agricultural usages the concept and process of charged-particulate conveyance by a non-uniform 3-phase traveling electric field as earlier proposed and developed by Masuda for industrial purposes [17] and later by Weiss [22]. A planar array of three series of equally spaced parallel electrodes, as in Figure 7, is energized by a 3-phase high-voltage source to create a periodic electric field having wave fronts progressing to the right in the phase-velocity direction. Charged

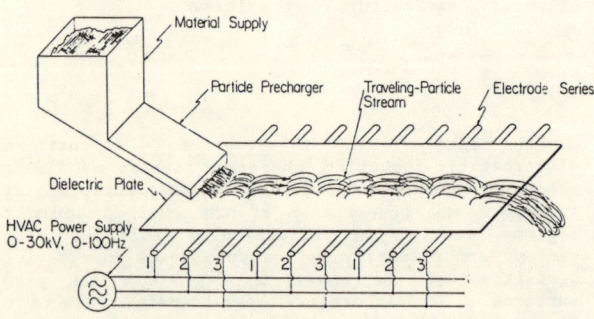

Fig. 7. Three-phase traveling electric-field electrode array for conveyance of agricultural particulates such as milled products and seed.

spraying. Over a $0 - 10^{12}$ ohm range of resistance in the grounding circuit, no significant reduction occurred in the 5-fold electrodeposition benefit. The study confirms that plastic-potted greenhouse plants should very satisfactorily facilitate the transient charge transfers required in electrostatic crop spraying.

Traveling Field Conveyance

There have been numerous successful applications of electric force fields for charged particulate deflection in various processes ranging from cell sorters to mineral enrichment [4]. These static deflection fields impart a component of velocity perpendicular to the major component which is generally imparted by gravitational or centrifugal forces. Separation of differing particulates along certain trajectories can be effected with subsequent collection at corresponding different locations (e.g., removal of rat and insect feces from small grain). Particle charging can be by any of the three discussed methods, and often the particle's charge-relaxation time characteristic is relied upon for controlling the particle's residual charge upon entering its trajectory and also its point of entry as it is spun from a rotating conductor. Difficult to separate agricultural particulates and seeds which may differ only in their electrical properties may thus be successfully sorted. Differing particle resistivity values, and especially their dependence upon the distinctive moisture sorption/desorption behavior of otherwise similar materials, become a prime factor on which electrical separation can be based. Rotating-drum electric force-field separators finding routine use in industry and agriculture are

particulates entrained within this traveling electric field can, under proper circumstances, be levitated and conveyed along the array in a contactless manner. Interchanging any two leads of the 3-phase input immediately reverses the direction of conveyance. For practical purposes the electrode array may be embedded several millimeters within a dielectric panel (e.g., cast acrylic) to initially contain the particulates and to triboelectrically charge them as taught by Masuda.

Computer analysis reveals individual particle trajectories to be cycloidal-like, typically at 1-3 cm elevation above the electric panel, and traveling along the panel at typically 3-8 cm/s [6]. This computer simulation also helps to clarify the effects which excitation frequency, the phase-lag of particle motion with respect to excitation voltage, and particle charge have upon both the magnitude and the direction of conveyance.

Laboratory-scale models of traveling-field devices have confirmed successful levitation and conveyance of agricultural particulates and seeds as large as 3 mm; operation at atmospheric relative humidities of 35-85% and corresponding seed equilibrium moisture contents has been verified [5]. A programmable 3-phase, variable frequency (5-300 Hz), variable voltage (0-20 kV peak), waveform generator has been developed specifically to facilitate ongoing experimental analysis of the effects which a wide range of electrical parameters have upon successful traveling-field conveyance and separation of agricultural particulates [3]. With this knowledge it is hypothesized that it will become possible to electrically "tune" traveling-field conveyance for transport and separation of individual components within various mixtures of agricultural particulates.

Summary

Electric droplet-force augmentation for efficient pesticide spray deposition has been accomplished, and commercial equipment is under development for pest control in routine crop-production use. On the basis of laboratory studies and computer simulations currently underway, electric force-field management of agricultural particulates *via* the traveling-field process offers much promise for a number of practical and otherwise difficult mass-transfer operations required in agricultural production and processing.

References Cited

[1] Cooper, S.C. and S.E. Law. 1987. Bipolar spray charging for leaf-tip corona reduction by space-charge control. *IEEE Trans*. IA-23(2):217-223.

[2] Cooper, S.C. and S.E. Law. 1987. Transient characteristics of charged spray deposition occurring under action of induced target coronas: space-charge polarity effect. *Proceedings of the 1987 Oxford University Conference on Electrostatics. British Inst. of Phys. Conf. Ser*. No. 85 (Section 1):21-26.

[3] Cooper, S.C. and S.E. Law. 1987. Three-phase variable frequency and waveform high-voltage power supply. *ASAE Paper No. 87-3516* (microfiche), St. Joseph, MI.

[4] Fraas, F. 1962. Electrostatic separation of granular materials. U.S. Dept. of Interior, Bureau of Mines Bulletin No. 603, Washington, D.C.

[5] Ganesh, M.J. and S.E. Law. 1988. Environmental and electrical effects upon levitation in traveling electric-field conveyance of agricultural particulates. *ASAE Paper No. 88-6065* (microfiche), St. Joseph, MI.

[6] Gan-Mor, S. and S. E. Law. 1987. Frequency and phase-lag effects upon transport of particulates by an ac electric field. *IEEE (IAS) Conference Record*: 1578-1584. Library of Congress No. 80-640527.

[7] Horvath, T. and I. Berta. 1982. Static Elimination. Research Studies Press, Chicester, England, pp. 3-6.

[8] Law, S.E. 1978. Embedded-electrode electrostatic-induction spray-charging nozzle: theoretical and engineering design. *Trans. of ASAE* 21(6):1096-1104.

[9] Law, S.E. 1983. Electrostatic pesticide spraying: concepts and practice. *IEEE Trans*. IA-19(2):160-168.

[10] Law, S.E. 1987. Basic phenomena active in electrostatic pesticide spraying. In " *Rational Pesticide Use*" (K.J. Brent and R.K. Atkin, Editors), pp. 81-105. Cambridge Univrsity Press, Cambridge, England. ISBN 0 521 32068 2.

[11] Law, S.E. and H.D. Bowen. 1966. Charging liquid spray by electrostatic induction. *Trans. of ASAE* 9(4):501-506.

[12] Law, S.E. and H.D. Bowen. 1985. Dual particle-specie concept for improved electrostatic deposition through space-charge field enhancement. *IEEE Trans*. IA-21(4):694-698.

[13] Law, S.E. and S.C. Cooper. 1987. Induction charging characteristics of conductivity enhanced vegetable-oil sprays. *Trans. of ASAE* 30(1):75-79.

[14] Law, S.E. and S.C. Cooper. 1988. Target grounding adequacy for electrostatic deposition of conductive pesticide sprays. *Trans. of ASAE* (in press).

[15] Law, S.E. and M.D. Lane. 1981. Electrostatic deposition of pesticide spray onto foliar targets of varying morphology. *Trans. of ASAE* 24(6):1441-1445 & 1448.

[16] Law, S.E. and M.D. Lane. 1982. Electrostatic deposition of pesticide sprays onto ionizing targets: charge- and mass-transfer analysis. *IEEE Trans*. IA-18(6):673-679.

[17] Masuda, S. 1971. Electric curtain for confinement and transport of charged aerosol particles. Conf. of Electrostatic Society of America, Albany, NY, 27 pp.

[18] Moore, A.D. 1973. Electrostatics and Its Applications. John Wiley and Sons, New York.

[19] Orr, C. Jr. 1966. Particulate Technology. The MacMillan Company, New York, pp. 144-168.

[20] Pohl, H.A. 1978. Dielectrophoresis. Cambridge University Press, Cambridge, England.

[21] Wark, K. and C.F. Warner. 1981. Air Pollution - Its Origin and Control. Harper and Row Publishers, New York, pp. 143-236.

[22] Weiss, L.C. 1982. Electrodynamic behavior of textile fibers. *Textile Research Jour*. 52(1):59-65.

[23] White, H.J. 1963. Industrial Electrostatic Precipitation. Addison Wesley Publishing Co., Reading, MA.

ARE AIR IONS BIOLOGICALLY SIGNIFICANT ?

Professor N.J. FELICI

Laboratoire d'Electrostatique et de Matériaux Diélectriques
C.N.R.S., B.P. 166 X - 38042 GRENOBLE Cedex (France)

Abstract

The possible effect of airborne electric charges on living beings have been considered since 1770. The contributions of Bertholon, Chizhevski and Krueger are critically examined. Chizhevski's contention that airborne superoxide anions are necessary for the maintenance of life appears utterly erroneous, while Krueger's effects seem genuine, though fairly weak in healthy individuals. The fact, however, that the superoxide anion has been recently recognized as a strong promoter of cancer and decay warrants a deeper investigation of the hidden effects of air ions on humans.

Historical Outlook

From its very beginning, electrical science was associated with biology and medicine. William Gilbert (1544-1603) was physician to Queen Elizabeth I of England. Sir Thomas Browne (1605-1682) who coined the word "electricity" (1646) was also a medical man.

In the following century, two momentous discoveries quickened the pace of investigations and made electricity fashionable : the friction generator (1740) by Bose (1710-1761) a professor at Wittenberg University, and the capacitor (1746) by Musschenbroek of Leyden. Both were popularized all over Europe by the influential Abbé Nollet (1700-1770) a member of the Academy of Sciences in Paris, who coined the phrase "Leyden jar". A flurry of experiments set in, and it was claimed that the electric shock could cure a number of ailments, from toothache to palsy. The illusory character of these cures, however, soon became apparent. The suffering patient was merely numbed for a time ; paralyzed limbs were caused to move, but a failing nervous transmission was not thereby restored.

A few years later (1752), the discovery of atmospheric electricity started another wave of excitement among scientists. The spectacular experiments on lightning showed Nature was playing with electricity on a grand scale, and this was a strong incentive to study the electric condition of the atmosphere. It was soon discovered that electrical activity is by no means restricted to thunderstorms, and more refined experiments showed it to be truly ubiquitous, even by fair weather. Thus, about 1770, scientists convinced themselves that the atmosphere was silently pervaded by electricity, and that living beings were, so to speak, immersed in an electric bath, whose electricity was penetrating their bodies by every pore, reaching into the inmost recesses of vital organs.

What a fascinating idea ! The very force that sparks the awe-inspiring thunderstorm was secretly at work in our heart and liver, for better or worse, while, unaware of Nature's sway we indulge in the vain pursuits of life.

Bertholon (1742-1800)

The first proponent of this theory, strongly reminiscent of the macrocosm/microcosm interaction of renaissance time, was the French physician P. Bertholon. Most distempers, if not all, were ascribable, according to him, to some electrical imbalance. More precisely, he viewed the exchange of electricity between the air and the body as essential for health and well-being.

In his work "On the electricity of the Human Body" (1780), Bertholon went a step further, claiming the very function of the lungs to be the absorption and excretion of electricity. This electricity, he believed, was carried to our organs or removed therefrom by the blood. A nice theory of respiration indeed !

Bertholon divised a number of clever experiments to prove his point. He subjected growing plants to an electric field, augmented by corona ionization. He also wetted his plants with a spray of electrified water, using an arrangement imagined thirty years before by Nollet. In his book "On the Electricity of Plants" (1783) he describes the beneficial effect of electricity, growth being accelerated and plants looking more vigorous.

With Bertholon the stage is set for the developments that took place a century later, when his ideas experienced a spectacular revival at the hands of A.L. Chizhevski (d. 1964) the great pioneer of air ionization.

Chizhevski's Nature and Deeds

I had the privilege to be introduced to Chizhevski's work by one of his French disciples, Dr. André Denier, a physician of La-Tour-du-Pin, in the Dauphiné province. In 1932, Denier travelled to Moscow to work with him, and entertained an important correspondence.

Chizhevski was an extraordinary man. Highly cultured, fluent in many languages, an incredibly prolific writer, he radiated enthousiasm, and possessed a magnetic power of persuasion.

He worked relentlessly to promote his ideas. He not only divised a number of clever experimental techniques and published hundreds of papers, but he also applied his results on a full scale to medicine, agriculture, and breeding, in spite of the difficult conditions prevailing in the Soviet Union after the Revolution.

He was no professional physician, but nevertheless proficient in biological disciplines.

The zenith of his career emcompasses the years 1920 to 1940. As early as 1919, he won the support of two influential dignitaries, A.M. Gorki and A.V. Lunacharski, and set up a research laboratory in the 3rd Medical Institute in Moscow. He had disciples all over the world, even in Germany and Japan, whose relations with the U.S.S.R. were then considerably strained.

After 1945, however, he suffered a grave setback. As we shall see, enthusiasm led him sometimes astray. He propounded a new interpretation of history, linking major events, like revolutions, with cosmic phenomena (e.g. sun spots). He so displeased the political theorists in Moscow, and was relegated to Moldavia, in the south of the Union, and, later on, to Central Asia. Nevertheless, he continued to write and publish. In 1960, four years before his death (he had been permitted to return to Moscow) came out his monumental work "Air Ionization in the National Economy" a stupendous book I never tired to read (1).

In this book, Chizhevski's central idea appears with stark cosmological simplicity. There are two antagonistic forces in the realm of aerobic life : positive and negative electricity. Negative electricity, in the form of the superoxide ion O_2^- is a universally beneficial factor, while all sorts of positive ions are detrimental.

Thus, Chizhevski's prescription for economic progress sounds marvellously simple : treat with superoxide ions whenever possible. We find in his book a number of applications of his method, together with detailed quantitative accounts of the results, always showing a favourable effect.

Medicine :
High blood pressure / Tuberculosis / Bronchial Asthma / Stomachal Ulcers / Duodenal Ulcers / Gynaecological Disorders / Healing of Wounds.

Hygiene :
Precipitation of airborne germs / Prevention of bacterial multiplication / Prevention of Professional diseases / Treatment of growing children.

Agriculture and Breeding :
Poultry / Swine / Sheep and Wool / Cattle and Milk / Rabbits / Bee-keeping / Salads / Beans / Cucumbers / Greenhouses.

Chizhevski on Ion Deprivation

In the 1930's, Chizhevski started another line of research. Instead of adding ions to the ambient air, he decided to remove the existing ones completely, and to submit animals to a "de-ionized" atmosphere. He knew filtering a gas (through cotton wool) removes its small ions entirely, in particular the superoxide O_2^-.

Animals (mice, rats, rabbits, guinea pigs) were kept in air-tight boxes with the necessary arrangements for food and removal of excrements, while filtered air was passed through the boxes by a pump. The results were impressive. In a matter of two or three weeks, the animals showed signs of illness, their condition deteriorated, and they eventually died. Postmortem examinations showed a number of characteristic disorders, both at macroscopic and histological levels (2).

Chizhevski did more. Applying to the letter one of Francis Bacon's criteria, he injected into the deionized air flow, downstream of the filter, negative or positive ions, by means of an electric corona. With negative ions, the animals survived in spite of filtering, while the injection of positive ions did not prevent their dying.

It thus seemed that Chizhevski had discovered an essential factor in the biology of respiration. According to him, oxygen was useless if not seeded with superoxide ions that acted as catalysts.

Criticism of Ion Deprivation Theory

Chizhevski's results were received with skepticism. Scientists did not care for an explanation of the mysterious role of superoxide ions. Moreover, as shown later on (3), devastating criticism could be levelled at Chizhevski's methodology. No doubt his filter eliminated all small ions from the air. He overlooked, however, that ionizing radiation permeates all bodies and restores the ionic equilibrium of air in a matter of minutes. Chizhevski's own data show that the renewal of the air in his boxes was slow and lasted a minute or more ; thus de-ionization was limited, if not insignificant.

In the 1970's the French Commissariat à l'Energie Atomique started a research program on this problem (3). By resorting to an adequate geometry, it was possible to renew the air the animals breathed every second of time. Thus, de-ionization was at least 95 % and presumably 99 %. After several weeks, the animals (white mice) were still in perfect health. Repeating the experiment gave consistent results. No animal died, nor gave any sign of impaired health.

At the Conference "Ions Atmosphériques et Pollution" (University of Pau, France, 1976), medical people pointed out that a similar de-ionization prevails in the "white rooms" used in hospitals to protect the diseased against airborne pathogens. No adverse effect was ever observed, even for residence times of several months.

How Chizhevski's experiments were flawed is impossible to tell at the moment. At any rate, the credibility of his work is thereby seriously tainted.

Air Ionization after Chizhevski

After his banishment from Moscow, Chizhevski was no longer permitted to travel abroad or to meet foreigners, and his scientific influence was much reduced. In the 1950's, some of his ideas were popularized by clever tradesmen who offered negative ion generators for domestic use. In the U.S.A., the Food and Drug Administration reacted strongly. Any claim to a medical effect was labelled "quackery" and ads for ions generators had to be formulated in careful terms (4). In France, however, regulations are not so strict, and negative ions are openly advertised as a cure for a number of ailments (5).

A respectable line of research, however, was launched in the U.S. by A.P. Krueger (6) who introduced the "serotonin hypothesis". Krueger focused attention on behaviour changes when the balance of + to - ions is upset. In normal air, there is a slight excess of + over -. It is easy, however, to create a large dominance of the ions of either sign. Krueger states that an excess of - ions is beneficial to behaviour, generating a feeling of well-being, favouring efficient work and reducing group tension, while + ions have the opposite effect (uneasiness, headache, aggressivity, errors). He related these observations to a change of the level of the neurohormone serotonin in the blood.

Some scientists confirmed Krueger's theory while others did not. J.M. Olivereau (1971) subjected rats to a stressing situation and measured the time they needed to escape by jumping. He found they reacted better when treated by - ions and worse with + ions, the relative differences in reaction time being small, however. Colonel Badré (7) studied the behaviour of seamen by double-blind methods to decide whether it was useful to ionize the air of submarines when on long patrols. He could not detect any difference in the seamen's ability to solve problems, etc.

Existing literature shows that although Krueger's effects cannot be denied, they are usually small, and sometimes undetectable. This may well be a mere consequence of homoeostasis in healthy individuals (8).

The Superoxide Ion as a Great Destroyer of Life

Anyone reviewing the matter about 1980 might have been tempted to say (with some exaggeration) "Much ado about nothing". In this decade, however, the progress of biochemistry cast a completely different light on the problem.

To-day, the superoxide ion O_2^-, together with other oxygen radicals ($OH^{\cdot}$, $O_2H^{\cdot}$) the hydrogen peroxide H_2O_2 and the singlet oxygen molecule, appears to be a most powerful reactant of biochemistry. Chizhevski's vision is thereby vindicated, but quite in the opposite sense. It is true that O_2^- plays a universal role, but by no means as a benevolent agency. It ranks high among the destroyers of life, the promoters of cancer and aging, the purveyors of decay and death no organism can escape.

Oxygen radicals are inevitable by-products of oxidative processes, and this explains why warm-blooded animals with high metabolic rates (mice) have such a short life-span, while humans and elephants live much longer, their metabolism being necessarily lower by mere size effect.

There is an enormous literature on the subject. I refer the reader to the review papers by Ames (9) and Cerutti (10).

It is interesting to notice that superoxide ions are generated by activated lymphocytes when destroying microbes (11). The bactericidal effect of negatively ionized air, claimed by Chizhevski, and others, appears quite plausible.

As concerns the inhalation of negative air ions, the problem of its innocuousness is clearly posed. Quite roughly, one may compare the energy of the superoxide anions to that deposited in living matter by ionizing radiation, since both types of damage are similar. It would seem that prolonged inhalation of the highest concentrations attainable (10^6 to 10^7 ions/cm^3) may be comparable to a dose of 1 to 10 rem/year which is by no means negligible.

Conclusion

Visionaries like Chizhevski are often useful. His naive hope of a natural panacea we must relinquish, but many results of his, however controversial, deserve further scrutiny.

The fact his cherished superoxide ion is destructive of life does not preclude stimulating or beneficial effects. The same can be said of almost every spice in our kitchen and of many drugs in our pharmacy, as shown recently by Ames and al. (12).

The recent breakthrough of biological science opens new vistas to experimentation with air ions. Let the better part of Chizhevski's spirit abide with us, for enthusiasm shall never die !

References

(1) Aeroionifikaciya v narodnom chozyaistve (in Russian) Gosplanizdat, 1960.

(2) Ibidem, chapter 4, pages 278 to 332.

(3) Comptes-Rendus Académie des Sciences Paris, 1983, tome 297, Vie Académique, page 12.

(4) M. SUN, Ion Generators : Old Fad, New Fashion. Science, 1980, vol. 210, p. 31.

(5) J. METADIER, Les Oxions, Editions Christian Godefroy, B.P. 9, 27760 La Ferrière sur Risle, France.

(6) A.P. KRUEGER and E.J. REED, Biological Impact of small air Ions, Science, 1976, vol. 193, p. 1209.

(7) R. BADRE and al., Revue Générale de l'Electricité, 1972, vol. 81, p. 240.

(8) Comptes-Rendus Académie des Sciences Paris, 1983, tome 297, Vie Académique, p. 15.

(9) B.N. AMES, Dietary Carcinogens and Anticarcinogens, Oxygen Radicals and Degenerative Diseases, Science, 1983, vol. 221, p. 1256.

(10) P.A. CERUTTI, Prooxidant States and Tumor Promotion, Science, 1985, vol. 227, p. 375.

(11) C.K. EDWARDS III, S.M. GUISSUDIN, J.M. SCHEPPER, L.M. YUNGER, K.W. KELLEY, A Newly Defined Property of Somatotropin : Priming of Macrophages for Production of Superoxide Anion, Science, 1988, vol. 239, p. 769.

(12) B.N. AMES, R. MAGAW, L.S. GOLD, Ranking Possible Carcinogenic Hazards, Science, 1987, vol. 236, p. 271.

THE EFFECTS OF THE ELECTRICAL FIELD ON PLANT GROWTH

Cywiński Kazimierz
Barmuta Piotr
Technical University Department of Electrical Engineering
Bialistok, Poland

1. Introduction.

Increasing food problems give rise to investigations concerning methods of seed selection and accelerating the process of plant growth. It has been found that the germination time and plant growth-especially in the earliest stage-depent to some extent on such physical factors as e.g. electrical field, infrared radiation or ultrasounds.

The autors of this paper investigated the influence of the electromagnetic field intensity /especially short-duration transitory/ evenescent of alternating field effect /f=50Hz/ and electrostatic fields with corona discharge on the germination time and an increase of plant substance. Seed-grain of wheat, oat, barley, seeds of fodder crops, hay and weed seeds were the subject of the investigation. The paper presents the effect of electrostatic separation of seeds and how electrostatic fields influence plant germination and growth [1,2,3,6].

The investigations have shown a water consumption increase and an activation of the respiration process in the seeds examined. It has been observed that in the process of electrostatic separation of the wheat seeds in shortly-activated electric field there is a change in the biopotentials of the cells responsible for growth, which accelerates the germination process. In comparison with the control sample, the influence of the electric field on growing wheat could be seen in an increased accumulation in its green substance of the elementary components essential to its proper development, such as nitrogen, potassium, calcium, sodium or phosphorus, which is a component of energy-rich compounds /ATP and ADP/ [4].

To sum up: electic field as plant growth stimulator can be used:
1/ in the electroseparation process [5],
2/ in the germination and growth process.

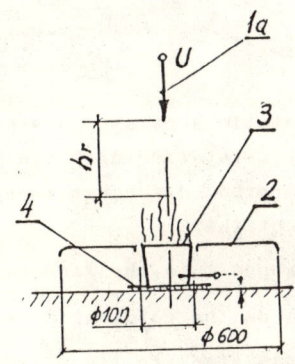

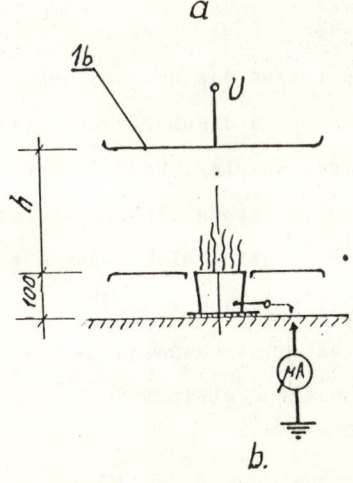

Fig 1. Laboratory set-up for germination:
a/rod-plane configuration, b/plane-plane configuration. 1-H.V. electrodes, 2-protection electrodes.

Electric fields and corona discharges influence on the seeds and grains in the separation process.

2. Material and methods.

A. Electric separation.

We examined among others seeds of lucerne, grains of oat, barley and wheat. The separation was made in electrostatic separators designed as shown in fig.2.

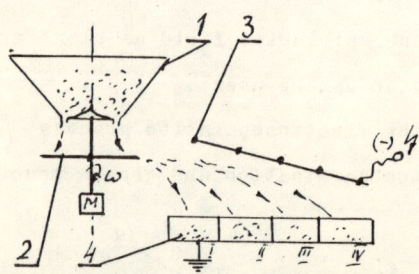

Fig.2 Electrostatic separator operation [6]. 1-feeder, 2-rotating disk electrode /RDE/, 3-corona discharge, 4-separate seeds container.

The seeds were put into the feeder-1. Next the seeds fell on the rotating disk electrode in the strong electric fields zone-3. In this field deflected seedsfell into the containers-4.

Seed separation was done at negative polarization of corona discharge at voltage value U=0kV /control sample/, U=40,60 and 80 kV. For each kind of seeds the proper rotational speed of the rotating disk electrode was chosen /ranging to 300 rpm/.
The effect of electroseparation was evaluated according to the distribution of mass and grain size.

In the electroseparation process the seeds of each obtained fraction and seeds obtained from the control test /U=0 kV/ were planted in specially prepared soil.

During the investigations the following aspects were evaluated:
1/ germination time of seeds,
2/ increase in the length of plant,
3/ changes in the coloring of plant,
4/ changes in the basic physical parameters of plant tissue.

Table I

Fraction	Germinating %			flax
	0 kV	40 kV	65 kV	80 kV
I	37	70	60	53,3
II	50	73,3	63,3	50
III	30	47	67	70

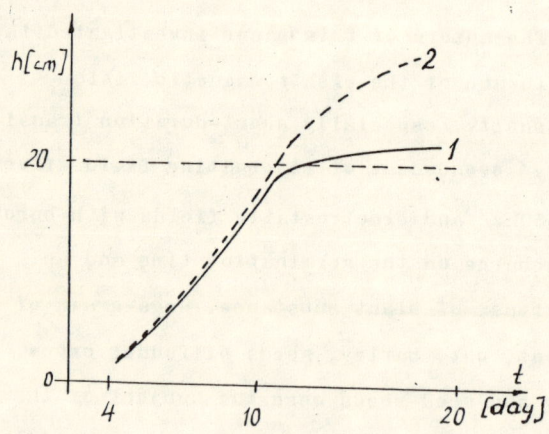

Fig 3. Germinating of electroseparated oat: 1-120 kV/cm, 2-0 kV/cm.

B. Electric field action upon the seed germination on and plant growth.

The seeds isolated in the electroseparation process were poted and part of them exposed to non-homogeneous electric field /f=50 Hz, U=60 kV/ as shown in fig.1. Different voltage cycles were applied as shown in fig.4.

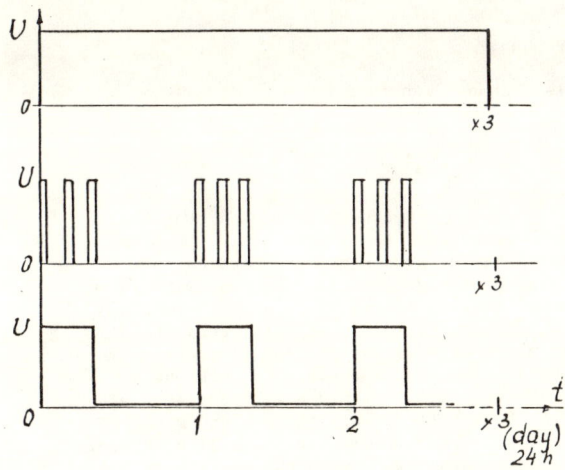

Fig.4 Examples of high voltage cycles.

In order to ensure homogeneous electric field the ring electrode was placed around the pot as shown in fig. 1a.

Corona discharge current flowing along the stalk was measured with an izolating transmission located under the pot.

Similarly to preceding scheme the electric field effect on the germination and plant growth was evaluated in relation to:
1/ germination time of seeds,
2/ increase in length of plant,
3/ changes on the surface of plants /f.e. colour/,
4/ changes in the basic physical parameters of plant tissue.

The current flowing the system: corona discharge-air-plants-soil was olso measured.

3. Results and conclusions.

In the electroseparation process the best results were obtained for oat grain.
In comparison with the control sample /U=0kV/ electroseparated seeds had:
1/ shorter germination time and increased of germination power,
2/ faster growth especially in the initial phase,
3/ differences in shape, thickness and coloring of stalk and leaves,
4/ changes in resistivity of plant tissue,
5/ effect of increased water transport from roots to leaves.

In conclusion we postulated that further investigations on electric field activation should be made [4,6], in order to obtain seeds greater germination power and more resistance to weather conditions, bacteria, mildew and others diseases.

References.

1. Taruškin W.J.:"Ocienka raznokačestwiennosti siemion s pomošč. el.polja".Viestn.s-ch nauki.1975 z s.116+122.
2. Šmigiel W.N.:"Razdielenije owsa i owsiuga v el.polje".Mieoh.i el.siel.choz.1973.09, s.10+12.
3. Harmond J.E.:"Cleaning by Electrostatic Separation".Agricutural Angin.1961 Jan. p.22+25.
4. Jasiński A.,Kilarski W.:"Ultrastruktura komórki".Wyd.Szkolne i Pedag.Warsaw 1987.
5. Czerwiński W.:"Fizjologia roślin".Warsaw PWN 1977.
6. Cywiński K.:"Elektrostatyczny separator do nasion i materiałów rozdrobnionych". Zgł.pat. Pol.Biał. 1988.

EFFECTS OF HIGH-VOLTAGE ELECTROSTATIC FIELD ON GROWTH IN PLANTS*

Ye Jia-ming

(Zhejiang University, Hangzhou)

Gao Tian-shun, Ge Lan, Zhou Hao-le, Zhang Jian-feng

(Northeastern Normal University, Changchun)

Abstract

The effects of high-voltage electrostatic field on plant growth and mitosis were studied on rye (Secale cereale), barley (Hordeum vulgare) and broad bean (Vicia faba). Experimental results showed that a moderate intensity of electrostatic field could promote the germination of seeds and growth of seedlings. But an excessive intensity of electrostatic field inhibited the growth of seedlings. The promotive intensity and the inhibitive intensity were different for the three species of plants. The variations of growth rate were correlated to the variations of mitotic cycle.

Introduction

There are many charged substances presented in living organisms, the distribution, arrangement and movement of these substances possess important living significances. If the intensity of environmental electrical field vary in a great enough degree, it may be possible to exert an influence on the charged substances of organisms, and as a result to exert an influence on the life activity. Authors had proved that electrostatic field with a high enough intensity can induce chromosome aberration in rye and broad bean root tip cells[1], and variation of chlorophyll content of wild soybean[2], and it can also exert an influence on the functional activities of animals[3,4]. Other researchists had also reported the effects of electrostatic field on living bodies[5-9]. These researches showed that electrostatic field is an efficient physical factor that can exert influences on manifold kinds of functional activities of living organisms. This paper reported the effects of high-voltage electrostatic field on the seedlings growth and mitosis of three species of plants.

Materials and Methods

As materials rye (Secale cereale), barley (Hordeum vulgare) and broad bean (Vicia faba) were selected.

Experimental field is an approximate uniform negative high-voltage electrostatic field between a pair of copper plate electrodes at a distance 200mm from each other, the possitive electrode was grounded. Direct-high-voltage was supplied by High-Voltage-Direct-Supply Model MZ-3B made in Department of Physics, Northeastern Normal University.

Soaked and about to germinate seeds were grouped randomly. Experimental groups were exposed to electrostatic field with different intensity and effective time: experimental rye group No.1 was exposed to 150kv/m electrostatic field for 6 hrs, group No.2 was exposed to 400kv/m field for 6 hrs; experimental barley group No.1 was exposed to 200kv/m electrostatic field for 2X8 hrs (with 14 hrs interval between each 8 hrs), group No.2 was exposed to 350kv/m field for 2X8 hrs; broad bean experimental group No.1 was exposed to 150kv/m electrostatic field for 8 hrs, group No.2 was exposed to 200kv/m field for 2X8 hrs, group No.3 was exposed to 350kv/m field for 2X8 hrs, and group No.4 was exposed to 400kv/m field for 8 hrs. The ambient temperature was about 20°C when the seeds were exposed to electrostatic field. Contrast groups were not exposed to electrostatic field. The culturing conditions were identical for experimental and contrast groups.

The germination percentage, length of seedlings and roots, fresh weight and dry weight of seedlings and the mitosis metaphase cell index were measured at different time after exposed to electrostatic field. The measure method of metaphase index was: root tips were taken off from plants and underwent the pretreatment with colchicine (0.1%) for 5 hrs, then it was fixed in Carnoy's fluid for 4 hrs; after this, the root tips were hydrolyzed in HCl (1N) for 12min then stained by iron alum haematoxylin; at last the specimen was pressed on the slide in acetic acid (45%) and observed under microscope, mitosis metaphase index were enumerated.

The data were statistically treated.

Experimental Results

1. Germination percentage of barley seeds:

To the 3rd day after the seeds were soaked, mean germination percentage of barley seeds of contrast groups was about 38%, but mean germination percentage of experimental groups was about 64%, there was very remarkable diversity between the two. The germination percentage of the two experimental groups that exposed to different intensity of electrostatic field little differed from each other. In the 4th day, the mean germination percentage of contrast groups was about 52%, and of experimental groups was about 78%, diversity was still very remarkable. Germination percentage of experimental groups rye was also higher than the contrast groups.

2. Length of seedlings:

Length of rye seedlings were measured at 5th day after exposed to electrostatic field. 30 seedlings were sampled randomly from each group, results of measurement were listed in Table 1.

As seen from Table 1, among the three groups, mean value of seedlings length of contrast group was the smallest one and its seedlings length were most irregular. Conversely, the group exposed to 400kv/m electrostatic field had the longest mean seedlings length and its seedlings length were most uniform. Results of t-test showed that the

---

* Projects Supported by the Science Fund of the Chinese Academy of Sciences.

Table 1. Length of rye seedlings

Group	Length of seedlings (cm)										Mean value
Contrast group	12.3	11.2	11.7	9.7	4.3	7.4	6.1	13.0	10.6	11.1	8.653± 3.031
	4.2	4.5	6.5	10.1	4.1	7.9	7.2	9.9	11.8	5.5	
	14.0	8.5	6.6	5.6	10.1	9.3	4.9	13.9	10.5	7.1	
150 kv/m group	11.0	6.4	6.9	9.8	4.0	8.0	10.2	8.0	11.6	10.1	9.926± 2.468
	10.0	10.7	12.1	13.7	10.4	14.8	6.0	9.2	13.2	10.0	
	9.4	12.7	11.2	8.9	10.7	10.0	7.6	7.0	11.0	13.3	
400 kv/m group	9.9	12.9	12.1	12.2	12.1	11.1	11.2	10.3	10.5	12.0	10.663± 1.674
	8.7	11.0	13.2	8.3	10.3	5.6	11.8	12.7	10.0	8.0	
	11.0	12.2	9.9	9.5	8.4	11.0	10.4	11.6	11.8	10.2	

Table 2. Length of broad bean seedlings (mean value, cm)

Time / Group	9th day	13th day	17th day	20th day
Contrast group	3.80	4.10	7.03	9.70
150kv/m group	6.32			
200kv/m group		9.48	16.92	20.93
350kv/m group		8.76	11.16	16.56
400kv/m group	2.48			

150kv/m group made little diversity to the contrast group (t =1.784<t =2.000, P>0.05); and the 400kv/m group made very remarkable diversity to the contrast group (t =3.178>t =2.660, P<0.01). The results showed that 150kv/m electrostatic field had little influence on the growth of rye seedlings, but 400kv/m electrostatic field could promote the growth truely.

Seedlings length of barley were measured at the 6th, 9th and 15th day after exposed to electrostatic field. The measurement results at 6th day were: mean length of seedlings of contrast group was 6.42cm, the 200kv/m group --- 7.69cm and the 350kv/m group --- 7.68cm. Results of statistical treatment showed that the two experimental groups both made very remarkable diversity to the contrast group (P<0.001). But to the 9th and 15th day, there were not remarkable diversity between the experimental and contrast groups.

From every experimental and contrast groups length of 60 seedlings of broad bean were measured, the results of measurement were listed in Table 2. Results of statistical treatment showed, length of broad bean seedlings of every experimental groups measured at the 4 times all made remarkable or very remarkable diversity to the contrast group (P<0.05 or P< 0.001). IN Table 2, 150, 200 and 350kv/m electrostatic field obviously promoted the growth of broad bean seedlings but the promotive effect of 350kv/m field was slightly less than the other two fields. 400kv/m field inhibited the growth of broad bean seedlings.

3. Length of roots:

Measurement results of root length of barley seedlings were listed in Table 3.

Table 3. Root length of barley seedlings (mean value, cm)

Time / Group	6th day	9th day	15th day
Contrast group	2.98	6.61	11.09
200kv/m group	2.52	6.07	8.88
350kv/m group	3.11	6.62	9.07

As seen from Table 3, root length of 200kv/m group in the three measurements were all shorter than the contrast group, and results of t-test showed the diversity was remarkable or very remarkable (P<0.05 or P<0.01). In the 350kv/m group, root length of the first two measurements were comparatively longer than the contrast group, but statistical results showed there was not remarkable diversity; in the 3rd measurement the root length were shorter than the contrast group and t-test result showed the diversity was very remarkable (p<0.01).

Measurement results of root length of broad bean seedlings were listed in Table 4.

Table 4. Root length of broad bean seedlings (mean value, cm)

Time Group	13th day	17th day	20th day
Contrast group	5.57	10.10	11.15
200kv/m group	11.27	14.57	13.33
350kv/m group	11.09	12.38	13.42

In Table 4, root length of broad bean seedlings in the three measurements were all longer than the contrast group. Statistical results showed that the root length of the two experimental groups in the first two measurements made very remarkable diversity to the contrast group ($P<0.001$), and in the 3rd measurement diversity was remarkable too ($P<0.05$).

4. Fresh weight and dry weight:

Fresh weight of barley seedlings were measured three time. At the 6th day after exposed to electrostatic field, mean fresh weight of contrast group was 136.94mg, 200kv/m group--155.14mg, and 350kv/m group--149.11mg. Results of t-test showed that the diversity between experimental groups and contrast group were remarkable ($p<0.05$). To the 9th and 15th day, there were not remarkable diversity between the experimental and contrast groups. These results were conformal to the variations of seedlings length.

Measurement results of broad bean seedlings fresh weight were listed in Table 5.

Table 5. Fresh weight of broad bean seedlings (mean value, g)

Time Group	13th day	17th day	20th day
Contrast group	2.96	3.54	4.22
200kv/m group	3.46	4.74	5.14
350kv/m group	3.24	4.24	4.73

Results of t-test showed that the fresh weight of broad bean seedlings measured at the three times all had very remarkable diversity to the contrast group ($p<0.01$). Variations of fresh weight were conformal to the changes of seedlings length.

Dry weight of experimental and contrast groups hadn't remarkable diversity.

5. Mitosis metaphase cell index:

The rye root tips were taken at the 17th hrs after exposed to electrostatic field for measurement of metaphase cell index. 10 root tips were observed under microscope from each group, and 500 cells were observed and counted from each root tip. The metaphase cell index was listed in Table 6.

Table 6. Metaphase cell index of rye root tips after colchicine treatment

Group	Root tip No. 1 2 3 4 5 6 7 8 9 10	Sum	Metaphase cell index
Contrast group	6 12 7 14 6 11 6 8 9 8	87	1.74%
150kv/m group	11 12 5 8 13 12 6 13 9 13	102	2.04%
400kv/m group	17 16 23 12 22 14 10 13 23 16	66	3.32%

As seen from Table 6, the metaphase cell index of contrast group was the smallest one. The index of 150kv/m group was bigger than the contrast, but results of t-test showed that the diversity wasn't remarkable ($p>0.05$). The metaphase cell index of 400kv/m group was the greatest and made very remarkable diversity to the contrast group ($p<0.01$). These results were conformal to the variations of seedlings length.

Metaphase cell index of broad bean root tips were measured at the 3rd and 12th hrs after exposed to electrostatic field. Among which, at the former 10000 cells were observed and counted from each group, and at the latter 5000 cells from each group. Measurement results were listed in Table 7.

As seen from the Table, in the measurements at the two times metaphase cell index of 150kv/m group were all greater than the contrast, and the index of 400kv/m group were all smaller than the contrast. Results of statistical treatment showed that the diversities were remarkable ($p<0.05$). These results were also conformal to the variations

Table 7. Mitosis metaphase cell index of broad bean root tips after colchicine treatment

Group	3rd hrs			12th hrs		
	Cell Number Observed	Number of Metaphase Cell	Metaphase Cell Index	Cell Number Observed	Number of Metaphase Cell	Metaphase Cell Index
Contrast group	10000	355	3.55%	5000	282	5.64%
150 kv/m group	10000	491	4.91%	5000	353	7.06%
400 kv/m group	10000	289	2.89%	5000	219	4.38%

of seedlings length.

Discussion

As seen from above experiments, high-voltage electrostatic field could really influence the seedlings growth of plants such as rye, barley and broad bean, the influences were fairly complex.

In the first place, the effects of different intensity of electrostatic field were different. For rye, 400kv/m electrostatic field promoted its growth and mitosis, but 150kv/m field made little influence. To the broad bean, 150, 200 and 350kv/m electrostatic field all promoted its growth, but the effects were not identical, for example, the promotive effect of 350kv/m field was less remarkable than the 200kv/m group; further more, 400kv/m electrostatic field made inhibitory effects on the broad bean growth. As is known to all, effects of manifold environmental factors on living organisms are correlated to its intensity or dore: too weak (subthreshold) factors make little or no influence on organisms, a moderate intensity of factors can stimulate or promote living activations commonly, but supperstrong factors always make inhibitive or injurious effects on organisms. Results of above experiments showed that the electrostatic field made effects on plants growth according to these rules.

Secondly, the effects of the same intensity of electrostatic field on different plants were not quite the same. As indicated above, 150kv/m electrostatic field obviously promoted the growth of broad bean seedlings, but it took no obvious effect on rye; 400kv/m electrostatic field remarkably promoted the mitosis and growth of rye, but it inhibited those of broad bean. These suggested that the sensitivity and tolerance of different plants to electrostatic field are not quite the same.

The experimental results also showed, it seems that the same intensity of electrostatic field could take different effects on different organs of plants. On barley, 200 and 350kv/m electrostatic field promoted the growth of seedlings but inhibited the root growth. It suggested that the reaction to electrostatic field possess organ specificity. But it is related to species of plant, the same as above intensity of electrostatic field took promotive effects on both seedlings and roots growth of broad bean.

In the experiments, measured variations of seedlings growth were conformable to the variations of mitosis cycle. For example, 150kv/m electrostatic field took no obvious effect on rye root tips metaphase cell index, correspondingly, the seedlings growth rate also hadn't remarkable variation. 400kv/m field remarkably promoted the growth rate of rye seedlings, at the same time the metaphase cell index of experimental groups were also obviously higher than the contrast. Metaphase cell index is a confidence index of the revolve rate of mitosis cycle (the greater the metaphase index the shorter the mitosis cycle). Experimental results from broad bean also showed the same correlation between growth rate and mitosis cycle. It suggested that the electrostatic field influences the plants growth by taking effects on the mitosis cycle.

References

1. Ye Jia-ming, Gao Tian-shun, Zhou Hao-le and Zhang Jian-feng: Effects of high-voltage electrostatic field on chromosome aberration in root tip cells of rye (Secale cereale) and broad bean (Vicia faba). Journal of Northeast Normal University, 1985, No.1, 61--66. (in Chinese)
2. Ye Jia-ming and Chen Min: Effect of high-voltage electrostatic field on chlorophyll content of wild soybean. Proceedings of 1987 annual conference on electrostatics, ESC (CPS), 290--294. (in Chinese)
3. Ye Jia-ming and Wang Rong-yi: Effect of high-voltage electrostatic field on the amplitude of R-wave of the electrocardiogram from albino mice. Journal of Northeast Normal University, 1985, No.3, 75--84. (in Chinese)
4. Ye Jia-ming and Wang Rong-yi: Acute experimental results of effects of high-voltage electrostatic field on animal organisms. Wait for publishing.
5. Zhang Li-ping, Zhang Chang-zhong and Ye Jia-ming: Effects of high-voltage electrostatic field on starch amylase and peroxidase and its isoenzyme activity of sprouting process of barley seeds. Journal of Northeast Normal University, 1987, No.2, 41--45. (in Chinese)
6. Murr L.E.: Mechanism of plant-cell damage in an electrostatic field. Nature (Lond.), 201, 1305-1306, 1964.
7. Murr L.E.: Biophysics of plant growth in an electrostatic field. Nature (Lond.), 206, 467--470, 1965.
8. Sidaway G.H.: Influence of electrostatic field on seed germination. Nature (Lond.), 203, 303, 1966.
9. Sidaway G.H. and G.F.Aspray: Influence of electrostatic field on plant respiration. Int. J. Biometeor., 12, 321--329, 1968.

EXPERIMENTAL STUDY ON PROMOTION OF PLANT GROWTH BY ELECTROSTATIC FIELD

Yan Li Dai Xiyao Li Yiaoling Ma Ancheng Meng Xiancai Wang Qingzhao

Anshan Research and Design Institute of Electrostatic Technology, Liaoning, China

Abstract

Natural electrostatic field is an essential environmental-physical factor for plant growth. In cooperation with some other environmental-physical factors, it can lead plant body to normal functions. In a proper electrostatic field applied, the activating energy of plant can be raised and certain critical conditions of reactions can be reduced. It can also promote the plant growth, improve its reproduction characters, enhance its disease-resistability and increase its yield. Therefore, its practical value is evident.

Introduction

The atmospheric electrostatic field has been proved to be an indispensable factor for normal growth of plant. If the plant is covered with a grounded metallic network so as to shield the plant from atmospheric-electrostatic field and ion current, the plant can not grow normally. On the contrary, a suitable electrostatic field applied can promote growth of plant.

Research of effects of electrostatic field on plant has received upper attention. Many experts [1-7] have done experimental research into the effects of electrostatic field on physiology, biochemistry, reproduction characters and ecology of plants and have acquired achievements to some extent. In this paper, we focus on introducing partial result of experiments employing electrostatic field to promote growth of plant and on discussing its practical value.

General Principle

From ancient times, green plant lives in the great electrostatic field of the earth with positive voltage of about 30 kilovolts between the ionosphere and the ground. In the long history, green plant develops an adaptability to their circumstances and makes use of environmental factors to promote their developing genetic traits. All these provide people with premier conditions to utilize electrostatic field effects on plant growth.

Environmental-physical factors have many sorts. The effect on plant results from mutual comprehensive action of the factors. It will not be admitted to consider merely the electric field without other environmental factors matching with it. It is known that there are environmental-physical factors such as magnetic field, weight field, universal gravitational field and so on, which match mutually with electrostatic field and cause effect on plant simultaneously. Under the normal conditions of other environmental factors, influences of electrostatic field on various plants follow certain regularity. The general tendency is that there is no obvious effect on plant with weak field intensity, certain promotions present with the increase of field intensity (Fig. 1); but the field will cause inhibition, variation and deadly effect with the intensity further increasing.

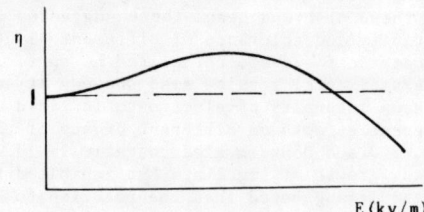

Fig. 1 The effect tendency of applied electrostatic field on plant

η --Effect of applied electrostatic field/control

Influence of electrostatic field on plant is complicated, and its effects may be roughly divided into the following three aspects:

1. Effect on botanical structure

The up-to-down natural electrostatic field establishes inside regular structures of plant body, such as arrangement of enzymes in the cell, overlapping state of chloroplast thylakoid, electric current moving along the external surface of the root in symmetrical manner, starch grain arrangement in the cell etc., which guarantee to sustain plant body in balance to perform various functions. For example, covered with a grounded thin metallic network, the plant is screened from electric field and ions so that the plant body is out of order in function to some extent: Its metabolism is imbalanced, absorption of mineral elements is reduced, proportion of accumulated elements is out of co-ordinate, chlorophyll content in leaves is reduced, disease-resistant and stress-resistant ability weakens, and growth slows down. In addition, as the downward-stretching root is set horizontally, the electric current moving symmetrically along the external surface of root becomes out of symmetry immediately and along the upper part of the root cap flows out from the top. It also flows along the lower part of the root cap, making the root to bend downward.

2. Supply of energy to cells

Electrostatic field energy can be accepted and employed by plant through orientation of its polar molecules, transportation of Ca^{24} and hormones, alteration of transmembrane electric potential, etc. This energy influences proton current and electron transmission of plant body, phosphorylation coupling with them, and growth and differentiation of plant cells. For instance, when a suitable electrostatic field is applied, cellular ATP content is raised significantly. Fig. 2 illustrates the alteration of ATP content in tomato plant with lengthening of time while the plant in an electrostatic field (field intensity is 80 kv/m).

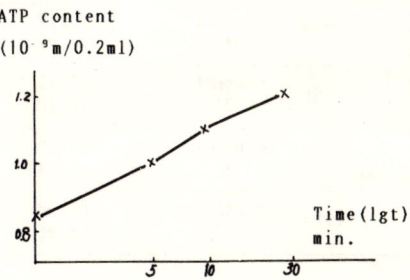

Fig. 2 Effect of electrostatic field on ATP content of tomato plant

3. Effect of reaction condition on plant

Electrostatic field has influences over electrolytic R-base of active sites of enzyme protein macromolecule. Change of enzyme activity influences the conditions required by plant physiological and biochemical reactions. In addition, it has effect on light compensation point of photosynthesis. If plant is placed in a suitable electrostatic field, it is found that there is still photosynthetic reaction even though light intensity is below the compensation point.

Methods and Results

In quest of optimum conditions of electrostatic field to promote plant growth, we have experimented on effects of multi-combination of field strengths and treating time on plants. By the analysis of laboratorial mechanism and field experiments, it is considered that the effect on the promotion of plant growth is obvious if the field intensity is in the range of 18-40 kv/m and the power supply maintains 4-6 hours per day. Partial results of experiments are shown as follow:

1. Effect of electrostatic field on growth of crop

The experiment was designed according to random combination, with repetition of three times and 10 plants sampled at random in each small zone. The investigation result is an average value of three repeats. It has been found that a properly applied electrostatic field can promote the crop growth (Table 1).

Table 1. Effect of electrostatic on the growth of crop

Investigation result / Crop treatment	Item	Height (cm)	Stem thickness (mm)	Leaves
Capsicum (Liaojiao No.1)	control	11.5	1.70	22.1
	Electric field	14.2	2.14	25.2
Cucumber (Jinyan No.2)	control	32.7	6.80	7.8
	Electric field	41.6	7.70	8.4
Corn (Shendan No.3)	control	38.5	15.9	8.1
	Electric field	42.1	17.7	8.7
Sorghum 1253	control	31.1	9.2	6.8
	Electric field	32.6	10.0	7.7

We also experiment upon the effect of time in electrostatic field on the growth of cucumbers. Thirty new pots were filled up with same type of soil and sowed in at the same time. The first group (10 pots) was not placed in the electrostatic field as the control, the second group was in the field for 10 days, and the third group was for 20 days (Table 2). We discovered that the growth situation in the field for 20 days was superior to that for 10 days which even excelled the first group not in the field (i.e. the control).

2. Effect on reproduction characters of crop

Growing in applied electrostatic field, the reproduction characters of plant are also affected. Corn (Shendan 3)

was taken as an example to demonstrate this point as shown in Table 3. Corn shoots developing in the electrostatic field exceed the control in morphological organs, also in plant height, stem stoutness, leaf-area coefficient, spike length and width. Their bald tip percentage of spike is lower than that of the control and the single yield was 124% of the control.

Table 2. Effect of time (days) in electric field on the cucumber developing situation

Investigation result\Item Treatment	Height (cm)	Stem thickness (mm)	Leaves
Control	15.2	6.3	4
In electric field for 10 days	17.3	6.7	4
In electric field for 20 days	19.3	6.7	4.4

Table 3. Effect of electrostatic field on reproduction characters of corn

Investigation result\treatment Item	Control	In electric field
Plant height (cm)	252.3	272.3
Stem thickness (mm)	2.23	2.37
Height of spike site (cm)	102.7	108.3
Leaf area coefficient	3.16	3.48
Spike length (cm)	18.45	19.06
Spike width (cm)	4.7	4.95
Lines of grain	15.8	15.3
Grains per line	36.7	41.3
Grains per spike	579.4	628.6
Percent of bald tip	11	8
Spike weight (g)	185.4	234.6
Grains weight per spike (g)	163.4	204
Grain weight percent (%)	87.5	87.11
Rachis thickness (cm)	2.8	2.9
Hundred grain weight (cm)	28.9	33.6

3. Effect of electrostatic field on crop yield

Partial data of field experiment in recent years are arrayed in Table 4. It may be seen that electrostatic field has effect of increasing production of crops to some extent. However, as factors influencing agricultural trial are quite complicated and subjected to various natural restrictions, the present consequence is only preliminary because in the process of our experiment between 1985 and 1986, heavy rain brought about disastrous flood and in rainy days, applied electrostatic field could not be supplied normally, which hindered regular experiment.

Table 4. Effect of electrostatic field on yield crop

Crop	Test time	Field strength (kv/m)	Average yield in small zone (kg)	Area (m^2)
Cucumber (Jinyan No.2)	1983	control 25	124.10 154.65	24
	1984	control 40	96.66 152.77	24
	1985	control 40	25.61 24.92	6
	1986	control 25	33.15 34.30	6
Capsicum (Liaojiao No.1)	1984	control 40	45.69 53.61	24
	1985	control 40	16.41 18.45	6
	1986	control 25	21.45 24.41	6
Tomato (Qiang-limishou)	1985	control 40	30.17 34.99	6
	1986	control 25	31.41 37.10	6
Corn (Shendan 3)	1983	control 25	15.20 19.24	12
	1984	control 25	12.58 13.96	12

Besides, the comprehensive laboratory of Liaoning Academy of Agricultural Science had analysed and examined the nutritional composition of fruit (cucumber, tomato) in two consecutive years. It is believed that there is no alteration in nutritional composition of crop growing in the electrostatic field.

4. Effect of electrostatic field on disease damage to crop

During the experiment, we found that crops in the electrostatic field zones had relatively lower incidence of disease than those in the control zone. The investigation data are shown in Table 5. The percentage of incidence of tomato Alternaria solani in the electric area is only 23.8% of the control. The percentage of incidence of

Capsicum mosaic virus is 14.6%. Both drop off evidently.

Table 5. Effect of electrostatic field on disease-resistant ability of crop

Investigation result / Treatment Item	Crop	Tomato (Qiangli-mishou)	Capsicum (Tongfeng 37)
Plant in small zone	control	102	105
	electric field	95	105
Plant with disease	control	9	41
	electric field	2	6
Incidence of disease (%)	control	8.82	39.04
	electric field	2.10	5.71
Type of disease damage	control	TAS	CMV
	electric field	TAS	CMV

5. Effect of electrostatic field on living percentage of grapevine cuttings

The experiment was contrived in contrast manner, with repetition of three times. On May 4, each zone has the same condition for cuttage and uniform number of Jufeng grapevine pieces planted. On September 26, we made investigation and saw that average percentage of living cuttings in the electric field areas was 56% superior to the control, and developing situation was better as well. At the same time, we investigated chlorophyll contents of leaves with a type HYL-1 chlorophyll meter, and learnt that the chlorophyll content in the electric field was 47% higher than the control. All these indicate that electrostatic field boasts investigating impact on both percentage living pieces and level of their metabolism.

Discussion

Electrostatic field has a series of effects on plant growth. However, scientific approach to the systematical regularity is a quite complex and massive task, and analysis of mechanism and culture experiment to be done simultaneously are needed. It is a fascinating project to make utilization of electrostatic field to promote the growth of plant and increase its yield. Nevertheless, since the experiment is affected by multi-objective factors, the formation of integral technique is not an easy work in any case.

Since electrostatic field accelerates plant growth, the physical parameters needed to define include electrostatic field strength suitable for the plant growth in various botanical development stages, the length of time to supply electricity in these stages, and the manner of supply, and so on. It has been learnt that the effect of electrostatic field on the growth of hydroponic crop and the increase of yield is comparatively stable. This is a result of fewer random disturbing factors and more controllable ones under hydroponic conditions.

On this account, this technique being applied to crop-culture in protective land, stable and prominent benefit may be achieved.

References

(1) 刘福全，1983；中国物理学会静电专业委员会学术交流会资料.
(2) 王秀文，1983；同上
(3) 白希尧，1984；静电气学会志 5,339—343 东京
(4) 白希尧等，1984；自然杂志 12,902—
(5) 马安成等，1985；辽宁省生物物理学术交流会资料.
(6) 叶家明等，1985；东北师范大学学报（自然科学版）1,61—66.
(7) 叶家明等，1985；首届全国环境生物物理学术会议论文摘要汇编.
(8) 贝时璋，1984；大自然探索 4,13—15.

EFFECT OF STATIC ELECTRICITY ON INITIAL GROWTH OF PANAX GINSENG*

Huang Shuzhen
  Changchun Normal College,P.R.C.
Yan Jixiang
  Jilin Agriculture University,P.R.C.
Xu Shaozeng
  Jilin University of Technology,P.R.C.

Abstract

Relationship between electrostatic field and ginseng growth was obtained by using electrostatic technique. Treatment of ginseng seeds with electrostatic field may shorten their duration of morphological after-ripening, increase probability of appearing multi-stem ginsengs. Positive effects of appropriate electrostatic field on initial growth and development of ginseng are confirmed.

1. Introduction

The technique of static electricity accelerating plant growth, sometimes called " electricity culture ", refers to promoting plant by static field or air ion flow. This technique may notably increase the plant yield and speed up the growth of plant. Since Giambatissta et al. suggested that ion flow may have effect on the growth of plant and carried out some experiments in 1775,many researchers have done experiments on Vicia faba,Pelargonium hortorum,Hordeum spp.,Triticum spp.,Zea mays, Avena sative,Paddy rice,and vegetable,etc.,obtaining some positive effects on accelerating growth and increasing yield of plants(1). In 1985, Wang Rongyi et al. performed an experiment on perennial plant Clivia miniata, almost doubled its growth speed.

Panax ginseng is a famous medicinal plant. Chemical analysis and pharmacological experiment demonstrated that Panax ginseng contains panaxin, panacene, carbohydrate panaxic acid and vitamin etc. (2,3). It has special tonic effect on human body,and is used in curing many diseases.In addition,it is effective in recovery of sportsmen from exhaustion.

But Panax ginseng grows very slowly. The seed of ginseng has the characteristics of long-term dormancy. The naturally ripened seed of ginseng is full of endosperm. Though the embryo forms cotyledon, and protoradicle, they are too small; it doesn't develop completely, the sowed seed doesn't sprout untill 18 months has passed.The developing process of an embyro from unsoundness to soundness is called morphological after-ripening,and artifically promoting the after-ripening of ginseng seed needs 90-120 days. After this process a seed still requires a period of low temperature before sprouting. This process is called physiological after-ripening. In a word, developing a seedling of ginseng from its seed needs two periods:morphological after-ripening and physiological after-ripening. Thus, no matter how early you sow ginseng seeds at the beginning of a year, you won't get their seedlings in the same year. Practice has proved that ginseng seed treated by hormone promoted its after-ripening. In this paper , the effects of ginseng seed treatment by static electricity on the after-ripening and the initial growth of seedling are studied.

2.Effect of static electricity on after-ripening of Panax ginseng seed

The Panax ginseng seed for experiments were provided by the medicinal plant laboratory of jilin Agriculture University. They were domesticated and naturally ripened, but didn't complete morphological and physiological after-ripening. We treated the seeds with static electricity, then notice its effect on seed after-ripening and probability of forming multi-stem ginseng.

Procedure:Experiments were carried out at room temperature,the light entering the room was adjusted by a screen window.After screening,rejecting the weak, injured and small ones from 5400 grain of seeds,we chose 3900 grains which were nearly in the same size. Random divided them into 2 groups of A and B, then subdivided each group into 5 equal parts.Every part with 390 grains was put into a plastic dish with 4 layers of gauze on its bottom, then covered them with 6 layers of gauze. Kept moist of all gauzes throughout the experiment to facilitate the seed sprouting, treated each dish with 5 different

-148-

intensities of high-voltage electrostatic fields.

For the group A a uniform electrical field was applied. Fixed a grounded plate electrode on the bottom of the plastic dish, put the aguze with evenly placed seeds on the grounded electrode. Another plate electrode with high voltage set over the dish horizontally. Supplied with electricity 3.5hrs. a day. Different intensities of electrical fields were obtained by different distance between electrodes. Covered the seeds with 6 layers of moist gauze when not being charged.

For the group B an ununiform electrical field was applied which was created by a line high-voltage electrode over the plastic dish and the same grounded plate electrode in the plastic dish. Chose 1.5mm copper wire as high-voltage electrode to avoid the possible disturbance of ions and gases(for example, ozone) made by discharge. The perion of charging was equal to that of group A.

The 2 groups were charged by the same high-voltage power supplier made by the Static Electricity Institute of Northeast Normal University.

Experimental results: The experiment was set about on Aug.23th,1984. A few seeds were found to begin splitting on Oct. 20th, till Nov. 28th the number of splitting was kept constant essentially. This period lasted 88 days. Under natural conditions this process needs 180 days. Even use controlled temperature and treatment with hormone it also needs 90-120 days. But by means of static electricity technique the shortest time for after-ripening is only 59 days, the longest is 88 days.

The experimental data for the 2 groups are listed in Tab.1.

Tab.1

Field intensity	split number Group A	split number Group B	unsplit number Group A	unsplit number Group B	split (%) Group A	split (%) Group B
2.08Kv/cm	280	198	110	192	72%	57%
6.20Kv/cm	195	175	195	215	50%	45%

From Tab.1 it can be seen that the high-voltage electrostatic field promotes after-ripening of seed and field intensity ranging from 2 Kv/cm has nearly same notable effects on after-ripening of seed. After morphological after-ripening the splitted seeds went on being treated with electrostatic field, trying to accelerate their physiological after-ripening.

For this purpose the seeds were put into the uniform electrostatic field with intensity 6.25Kv/cm on Nov. 18th. A grain of seed began to sprout on Dec. 18th, till Dec. 27th, the rate of sprouting was nearly constant. This period lasted 40 days. This experiment demonstrated that the whole after-ripening of Panax ginseng seed treated with electrostatic field only needs 120 days, which is 50-60 days less than that of ordinary treatment.

On Nov. 27, we chose 10 grains of sprouted seed at random from each of 5 plastic dishes and sowed them in 5 flower basins. The first ginseng seedling appeared at Jan. 3rd,1985, then the seedlings were grown in uniform electrostatic field. Chose 5 seedling at random from each basin 50 days later, counted up the numbers of single-stem, double-stem and multi-stem seedlings, the results are shown in Tab.2.

Tab. 2

stem number	single-stem	double-stem	quadri-stem
seedling number	19	5	1
percentage	76%	20%	4%

It can be seen from Tab.2, the Panax ginseng seeds can sprout by treatment with electrostatic field without low temperature process. In addition, the ratio of multi-stem ginseng increases from some thousandths under ordinary condition to 15-25%. Therefore, electrostatic treatment effectively accelerates the initial growth and developmemt of Panax ginseng. Developing a quadri-stim ginseng seedling from seed directly is unprecedented(see Fig.1)

Fig.1 Photo of quadri-stem panax ginseng

3. Effect of electrostatic treatment of seed on speed of germination

Chose experimental Panax ginseng seeds of Jilin Agriculture University which had completed normal after-ripening and divided them into groups at random, put into different intensity electrostatic fields for relatively short time (less than 24 hrs). The applied voltages of each field are 0, 20, 60, and 80 Kv. Then sowed the seeds into different basins according to different treating voltages. Ginseng soil was supplied by Jilin Agriculture University. Management was normal. Took down the sprout date of each ginseng plant in each basin, the results are shown in Fig.2

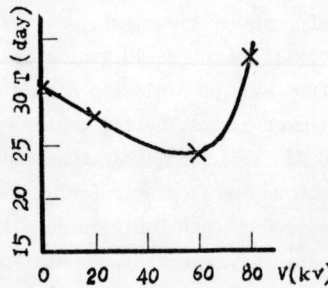

Fig.2 Sprout duration of Panax ginseng plant versus externally applied voltage

Fig.2 showed that under 60 Kv condition it took in average 24 days (from March 7th,1984 to March 30th,1984) for seeds to sprout from sowing whereas the counterpart control needs 30 days. Seeds treated with electrostatic field germinated 6 days earlier than non-treated seeds,i.e., the former needed 20% less time for germination.

4. Effect of electrostatic treatment of ginseng on its root mass

The ginseng seedlings which were cultivated in the above mentioned basins were treated with different intensity fields of 0, 0.39, 0.45, 0.60, and 0.75 Kv/cm according to the sequence of field intensity with which the seeds were treated. Thick wire was used as high-voltage electrode and short distance between electrodes was adopted to obtain powerful ununiform electrical field without corona discharging and resulted ions and ozone. The system was charged 3.5 hrs every day. We went on experimenting untill the ginseng plants withered and didn't grow any more. Then the relationship between field intensith and ginseng root mass was drawn and is shown in Fig.3.

Fig.3 indicates that the root mass of Panax ginseng which is treated with suitable negative electrical field is 27% more than that of the counterpart(controls).

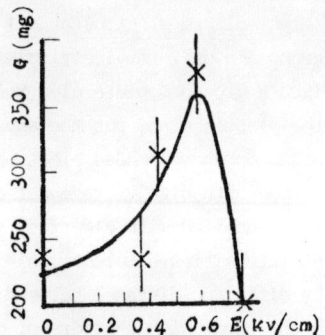

Fig.3 Panax ginseng root mass versus field intensity

4. Summary

One grain of Panax ginseng seed ususlly only grows into one single-stem seedling. If ginseng seeds are treated with gibberellin, a few of them may grow into double-stem seedlings. However, if they are treated with high-voltage electrical field, their after-ripening period may be shortened, their dormancy may be broken, the double-stem rate increases.especially, quadri-stem ginseng appears,and the root mass increases. Therefore,static electricity technique provides new means of cultivating multi-stem ginsengs and raising the yield of ginseng plant.

*The Project supported by Foundation of Chinese Academy of Sciences.

References

1. H.A.Paul and G.W.Todd,"Electroculture for Crop Enhancement by Air Anions", International Journal of Biometeorology, 1980.
2. Li Yinqin et al.,"Study on Polysaccharide of Ginseng Stems and Leaves", Transactions of Materia Medica, 19(10),764,1984.
3. Panax Ginseng, Jilin People's Press,pp.8,1985.

THE INFLUENCE OF HV ELECTROSTATIC FIELD ON THE ROOT SYSTEM OF PLANTS AND ABSORBING NUTRITIVE ELEMENTS

Wang Rongyi Wang Xiuwen
(Dalian University of Technology)

ABSTRACT

The experiments have shown that HV electrostatic field has a specific influence on the growth of plants. The result has shown that the growing rate of plant root system treated with positive HV electrostatic field is faster than that of contrast groups, and the growth of those treated with negative HV electrostatic field is slower than the contrasts. The result has also shown that HV electrostatic field has some influences on absorbing nutritive elements of the plants.

I. The Influence of HV Electrostatic Field on Root System of Plants

Material and Method

The materials used in these experiments were Allium sativum, Allium fistulosum and Cliva miniata. Allium sativum and Allium fistulosum were cultivated with water. Set up the thin copper mesh on the mouths of beaker filling with tap water. Put bulbs of Allium Sativum and Allium fistulosum on the mesh. Six groups were used in each experiment, Two groups were used as contrast, two were treated with positive HV ES field and two were treated with nagative HV ES field. The method of applying ES field was as follows, The low voltage terminal of HV DC Power Supply was grounded and connected with the coppernet on the mouths of the beakers, and high voltage terminal of the power supply was connected with a thin mesh electrode which was set up 15-20cm over the plants. Both the applied positive and negative HV were 11-22 KV.

When folwerpots cultivated cliva minita was treated with ES field, the metal electrode putting on the bottom of the flowerpot was grounded and another electrode was netted with thin mesh which was set up 7- 50 cm over the plants. The applied HV was 14-23 KV.

Results and Discussion

1. The influence of HV ES field on the growth of Allium sativum root system

The experiment was carried out respectively in Jan. 9 and Feb. 25. The growth of root system was observed at regular intervals. The result is as table 1.

Table 1. The Influence of HVES Field on the Growth of Garlic Bolt Root System

Planted on Jan. 9, 1984

Groups	Jan. 18 Roots Average Length	Jan. 18 Chromaticity	Jan. 21 Roots Average Length	Jan. 21 Chromaticity	Feb. 10 Roots Average Length	Feb. 10 Chromaticity
Contrast Groups	3.9 Cm	Plate Yellow	7.9 Cm	Plate Yellow	12.1 Cm	Plate Yellow
Negative HV Groups	3.7 Cm	Rust Yellow	7.1 Cm	Rust Yellow	11.7 Cm	Rust Yellow
Positive Hv Groups	4.0 Cm	Bright White	9.5 Cm	Bright White	14.6 Cm	Bright White

Planted on Feb. 25, 1984

March 5 Roots Average Length	Chromaticity	March 18 Roots Average Length	Chromaticity	March 28 Roots Average Length	Chromaticity
6.2 Cm	Slight Yellow	9.7 Cm	Slight Yellow	13.0 Cm	Slight Yellow
6.0 Cm	Rust Yellow	9.1 Cm	Rust Yellow	12.5 Cm	Rust Yellow
6.8 Cm	Bright White	12.1 Cm	Bright White	14.1 Cm	Bright White

From table 1. We can find that both positive and negative HV ES fields have specific influence on growth of Allium sativum root system, The Allium sativum treated with negative HV ES Field has the roots with different length, and average length is slightly shorter than that of contrast groups, the surface of the roots is short of lustre, coloured yellow, and root tips brown; But the Allium sativum treated with positive HV ES field has the roots with uniform length, and average length is longer than that of contrast groups, its roots are white and lustrous. This result indicates that positive HV ES field promotes the growth of Allium sativum root system, but negative HV ES field has slightly inhibitory effect.

2. The influence of HV ES field on Allium fistulosum root system

The experiment method was the same as that of the Allium sativum. The result is as table 2.

In table 2, we can find that the influence of HV ES

field on Allium fistulosum root system coincides with the experimental result of the Allium Sativum. The positive HV ES field has some promotive effects in view of the average length uniformity and lustre of the roots. But negative HV ES field has some inhibitory effects.

3. The influence of HV ES field on the growth of Clivia miniata root system

Table 2. The Influence HV ES Field on the Growth of Allium Fistulosum Root System

Planted on Fed. 25, 1984

Groups	March 18		March 28		Apr. 5	
	Roots Average Length	Chromaticity	Roots Average Length	Chromaticity	Roots Average Length	Chromaticity
Contrast Groups	5.8Cm	Slight Yellow	7.3Cm	Slight Yellow	8.5Cm	Slight Yellow
The Group of Negative H-V	5.3	Rust Yellow	7.3	Rust Yellow	7.9	Rust Yellow
The Group of Positive H-V	6.2	Bright White	9.1	Bright White	11.8	Bright White

In the experiment, the Clivia miniata consists of two groups, The first eight of them were plants over 2 years, 2 were treated with negative HV ES field over a long period of time, 2 were treatd with positive HV ES field over a long period of time, and 4 were contrasts; The second was the seedling over 3 months, 10 pots, each with 10 plants, 2 pots were treated with positive HV ES field, 6 were treated with negative, and 2 were contrasts. The experimental result is shown in table 3 and table 4.

We can find that both the grown plant and the seedlings treated with positive HV ES field of the Clivia miniata root system is more luxuriant than that of contrast groups, The root system is thick and uniform, colouring milky white, with more roots and longer average length, and has luxuriant root hairs. The effect of negative HV groups is worse than contrast groups. Therefore, in our experimental condition, the positive HV ES field has a promotive effects on the root system of plants, and negative HV ES field has a slightly inhibitory effects.

Table 3. The Influence of HV ES field on Clivia miniata Root System

Average Root Number of Plants

	over 1 year	over 2 years
Contrast Group	13	27
Negative HV Group	15	26
Positive HV Group	18	34

Table 4. The Influence of HV ES Field on the Root System of Clivia Miniata Seedlings

Groups	Everage length of Roots (Cm)
Contrast Group	9.87
Negative HV Group	9.48
Positive HV Group	11.49

II. The influence of HV ES field on the absorbing nutritive elements of plants

Materials and Method

Put the Allium sativum in 6 different sorts of nutritive solution and HV ES field was applied. The basic or common nutritive elements were 252 ml water with KH2PO4 of 0.68 g (P, 0.155g, K, 0.195g). The 6 sets of nutritive solution were as follows,

1. $KH_2PO_4+MgSO_4$ 2. $KH_2PO_4+MnSO_4$ 3. $KH_2PO_4+CuSO_4$
4. $KH_2PO_4+FeSO_4$ 5. $KH_2PO_4+ZnSO_4$ 6. H_2O (Distilled Water)

Each set was divide into 3 groups which were put in glass beakers, and the liquid in the beakers was grounded. Apply HV field of 14-30KV with copper wire which was set up over the plants, the voltage and distance were adjusted to field strength of 1KV/cm. The 3 groups were positive HV group, negative HV group and control group. We energized continually 29 days from March 24 to Apr. 21. (with some intermissions), performing water culture experiments, measuring the pH value, detecting the relative content of different elements in the solution and observed the colour change of the solution.

Result

After applying HV ES field 29 days, remainder elements in the containers were measured. The result is as Table 5 and Table 6.

Table 5 The condition of Absorbing Nutritive Elements

Element composition	Relative Content		
	Positive HV	Contrast	Negative HV
Mn	0.95	6.0	3.2
Cu	2.7	2.1	2.5
Zn	1.94	1.4	0.55

Table 6 The pH Value of Solution

pH Value of Element	Before Experiment	The pH value of solution after 29 Days		
		positive HV	contrast	negative HV
KH2PO4+MnSO4	4.4	6.4	5.4	5.5
KH2PO4+CuSO4	3.8	5.4	5.4	5.8
KH2PO4+FeSO4	3.9	6.0	5.2	5.4
KH2PO4+ZnSO4	4.2	5.0	6.2	5.4

The colour change of solution is as table 7

Table 7.

Elements	Applying Positive H-V				
	March 24	Apr. 1	Apr. 7	Apr. 12	Apr. 21
KHPO4+MgSO4	Colourless	Milky White Bright	Milky white Bright	Slight yellow	Slight yellow
KH2PO4+MnSO4	Colourless	Milky white	Rusty yellow	Dark Rust yellow	Dark Rust yellow
KH2PO4+CuSO4	Pale Blue	Pale BlueGreen Transparent	Pale BlueGreen Transarent	Pale BlueGreen Transarent	Blue Dark Green
KH2PO4+FeSO4	Pale Green	Milky white-Bright Transparent	Milky white Transparent	Rust yellow	Rust yellow
KH2PO4+ZnSO4	white Little Precipi- tation	Milky white Transparent	Milky white Transparent	Milky white	Milky yellow
H2O	Colourless	Rust yellow	Dark Rust yellow	Clear White	Rust yellow Transparent

Contrast

Apr. 1	Apr. 7	Apr. 12	Apr. 21
Milky White	Rust yellow	Rust	Milky yellow
Rust yellow	Slight yellow Milky White	Rust yellow	Glossy yellow
Pale Blue- Green	Pale BlueGreen	Pale Blue- Green	Blue
Milky White (Slight yellow)	Milky White (Slight yellow)	Yellow- Brown	Black- Brown
Slight yellow	Yellow Turbid	Yellow	Milky yellow
Slight yellow	Slight Yellow	Slight yellow	Rust yellow

Applying Negative H-V

Apr. 1	Apr. 7	Apr. 12	Apr. 21
Rust yellow	Black Rust yellow	Rust yellow	Milky yellow
Clear Bright	White Bright	Slight yellow	Slight yellow
Blue- Green	Blue- Green	Blue- Green	Blue- Green (Pale)
Milky white	Milky white	Milky white Turbid	Inky
Milky white	Milky white	Milky white	Milky yellow
Milky white	Milky white	Pale yellow	Pale yellow

Comparatively unitary nutritive elements were added in the distilled water in order to analyse conveniently. The growth of the plants were depended on the nutritive elements stored by themself, because of the concentration of solution was higher, the growth of the plants was not good enough. If the growth were better, the difference of absorbing various elements would be more obvious otherwise. As to the growth of the root syetem, those applied with HV ES field were better and colored white ; the grouth of those applied with negative HV ES field were poor and colouring dark. The contrast group is the middle.

The results have shown that applying HV ES field has some influences on absorbing nutritive elements of the plants, and apply different electric field has different influence on various elements. It also has a specific influence on the pH value of nutritive solution, but the influence isn't obvious.

Observing the colour of the solution has shown that different polarity of HV ES field has extremely different effect on the colour of solution.

III. Result and Discussion

In this paper, we can bring about the following two results,

1. HV ES field can effect the growth of root system of plants. Applying positive HV ES field can promote the growth of root system of plants, and applying negative HV ES field has some inhibitory effects on growth of root system.

2. Applying HV ES field has some influences on absorbing nutritive elements of the plants. Different electric field has different effect on various elements.

These two results indicate the following new prospects,

1. Applying positive HV field with proper amount can promote the growth of roots of such plants as ginseng and carrot and increase their output.

2. Using HV ES field to control the content of trace elements and the nutritive composition of plants. It would improve the nutrition and savour of the vegetabe and fruit, it can control the content of effective composition of medicinal herbs.

3. The colour change of solution in above experiments is an interesting and unexpected finding. Applying positive or negative HV ES field changes the colour of nutritive solution of water-cultured plants.

It is obviously different that applying HV ES field with different polarity would affect the chemical reaction and material exchange between plants and solution. This evidence provides new possibility and approach of using ES technic to affect the chemical reaction.

References

1. H. A. Pohl and G. W. Todd, Electroculture for crop enhancement by air anions. Int. J. Blometeor. (Reprint from 1980)

2. 岛山英雄，野外植物の生体电位.静电気学会誌 1982; 6(5),276-284

BIOLOGICAL POTENTIAL IN PLANT AND EXTERNAL ELECTROSTATIC FIELD*

Xu Shaozeng
Jilin University of Technology, P.R.C.

Huang Shuzhen
Changchun Normal College, P.R.C.

Dong Zhanhua
Jilin University of Technology, P.R.C.

Abstract

In the paper authors analyse the curve of biological potential of green bean seeding hypocotyl with variation of external electrostatic field which appears not agree with physical law and conclude that an adjusting funtion inside biological body is the main impetus to promote plant growth. Therefore, mutual contradicted experimental phenomena of biological effects caused by static electricity which have been observed so far may be explained in a unified way.

1. Introduction

The effects of static electricity on plant growth have been widely investigated experimentally at home and abroad since the end of 18th century. Many experiments showed both positive and negative electrostatic fields and both cation and anion flows promote plant growth under appropriate conditions. Like any other external factors, static electricity also has its corresponding doses for promoting, inhibiting and destroying the plant growth(1-4). The purposes of the work is to search a representative physical parameter and its correlation with threshold values concerned, and by means of research on the parameter to obtain a procedure to determine the appropriate dose for promoting plant growth.

2. Experimental investigation on biological of plant

2.1 Preparation of samples

We chose uniform and plump grains of green bean seeds and grew them in a flower basin at room temperature(25+5°C). 6-8 days after seeds sprouted, cut the whole sprout hypocotyl to be used in experiments.

2.2 Selection of electrode

We used very fine metal wire(ϕ0.4 mm) as electrode to insert into the sample to measure the electric potential. In order to reduce the error caused by chemical electric potential occured between the electrode and sample, the time of insertion was kept as short as possible. many experiments showed, repeatability and stability of the wire electrode are superior to other types of electrode such as siliver-mercury electrode, etc.

2.3 Obtainment of electrostatic field

In order to obtain a uniform electrostatic field, we used two pieces of parallel metal plates as electrodes with large area of electrodes and small distance between them. The sample is put on an insulated stand in the center area of electrode system. The measuring apparatus is insulated against the ground and the probe wire is led into the electrostatic field along the iso-potential plane.

2.4 Experiments on relationship between biological potential and external electrostatic field

The biological potential of sample varies with external electrostatic field and gradually recovers after removing the field. The shape of relation curve between them is similar to that of magnetic hysteresis loops. Therefore, the sections of the curve at field intensity $E>0$ and $E<0$ have to be measured by using two sets of samples separately. Experimentally obtained electrification and restoration characteristics of biological potential are shown in Fig.1-a and 1-b respectively.

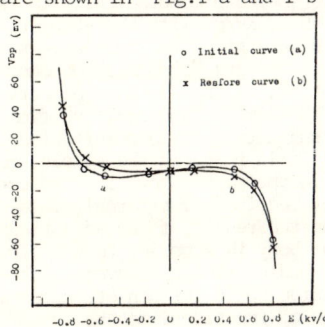

Fig.1 The apparent biological potential of green bean hypocotyl versus external electrostatic field

3. Results and discussions

3.1 Concept of apparent biological potential

Without external electrostatic field, the measured average value of biological potential of the sample cotyledon relative to the root is

$$V_{bpo} = -8.64 \pm 0.54 \text{ mV}$$

Under external field the biological potential V_{bp} consists of three parts. One is initial biological potential V_{bp1} which is produced by the adjustment inside the V_{bp2} biological body of the sample under the action of external field. The action of V_{bp2} consists in overcoming the change caused by external field, the directions of V_{bp1} and V_{bp2} are contrary.

So, the potential value measured can be expressed as

$$V_{bp} = -|V_{bpo}| - |V_{bp1}| + |V_{bp2}|$$

which may be defined as apparent biological potential.

From Fig.1 it can be seen that the electrification

characteristic of the sample biological potential is different from the recovery characteristic. This difference implies that the plant organism is more or less injured (or fatigued) under the action of over electrification, it needs some time and certain conditions to recover. The coincidence of the asymptoties of two curves indicates that the additional potential V_{bp2} produced by the self-adjustment in the biological body may be neglected when the value of external field increases to a certain level. Then V_{bp1} plays a major role in V_{bp}.

In order to analyse in details the negative electrical field part of the curve a, this part is reproduced with enlarged ordinate scale in Fig.2.

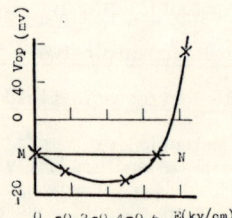

Fig.2 Negative electrical field part of biological potential

According to physical law the potential of the sample should increase monotonically in the interval of 0 to -0.65 Kv/cm with the absolute value of negative field intensity, but in fact, as we can see from Fig.2, it appears "abnormal" descent in the curve segment MN. In the E>0 part of curve a, we can also see the similar "abnormality" from Fig.1.

This fact suggests a certain sort of action of nonelectrical field force in the sample, making ions or dipoles in the sample move in the opposite direction to that required by external field. This is supposed to be the result of the adjusting action in the biological body is superior to the others.

Biological bodies manage to overcome the influence from outside world through strengthening the normal action of functional metabolism of organic tissues and organs, this segment corresponds to the range of electrostatic field intensity promoting plant growth. Therefore, physical and chemical factors such as positive and negative electrical field, alternating electrical field, and cation and anion flow below a certain threshold value may stimulate the self-adjustment of plant body to promote their growth.

With the increasing of absolute value of external electrical field, self-adjustment appears weaker and weaker, but the adjusting function still exists and compensates, postpones and resists the change caused by external electrical field. Plant yield may not increase although the intensity of electrical field is relatively strong now. The segment NP of the curve is the transition segment from promoting growth to inhibiting. So it is not true that the stronger the field intensity is, the greater the effect promoting growth of plant.

Above point P, the adjusting ability in plant becomes negligible in comparison with the action of externally applied electrical field, the change of the apparent biological potential in sample is in accordance with the physical law: V_{bp} increases linearly with the increase of the absolute value of electrical corresponds to inhibiting and deathful stage. Our results are in good agreement with those obtained by Murr et al.(2,3).

3.2 Biological potential and total mass

The biological potential of plant and its total mass change under the action of electrical field. Let us analyse these changes and then search for the relation between biological potential and total mass. The relations reported by L.E.Murr between the percentage of the mass increment of the frond of yellow bush beans and the intensity of alternative electrical field is shown in Fig.3(5).

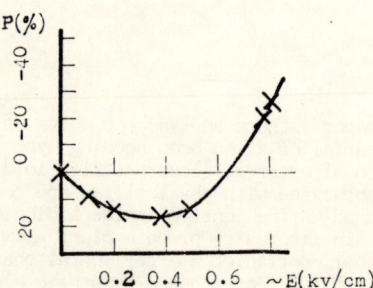

Fig.3 Response curve for frond of yellow bush bean in an electrostatic field

In Fig.3, E>0 and E<0 parts of the curve a are symmetrical about the zero field point. Only E>0 part is reproduced in Fig.4 for convenience. Experiments have confirmed that positive and negative electrical fields have the similar effect for the plant growth, so we can compare the curves in Fig.3 and 4. In Fig.4 the change of biological potential is used as ordinates, but the

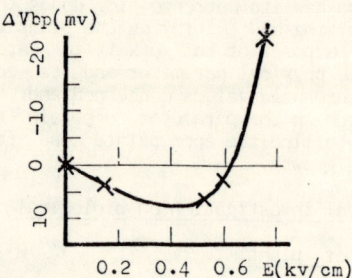

Fig.4 Response curve for phaseolus radiatus in an electrostatic field

percentage of the frond mass increment is used as ordinates in Fig.3. These two ordinates may be compared each other.

The curve in Fig.3 indicates that the mass increment of frond of yellow bush beans gradually increases with the increase of the intensity E of alternative field and reaches a maximum at E= 0.35 Kv/cm, then gradually reduces with the increase of E and drops to 0 at E≈0.6 Kv/cm. Similarly in Fig.4, the maximum of biological potential increment corresponds to E= 0.465 Kv/cm, and drops to 0 at E≈0.6 Kv/cm. In short, curves in Fig.3 and 4 have similar form with almost the same characterized values. The experiments performed by authors with the araliad perennial panax ginseng led up to similar results. In summary, our investigations on the biological potential of plant samples provided clues of searching for the optimum for promoting plant growth range of electrostatic field intensity.

*The project supported by Foundation of Chinese Academy of Sciences.

Acknowledgement

The advices and encouragements from professor Liu Xun jun of Jilin University of Technology are gratefully acknowledged.

References

1. H.A. Paul and G.W. Todd, "Electroculture for Crop Enchancement by Air Anions", International Journal of Biometeorology, 1980.
2. Huang Shuzhen, "Effect of High-Voltage Electrostatic Field on Growth of Green Bean Seedlings", Journal of Changchun Normal College, Natural Science Edition, No.1.,pp.82-84,1985
3. L.E. Murr, "Plant Growth Response in a Simulated Electric Field Environment", Nature,Vol. 200,No. 4905,pp.490-491,1963.
4. Eiyu Shimayama,"Biological Potential of Wild Plants", Journal of Society of Static Electricity, 6,5, pp.276-284,1982.
5. L.E. Murr,"Plant Growth Response in Electrostatic Field", Nature, Vol.207, No.5002, pp. 1177-1178,1965.

SEPARATION AND BIOSTIMULATION OF SOYBEANS USING HIGH-INTENSITY ELECTRIC FIELDS

R. Morar, Al. Iuga, L. Dascalescu, V. Neamtu
Polytechnical Institute of Cluj-Napoca, str. E, Isac 15, Romania 3400
I. Munteanu, Agricultural Research Center, Turda-Cluj, Romania

Introduction

Agricultural applications of electrostatics include seed cleaning [1,2,3,4,5] and biostimulation.[6,7] ELSEP laboratory electroseparator (Fig.1), conceived by the authors, has been used to study these processes.

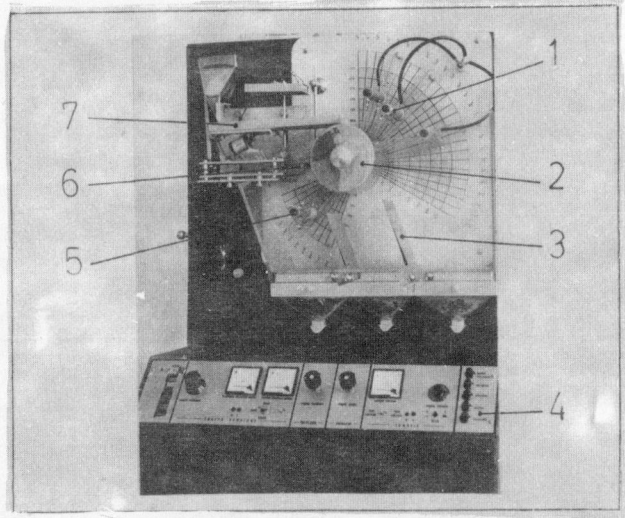

Fig.1. ELSEP laboratory separator; 1-active electrodes; 2-earthed electrode; 3-splitter; 4-control panel; 5-wiper electrode; 6-brush; 7-vibratory feeder.

The preliminary results obtained with soybeans are presented in this paper.

Ion bombardment

The most common form of electrostatic separation is based on charging the particles by ion bombardment (Fig.2).

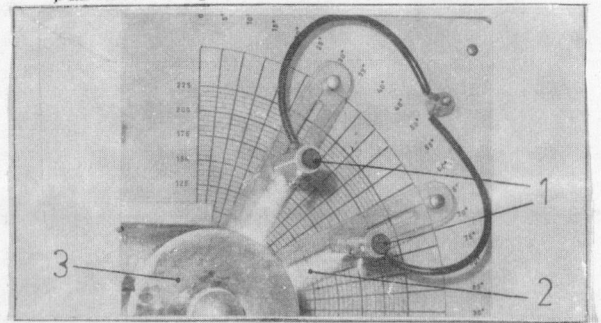

Fig.2. Ion bombardment; 1-wire-type corona elctrodes; 2-corona discharge; 3-earthed rotating roll-electrode.

The granular material (seeds and impurities) is fed onto an earthed rotating roll and charged by the corona electrodes. The conducting particles loose their charge to the grounded surface and are thrown free from the roll by the centrifugal force. The charged non-conducting seeds are electrostatically pinned to, and move with the surface of the roll.

The conductivity of the seeds depends on their moisture content. Therefore, dead shrivelled seeds, which are characterized by a lower moisture content, act like poor conductors, and can be separated from the normal ones.

Shape separation

The electroseparability characteristics of materials depend not only on their conductivity, but also on the shape of the particles. The faltness coefficient, K, is determined by the ratio of particle length L to thickness T (Fig. 3a). F.S. Knoll and J.B. Taylor[8] stated that effective separations can be carried out if the ratio of flatness coefficents K_S/K_B is greater than 2.

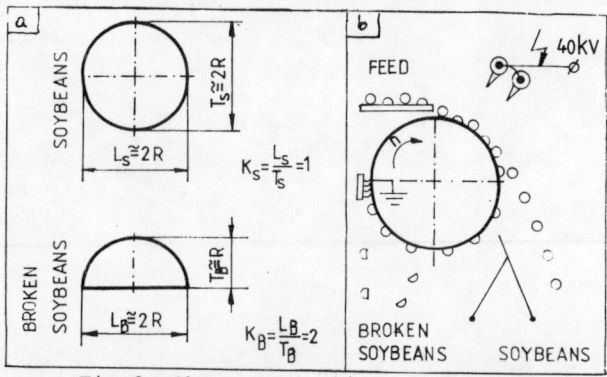

Fig.3. Shape separation; a-flatness coefficient K calculation; b-principle of operation; K_S, K_B-soybeans and broken soybeans flatness coefficients, respectively.

Broken soybeans are characterized by a flatness coefficient $K_B = 2$. With regular soybeans, the value of the same coefficient is $K_S = 1$. It can be concluded that removing of broken soybeans is achievable by electroseparation.

A simplified physical model has been used to evaluate the electric image forces

F_i which pin the particles on the surface of the roll.

Damaged soybeans (Fig.4b) are almost seven-fold more attracted to the earthed electrode than the good ones (Fig.4a):

$$F_i = Q^2 : \{4\pi\varepsilon_0 [2(R+\tfrac{2}{\pi}R)]^2\} \quad (1)$$

$$F_{iB} = Q^2 : [4\pi\varepsilon_0 (2\times\tfrac{2}{\pi}R)^2] \quad (2)$$

$$\frac{F_{iB}}{F_i} = \frac{[1+2/\pi]^2}{(2/\pi)^2} = \frac{(\pi+2)^2}{4} = 6.6 \quad (3)$$

index B being used for broken soybeans.

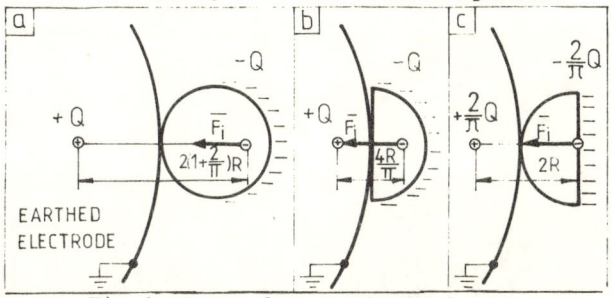

Fig.4. Image force calculation:
a-soybeans; b,c-broken soybeans.

The position of non-spherical fragments of beans on the surface of the roll affect the magnitude of the image force (Fig.4c):

$$F_{iB} = (2Q/\pi)^2 / [4\pi\varepsilon_0 (2R)^2] \quad (4)$$

The trajectories of such particles are quite similar to those of regular soybeans, because:

$$F_{iB}/F_i = (2/\pi)^2 [1+(2/\pi)]^2 = 1.08 \quad (5)$$

Therefore, although the vibratory feeders employed with electrostatic separators usually prevent the situations shown in Fig.4c, it could never be obtained 100% separability of broken soybeans from good ones.

Experimental results

The researches have been accomplished on ELSEP laboratory separator, provided with an original high-voltage generator SIT 75 (d.c.,negative polarity, 0...75 kv peak value). Samples of actual harvested ALTONA soybeans have been electroseparated, using the flow-sheet in Fig.5. The patent-pending technology elaborated by the High-Intensity Electric Fields Research Laboratory of the Polytechnic Institute of Cluj-Napoca ensures purities of the clean product superior to those of mechanically separated soybeans (Table 1).

Table 1. Analysis of ALTONA soybeans products (flowsheet in Fig.5)

Product	Purity %	Weight%
Harvested (Feed A)	75.65	100
Rejected-sep.1 (B)	16.48	22.4
Rejected-sep.2 (D)	52.65	8
Rejected-sep.3 (F)	81.93	4.6
Clean-separation 3	98.45	65
Mechanically separated	97.25	-

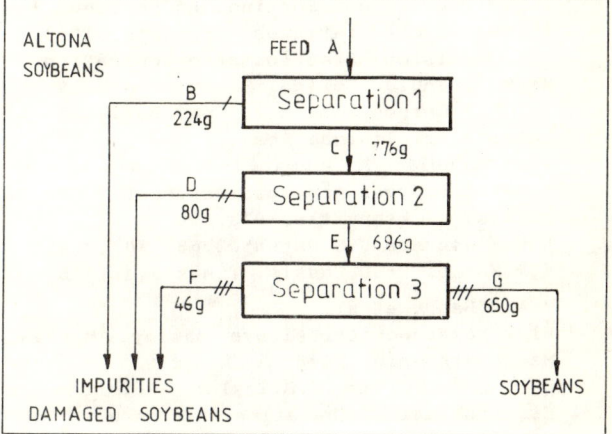

Fig.5. Flowsheet of soybeans electroseparation.

Bioeffects

The soybeans passing between the active electrodes of the electroseparator have been subjected 0.5 s to a high-intensity electric field (8 kv/cm). The poor conductivity of the beans prevented the conveyance of the aquired negative ions. The soybeans remain electrically charged after dropping off the carrier roll. These phenomena, as well as the ozone generated by the corona discharge, seem to exert a biostimulating action on soybeans. The germinative power of these beans has been laboratory determined at Turda Agricultural Research Center. There have been evaluated gains of up to 25% over untreated soybeans. The bioeffects could be enhanced by using low roll-velocities and extended corona fields.

Conclusions

Impurities, as well as shrivelled and broken soybeans could be easily removed from harvested soybeans by electrostatic separation. The product of a three-stage electroseparator is better than that of a standard seed conditioning station. The biostimulation achieved by passing the beans through a high-intensity electric field represents an additional advantage of electroseparation technologies.

Further research aims at a cost effective special design of an industrial corona discharge separator for soybeans conditioning.

References

1 D.M.Tombs. Seed Sorting. La physique des forces electrostatiques. Grenoble, 1960.
2 O.C.Ralston. Electrostatic Separation of Mixed Granular Solids. Elsevier,N.Y.1961.
3 * * * Carpco Inc. H.T. Gen. Bulletin-75 Foodstaffs. Tipical Results.
4 F.S.Knoll. A Summary of Carpco's Non-Mineral Process Applications.Part III:Foodstaffs. Jacksonville,1973, pp 6-7.
5 H.B.Johnson.U.S.Patent 2,687,803, 1954
6 R.Morar. Dr.D.Thesis. Timisoara,1976.

7 A.M.Basov et al. Ělektrozernoočistitel'nye mašiny. Moskva, Masinostroenie, 1968.
8 F.S.Knoll and J.B.Taylor. Advances in Electrostatic Sparation. Minerals and Metallurgical Processing, May 1985,106-14.

PHYSIOLOGICAL AND BIOCHEMICAL EXPERIMENTS IN ELECTROSTATIC TREATED SEEDS

Bai Xiyao, Ma Ancheng, Ma Jingrun, Li Xiaoling, Yan Li, Wang Qingzhao
Anshan Research and Design Institute of Electrostatic Technology
P.R.China

Abstract

This paper discusses the effects of electrostatic field on seeds. Electrostatic treatment may enhance seed vigor by influencing processes of free radicals and stimulating the activity of protein and enzyme. The paper also covers the experiments about the influences on conductivity, vigor indes, ATP content, dehydrogenase, and amylase of seeds.

Electrostatic treatment of seed causes a series of its physilogical and biochemical changes. Here we present our experimental studies on physiological and biochemical indexes, conductivity, vigor index, ATP content, activities of amylase and dehydrogenase in seeds, which indicate seed vigor.

1. Experiment of conductivity of seed

The cell membrane with normal function of seeds of high activity is not only an interficial film separating its interior from exterior, but also an important place for the material cycle and information-conveying and is related to a series of important life phenomena, such as cell identification, hormone effects, development and differentiation[1]. In recent years many reports claimed that there exists a inverse correlation between seed vigor and conductivity. But Xu Benmei's[2] report shows that after seed is slightly damaged, the conductivity is a little bit less than the control, and has no direct relation with the membrane permeability. In this case though membrane structure is altered, change in membrane permeability has not been realized. Koziowski's[3] experiments point out: all the contents in the leached liquid from seeds, including inorganic salts, amino acid, nucleotide and monosaccharide are of small molecules, whose movement from high concentration to to low concentration through the membrane closely relies on the concentration gradient, probably not on permeability change of the membrane. Actually, even for the seed with the richest vigor, carbohydrate and salt would permeate with the leached liquid. The conductivity of the seeds with high vigor due to the difference of metabolic action would increase correspondingly, which is in positive correlation. The wording of " seeds with high vigor generally being of low conductivity " already fails to give a full description to the conductivity items of seeds at various activety levels. Actualy, relationship between vigor and conductivity is rather complicated and related to various factors. The increase in potential of membrane induced by applied electrostatic field has effect upon conductivity of seed[4].

The soyabean seed of Tiefeng 18 for experiment is selected and seed weight is definitely weighed on balance. Rinse the seed and containers with tap water and distilled water for several times. Dry up the seed with qualitative filter paper and put them into a beaker. Pour in 100 ml distilled water and then keep it in an incubator at a constant temperature of 25 °C for 4 hours to soak the seed, then the measurement is made. With untreated seeds as a control, 8 electrostatic treatments are made, each of which repeats twice. The conductivity is taken from the average of two repetitively measured values. The functional relations between conductivity of seed and electrostatic field strength, treating time etc. are shown in Fig.1. From the curves in Fig.1, it is seen that the conductivity of seed treated in either positive or negative electrostatic field is higher than that of the seed for the control. Treatment in a positive electrostatic field of 200 kV/m creates maximum conductivity, about 56--86 percent higher than that of the seed for the control. The seed treated in a negative field of 200 kV/m acquires a conductivity of over 22% higher than that of the control. The conductivity of seed treated in positive field is higher than that of seed treated in negative field. The three curves in the figure show that along with the elongation of the soaking time of the seeds, the more the permeated things, the greater the conductivity increases. Of course there exists a threshold.

As shown in Fig.2 with the increase of soaking temperature, the material exchange between seeds and deionized water speeds up and the conductivity increases. When the soaking temperature rises from 20°C to 24°C, the conductivity increases by 80 % to 100%. This proves that the influence of electrostatic field strength and treating time is less than that of soaking temperature. The measurement of conductivity should be made strictly at identical temperature.

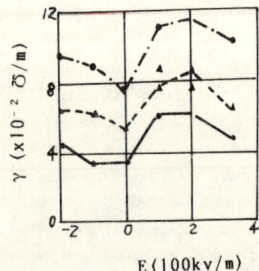

Fig.1 Curves of relation between seed conductivity and field strength
Note: Soaking 4 hours, treating 0.17 hour; Soaking 8 hours, treating 0.17 hour.

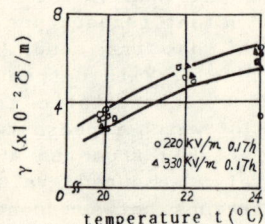

Fig.2 Curves of relation between treated seeds conductivity and soaking temperature

Note: Field   Time   Symbol
      strength
      200 kv/m   0.17h
      330 kv/m   0.17h

2. Experiment in seed vigor index

Experiments in seed activity index have been carried out over seeds of capsicum, wheat, corn and tomato. The sprouting rate is the average of 4 repetitions, for each of which 100 grains of seeds are used. The impetus of growth is expressed by root length and height of the seedlings. Random sampling is made. Each measured value is an average of four repetitions, each with 10 seedlings. The results of the experiments are shown in Table 1, from which it is seen that the reduced activity index of capsicum, after treatment in 176.5 kV/m field, increases by over 61.1%, and tomato index increases by over 76%. The reduced activity index of wheat increases by 18.7% after in 200 kV/m field, and corn index increases by 13.6% after in -200 kV/m field. The seed activity indexes after treatment in electrostatic field generally increase in a range of 10 to 120%. The activity indexes of aged seeds of low activity increases more obviously than that of high activity. In Table 1, the reduced activity index of tomato seed of 1982 increases by 46.3% than that of 1983. It follows that the treatment to old and aged seeds results in obvious and greater increase of seed activity.

3. ATP content experiment in germination

ATP is a high energy compound needed by all living organisms. Various physiological and biochemical action become active during germination. A large quantity

Table 1. Experimental data on activity of electrostatically treated seeds

seed	field strength (kv/m)	concentration of ions (ion/cm^3)	time (hour)	sprouting rate (%)	root length (cm)	plant height (cm)	GS* value	activity increase (%)
Capsicum	0	0	0	35	2.12		74.2	0
(Liaojiao	176.5	14x10^6	0.17	43	2.78		119.5	61.1
No.1)	176.5	14x10^6	0.08	50.3	2.42		121.7	64.1
Wheat	0		0	95.3	1.68		160	0
	200		0.17	98	1.94		190.1	18.7
Corn	0		0	98.7	1.98		195.4	0
	-200		0.17	100	2.22		222	13.6
Tomato 1982**	0	0	0	26.7		2.74	73	0
	176.5	14x10^6	0.17	48		3.4	163.2	123.0
Tomato 1983***	0	0	0	48.25		5.8	279.9	0
	176.5	14x10^6	0.5	55.25		8.95	494.5	76.7

Note: * GS Value stands for reduced vigor index which is the product of sprouting rate G and growth situation S[8].
     ** Tomato seeds, yielded in 1982 and experimented in 1984.
     *** Tomato seeds, yielded in 1983 and experimented in 1984.

of energy is required for the adjustment of biological synthesis and energy-absorbing reactions. Since organic material release energy by biological oxidization, ATP content is the key lever for the energy conversion and can be used as an index for seed activity. ATP content in germination and seed activity are in positive correlation[5]. The driving power for the synthesization of ATP increases with the gain of membrane potential induced by applied electrostatic field[6].

Treat the rice seeds (Liaoyou No.1) in electrostatic field with strength of 33 to 267 kV/m for 0.17, 0.5 and 1 hour respectively. Each treatment is made in three repetitions with a group of untreated seeds for control. Sampling is made randomly for the three repetitions. The measurement is made by means of luciferase and results are shown in Table 2 and in Figure 3. Table 2 shows that the ATP content increases by 76.7 to 216.7%. The increase is obvious after electrostatic treatment. Table 2 and Figure 3 show that the ATP content difference between treated seeds and untreated seeds(i.e.the control) is in positive cooperativity. ATP content of treated seeds is far higher than that of the untreated for control.

The curve of relation between electric field strength and variation of ATP content in seed is shown in Fig.3. It is seen that in low strength (< 100kV/m) region, no obvious ATP content increase appears even the treatment lasts a very long time. The strength ranging from 100 kV/m to 200 kV/m gives an obvious increase of ATP content. The optimum strength of electrostatic field for treatment of seed is aroud 166 kV/m. ATP content tends to

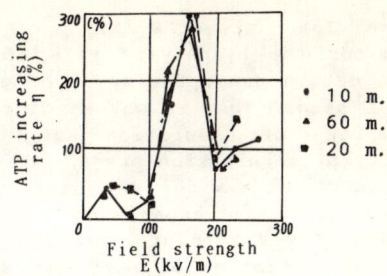

Fig.3 Curves of relation between field strength and ATP content increase rate

decrease when strengh is greater than 167 kV/m. The three curves in Fig.3 indicate that different treating time (0.17, 0.5, 1 hour) gives little influence to ATP content and results in almost no difference in ATP content. It is to say that the seeds are very sensitive to the strength of electrostatic field. Under optimum conditions of 166 kV/m field strength and 0.17 hour of treating time, ATP content of seeds increases from original 1×10^{-10} mole/0.2ml to 4×10^{-10} mole/0.2ml by a net increase rate of 300%. To treat the seeds under suitable electrostatic conditions would remarkably increase ATP content of seed in germination period, thus the activity of seeds may be increased.

4. Experiment of amylase variation in germination

Germination of seed relies on nutritional material stored in seed and energy supplied in the degradation of starch. The degradation of starch also provides energy and raw material for the construction of biological structure. The starch hydrolyzes into maltose at the catalyzation of

Table 2. Experimental data of ATP content of treated seeds

field strength (kv/m)	ATP content($\times 10^{-10}$ mole/0.2ml)/ increase rate(%)		
0	3/0	3/0	3/0
33.3	8.0**/166.7	8.5*/183.3	9.5*/216.7
200	5.3**/76.7	5.5**/83.3	5.3*/76.7
treating time(hr.)	0.17	0.5	1

Table 3. Experimental data of amylase general activity of corn seed

field strength(kv/m)	general amylase activity (mg/grains)/increasing rate(%)			
0	209/0	220/0	311/0	221/0
-100			407**/30.9	333**/50.7
-200	278**/33	307**/39.5		
-267	291**/39.2	298**/35.5		
treating time	0.08	0.17	0.5	1

amylase. Intensity of activity of amylase determines the hydrolysis rate of starch and in cetain sense the speed of germination of seeds. That is why we discuss the regularity of variation of amylase activity at germination stage.

Seed of M17X wide leaf Lujiu corn is employed for the experiment with field strength of -67 kV/m and treating time of 60 minutes. Samples of seed are selected randomly. Each treatment and the control is made up of three repetitions. The treated and untreated seeds are put to germinate in regular way at the same time. After 12 to 48 hours, measure the amylase activity. In the experiment we use Bemfeld method in measurement of amylase activity. The result by this method is general activity for amyulases α, β and Q. Measure the quantity of maltose converted from starch in the hydrolyzation of amylase of corn seed (embryo) at 30 °C within 5 minutes. Define the amylase activity with measured mg/grains value. The amylase activity increase is extremely notable after the seeds being treated in a negative electrostatic field. As shown in Table 4, there is a notable difference of amylase between treated and untreated seeds. An increase of 31--51% of amylase results from the electrostatic treatment of seed. The functional relation between amylase activity of seed and the strength of electrostatic field is shown in Fig. 4. The curves in the figure show that the treating time (0.08--1 hour) gives little influence to seed amylase; but with the field below 200 kV/m, the amylase activity increases along with the increase of the field strength but decreases as the field strength exceeds -200 kV/m.

5. Experiment of dehydrogenase activity variations during getmination

In the reaction of biological oxidization, dehydrogenation plays an important role. Under certain conditions, the

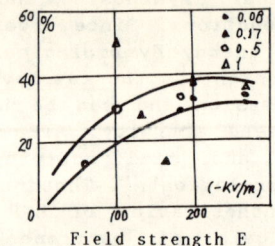

Fig.4 Curves of relation between field strength and amylase general activity

Note: Treating time (h)
   --0.08;   --0.17;   --0.5;   --1.
Field strength E   Amylase general activity increasing rate η

dehydrogenation always takes place with the presentation of dehydrogenase. The measurement of dehydrogenase activity can reflect quantitatively the biological oxidization level. Dehydrogenase can be employed not only to reflect sprouting potential of seeds, but also to appraise the vigour level of seed. It is capable of measuring seed activity exactly, quickly and directly[7]. Seed activity and dehydrogenase activity are in positive correlation[8].

Seed of M17X wide leaf Lujiu corn are used in the experiment with field stregth of 67 to 267 kV/m and treating time of 5 to 60 minutes. Samples of seed are selected randomly. Each treatment and its control is made up of over 5 repetitions. The treated seeds and untreated seeds (the control) are put in an incubator to germinate at a temperature of 35°C. Measure the dehydrogenase activity of 12 groups of seed by means of TTC quantitative method 24 hours after germination. The electrostatic treatment for seed increases seed dehydrogenase activity by 15 to 50%, as shown in Table 4. It shows different result of the increase of dehydrogenase activity. The optimum field strength ranges from 150 to 190 kV/m.

Table 4. Experimental data of dehydrogenase activity electrostatically treated seeds

treating time(hr.) dehydrogenase activity(mg/ml)/increasing rate(%) field strength(kv/m)	0.08	o.17	0.5	1
0	17.4/0	17.3/0	17.1/0	16.0/0
100			21.6**/26.3	24.0*/50
200	20.0**/14.9	20.3**/17.3		

According to the Table, the treating time has little influence on the dehydrogenase activity so that the seeds might be treated for about 5 to 10 minutes in electrostatic field.

Conclusion

The electrostatic treatment for seeds (vegetable and grain) under appropriate conditions would promote germination of seeds, increase seed activity by 40% and strengthen resistability of seed to adversity. The treatment features consist in low cost, less energy consumption, easy operation and remarkable increase of yield. It is really a mew technique of great practical use.

References

(1) 陈秀楚：生物参考资料，科学出版社，1 3 (1 9 8 1) 2 4 7
(2) 徐本美：种子，1, (1 9 8 3) 1 8
(3) T.T.Koziowoki: Seed Biology, 1 (1972) 29
(4) Bai, Xi Yao et al, Proc. of Electrostatic Soc. Japan 5 (1984) 339
(5) 徐本美：种子世界，7 (1 9 8 5) 3 0
(6) 白希尧：自然杂志，7 (1 9 8 4) 9 0 2
(7) 傅家瑞：种子生理，科学出版社 (1 9 8 5) 3 8 4
(8) 徐本美：种子，2 (1 9 8 2) 1 2.

STUDY ON BIOLOGICAL EFFECT OF SEEDS IN ELECTROSTATIC TREATMENT

Bai Xiyao   Yan Li   Chen Zuoli   Li Xiaoling   Ma Jingrun   Wang Qingzhao

Anshan Research and Design Institute of Electrostatic Technology, Liaoning, China

Abstract

The vital factor index, specific conductance (positive correlation), ATP in germinated seeds, starch enzyme and dehydrogenase activity of seeds obviously increase when they are treated in appropriate electrostatic field or ion shower. Seed activity increased by around 40%; disease resistability and adversity resistability are enhanced; crops yield increases by 5-20%. Method of electrostatic treatment for seeds and physiological, biochemical and biological effect experiments are discussed in this paper.

Seeds are among the most fundamental productive materials in agricultural and forestry production. Influence of seed vigor exists over the whole living period of plant. Seeds of superfine quality are a necessary condition for increasing yield and seeds of high vigor would increase yield by 5-40% more than those of poor vigor. It shows that to search for a better vigor of seed is very important and directly linked with the development of national economy.

Electrostatic field and its physical phenomenon have a wide range of influence over plant seeds [1-4]. Electrostatic field and ionic current affect activity of free radical of seeds, activity of protein and enzyme, even the biological oxidation and reduction during electronic transmission.

Method of Electrostatic Treatment for Seeds

The treatment for seeds in a electrostatic field of certain strength and certain ion current would increase seed vigor and enhance the growth and development. The experimental apparatus for the electrostatic treatment for seeds is shown in Fig. 1.

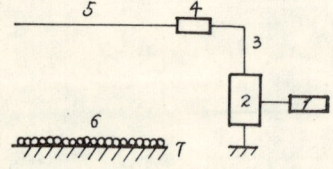

Fig. 1 Experiment apparatus of electrostatic treatment for seeds

Units 1 and 2 make up a high voltage generator with an output of 40-200KV, 2-10mA. High voltage is connected to the corona wire (5) via a protective resistor (4) with high voltage cable (3). The corona wire and "earth" pole (7) form a 100-400KV/m electrostatic field and produce a 10^6-10^8 air ions/cm³ with ozone, nitrogen oxide, etc. In the mean time, seeds (6) with a maximum thickness equaling two grains of seed are placed onto the "earth" pole to be "showered" thoroughly by the electrostatic field and ion current. The electrostatically treated seeds had better to be sowed within 1-3 days or 10 days at most, otherwise it will lead to a poor result.

Corona wire is made from 0.6mm tungsten alloy, the "earth" pole from special semiconductive or conductive material. Distance between corona wire and seeds generally ranges from 200 to 400mm.

Experiment and Result

1. Effect of electrostatic field on germinating and growth situation

Seeds of tomato (Qianglimishou), cucumber (Jinyan No.2) and capsicum (Liaojiao No.1) were treated in an appropriate electrostatic field (100-250KV/m) for 0.17-1 hour. The sprouting tendency of the treated seeds was increased by 14.5-79.8%, 29.3-36.3%, 22-43% respectively as compared with groups of untreated seeds. It showed that the speed and degree of size-uniformity of sprouting of electrostatically treated seeds have been increased in germination period. But there is a threshold in both field strength and treating time. For instance, the sprouting tendency of tomato seeds tends to decrease when treated with a field strength of 400KV/m and a treating time of 2 hours.

From Table 1, it can be seen that the plants of capsicum, tomato, cucumber, corn, rice, bean and Chinese cabbage whose seeds were appropriately treated under electrostatic condition, grow relatively well in height, thickness of stalk, length of root, and number of leaves. Difference in growth situation of plants of capsicum and tomato is obvious. During the electrostatic treatment, the electrostatic field and ion current act on organisms inside the seed, ozone and nitrogen oxide stimulate plumule in dormacy condition and thus speed up the emergence as shown in Table 2. The emergence of electrostatically treated tomato seeds takes place one day earlier.

Table 1. Experimental data of electrostatic effect of growth enhancement on seeds

Experiment			Treatment condition			Morphological coefficient of plant							
Place	Year	Species	Strength of field (kv/m)	Concentration of ion (ion/cm³)	Time (h)	Plant height (cm)	Stem thickness (mm)	Leaf number	Stem fresh weight (g)	Fresh leaf weight (g)	Root length (cm)	Root fresh weight (g)	Plant fresh weight (g)
Trial Farm of Anshan Institute of Electrostatic	84	Capsicum (Liaojiao No.1)	0 +176.5 +176.5 +353.3	0 +14x10⁶ +14x10⁸	0 0.08 0.17	0.94 1.1 0.9 0.82	2.14 2.44 1.98 1.98				2.12 2.42 2.78 2.42		
	85	Capsicum (Tongfeng 16)	0 +200	0 +14x10⁶	0 0.17	13.4 19.0	2.8 2.9	8.5 9.5			6.4 6.8		
	86	Capsicum (Tongfeng 37)	0 +200 -200	0 +14x10⁶ -14x10⁶	0 0.17 0.17	9.0 14.85 9.9	3.3 4.4 3.8		1.36 2.31 1.32	1.92 2.28 1.67	6.65 8.75 8.9	1.70 1.86 1.44	4.98 6.45 4.43
Jiupu Farm in Anshan	86	Capsicum (liaojiao No.1)	0 -200 +200	0 -14x10⁶ -14x10⁶	0 0.17 0.17	19.75 21.17 20.74	4.5 5.0 4.6						
Trial Farm of Anshan Institute of Electrostatic Technology	84	Tomato (Qiang)	0 +200	0	0 0.17	12.95 16.75	2.62 2.64	3.6 4.45			4.25 5.18		
	85	Tomato (Qiang)	0 +200	0	0 0.17	22.8 52.3	3.22 4.55	7.4 9			8.6 12.1		
	86	Tomato (Qiang-limi)	0 +200 +200	0	0 0.17 0.17	15.7 18.7 27.9	3.4 4.11 4.37		1.32 2.39 3.75	1.85 3.45 4.40	7.65 10.1 11.75	0.54 1.23 1.10	3.71 7.15 9.25
Jiupu Farm in Anshan	86	Tomato (Qiang)	0 +200	0	0 0.17	3.92 5.28	0.83 1.04		0.02 0.025	0.01 0.03	1.86 2.44	0.01 0.02	0.04 0.075
Trial Farm of the Electrostatic Institute	84	Cucumber (Jin 2)	0 +200	0	0 0.33	55.37 62.0	8.72 9.79	10.8 11.8					

Table 2. Experimental data of electrostatic effect on seed emergence speed

Treatment condition		Time of interval (day)			Difference (day)	
Field strength (kv/m)	Concentration of ion (ion/cm²)	Time (h)	Sprouting through earth	Emergence	Sprouting through earth	Emergence
+200	+14x10⁶	0.17	6	8	1	1
-200	-14x10⁶	0.17	6	8	1	1
0	0	0	7	9	0	0

2. Disease resistability of seeds increasing due to electrostatic treatment

Seeds treated electrostatically together with group of seeds untreated for control are sowed into earth with germs for experiment. The results are shown in Table 3. Emergence rate of treated tomato and cucumber seeds has been increased by 15.6% and 13.4% respectively as compared with untreated groups and the emergence of treated seeds takes place one day earlier. The death rate of untreated seeds is 23.6% and 8.8% higher respectively than that of treated ones. The disease resistability of treated seeds is higher than that of untreated ones.

Disease resistabilities of electrostatically treated seeds and groups of untreated seeds for control both sowed in vegetable land are shown in Table 4. The death rate of treated capsicum seeds due to virus decreases by 22.1-44.4% as compared to the untreated group. The death rate of treated tomato seeds due to Alternaria solani decreases by 14.6-52.1%. No smut has been taken place in an enlarged experiment for electrostatically treated sorghum seeds in Dagushan township, Anshan city. This is due to the strengthened activity and disease resistability of electrostatically treated seeds, the highly concentrated oxidizing agent-ozone and nitrogen oxide produced from strong corona discharge during the electrostatic treatment for seeds have strong bacteria-killing effect and are 15 to 30 times faster in bacteria-killing rate than the chlorine [3]. Bacteria and viruses on surface of seeds would be killed during electrostatic treatment, which leads to increase of emergence rate and survival rate of seeds and a reduction of disease of plant in farmland.

Table 3. Experimental data of resistability against disease of electrostatically treated seeds

Virus disease	Seed name	Treatment condition		Emergence rate (%)	Death rate (%)
		Field strength (kv/m)	Time (h)		
damping-off	Cucumber (Jinyan No.4)	0	0	59.4	42.1
		200	0.33	75	33.3
witherness	Tomato (Qiangli-mishou)	0	0	86.6	26.9
		200	1	100	3.3

Table 4. Practically measured data of disease resistability of electrostatically treated seeds in vegetable field

Name of disease	Treatment condition		Death rate (%)	Difference (%)
	Field strength (kv/m)	time (h)		
Capsicum virus disease	0	0	71.75	0
	+200	0.17	39.78	-44.4
	-200	0.17	55.78	-22.1
Tomato Alternaria solani	0	0	52.17	0
	+200	0.17	25.0	-52.1
	-200	0.17	44.56	-14.6

3. Influence of electrostatic treatment for seeds upon yield of grain

The electrostatic treatment for seeds increases the sprouting rate; results in early emergence, size-uniformity and strong seedlings; increases growth situation of plant; enhances the resistabilities against disease and stress to some extent; and thus relatively increases the crop yield. The experiment is done in groups in random order with three repetitions for both treated and untreated groups. The data are processed for variance analysis and difference significance test on microcomputer. The statistic data show that the yield increase rate generally ranges from 5 to 20%. This is of tremendous practical value for agricultural production. The practice in the experimental base for spreading of the technology in 1987 gave a good example for this. The spreading test made in Aji town of Tieling County in 8,300 mu (15mu=1 hectare) showed a per mu yield increase of 36.9kg and a total increase of 306,179.5 kg. The test made in Pailou Town of Haicheng City in 2,850 mu for sorghum also showed an average yield increase of 46.7 kg per mu and a total increase of 133,183 kg. It is proved that only with the electrostatic treatment for seeds, the crop yield could be increased under same surroundings. **Table 5 shows the statistic data of crop yield.**

Fruit composition of plant of electrostatically treated seeds shows no evident difference as compared with group of untreated seeds for control.

Conclusions

Physical parameters selected from a great number of physiological and biochemical experiments and farmland tests are as follows:

1. Suitable strength of electric field is 100-200KV/m, the field polarity is determined according to characteristics of the seeds themselves.

2. Appropriate time of treatment for seeds is 0.08-0.17 hour.

3. The seeds to be treated in an electrostatic field should be placed in a thin layer with a maximum thickness of not more than the size of three grains of seeds.

To treat seeds (vegetables tomato, green pepper, cucumber, egg plant, kidney bean, Chinese cabbage; grain: corn, rice, sorghum, soy bean, peanut) under appropriate electrostatic conditions could promote the sprouting of the seeds, increase the seed vigor by 40%, strengthen the resistability of the seeds against disease and stress, increase the crop yield by 5-20% with a cost of only 5 cents per each 1 kg. The technique of electrostatic treatment for seeds features in low cost, less energy consumption, easy operation, obvious increase of yield and no pollution to seeds and environment. It is really a new technique of great practical value and popularization sense.

Table 5 Effects of Electrostatic Treatment of Seeds on The Productions

Trials Species	Year	Treatment conditions			Yeild	
		field strength (kv/m)	duration (h)	trial area (m²)	average yield (kg)	increase rate (%)
Cucumber (Jing Yan No.2)	1983	200	0.33	12	231.00*	40
		0	0	12	165.6	0
	1984	200	0.33	24	354.9	5
		0	0	24	337.6	0
	1985	200	0.17	6	105.2*	29
		0	0	6	81.6	0
	1986	200	0.17	6	66.1*	21
		0	0	6	54.5	0
Tomato (Qiang Li)	1985	200	0.17	6	73.3*	29
		0	0	6	56.7	0
	1986	200	0.17	6	54.1	12
		0	0	6	48.2	0
Kidney beans (double season bean)	1985	200	0.17	6	26.0**	24
		0	0	6	20.9	0
Green Papper (22-1)	1984	-200	0.08	12	95.2	5
		0	0	12	90.7	0
Green Papper (Tong feng 37)	1986	-200	0.17	12	55.7*	19
		0	0	12	46.8	0
Potato (Gangou)	1984	-400	0.17	12	61.0*	15
		0	0	12	53.1	0
Potato (Dong Nong 303)	1985	-200	0.17	6	51.2	14
		0	0	6	45.1	0
Maize (Shen Dan 3)	1983	200	2	24	45.7*	18
		0	0	24	38.6	0
	1984	200	2	12	23.8*	9
		0	0	12	21.8	0
Soy bean (Tiefeng 18)	1983	200	2	24	13.7*	19
		0	0	24	11.5	0
	1984	200	0.17	12	9.13*	16
		0	0	12	6.14	0
Peanut (Haihua No.1)	1984	-200	0.17	24	32.7	6
		0	0	24	30.9	0
	1985	-200	0.17	6	2.74	15
		0	0	6	2.39	0

Reference

(1) 黎先栋等：高压静电场对微生物的影响及其在农业中的应用，生物化学与生物物理进展，3, 36～39, 1986
(2) 白希尧等：静电界・イオンによる农作物の成长促进效果に关する实验，静电气学会志，5, 339, 1984
(3) 白希尧等：静电技术在农业中的应用，自然杂志，7, 902, 1984
(4) 唐建军：电场对水稻种子萌发和幼苗生长影响，种子，5--6, 68～72, 1986

STUDY ON THE TOTAL ACTIVITY OF AMYLASE OF SPROUTING MAIZE SEEDS AFFECTED BY TREATMENT OF ELECTROSTATIC FIELD

Ma Jingrun   Zhang Bo

Anshan Research Academy of Electrostatics Technology, Liaoning, China

Abstract

By use of electrostatic field, ion atmosphere (ion-wind) the physical factor for maize seeds treating stimulates the "activation" of tested seeds to promote the total activity of amylase of sprouting seeds. The contrast comparison shows gain of 12.83-120.48%.

By the changing rule of the total activity of amylase, we find out the suitable condition for maize seeds treated in the electrostatic field. Usually, they are within the effective range of the electrostatic field producing biological effect under the condition of High Tension Voltage for short time or Low Tension Voltage for long time. Under the same field strength and treating time, the more prolonged budding time, the more the total activity of amylase promoted. After 12 hours, the longer the time, the bigger the difference too.

The increase of the total activity of amylase denotes that the activity of seeds is promoting. This facilitates the promotion of germination tendency, rate of emergence, seedling growing tendency and plant production.

Introduction

In the vast world environment, there live myriads of various kinds of organisms. These living organisms have close relation with every kind of factors in the environment. Besides the chemical factors, there are physical factors such as light, heat, sound, electricity, force, ionization, radiation, electromagnetic field, etc. According to certain biological rules about these factors, these living organisms live and develop. We use electrostatic field as a kind of physical factor for treating maize seeds. The electrostatic field induces "activation" of dormant seeds by changing the molecular structure and function of the cell organization in the physiological and biochemical metabolic process, and exciting the protein, enzyme, fat, sugar, starch, etc. of the organisms to produce biological effects.

Under the action of electrostatic field, different seeds produce different biological effects, but total physiological and biochemical metabolic functions in budding increase gradually, and the activity of enzyme system increases as well. At present, the enzyme activity is also one of the important indices for estimating the quality of the seeds. For this reason, we emphatically tested the degradation and composition of starch in sprouting after the treatment of the maize seeds in the electrostatic field. The starch hydrolyzes into maltose with the aid of the catalysis of the amylase. High or low level of activity of the amylase determines the degree of the starch degradation, and thereof we can discuss the changing rule of the amylase activity and seek suitable condition for treating maize seeds in the electrostatic field.

The activity of the amylase of grain seeds is the highest. In this paper, we selected maize seeds for the test. Amylase is usually divided into α-amylase, β-amylase, Q-amylase, etc. Different kinds of amylase have different specific properties, and therefore, measurement must be taken separately, according to its properties. But in this test, we measure the total activity index of the amylase and use it to represent the promoting condition of the maize seeds activity.

Material and Methods

1. Testing principle

The amylase hydrolyzes the starch in seeds, and the hydrolysate is maltose. The maltose has the ability to reduce 3,5-bi-nitro-salicylic acid to salicylic acid of 5-nitro,3-amino (sub) revelation group. The depth of its colour is proportional to the amount of produced maltose. From this we can succeed in calculating the content of maltose. The total activity of the amylase is usually expressed in terms of the weight (in mg) of the maltose. The reaction equation is as follows:

$$\underset{O_2N\ \ NO_2}{\underset{OH}{COOH}} \xrightarrow{Reduction} \underset{O_2N\ \ NH_2}{\underset{OH}{COOH}}$$

2. Instruments and reagents

(1) Main instruments
   (i)   Type 721 spectrophotometer;
   (ii) LRH-150B-G organisms culture box;
   (iii) Constant temperature water-bath;
   (iv) Type 80-1 sediment separator.

(2) Main test reagents
   (i)   pH 5.6 sodium citrate buffer solution;
   (ii) 1% starch solution;
   (iii) 0.1% standard maltose solution;
   (iv) 1% 3,5-bi-nitro-salicylic acid.

3. Methods

(1) Treatment of maize seeds

We choose the maize seeds of assortment Mo 17*Wide Leaf Lu 9 (莫17*宽叶旅9) at random. By use of orthogonal test design, preliminary tests are conducted and relatively proper treatment conditions are found. Then repetitive tests are further conducted. Intensity of the electrostatic field is chosen to range -60 -267KV/m. The concentration of negative ions ranges 10^4-10^6/cm^3. Seeds are treated in the electrostatic field for 5-60 min (see Table 1). Soak every set of seeds treated under different intensity of electrostatic field and for different periods and the contrast set and put them into an organism incubator at 35°C for 12-48 hours for sprouting separately by conventional method.

(2) Extraction of the amylase solution of maize seeds and the measurement of total activity

Take plumules from budded maize seeds, grind uniformly with pH5.6 sodium citrate buffer solution, fix the volume to 5 ml, then centrifugalize (3000 revolutions per 12 minutes) to get amylase solution.

Mix 0.5 ml of amylase solution with 1% starch solution uniformly, and put it into 30°C water-bath to catalyze and to hydrolyze for 5 minutes, and then the maltose is transformed.

Add 2ml of 1% 3,5-bi-nitro salicylic acid into the transformed maltose solution, shake uniformly, and put it into 100°C boiling water for exactly 5 minutes, then cool it down, then use heavy distilled water to dilute it to 10ml, and shake uniformly.

Use type 721-spectrophotometer (520 nm wavelength) to measure light density value.

(3) Calculation of total activity of amylase of maize seeds (see Table 2)

According to the ratio of the solution density to light absorption value, we use 0.1% standard maltose light density value. The formula is as follows:

$$C\ sample = \frac{C\ standard\ sample}{A\ standard\ sample} * A\ sample \qquad (1)$$

We set: At 30°C within 5 minutes, the amylase of plumule of every maize seed is able to catalyze and to hydrolyze to transform into maltose, which is the total activity of the amylase. It has the following formula:

$$T = C\ sample * N * V \qquad (2)$$

In formulas (1) and (2):
  C standard sample: standard sample solution density (0.1%);
  A standard sample: standard sample solution light absorption value;
  C sample: sample solution density;
  A sample: sample solution light absorption value;
  V: total volume (5ml) of sample solution of extracting amylase solution;
  N: amylase solution diluting times (20) of the sample solution;
  T: amylase total activity (unit: mg per 3 maize seeds).

Discussion and Analysis of the Testing Result [3,4]

From Table 1 and Fig. 1, we get the result that through the effective electrostatic field and ion-atmosphere (or ion-wind), the treated maize seeds compared with the contrast get a total activity gain of the amylase to be 12.83%-120.48%.

Table 1. Statistics and comparison of the total activity of maize seeds at different electrostatic field treating conditions with the contrast's

SN	Testing Condition	Amylase Activity(mg/grains)				t Test and others
		Electro-static field	Con-trast	Diffe-rence	Gain (%)	
1	-183KV/m 5 min.	138.88	106.94	31.94	29.87	t=-1.783 P>0.1
*2	-200KV/m 5 min.	278.38	209.21	67.17	33.06	t=-3.194 P<0.01
*3	-267KV/m 5 min.	291.14	215.68	75.46	34.99	t=-5.113 P<0.01
*4	-200KV/m 10 min.	306.79	220.38	86.41	39.21	t=-3.315 P<0.01
*5	-267KV/m 10 min.	298.61	218.51	80.10	36.66	t=-3.121 P<0.01
6	-67KV/m 30 min.	82.50	73.12	9.38	12.83	t=-0.670 P>0.1
*7	-100KV/m 30 min.	407.44	310.69	96.75	31.14	t=-3.558 P<0.01
8	-133KV/m 30 min.	155.55	70.55	85.00	120.48	t=2.483 P>0.1
*9	-100KV/m 60 min.	333.33	220.95	112.38	50.86	t=-4.628 P<0.01
10	-135KV/m 60 min.	109.72	85.40	24.32	28.48	t=-2.598 P>0.1
11	-233KV/m 60 min.	70.66	86.66	-16.00	-18.46	t=4.865 P<0.05

SN = serial number
* expresses very notable

Table 2. Example of Calculation on total activity of amylase in maize seeds

1	2	3	5 *C	4 N	Total Activity of Amylase(mg/grains)						*SS	Other
					Electrostatic field			Contrast				
2					1	2	3	1	2	3		
	(1)		A		0.12	0.155	0.21	0.12	0.15	0.125	0.185	
			Ā T			0.162 87.57			0.132 71.35			
	(2)		A		0.29	0.22	0.11	0.175	0.12	0.225	0.18	
			Ā T			0.207 115.00			0.173 96.11			
	(3)		A		0.24	0.215	0.30	0.22	0.205	0.225	0.215	
			Ā T			0.252 117.21			0.217 100.93			
			A		1.30	1.28	1.10	1.05	0.76	0.75	0.22	
			Ā T			1.23 559.09			0.85 386.36			
			A		1.25	1.30	1.00	0.95	0.845	0.905	0.23	
			Ā T			1.18 513.16			0.90 391.30			
	5 Σ n=1		T T̄			1391.91 278.38			1046.85 209.21			

*SS=standard sample *C=content
1.Number; 2.Treating condition; 3.Content; 4.Testing data; 5.Contrast; 6.Term

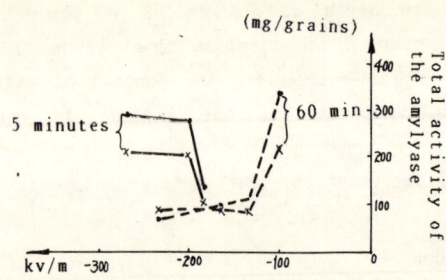

Fig. 1. Strength of the electrostatic field
Note: Under the electrostatic field condition of the same time but not the same voltage, the comparison between the budding in the electrostatic field [·] and the total activity of the amylase of the contrast [x] of the treating seeds.

The optimum conditions of the electrostatic field are:

-200KV/m/5 minutes, -267KV/m/5 minutes, -200KV/m/10 minutes, -267KV/m/10 minutes, -100KV/m/30 minutes, -100KV/m/60 minutes.

The total activity of amylase changes in accordance with this tendency: In the effective electrostatic field, under either short time (High Tension Voltage) or long time (Low Tension Voltage) treating condition, the effect is notable or obviously notable. Within effective range and under certain short time treating condition, as the Voltage increases, the total activity of the amylase increases gradually; but if treating time is long, when the Voltage increases, the total activity of amylase is lowered gradually instead. Finally total activity of amylase is inhibited by electrostatic field.

Under certain effective electrostatic field strength treating condition, the activity of the amylase of maize seeds increases as the sprouting time prolonged. After 12 hours, the difference between the treated sample and the contrast sample is notable [4] (see Table 3 and Fig. 2).

Comparing the contrast (untreated) to the sets of seeds treated with different "electrostatic dose", the difference in activity of the starch amylase in germination of the seed is remarkable if the dose ranges between 1000KV/m/min. and 6000KV/m/min. Define the starch amylase of the contrast as 100%. That of the treated maize seed may increase to 128.48-220.48%. The effect becomes less remarkable if the dose is below 915KV/m/min. If the dose is over 7980KV/m/min., the activity of the starch amylase is inhibited and declined or the seeds are even stifled (see Fig. 3).

Table 3. Statistics and comparison of the total activity of the amylase in maize seeds at different sprouting time under the same electrostatic field treating condition and the contrast.

SN	Testing Condition	Amylase activity (mg/grains)			t test and other
		Electrostatic field	Contrast	Difference	Gain (%)
2-1	-200kv/m 5minutes	87.57	71.35	16.22	
2-2	"	115.00	96.11	18.89	
2-3	"	117.20	100.93	16.27	
2-4	"	513.04	391.30	121.74	
2-5	"	559.09	386.36	172.73	
∑T		1391.90	1046.05	345.85	
T̄		278.38	209.21	67.17	33.06
3-1	-267kv/m 5minutes	101.53	70.25	31.28	
3-2	"	313.60	211.40	92.20	
3-3	"	458.30	355.40	102.90	t=-5.113 p<0.01
∑T		873.43	647.05	226.38	
T̄		291.14	215.68	75.46	34.99

*SN--serial number

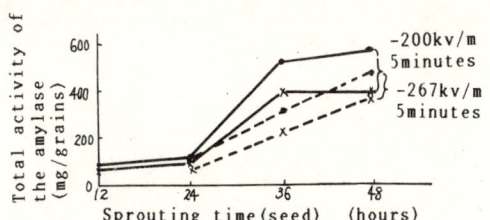

Fig. 2 Relationship between total activity of the amylase and the sprouting time of the maize seeds. ([•]:treated with electrostatic field; [x]: untreated, the contrast.)

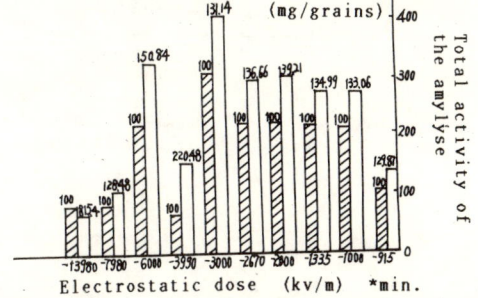

Fig. 3 Comparison of the total activity of the sprouting seeds treated with different electrostatic dose (KV/m/min.) and that of the constrast. (: for the treated, : for the contrast. The value for the contrast is defined as 100%.)

In short, by the action of the electrostatic field and ion atmosphere (or ion-wind), the maize seed changes its total activity of amylase in the sprouting process. The main effect is field strength and the time effect follows the field strength. In addition, the superposition of the electrostatic field with appropriate polarity on the "organisms electrostatic field" of seeds in equilibrium makes them polarized, and stimulates or strengthens the seeds "organisms electrostatic field" gradually. The "micro-electrostatic field" is especially constructed by the cytoplasm membrane of the organisms. It causes the macromolecules in the cell membrane of the organisms, especially the molecules of protein rearranged, i.e. by the change from original gathered condition —— dispering condition —— getting together again. This causes the dormant seeds "activated", performing metabolism, increasing the absorption of water and nutrition and transform quickly. It enhances the strength of respiration, which speeds up the physiological and biochemical reactions, and accelerates the process of the sprouting of seeds.

Electrostatic field promotes the activity of all kinds of enzyme in the cells of seed in various degrees. In this paper, the total activity of amylase of the maize plumula ascertains this fact (see Table 1). The promotion of the total activity of the amylase [3] suggests that by the processes of catalysis and hydrolysis, the promoted capacity of the starch transforming to maltose is enhanced. At the same time, in the process of metabolism, the maltose is used as energy sources of producing ATP and reduction type NADPH. It promotes the sprouting, the rate of emergence and seedling's growth and increases the yield of the plant (5-20%).

In general, the electrostatic field, ion atmosphere (or ion-wind) treated maize seeds (or other seeds) is an effective technical step for increasing the production. It gives deep effect on and immeasurable contribution to the agricultural production.

References

1. Yan Gouguang, Wang Fujun, ed., Agriculture Apparatus Analyse, 1982.
2. Bai Xiyao et al., Electron Techniques, pp32-35, Application of Electron Techniques to Agriculture, 1984.
3. C.A. Price, Molecular Approaches to Peant Physiclogy, McGraw-Hill, 1970.
4. A.M. Mayer and A. Poljakoff-Mayber, the Germination of Seeds, 1978.

A PRELIMINARY STUDY OF THE BIOLOGICAL AFTER-EFFECT ON CROP SEEDS TREATED BY UNIFORM ELECTROSTATIC FIELD

Liang Yun-zhang, Shi Xin-min, Yang Ti-qiang

Electrostatic Laboratory, Inner Mongolia University, Huhehot, China

Abstract

Serial macro phenomena resulting from the treatment of beet seeds by uniform electrostatic field reveal the existence of a biological after-effect, and the micromechanism for such process is preliminarily studied on the basis of X-ray diffractometry.

I

Today, the application of electrostatic technology is utilized in different areas, one of which is the biological one. The technology has made a gratifying progress in biological engineering [1,2], and the electrostatic biological effect is also a fascinating field to study.

The electro-biological phenomena are used to study the effect of a biological body, including microbes, plants, animals and man which are acted on by an electrostatic field.

There are two different modes, uniform and nonuniform, by which electrostatic field can act on biological body. The nonuniform field, electric corona field, acts on the biological body by positive and negative ions in the air which are produced by a high voltage. This is called ionic biological effect. There are many reports on this test [3].

As for the crop, it was reported that after the crop seeds were treated with the corona field, marked effects on plant organs, disease resistance and output were observed confirming the biological effect made by the electric corona field [4]. During the growing period, it was also shown that a certain density of negative ions could produce influence on the crop [5]. Little information, however, is available in the literature on the biological effect after the uniform electrostatic field being applied on the crop.

II

An observation has been made on the germination percentage, germination potential and growth potential for beet seeds and barley seeds treated in uniform field, and also the output, sugar content and quality examination has been made for the root tuber of beet. It is found that the electric field strength and the treatment time are the main factors influencing the seeds by uniform field when an adequate electric strength and treatment time are taken, the germination can be increased by 12.6% in the optimal treatment.

Another observation is made on the beet seeds treated in a small field area. It is found that the weight of 100 seedlings has increased by 90%, the height by 10.8%, and the number of green leaves by 19.7%. Having tested repeatedly seven times over 3 years in a small field area, we have found that the average sugar content has increased by 0.9 degree(*) and the output per mu (15 mu = 1 hectare) by 11%. The comparison of exemplary tests on 13 small pieces of cropland indicates that sugar content increases by 0.25 degree on an average, the output per mu by 7.66%, the sugar content per mu by 13.9%.

After treatment of barley seeds with different strength of electrostatic field, and different time, it is also found that their germination percentage, germination potential and growth potential are apparently higher than those of the control groups. It is determined by the small field tests that there is a 3% increase in output of barley seeds, when subject to optimal treatment, and that the protein content in the barley (generally 11-12%) is raised by 3.7%.

We can see from the example that germination percentage can be raised, the growth promoted, the output increased and the quality improved after the crop seeds have been treated in the uniform electrostatic field with the adequate strength and time. It is proved that when a uniform electrostatic field is applied to the crop seeds, a significant biological after-effect can be expected.

Determination of enzymic activity by some authors [6] from the biochemical point of view revealed that soaked barley seeds treated by the uniform electrostatic field exhibited higher activity in the amylase, peroxidase, isozyme and more content of α-amino nitrogen than that of the control groups. The content of starch, however, is less.

Comparison between treated and untreated seeds has been made with D/MAX-III X-ray diffractometer (made in Japan) based on the structural point of view.

1) The field strength is 1 Kv/cm (marked No.1), 2 Kv/cm (marked No.2) and the treatment time is 4 hours for the barley seeds. We mark the control as No.0. The result of Fe marked K_α diffracted ray is given on the conditions of the Fe target, tube voltage 40Kv and the current 20mA (Fig.1).

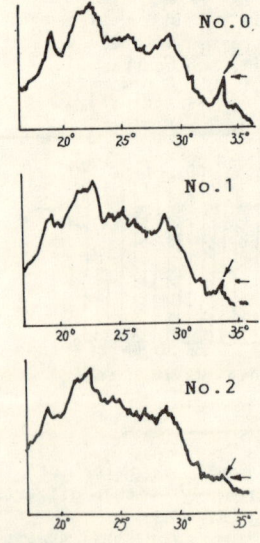

Fig. 1 Barley seeds diffraction curves

We can see that the diffraction peak varies with the different field strength at $\theta = 33.6°$. The diffracted curves show that the crystal structure of barley seeds have changed to some extent.

2) The treatment time was 10 minutes (marked No.1), 4 hours (marked No.2), 12 hours (marked No.3) and the field strength was 4Kv/cm for the beet seeds. We mark the control as No.0. The result of the marked K_α diffracted ray is given on the

conditions of the Fe target, tube voltage 40Kv and the current 20mA (Fig. 2).

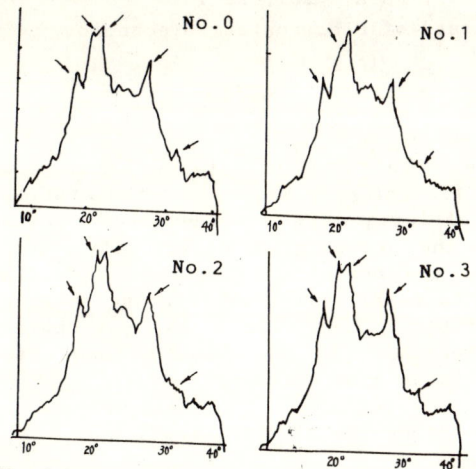

Fig. 2 Beet seets diffraction curves

It can be seen from Fig. 2 that the peak value varies apparently at $G=33.6°$. It is shown that the crystal stucture of beet seeds varies with the different treatment time.

Comparing the above results with the macroeffect, we can see that there is a relationship between the varying peak value and the macroeffect. Variations in the diffractogram can be expected as an important criterion for the macroeffect by further studies.

III

Electrostatic field of a material has energy. The embryo of crop contains a lot of high-energy mutients, for example, fat, protein carbohydrate, phosphoric acid and the physiologically active substances containing vitamins, enzymes and plant hormones which are physiologically needed. When the seeds are exposed to the uniform field, they are affected by it. Thus, the influence produces a series of macroeffects. The catalysis of enzyme is of importance to plant growth. The influence of the electric field increases the enzymic activity [6].

The activity is only in certain sites in macromolecules. Forming the activity centre, the structure is necessary for apoenzyme [7]. By the action of uniform electric field, the macromolecule of apoenzyme may be polarized and structure of the activity centre will be changed. The change improves the enzyme activity so that they produce a series of biological macroeffects. The results indicated by X-ray diffractometry may be regarded as evidence.

IV

We briely report the preliminary work of uniform electric field on biological after-effect. The work is, however, initiative and therefore further study is needed.

References

(1) 水野 彰 （日）静電気学会誌 10(1986), 6, 511
(2) 浦野直人 （日）静電気学会誌 10(1986), 6, 517
(3) 康智遥 自然杂志 5 (1982), 17, 848
(4) 白希尧等 静电学术报告会论文集 (37), 中国物理学会 静电专业委员会, 295
(5) 文荆江 静电及其应用, 科学出版社 (1978)
(6) 叶家明等 中国静电学术交流会论文集 (1985)
(7) 阎隆飞 李明启 基础生物化学, 农业出版社, 1985, p115-118

A HIGH-PRESSURE SHOCK STERILIZING TECHNIQUE

Bai Xiyao, Chen Tianming, Wang Juan, Fu Mingzhi, Xu Yu
Anshan Research and Design Institute of Electrostatic Technology
P.R.China

Abstract

The use of tremendous pressure shock wave generated in the course of high-voltage pulse discharge in liquid kills bacteria in it at ambient temperature. People can enjoy drinks with original taste and colour of fruits and vegetable juices and vitamins undamaged. This paper states the mechanism of sterilization, experimental process and results.

Introduction

Presently, fruit and vegetable juice-based drinks have been developed into one of daily necessities in commercial production. Therefore, how to improve the quality of these drinks and reduce their costs becomes a very important problem. At present, sterilizing technique is the main factor which determines the quality of these drinks. The available sterilizing techniques include the high temperature, ultraviolet light, ultrasonic and microwave sterilizing techniques, and the bacteria filtering technique with filter films. Usually, the high-temperature sterilizing technique is widely adopted in consideration of productivity and cost. In order to minimize the effect of the high-temperature sterilization on the quality of these drinks, the conventional indirect heating sterilizing technique has been developed into direct high-temperature sterilization and pasteurization. By use of these techniques, juice can be heated up to higher than 100 °C for the heating speed has been heightened from 3K/s to 150K/s. Despite of this, heating would lower the quality. High heating speed does nothing but only reduce the harmful effects to a comparatively low degree. High heating temperature results in reduced quality and spoils its taste, fragnance, colour and vitamines.

Scientists are seeking a new technique of ambient temperature sterilization. Scientists from U.S.A and U.S.S.R use the ultrasonic sterilizing system. Japanese scientists Mizunosho et al kill the saccharomyete in pure water (without ions) with extremely short high-pressure impulse (20kv). Real fruit and vegetable juices contain a large number of ions and fruit acids. Their resistivity is much lower than that of pure water, and the sterilizing efficiency is considerably decreased with the decrease of resistivity. Therefore, the ambient temperature sterilization is fairly difficult for the fruit and vegetable juices with a low resistivity. People have long been seeking new effective techniques of sterilization.

Mechanism of Sterilization

By use of high-voltage (5--36kv) pulse (pulse duration 1 μs), the main gap of liquid in a pot discharges intensely with an instant power of millions kw. Because the fruit and vegetable juices are incompressible, the instantaneous shock force reaches 10^5 pa. The electrical energy is converted directly into impulse pressing energy carried by a powerful shock wave transmitted at a speed of 1000--5000 m/s. The shock applies a pressure of 10^3 kg/cm² to the bacteria and kill them by breaking the cell membranes, crushing the bacteria or resulting in forced internal vibration within the bacteria.

Now the question is how to use limited amount of energy to produce possibly highest efficiency of sterilization, i.e. to generate possibly most powerful instantaneous shock so as to apply it the sterilization.

The amplitude and front rise rate of current pulse produced by discharge in a liquid determine the intensity of the electrohydraulic effect.

The maximum amplitude of the current pulse of discharge

$$I_m = U_a \sqrt{\frac{C}{L}}$$

The maximum rise rate of the pulse of discharge

$$\left.\frac{di}{dt}\right|_m = \frac{U_a}{L}$$

As shown in the formulas, the maximum amplitude I_m of the current pulse of discharge depends on the capacitance C of the capacitor, the critical breakdown voltage U_a of the gap and the inductance L in the discharge circuit. I_m is proportional to U_a and the square root of C, but reversely proportional to the square root of L. The maximum rise rate of the current pulse

$\left.\frac{di}{dt}\right|_m$ is independent of the capacitance C and the total load resistance. It is only proportional to L. L tends to damp the change of the current pulse. Once the capacitor is seleted, U_a and C are determined. In order to increase the amplitude of the current pulse, it is very important to reduce the inductance in various parts of the circuit. The structural form of a circuit determines the value of inductance in the circuit. Therefore, much attention must be paid to the structure in designing a discharge circuit so that the inductance in the circuit can be reduced.

Charging voltage Uc creates the greatest effects on the intensity of the electrohydaulic effect.

As shown in the formulas, the electric field energy Wc stored in the capacitor and the main gap discharge power N is proportional to the square of the critical breakdown voltage U_{cK} across the gap. The maximum amplitude of the discharge pulse current I_m and the maximum rise rate of the discharge pulse current $\left.\frac{di}{dt}\right|_m$ is proportional to U_a. Therefore, the increase of U_a is the most economical and effective method for improving electrohydraulic effect. Nowadays high-voltage techniques are developing rapidlly. High-voltage supply and insulation become easier than ever before, and the costs are reduced significantly.

By using this technology, bacteria in the tap water, fruit and vegetable juices and drinks containing ions and fruit acids can be killed at ambient temperature with their colour, taste, vitamines and other nutrients unchanged. This technology can provide users with plenty of drinks of natural fruit and vegetable juices with original colour, taste and vitamin content.

Experiment and Result

The experiment facility is illustrated in Fig.1. Industrial water and its solution are filled in a φ 400mm spherical container, with the colibacillus, acetobacter and other bacteria as test samples. The main gap between the electrodes is 40mm. The pulse current is applied on the main gap discharge electrodes through the high-voltage switch. The test liquid is poured into the container through the inlet at the top, and drained away through the valve at the bottom. Before and after the experiment, a very small amount of test liquid was taken and spread on the malt-agar culture medium. The colibacillus, acetobacter and other bacteria had been cultured at 38 °C for 24 hours to evaluate the sterilization efficiency in terms of resultant number of colonies (average value of 3 samples).

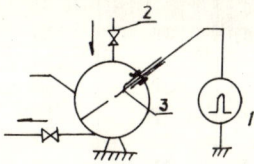

Fig.1 Scheme of experimental sterilizing facility by discharge in liquid. 1-High-voltage pulse supply; 2-Valve; 3-Discharge electrode; 4-Sterilization container.

Pressure shock wave produced by high-voltage pulse is in close relationship with the resistivity, Ph value, etc. of the liquid. Experiment data of the intensity of the pressure shock wave, sterilizing efficiency for the liquid and the resistivity are shown in Table 1, and their functional relation is shown in Fig.2. When the resistivity of the liquid increases, the energy required for each unit sterilized volume decreases significantly. When the resistivity decreases, the energy required increases significantly. The functional slope is -0.019 J/1kΩ.

Table 1. Sterilizing energy of various bacteria-bearing liquid

Bacteria-bearing liquids	Resistivity (kΩ·cm)	PH value	Required energy when sterilizing efficiency is 100% ($\times 10^2$ J/l)
Pure water without ions	≥10000		
Tap water	45-600	6	2.4
Tap water containing 10% sugar	80-100	4-5	7.2
Sweet sorghum juice	15-30	4-5	1.2

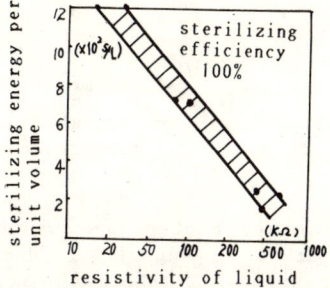

Fig.2. Functional relation of the resistivity of liquids and the required energy in sterilizing when the sterilizing efficiency is 100%

When the colibacillus suspended in the

Table 2 Sterilization data for the cultured coli bacillus in industrial water

Date	Liquid temperature (°C) Before experiment	Liquid temperature (°C) After experiment	Resistivity of liquid (kΩ·cm)	PH value	Number of pulses	Electric energy of pulse (J)	Sterilizing efficiency (%)
12/10-86				6	30		99.9
13/11-86	36	36	465	6	30		99.9
24/11-86	16	16	501	6	30		99.9
16/4 -87	30	30	420	6	30	4050	99.9
1/6 -87	32	32	580	6	30	3524	98
3/6 -87	32	32	610	6	30	2586	98
9/6 -87	25.5	26	590	6	10	2586	97
9/6 -87	25.5	26	590	6	20	2586	98.7
9/6 -87	25.5	26	590	6	30	2586	100
9/6 -87	25.5	26	590	6	30	2586	100
9/6 -87	25.5	26	590	6	50	2586	100
9/6 -87	25.5	26	590	6	60	2586	100
9/6 -87	25.5	26	590	6	70	2586	100

tap water as a test sample, the functional relation of the sterilizing efficiency and the energy consumption per unit volume is as shown in Fig.3. The amplitude of voltage pulse is 45 kv, the pulse energy per unit volume 60 J/l, the PH value of the water 6, the resistivity 590 kΩ, and the temperature 26 °C before and after the experimtnt. In Fig.3, the relation between the energy consumption per unit volume and the sterilizing efficiency appear to be an exponential-typed function. As shown by the curve in Fig.3, the sterilizing efficiency is 100% when the energy consumption per unit volume is greater than 240 J/l, with the countless colibacillus (diluted by 100 times) being reduced to zero. As shown in Table 2, this efficiency is stable. Since October 1986 the reproducibility of colibacillus sterilizing efficiency has been very good, and the sterilizing efficiency has been stabilized above 98%.

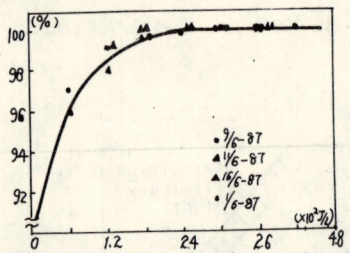

Fig.3 Functional relation of sterilizing efficiency and the sterilizing energy per unit volume

When the mixed bacteria suspended in the tap water or syrup (10%) and the acetobacter suspended in the sweet sorghum syrup are used as tesr samples, the experimental data of the sterilizing efficiency and energy consumption are shown in Table 3. When the pulse discharge energy per unit volume is 900 J/l, the sterilizing eficiency for mixed bacteria is 100%. When the energy is 1200 J/l, the sterilizing efficiency for acetobacter is 100%.

Table 3. Experimental data of sterilization for mixed bacteria and acetobacter

Energy consumption for sterilization (×10² J/l)	Mixed bacteria (dilution×10²) (colonies)	Acetobacter (dilution ×10⁴) (colonies)
0	6×10⁴	440×10⁴
1.8	2×10⁴	
3.6	1×10⁴	
5.4	6.6×10³	
6.0		120×10⁴
7.2	3.4×10³	
9.0	0	4.4×10⁴
12	0	0
15		0
18		0

The energy consumption for this technique is less or equal to one thousandth of the energy consumption for high-temperature sterilization. It is far less than the energy consumption for ultrasonic sterilization, microwave sterilization, etc..

Pulse discharge in liquids does not damage vitamine C. The original vitamine content can be retained after experiment, as shown in Table 4.

Table 4. Vitamine C content

Discharge energy(J/l)	Resistivity (kΩ·cm)	PH value	Vitamine C content(mg/ml)
0	500	6	0.03
300	500	6	0.03

The designed technological procedure is as follows:
Material selection→ Washing→ Crushing→ Squeezing→ Filtering→ Settling→ Sterilizing→ Sterile filling→ Finished product.

Conclusion

The pulse discharge in liquids can produce tremendous mechanical energy, which has a sterilizing effciency of 100%, with a very low energy consumption. It can kill bacteria at ambient temperature, and does not affect the taste, colour, vitamine and ingredients of fruit and vegetable juices and drinks. People can enjoy the original taste of wild fruits, orchard fruits and vegetables with vitamine C undamaged, but without any need of additives. This technology can be used not only for sterilization of drinks and condiments, but also for waste water treatment. However, in this case there is a high noise level. After silencing treatment, the noise level can be reduced to below 80db.

Reference
(1) Bai Xiyao et al: Nature, 10,3(1987) 190-194
(2) Akira Mizuno et al, Proc. of 1986 Annual Meeting of the Institute of Electrostatics Japan, (1986) 405-408
(3) Zhang Lingfei et al, Electronic Science and Technology, 8,16,177(1986) 28-31
(4) Du Peng, Food Industry Science and Technology,3(1987) 37-42

OBSERVATION OF CHANGES OF SOME BIOLOGICAL INDICES IN PATIENTS WITH OCCUPATIONAL LEUKOPENIA BEFORE AND AFTER THE NEGATIVE AIR ION THERAPY

Li Weimin, Li Anbo, Zhang Jimei
The Faculty of Public Health, Xi'an Medical University
Xi'an City, P.R.China

ABSTRACT

To evaluate the exact biological effects of negative air ions, we conducted a clinical trial study on occupational leukopenia in double-blind fashion with placebo. 5-HT in blood; catecholamines, 17-KS, 17-OH in urine; ECG; IgG, IgA, IgM, E-RFC, LCT; hematological indices etc. were determined before and after treatmemt. Only 17-KS and catecholamines in urine were decreased after negative air ions treatment in patients with benzene poisoning. Other indices of the patients had no significant changes either in group treated by batyl-alcohol or in the control groups.

Our results indicated that both the negattive air ion and batyl-alcohol were not effective for the patients with occupational leukopenia. To investigate the more effective drugs and new therapeutic method were suggested.

INTRODUCTION

Despite of the fact that investigations into the air ions has been in progress for more than a hundred years, the subject on the biological effects of air ion remains inconclusive. From the available data, we know that the previous studies were mainly the experimental studies on the effects of air ion on animals and plants, while the systemic clinical observations were in the minority and only a few studies were on the effcts of air ion on the occupational disease. In consideration of that we selected some patients with occupational leukopenia for the clinical observation of the negative air ion inhalation. By observing carefully the changes of some biological indices (hemogram, immunity, neurotransmitters, endocrine hormones, ECG, etc.) of the patients before and after the negative air ion therapy(NAIT), we tried to evaluate further the biological effects of negative air ion, to look for a new way to prevent and treat occupational leukopenia, and to find out some mechanism of NAIT during the treatment.

CLINICAL DATA

1. Grouping and Schemes of Therapy

All the patients participating in the therapy were divided into four groups. Group 1 and group 2 were the hospitalized patients with mild chronic benzene poisoning who were diagnosed according to the China National Diagnostic Criteria, observed in random order in double-blind fashion with placebo, group 3 and group 4 were the outpatients with occupational leukopenia caused by other factors (high frequency, X-ray and chemicals), treated in random order. Table 1 shows the characters of the patients and the scheme of therapy.

Table 1 The characters of patients and the scheme of therapy

Groups	Cases	Male	Female	Age	treatment
1	12	4	8	44.3	active ioniser(No.39) and placebo
2	12	4	8	41.0	placebo ioniser(No.40) and batyl-alcobol
3	12	3	9	38.9	active ioniser (AS-110 model)
4	12	3	9	37.6	no treatment as a control

Both No. 39 and No. 40 air ions generators were produced by the Beijing Medical Apparatus Factory, with the same fashion and wind speed. The negative air ions concentration of generator No. 39 is 500,000 ions/cm (detecting distance: 30 cm); while generator No. 40 is zero. Group 1 and group 2 were treated with negative ions generators for 40 minutes, twice daily and six days weekly. Placebo and batyl-alcohol (produced by Shanghai Yanan Pharmaceutical factory) were administered 50 mg three times daily by oral. All the patients were told that they were treated with batyl-alcohol and negative air ions. During the treating period, a doctor was specially responsible for grouping and treating, and the author was responsible for observing the therapeutic effects. Group 3 were treated only with negative air ions (As-110 negative air ions generator, concentration: 1200,000ions/cm) for 40 minutes once a day. Group 4 were not given any treatment as a control. The observation period for all groups lasted a month. During the period, the climatic conditions of the therapeutic room were monitored and recorded. The basic air ions

Table 2 The changes of hematological indices of the patients (mean value)

Groups	WBC before	WBC after	RBC before	RBC after	HCT before	HCT after	HB before	HB after	PC before	PC after
1	3905	4149	361	361	33.1	32.9	11.5	11.5	9.8	11.2
2	3764	3991	364	349	32.1	31.9	11.3	11.1	9.5	10.3
3	3933	3786	408	416	34.0	36.6	12.3	13.0	9.9	9.7
4	3950	3650	394	402	34.6	35.1	11.9	12.5	10.0	9.8

concentrations in the therapeutic room were 300 positive ions/cm and 250 negative ions/cm, which were detected by DLY-IA model atmosphere ions monitor.

2. The Indices and Methods

The indices of observations include symptoms and signs; ECG; white blood cell count (WBC/mm^3), red blood cell count (RBC 10^4/mm^3), Hct(%), mean red blood cell volume (MCV, μm^3), Hb (g/dl) and platelet count (PC 10^4/mm^3); serotonin in blood (5-HT μg/ml, detected with improved Curzon spectrum assay); catecholamines (CA), noradrenaline (NA) and adrenaline (Ad) in urine (method of 3-hydroxyindole); 17-hydroxysteriod (17-OH) and 17-ketosteriod (17-KS) in urine; serum immunoglobulines IgG, IgA and IgM (single immunodiffusion), lymphocyte transmission rate (LCT) and E-RFC. The blood indices were detected with M_5-electric blood cell counter; 5-HT in blood and catecholamine in urine were detected with MPF-4 spectroflourometer. The blood indices were detected twice weekly; the other indices were done once before and after the treatment respectively.

RESULTS

1. Symptoms and Signs

The symptoms and signs (amnesia, headache, insomnia, weakness and restlessness) of each group were analysed. There were no significant differences before and after the treatment, neither between the treatment and control groups.

2. Electrocardiogram (ECG)

ECGs were recorded before and immediately after the patients' inhalating the air ions for 40-60 min. and after a month's therapy. They were analysed carefully, including P-P, Q-T and P-R intervals and the amplitude, etc. There were no significant differences in ECGs before and after NAIT, and the changes of ECG were irregular.

3. Hematological Indices

When the hematological indices of each group were analysed, there were no significant differences (P>0.05) before and after the treatment; neither between group 1 and 2, group 3 and 4 (Table 2).

4. 5-HT in Blood

There were no significant differences (P>0.05) in blod 5-HT levels in group 1 and 2 before and after the treatment. However after the treatment, the levels of 5-HT in group 3 and 4 were reduced (P<0.05). Nevertheless, there were no significant differences (P>0.05) between group 3 and group 4. See Table 3:

Table 3 The changes of 5-HT in blood (ng/ml)

Groups	Cases	Before (X±SD)	After (X±SD)	P
1	11	129.3±29.6	140.4±27.1	>0.05
2	8	130.9±26.3	123.5±21.1	>0.05
3	12	114.6±25.6	89.6±23.6	<0.05
4	12	124.5±24.9	101.5±24.2	<0.05

5. Catecholamine in Urine

Only on group 1 and 2, the catecholamines in urine were detected. The result showed that the levels of catecholamines and adrenaline in urine in group 1 decreased after NAIT. However, the urinary catecholamines levels in group 2 were no significant differences (P>0.05) before and after NAIT. See Table 4:

Table 4 The levels of catecholamines in urine (μg/24h)

Groups	Cases	CA before	CA after	NA before	NA after	AD before	AD after
1	9	39.16	26.57*	29.28	21.85	9.88	4.72*
2	9	23.74	24.90	17.18	19.42	6.56	5.48

Note:* Comparison before and after NAIT P>0.05

6. 17-OH and 17-KS in Urine

Only were the 17-OH and 17-KS in urine of the patients with benzene poisoning examined. The results showed that the levels of 17-KS in urine in group 1 decreased after NAIT, but the other indices were not changed. See Table 5:

Table 5 The changes of 17-OH and 17-KS in urine (mg/24h)

Groups	cases	17-OH (X±SD) before	17-OH (X±SD) after	17-KS (X±SD) before	17-KS (X±SD) after
1	8	7.7±2.8	8.5±3.1	11.3±3.1	7.1±2.1*
2	10	7.8±3.6	6.3±3.7	7.8±3.6	8.7±5.0

Note: * Comparison before and after NAIT P<0.05.

7. The Immunological Indices

Only were the immunological indices of group 1 and 2 examined. The results showed that there were no significant differences (P>0.05) before and after NAIT.

DISCUSSION

The argument on the exact biological effects of negative air ions has been waging. Krueger A.P. et al held that the negative air ions can stimulate the growth of animals and plants, kill microorganisms, speed the biological oxidation of tissues and cells, reduce the amount of 5-HT in brain and influence further the functions of every system of the organism. But Proling A.P., Anderson I et al took the opposite point of view. They considered that the air ions take hardly significant biological effects, because it accounts for few of air molecules of atmosphere. In recent years, some writers have used the critical design and advanced methods to repeat some classical experiments. Most of the results obtained were not consistent with the previous results. They considered that the biological effects of negative air ions reported previously might be the action of ozone and superoxide anion radical, or the false positive results caused by the factors such as uncritical experimental design. Some workers conjectured that after negative air ions enter the bodies by inhalation, they would take biological effects by stimulating the levels of interferons, endorphins and enkephalins. In our studies, the indices such as symptoms, signs, ECG, hemogram, endocrine, etc. were monitored critically in double-blind fashion. The results showed that there were no significant changes of the indices between treatment group and the control group, and there were no significant differences before and after therapy among each group either. The blood 5-HT levels in group 3 after the treatment were decreased, while so did in group 4. It showed that the decrease in 5-HT in blood may be caused by other factors instead of the negative air ions.

The effct of air ions on adrenal gland are still not clear. Some workers considered that the biological effects of the negative air ions resemble the ones of mineralcorticoid, and the effects of positive air ions resemble the ones of glucocorticoid. Sulman F.G. et al reported that there were no significant changes of 17-OH, 17-KS and catecholamine in urine of human being after negative air ions inhalation, while Milcu S.M. et al reported that the 17-KS dropped after inhalation of air ions. Our results were basically similar to that. As we know, catecholamines in urine not only represents secreting of adrenal medulla but also is the index which reflects the tonicity of sympathetic system. Some workers reported that the inhalation of the negative air ions can increase the tonicity of parasympathetic nervous system, reduce that of sympathetic system and take tranquil effects on human and animals. In our studies, we found that the level of catecholamines in urine in group 1 reduced after NAIT. Perhaps these changes were due to the decreasing in tonicity of sympathetic nerve system. We all know that many factors can affect values of indices above. So the changes of 17-OH, 17-KS and catecholamines in our experiment were not certainly caused by NAIT. It is necessary to study further the relationship between air ions and the functions of adrenal.

In brief, only 17-KS and catecholamines in urine were decreased after NAIT in the patients with benzene poisoning. The other indices of the patients had no significant changes, either in group treated by batyl-alcohol or in the control groups.

Our results indicated that both the negative air ions and batyl-alcohol were not effective for the patients with occupational leukopenia. To investigate the more effective drugs and new therapeutic methods were suggested.

REFERENCES

1. Kotoka,S., Effects of air ions on microorganis and the other biological materials, CRC. Critica Reviews in Microbio.,6;108,1978
2. Krueger, A.P., and Reed,E.J., Biological impac of small air ions. Science,193;1209,1976
3. Proling,A.P.; Are artificial air ions biological relevant climate factor? Int.J.Biometeor, 29(3);233,1985
4. Norgrady,S.M., et al,Ionisers in the managemen of bronchial asthma. Thorax 38(12);912,1983
5. Anderson,I., Effects of natural and artificiall generated air ions on living organisms Int.J.Biometeor, 16. Suppl in Press,1971
6. Goheen, S.C.,et al; Ozone produced by coron discharge in the presence of water. Int.J.Bioметerc 28; 157,1984
7. Wehrer,A.P., Stimulation of interferons an endorphins/enkephalins by electroaerosol inhalation Int.J.Biometero.28(1); 47,1984
8. Li, Anbo, The Reviews of Labour Hygiene.p136 Publishing House of People's Health. 1983
9. Sulman, F.G., Absence of harmful effects c protracted negative air ionisation,Int.J.Bioмeterc 22; 53,1978
10. Milcu,S.M.,et al, Air ions and the urinary 17 ketosteroids, REV.ROUM. MEDENDOCRINOL. 14(3);207,197

EFFECT OF NEGATIVE IONS ON SPONTANEOUS AND CYCLOPHOSPHOMIDUM-INDUCED MICRONUCLEUS FREQUENCY IN MOUSE BONE MARROW POLYCHROMATOPHILIC ERYTHROBLASTS

Xie Lin*, Yin Jinghua*, Wang Yongchao**, Song Yapong*
*Beijing Research & Application Centre for Air Ions;
**Biology Department ;
Beijing Normal University, Beijing, China

Abstract

In this paper, the micronucleus frequency in mouse bone marrow polychromatophilic erythroblasts (PCES) is used as a criterion to explore whether negative ions would have cytotoxicity and would protect mice from the damage induced by a radiomimetic chemical--cyclophosphomidum (CP). The results showed that negative ions accelerate the growth and decevelopment of normal mice and don't result in insidious or chronic damage and that negative ions have a protective function on chronic damage induced by CP.

Introduction

Scientists have known that there are good biological and medical effects of negative ions on plants, animals and human beings, but have not known whether a long-term exposure to negative ions would induce insidious or chronic genetic damage. In addition, some studies have showed that there is a significant protective function of negative ions on acute and chronic damage induced by 60 Co -ray (1,2), but no report has shown whether negative ions would protect animals from a radiomimetic chemical cyclophosphomidum (CP). In this paper, micronucleus frequency in mouse bone marrow polychromatophilic erythroblasts (PCEs) was used as a main criterion to explore these two questions.

Materials and Method

Animal Rooms The two rooms and all equipment used in this experiment were under identical conditions except for air ions and were made with materials which don't absorb negative ions. In Room N, there were four mouse cages with AS-140 type negative ions of oxygen generators (corona effect), the concentration of negative ions was $10-10$ ions/cm at the floors of the cages. In Room C there were four cages with the corresponding ventilators in order to keep the same airflow as the negative ions. During the experiment, the generators and ventilators worked continuously, temperature was kept at 22 2C, relative humidity was 44 5%, airflow was 6.5m /min, ozone concentration was lower than 0.012ug/1 in both rooms.

Animals Eighty male mice of Kunming strain (about 13.9g body wt) were divided randomly into eight groups according to their body weight. Mice in N1-N4 groups were put into Room N and mice in C1-C4 groups were put into Room C. Normal diet was given.

Sample Preparations After the beginning of the experiment, each mouse in N1-N2 and C1,N2 and C2,N3 and C3 groups was given a CP injection with 145mg/kg body weight at 15th day, 30th day and 60th day respectively. Six hours later, they were killed, the bone marrow cells in their femora were collected, cell suspensions were made and centrifugalized with 1000 rpm for five minutes, then cell slides were made and dyed with Giemsa . The mice in N4 and C4 groups weren't given CP injections and were killed and sampled at the 60th day.

Examination and statistics 1000 PCEs on one slide were examined and the number of the PCEs which contain one or more micronuclei was counted under microscope, then micronucleus frequency in PCEs was calculated and treated by bilateral t-test. Some researchers have pointed out that the spontaneous micronucleus frequency in mouse PCEs is 0.1-0.3% and others have shown that it is 0.258+0.04% and that more than 0.5% is abnormal.

Body Weight And adrenal Gland Index During the experiment, mice in various groups were weighted four times. The adrenal glands were taken from ten mice in N4 and C4 groups and weighted by a photoactive balance when their bone marrow cells were collected, then the percentage of adrenal weight to body weight was calculated.

Results

Body weight Table 1 shows that the mice exposed to negative ions are significantly heavier than those in control groups at 30th day and 60th day (P<0.01). This indicated that negative ions accelerate the growth and development of mice.

Table 1. The average body weight of mice in two rooms

Time after experiment (day)	N room		C room	
	mean value	SD/X (%)	mean value	SD/X (%)
1*	13.93**	12.87	13.91	11.58
15	28.22**	2.89	25.98	10.78
30	35.58***	8.22	28.32	12.22
60	36.98***	6.93	32.76	3.15

*   20 mice in each group, 10 mice in other groups
** t test, P> 0.05
*** P<0.01

Frequency of Spontaneous Micronuclei in PCEs

Table 2. shows the micronucleus frequency in PCEs of mice in N4 and C4 groups 60th day after the beginning of the experiment .It is known that there is no significant difference between N4 and C4 groups (P>0.05).This result indicated that the exposure to negative ions for 60 days doesn't induce the genetic damage, e.g. chromosome break.

Table 2. The micronucleus frequency in PCEs of mice in various groups at different times

time after experiment (day)	number of mice	mean value(%)	SD/$\bar{X}$ (%)
15 N1	10	2.31*	14.35
C1	10	2.40	11.11
30 N2	10	1.29**	13.89
C2	10	2.18	6.04
60 N3	10	0.37**	44.23
C3	10	2.94	13.34
N4	10	0.30*	27.22
C4	10	0.38	32.35

* $P>0.05$
** $P<0.01$

Mice in N4 and C4 groups weren't given CP injections

Frequency of Micronuclei in PCEs Induced by CP

It can be seen from Table 2 that the micronucleus frequency in PCEs of mice in N1 group is not significantly different from that in C1 group ($P>0.05$) at 15th day; that the micronucleus frequency in PCEs of mice in N2 group is 41 percent lower than that in C2 group ($P<0.01$) at 30th day; and that the micronucleus frequency in PCEs of mice in N3 group is reduced to normal level at 60th day. This indicates that negative ions protect the mice from CP and that longerterm exposure to negative ions seems to be more efficient.

Index of Adrenal Gland

Table 3 shows the indexes of adrenal glands of mice in two groups at 60th day after the beginning of the experiment. The indexes of adrenal glands of mice in N4 group are significantly higher than those in C4 group ($P<0.01$). The result is the same as the result from rats.

Table 3. Index of adrenal glands in two groups (%)

group	number of mice	mean value	SD/$\bar{X}$ (%)
N4	5	0.01264	12.43
C4	5	0.00990	3.71

60d after beginning of the experiment.
$P<0.01$

Discussion

It has been proved that micronuclei are involved in chromsomal damages. After chromosomes are injured by physical or chemical factors, acentric chromatids or chromsomal fragments stay in the cytoplasm during the mitotic anaphase, then form one or more subnuclei, called micronuclei, in the cytoplasm after the mitotic telophase. Micronuclei exist in nucleate or non-nucleate cells of bone marrow and peripheral blood. However, only the examination of micronuclei in bone marrow PCEs becomes a standard procedure because the micronuclei in bone marrow PCEs are more easily distinguished. J.A. Heddle has pointed out that the examination of micronucleus frequency in PCEs is an important criterion to study cytogenetic damage (5).

CP is an alkylating agent. Its structural formula is

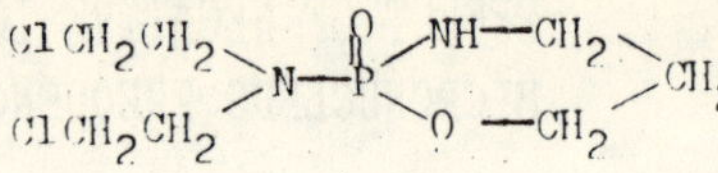

$$\begin{array}{c} ClCH_2CH_2 \\ ClCH_2CH_2 \end{array} N-P \begin{array}{c} O \\ NH-CH_2 \\ O-CH_2 \end{array} CH_2$$

CP itself is not active. After entering the body, CP is transformed into 4-hydroxycyclophosphomidum and aldehydicphosphomidum by cytochrome p450 mixing functional oxidase in liver mitochondria, then transformed into inactive 4-ketocyclophosphomidum and carboxylphosphomidum by some enzymes in liver, kidney or other tissues. However, in those fastgrowing tissues such as hematopoietic organ, gastrointestinal tract, genital cells and cancer cells, there is a shortage of these enzymes, so that 4-hydroxycyclophosphomidum and aldehydic-phosphomidum can't become inactive compounds, Aldehydic-phosphomidum is not stable and can be decomposed into a strong-toxic alkylating group-phosphami nitrogen mustard (Fig. 1). Phosphamic nitrogen mustard acts on cell components such as protein, enzyme and nucleic acid through its alkylating group, but its lethal effect is due to its effect on DNA.

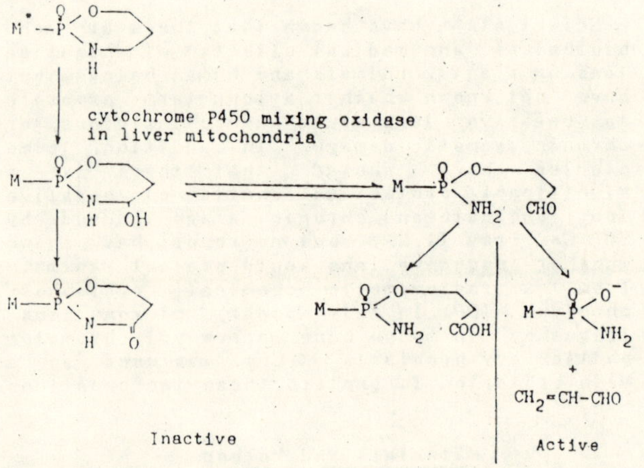

Fig. 1 CP metabolic path after entering body

It has been known that alkylating group can cross-connect with DNA or between protein and DNA, which will result in depurine from DNA. Depurine can cause mistakes in genetic codes or breakage of DNA chains. In addition, the phosphatic group on DNA chain can be alkylated and form phosphoglyceric bond, which will break DNA chains. Normal cells have endonuclease and exonuclease which can excise damaged DNA, then the damaged parts can be repaired if the dosage of CP and the sensitivity of cells is lower (6).

It was observed in this experiment that there is a signifcant protective function of negative ions on CP-induced genetic damages. The protection of negative ions from CP might occur in the following two processes: 1. CP metabolic process after entering body; 2. DNA reparation after it is injured. Table2 showed that the different exposure time to negative ions changes the protection of negative ions from CP. For instance, the exposure to negative ions for 15 days does not change micronucleus frequency, while the exposure to negative ions for 30 days and 40 days reduces micronucleus frequency by 41% and 82% respectively. It seems to indicate that the protection of negative ions from CP occur in the reparation process rather than in the CP metabolic process.

DNA reparation first depends on the genetic component of organisms, but the internal and extraneous conditions under which organisms exist are important to DNA reparation. Many studies have shown that negative ions can improve the physiological state(8-15), and it has been observed in this experiment that negative ions are advantageous to the growth and development of mice, so negative ions should increase the probability of DNA reparation. In addition, DNA exists in mammals in the form of chromatin. Histone in chromatin plays an important role in preserving high construction of chromatin and inhibiting replication, transcription and reparation of DNA chains. It has been proved that catecholamine hormones can evoke a series of reactions through their acceptors, for instance, removing the repression of histone and untying DNA helices(16.17), which will activate DNA reparation if it has been injured. Hansell and Frey pointed out that negative ions accelerate the secretory function of the adrenal cortex and medulla(18,19), and our study also showed that negative ions enhance the adrenal index of mice, indicating that negative ions raise the secretory level of the adrenal gland. This is advantageous to reparation of damaged DNA.

The reparation of mammal DNA needs a series of enzymes such as restriction endonucleases, DNA polymerases, DNA ligases and so on. It has been known that the complexes of steroid hormones and their acceptors can combine with relevant repressible proteins and separate them from their control sites, which reults in the occurence of transcription As negative ions induce adrenal cortex to secrete steroid hormones, DNA transcription will be activated and the enzymes for reparation might be produced.

The higher secetory function of the adrenal gland raises the strain ability of organisms and reduces damage induced by the factors such as cold, wounds, radiation and chemicals. In this experiment, it was observed that the relationship between the protection of negative ions from CP and the increase in adrenal secretion is homogeneous, and it was suggested that negative ions might help reparation of DNA damages induced by CP through adrenal hormones. However, there is a want of direct proofs. We do not know, for instance, whether adrenal hormons would activate the expression of the specific genes concerning reparation and raise the activity of the specific enzymes concerning reparation. These problems need to be explored in future.

Reference

1. Xie Lin, et al, Chinese J. Radiol. Med. Prot. 6(2) p. 73-76. 1986.
2. Guo Xuecong, et al, Acta Genetica Sinica. 13(6) p. 470-475. 1986.
3. Wu Heling, et al, Method and Technique in Genetics. Beijing. p. 222-223. 1983.
4. Huang Xingyu, et al, Examination of Environmental Chemical-Induced Aberration and Cancer, Zhejing. p. 218-235. 1985.
5. Heddle, j. A, et al, Mutat. Ras. 123 p. 61. 1983.
6. Shanghai Medical Industry College, Anti-Tumour Drugs. p. 12-33. 1985.
7. Stent, G. S, Molecualr Genetics. 1971
8. Slote, L, Int. Conf. on Ionization of Air. p. 20-25. 1961.
9. Kruegue, A. p, Int. Conf. on Ionization of Air. p. 17-19. 1961.
10. Lotmar, R, Int. J. Biometeor. 4. p. 323-327. 1972.
11. Bhartion, Int. J. Biometeor. 1. p43-52. 1978.
12. Hinsull, S. M, Int. J. Biometeor. 4. p. 327-329. 1981.
13. Sulman, F. G, Int. J. Biometeor. 1. p. 53-58. 1978.
14. Wehner, A. P, Int. Conf. on Ionization of Air. p. 15-16. 1961.
15. Hansell, C. W, Int Conf. on Ionization of Air. p. 26-30. 1961.
16. Beijing Medical College. et al, Biochemistry. Beijing. p. 280-283. 1978.
17. Li Zhengang, Molecular Genetics. Hefei. p. 42-43. 1985.
18. Hansell, C. W, Int. Conf. on Ionization of Air. p. 26-30. 1961.
19. Frey, A. H, TRE. Trans. Bio. Med. Electronics, p. 12-16. 1961.
20. Bacq, Z. M: 1961. Shen Shumin et al, Fundamental Radiobiology. Beijing. p. 283-285. 1965.

TREATMENT FOR TISSUE INJURY BY USE OF MICRO-POROUS POLYMERIC ELECTRET FILM

Sun Caomin, Cui Mi and Zhu Hesun
Beijing Institute of Technology
Beijing, P O Box 327, China

Abstract

The electret is made of micro-porous polymeric film. Its both sides are uniformly negatively charged with a mass charge density about -8×10^{-8} c/g. Applied to the injured and painful area, it can have curative effects of invigoration of blood circulation and elimination of blood stasis. Animal experiments and clinical observations indicate that the micro-porous polymeric film can cure both acute injury and relieve chronic pain of soft tissues with an efficacious rate of about 90%. It can also accelerate the union for fracture. It shows that the electrect has a broad applicable prospect in biomedical engineering.

Introduction

It can be traced back to late 1960's, that many scientists, such as S. Mascarenhas (Brazil), Yasuda (Japan) and Fukada (Japan), began to investigate the phenomena in bioelectret. They discovered that many bio-materials may enter into an electret state, preserving electric charge and permanent polarization. Later, many bio-polymers, such as collagen, protein, DNA, RNA, some enzymes and blood vessel were proved to have electret effects.

In early 1970's, experimental research in healing of wound by stimulation of external electromagnetic field came forth. Professor Bassett of Coulumbia University inserted a pair of platenum needle-shaped electrodes to the both ends of a fractured bone. The stimulation by less than 20 µA current can accelerate the formation of callus near the negative electrode and union of the fracture. Afterwards he replaced the inserted needles by Helmholtz coil which generates a changing electric field by a changing magnetic field and attained a similar result. A physical research from Purdue University discovered that there is an obvious volatge difference between the wounded and the normal areas of derma of mammal. It may be as high as 200 mv/mm. An electric current of about 1 µA flows into each millimeter of the perimeter around the wounded area. Furthermore they proved that spontaneous healing has relation to electric vectors and external electric field may accelerate wound healing. Japanese Professors Inoue and Fukada used teflon electret film or PMLG (r-Methyl-L-Glutaminates) electret film to bind up the fractured part of a rabbit femur. Week by week, examinations by roentgenogram showed that callus formed in two weeks and rigid callus had come onto being after four weeks.

Having analysed the above results, we considered that since there is an obvious voltage between the wounded area and the normal area of an organism and a negative electrode set at the wounded area has contribution to the wound healing, why not we adopt the simpler and easier way by applying a sheet of negatively charged electrect to the injured and painful area of the soft tissue to cure the injury? Moreover we wonder if we can utilize the polarizable property of bio-electrects by wrapping a sheet of electrect with negative charge over the skin to advance union of a simple fracture. We began the research from 1982. Numerous animal experiments and clinic tests for four years showed that the artificial electret film can cure various acute and chronic injuries of soft tissues and accelerate union of simple fractures so long as it preserves appropriate amount of negative charge and is closely applied to the skin of the injured area.

Micro-Porous Electret Film for the Treatment

There are many materials of which electret can be made. Electret for medical use should be: (1) nontoxic and incapable of causing allergic dermatitis; (2) capable of ventilation over the skin of the injured area and keeping normal blood circulation; (3) having enough capacity to store charge and appropraitely large dielectric constant; (4) available abundantly.

Having optioned the samples, three kinds of microporous electret films shown in Table 1, are taken into consideration, their properties are listed in Table 2 & 3

Animal Experiments

The experimental objects are normally healthy rabbits. The goal is to examine the curative effects of the micro-porous films through contrasting experiments. The experimental work is to compare the sectiones of the normal tissue, the injured tussue, the injured tissue after treatment by external application of normal electret film and the injured tissue after treatment by inadequately charged contrast electret. The trunk or limbs of the rabbit is injured with an impact of 2 5 kgm/s impulse, the injured area being 10x10 cm^2. Two hours after the injury, take the sample and separately apply a normal

Table 1 Appearance of Microporous Electret Film

No.	appearance	size of pores	dielectric constant ε	toxity and side effects	usual uses	material	clinic-curative effect
A	transparent film	~1μ	~2.1	no	medical	teflon	good
B	adhesive-bonded cloth by superfine fibre	~25μ	~1.5	no	medical and dust filtration	P.P.	good
C	milky white film	5~20μ	~5	no	medical	nitro-acetate fibre	good

Table 2 Properties of the Micro-porous Electret Films
(measured at 23°C and R.H. 50%)

No.	surface potential (v)	surface resistivity (Ω)	volumeric resistivity (Ω cm)	dielectric constant ε	stored charge 10^{-8} c	mass charge density 10^{-8} c/g	surface charge density 10^{-6} c/m^2	sample area cm × cm
A	-8000			2	4.25		5.86	6.6x11
B	-650	2.04x10^{16}	4.86x10^{17}	1.2--1.5	4.03	8.06	4.03	10x10
C	-1000			5	5.0		3.36	8.7x17.5

Table 3 Decay properties of the micro-porous electret films

No.	initial surface potential (v)	after starage for 3 months	after storage for 6 months	after storage for one year	after storage for two years	after storage for 3 years
A	-800	-800	-800	-780	-780	-780
B	-680	-680	-600	-600	-600	-600
C	-1000	-1000	-1000	-980	-980	-980

electret film and a contrasting electret film to two different injured rabbits. The results of examinations are shown in Table 4 and the pictures to show the changes are in Fig.1 (see the color pictures).

The results indicate that: (1) After application of the electret film, a long period of arterial hyperaemia arises at the injured area so that more oxygen and nutrients can be brought to the tissue and hydrops can be eliminated faster; (2) Inflammatory reaction is reduced owing to the application of the electret film.

Fig.1(the color pictures) Sectiones of the Tissues
a. section of a normal tissue ;
b. the section two hours after injury (edema and exudation of inflammatory cells);
c. 48 hours after application of normal electret film (blood vessel hyperaemia and reduced inflammatory cells);
d. after 48 hours without application the film (serious edema and increaed inflammatory cells);
e. 48 hours after appliation of inadequately charged electret film (deficient hyperaemia in blood vessel and increased inflammatouy cells).

Table 4

test items	degree of edema	blood vessel hyperaemia	exudation of inflammatory cells
apply normal electret film	light	remarkable	few
apply contrast electret film	serious	insufficient	many
no application of the film	serious	insufficient	many

Clinical Application

Between 1983 and 1987, over 700 cases were observed clinically in Xiehe, Friendship, The Chaoyang Red Cross and PLA-General Hospitals in Beijing. For treatment of injury of soft tissues, the total effective rate reaches 90% as shown in Table 5.

There is not any allegic reaction and side effect in the 728 cases cited above. The patients felt warn and comfortable when the film is applied. Pain can be relieved in 5 min. at the fastest or in 24 hours at the slowest. Generally for acute injuries, swelling vanishes and symptom abates in 48 hours. It is really a means more convenient for use than others.

Table 5

states of illness	total cases	healed	remarkably effective	effective	no effect	total effective rate(%)
acute injury of soft tissues (sprain, bruise etc.)	186	32	84	54	16	91.4
chroni injury of soft tissues(arthritis, tenosynovitis periarthritis of shoulder, pains in waist and leg)	311	27	110	135	39	87.4
mamma swelling and pain	173	3	54	101	15	91.3
simple fracture	58	10	48	0	0	100
total	728	72	296	290	70	90.4

Exploration of the Mechanism

We have not found any existing works on the mechanism of curing injury of soft tissues and fracture with electret films. Here we are trying to approach the subuect.

Various soft tissues, e. g. muscles, tendons etc, appear to be bio-electrects with piezo-electric, pyro-electric and charge-storage effects. When they are damaged, their polarized states are disturbed. For the polarization of normal cell membrane, the outer appears positive, while the inner negative. After injury, potassium ions (K^+) in the cell membrane are reduced and the membrane potential decreases. Furthermore scatter of K^+ and injury current would cause inverse piezo-electric and inverse pyro-electric effects, which affect nerves and induce local ache or heat generation. The voltage between the normal area and the injured area of the soft tissue can be reduced by applying the electret film to the injured area. (Fig.2)

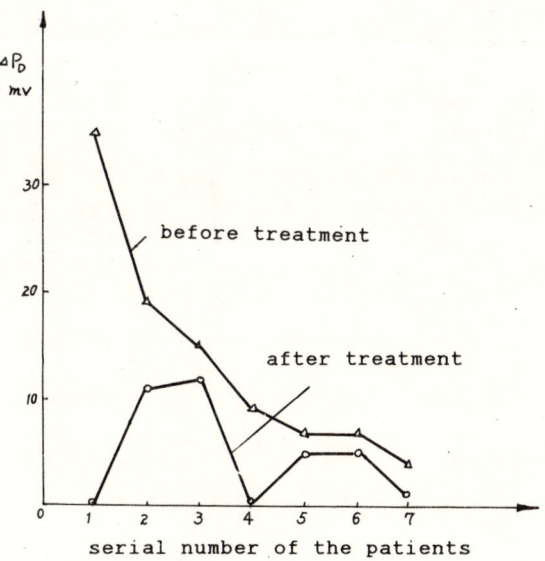

Fig 2 changes of injury potential

As for the changes in injured soft tissues and due to application of the electret film to injured area, we may make an explanation in terms of tri-capacitor model on the basis of the animal experiments and the clinical observations.

Assume that there are three capacitive states in a soft tissue. C_1 is a crystalline electret. C_2 is an amorphous electret, C_3 is an electret of condensed phase (Fig.3). C_1 is an electret easily to be activated. Normally they are in equilibrium. As the tissue is in motion, by piezo-electric effect C_1 charges C_3 and C_2. In fact C_2 and C_3 also charge C_1.

As it is injured, there appear obvious voltages between C_1 and C_2 and C_3, named "injury voltage", causing disequilibrium and injury current. When a negatively charged electret film is applied to the injured area, it can restore the vigor of C_1 and generate an electric field E_1. Because C_1 charges C_2 and C_3, vigores in C_2 and in C_3 are also increased. The field of the whole organism is increased to E_2 and its spontaneous restorative ability is also strengthened, causing wound healing and reduction of injury voltage.

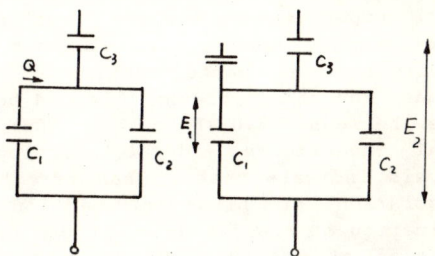

Fig 3 Tri-capacitor model

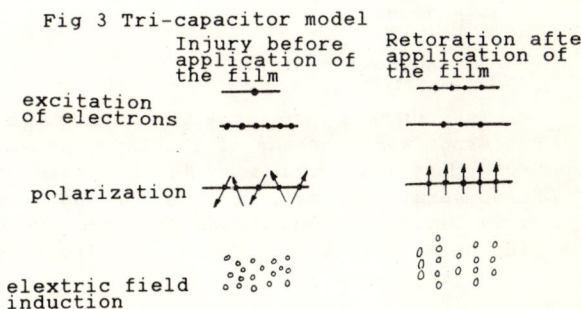

Fig 4 changes in electric properties as the film is applied

Moreover it is inferred that after injury of the tissue, the excited electrons in its bio-molecules are reduced and its polarization is distroyed so that both its own internal field inductive ability and its resistance are reduced. With the electret field applied, an additional field is exerted, causing increase of excited electrons and restoration of its polarization and internal field inductive ability. This leads to recovery of the tissue. Therefore it is a treatment to strengthen the spontaneous restorative ability and no side effects may appear.(see Fig.4)

In the light of the animal experiments and clinical observations we consider that the negatively charged micro-porous electret film functions as follows:

a. It functions as a negative electrode to repair the injured organism without an electrode needle inserted.

b. It provides a compensational potential to reduce the injury voltage and restore the injured organism to have the normal electric characteristics.

c. It also functions as a weak-current stimulation and advances wound healing.

Based upon above inferences we have also cured angina pectoris, phlebitis, shingles etc. by this means.

Conclusion

Four years' clinical observations and animal experiments indicate that negatively charged micro-porous electret film is effective in wound healing of soft tissues. The charge of the film had better range between -5.0×10^{-8}c and 10×10^{-8}c. Animal experiments and histocytological analysis indicate that it has effects of invigoration of blood circulation and elimination of blood stasis, it can reduce exudation of inflammatory cells and has functions of dephlogistication and delumescence.

Acknowledgement

The authors are greatly indebted to Professor Wan Minsheng of The Chaoyang Red Cross Hospital, Professor Ma Chengxuan of PLA-General Hospital, Professor Liang Dong of Ji Shuitan Hospital and Professor Zhao Ping of Hiehe Pospital for their effective supports.

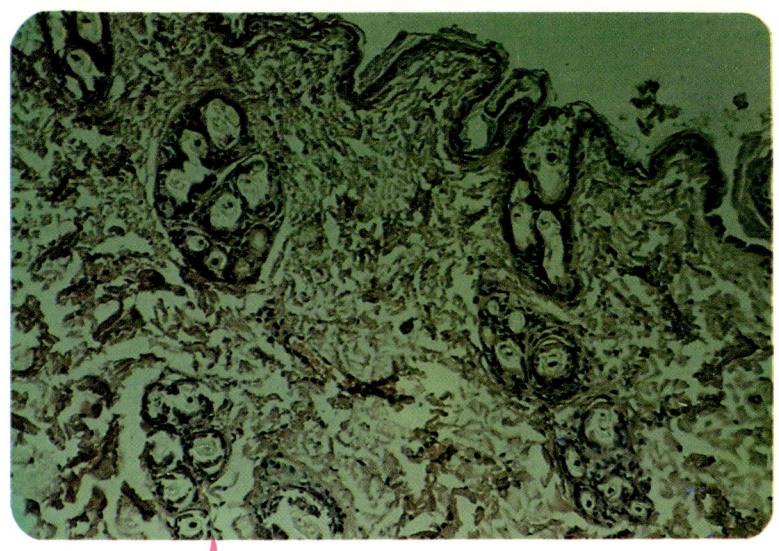

Fig. 1

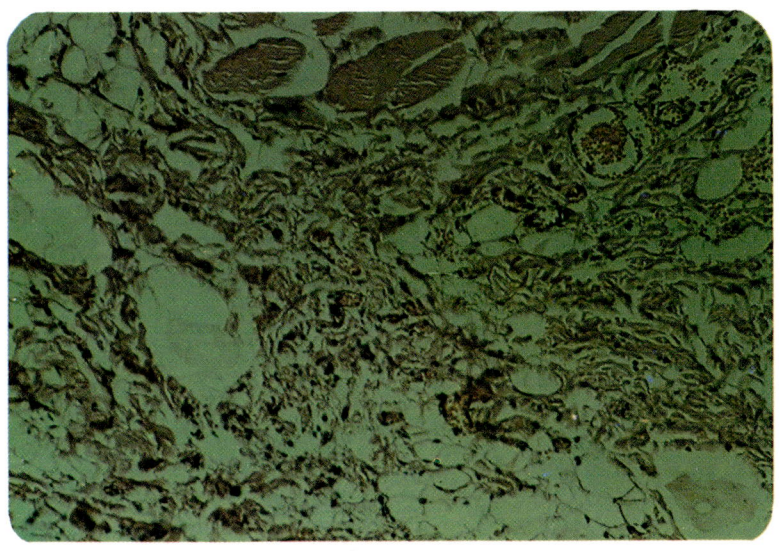

Fig. 2

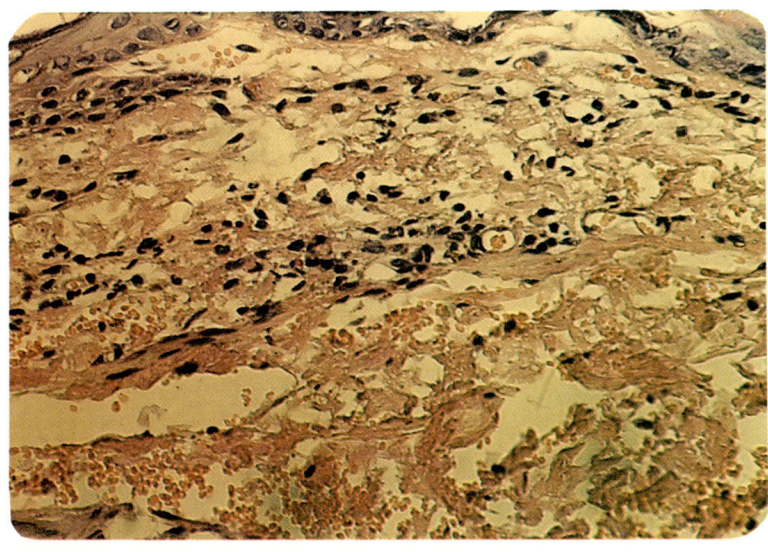

Fig. 3

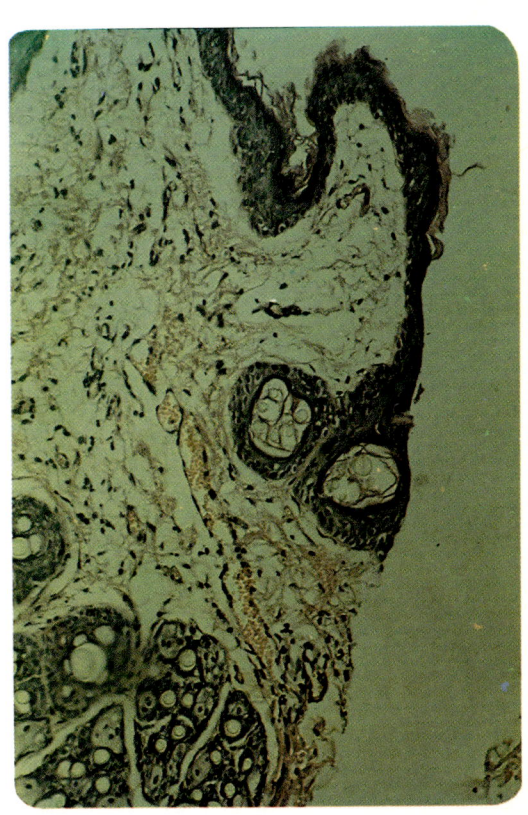

Fig. 4

Fig. 5

ELECTRICALLY STIMULATED BONE GROWTH BY ELECTRET FILM

Zhang Guoguang, Wang Shoutai and Zhang Hekang

Applied Chemistry Department, Shanghai Jiaotong University

Abstract

A new plasma injection method is used to obtain perfluoro-ethylene-propylene copolymer (FEP) electret film. Since 1982 we have been in collaboration with Shanghai Central Railway Hospital, Zhong Shan and Ruijin Hospitals for clinical studies and animal experiments. Healing processes for rabbits fracture are histologically and mechanically examined and, moreover, investigated by x-ray and trace elements determination. There are more than four hundred fracture cases. Both clinical studies and animal experiments show that FEP electret film may promote the healing effect on fresh fracture, delayed union and non-union cases.

1. Piezoelectric Phenomenon of Bone

All through the ages, it was believed that a callus appeared only because of fracture. Küntscher (1) and others, however, proved that callus formation could be induced by mechanical, thermal or chemical irritation. This phenomenon is known as "callus without fracture". They drove two plastic nails through both ends of an intact rabbit femur, and metal spring are hooked externally between the nail to exert axial pressure along the bone (Fig.1). Uptake of isotope P^{32} becomes the greatest on the spot where the callus is formed after injection of the isotope into the experimental animals. Whereas the other regions of the same femur and the opposite half of the animal takes up P^{32} minorly. This indicates that the mechanical irritation promotes the metabolism of bone cells, resulting in the proliferation of the periosteum to form new bone (2). From an electrophysiologic point of view, active regions in biological tissues are always

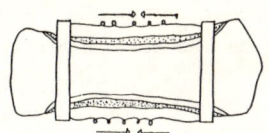

Fig.1 compression force applied parallel to the born axis. Darkened areas represent subperiosteal new bone formation.

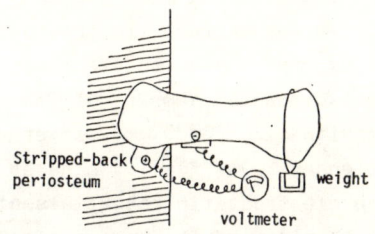

Fig.2 Drawing depicting bone fixed on one end with weight applied to the other end. Region under compression is electronegative.

electronegative compared to resting regions. In 1953 Iwao Yasuda experimentlly confirmed this phenomenon. He fixed one end of a long tubular bone to a stationary stand while a weight was exerted on the other end of the bone to bend it (Fig.2). Voltmeter readings showed that region under compression is electronegative as compared to the non-stressed portion. In contrast, regions under tension were electropositive. When mechanical vibrational wave was applied, from oscilloscope it is observed that cycle of variations in potential is synchronal with the mechanical vibrational wave. This is a quite piezoelectric effect of bone. Bones boiled in saline solution for 10 minutes or the dried bones still exhibited piezoelectricity. Therefore, piezoelectricity of bone has no relationship with bone viability. On the basis of these experiments Iwao Yasuda proposed the following scheme: Force→Electricity →Callus Formation. For other sorts of stimulation, they are first converted to electrical energy, and then act on the cells to cause them to proliferate and form the callus. Since the amounts and polarity of electricity can be controlled easily, electrically stimulated bone growth techniques has been developed quickly.

2. Plasma-injected F_{46} Electret Film

We apply the charge injection by plasma. In order to have effective medical treatment the film electret must possess fixed charge. We use surface electrostatic potentials to express charge in electret in volt. For two hundred samples of FEP electret film with size 20*20 mm², after plasma injection at 2000v for 1.5 minutes, we measure the electrostatic potentials at surfaces of both sides of the sample after storage of 2,3,75,100 and 160 days respectively. We use simple linear unbiased estimate and Kolmogorove test. Surface potentials obtained obey Weibull distribution ($\alpha = 0.2$). In table 1, we list parameters of Weibull distribution. Distribution curve is shown in Fig.3. As time goes on, characteristic value η gradually decreases and approaches to about 70v. The value of shape parameter m is between 1.95 and 2.23 with small variations. TSC spectra of film electret are shown in Fig.4, and only one peak appears in the TSC spectra. The temperature of main peak is between 150° and 190°C, indicating that F_{46} film electret possess better stability.

3. Curative Effect of F_{46} Electret Film

Femurs of forty healthy rabbits raised by Shanghai Central Railway Hospital (3) were cut off for the experiment. And then radiuses of fifty five female rabbits of New Zealand origin were cut short by 3mm for study in Shanghai Zhong Shan Hospital. Of the fifty five animals, forty were self control, thirteen were control of variant, and the two others were wrapped with

Table 1 Parameter of Weibull Distribution

Storage Time (day)	2	30	75	100	160
Shape Parameter m	2.23	1.95	2.23	2.15	2.06
Scale Parameter n	105	74	73	71	64

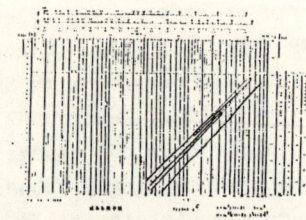

Fig.3 Weibull distribution curve of surface electrostatic potential under different storage time.

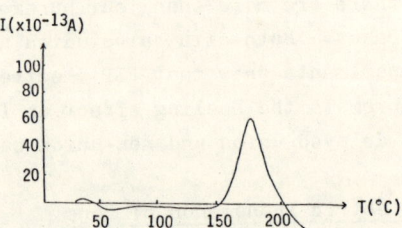

Fig.4 TSC spectrum of electret film
Ampde voltage 2Kv
Operating time 2 minute

electret film at both sides of the radiuses. Electret film or non-injected film was changed once a week. Under double blind method the animals were subsequently sacrificed 1,2,2.5,3,3.5,4 and 6 weeks after the operation respectively for investigations by x-ray, histology, biomechanics and atomic spectroscopy with consistent results. On electret film side, callus appears after two weeks and rapidly increases between 2.5 and 3.5 weeks, at the end of four weeks the fracture healed up. On non-injected regular film side, callus begins to show after 2.5 weeks and increases slowly, the fracture does not heal up until 6 weeks.

Shanghai Central Railway Hospital, Zhongshan Hospital and Ruijin Hospital etc. engaged clinical study. [4,5,6] There are more than four hundreds cases of fracture. Both clinical study and experiment show that FEP electret film promotes healing of fresh fracture, delayed union and non-union. The electret film can shorten healing period of fresh fracture to about two thirds.

4. Mechanism of Electrically Stimulated Bone Growth

Direct effect: (7,8,9,10)

A direct effect must activate the cell's cyclic-AMP system, that is, must activate the intracellular or second messenger which triggers various enzyme systems to bring about specific physiological responses of the bone or cartilage cell, therefore induced osteogenesis.

Indirect effect: (11,12,13)

Effect of electrical stimulation on the cell's microenvironment is that the local tissue oxygen tension is decreased and the local tissue environment becomes more alkaline. Both is favorable to osteogenesis.

5. Conclusion

Electret film possess greater curative effect for delayed union and non-union. It is easy to operate. In contrast with animal experiment, it can shorten healing period of fresh fracture to about two thirds. Through local effect, it promotes healing of fracture. Healing mechanism is pending for further study.

References

(1) Küntscher, G. Zentr. Chir. 68:857, 1941
(2) Yasuda I. Clin. Orthop. 1977;124:53
(3) Shen Jun Hui, Chinese Orthop. 1985;5(6):363
(4) A Booklet of Directions for Broken-bone Curing Film, Jiaotong University, 1986.
(5) Chen Zhongwei, Trauma Magazine, 1986;2(1):1
(6) Zheng Chunbo, et al. Acta Universitatis Medicinalis Secondae Shanghai, 1987;7:4
(7) 深田荣一 静电气学会志 1980;4(1):55
(8) 井上四郎 静电气学会志 1980;4(5):290
(9) Brighton Ct. Clin. Orthop. 1977;124:106
(10) Wang Shou Tai, External Medicine 1985;4:195
(11) Hoppenstall: Fracture Treatment and Bone Healing P77, Saunders, Philadelphia 1980.
(12) Brighton CT. Clin. Orthop. 1975;107:277
(13) Brighton CT. J. Bone Joint Surg.(AM) 1969,51:1383

ACUTE EXPERIMENTAL RESULTS OF EFFECTS OF HIGH-VOLTAGE ELECTROSTATIC FIELD ON ANIMAL ORGANISMS

Ye Jia-ming

(Zhejiang University)

Wang Rong-yi

(Dalian Industrial College)

China

Abstract

Results of acute experiments showed that electrostatic field with efficient intensity induced manifold functional variations in experimental animals: respiratory rate of rabbits increased, arterial blood pressure of rabbits increased or decreased, amplitude of R-wave in ECG of rabbits varied, force of myocardial contraction and heart rate of toad hearts in vitro and those in vivo varied,
rhythms and amplitude of EEG waves of albino rats and rabbits increased higher-voltage electrostatic field resulted in abnormal EEG waves.

Introduction

History of development of sciences and technology had proved repeatedly that the apply and popularization of many new techniques always bring some effects on the environment of mankind. Recently, high-voltage direct current transmission, electrostatic dust removing and other electrostatic techniques have been applied more and more than ever, so scientists are duty-bound to find out the effects of high-voltage electrostatic field on human and animal organisms.

The effects of electrostatic field on living organisms were studied not enough[1-8], no systematical theoretical elaboration has been developed, and so far there is not unanimity of views on whether the electrostatic field can make injurious effects on human and animal organisms. This paper reported the results of acute experiments on effects of high-voltage electrostatic field on animal organisms.

Materials and Methods

85 rabbits, 10 albino rats and 82 toads were used in the experiments, among them, rabbits were used for respiratory movement, arterial blood pressure, ECG and EEG experiments, albino rats were used for EEG experiments and toads were used for experiments of heart beat.

Rabbits used for respiratory movement experiments were not anesthetized and were fixed prostrately by belts and cloth cover, its respiratory movement was recorded by self-made respiration sensor. Rabbits used for blood pressure experiments were anesthetised by urethane (ethyl carbamate) intravenous injected, and heparin was intravenous injected for blood coagulation-proofing, arterial blood pressure was recorded by arterial carotid intubation. Experimental toads were grouped in two groups: in Group No.1, the hearts in vitro were infused with Straub's method, and in Group No.2,

---

* Projects Supported by the Science Fund of the Chinese Academy of Sciences.

** Members taken part in experiments: Gao Lan, Hui Xiu-juan, He Li, Li Zi-wei, Zhang Xin-li.

the hearts in vivo, toads were anesthetized by urethane injected subcutaneously and then its abdominal wall was anatomized and the heart was linked to mechanicelectrical transducer for recording the heart beat. In the above experiments Two-channel Physiological Recorder Model LMS-2A made in Chengdu Instruments Factory was used for recording. Some rabbits used for ECG experiments were motion-prevented by muscular relaxor, and the others were local-anesthetized by novocaine in the tip of limbs, the electrocardiosignals were leaded by needle electrodes and ECG of standard lead II was recorded by Electrocardiograph Model XDH-2 made in Shanghai Medical Instruments Factory. Albino rats used for EEG experiments were anesthetized by ether inhalation or pentobarbital sodium injected peritonially, rabbits used for EEG experiments were motionprevented by muscular relaxor or local anesthetized, then the animal's skull top skin was dissected and needle electrodes were put into os frontalis, os parietale and os occipitalis, cortical EEG (electrocorticogram) was recorded by Electroencephalograph Model ND-82B made in Shanghai Medical Electronic Instruments Factory.

The positive plate electrode for applying electrostatic field was grounded, experimental animal was set upon this electrode, the negative high-voltage electrode (plate or needle) was hanged above the animal. Direct-high-voltage was supplied by High-Voltage-Direct-Supply Model MZ-3B made in Department of Physics, Northeastern Normal University.

Experimental Results

1. Effects of high-voltage electrostatic field on respiratory movement of rabbits:

The respiratory rate of 25 experimental rabbits before exposed to electrostatic field showed individual difference, but respiratory movement of all the animals were regular, its rate and amplitude hadn't obvious spontaneous fluctuation.

214 experiments of waking rabbits exposed to electrostatic field were made, intensity of applied electrostatic field were 100 to 450kv/m, exposed duration 60 or 120sec per time with 5-10 min interval between each experiments. Results showed that respiratory movement of waking rabbits were very sensitive to electrostatic field: respiratory rate obviously increased after beginning of action of high-voltage electrostatic field with a short latent period (less than 1 sec in some rabbits). Commonly, the respiratory rate increased most remarkably at the first 10-20 sec of electrostatic field action, then the respiratory rate slightly decreased but still higher than the contrast value until the end of electrostatic field actions. In some rabbits during the period of electrostatic field action the respiratory

rate increased again after its decrease. Commonly, the respiratory rate of first 30 sec of electrostatic field action exceeded the contrast value by several tens percent, in a few rabbits respiratory rate increased by about 10%, and in several rabbits the rate increased by a factor of almost 100%. Results of statistical treatment showed that the respiratory rate of waking rabbits during electrostatic field actions made very remarkable diversity to the contrast value ($P<0.001$).

In the experiments, electrostatic field from 100kv/m to 450kv/m all effectively induced the respiratory rate increase. Within this scope of intensity degree of respiratory rate increase didn't directly correlated to the intensity of applied field.

After the end of electrostatic field actions, commonly in a duration of about 10 sec the respiratory rate of rabbits recovered to near the initial value.

During the electrostatic field actions of the above intensity the amplitude of respiratory movement of experimental rabbits didn't variate obviously.

17 rabbits from the 25 experimental animals were anesthetized by urethane injected peritonally after performing the experiments of exposed to electrostatic field at waking state, then they were exposed to the same electrostatic field at anesthetic state. Intensity and action duration of electrostatic field were the same as above, 126 experiments were made.

Respiratory rate and amplitude of anesthetic rabbits were obviously lower than waking state. During the actions of above intensity of electrostatic field, respiratory movement of anesthetic rabbits almost didn't variate: from the 17 rabbits, except the only one that was anesthetized too lightly, respiratory rate of the other 16 rabbits made no remarkable diversity to the contrast ($P>0.05$). Amplitude of respiratory movement didn't variate commonly, only in some anesthetized rabbits a deep breath was made at the beginning of electrostatic field actions. Removing of the electrostatic field didn't influence the respiratory movement of anesthetized rabbits.

2. Effects of high-voltage electrostatic field on arterial blood pressure of anesthetic rabbits:

Before exposing to electrostatic field arterial blood pressure of 20 experimental rabbits were stable on the whole. In the recorded blood pressure curves, first grade waves (heart beat waves) and second grade waves (respiratory waves) were very clear, third grade waves were seldom seen. Blood pressure value of most rabbits were 80 to 100 mmHg.

Duration of electrostatic field actions were 60 sec per time with 5 to 10 min interval. Intensity of electrostatic field lower than 50kv/m didn't induce blood pressure variation of anesthetized rabbits. 100 to 350kv/m electrostatic field commonly induced variation of blood pressure. In some deep anesthetized rabbits responses arised until the electrostatic field rise up to or over 200kv/m.

Reactions of different individual were not quite the same evidences were as follows:

(1) Electrostatic field induced blood pressure increase. In some rabbits, after the beginning of electrostatic field action, blood pressure increased by 15 to 20 mmHg with a latent period of several seconds, and kept this high blood pressure level during the electrostatic field acting period. Blood pressure recovered to the initial value in a duration of several minutes after the end of electrostatic field actions. In a few of rabbits only arised a short time blood pressure increase during the period of electrostatic field actions. In other one or two rabbits, during the field actions period, blood pressure increased and decreased alternately.

(2) Electrostatic field induced blood pressure decrease with a latent period of several seconds or over than ten seconds. Pattern of the blood pressure decrease were as follows: sustained decrease, once short time decrease and multiple decrease.

(3) Electrostatic field induced blood pressure fluctuation with a latent period of several seconds. The blood pressure increased to a level higher than the initial highest value and decreased to a level lower than the contrast lowest value, fluctuation was shown during the whole period of electrostatic field actions.

3. Effects of high-voltage electrostatic field on ECG of rabbits:

All contrast ECG of 20 experimental rabbits were sinus cardiac rhythm, most of the QRS-waves-group were RS-mode, and a few were qR-or Rs-or R-mode. Abnormal ECG hadn't been seen.

Intensity of applied electrostatic field were 100 to 300kv/m, active duration were 60 sec per time with interval of 5 min. In most of 81 experiments, electrostatic field induced variations of R-waves of ECG of experimental rabbits, among which, amplitude increased--50 times, amount to 61.73% of the total experimental times; amplitude decreased--28 times, amount to 34.57%; amplitude didn't variated--3 times, amount to 3.7%. Results of statistical treatment showed that R-waves increase or decrease both made obvious diveristy to the contrast value.

4. Effects of high-voltage electrostatic field on heart beat of toads:

41 toad hearts in vitro were exposed to electrostatic field for 30 sec per time with interval over 5 min. Experimental results showed that during 660kv/m electrostatic field actions and after its remove force of myocardial contraction in vitro made no remarkable diversity to the contrast value. But during 1000kv/m and 1330kv/m electrostatic field actions, myocardial contraction force were obviously inhibited, results of t-test showed that diversity were very remarkable ($P<0.01$). After removal of electrostatic field, myocardial contraction force recovered gradually and 1000kv/m group recovered faster than 1330kv/m group.

High-voltage electrostatic field also inhibited the heart rate of toad hearts in vitro. During 660kv/m field actions heart rate had been obviously lower than the contrast ($P<0.05$), after remove of the field, the heart rate recovered very fast. Inhibition on heart rate of 1000kv/m and 1330kv/m electrostatic field were more strong, and the heart rate hadn't recovered at the period for 1 min. after removing of the field but decreased furthermore.

41 hearts in vivo were exposed to electrostatic field for 60 sec per time with over 5 min interval. Applied 660,1000,1330 and 1660kv/m field induced increase of myocardial contraction force, and made very remarkable diversity to the contrast ($P<0.001$). After removal of 660kv/m field, myocardial contraction force recovered quite fast; but after removal of the field in the other three groups myocardial contraction force decreased: not only very obviously lower than the force during the period of electrostatic field actions, but also very remarkably lower than the contrast value. It suggested that the action of strong electrostatic field on myocardial contraction force of toad hearts in vivo possess negative after-effect.

All of above four intensity of electrostatic field induced heart rate decrease in the heart in vivo. At the first one minute after removing of field, heart rate of all the four groups were still very remarkably lower than the contrast value; in the 1330kv/m group, heart rate were even very obviously lower than the heart rate during field actions. Those of the other three groups were lower than the latter but hadn't obvious diversity.

5. Effects of high-voltage electrostatic field on EEG of albino rats and rabbits:

The cortical EEG (electrocorticogram) of 10 anesthetized albino rats before exposed to electrostatic field varied from the anesthetic depth: during light and moderate anesthetization the EEG were typical θ-waves commonly, and in the EEG of some animals appeared spindly sleeping-waves; during deep anesthetization EEG were typical δ-waves.

Applied 300kv/m electrostatic field on the background of θ-waves EEG, the field action induced rhythm and amplitude of waves increase by about 50% in most albino rats. In a few animals the field also induced appearance of bursts of high-amplitude peak-waves, its amplitude were over twice as large as the background-waves, its frequency were about 15Hz and every burst lasted 1.5 to 2 sec. In a few experiments at the beginning of electrostatic field actions arised these high-amplitude peak-waves, then transformed into epileptoid-waves of composite spina-slow-waves mode.

On the condition that the background EEG were spindly sleeping waves (composite fast-slow-waves) electrostatic field made the fast-waves-phase of composite waves prolonged by twice or several-fold, its amplitude increased too.

When the background EEG were low-rhythm δ-waves, electrostatic field made the EEG of most animals transforming into θ-waves, and those of a few rats transforming into spindly sleeping-waves, in a few experiments arised K-composite-waves.

The electrocorticogram of 20 nonanesthetized rabbits before exposed to the electrostatic field were typical α-waves with frequency of 6 to 12 Hz. In most of experimental rabbits there were bursts of moderate-fast-waves (13--16Hz) placed in between the α-waves with slightly lower amplitude. β, θ and δ-waves were very seldom seen, only in one or two animal there was short-time burst. Abnormal EEG waves hadn't been seen commonly.

When exposed to 60, 90, 120---270kv/m electrostatic field the variations of waking rabbits EEG were increase of rhythm and amplitude, among which:

In most of rabbits which background-EEG mainly were α-waves the electrostatic field induced increase of α-waves rhythm and amplitude. In some experiments the rhythm of α-waves increased but its amplitude slightly decreased, it suggested that the degree of asynchronization of cortex cells activity increased.

In 7 rabbits which background-EEG had not moderate-fast-waves electrostatic field induced bursts of moderate-fast-waves, but the threshold differed from individuals. In the other 13 rabbits which background-EEG possessed moderate-fast-waves bursts, electrostatic field induced increasing of the times of bursts and rhythm of moderate-fast-waves, waves amplitude increased in most of the rabbits but slightly decreased in the others.

From 18 rabbits which background-EEG didn't possess β-waves, electrostatic field induced β-waves bursts in one half of them. In the two rabbits which background-EEG possessed β-waves, electrostatic field induced increasing of the times of β-waves bursts.

60 and 90kv/m electrostatic field didn't induce abnormal EEG waves. 120 to 270kv/m electrostatic field resulted in abnormal EEG waves (peak-waves, high-amplitude-spina-waves, peak-waves-bursts, spina-waves-bursts, epileptoid high-amplitude composite spina-slow-waves, etc.), appearance frequency of abnormal EEG waves tended towards increase with the intensity of applied electrostatic field.

Discussion

Results of experiments showed that during the actions of an efficient intenisty of electrostatic field, the respiratory rate, arterial blood pressure, ECG, myocardial contraction and EEG all obviously varied, it proved that high-voltage electrostatic field is an effective environmental physics stimulus factor which can induce variations of manifold functional activities of animal organisms.

Experimental results suggested that the high-voltage electrostatic field stimulated nervous system strongly. For exmaple, respiratory rate of waking rabbits increased remarkably during the electrostatic field actions but those of anesthetized rabbits almost didn't been influenced. In the EEG of albino rats and rabbits during electrostatic field actions the waves rhythms and amplitudes increased commonly, it suggested the cortex cells activity were stimulated. Abnormal EEG waves appearance suggested that the high-voltage electrostatic field as an environmental physics factor at least is a too strong stimulus that make the animals feeling uncomfortable. The body responses of albino mice in a high-voltage electrostatic field that were seen in another experiment is an exmaple too: after the beginning of electrostatic field actions the mice were deeply agitated, run for escape, jumped and shouted loudly. If the negative high-voltage electrode was a needle electrode that made the electric field unhomogeneous, most of the mice would move to the place with weakest intensity of electrostatic field in the container, then curled up at this place, some of them would lick the paws and scratch face over a long time; some mice would lie prostrate and its hair were lodging. If the two electrodes both were plate electrode, the mice would run and escape for a short time, then gather together and lie prostrate, this state would continue for 1--2 hrs or more.

High-voltage electrostatic field also can directly influence some organs, for example the myocardial contraction force and heart rate of toad hearts in vitro were inhibited. This results were comform to the results of author's former research that showed the amplitude of R-waves of ECG of anesthetized albino mice were inhibited by electrostatic field. During electrostatic field actions the myocardial contraction force of toad hearts in vivo increased, R-waves amplitude of most of waking rabbits ECG increased and the rabbits arterial blood pressure varied differently, these results showed that when the electrostatic field acted on intact animal organisms it had both direct effects on some organ systems and indirect effects on these organ systems through stimulting nervous system.

As stated above, electrostatic field can induce increase of respiratory rate, fluctuation of blood pressure, increase of ECG and myocardial contraction activity, increase of cortex activity even appearance of abnormal EEG waves, these suggested that the animal organisms had get into stress state. These showed that high-voltage electrostatic field may be an environmental physics factor causing harmful to animal organisms if its intensity rised up to or over a limit. So it is necessary to investigate thoroughly the property and degree of effects of high-voltage electrostatic field on animal organisms through further experiments especially chronic experiments.

References

1. Ye Jia-ming and Wang Rong-yi: Effect of high-voltage electrostatic field on the amplitude of R-wave of the electrocardiogram from albino mice. Journal of Northeast Normal University, 1985, No.3, 75--84.
2. Ye Jia-ming, Gao Tian-shun, Zhou Hao-le and Zhang Jian-feng: Effects of high-voltage electrostatic field on chromosome aberration in root tip cells of rye (Secale cereale) and broad bean (Vicia faba). Journal of Northeast Normal University, 1985, No.1, 61--66.
3. Zhang Li-ping, Zhang Chang-zhong and Ye Jia-ming: Effects of high-voltage electrostatic field on starch amylase and peroxidase and its isoenzyme activity of sprouting process of barley seeds. Journal of Northeast Normal University, 1987, NO.2, 41--45.
4. Skorobogatova A.M., G.M.Tarasova, A.V.Soloviev, V.G.Plotnikov, L.B.Plotnikova, and M.N.Yakovleva: The effect of a static electric field on the body of humen and animals. Gig. Sanit., 6, 20--24, 1976.
5. Hinkle L., C.D.Maccaig and K.R. Robinson: The direction of growth of differentiating neurons and myoblasts from frog (Xenopus laevis) embryos in an applied electric field. J. Physiol. (Lond.), 314, 121--136, 1981.
6. Senatra A.D., D.A.Perego and G.Giubilaro: Biological effects of low-level, very low frequency (VLF) electric field on the blood sedimentation rate. Int. J. Biometeorol., 22, 1, 59--66, 1978.
7. Drozdz M., E.Kucharz and K.Ludyga: Effect of longtern exposure to highenergy electric fields on some indexes of connective tissue metabolism in guenea pig. Rev. Roum. Biochim. 17, (1), 27--30, 1980.
8. Kozyaryn, I.P., I.A.Mykhalyuk and L.D.Fesenko: Effect of electric field of commercial frequency of different intensity on balance and metabolism of copper, molybdenum and iron in laboratory animals. Fiziol. Zh. (Kiev), 23, 3, 369--373, 1977.

Proceedings of International Conference on Modern Electrostatics

III. Novel Electrostatic Techniques
and
the Prospects of Industralization

CORONA-ELECTROSTATIC SEPARATION PROCESSES FOR RECOVERY OF CONDUCTING AND INSULATING MATERIALS FROM INDUSTRIAL WASTES

L. Dascalescu, R. Morara, Al. Iuga, V. Neamtu
Polytechnical Institute of Cluj-Napoca, str. E. Isac 15, Romania 3400.
I. Csorvassy, Electromures Company, Tîrgu Mures, Romania

Introduction

The recent advances in electrostatic separation technologies are demonstrated by the multitude of industrial applications, based on either electrophoresis or dielectrophoresis - processes involving charged and uncharged particles, respectively /1/. A typical example of electrophoretic application is the recovery of metals and reusable plastics from wire and cable scrap /2/.

The theoretical and experimental studies performed by the High-Intensity Electric Fields Research Laboratory of the Polytechnic Institute of Cluj-Napoca aimed at improving metal/insulation separation technologies, by using the most adequate types of electrodes and determining the optimum operating parameters (high-voltage levels, feeding rate, roll velocity, etc.).

Materials

Large amounts of metals and plastics accumulate as either preapplication or post-application scrap /3/, involving specific recycling technologies. Preapplication scrap includes only material from plants where electric wires are fabricated or compounded into a final product (electrical machines, control panels, etc.). This scrap, clean and homogenous, is easier to process than post-application scrap (such as damaged windings of electrical machines), which is usually a mixture of different sorts of metals and plastics, contaminated with dust or other waste.

Most of metal/insulation separation technologies require preliminary scrap-granulating operations, in order to carry out mechanical component disassociation. Since the specific weight and the shape of metallic granules differ from the characteristics of non-conducting particles (Table 1), precipitation from solution and pneumatic separations are possible, but neither of these methods succeeds in recovering reusable plastics free of metallics.

MATERIAL	Specific weight $\gamma [g/cm^3]$	Relative permittivity $\varepsilon_r [-]$	Volume conductivity $\sigma [S/m]$
Cu	8.9	∞	$56 \cdot 10^6$
Al	2.7	∞	$35 \cdot 10^6$
PVC	1.35	3...7	$10^{-9}...10^{-14}$
PE	0.94	2.3	$10^{-13}...10^{-16}$
Rubber	0.9...1.3	2.7...7	$10^{-11}...10^{-15}$

Table 1. Physical characteristics of the components of wire & cable scrap.

Difficulties arise also in separating of fine metallics from stranded wire. This paper illustrates the advantages of electroseparation technologies, which ensure higher qualities and productive rates, as compared to conventional methods of waste treatment.

Improved corona-electrostatic separation techniques

It has been demonstrated by Y. Tonoya and Y. Nakamura (cited in /4/) that corona charging (Fig.1) is ten-fold more efficient than triboelectrification, and that induction charging has the lowest charging capability.

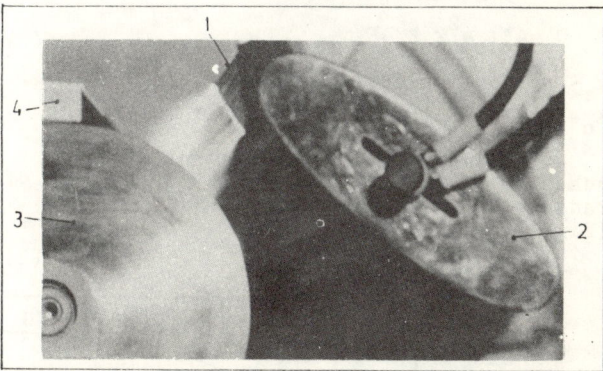

Fig.1. Ion bombardment; 1- needle-type corona electrode; 2- electrostatic electrode; 3- earthed rotating roll electrode; 4- vibratory feeder.

Wire-, blade-, and needle-type electrodes have been tested /5/ in order to achieve the most effective ion bombardment. Needle-type electrodes (Fig.2), which resist better to mechanical shocks and vibrations, as well as to spark discharges, are recommended to be used with industrial electroseparators.

Fig.2. Types of electrodes in roll-type electroseparators; 1- corona; 2- rotating roll; 3- corona, for industrial use; 4- electrostatic, for industrial use; 5- electrostatic.

Preliminary tests /6/ indicated the opportunity of combining the pinning effect of the corona-electrode on plastics and the attraction forces developed by the electrostatic electrodes on metallic particles. Therefore, an extended electrostatic field has been generated by using several types of tube-electrodes (Fig.2). Finite-element computation /5/, electrolytic tank modelling /7/, and experimental measurements enable an accurate plotting of the electric field in the electroseparator (Fig.3).

Metal/insulation electroseparability

The forces acting on the particles in a roll-type electroseparator are (Fig.4):

- the electric field force:

$$F_Q = Q_s E_o , \quad (1)$$

where Q_s is the saturation charge of spherical particles with radius a, conductivity σ and relative permitivity ε_r:

$$Q_s = 4\pi\varepsilon_o a^2 k_s k_{\varepsilon f} E_o , \quad (2)$$

k_s - saturation coefficient, $k_{\varepsilon f} = 3\varepsilon_r/(\varepsilon_r+2)$,
E_o - field on the surface of the roll;

- the image force, which ensures the adhesion of the low-conductivity particles of radii a on the surface of the earthed roll (Fig.4):

$$F_i = Q_s^2 : [4\pi\varepsilon_o(2a)^2] ; \quad (3)$$

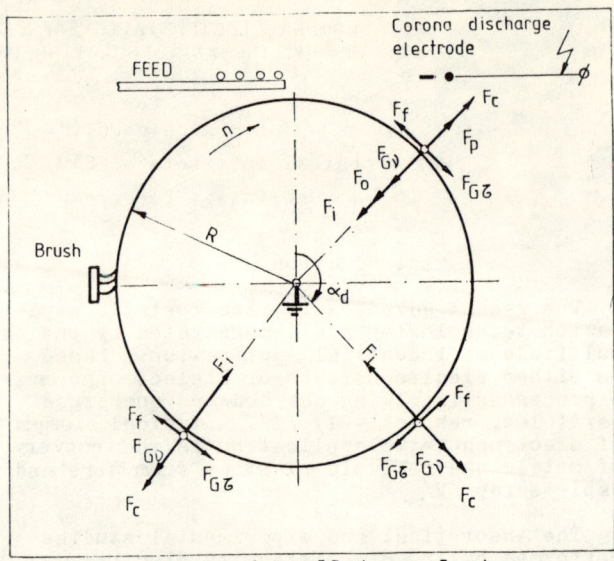

Fig.4. Forces in roll-type electro - separators.

- the centrifugal force, which determines the departure of the particle from the roll:

$$F_c = m\omega^2 R = \frac{16\pi^3 a^3}{3} n^2 R \gamma , \quad (4)$$

where: n[rot/s] - the velocity of the R radius roll, m[kg] - the mass of the spherical particles of specific weight γ [kg/m³];

- the gravitational force F_G ;

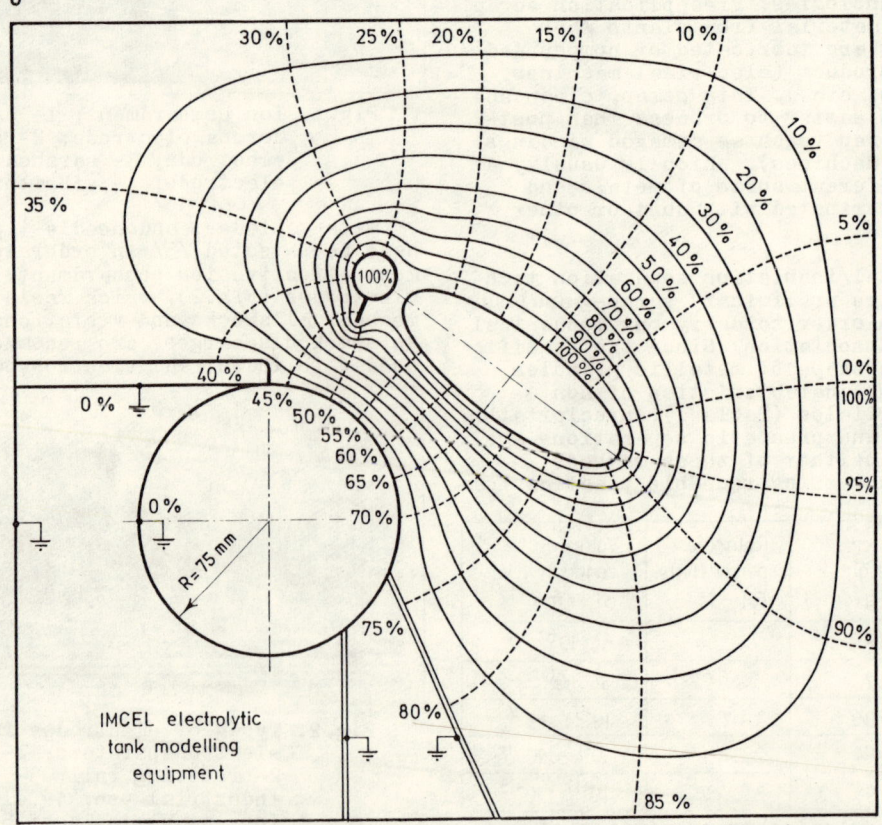

Fig.3. High-intensity electric field of ILES-1 laboratory electroseparator.

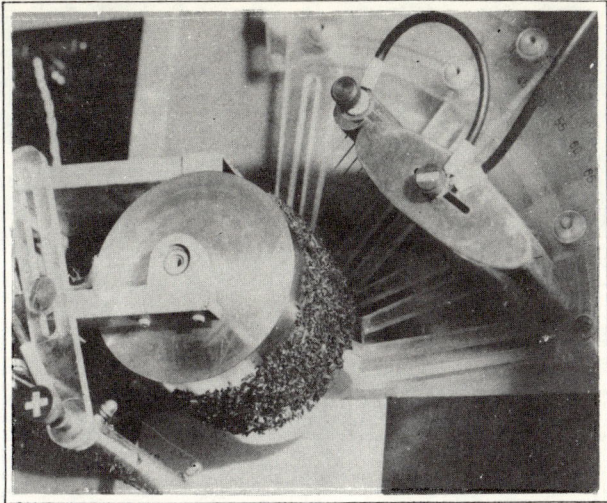

Fig.5. The pinning effect of the electric image forces acting on PVC granules.

- the non-homogenous field force F_p.

The departure angle α_d represents the solution of the equilibrium equation of the radial forces /8/:

$$F_Q + F_i - F_p - F_c + F_G \cos \alpha_d = 0, \quad (5)$$

or, if F_p is ignored:

$$\cos \alpha_d = \frac{\omega^2 \cdot R}{g} - \frac{3\varepsilon_0}{4g} E_0^2 \frac{k_s k_{\varepsilon f}}{a \gamma} (k_s k_{\varepsilon f} + 4). \quad (6)$$

A Commodore personal computer has been employed to simulate PVC/copper, PVC/aluminium and rubber/copper separation (Fig.6), for various granulometric characteristics and different operating conditions (γ, E_0, k_s).

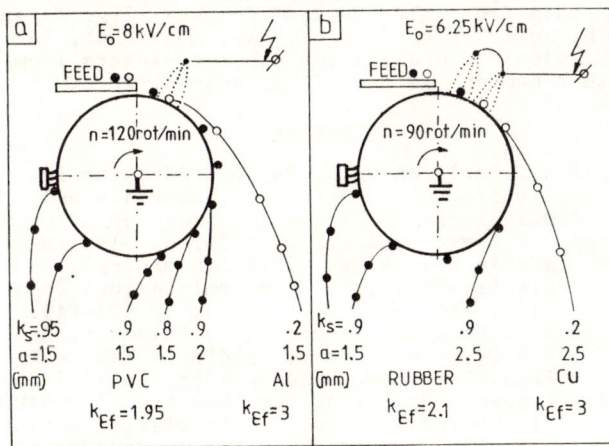

Fig.6. Computed separation trajectories; a- Al/PVC; b- Cu/rubber.

The calculated parabolic separation trajectories have been experimentally confirmed (Fig.7).

Computer simulation suggested the optimum particle size and roll velocity, as well as the necessity of using two corona-electrodes with larger rubber granules (Fig.8).

Fig.7. Experimentally-determined trajectory of aluminium particles in a corona electrostatic separator.

Experimental results

ILES-1 laboratory electroseparator, conceived by the authors, has been provided with two high-voltage generators: SIT 60 (d.c. negative polarity, 0...60 kV peak), and SIT 50 (a.c., 50 Hz, 0...50 kV peak), supplying the active and wiper electrodes, respectively.

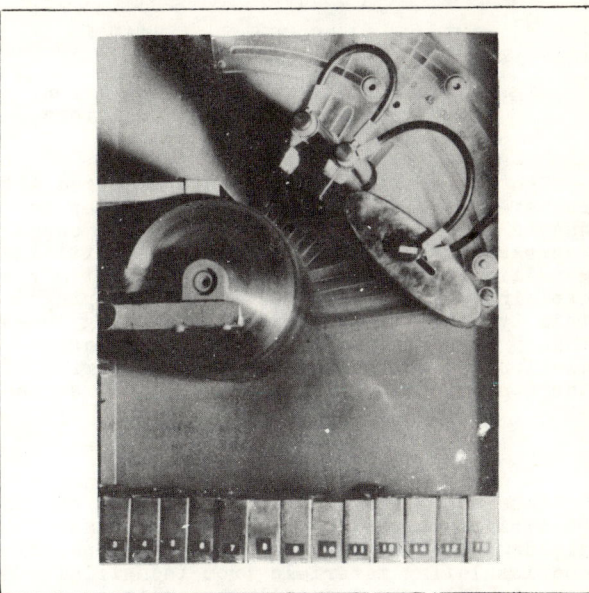

Fig.8. Electrode arrangement for copper/rubber separation.

The separation tests have been carried out on samples of actual industrial residues, offered by "Electromureș" Company of Tîrgu Mureș. The experiments have pointed out the efficiency of corona-electrostatic methods in separating metal/insulation granular mixtures. It has already been shown by Dascalescu et al. /6/ that ILES-1 can recover at a purity beyond 97 % more than 90 % of the metal contained in a 62.5 % Cu , 37.5 % PVC granular feed.

Almost 90 % of the aluminium in a 44.2 % Al, 55,8 PVC chopped-wire waste has been recycled at a purity up tp 98.5 % employing a single step of separation (Fig.9). The 2.4 % impurities of aluminium in the PVC concentrate could be diminished by a second operation of "cleaning", using high-intensity electric fields.

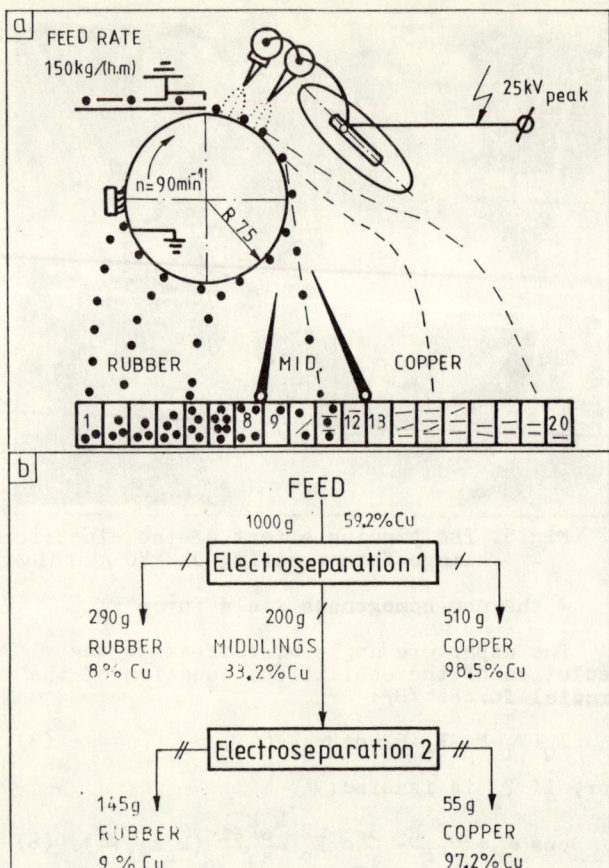

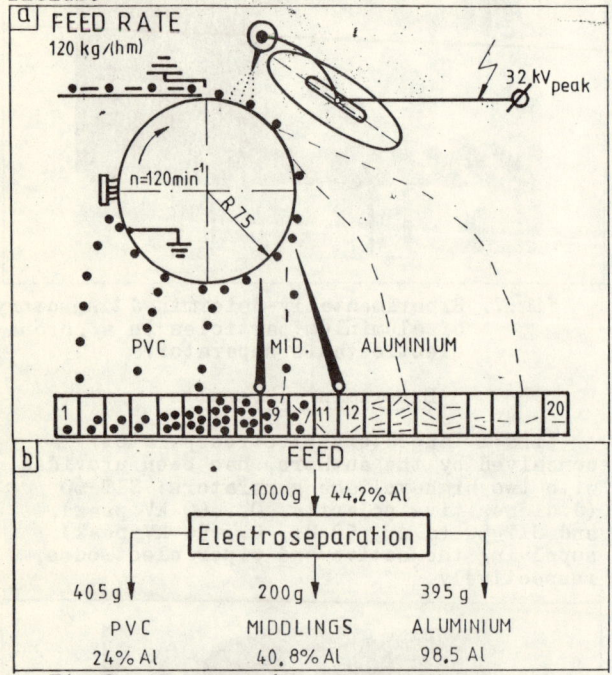

Fig.9. Aluminium/PVC electroseparation;
 a- optimum operating conditions;
 b- laboratory results.

Two steps of separation were required with rubber-insulated cable wastes, in order to ensure a 97 % purity of the recovered copper, representing more than 90 % of the initial metallic content (Fig.10). This result may be explained by the difficulties of chopping this type of wires. The granules being longer than 5 mm, complete copper-rubber disassociation was not achievable and the purity of the non-conducting fraction could not surpass 92 %.

Conclusions

Laboratory and pilot-plant tests prove the possibility of using corona-electrostatic fields to recover several sorts of conducting and insulating materials from industrial wastes. The experimentally-established design criteria have been used for developing ELSIM-series industrial electroseparators, fabricated by "Electromures" Company of Tîrgu Mures.

A wider range of applications /9/ are presently under test.

Fig.10. Copper/rubber electroseparation;
 a- optimum operating conditions;
 b- laboratory results.

Acknowledgements

This paper has been partly supported by grants from "Electromures" Company, Tîrgu Mures. The assistance of Eng.V.Duka, Eng.V.Dub, Eng. M.Radovici, and all the other co-wokers from that company is gratefully acknowledged.

References

/1/ F.S.Knoll and J.B. Taylor. Advances in Electrostatic Separation. Minerals and metallurgical Processing, May 1985, 106-14.
/2/ J.E.Lawver and W.P.Dyrenforth. Electrostatic Separation. In:A.D.Moore(Ed). Electrostatics and its Applications.John Wiley and Sons, N.Y., 1973, pp.221-249.
/3/ W.L. Hawkins. Recycling of Polymers. In: M.B.Bever(Ed). Encyclopedia of Mat.Sc.& Engn.,Pergamon, Oxford, 1986, pp.4127-30.
/4/ K.Asano. Review of the 1980 Annual Meeting of the Institute of Electrostatics Japan. J.of Electrostatics, 9(1981), 367-384.
/5/ Al.Iuga. Dr.D.Thesis. Iași. 1984.
/6/ L.Dascalescu et al. Copper Recovery from Wastes in Electrical Machines Industry. Proc. BICEM'87, Beijing, China.
/7/ Al.Iuga et al. Modelling of High-Intensity Electric Fields in Industrial Separators. Modelling,Simulation & Control-A,1989.
/8/ R.Morar et al. Numerical Simulation of Electrostatic Separation of Materials. Modelling,Sim.&Control-A,/3,1987, 57-64.
/9/ Bulletin 683.Bur.of Mines,U.S.A., 1985.

ELECTROSTATIC PARTICLE LEVITATION ON A QUADRUPOLE SYSTEM

B Makin, BSc, PhD, CEng, FIEE, FInstP and David Hu, BSc

University of Dundee,
Department of Applied Physics and Electronics & Manufacturing Engineering
Dundee, Scotland

Abstract

The Electrostatic levitation of charged particles has been stated by Earnshawn Theorem [1] that it is impossible to have a state of static stable equilibrium. Thus levitation can only be achieved dynamically or by using active feedback control techniques. An a.c. low frequency five electrode quadrupole levitation system has been set up and levitation of non-conducting liquid droplets from 100 - 800 μm has been achieved. A 3d numerical solution is used to analyse the confinement force and its stability. The charging and launching process of single particles is similar to the spray process of a non-conducting liquid. The charge to mass ratio of a levitated particle is calculated and compared with the Reyleigh and Maxwell charge and it is found that it is well below these limiting values.

Introduction

The first introduction of an a.c. quadrupole was made by Weuker et al [2]. The electrodynamic behaviour of a charged particle in such an experimental rig can give rise to many applications, such as mass spectrometer [3-5], confinement and transportation of laser fusion targets [6-8], and also as a strong focusing lens in linear accelerators [9].

Levitation of solid and liquid particles is of great interest for experimental and industrial applications. There are a few kinds of levitation methods in ground environment (1-g gravity). The Jet Propulsion Laboratory [10] has conducted a series of experiments and analysis in 1-g and micro or zero-g environments for the purpose of containerless material processing. The first accoustic levitation experiments carried out on space lab 3 in 1985 by JPL were very successful. In accoustic levitation, 3 sound sources are used to generate a stable minimum but these experiments have to be carried out at atmospheric pressure. Electrostatic levitation with active feedback control using two orthogonal TV cameras to determine the position of the centre of mass have been tested on both a 1-g environment and in micro gravity tests on board aeroplanes. A successful combination of accoustic/electrostatic control has also been obtained in 1-g tests where levitation was obtained in a vertical anti-gravity electrostatic field. The accoustic component consisted of two orthogonal sources arranged in a horizontal plane at the levitation position. By careful control of the phase difference in these sources it was possible to generate a torque sufficient to rotate the particle in either direction or bring it to rest. Water droplets of diameter 3 mm were levitated which is a significant achievement. The laboratory have also observed crystal growth in the same rig.

I.J. Lin [11] has summerised all the levitation methods and concludes that the general condition for all types of levitations in a conservative field is that

$$\text{grad } \phi = 0 \quad (1)$$

and

$$\nabla^2 \phi > 0 \quad (2)$$

where ϕ is the scaler vector of the field which decides the force of equation

$$\overline{F} = -\text{grad } \phi \quad (3)$$

For an electrostatic field if it is divergent-free, Laplace equation is satisfied everywhere, i.e.

$$\nabla^2 \phi = 0 \quad (4)$$

which is inconsistant with equation (2). So the Earnshawn Theorem can be described as levitation in electrostatic field is never in stable equilibrium. For a time variant field, the $\nabla^2 \phi$ is no longer zero. The governing equation should be Maxwell's equation.

$$\nabla^2 \phi = \frac{1}{c^2} \frac{\partial^2 \phi}{\partial t^2} \quad (5)$$

Through carefully choosing the frequency and a.c. voltage, an electrical dynamical equilibrium can be set up.

A novel equipment based on the a.c. quadrupole is to use a low frequency source to levitate small liquid charge droplets of various kinds. The quadrupole levitator consists of 5 electrodes the bottom one supplied with d.c. voltage and four bars forming a quadrupole to give a lateral confinement force. In this paper, we present the analysis with some experimental data obtained on this system. Without using feedback control, a very stable levitation can also be achieved in this system. JPL has spent 8 years in developing its feedback control techniques. Comparing this, the electrodynamic method of levitation has still many interests to be uncovered.

Equation of Motion in an A.C. Quadrupole: Stability Criteria

The standard equation of motion within an a.c. quadrupole for 2D hyperbolic electrodes is governed by Mathieu's equations [12] which can be written as:

$$\ddot{W} + (a - 2b \cos 2u) W = 0 \quad (6)$$

W represents x, and y co-ordinates separately and where the quadrupole is supplied with voltage
$\phi = U - V \cos (\omega t)$

then

$$a = a_x = -a_y = 4qU/m\omega^2 R_o^2 \quad (7)$$

$$b = b_x = -b_y = 2qV/m\omega^2 R_o^2 \quad (8)$$

$$u = \frac{1}{2} \omega t \quad (9)$$

Where U is the d.c. voltage, V is the peak a.c. voltage, q is particle charge and m is the particle particle mass, $2R_o$ denotes the nearest separation between two opposite hyperbolae electrodes.

The Mathieu equation has a stable solution, which reflects the stable confinement of the particle only if parameters a and b satisfy certain relations. According to these relations, the Mathieu stability chart can be drawn in a - b plane and is shown in Fig. 1. The shaded areas are the stable zones which refer to different values of a and b. Thus only the interesting shaded area will be considered and half the stability plot is shown in Fig. 2 where the liner of constant β_x and β_y represent parametric characteristics of constant particle frequency [4,5].

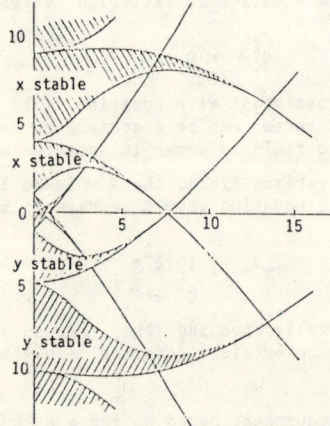

Fig. 1. Mathieu stability diagram.

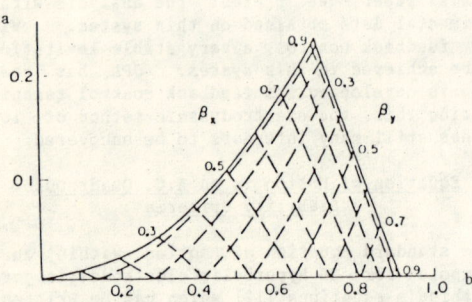

Fig. 2. Stability Chart in both x and y directions.

The difference between a mass spectrometer and a levitator is that the former adjusts the a.c. and d.c. voltage to give maximum resolution of q/m ratio, and in the later the voltages are selected to contain particles with a wide range of q/m. For levitator, a can be made equal to zero by removing the d.c. voltage, i.e. U = 0. Hence according to Fig. 2, b varies from 0.0 to 0.908. Choosing a frequency of 50 Hz with U = 0, V = 2.5 kV and R_o = 10 mm, the q/m ranges from 0.0 to 1.79×10^{-3} C kg^{-1}. Although the stability chart was obtained for an ideal hyperbolic quadrupole, it is still verified by our experiment that all our levitated particles are located in the stable area, as their charge/mass ratio is less than the maximum value.

Numerical Solution

In order to obtain the q/m ratio more accurately, a finite element program has been used to compute the 3-D field of the quadrupole system. The electrode arrangement is where the electrodes are cylindrical rods of radius R.

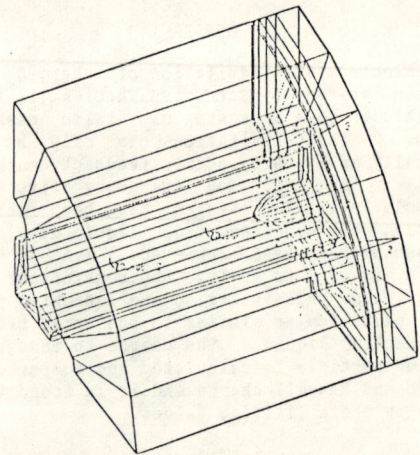

Fig. 3. Finite Element Mesh Geometry.

Fig. 3 is the 3-D finite element mesh geometry for a quadrupole levitator using approximately 4600 nodes. We solve Laplace equation within this area, because the frequency is quite low, we ignore the time variant voltage and set the a.c. bars as zero voltage. Selecting fixed known voltages to the electrodes the computing zone is extended to 3-4 times larger than the confinement zone. The boundaries can be selected with either $\phi = 0$ or $\frac{\partial \phi}{\partial n} = 0$, where $\frac{\partial \phi}{\partial n}$ denotes the normal derivative of the potential. The potential and field in the z-direction (the central vertical axis) are shown in Fig. 4 and Fig. 5 for R_o/R = 2.0 and 3.0. The z origin is taken at the centre of the lower hemispherical electrode. For a particular geometrical configuration at a certain voltage the electric field E_z can be obtained at a specific levitating height, and then we can calculate the q/m ratio from the equilibrium condition.

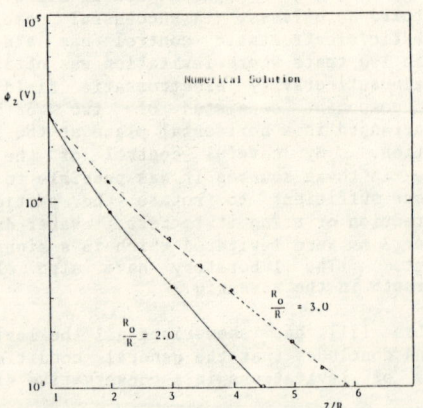

Fig. 4. Distribution of Potential on z-axis

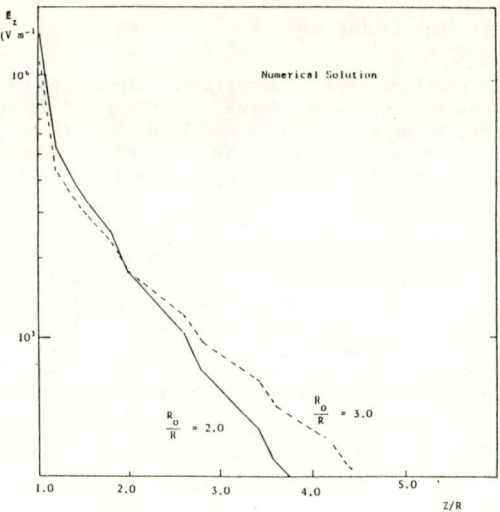

Fig. 5. Distribution of E_z along the z axis.

$$mg = qE_z \qquad (10)$$

One interesting observation from Fig. 5 is where R_o increases from small to large values there is a crossover in the E_z field. This is clearly seen for the two values of R_o/R = 2 and 3. Further analysis of stability by numerical method for the real 5 electrode levitator is still being investigated and will be reported later.

Fig. 6. Experimental quadrupole levitation system.

Experiment

Fig. 6 is a photograph of the experimental quadrupole levitation system. It consists of 5 cylindrical electrodes each of radius R = 4.5 mm. The bottom electrode is supplied with d.c. voltage and the bars forming the a.c. quadrupole have variable spacing R_o. Each opposing rod is interconnected and an a.c. supply is used with a phase difference of 180°. Charging and launching of the charged liquid particle is through a small capillary of Φ1 mm within the bottom electrode. To raise the electric field to improve the charging efficiency, a sharp needle of diameter 0.1 mm is located at the opening of the capillary. The launching process is through electrostatic spraying from a Taylor cone formed by the electrostatic field. Fig. 7 shows the video image of the whole launching process. Experiments were taken using two non-conducting oils; silicon oil and glycerol. Two spraying voltages for these liquids are 4.1 kV and 3.83 kV respectively.

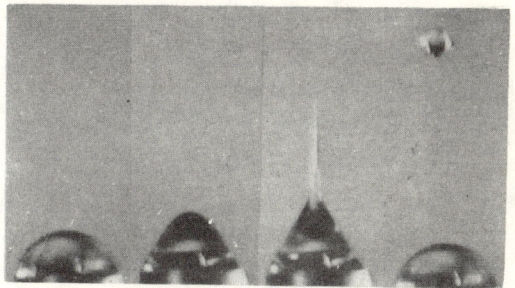

Fig. 7. Launching process of a levitated Particle.

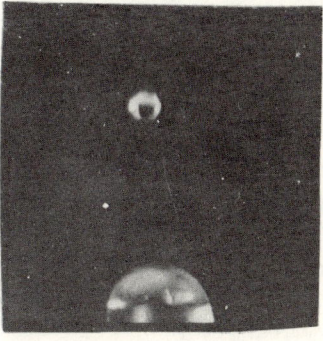

Fig. 8. Levitated Particle of Φ300 μm.

Fig. 8 shows the video image of a particle of diameter 300 μm in the quadrupole system. There is an image processing system which also digitises the image for measurement purposes. The particle size can be measured directly from the display. The experimental rig uses an a.c. supply generated from a microcomputer which can give frequency variation from 6 - 279 Hz and voltage from 0 to 5 kV p-p.

Fig. 9 shows experimental data of V_{dc} against levitating height for different sized particles.

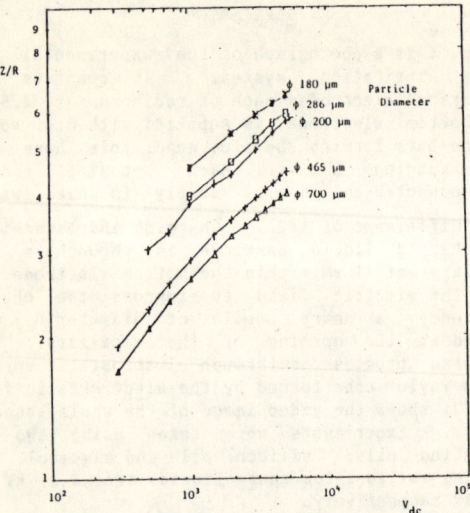

Fig. 9. Experimental Data of Z/R vs V_{dc}.

Fig. 10 shows the experimental data of R_o against levitating height for particle diameter 400-600 μm.

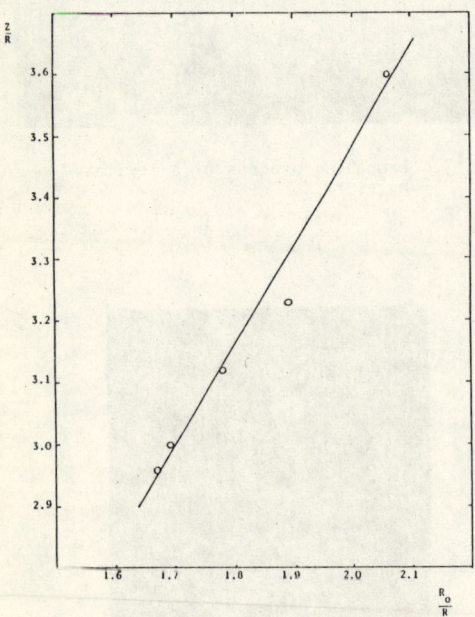

Fig. 10. Levitating Height (Z/R) with Quadrupole Spacing (R_o/R) particle size φ400 ~ φ600 μm V_{dc} = 3 kV.

Discussion of Results

Using the theoretical calculation of the electric field shown in Fig. 5 it is possible to determine the charge mass ratio of a particle reported in the experimental section. For a levitation particle of diameter 465 μm the equilibrium position was 13.5 mm above the lower electrode which corresponds to z = 18 mm measured from the centre of the injecting electrode at a voltage V_{dc} = 3 kV. The computed electric field (E_z) is 2.4×10^4 V m^{-1} and the q/m ratio from equation 10 is 4.1×10^{-4} C kg^{-1}.

An alternative method to determine the charge/mass ratio is from the expression:

$$\text{Levitating Force} = \frac{q \cdot RV}{z^2} = mg \quad (11)$$

This relation uses a simplified expression from the Coulomb repulsion force of two point charges where the lower electrode is modelled as a sphere of radius R with induced charge $4\pi \varepsilon_o RV$. Using the same conditions as above the charge/mass ratio is 2.4×10^{-4} C kg^{-1}.

This is a very satisfactory agreement as the assumptions used in the Coulomb force repulsion are quite significant and there is a major problem identifying the true origin of z in this model. It is noted in Fig. 9 that the relationship of z with V is not constant for all particles or for different levitating zones. The Coulomb model assumes $z \propto V^{0.5}$ but the general tendency for larger particles (φ 700 μm) at reduced voltages is like $z \propto V^{0.3}$ and for smaller particles of diameter 180 μm the dependency is more like $z \propto V^{0.25}$.

The crossover aspect observed in Fig. 5 from the numerical calculations has again been confirmed from experimental observations. Increasing R_o/R from 1.5 to 2.0 in Fig. 10 results in levitated particles of diameter 400 - 600 μm increasing the levitation height z/R from 3.0 to 3.6. The height vs. electrode spacing is linear for a constant voltage of 3 kV and this can only occur from an increase in the E_z field.

From the stability considerations for a typical quadrupole operation of 2.5 kV at 50 Hz the maximum q/m is 1.79×10^{-3} C kg^{-1}. Using the Coulomb method to estimate charge/mass ratio, the experimental data is presented in Fig. 11 showing the typical values are between 2×10^{-4} and 4×10^{-4} C kg^{-1}. For comparison the limiting expressions derived by Maxwell & Rayleigh are also shown in Fig. 11 and it is confirmed that the particles should be stable and they have only approximately 15% of the limiting Rayleigh charge.

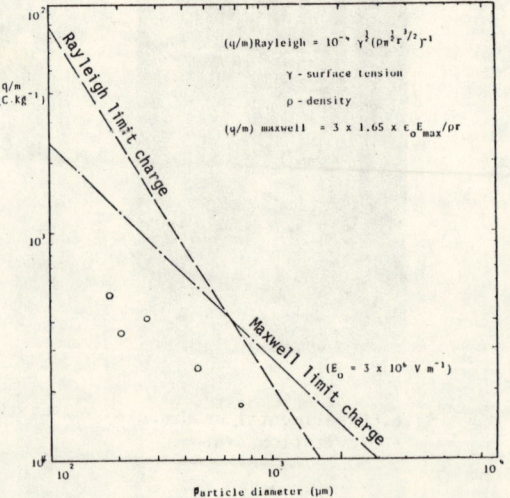

Fig. 11. q/m ratio for different particle diameters at a levitating voltage of 3 kV$_{dc}$ and 2.5 kV$_{ac}$.

Conclusions

1. Levitation in an a.c. quadrupole is very stable in an atmospheric environment for particles up to 700 μm diameter with a q/m ratio less than 10^{-3} C kg^{-1}.

2. The numerical solution of the 3-D geometry using a finite element model can simulate the experimental rig. The relative tollerance for the solution of the field is less than 0.01 which gives a good approximation.

3. The Coulomb method of estimating the q/m ratio is approximate and is less than the value obtained using the numerical computed field variation along the z axis.

4. The derived values of q/m ratio for a range of levitated particles is approximately 20% of the Rayleigh limiting value.

5. The influence of increasing the quadrupole electrode spacing at constant levitating voltage is to increase the levitation height. This effect is similarly observed in a crossover effect in the E_z field.

Reference

1. MAXWELL, J.C., "A Treatise of Electricity and Magnetism", Dover Press, New York 1954, art. 116.

2. WUEKER, R.F. et al, J. Appl. Phys., Vol. 30, N3, (1959), p 342.

3. DAWSON, P.H., "Mass Spectrometry and its Applications", Elesevier, Amsterdam, (1976).

4. DAWSON, P.H. et al, "Advances in Electronics and Electron Physics", (1969) 27, p 59.

5. DAWSON, P.H. et al, Ibid., Vol. 53 (1980), p 153.

6. MASUDA, S., Conference Record, IEEE/IAS (1980). Annual Meeting, p 1005.

7. IZAWA, Y., J. Vas. Sci. Technology A3 (3) (1985), p 1252.

8. JOHNSON, W.L., and HENDRICKS, C.D., Conference Record IEEE/IAS (1980), p 1011.

9. CRANDALL, K.R., Los Alamos Nat. Lab. Report 1A-9695-MS (1983).

10. RHIM, W.K., Proceedings IEEE/IAS Conference (1986) p 1338.

11. LIN, I.J., J. Electrostatics, Vol. 15 (1984) p 53-65.

12. McLACHLAN, I., "Theory and Application of Mathieu Functions", Oxford, (1947).

13. HENDRICKS, C.D., J. of Colloid Sci., Vol. 17, (1962) p 249-259.

EXPERIMENTAL STUDY ON A NEW TYPE ELECTROSTATIC SEPARATING DEVICE

Dou Weiguo, Hao Guanghui, Zhang Fu, Liu Binjiang
Inner Mongolia College of Agriculture and Animal Husbandry
Huhehaote City, P.R.China

Abstract

An electrostatic separating method with the free-falling of corona precharged particles through high electrostatic field is presented in this paper. The author has performed a number of experiments on electrostatic separating of shell from meat of peeled off sunflower seeds. Based on these experiments, the influence of voltage, electrode span, speed of ionic wind, configuration of electrodes, corona wire and the fluctuation of output voltage of power source on separating effectiveness, has been discussed in detail.

1. Introduction

The experimental study on electrostatic separation of agricultural stuff shows that a fairly satisfactory separation result can be achieved by using the separation method of free-falling corona charged particles, so long as the ratios of surface area to specific mass between the two kinds of particles differs significantly[1][2]. The purpose of this article is to analyze and discuss how the factors, such as electrode span, applied voltage, speed of ionic wind, corona wires and the sort of power sources etc, affect the separation result. The problem that the fine broken bits of shell of sunflower seed return back to the negative electrode after losing the free charge near the earthed electrode, is solved to improve the separating efficiency.

2. The method and device of separation

The experimental device is showed in Fig.1. The feeding rate, voltage of power source, electrode span are all adjustable

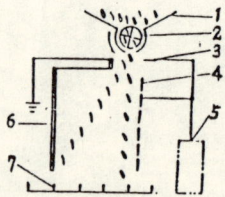

Fig.1 The schematic drawing of separator
1.container 2.feed hooper 3.prickle electrode 4.wire electrode 5.high voltage D.C. power source 6.earthed electrode 7.catch bin

We have experimentaly found that the primary cause which affects the stability of separating effectiveness is the backward motion of the fine broken shell bits of sunflower seed [1].

The improved discharging electrodes consist of prickles electrode for the precharging and a corona wire used in the separation region to recharge the backward shell bits as shown in Fig.1. The satisfactory separation of sunflower seed from its shell has been obtained with this kind of electrode configuration. The experiments carried out in this paper consist of two parts. The first one is to analyse the dependence of separation on the applied voltage, the configuration of electrodes etc, through the orthogonal experiments. The second is to study the output fluctuation effect of the negative high voltage power on the separation quality by double-factor experiments.

3. The experimental scheme and result analysis

(1) Orthogonal experiment and result

Based on the number of factors to be considered and the levels to be taken, the experimental procedure of the orthogonal table of $L(4,2^4)$ is adopted. The disposition of experimental factors and the levels are given in Table 1. The scheme of orthogonal experiment and the numerical analysis of data are listed in Table 2. The variance analysis of data in orthognal experiment is given in Table 3.

(2) Discussion of experimental result
(a) The disposition of electrode span and voltage

The result of variance analysis gives $F_R > F_{0.05}(3.2)$, that is to say, the different disposition of voltage and electrode span affects the experimental index notably, we can get $K_{R3} < K_{R1} < K_{R4} < K_{R2}$ (25.7 < 27.9 < 32.55 < 37.2) from Table 2. This shows that the percentage of shell in sunflower meat is relatively low and the separation quality fairly good when the voltage of power is set to 40 kv and the span between positive electrode and negative electrode is 6 cm. The reason is that the surface charge density of particles is increased as the field strength gets stronger, and there is existes a denser negative ions in electric field [3][4].

Table 1. Experimental designing for electrostatic separating

level experimental factor	1	2	3	4
A: volt/electrode span (kv/cm)	A_1: 50/9	A_2: 33/6	A_3: 40/6	A_4: 45/9
B: electrode constitution (negative electrode--positive electrode)	B_1: corona wire--zinc coated steel plate	B_2: corona wire--polished aluminum alloy plate		
C: constructional disposition	C_1: type 1	C_2: type 2		
D: feeding rate (kg/h)	D_1: 160	D_2: 220		
E: sort of power source	E_1: 50 Hz half-wave rectified	E_2: 1000 Hz with voltage doubling rectifier		

Table 2. A scheme of orthogonal experiment and the numerical analysis of data

factor number of experiment	A voltage/ electrode span	B electrode constitution	C construc- tional disposition	D feeding rate	E type of power	experimental index Y_i percentage of shell contained in meat %
1	A_1	B_1	C_1	D_1	E_1	10.10
2	A_1	B_2	C_2	D_2	E_2	17.80
3	A_2	B_1	C_1	D_2	E_2	10.70
4	A_2	B_2	C_2	D_1	E_1	26.50
5	A_3	B_1	C_2	D_1	E_2	17.30
6	A_3	B_2	C_1	D_2	E_1	8.40
7	A_4	B_1	C_2	D_2	E_1	23.40
8	A_4	B_2	C_1	D_1	E_2	9.15
K_1	27.90	61.50	38.35	63.05	68.40	$R = \sum_{i=1}^{n} Y_i = 123.35$
K_2	37.20	61.85	85.00	60.30	54.95	$P = \frac{1}{n} R^2 = 1901.9$
K_3	25.70					$W = \sum_i Y_i^2 = 2236.72$
K_4	32.55					
Q_j	1941.12	1901.92	2173.93	1902.85	1924.52	
S_j	39.22	0.02	272.03	0.95	22.62	

The parameters relavent to the material to be separated:
weight of thousand grain-- 128 g; moisture content--8.5%; ratio of weight of shell to meat--48:52

Table 3. A variance analysis of data on electrostatic separation

variance source	sum of square deviation	freedom	sum of means square deviation	F ratio	dominance
factor A	$S_a = 39.22$	3	13.07	26.95	*
factor C	$S_c = 272.03$	1	272.03	560.89	* *
factor E	$S_e = 22.62$	1	22.62	46.64	*
factor B	$S_b = 0.02$	2	0.485		
factor D	$S_d = 0.95$				

$F_{0.05}(3,2) = 19.16$; $F_{0.01}(3,2) = 99.17$; $F_{0.05}(1,2) = 18.51$; $F_{0.01}(1,2) = 98.50$

Due to the bigger difference between the seed surface and the shell surface, there is a satisfactory separation.

(b) Construction disposition

Two different types of construction design are used in the device. The ionic wind in type 1 is stronger than that in type 2. There is $F_c > F_{0.01}(1,2)$ from Table 3, which means that the experimental index is affected obviously by changes of construction design. It can be seen in Table 2 that $K_{c_1} < K_{c_2}$ (38.35<85), which indicates the separating effectiveness is satisfactory.

(c) The types of power source

There is $F_E > F_{0.05}(1,2)$ as shown in Table 3, the alternation of power sources affects the separating index apparently. $K_{E_2} < K_{E_1}$ (54.95<68.4) as shown in Table 2, a fairly good separation result can be obtained by using 1000 Hz high volt power with voltage doubling rectifier.

(d) Feeding rate and disposition of electrode

In Table 3, the values of S_B and S_D are both much smaller than that of S_A, S_C and S_E. This means that the variations of the two factors, electrode diposition and feeding rate, cause less fluctuation of separating index, so these two factors are regarded as subordinate ones, S_B together with S_D is to be dealt with as experimental error in the data processing.

(3) The affect of voltage fluctuation of power source on separation quality

To study the effect of voltage fluctuation of power source on separation efficiency, the experiments has been performed with three different types of high voltage sources. Under remaining the other factores unchanged, the experimental results are shown in Fig.2.

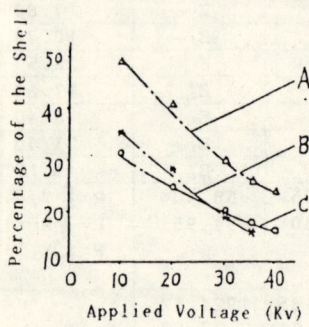

Fig 2 The graphical relationship between separation efficiency and the applied voltage with different power sources. Curve A, B and C correspond to the 50 Hz source with half-wave rectifier, 50 Hz source with half-wave rectifier and capacitive filter, 1000 Hz source with voltage-doubling rectifier respectively.

As it is shown in Fig.2, the percentage of shell contained in the meat of sunflower seed is comparatively high for 50 Hz half-wave rectified power supply (case A), while it is relatively low for 50 Hz half-wave rectified power supply with capacitive filter (case B) and if the 1000 Hz voltage-doubling rectified power is used, the percentage of shell in sunflower seed meat will be reduced apparently (case C), but it is slightly different from case B.

Analysis shows that the output voltage of case A is not smooth enough and continuous, there is a larger ripple factor ($s=1.57$), while for case B, it is improved due to a capacitive filter and the ripple factor is decreased to $s=5 \times 10^{-3}$. The output for case C is smooth and continuous, the ripple factor is also small $s=6.7 \times 10^{-3}$. This indicates that the reatifier and filter used in the negative high voltage sources have a significant effect on the separation qulity. With a small fluctuation high voltage source, there would exist both stable electrical field and ionic wind in the separation region, which cause sufficiently charging of particles and high separation efficiency.

4. conclusion

a. The exeriments has shown that the electrostatic separation methode of free falling corona charged particles works well so long as the ratios of surface area to specific mass of two kind particles differs from each other obviously.

b. The separation electric field strength, the negative ion density and the speed of ionic wind are three main factors which affect the result of separation. A quite good result of separation could be achieved by optimizing the factor levels on the basis of experiments.

c. The fluctuation of output voltage in a high volt D.C power source has a significant effect on electrostatic separation quality. The separating result would be satisfactory if the power source is used with effective rectifier-filter. In other words, the electro-static separation efficiency will be improved if the ripple factor is reduced.

References

[1] Dou Weiguo, Oorgandale, Liang Yunzhang, Shi Xinmin, An experimental study on an electrostatic separation device, Transaction of CSAM No.4.1987

[2] Duo Weiguo, A trajectory analysis of precharged particle travelling through a high volt electrostatic field, Transaction of Inner Mongolia College of Agriculture and Animal Husbandry, No.1.1987

[3] Jian Yifu, "Electrostatic Handbook", Science and Technical Publishing House, 1981

[4] A.D.Moore, Electrostatic and Its Application, 1973

STUDY ON DIELECTROPHORETIC OIL PURIFICATION TECHNIQUE

Bai Xiyao, Huang Meixiang, Ren Dianfa, Ning Wu,
Yan Shizhong, Man Shuling

Anshan Research & Design Institute of Electrostatic Technology
P.R.China

Abstract

The electric field for dielectrophoretic oil purification is filled with fibre-shaped dielectrics with very small radius of curvature so that a stable high gradient field is formed. The impurities, dirts and water droplets polarized move towards the intense field and adhere to the surface of the dielectrics. The contaminated dielectrics can be self-regenerated to a certain extent. The article discusses the mechanism of dielectrophoretic oil purification and various factors which affects the purification efficiency.

Introduction

Oil is contaminated by some impurities and dirts, such as water, small quantities of fibre powder, scraps of insulating cardboard, powders of rust, resin, carbon, lacquer and metals etc. during processing, feeding, transportation and in the course of its operation. The impurities and water droplets suspending in the oil and ions generated by them may ruin the insulation of the oil. They may cause hydraulic systems clearance block, choke, poor lubrication and intensified abrasion. Small amount of water with salts solved in it disperses in the oil and erodes and rusts the equipments.

Early 1960s, many scientists tried to apply the theory and technique of electrostatic precipitation to oil purification, using dielectric wavy board as collectors[1-3]. In this test rig, the dispersed impurities and water droplets remote from the collecting electrodes will be impossibly cleared off. It takes long time or large collecting electrode area to gain satifactory purification. That leads it to be impractical. In recent years, barbed electrodes are used to intensify the electrostatic field strength[4,5], or dielectrics with small radius of curvature are filled in the uniform or nonuniform electrostatic field to form a high gradient electrostatic field, producing a dielectrophoretic oil purification effect[6]. Because the collecting electrode forces suspending impurities and water droplets to adhere to dielectric surfaces, the dielectrics are polluted so that the intense electrostatic field is destroyed and loses its precipitation effect. In this case, captured impurities may re-disperse, causing frequent flashover and short-circuit between electrodes and relatively large consumption of power. Thus the efficiency of oil purification is reduced or even lost thoroughly. Thus frequent replacement of the dielectrics is needed. Besides, the impurity contents of the oil to be treated must be still limited to a relatively low level.

This paper present a newly developed dielectrophoretic oil purification technique. In the dielectrophore (a thick dielectric body) in the oil purification tank is filled with thin dielectrics with very small radius of curvature. In the vicinity of the dielectric surface, a highly intense gradient electric field is stably formed without influence of surface contamination. The dielectric surface has ability to regenerate itself. With this technique, removal rate of water and other impurties can be above 90% and insulating strength of purified new transformer oil can reach 50kv/2.5mm through one cycle of operation. The technique has advantages of high efficiency and low consumption.

Experimental Results and Discussion

The technological process and equipments of dielectrophoretic oil purification are shown in Fig.1. A dielectrophore (a dielectric body) is set in an oil purifying tank and placed in an electric field between the corona electrode and ground. When a high voltage acts on the electrode, a non-uniform high gradient electric field is produced in the vicinity of dielectric surfaces. A pump supplies oil from original oil tank to the oil

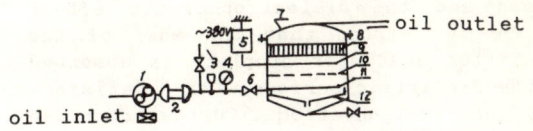

Fig.1 Sketch of oil purification of stable gradient electric field
1. oil pump; 2. primary filter; 3. oil pressure detector; 4. manometer; 5. high voltage power source; 6. valve; 7. oil purifying tank; 8. dielectrophore; 9. corona electrode; 10. flow distributor; 11. oil inlet nozzle; 12. waste oil chamber

effect

purifying tank through a flow distribution pipe. When oil passes through the dielectrophore, impurities and water drops adhere to the dielectric surfaces and stay there. Hence, the whole process of oil purification is ended. Generally, the insulating strength meets or exceeds the requirements of standards through one or several cycles of purification. The experimental data provide necessary parameters for design of technological process and equipments.

1. Relation between applied field strength and efficiency of purfication

The tested material is transformer oil No. 25. Oil flow rate is 360 l/h. Linear speed is 1.2 mm/sec. Thickness of dielectrophore is 60 mm. The dielectrophore is filled with dielectric fibre. The relations between field strength and insulating strength of oil are shown in Fig.2:

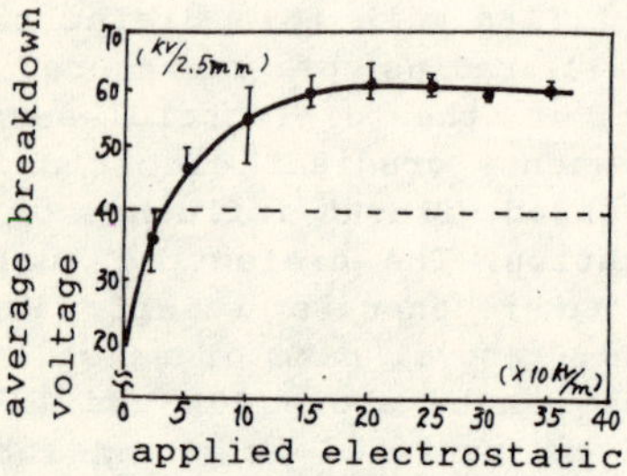

Fig.2 Curve of relation between applied electrostatic field intensity and strength of oil insulation

It can be seen from the figure that the field strength acting on the dielectrophore made from fibre materials should be greater than the knee point 150 kv/m, but it is proper to select 200 kv/m. In this case the breakdown voltage of oil would be above 60 kv/2.5mm. This shows that only when the exerted electric field intensity is sufficiently high, a high gredient field which meets the requirements of oil purification can be formed and the dielectrophoretic effect can be so strong that the most of the impuritics and water droplets is absorbed by the dielectric surface with a satisfactory purification effect. Further increase of external voltage does not lead to enhancement of purification efficiency, while resulting in problems of insulation and power consumption.

2. Relations between thickness of dielectrophore, speed of oil flow and efficiency of purification

Test material is transformer oil No. 25. Electrostatic field acted on the dielectrophore is 200 kv/m. Linear speed is 1.0 mm/s and the flow rate is 360 l/h. Under these conditions, the insulating strength of the tested oil increases linearly with the increase of the thickness(L) of the dielectrophore. The thicker the dielectric body, the more the probability for impurities and water to be absorbed. However there is a limit of the thickness beyond which insulating strength tend to become level.

U-V line in Fig.3 shows the dependence of insulating strength U(in volt across a gap of 2.5 mm) on the linear speed V(in mm/s) at which the oil passes through the dielectrophore with its thickness equal to 60 mm. As the linear speed V increases from 4 mm/s to 24 mm/s, the insulating strength is almost constant. Therefore, purification can be conducted at properly high speed.

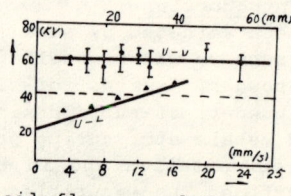

oil flow speed V through the dielectrophore

Fig.3 Curve of relation between flow speed, thickness and insulating strength

3. Experiment in removal of impurities and water droplets in oil

Table 1 shows the experimental data on 25# transformer oil purified by the stable high gradient electric field at an external electrostatic field intensity of 110-220 kv/m. From the data, it can be found that the removal rate of various impurity particles in the oil through one cycle of operation of equipment Model WYJ-10 designed according to this technique is above 95%. The purification precision can reach 0.1μ, rank 2-3 of NAS(1638). The effect of purification of minute impurities is especially outstanding.

Take waste 100L 20# mobile oil as test material. The applied electrostatic field intensity is 320 kv/m, linear flow speed is 1.32 mm/s, flow rate is 600 l/h and the oil temperature is 55°C. The oil is treated in cyclic operation for 0.3h. The impurity contents are reduced from 0.039% to 0.0027%, removal efficiency equal to 93%. When the oil is treated in cyclic purifitation for 0.5h, impurities of 5-10 μm are reduced from 1119375 particles/100ml to 14670 particles/100ml, impurities of 10-15μm are reduced from 104630 particles/100ml to 1440 particles/10ml, and the removal efficiencies are 98.7% and 98.6%. The precision of purification is increased from greater than rank 12 of NAS (1638) to less than rank 6.

Table 1 Data of particle size on oil purification

particle size of impurities (μm)	particle concentration of impurties (particles/100ml)		
	original oil	oil purified once	removal efficiency (%)
5-10	96012	1655	98.3
10-15	20353	535	97.4
15-25	9075	268	97.0
25-50	5000	93	98.1
50-100	790	33	95.8

Removal efficiency $\eta = (N_0 - N)/N_0$; N_0 --number of particles before purification; N--number of particles after purification.

New transformer oil can have its breakdown voltage raised from 20 kv/2.5mm to above 50 kv/2.5mm through one cycle of purification at temperative 10°C-70°C. Old oil with breakdown voltage over 20kv/2.5mm can raise its value to above 45 kv/2.5mm through one or several cycles of operation. Trace water content can be lowered under 25 ppm through purification at relative humidity 50%. Table 2 gives the experimental data.

Table 2 Experimental data of purifying transformer oil No.25 by dielectrophoresis

applied electrostatic field intensity (kv/m)		110		
environmental temperature (°C)		25		
environmental humidity (%)		74		
capacity of oil purification (l/h)		600		
operating time of purification (h)		one cycle		
breakdown voltage (kv/2.5mm)	before purification	21	23	33
	after purification	64*	64*	64*
	increment	43	41	31

Note: Symbol * denotes that oil in cup has not been broken down but air outside the cup is broken down at the voltage.

This technology has a very strong ability of removing water in oil. It can remove water in oil with water content more than 20%. The rate of water content of transformer oil No.25 is reduced to 0.063% after purifying once. As seen in Fig.4, the efficiency of water removal is up to 98.2%. After purifying for 40 hours, the rate of water content is reduced to 0.004%, and the efficiency of water removal is 99.8%. In purifying oil with high water content, the dielectrics have ability of self-regeneration. The waste water is drained continuously through a drain valve. If the permissible value of water content is greater than 50 ppm, the dielectrics may not be replaced. If the water content is below 40 ppm, the dielectrics need certain drying treatment.

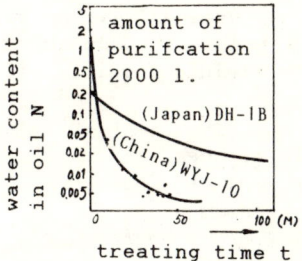

Fig.4 Curve of precision and efficiency of removing water from oil

The result of oil purification by dielectrophoresis is greatly affected by oil temperature. The functional relation is shown in Fig.5. It is seen that the temperature has great effect on oil purification. With the increase of oil temperature, the content of impurities is reduced. An evident turning point is at 50°C. The impurity content is dropped down quickly below 50°C, and very slowly above 50°C. Generally, it is suitable to filter oil in the safe region at 55°C. If temperature is higher, the oil is oxidized easily; if lower, the viscosity is too great to allow good oil filtration.

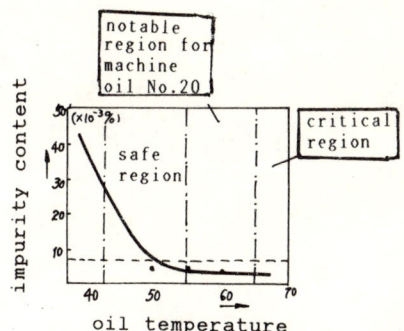

Fig.5 Curve of relation between oil temp. and impurity content

Conclusion

The experimental data indicate that the technology of oil purification with stable gradient electric field is feasible and ideal, and has a great potential and

need to develop promptly. The feature are as follows:
1. The efficiency and precision of oil purification is greatly raised. The collecting efficiency of impurities after purifying once or several times is greater than 90%. Power loss, cost and oil loss are reduced. It is no need to add adhesion agent, and it can remove ultramicro (0.02μm) impurities.
2. It has the ability of self regeneration. The dielectrics can be cleaned and regenerate by itself after being contaminated.
3. It can be used to purify transformer oil, kerosene, lubrication oil, hydraulic fluid and edible oil.

References

[1] A.Sasaki: Filtration & Separation, Nov./Dec.(1984) p407-410
[2] 静电气学会编：静电气ハンドブツク, p591-596, オーム社 (1981)
[3] 香山正晃：特公昭 57-81851
[4] 下田诚：特公昭 56-163713
[5] 香山正晃：特公 (A)昭 57-10359
[6]. J.I.Lin.: Coal gold base minerals of Southern Africa, 46-July(1982)

ELECTROSTATIC SEPARATION OF FINE-GRADED TALCUM POWDER MINERALS

Wang Jianzhong Bai Xiyao Liu Yanchun Yan Shizhong Man Shuling
Anshan Research and Design Institute of Electrostatic Technology China

Abstract

This article introduces electrostatic separation of talcumpowder of the diameter range of 0-0.074μ and contributes a technique by which fine talcum particles can be effectively charged and separated. this technique also provides an effective means to separated fine-graded minerals with advantages of saving energy and no pollution.

Introduction

Talcum is a kind of a non-metallic hydrated minerals of magnesium silicate. After being disintergrated, its monomer has very small size and breaks easily. In the process of grinding, it is easy to be overground and become muddy. Talcum and its associated impurity-magnesite are all non-metallic minerals with insignificant difference in electricproerty. As the talc mineral is treated with a conventional electrostatic separator, the powder would agglomerate and be poorly charged owing to its too fine a size and too small a difference of electric property. In this experiment, single-layered fluidized bed and electrodes capable of generating high concentration of ions are utilized so that problem in inter-agglomeration of the over-fine particles is solved effectively and separation is realized because the paricles can be highly charged in out system.

Fundamental Principles

Electrostatic separation of talcum from its impurities is based upon their different electric properties and therof different forces exerted on the particles in the electric field. The voltage and the polarity of the high voltage electrod are determined by the rations of conductivity and rectification characteristics of the different contents. Electrification of particles includes three kinds of processes. Firstly, charge on electrified particles by electric conduction and induction is determined by conductivity of the materials and duration of contact of the particles with the elctrode. Secondly, corona chrging is the essential process for electrifying the particles. The charge captured by the paricles has relation to their surface characteristics. Thirdly, tribo-charging mainly happens in the process of fluidization. Particles with different surface properties may obtain different charge. In electrostatic separation, the electric properties and aurface characteristics of different particles are utilized with the aid of the three charging techniques so that different materials can obtain different quantities of charge and separation can be attained in an electrostatic field.

Experimentals

Fig.1 shows the function diagram of the experimental divice. Compressed air passes through a dryer and enteres the separation tank, within which a high voltage electrode is set up. A grounded electrode operates as a collector, located near the tank.

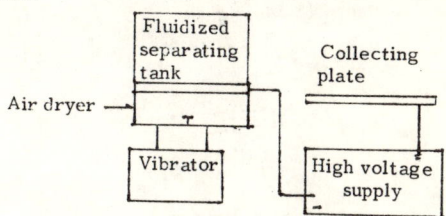

Fig.1 Experimental device for electrostatic separation of talcum powder

Test material is graded talcum powder with a diameter range of 0-74μ. In order to optimize the separation, effects of many factors on the process should be investigated, i.e. ambient temperature and humidity, flow rate and pressure of the compressed air, feeding rate of raw material, structure, polarity and voltage of the high tension electrode and some parameters of the test material.

1. Effect of compressed air on the seprartion
Pressure used for fluidization of the talcum powder ranges 1-4kg/cm^2. Fig.2 shows that the pressure does not cause any effect on the grade of the

product.

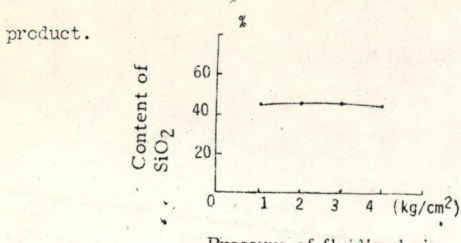

Fig.2 The relationship between pressure of fluidized air and grade of talcum powder through one cycle of separation

2. Polarity and voltage of the high tension electrode

The contact between the particles and the matallic electrode may be non-ohmic. For example contact between magnesite and the electrode has a property of negative rectification. Fig.3 shows that negative polarity is better than positive for rejecting SiO_2.

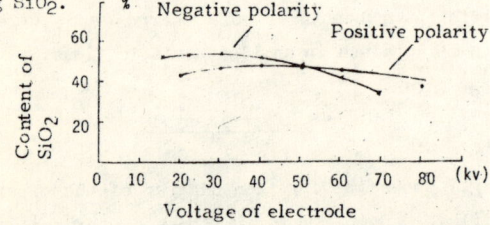

Fig. 3 The relationship between polarity of electrode and the grade of talcum through one cycle of separation

3. Structure of high voltage electrode

Fig.4 shows that three layer closed-type electrode is the best for separation.

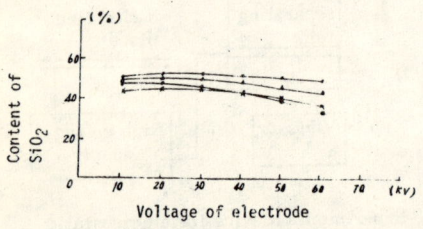

• One layer colsed-tupe electrode ▲ Two layer closed-type electrode
  Three layer closed-type electrode ♦ Four layer closed-type electrode

Fig.4 The relationship between structure of electrode and the grade of talcum through one cycle of separation

4. Voltage of the high voltage electrode

Determination of the voltage closely relates to the conduction properties of the different components. Too low a voltage may lead to insufficient charging of the particles. If the voltage is too high, the dielectric particles become conductive. That also results in poor separation. Our experiments indicated that 30kv is the optimum(see Fig.3).

5. Separation under optimum conditions

In this experiment talcum powder is devided into five size grades. The optimum conditional parameters vary with different grade. Table 1 shows the results obtained under their respective optimum conditions.

Table 1 Test results obtained under the optimum conditions

Size of material in terms of screen mesh	Grade of concentrate ore after one cycle of separation				
	SiO_2	CaO	MgO	Al_2O_3	Fe_2O_3
-200+280	61.28	0.17	31.60	0.15	0.04
-280+300	59.94	0.27	31.60	0.14	0.04
-300+400	58.65	0.22	31.40	0.14	0.04
-400+0	48.00	0.97	33.90	0.16	0.08

Conclusion

In this experiment the problem in agglomeration of talcum powder has been solved by use of fluidized bed technique. Besides, in view of insignificant difference of electric properties between talcum and magnesite, an electrode system is meticulously designed. Under the optimum conditions, negatively charged talcum particles fly out of the tank and are precipitated on the collecting plate under the action of the electric force and other forces, while the particles of magnesite can not because of insufficient charge on them. In the range of size of -200+280 screen mesh, the third class talcum can be raised to the first class through one cycle of electrostatic separation.

References

[1] Burgess K. I., Inculet I. I. and Borgougnau M. A. "Ore Beneficiation in A Fluidized Bed by Means of An Electric Field" Chem. Engng. Prog. Chem. Ser., 66 No. 105, 1970, 236-42.

[2] 杨津基， 气体放电，中国科学出版社

[3] КАПЦОВ Н. А. ЭЛЕКТРИЧЕСКИЕ ЯВЛЕНИЯ В ГАЗАХ И В ВАКУУМЕ. М. ГОСТЕ-ХИЗЦАТ 1950

INDUSTRIAL PILOT TEST OF ELECTROSTATIC PURIFICATION TECHNOLOGY FOR DRY-DISTILLATED SHALE GAS

Wang Yan, Wang Shouzhong, Cai Runfu, Wang Yanming
Anshan Research and Design Institute of Electrostatic Technology
Dai Mingquan
Fushun No.2 Refinery   P.R.China

Abstract

This paper describes properties of dry-distillated shale gases, industrial pilot test setup for pertinent electrostatic purification (oil stripping and dust/ash removing), technological procedures and basic parameters concerned. And also the safety supervision of gas purification are discussed, includeng V--I characteristics, electrostatic purification efficiencies and significances of experiments under discussion.

Specific Characteristics of Dry Distillated Shale Gas and Electrostatic Purification Devices

1. Special characteristics of dry distillated shale gases

At Fushun №. 2 Refinery, the electrostatic purification device is arranged at upstream part of ammonia sulphate column and downstream part of wash-over column. And such unusual arrangement gives the gas the following special characteristics compared with the ordinary gas of our country and the dry-distillated gas from foreign countries.

Greater gas rate, about 11×10^4 Nm^3/h, which is 4-5 times as high as ordinary gas rate; higher oil content, about $37.74 g/Nm^3$ which is 16-20 times as that of ordinary gases (oil solidifying point: $33°C$; work temperature for purification: $70°C$, specific resistivity 5.99×10^7 Ω cm); huge water content, up to 317.2 g/Nm^3, dust content 309.2 mg/Nm^3 (specific resistivity is 9×10^6 Ω cm at 70 $°C$). Composition of the gas: CO_2 19.2%, CO 3.4%, C_nH_n 0.8%, CH_4 3.6%, H_2 8%, N_2 64.3%, O_2 0.6%, Vapour of petrol 29 g/Nm^3, Ammonia 5-6 g/Nm^3 Calorific value 700-800 $kcal/Nm^3$.

2. Features of electrostatic purification devices

In order to meet the special requirements for shale gases, the purification devices adopt profiled plate-wire geometry and two-field horizontal cylindrical structure. This design is characteristic of both the vertical tubular-typed oil arresters and the horizontal-typed dust precipitators.

Horizontal type has greater effective field cross-sections; larger effective area for oil and dust arresting; ability to treat large volume of gas effectively; possibility to improve purification efficiency by introducing more field; easiness to be manufactured and installed and less steel consumption.

Its shortages are less current density than vertical tubular oil arrestor and lower gas velocity (within 0.6-1.2 m/s).

For cylindrical-typed device, uniform stress exerted on the shell results in no stress concentration due to thermal deformation, uneasy occurrence of cracks and good sealing. So it is suitable to treatment of explosive, infammable and high pressure and temperature gas. However the utilization ratio of precipitation space is not so sufficient as that of rectangular ones' and its installation is not so easy to be conducted. Instlation chambers have special structures for reliable moistureproof and anti-fouling.

Vertical-typed oil arrestors prevail both domestically and abroad owing to limited amount of shale to be treated and low oil content (<3 g/Nm^3). The conditions are basically different from ours.

3. Technological process of industrial pilot experiment

To meet the practical requirements of production, the experimental technological process is made as shown in Fig.1.

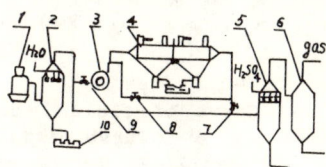

Fig.1 Illustration of electrostatic purification process. 1. Dry distillation furnace. 2. Washing tower. 3. Blower 4. Electrostatic purifier. 5. Ammonia sulphate tower. 6. Cooling column. 7,8 and 9 Valve, 10.Oil pool

An experimental device is set on the main pipeline after washing tower and before ammonium sulphate tower. Two joints of φ700 are prepared beforehand for connection with valve 7 and 9. A certain amount of gas is extracted from the main pipeline (negative pressure 1950 p_a) and sent to the electric purifier 4 with a relay blower 3. After the gas is purified it is returned to the main pipeline.

4. Safety supervision of experiment

Safety is a thing of essence since the electrostatic sparks give great dangers. The only controllable factor for safety is the oxygen amount contained in gas. The dry distilled gases at Fushun №.2 Refinery has the lowest explosion limit of 4.6%. Hence, the oxygen amount contained when purified electrically must be <1.6%. Apart from regular measurements, oxygen contained amount is tested any time as necessary. Wheb the oxygen amount exceeds the limit, worning is given automatically or power supply is cut off automatically.

Basic Experimental Data and Analysis

The differences between design values and actually measured values are shown in Table 1. The following is devoted to analysis of the test.

1. Difference in V-I characteristics

In the process of gas purification, the V-I curves vary significantly. I-V curves before gas entering & after gas entering and when the two fields are powered simultaneously & seperately are shown in Fig.2.

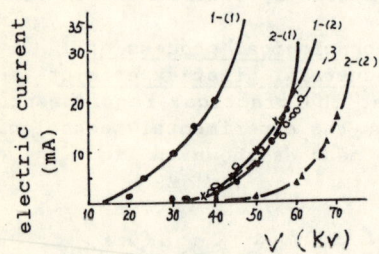

Fig.2 V-I Curves
1-(1) V-I Curve of field 1 before loaded with gas;1-(2) V-I Curve of field 2 before loaded with gas;2-(1) V-I Curve of field 1 after loaded with gas; 2-(2) V-I Curve of field 2 after loaded with gas; 3:V-I Curve of field 2 with both field 1 and 2 energized and loaded with gas,other conditions being equal to 2-(1).

(1) Changes in V-I characteristics of field 1 and 2 separately energized due to loading them with gas are different from those for ordinary ESP. The reason is as follows. After the field is loaded, crude oil carried by the gas is entrapped over the surface of corona wires and weakens the emitted ionic current at the same applied voltage. However, the crude oil layer obstructs the onset of sparkover when operation voltage is increased. That leads to a stable performance of the precipitation at relatively high voltage. In consideration of proportionality of driving speed to voltage squared, high voltage is in favour of purification.

(2). The V-I characteristics of field 2 with both field 1 and 2 energized and loaded with gas appears as curve 3. In comparison with that of individually energized field 2, it is found that current in field 2 is remarkably increased as both field are switched on electrical-

Table 1. Design values and actually measured values of basic technical data

Effective cross section of field (m^2)	1.2 (accounting for 70% of entire shell cross-section)			
Field number	2			
Entire area of precipitation electrodes (m^3)	62			
Entire length of fields (m)	7			
Gap length between electrodes (mm)	30			
Corona wire shapes	1 RS prickle 2 star-shaped			
Precipitation electrode	C480			
Basic technical data	Design value		Actually measured value	
Processed wind capacity (m^3/h)	3456-8640		4500-5000	
Wind speed (m/s)	0.8-2		1.05-1.15	
Migration velocity (m/s)	0.8		0.058	
Residence time of the gas (s)	8.75-3.5		6.67-6.1	
Oil arresting rate (%)	96		93.2	
Dust precipitation efficiency (%)	96		~100	
	Field 1	Field 2	Field 1	Field 2
Voltage (kv)	35-40	40-45	54-58	55-60
Current (mA)	12.4	12.4	13.5	15
Current density (mA/m^2)	0.4	0.4	0.44	0.48

ly. The reason is considered to be that the charge carriers generated in the gas in field 1 is transferred into field 2 so that the current in field becomes larger than that of individually energized field 2.

Electrostatic Purification Efficiency

In the test gas flow rate is 4600-5000 m^3/h and the linear velocity is 1.05-1.15 m/s. Table 2 shows the I-V value adopted and efficiency distribution measured is given in Fig. 3. The average efficiency is 93.2%.

Table 2. V-I relations of the fields

		V (kv)	I (mA)
Field 1	range	55-57	12-15
	averaged	56	13.5
Field 2	range	56-59	13-18
	averaged	57.5	15.5

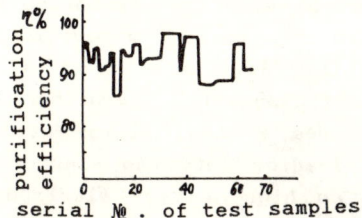

Fig.3 Distribution of electrostatic purification efficiency

One more thing about purification efficiency needs to be explained. Crude oil content in the gas at the outlet is collected by two stages of wind and water cooling. The specific gravity of the sample is only 0.7 g/cm^3. In fact, most part of the collected oil is light oil, which is not crude oil we want to precipitate with the device. The collected sample is preserve in open air for 24 hours and 70% of it evaporates. The residue is used to calculate the purification efficiency, but its weight is more than that of the actual penetrated crude oil. We believe that real efficiency must be higher than that listed above. As a calateral evidence of inference. the dust, uniformly mixed up with the oil, has been precipitated nearly at an efficiency of 100%.

Technical/Economic Effects for Electrostatic Purification

Electrostatic purification of dry distillated shale gases will bring along great social and economic efffects. It may yearly reclaim 10090 ton of crude oil worth about 4.50 million yuan.

The elinimation of produced sulfonated oil will not only reduce the loss of crude oil and sulfuric acid, but also avoid contaminating water resource, soil and atmosphere as well as the choking of pipes and nozzles. The reduced ammonium sulfate processing time and technological steps will lower the cost of chemical fertilizer and improve its quality.

Conclusions

The industrial pilot test has proved that the electrical purification technology for dry distillated gases is safe, feasible, successful and applicable to the future commercial plants.

ELECTROSTATIC SEPARATION OF FIBRE AND METAL FROM RUBBER POWDER

Liu Yanchun, Yan Shizhong, Ren Dianfa, Man Shuling
Anshan Research and Design Institute of Electrostatic Technology
P. R. China

Abstract

This article introduces a method of separating electrostatically fibre and metal from rubber powder bits. This method resolves a series of problems such as high power consumption, environmental pollution and complication of process in production of reclaimed rubber. Roller electric separator with structurally improved electrode is used to separate electrostatically the rubber raw material ranging from 0 to 0.8mm in granular size from impurities such as fibre, metal bits and so on. More efficient separation has been achieved.

Introduction

The available production processes of regenerating rubber can not yet cope with the production development. The raw material for regenerated rubber includes old tyre and used rubber sole. It contains fibre, iron, copper etc. At present, the production of regenerated rubber widely employs combined method of magnetic separation and pneumatic concentration based on the weight difference of the rubber powder and the fibre. However, a part of the rubber powder can not be separated from the fibre due to the air flow. The separation is thus comparatively poor. Besides, this production process is too complicated, the energy consumption is rather high and it causes environmental pollution too. The present method is based on the electrical difference of the component parts to be separated. As the electrical difference between the rubber and the fibre is very slight, it is difficult to use conventional method to achieve satiafactory separation.

In view of the situation mentioned above, we adopted a cylindrical electric separating device of with a static induction electrode placed above high voltage electrode. By using this device impuities can be separated from the rubber with small energy consumption. Moreover, the device is attached with a dust collector, consequently it does not pollute the enviroment. By now we can clearly see this is a very effective method in the production of regenerated rubber.

Test device

A cylindrical electric separating machine is used in this test. A schematic diagram is shown in Fig.1. Here 1 is the adjustable electromagnetic vibrating feeder, which assures a single layer of the raw material feeding into the electric field region. The high voltage electrode 2 is composed of three corona wires connected to negative high voltage d.c. power source, which causes the corona discharge to the earthed cylinder 4 (with a diameter 400 mm). The metallic inductive electrode 3 is placed above the corona electode. Separation plates 6 and 7 are mounted under the cylinder. The plates could be regulated forming different angles. 8,9,10 are the material hoppers.

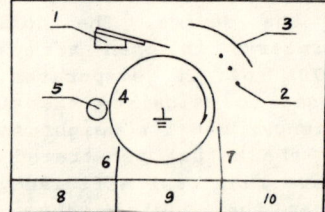

Fig 1 Schematic diagram of electric separating machine.

Basic Prinsiple

Electrostatic separation of fibre and metals from rubber is based on the difference in either electric properties of the component parts or the electric forces on them. When the prepared mateials (grains of certain size) are fed through the feeder, they are distributed evenly on the surface of the cylinder 4 forming single layer and are charged by corona discharge. At the same time the electrode 3 is positively charged by induction. Due to the conductivity difference between rubber, fibre and metal, there exists different leakage currents between them and the cylinder. The final charge deposited on these particles is controlled by their electric conductivity, bringing about different acting forces as shlown in Fig. 2.

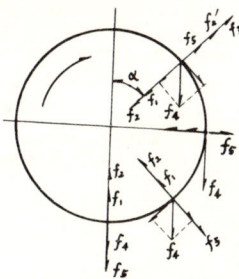

Fig 2. Force diagram analysis in the materials separation. f_1: attracting force between the particles and the cylinder, f_2: Coulumb force on the particles; f_3: acting force due to the uneven electric field; f_4: gravitational force on the particles; f_5: centrifugal force on the particles.

While the cylinder is in operation the great attractive force between the rubber powder and cylinder due to its large charge leads to $f_1+f_2 > f_5 + mg\cos\alpha$ Thus the rubber powder will be attracted on the surface of the cylinder until it is brushed out of it, and drops into hopper 8 or hopper 9. But for the metallic iron and copper, the remaining charge is so small that there is $f_5 + mg\cos\alpha > f_1 + f_2$, they would fall into the hopper 10 under the action of resultant force of the gravitational and centrifugal force. For the longer and thinner fibre, due to the uneven electric field, it will be attracted to the stronger field region and drop tnto hopper 10, while for the small fibres or the nap, owing to its light weight, they could be attracted to the immediately to the inductive electrode. In this way the separation of the metal from the rubber powder could be accomplished.

Experiments and Results

Test materials: pulverized old rubber sole; pulverized pure rubber, size range 0-- 0.8 mm (0-- 20 mesh).

Main compositions: rubber, fibre and iron and copper filings.

Because of the great difference of the electrical properies between the rubber and metals, they could be easily separated from each other. The later part of this paper is concentrate on the separation of fibres from rubber powder. The main parameters related to the separation are as followes:

(1) atmospheric humidity and temperature;
(2) size range of the materials;
(3) rotational speed of the cylinder;
(4) both polarity and value of the applied high voltage;
(5) construction of the electrodes.

1. Effect of the applied voltage polarity on the separation

Generally speaking, the separation efficiency depends on the difference in the charge deposited on the components. From Table 1 which shows the contrasting test results with both positive and negative corona discharge, we can see that a satisfactory result could be obtained with negative corona discharge under certain conditions. This is because there is a large corona current density when a negative voltage is applied to the electrode.

Table 1. test results

test results	fibre content before separation(%)	fibre content after separation(%)
positive	8.0	1.04
negative	8.0	0.33

* test condition: rotational speed: 60 rpm; voltage of electrode: 30 kV; construction of electrode: three corona wires, one inductive electrode; gap of electrodes: 70 mm.

2. Optimizing the distance between electrodes, applied voltage and the rotational speed of the cylinder

From the results obtained we can see that the three parameters mutually affect and restrict each other. For example, when given a certain discharge electrode array, the deposited charge is controlled by the applied voltage and the rotational speed, which affect the separation efficiency. With our repeated experimental results, the optimized distance between electrodes, the applied voltage and the rotational speed are 70 mm, 30 KV and 60 rpm respectively for the powder with diameter ranging from 0 to 0.8 mm.

3. Effect of the induction electrode on the separation

Even with optimized parameters suggested above, a satisfactory separation could still not be obtained because of the charge loss of the shorter fibres or the nap on the cylinder and its droping down to the hopper 9. In order to overcome this problem the inductive electrode 3 is proposed to be placed above the corona discharge electrodes to attract the shorter fibres. The contrasting test results are shown in Table 2.

Table 2. Test result with different type of electrode

test results type of electrode	fibre contents before separation(%)	fibre content after separation(%)
without inductive electrode	8.0	2.95
with inductive electrode	8.0	0.33

* test conditions: rotational speed: 60 rpm; voltage of electrode: 30 kV; corona pole: three corona wires; polarity of electrode: negative; gap of electrodes: 70 mm.

4. Satisfactory test results

Using the optimized parameters and the inductive electrode, a satisfactory separarion of fibre from powdery rubber has been obtair d. See Table 3.

Table 3: Test result under optimized conditions

separating times	test results	
	fibre content before separation(%)	fibre content aftert separation(%)
1 time	8.0	1.04
2 times	1.04	0.33

* test conditions: rotational speed: 60 rpm; voltage of electrode: 30 kV; construction of electrode: three corona wires, polarity of electrode: negative; gap of electrodes: 70 mm.

Conclusion

From the tests it is found that by using an improved electrode the electrostatic separation method could be successfully used to separate the fibre from the rubber powder ranging from 0 to 0.8 mm in size, and the content of fibre could be reduced from 8.0% to 0.33%.

Reference

<1>. Electric Separation of Minerals. Publishing House of Metallurgical Industry (1982).

USING ELECTROSTATIC BONDING ON SEMICONDUCTOR DEVICES TECHNOLOGY

Xie Baixing

Kunming Institute of Physics, Kunming, China

Abstract

This paper describes an electrostatic bonding method. When a D.C. voltage and heating are applied on the glass (it includes a certain quantity of B_2O_3) and semiconductor, the mobile positive ions in the glass will move towards the interface, the electrostatic coulbomb acting force will come into being and the bonding of the glass-semiconductor wafer (such as Ge, Si etc.) will be carried out. Because the bonding process is finished by using space charge attraction, it is a kind of tight seal with low energy consumption, non-pollutant, non-destructive. It is very suitable to semiconductor devices technology.

Introduction

Electrostatic bonding is a kind of bonding method with non-destructive, non-adhesive and low energy consumption. Whenever an external electrostatic field and heating are applied on bonding mediums (such as glass etc.), interface so as to make the mobile positive ions in the glass move towards the cathode through a distance, the region adjacent to the anode acquiring a space charge of immobile ions, it is said to have done work which in turn generates the electrostatic bonding force.

Test

Basic principle and acting force

Electrostatic bonding is accomplished in both sides sealing. One side is a metal or semiconductor, the other side is a glass, ceramic or insulator(or other special material), as shown in Fig. 1. The mobile positive ions in the glass move towards the cathode and become largely neutralized. At the same time, the region adjacent to the anode becomes depleted of positive ions and acquires a space charge of negnative ions (i.e., the bonded charges). In order to increase the conductivity of glass, the glass temperature is raised with heating which causes the rapid build-up of the space charge (polarized region). The electrostatic attraction force between semiconductor and glass surface pulls the parts together.

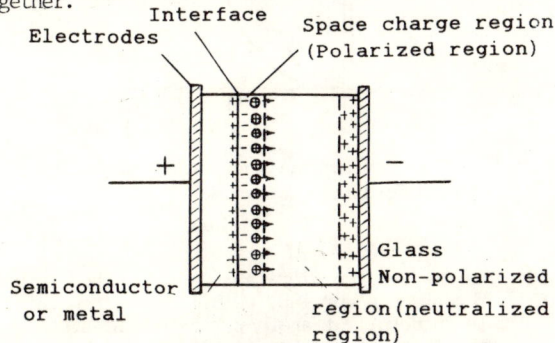

Fig.1 Charges distribution of semiconductor-glass interface by applying a dc voltage

Supposing the positive ions in the glass movethrough a distance r', the work w done by external electric field E can be expressed as

$$W = 1/2 \iiint_v \rho(r')V(r')d^3r' \quad (1)$$

Where the volume v is taken optionally. V is the potential at point r'. ρ is charges density. According to Poisson Equation, $\rho(r') = div(\varepsilon_0 E)$, substituting it into Eq(1). We obtain

$$W = 1/2 \iiint_v \varepsilon_0 \, div EV(r')d^3r' = 1/2 \iiint_v \varepsilon_0 E^2(r')d^3r'$$

Due to the acting force F is equal to the work done per unit distance, the electrostatic acting force is written by $F = 1/2 \varepsilon_0 E^2_{gap}$ Where the ε_0 is the air dielectric constant

$$V_{gap} = V_{applied} - V_{polarized\ region}$$

$$E_{gap} = \frac{V_{gap}}{X_{gap}}$$

X_{gap} is the distance of the air gap. When the X_{gap} is reduced to very small, the E_{gap} will become very large.

Bonding between Ge, Si and domestic glass

A number of domestic glasses have been successfully bonded to Ge,Si and other semiconductors with expansion coefficients between $4 \cdot 10^6$ and $6 \cdot 10^6$/C. The thermal properties can be matched to those of commercial glasses.

In our experiments, two types of domestic glasses with brand marks of DM-305 and DM-308 are well bonded, in which contains more than 20% boron oxide. The bonded parts are heated in boiling water for ten minutes without interface falling-off, this is a satisfactory bonding. Well-bonded parts have no interference fringes or bubbles. Inspection of bonded interface with scanning electron microscope(SEM) is shown in Fig.2

Fig. 2 is a SEM photo of Ge-glass bonded interface which shows Ge and Si content variation in diffused interface region. This photo alos indica-

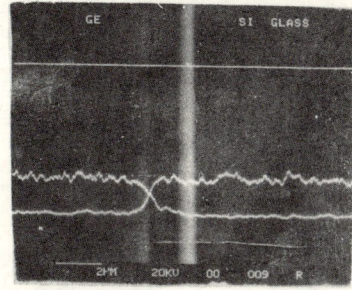

Fig.2 Variation in Ge, Si content at the interface.

tes that a slight molten layer may be existed in the interface.

Ions current and temperature variation

The value of positive ions current depends on applying voltage, temperature, contact area and boron content of glass. Ions current of Ge-glass No. 8 during sealing process is recorded as shown in Fig 3. The ion current must be controled properly, in order to have a successful bonding, Generally the well-bonded current is about 30-100μA

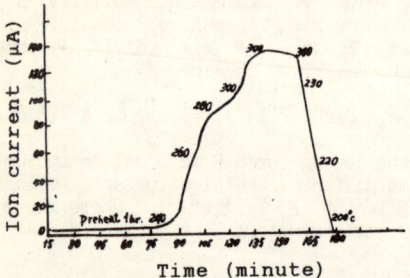

Fig 3 Ion current variation in Ge-glass No.5 during sealing process

Applications of electrostatic bonding to semiconductor devices technology

We put electrostatic bonding into the practice as following four respects:

1. Fabrication of large-area photosensitive field emission arrays

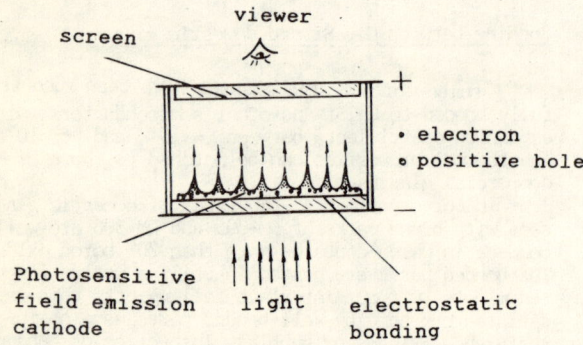

Fig.4 Photosensitive field emission water image tube by proximity focussing

The solid image tube shown in Fig.4 illustrates the very small tip emitters which are fabricated from p-type silicon. It is a proximity focus wafer tube consisted of photosensitive field emission cathode and screen anode. The key process is etch-undercutting a precision oxide pattern which is delineated on P-Silicon wafer with thickness of 25-40μm by photolithography. The silicon emission cathode wafer must be first bonded to glass for mechanical rigidity. Also the bonded interface can resist from strong acid. Through the tilting table of selective etching method then sharply-pointed dense emitters are formed. The emitted current density J can be written as

$$J = 1.54 \cdot 10^6 \cdot \frac{F^2}{\phi} \cdot \exp\left\{-6.83 \cdot 10^7 \cdot \frac{\phi^{3/2}}{F}\right\}$$

Where F is the external electric field at the tip (v/cm), and ϕ is the work function (ev), F is given by

$$F = \beta \, V_{A/d}$$

β is the field enhancement factor and $V_{A/d}$ is the average electric field between anode and cathode. The factor β is expressed as follows

$$\beta = 0.36 \frac{h}{r}$$

Where h and r are height and radius of tip emitters respectively.

The cut-off response wavelength of silicon cathode is about 1.1μm.

2. Fabrication of cold field emission cathode

With the same process as above, and the field emission cathode is made from n-type silicon, the array current density may be up to 1/4 A/cm². However the cathode-anode separation of 0.014cm is more difficult to be reached due to the destructive arcing.

3. Fabrication of resist high and low temperature IR filter

Ge, Si etc. slices can be sealed to the glass ring or fiber plate by electrostatic bonding method to form IR filter. It is a good IR filter because the bonded interface can resist heat and freeze. Such filter can work for all weathers (or corrosive environment).

4. Using electrostatic bonding to form p-n junction

The bonding process is also suitable for two semiconductor surfaces which are firstly polished. The parts are then assembled and heated to one temperature. The DC voltage is connected to the semiconductor-semiconductor (one side is high resistivity, and the other is low) assembly. In our work, two surfaces has been sealed to form a heterojunction. It seems to possess all p-n junction characteristics. By suitable choice of two semiconductors and DC voltage or temperature, it is possible to produce p-n heterojunction. A current-voltage characteristic plot is shown in Fig.5.

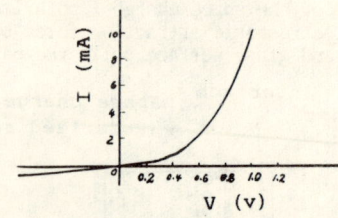

Fig.5 I-V curve of heterojunction made by electrostatic bonding

The new p-n junction fabrication method is now under research in our lab. We will report on this articale in detail at another meeting. As stated above the electrostatic bonding appears to have considerable theoretical meaning and practical applications. It is worth further researching.

TECHNOLOGY DEVELOPMENT AND MECHANISM RESEARCH OF ELECTROSTATIC SPINNING

Zhang Qu

China Textile University, Shanghai, China

Abstract

This paper is mainly composed of two parts: Firstly, it elucidates the basic principle of electrostatic spinning and its developments in the world, describes the history of electrostatic spinning research in China, the processing procedure, the scale on which the industrial production develops and the features of the products, makes a summary of the main advantages and the existing problems of the new procedure, and discusses the main stream of its developments. The second part of this article is a review of the recent research work in China, including the spatial distribution of electric field, the measurement of the electrostatic parameters of fibre, the electrification of fibre, electrostatic force acted on the fibre, the motion pattern in the electric field, and the theory related to the electrostatic spinning.

Introduction

Electrostatic spinning is a kind of open--end spinning. This processing method put the fibres under the high voltage electric field after opening, the force acted on the fibre by electric field arranges, straightens the fibres which then converge to a strand and spin into yarn.

The thought of electrostatic spinning was brought forward by Kennedy in 1940s. Afterwards, a great number of research work, patents and theoretical papers were published in more than a dozen countries, with America, Soviet Union and Japan among them. But, up to now, no report of successful research and industrial production has been heard. China began this research in the late 1950s. Through persistent efforts, improvements of the processing procedures, the design, making and adjusting of the model machine, a production workshop of comparatively large scale was built up in 1980 and it has run normally since then.

Processing Procedure

Strands of fibres are fed through the rollor, opened by the combing rollor. The individual fibres are transported into the high voltage electrostatic field, straightened under the electrostatic force, ranged in parallel to the electric force line, and converged to the central line of the main electric field. And so an open--end strand of fibre is formed which is twisted by the rotor and drawn out by the drawing--off rollor.

Processing Technology Specification

Unit Output: 70-80 kg/hr. (cotton yarn of No.20)
Count-Strength Product: 1800--2100
Yarn Evenness: 1st grade
Neps and Trashes: 1st grade--excellent
End Breakage: 30-60 piece/millenary spindle hr.
Finished Products Rate: about 96%
Package Volume: 1.2--1.5 kg

Processing Features

High Output: spinning speed: 40-45 m/min. (yarn No.20)
Bigger Package: package with weight about one kilogram can be directly wound. therefore, the doffing cycle can be prolonged.
Shorter Procedure: roving and winding procedure are not needed.
Energy Saving: consuming 8--10% less energy compared with the conventional ring spinning processing.
Less Dust Content: dust content of the air in the workshop: 0.8--1.2 mg/m^3.
Noise: <85 dB

This procedure has many advantages, it does not demand very much from the distribution of cotton, there are fewer neps and trashes in yarn, woven fabric has a higher fastness, higher abrasion resistance, and a brighter color, and it improves the strength of the knitting fabrics. The disadvantage of electrostatic spinning is that the strength of the yarn and the finished product rate are not very high. The adaptation to the change of fiber is not very satisfactory.

Products

Using the electrostatic spinning, we can spin cotton yarn of number 10-60 and various blending yarn of number 10-32. With these yarn, we can produce a variety of woven fabric as follows:
Loom Weaving: serge, plain cloth, flannel twills, creppella.
Knitting Fabric: sailor's striped shirt, jersey flannel, gloves, hosiery.
Warp-knitting: acrylic high bulk colored T-shirt, mesh table cloth.
Terry Cloth: sheet, towel.
Special Product: core-spun yarn, hand knitting yarn, bunchy cloth, burnt-out polyester-cotton fabric.

Among the above, some of the cotton, ramie-cotton, polyester-cotton products have been exported to foreign countries, which have won many praises.

Theoretical Analysis

In recent years, with the improvement of technology of electrostatic spinning, a series of theory concerned have been studied in which the distribution of the electric field and behaviour of fibre are introduced briefly as follows:

Distribution of Electric Field

Electrostatic spinning works depending on the effect of the electric field on fibre. Therefore, the intensity and distribution pattern of the high-voltage electric field play important role on the normal performance of spinning and improvement of the quality of yarn. The electric field distribution in two or three dimension space of different electrode system has been calculated successively with various numerical methods, such as finite element method, analogue charge method, boundary element method etc. aided by computer. Besides, solid simulation method is also adopted to measure the distribution of the field. Comparison of the intensity of the field in every point and the distribution of force line has provided the basis to design a reasonable electrode system and spinning equipment. Research and practice show that due to the existance of discharge in the field and ion exchange, the funtion of the field for spinning not only relates to the electrode shape and its distance, but also concern with the polarity and the material of the electrode.

Charge of Fiber

The charge transferred to fibre in the static electrification process can be contact charge, induction charge, ionization or electrode charge.

Contact charge is generated by contact, friction and separation of fibre with metal, polymer objects and air flow before fibre enters the field.

The amount and polarity charge are concerned with material, surface of objects, pressure, speed and the ambient temperature and humidity. Fibre shows a certain conductivities because of water, impurity and oil on it. After fibre enters the field, charges of different sign will pile up at the two ends of the fibre. Thus, induction charge is formed. The quantity of charge is concerned with electrostatic parameters and the charging time after it enters the field. Its charging pattern is

$$Q=Q_0+(Q_s-Q_0)*exp(-t/t_s)$$

where Q_0 is the initial charge of fibre when it enters the field. Q_s is the saturation charge on one end of fibre; t is the time constant of fibre. Theoretical study and techorological practice prove that in the spinning process when fibre moves to the area of high intensity of the field and its charge reaches the limited value, it will generate ion discharge at the end of fibre. The value of Q_i is determined geometry, electrostatic parametres of the fibre and applied field intensity etc. The result of the discharging will cause the change of the quantity and polarity and lead to the changes of electric field force, speed and direction of fibre motion. One of the characteristics of ion discharge is pulsation. This can be tested by the discharge current. It can also be confirmed by the radio frequency interference with several MHz.

The air flow and the field force make a small part of fibre gain higher motion speed, reach and contact the surface of electrode. At this time, the electric charge at the near end discharges with the electrode until the attraction with electrode change into repulsion so that the fibre seperates from electrode.

Among four kinds of charging patterns described above, corona discharge is the most important. It has much more charges than that produced by friction and induction. As to the polarization charge, it is smaller and can almost be neglected.

Electrostatic Parameters of Fiber

The charge of fibre, the electric force acted on the fibre and its motion are all closely related to the electrostatic parameters, including conductivity and dielectric constant.

The mechanism of the conductivity of fibre is mainly ionic conductivity, which is formed by ionization of water and impurity under the electric field. One of the outstanding characteristics of ionic conductivity is that it not only has charge motion, but also include motion of carrier itself. Fibre material is a soft, porous and elastic object which has no fixed length, section and shape, thus its conductive property is not indicated in terms of volume or surfase resistivity, instead mass resistivity is often used, i.e.

$$\rho_m = R*w/L^2 \quad (ohm*g/cm^2)$$

where R is resistance measured; w and L are weight and length respectively. The measured result shows that the conductivity of different fibre has great diversity. Generally speaking, resistivity of natural fibre usually lower than that of synthetic, its value varies greatly with humidity (moisture of fibre or R.H. of surrounding), this is caused by amount of porosity of it and affinity with water. The experiment also indicates that the value of conductivity is more or less related to property and amount of impurity contained in fibre.

Dielectricity of fibre is described by its relative dielectric constant ε_f. Because fibre materials generally contain a certain amount of air, the measured value actually is the dielectric constant ε_m of the mixture of fibre and air, while vacuum (air) dielectric constant is a certain known value. From the expression

$$\varepsilon_m = \varepsilon_f * F + \varepsilon_o * (1-F)$$

the dielectric constant of fibre can be calculated, where F is filling-in coefficient of sensor.

Frequency and moisture are most important ones among all factors affecting the dielectric constant. Because relaxation time needed to polarize fibre, polarization effect for different frequency may be different. The higher the frequency is the weaker the polarization effect is, and in the electrostatic field it is the most effective. Due to orientational alignment of polarized water molecule in the direction of the field, the dielectric constant of the fibre in low frequency will increase with increased moisture.

Force Acted on Fiber and Motion of Fibre in Electric Field

Electrostatic balance and strainometer are used to measure the force acted on the fibre in the electrostatic field. In fact, the charge on the fibre relates to its position and time entering the field. The acted force and motion of the fibre are a dynamic process. So after considering Coulomb force, image force, air flow force force and gravity and torque, the motion equation of fibre in the two dimension field are derived:

$$m\ddot{x} = \sum F_{xi} + F_{xf}$$
$$m\ddot{y} = \sum F_{yi} + F_{yf} - F_G$$
$$J\ddot{\phi} = \sum M_i + M_f$$

where m and J are mass and moment of intertia respectively; F_i, F_f and F_G are electric, flow and gravitational force respectively; M_i and M_f are torques acted on the fibre by electric and flow field respectively.

A computer program is made to get numerical solution under given conditions, while the motion trace of fibre is also obtained by stroboscope and both are checked with each other.

By using the above method, the force acted on the fibre, motion speed and direction of the fibre at different position in space can be got. Technological conditions such as the variety of fibre, its length, voltage, humidity etc. which affect the fibre motion have also be studied.

The study of mechanism above mentioned has provided theoretical basis for the physical nature of electrostatic spinning and the way to control effectively motion of the fibre flow and improve its technology.

References

[1]. Oglebsy, s. and Thomas: U.S.Patent 2468827(1949)

[2]. Amoto,R.S.:Electrostatic Spinning Appartus U.S.Pat.3665695(1972)

[3]. K.J.Binns,P.J.Lawrenson: Analysis and Computation of Electric and Magnetic Field Problem,Oxford,Pergamon Press,(1973)

[4]. Morton, W.E. and Hearle J.W.S.: Physical Properties of Textile Fibres, The Textile Institute,(1975)

[5]. Dogu, I. : The Fundamentals of Electrostatic Spinning. part 1,2,3, Textile Res.J.,45,46,47(1975,1976,1977)

[6]. K.Q.Robert, A.Baril: Fortschritte bei der Erforschung des Spinnens, Chemiefasern/Tectilindustrie,Dez.(1984)

[7]. Zhang Qu and Xu Ying: Motion Pattern of the Fibre in Electric Field, Journal of CTU Vol.13,(1987)

[8]. Попков, В.И. и Глазов, М.И. Кинетика зарядки и динамика волокон в электрическом поле, Канд. дисс., Автореф Т 159 (1980)

THE BASIC PRINCIPLE OF ELECTROSTATIC POWDER COATING

Zhou Shiyue

Guiyang Electrical Machinary Plant
P.O. Box 7, Guiyang, Guizhou, China

Abstract

This article studies the basic principle of electrostatic coating by polarizable high molecular powder in an electrostatic field provided by electrostatic generator. Experiments prove that the electrostatic charge gained by molecular powder in an electrostatic field is not a free charge, but a binding one. The writer puts forward the new theory that " the physical effect of electrical dipolar polarization occurs in the high molecular powder in an electrostatic field ".

Text

1. Attributes of charged powder in electrostatic field

Currently, the ' corona discharge ' theory has been introduced as the basic principle of powder electrostatic coating at home and abroad. The core of this principle is that the powder in space will be charged by the free electric charge and become an ionic powder. The ion-type electroadsorption will take place as soon as this charged powder lies closely to the grounded metal parts.

It is discovered by the author that some phenomena occured in powder coating with high voltage electrostatic generator are in contradiction with current theory concerned. For example, an armature iron core, 54mm long with a diameter of 50mm, moves up and down through the liquidized powder in an electrostatic liquidized-bed to bring the surface of its slots into full contact with the charged powder. We find that there is a few powder adsorbed in the slots and more powder is adsorbed, but not uniformly distributed over the ends of it; insulation materials(silastic, laminated bakelite plate, ect.) can also adsorb electrostatically a thick layer of powder; charged powder has a dual characteristic: repellency and attractiveness.

As illustrated in fig.1, an iron plate of 50 mm x 100mm is plced in the electrostatic fluidized bed in order to be coated with expoxy powder under negative high voltage of 60 kv. It is neither grounded nor contacted with metal while the plate is coating in order to prevent the charge to run away from iron plate. The weight of the powder to be coated on

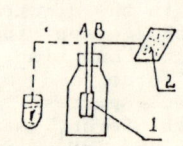

Fig.1
1. electroscope
2. iron plate

the single face of the specimen is about 7 g. The aluminum foil in the electroscope shows no action while the plate is connected to the terminal B. Then we moisten the powder with alcohol containing 60-70% of salt water and make it dissoved and linked together for decreasing its rather high electrical resistiving, and the the aluminum foil shows no action again. Now the charged powder is picked up and put into a beaker(refer to fig.1). We stir the powder with a lead connected to the terminal A of the electroscope, and no action of the alumnium foil is observed.

These experiments indicate that no free charge is transfered to the powder by corona discharge.

A comparison of the quantity of electricity between charge and friction charge has been made by a Japanese scholar[1]. The charged quantities of electricity in saturate status are not just the same between these two conditions.

According to the relation between the charged quantity of electricity and the particle diameter of powder, Panthenier Morean-Hanot formula can be used for corona charge:

$$q = 12 \frac{\varepsilon}{\varepsilon+2} \pi \varepsilon_0 r^2 E$$

This formular is applied to maximum charge gained by a theortical spheric dielectric particle in electric field, but can not be applied to friction charge. It indicates that there is a difference between these two charge mechanism. The charged quantity of electricity in the formula is not only related to E and r, but also is dependent closely upon the dielectric coeffcient. So the electrostatic character can not be considered separately from the physical effect of electric dipole polarization.

The materials generally used in electrostatic coating such as expoxy, polyster, ect. and resin powder are categorized as polar polymer solid materials. In these materials, there is an electric moment in the valence bond of the molecular bonds, and polar redicals on its branched chain. Therefore, the sway polarization of the elastic-linked polar molecule under the acting of the added electric field is the main factor in producing the electric adsorption of the powder.

When the field strength reached 10^5 v/cm, the mean electric moment of particles directly propoetional to field strength (the field strength in general electrostatic coating is in the order of 10^3 v/cm). We can make a conclusion that the degree of dipole powder polarization is proportional to the field strength. One of the important conditions to guarantee the uniformity of coating layers is the control of the consistence of the degree of powder polarization and the uniformity of the field strength on the various parts of the workpiece to be coated.

2. Verification of powder polarization by several experiments

(1) As shown in fig.2, a negative pole in the electrostatic fluidized-bed is hung in the space of the bed and kept out of contact with expoxy insulation powder. The surface resistivity ρ_s of powder is $10^{13} \Omega$, so the charge run away from powder is slowly because the ρ_s of the powder is very high. Therefor, the powder can obtain electrostatic adsorption long time (over two hours) after it is electrostatically adsorbed on the surface of a workpiece. The powder is not liquidized since no air is supplied at this time. Then connect the negative high voltage and make the corona discharge of the pole not affect the powder in the interior part of the powder layer. Cut-off the high voltage after in-

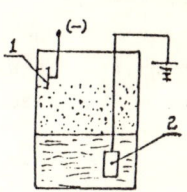

Fig.2
1. pole
2. test rod

putting it for about 1 second and supply the liquidized-bed with 0.01 Mpa pressure to make the powder to be sligthtly liquidized, and then quickly submerge the test-rod into the powder. The surface of the rod now immediately adsorb the powder, and the smaller the clearance between the rod and the pole, the more powder is adsorbed. If we ground the pole several seconds after cut-off the high voltage and then make the powder liquidized, the powder adsorption effect is greatly decreased while submerging the rod in powder.

(2) In the same apparatus in fig.2, supply the liquidized-bed with 0.05 Mpa pressure first, and cut-off the negative high voltage after inputting it about one second, and after a short delay of 60 seconds, submerge the test rod into the powder. We can see that the powder adsorption become apparent. If rotate the submerged rod in the powder, a certain thickness of powder can still be remained on its surface. In case that we immediately ground the pole several seconds after cut-off the high voltage and submerge the rod into the powder after 60 seconds, the powder adsorbed can be cleared from its surface.

(3) The pole in the fluidized-bed is placed on the bottom of the powder, as shown in fig.3, so the powder can make a full contact with the pole in order to obtain the free charge. Now we supply the liquidized-bed with 0.04 Mpa pressure, and cut-off the high voltage after inputting it for two minutes to

make the powder completely charged assuring that the powder flown in the bed has gained the full electrostatic characteristic. When the test rod is dropped down and reaches the surface of the liquidized powder layer, the adsorption powder of the test rod is rapidly increased. Even though after we cutoff the high voltage for 60 seconds, the rod still has the adsortive powder. But if we cut-off the high voltage several seconds and ground the pole simultaneously and then move the rod to reach the surface of the powder closely, the adsorption powder of the rod disappeared obviosly.

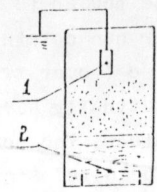

Fig.3
1. test rod
2. pole connecting with the negative high voltage

A conclusion can be made from above the three experiments that in an electrostatic field the electro-adsorption of powder charged with free charge is negligible, and the effect of corona discharge is not apparent. It is the electrostatic adsorption effect due to powder polarization induced by electric field that plays an important role in the powder electrostatic coating, and the significant action of the residual voltage of the pole for maintaining the powder polarization characteristic (so as to maintain the eletroadsorption characteristic of the powder) is verified.

(4) Exchange the polarity of the spray gun and the workpiece in an electrostatic spraying apparatus. The workpiece is connected to the negative high voltage and the gun ground (positive polarity). By such arrangement, the workpiece can adsorb powder excellently as if the powder becomes particles with positive charge. Both the spray gun and the workpiece are connected with the negative high voltage simultanously.

The workpiece can adsorb powder as before. In fact, in all electrostatic spraying it is the electric pole of the spray nozzle that adsorbs the charged powder first.

3. Stress analysis after polarization of powder in an electrostatic field

The loading condition of powder in electrostatic coating a metal plate is shown in fig. 4 and fig.5a. Compute the magnitude of the electrostatic force F to be resisted by the dipole powder at point A at a distance R from the plate. Only the attractive electrostatic force Fz along the z axis of the plate is considered, and $F = \Sigma F_z$ (gravitational force is neglected).

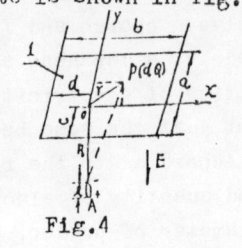

Fig.4
1. metal plate

Through computation, we have derived an equation to find out the magnitude of the attractive electrostatic force applied on the powder in an electrostatic field. The derivation of these computation is omitted. The formula is listed below.

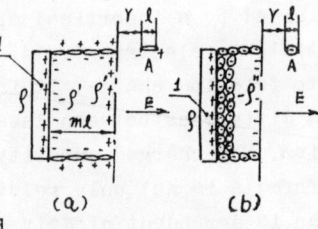

Fig.5
1. metal plate
ℓ. length of dipole powder
ρ. density of induced charge
ρ' and $-\rho'$. density of polarized charge
$-\rho''$. free charge density of powder

$$F = F_{z1} + F_{z2} + F_{z3}$$
$$= (K_1-K_2)\left[\operatorname{arctg}\frac{\varepsilon}{a\cdot\alpha} + \operatorname{arctg}\frac{\eta}{(a+n\ell)\cdot\beta}\right.$$
$$\left. - \operatorname{arctg}\frac{\eta}{a\cdot\beta} - \operatorname{arctg}\frac{\varepsilon}{(a+n\ell)\cdot\alpha}\right]$$
$$+ K_2\left[\operatorname{arctg}\frac{\varepsilon}{(a-nm\ell)\cdot\alpha} + \operatorname{arctg}\frac{\eta}{(a-n(m-1)\ell)\cdot\beta}\right.$$
$$\left. - \operatorname{arctg}\frac{\eta}{(a-nm\ell)\cdot\beta} - \operatorname{arctg}\frac{\varepsilon}{(a-n(m-1)\ell)\cdot\alpha}\right]$$
$$- (K_1-K_2)\left[\operatorname{arctg}\frac{\theta}{a\cdot\gamma} + \operatorname{arctg}\frac{\phi}{(a+n\ell)\cdot\delta}\right.$$
$$\left. - \operatorname{arctg}\frac{\phi}{a\cdot\delta} - \operatorname{arctg}\frac{\theta}{(a+n\ell)\cdot\gamma}\right]$$
$$- K_2\left[\operatorname{arctg}\frac{\theta}{(a-nm\ell)\cdot\gamma} + \operatorname{arctg}\frac{\phi}{(a-n(m-1)\ell)\cdot\delta}\right.$$
$$\left. - \operatorname{arctg}\frac{\phi}{(a-nm\ell)\cdot\delta} - \operatorname{arctg}\frac{\theta}{(a-n(m-1)\ell)\cdot\gamma}\right]$$

where:
$A_1^2 = (a-c)^2 + D^2$ $A_2^2 = (a-c)^2 + d^2$
$B_1^2 = c^2 + D^2$ $B_2^2 = c^2 + d^2$
$D = b - d$ $n = \dfrac{a}{R}$
$\alpha = \sqrt{a^2 + n^2 A_1^2}$ $\beta = \sqrt{a^2 + n^2 B_1^2}$
$\gamma = \sqrt{a^2 + n^2 A_2^2}$ $\delta = \sqrt{a^2 + n^2 B_2^2}$
$\varepsilon = n^2 \cdot D(a-c)$ $\eta = n^2 \cdot c \cdot D$
$\theta = n^2 \cdot d(a-c)$ $\phi = n^2 \cdot c \cdot d$

m – Number of adsorbed powder layer

4. Discussion

(1) Comparision of polarized electrostatic force and electrostatic force induced by corona discharge.

The polarized electrostatic force applied on the powder at A is shown in fig.5a.

$$F_p = (F_{\rho} + F_{\rho'}) - F_{-\rho'} \quad (1)$$

The electrostatic force induced by corona discharge on the powder at A is shown in fig. 5b.

$$F_c = F_{\rho} - F_{-\rho''}$$

Since the center of positive charge of the neighbouring polarized powder in the powder layer almost coincides that of negative charge, $-\rho'$ merely indicated the intensity of negative charge of the powder adsorbed closely near the workpiece surface. So,

$$F_{-\rho''} = mF_{-\rho'}$$

and

$$F_c = F_{\rho} - mF_{-\rho'} \quad (2)$$

We can see that $F_p > F_c$ by comparing equation (1) with equation (2). In practice, the above deduction is verified by the phenomenon that the powder adsorption ability of high pressure spray gun is really stronger than that of friction electrostatic spray gun.

Following the increasement of the number m powder layer, F_c is approaching to zero. That is, there is a gradually enhancement of so-called 'self-limiting effect' of the thickness of coating layer. But, in fact it is not apparent in the powder electrostatic coating. When the powder is adsorbed by the workpiece surface, the adsorption rate is not reduced regularly due to the increasement of the thickness. However, this phenomenon accords comparatively with the adsorption characteristics of the polarized electrostatic force.

(2) The mutual induction among the gaseous powder can be neglected since the distance between the dipole particles r is sufficient large and the electrostatic force applied each other is very weak. The flying-up of the powder is accelerated under the negative high voltage. The distance between liquid state powders is small, so the probability for $r \approx \ell$ or $r < \ell$ is larger. Hence, there is an apparent dipole inductive directional effect. Under the weak air flow and strong negative high voltage condiction, the dipole powder will be adsorbed each other in a front versus and manner. Especially for the powder with a larger ε value or after the affection with damp, this action is more apparent. It causes the liquid state powder to be contracted and results in the occurence of the phenomena of blocking-up the hole and connecting the slots with the powder in coating.

(3) Although the insulative materials can not conduct electric current but similar physical effect of powder dipole polarization also takes place over these surface in an electrostatic field. The bound charge directionally arranged and the powder are adsorbed each other. The status of such adsorption depends upon the characteristic and behaviou of the bound charge in the field.

(4) The ununiformity of layer is caused by two factors: the ununiformity of the electrostatic field distribution and the ununiformity of powder density. The ununiformity of powder layer can not be thoroughly eliminated by prolonging the operation time, because the polar electrostatic adsorption is a continuous process.

Reference :

1 The friction charging of polymer powder. " Metal Surface Technique " V.29, No.9

STUDIES ON ARTIFICIAL CURING OF WHITE SPIRIT BY MEANS OF ULTRAHIGH VOLTAGE (UHV) ELECTROSTATIC FIELD

Guan Xiaosheng (Liaoning University)

Sun Qingwen
Guo Shichen (Shenyang Laolongkou Distillery)

China

Abstract:

Studies on artificial curing of white spirit (distilled from fermented sorghum or maize) is deemed as an important subject both at home and abroad. This article presents a new method of curing white spirit by means of electrostatic field. The technique appears to be promising for industrial production. At the same time, the discovery of this technique provides a new approach to catalytic chemical reaction and bears certain theoretical value for developing chemical catalysis theory.

A. PREFACE

In white spirit production, the new product freshly obtained from distillation is generally characterized with sharp pungency, nose irritation and strong stimulation. To produce product of superquality with mellow taste, elegant colour and lustre, and uniform and mildness, certain storage period has to be specified in production process. Storage period for some spirits of famous brand may be as long as one to three years, even five to seven years (such as China's Mao Tai), and the French Brandy Coneck, one of world's three wellknown wines, has often been stored as long as twenty-five years.

For the above, the breweries and distilleries have to be equiped with storehouses and containers of large volume. This not only occupies large area of land and consumes high cost but also results in heavy loss due to volatilization of spirit during long period of storage, which affects negatively the capital circulation.

Now study on artificial curing of white spirit has been carried out both at home and abroad. Methods being adopted are supersonic wave, microwave, ultraviolet, and laser techniques or radiation by radioactive isotopes, but none of these methods have been used for industrial production because of their small amount of processing, high energy cost and poor stability.

The present study is conducted on the basis of solving the above difficulties in brewery industry. Tests show that electrostatic method plays a catalytic role to chemical reactions of concern, apart from accelerating the physical processes in spirit curing.

B. MECHANISM OF ELECTROSTATIC CURING

1. Physical change

Physical change in curing process is first of all the variation in the degree of freedom of spirit molecules. The greater number of spirit molecules with higher freedom exists, the stronger the stimulation is. While in natural curing, larger groups of associated molecules are formed between spirit molecules and water molecules gradually, and the degree of freedom would derease with increasing extent of association of spirit molecules and, hence, giving forth a taste of mildness. The action of strong electric field forces the polarized molecules of spirit, water and etc. to line up along electrical field and enables them to form associated groups within shorter time. Moreover, the electrostatic treatment accelerates the volatilization of gases with low-boiling-point and liquids such as vulcanized hydrogen, acrolein and etc. Change caused by elctrostatic field is the major reason for the wine to become mild and reduce to great extent its stimulus of freshly distilled products.

2. Chemical change

The chemical changes in curing white spirit are mainly esterification redox reaction and condensation, which enable compositions such as alcohol, acid and aldehyde in spirit to reach new equilibrium. The main reactions are:

a. The esterification reaction:

Esterification reaction that results in ester by alcohols and acids expressed as:

$$RCOOH + R'OH \rightarrow RCOOR' + H_2O$$

resulting in the increase of esters and the slight decrease in acidity and spirit content.

b. Redox reaction:

Alcohol produces aldehydes through oxidation: $RCH_2OH \rightleftharpoons RCHO + H_2O$. Aldehyde produces acid through oxidation: $RCHO \rightleftharpoons RCOOH$, resulting in the decrease in spirit content.

c. Condensation reaction:

$$2R'OH + RCHO \rightarrow RCH(OR')_2 + H_2O$$

rearrangement of alcohol and aldehyde, reduces the stimulus of spirit and increases fragrance of spirit.

Aside from enabling polarized molecules to arrange regularly, the presence of strong electric field enables the ions or other charged clusters or ion groups in spirit to obtain huge energy and even breaks somebonds by the energy. After the newly emerged ions have obtained sufficient energy from electric field, they may go ahead to break off other chemical bonds and, at the same time, their probability of effective collision with other ions or ion groups (the collision that may cause chemical reaction is called effective collision) is remarkably increased because they have obtained energy from electric field. The presence of strong electric field, therefore, speeds up to a great extent the chemical reaction that can occur under natural storage conditions, that means the presence of strong electric field has played the role of accelerating chemical reactions.

C. EXPERIMENTAL SYSTEMS

The system is composed of high-voltage electrostatic generator, controllor, electric corona wire and special container. Schematic illustration of the experimental system is given as follows:

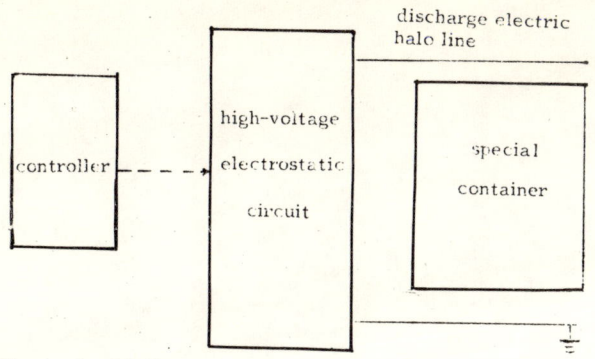

Figure 1. Schematic diagram of the experimental system

D. EXPERIMENT METHOD

Pour the white spirit to be treated into special container, switch on controller after the spirit is uniformly stirred, and adjust the output of high voltage. According to different white spirit, the required electric field strength and treatment period should be selected appropriately.

Range of changes in electric field strength·

2kv/m ∼ 400kv/m,

Range of changes in treating time:

1 sec. ∼ 16hrs.

Amount treated each time: 0.1l 500l. Sampling or composition analysis of the treated spirit can be conducted 20 days after static storage of the sample.

E. SAMPLING AND RESULT OF TEST

a. The conclusions of appraisal by experts (members of the National Wine Appraisal Committee) to the new spirit properly treated with electrostatic technique (the white spirit is CHENNIANG brand (Long-cured wine) by Shenyang Laolongkou Brewery) are as follows:

The new spirit properly treated by electrostatic technique is colorless and limpid, its flavour is fragrant and mellow with slight sweetness, improved softness, clean odd amount and lingering long flavour.

No restoration to original taste has ever occurred to the electrostatic-treated sample after it was stored for three to six months.

The primarily electrostatic cured sample tastes as the same in quality as compared with sample nataurally stored for five months.

b. Refer to Table 1 for results of chemical composition analysis of samples 14-0 (control sample) and 14-6 (treated sample) by liaoning Provincial Foodstuff Research Institute and Liaoning Provincial Hygiene Monitoring and Inspection Institute:

From Table 1, it could be seen that the changes of trace compositons follow up the law as follows.

a. Sulphid and acrylaldehyde are the main composition that make the newly-cured spirit taste pungent. Content of acrylaldehyde reduced remarkably after electrostatic treatment: from 5.60 g/100kg to 4.2 g/100kg, reduceed by 25%. Content of sulphide was checked out only in trace amount.

b. Miscellanous alcohol oils was reduced by 52% from 0.097 g/100mL to 0.050 g/100mL, which improves the quality of spirit.

c. Acetal in white spirit is a trace composition that makes spirit freshly fragrant and mild, which increases from 100.04mg/100mL to 102.9mg/100mL, 3% increasing after electrostatic treatment.

d. Content of isoamyl alcohol in spirit after electrostatic treatment was reduced from 31.96mg/100mL to 30.02mg/100mL, by 6%, which tastes puckery to the tongue and, it is obvious that this is one of reasons why the spirit quality is improved.

e. Ethyl hexanoate is the main fragrant body that characterizes the spirit of strong fragrant. It increases from 32.33mg/100mL to 34.62mg/100ml, increased by 7%, that enhances flavour of white spirit.

f. Ethyl acetate in white spirit gives out

Table 1. Changes of Main Trace Components that Give out Flavours and Tastes:

Trace Component	Calculated	Result		
		14-0	14-6	change in %
1 Acrylaladehyde	g/100kg	5.6	4.20	reduced by 25%
2 Sulphide		trace amount	trace amount	
3 Miscellaneous alcohol oils	g/100L	0.097	.050	reduced by 52%
4 Acetal	mg/100mL	100.04	102.91	increased by 3%
5 Ethyl acetate	mg/100mL	32.33	34.62	increased by 7%
6 Ethyl acetate	mg/100mL	240.16	250.43	increased by 4%
7 General esters	g/L	5.10	5.29	increased by 4%
8 Diacetyl	mg/L	100.4	94.6	decreased by 6%
9 Isoamyl alcohol	mg/100mL	31.96	30.02	decreased by 6%

fruit flavour and increases from 240.16mg/100mL, to 250.43mg/100ml after treatment, increased by 4%.

g. Most esters give out fruit fragrance which is the main composition in white spirit to form fragrance and increases from 5.10g/L to 5.29g/L, increased by 4% after treatment.

h. Diacetyl in white spirit is composition that gives out flavour and fragrant smell, and reduces from 100.4mg/L to 94.6mg/L, dereased by 6% after treatment.

i. Electric conductivity of white spirit after electrostatic treatment tends to increase, from $19.0\ \mu\Omega^{-1}$ to $19.5\ \mu\Omega^{-1}$, which is similar with the changing law in natural curing process.

F. CONCLUSION

a. Both the sensing quality of the electrostatic-field-treated spirit and the changing tendency in its trace compositions that give forth fragrance and falvour basically coincide with the changing law of the naturally-cured white spirit, based on both experts' appraisal and chemical analysis.

b. No restoration to its original taste has ever occurred to the electro-statically-treataed samples that have been stored for three to six months and re-appreciated.

c. It is desirable tha huge economic and socail effects will be gained if this technique is put into production because it features, as said above, with simple device, low energy consumption and larger amount of treatment per batch, so it has a broad prospect for popularization in use.

G. DISCUSSION

a. Amount of batch treatment should be increased, to 1000L and more, if possible.

b. Current experiments have been conducted only to medium-priced white spirit such as strong-fragrant-type and fresh-flavour-type od spirit. Artificial curing experiment should be conducted to luxury spirit and high-quality spirit.

c. To obtain optimum effect, other physical methods should be studied along with electrostatic treatment as a main method.

d. Theoretical research should be further conducted to reveal the mechanism of artificial curing together with the above said experiment and study.

e. Scope of experiment should be expanded. Apply this method of catalystic chemical reaction to other chemical reactions that are valuable but hard to realize, so as to open a new field of electrostatic chemistry

EFFECT OF ELECTROSTATIC FIELD ON LIQUID SURFACE TENSION

Xian Fusheng, Zheng Jiaqiang
Jiangsu Institute of Technology
Jiangsu Province, P.R.China

ABSTRACT

The variation of liquid surface tension under the action of electrostatic field is analyzed on the basis of charged adsorbability and liquid surface molecule orientation, then a mathematical model is established through experiments and analysis.

Introduction

Based on the decrease in the liquid surface tension in the electrostatic field, many new techniques have been developed, such as electrostatic painting, spraying and liquid purification. It is shown that surface tension is a major resistance for electrostatic atomization. The authors intend to measure the variation of liquid surface tension in electrostatic field and to analyze the regular data pattern and establish the mathematical model providing the quantitative basis for the electrostatic atomization and other related studies.

Theoretical Analysis

1. Liquid superficial layer molecularity and its surface tension.

The resultant force of any molecule acted by the other is zero in the interior of liquid, but as the distance between the molecule and the liquid surface is less than molecule action radius r, one part of the molecule action ball will be beyond the liquid surface. The surrounding gas density is very low and its action may be negligible. Therefore the resultant force acting on the superficial layer molecule will be directed towards the inner liquid and exerts an inward molecular pressure. The movement of a molecule from inner to outer layers must work against the moleculer pressure, and thus increases the potential energy of the molecule. For maintaining the stable balance with minimum potential energy the molecules of superficial liquid layer tend to squeeze into the interior of the liquid.

The interactions of molecules between superficial layer and interior of the liquid is shown in Fig.1. The greater the inter molecuar force, the larger the surface tension is:

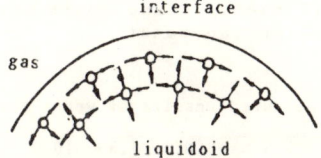

Fig.1 Interactions of molecules between Superficial layer and interior of the liquid

2. Liquid superficial adsorption and molecular ordered orientation in electrostatic field.

Most of liquid molecules have dipole moments and will be oriented at their phase interface. The concentration difference phenomenon between superficial layer and liquid phase is called adsorption.

For low concentration solution and gas interface, Gibbs adsorption isotherm can be written as follows.[2]

$$\sigma = \sigma_0 - r_2' RT \qquad (1)$$

Where σ_0 and σ are surface tensions of pure solvent and solution respectively. r_2 is the solute superficial surpus, R is a gas constant, T is Kelvin temperature.

The effect of electrostatic field on the liquid superficial adsorption is also analyzed and we know:

$$\frac{d(\Delta n)}{d\psi} \begin{cases} =0 & \psi=0 \\ <0 & \psi<0 \end{cases} \qquad (2)$$

Where n is a varied value of partial ion concentration of charged liquid
ψ is the electric potential at a certain point from the liquid surface.

From Equation (2), We know that for the charged surface of the conducting liquid its partial ion concentration increases with then the liquid surface tension will decrease from Gibbs' adsorption isotherm and Equation (1). But from the results of experiments on superficial layer adsorption saturation, the concept of the molecule ordered orientation at the surface is considered. The ordered orientation will be more remarkable as the superficial adsorption capacity increases. This conclusion can be analyzed from the statistic theory of dipole orientation in the external electric field.[4]

If there is no external electric field, collision makes every liquid molecule keep the isotropic statistical orientation. If there is an external electric field, every molecule tends to cause its orientation direction parallel to the external electric field while the surface tension acts against this tendency, so liquid molecules system finally tends to be a new balance. Because the number of the dipoles along the external electric field direction is more than that in the opposite direction, liquid molecule system becomes a weak anisotropic one. Suppose dipole orientation direction makes and angle θ with the electric field direction, from Maxwell-Boitzman statistics, there is a Langevin function in an external electric field:[4]

$$\cos\theta = L(E/T) \qquad (3)$$

Where E is the external electric field strength.

Its functional curve is shown in Figure 2. Obviously, as the E/T increases, $\cos\theta \to 1$ that is, when the dipolar electric layer forms in the external electric field, the orientation of the electric field prevails against the

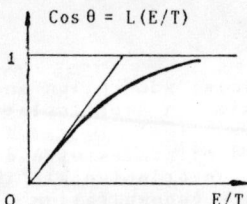

Fig.2 Langevin functional curve

disturbance of various factors (such as temperature etc) and causes the dipoles' direction along with the external field direction. Then two external poles of the dipole are situated in the liquid phase. The immovable negative pole comes into close contact with the interface (for the negatively charged liquid surface) and the external positive pole is dispersed in the liquid molecule layer at a distance from the surface. Therefore, after the surface becomes the saturated adsorption layer, the molecule single layer will vertically be ranked, as shown in Figure 3. such molecule ordered orientation

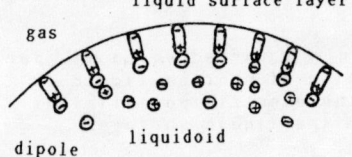

Fig.3 Interface adsorption dipolar electric layer diagram.

decreases the interaction force between molecules, and reduces the work which is required by moving the molecules or ions into superficial layer. Then from the analysis in Fig.1, the surface tension will decrease. Therefore, we can know that decreases with the increased external electric field strength, ordered orientation is more remarkable, and the value of suface tension is smaller.

Experiments and Results

A schematic illustration is shown in Fig. 45. In order to measure the surface tension conveniently, the instrument must be well insulated when the liquid is applied with high-voltage. Through experiments the insulation film is placed between the liquid testing cup and the adjustable base. The lifting ring is connected to boom with the insulation boom-head. And the electrostatic voltage is uniformly applied around the liquid cup by three silver-clad-needle electrodes through a high-voltage wire. Then the liquid cup, liquid and lifting ring are exerted by electrostatic voltage, while the complete set of the instrument well insulated.

The difference $\Delta\sigma$ of liquid surface tension before and after exerting electrostatic voltage is taken as the criterion and is measured at different electrostatic voltage.

Table 1 shows the measured results of tap-water with a conductivity of $2.8\times10^2 \mu v/cm$ at the environmental temperature of 29 °C and relative humidity of 95%.

Table 1 The measured results of tap-water

V(kv)	0	1	2	3	4	5	6	7	8	9
σ (dyn/cm)	0	0.1	0.2	0.4	0.8	1.2	1.9	2.2	2.6	3.7
V(kv)	10	11	12	13	14	15	16	17	18	18.5
σ (dyn/cm)	4.0	5.4	6.2	7.2	8.0	8.5	10.0	12.4	13.8	14.6

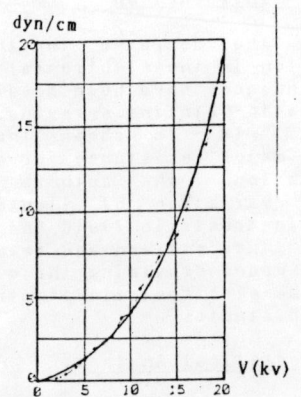

Fig.4 $\Delta\sigma$-V curves

Regression curve and experimental data are shown in Figure 5. Satisfactory fitting is obtained. Its mathematical model is determined as:

$$\Delta\sigma = 1.45(e^{0.133V} - 1) \quad (dyn/cm) \quad (4)$$

where V is voitage (in kv)

The second test liquid is sodium silicate Na sio, also called as water-glass. Its measured surface tension is 65.3 dyn/cm, and conductivity $1.7\times10^4 \mu v/cm$. Environmental temperature 31 °C and relative humidity 95%. The measured results are shown in Table 2.

By regression, these data can also be related as

$$\Delta\sigma = A(e^{BV} - 1)$$

but its coefficients A and B are not the same as those for the tap-water.

Table 2 The measured results of water-glass

V(kv)	0	2.5	5.0	7.5	10.0	12.5	15.4
σ (dyn/cm)	0	0.4	1.9	3.8	4.0	9.4	11.4

Fig.5 shows the measurements of oil-water interfacial surface tension in electrostatic field. A combination of 25ml tap-water and 25ml NO66 gasoline is used as the test liquid. The measured gasoline surface tension is 25.6dyn/cm, and the conductivity is 0.013μv/cm

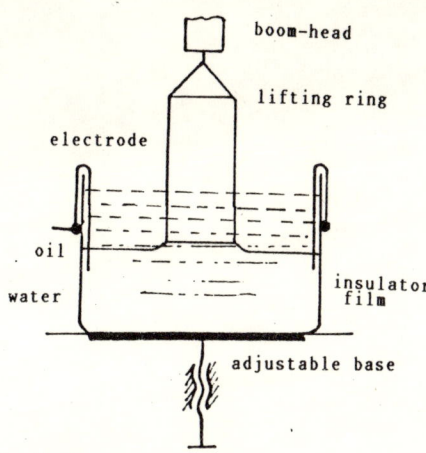

Fig 5 Measured area of gasoline-water interfacial surface tension

While measuring interfacial tension, the ring is pulled out from water into gasoline. The measured gasoline-water interfacial surface tension with no electrostatic field applied is 25.1dyn/cm at 25°C. Then suppose the interfacial surface tension difference with no electrostatic field applied is zero and the measured results are listed in Table 3.

Table 3 The measured results of gasoline-water interfacial surface tension

V(kv)	0	2.5	5	10	15
σ (dyn/cm)	0	1.6	2.5	4.8	7.6

When the voltage further increased (up to 17.5KV) strong fluctuation occurs in the interface and a small amount of water is attracted by the ring. This will affect the accuracy in measurement. The phenomenon may be caused by the zero interfacial surface tension and the greater conductivity of water than gasoline. Therefore the ring repels and attracts water alternatively near the interface and no interfacial surface tension can be measured. However from Table 4. it is found that the interfacial surface tension difference almost linearly increases with the external voltage, and the regression equation is

$$\Delta\sigma = 0.506V \quad (dyn/cm) \quad (5)$$

Conclusion

1. In the electrostatic field, becaues of charged adsorbability and liquid superficial molecule ordered orientation, the liquid surface tension decreases and its falling value increases exponentially, that is:

$$\sigma = A(e^{BV} - 1)$$

2. From the measurement of the surface tension change in tap-water and water-glass in an electrostatic field, it is found that the change in liquid surface with external electric voltage is related to its conductivity. This is because the smaller the conductivity, the weaker the action of the liquid as the charge carrier. And the charge are uneasily gathered on the surface and the surface tension is little changed.

3. oil-water interfacial surface tension is approximately linear with the external electrostatic voltage ($\Delta\sigma$ =kV). That is oil-water interfacial surface tension difference increases linerly with the external electrostatic voltage. Therefore it is possible that greasy dirt may be washed away by electrostatic washing with the aid of mechanical force.

References

(1). Liang-run Gao. Fu-shengXian: Electrostatic spraying Theory and Its testing Technique. For presentation at the 1986 winter Meeting ASAE.
(2). Fu-sheng Xian, Chundu. Wu: The effect of static Electricity and physical properties of liquid on spray Atomization.Trans of the chinese society of Agricultural Machinery, Vol 18, No1. 1987.
(3). Jilin University and sichuan University: physical chemistry and colliod chemistry, Higher educational publishing house, 1980.
(4). R.Coelho: physics of Dielectries for the Engineer, Elsevier Scientific Publishing Company, 1979.

* This paper is advised by Professor Liang-run Gao.
  Jiangsu Institute of Techunology, China

IV. Fundamentals of Electrophotography

Proceedings of International Conference on Modern Electrostatics

COLOR COPY MACHINE

Toru Takahashi

Copier Devel. Div. Canon Inc. Shimomaruko 3-30-2, Ohta-ku, Tokyo JAPAN

Abstract

The history of electrophotographic color copies will be briefly summarized. The number of units installed in the world was less than 1,000 in '86. But, in '87, Canon produced 6,000 units and probably 10,000 units in '88. The reason of such revolutional change is analzed. Image quality improvement is the main factor, which is realized with a combination of electrophotography, digital imaging technology and computer image processing software.

History

A history of electrophotographic color copier products is shown in Table 1.

```
1969  3M Color-In-Color    ZnO-EF, Dye transfer
1973  Xerox 6500           Se, Solid roller transfer
1977  Ciba Geigy CC-001    Silver Haraid
1978  Canon NP Color       CdS, Screen transfer
1983  Canon NP Color T     Tone control process
1985  Ricoh Color 5000     Se
      Konica Color 7       Silver Haraid
1987  Canon Laser Color Digital
1988  Kodak Color Edge Flush exp., OPC belt
```
Table 1. Color copier products

Canon's history of color copier development will be explained.

Color Laser Copier

Canon introduced the Color Laser Copier into the market in January 1987, hereafter referred to as the CLC. This is a combination of a CCD color scanner unit and color laser printer as shown with Fig. 1.

An original on the copyboard glass is scanned by an optical system including the CCD. Here, one of the world's top level sensors, which was newly developed, is used. The analog signals from the CCD are sent to the image processing unit. Here, analog signals are converted into digital signals and processed by computers. The information is sent to the laser unit in real-time. The laser beam hits the surface of the photosensitive drum, and

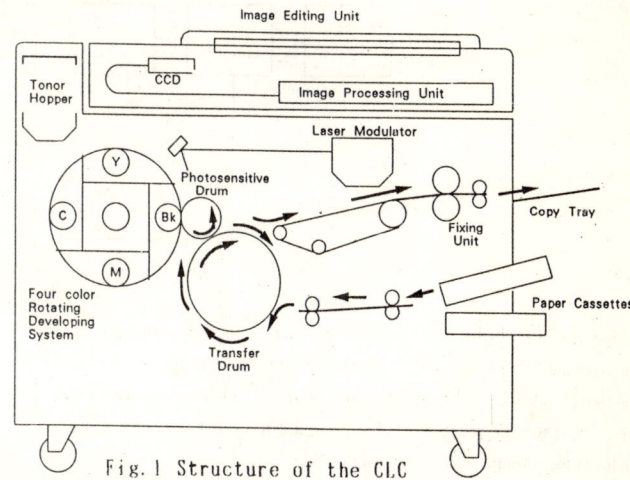

Fig. 1 Structure of the CLC

after going through the usual electrophotographic color processing, reproduces into copies. The difference from the usual processing is that in this case, black is used as well as cyan, magenta and yellow. Canon's highest LBP technology is applied here. Table 2 is the summary of the specifications of the CLC.

```
Original size       A3
Copy size           A3
Copy speed          5 (A4) copies/minute
R&E                 50%-400%
Resolution          400dpi/200dpi
Tone reproduction   >64 levels
```
Table 2. Specification of the CLC

Image Processing System

Fig. 2 is an outline of the signal processing system of the color image scanner. The analog signals of R, G, B from the color CCD sensor composed of five CCD chips are amplified and converted into 8 bit digital signals in the A/D conversion unit. The 8 bit digital image signals for each color of 5 channels are synthesized into one line along the direction of the primary scanning line in the color separation/5-channel synthesis unit.

Non-uniformity of the image data in one line caused by the difference of sensitivities and

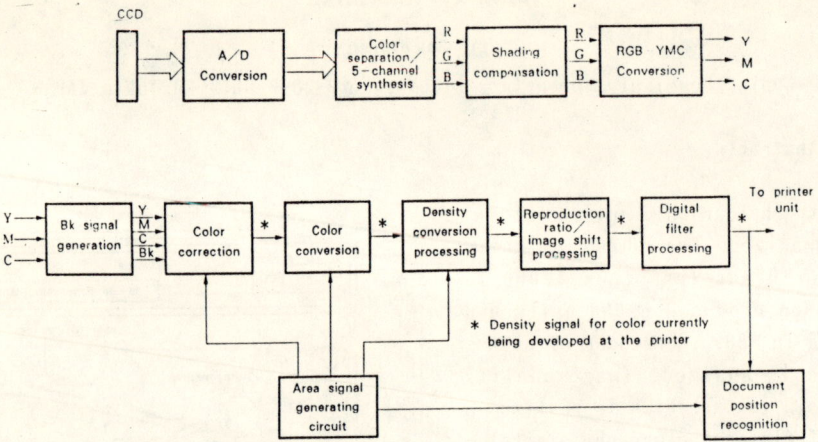

Fig. 2 Image Data Flow of the CLC

the difference of color separation between each element of the CCD. Non-uniformity of exposure level along the line of the primary scanning, is compensated for each color at the shading-compensation unit.

Thus uniformed R,G,B digital video data are converted into density signals corresponding to Y,M,C toner density for each. The signals are then transmitted to the Bk signal generation unit, the under color removal unit (UCR), and the the color correction unit. Here, a Bk signal is generated from the three color signals (YMC).

The transmission characteristics of the blue, green and red filters on the CCD sensor and the reflection characteristics of the cyan, magenta and yellow toners are corrected in the color correction unit. In the color conversion unit, image processing such as designated color conversion, painting, framing and blanking are done according to the operator's instructions. When the digital image signal is transmitted to the density conversion processing unit, the level of the signal is converted according to the information from the operation panel (copy density, color balance etc.) before the signal is output to the printer.

The CLC also has the reduction/enlargement unit, image shift unit, digital filter processing unit and document position & size recognition unit. There is also area data unit for the image editing. Due to the development of exclusive LSIs, the various image processes mentioned above are accomplished in real time. Thus, high speed, high image quality and a variety of intelligent functions are available.

Image quality

The imaging characteristics (D_o-D_p curve) of Canon NP Color, Canon NP Color T and CLC are shown in Fig. 3. The tone reproduction of model T was an improvement from the preceeding model with a special electrophotographic process. But it was not enough for users. As for the CLC, this curve is designed to fit to users requirement with the computer-controlled image processing system as explained. We can get the ideal curve freely. Other imaging characteristics will be shown by slide projecter.

Thus we have got a free control method for the electrostatic imaging field and we can expect the color copier industry to expand into the future.

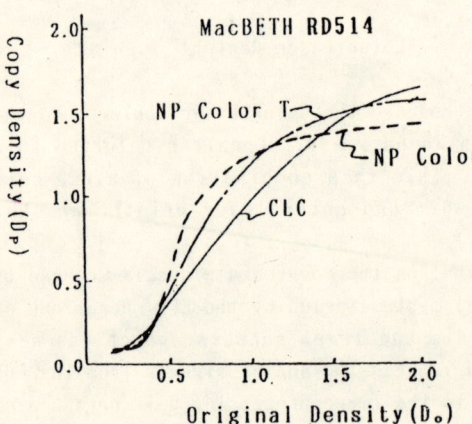

Fig. 3 $D_{original}$-D_{print} curve of the CLC

NOVEL ORGANIC PHOTORECEPTOR USING ORGANOPOLYSILANES

Kenji Yokoyama, Masaaki Yokoyama

Faculty of Engineering, Osaka University,
Yamadaoka, Suita, Osaka 565, Japan.

Masaie Fujino, Nobuo Matsumoto

NTT Basic Research Laboratories,
Musashino, Tokyo 180, Japan.

Abstract

The hole drift mobility and hole carrier injection from organic pigments for two typical organopolysilanes, i.e., poly(methylphenylsilane) $(SiMePh)_x$, and poly(methylpropylsilane) $(SiMePro)_x$, were investigated aiming an application to the photoreceptors of electrophotography. Hole injection from the photoelectrode consisting of several organic pigments was demonstrated in the surface potential photo-decay measurements for the layered photoreceptors, suggesting the possibility for the practical use. Actually, the printing test using a commercially available laser beam printer gave the fairly good quality of the images.

Introduction

From the requirements of higher sensitivity and practical standpoints, the majority of organic photoreceptors of electrophotography in use today, has the layered structure composed of the charge generation layer (CGL), generating charge carriers by photo-irradiation, and the charge transport layer (CTL), transporting the charge carriers injected from CGL. The materials for the CTL were primarily the polymers such as poly-N-vinylcarbazole (PVK) or its complex with 2,4,7-trinitro-9-fluorenone (TNF). These polymers are of the low drift mobility of only $10^{-7} \sim 10^{-6}$ cm^2/Vsec. The recent progress in CTL materials have provided the different type of polymeric materials in which low-molecular-weight charge transporting compounds such as aromatic amines and hydrazone derivatives are molecularly dispersed in inert polymer binder for the film forming. From the practical viewpoint of easy processing, however, the one-component polymer materials having high mobility would be preferred.

Organopolysilanes (Fig. 1) are the amorphous polymers characterized by their saturated Si backbone with two organic side groups (aryl or alkyl groups) attached to each Si atom. These polymers show the hole drift mobility of $\sim 10^{-4}$ cm^2/Vsec at room temperature,[1] which is the highest drift mobility in amorphous polymeric materials. It is very interesting that the charge carrier transport through Si-Si σ bond in organopolysilanes and the charge carrier injection from organic photoelectrode to the polymers are compared with molecularly dispersed charge transport system.

In the present work, the temperature dependence of the hole drift mobility in two typical organopolysilanes, i.e., poly(methylphenylsilane) $(SiMePh)_x$ and poly(methylpropylsilane) $(SiMePro)_x$, and hole carrier injection from organic pigments were investigated aiming an application to the photoreceptors of electrophotography. The differences in carrier transport and injection between molecularly dispersed system and organopolysilane polymer system are discussed. In addition, the printing test using the layered photoreceptor consisting of the combination of organopolysilane and organic pigment is demonstrated.

$R = C_6H_5(Ph)$
$n\text{-}C_3H_7(Pro)$

Fig. 1

Experimental

Organopolysilanes used in the present work (Fig. 1) were synthesized by the reaction of diorganodichlorosilanes with sodium metal in toluene at the temperature above 100℃.[2] The low-molecular-weight charge transporting compound used for comparison is N,N,N',N'- tetrakis (3-methylphenyl)-1,3-diaminobenzene (PDA, Fig. 2), which gives the excellent hole mobility of $\sim 10^{-5}$ cm^2/Vsec in the molecular dispersion of polycarbonate resin (1:1 in wt) and exhibits the ionization potential similar to (SiMePh)$_x$.

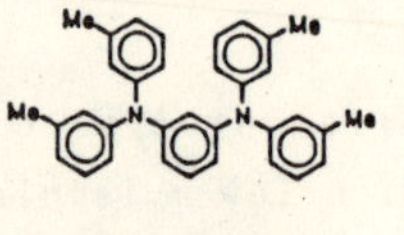

PDA 5.68eV

Fig. 2

The organic pigments for the CGL materials as photoelectrodes are listed in Fig. 3 together with the ionization potentials measured by a Surface Analyzer (Rikenkeiki Co., Ltd.) which is one version of UPS under atmosphere. These are some of metal and metal-free phthalocyanines (PbPc, TiOPc, H$_2$Pc), bisazo and trisazo pigments[3] (azo-A,B,C), and perylene pigment (Pery). The ionization potentials of the present organopolysilanes, (SiMePh)$_x$ and (SiMePro)$_x$, were obtained to be 5.69 and 5.82 eV, respectively, measured by the same apparatus.

Photoreceptors prepared have the layered structure consisting of the carrier generation layer (CGL: 0.5μm) on an aluminum substrate which contains respective organic pigment (50 wt%) dispersed in polyvinylbutyral resin and the carrier transport layer (CTL: ~7μm) of the organopolysilanes cast from toluene solution. For the mobility measurements, CTL single-layered samples were provided with semi-transparent Au electrode by vacuum deposition on the film surface to make a sandwich-type cell.

Hole drift mobility was measured by the usual time-of-flight technique using N$_2$ gas laser (337 nm, pulse duration: 3 ns, 50μJ).

The surface potential photo-discharge was measured by an Electrostatic Paper Analyzer EPA-8100 (Kawaguchi Electric Works Co.,Ltd.) in static mode.

Phthalocyanine pigments(MPc)

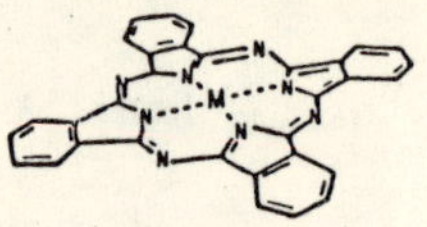

M = Pb 5.28eV
    TiO 5.38eV
    H$_2$ 5.40eV

Azo pigments

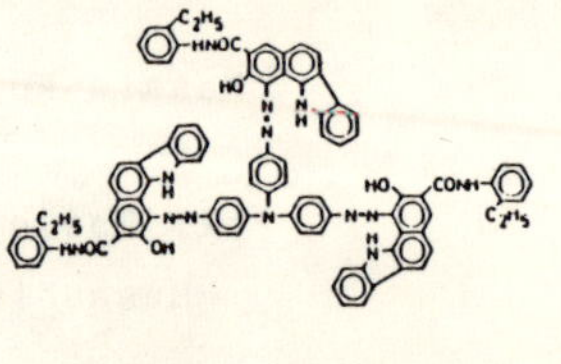

azo-A 5.62eV 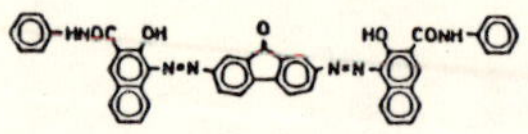   azo-C 5.92eV

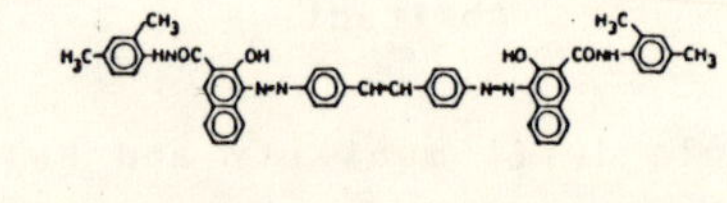

azo-B 5.74eV

Perylene pigments

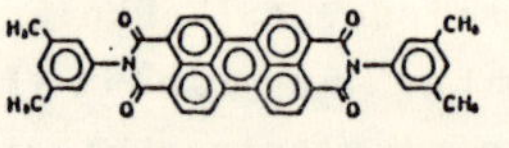

pery 5.83eV

Fig. 3 Several organic pigments used and their ionization potentials

Results and Discussion

1. Charge Transport Characteristics of Organopolysilanes

In Fig. 4 are shown the typical transient photocurrents. Only hole transport was observed with well-defined transit time due to the movement of hole carrier packet produced by pulse light irradiation on positively biased electrode. The temperature dependence of hole drift mobilities of (SiMePh)$_x$ and (SiMePro)$_x$ are shown in Fig. 5 in the temperature range of -50~+60℃. Both polymers exhibit the high mobility of ~10^{-4} cm^2/Vsec in the vicinity of room temperature even in a low field of 4 x 10^4 V/cm. Such high mobility in (SiMePro)$_x$ which has no π-electron pendant groups, clearly indicates that charges are transported through the Si-Si σ bond of polymer backbone. An Arrhenius plot of 1/T vs. logμ in (SiMePh)$_x$ having the glass transition above 200℃, gave a good straight line in the present temperature range, while the mobility of (SiMePro)$_x$ leveled off just above the glass

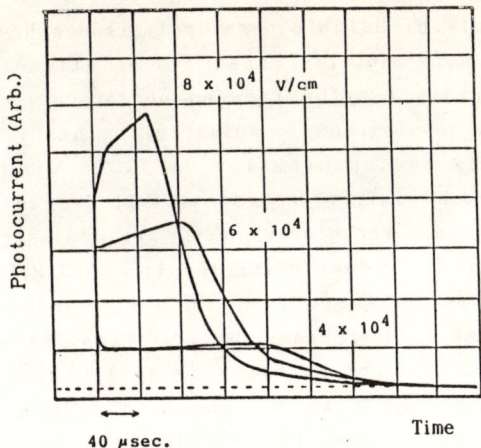

Fig. 4 Typical transient photocurrents of (SiMePh)$_x$.

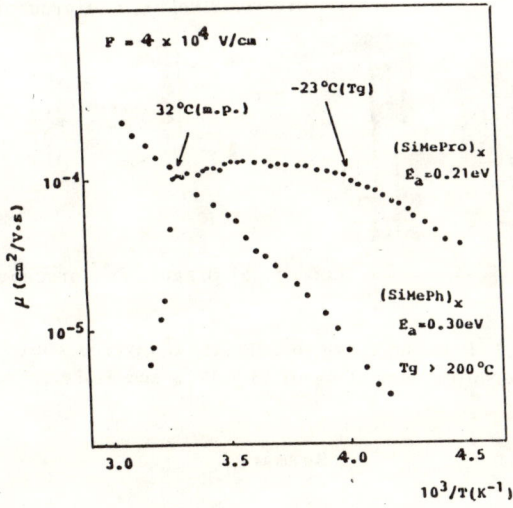

Fig. 5 Temperature dependence of the hole drift mobility of (SiMePh)$_x$ and (SiMePro)$_x$.

transition of -23°C and showed a drastic decrease at the melting point of 32°C. This seems to be due to the degradation reaction by photon or electric field above 32°C, since there could not be observed sufficient photocurrent in the measurements after cooling or a few days later. From this point, (SiMePro)$_x$ having low melting point is not suitable, and the higher melting point materials are needed for the practical use. The activation energies determined from the slope of the straight line were 0.30 and 0.21 eV for (SiMePh)$_x$ and (SiMePro)$_x$, respectively. These low activation energies are the desirable characteristics for CTL materials of layered photoreceptors.

2. Photo-induced Discharge Characteristics of Layered Photoreceptors

In Fig. 6 are shown some of the photo-induced discharge curves in several combinations with organic pigments from the surface potential photo-decay measurements. The values of half-decay exposure are listed in Table. The most excellent photo-discharge was observed for the photoreceptor consisting of TiOPc and (SiMePh)$_x$ (Fig. 6(a)) and their half-decay exposure was 1.4 lux·s, indicating the possibility of new type of organic photoreceptor.

In general, it is a common knowledge that the ionization potential of organic pigments in CGL should be larger than that of the CTL materials for effective hole carrier injection without energy barrier from CGL to CTL. In this sense, the TiOPc/(SiMePh)$_x$ combination has an energy barrier of about 0.3 eV

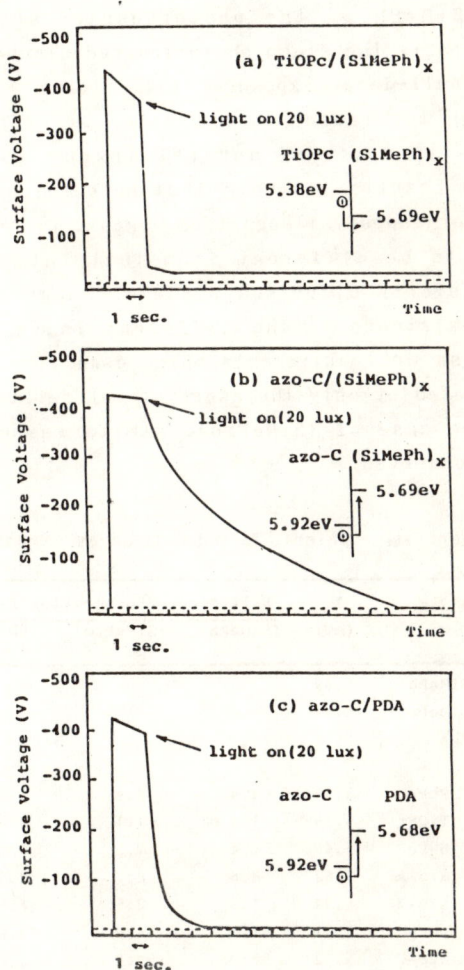

Fig. 6 Surface potential photo-decay curves of layered photoreceptors.

for hole injection, nevertheless the effective hole injection was observed with low residual potential of only -10~ -20 Volts surprisingly. Similar discharge curves were observed in other photoreceptors having energy barrier. On the other hand, however, in the case of the combination with azo-C (Fig. 6(b)), which has an ionization potential of 5.92 eV and can be considered to have no energy barrier for hole injection, the half-decay exposure was mysteriously as much as 56 lux·s in spite of the complete photo-discharge Pery pigment with similar ionization potential (5.83 eV) also gave a large value similar to azo-C. The complete photo-discharge with no residual potential seems to imply no energy barrier between CGL and CTL. As shown in Fig. 6(c), however, if we used the molecular dispersion of PDA in polycarbonate resin (1:1 in wt) which have the same ionization potential to that of $(SiMePh)_x$, the photoreceptor showed reasonably the steep photo-induced discharge with half-decay exposure of 2.6 lux·s. Although the reason is not clear at present stage in spite of several inspections, these results indicate that hole injection into organopolysilanes form organic pigments seem to be different from that into the molecularly dispersed systems, suggesting the existence of the different injection process or requirements which cannot be interpreted by only the energy level relation. In the case of $(SiMePro)_x$ similar aspects were observed.

Table Ionization potentials and half-decay exposures

Sample	I_p (eV)	Half-decay exposure(lux s)		
		$(SiMePh)_x$	$(SiMePro)_x$	PDA
$(SiMePh)_x$	5.69			
$(SiMePro)_x$	5.82			
PDA	5.68			
PbPc	5.28	4.4	4.4	5.2
TiOPc	5.38	1.4	1.1	4.2
H_2Pc	5.40	6.4	7.2	6.5
azo-A	5.62	3.8	2.7	2.5
azo-B	5.74	1.8	2.5	1.6
pery	5.83	41.3	96.0	13.6
azo-C	5.92	56.0	59.9	2.6
Se	5.92	4.1	7.8	7.3

3. Application to Laser Beam Printer

Solvent-Soluble organopolysilanes having high hole mobility appear to be attractive materials for the layered photoreceptor of electrophotography. Actually, as described above, the combination of TiOPc pigment provides fairly good sensitivity as an organic photoreceptor. Here, we demonstrate the printing test using the $TiOPc/(SiMePh)_x$ layered photoreceptor on a commercially available laser beam printer. Printing test was carried out by sticking small piece of the hand-made sample sheet on a drum. A printing example is shown in Fig. 7.

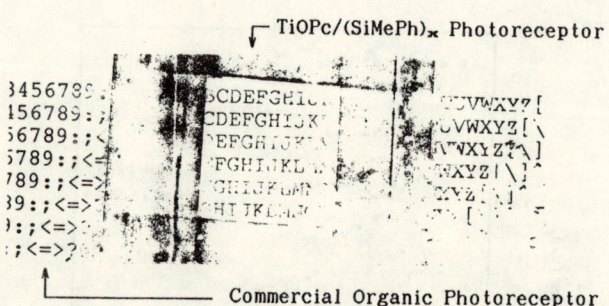

Fig. 7 Printing image in LBP with a layered photoreceptor consisting of $(SiMePh)_x$ and TiOPc.

Summary

In the present work, the hole drift mobility and hole injection from several kinds of organic pigments for two typical type of organopolysilanes were investigated. A layered organic photoreceptor in the combination of $(SiMePh)_x$ and TiOPc pigment gave fairly good sensitivity and good quality of printing image, indicating organopolysilanes to be a promising material for the practical use in electrophotography.

References

1) R.G. Kepler, J.M. Zeigler, L.A. Harrah, S.R. Kurtz, Phys. Rev. B, 35, 2818(1987); M. Stolka, H.-J. Yuh, K. McGrane, D.M. Pai, J. Polym. Sic. Polym. Chem., 25, 823(1987); M. Abkowitz, F.E. Knier, H.-J. Yuh, R.J. Weagley, M. Stolka, Solid State Commun., 62, 547(1987).
2) R. West, J. Orgnomet. Chem., 300, 327(1986).
3) These azo pigments were kindly supplied from Ricoh Company Ltd.

TRANSPORT PROPERTIES OF COPPER PHTHALOCYANINE

JIN Xiang-feng, ZHOU Shu-qin,
WANG Yan-qiao, QIU Jia-bai and REN De-yuan
(Inst. of Chem., Chinese Academy of Sciences, Beijing, China)

ABSTRACT

The temperature and field dependence of the traps parameters in the α-CuPc evaporated films and α-CuPc/PVK solution films have been studied by using of isothermal decay current (IDC) and I-V measurement. A method of characterizing the Gaussian distributed traps in IDC was presented. In the I-V experiments, Ohm's law was observed below the field of 1×10^4 V/cm; at higher voltage both of dark and photo conductions became non-Ohmic and met Poole-Frenkel effect.

Introduction

Copper phthalocyanine (CuPc) is a semiconducting photo-conductors. The studies of α-CuPc evaporated film and α-CuPc/PVK dispersed film is now of considerable interest[1-5].

Organic solids have impurities and structure defects which act as traps to the charge carriers moving in the solids. Thus the carrier transport properties of an organic photo-receptor are usually controlled by a succession of trapping and detrapping processes occurring during the carrier transit[6-7].

Various experimental methods for the carrier trap characterization, the trap energy E_t below the conduction band, the density of trap states N_t, carrier capture cross-section of the trap and the frequency factor ν for thermal escaping from the trap, have been developed[8]. By far the most important parameters are the density and energy of trap states and their distribution, if they are not discrete, particularly in polycrystalline and dispersive amorphous material. The application of steady-state space-charge-limited currents (SCLC) to the case of Gaussian distributed traps is considerably complicated[9]. As compared to the method of thermally stimulated conductivity (TSC) the isothermal decay current (IDC) method is much simpler experimentally and it also eliminates any artefacts connected with temperature gradients across sample, linearization of the heating rate, and trap parameters changing with temperature etc. The analysis of IDC characteristics as a means of determining the trap parameters of organic solids is superior to that of TSC method.

This presentation deals with the application of IDC experiments on the Gaussian distributed traps to the thin films of α-CuPc evaporated films and α-CuPc/PVK dispersive films.

Experimental

1. α-CuPc samples

The sample of α-CuPc is made from the mixture of o-benzonic dianhydride (with $(NH_4)_2MO_7O_{24} \cdot 4H_2O$), by fusing and sublimating at 570°C and 2 torr. The crystal of the sample was α-form determined by X-ray diffraction (Fig.1).

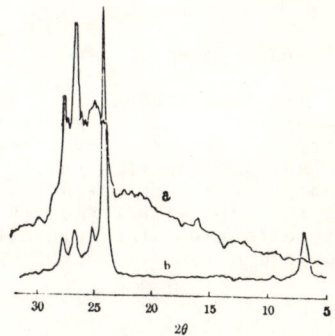

Fig.1 X-ray diffraction of α-CuPc crystalline

2. Preparation of photo-conductive films

Two kinds of samples were prepared: α-CuPc evaporated film was made into sandwich cell with Au and SnO_2 transparent conducting layer on a glass substrate as the bottom electrode. The top Au electrode of 0.13cm² was vacuum evaporated onto the CuPc film. The thickness of the α-CuPc film was 4 μm.

The film of α-CuPc dispersed in PVK was made in the form of sandwich: SnO_2/CuPc-PVK/Au. The ratio of CuPc to PVK was at 1:10 (w/w), The film thickness was determined by the use of interference microscopy in 2 to 5 μm. The semi-transparent top electrode (Au) of 100 to 150 Å was deposited on the α-CuPc/PVK in a vacuum evaporator (Beijing Instrument Factory, DM200).

3. IDC measurements

The experimental setup is schematically shown in Fig.2. In the IDC measurements, the sample(2) was illuminated by W-light source (58mW/cm²) at a definite temperature and applied electric field to fill the traps by photogenerated carriers. As steady state was attained(in 60 seconds), the light source(1) was cut off by shutter. Isothermal decay current from the thermal release of trapped carriers were measured by a Keithley model-642 electrometer(3) or log-amplifier(4). Data acquisition(5) and processing were controlled[10] in real time by IBM PC/XT micro-computer(6). Measurements could be done from 223 to 363 K.

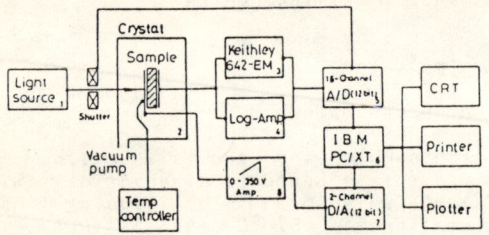

Fig.2 Block Diagram of IDC and I-V Measurements

4. I-V measurements

In I-V measurements, the applied triangular voltage was generated from a computer through D/A converter(7) and a linear voltage amplifier(8). The voltage increment on the sample was 0.02 volt per step. The rate of voltage increase was adjustable to match the relaxation time of the measured material.

Theory

Under high applied field the released carrier can be removed by the field before they recombine[11]. For a Gaussian distributed traps[12], We got the expressions as the following [13]:

$$\ln(It) = C_1(\ln t)^2 + C_2(\ln t) + C_3 \quad (1)$$

$$C_1 = -(kT/\sigma)^2 \quad (2)$$

$$C_2 = 2C_1[\ln\nu - E_m/kT] \quad (3)$$

$$C_3 = \ln\{-(qALH_t/2)(-C_1/\pi)^{1/2} \exp[(C_2/2)^2/C_1]\} \quad (4)$$

where I is the decay current, q is electronic charge, k is Boltzmann's constant, T is absolute temperature, E_m is the energy level at which the density of trap state shows a maximum, H_t is the total density of states, and σ characterizes the width of distribution.

The constants of C_1, C_2 and C_3 could be evaluated by least square fitting. And, when IDC curve are measured at several temperatures isothermally, the parameters of Gaussian distribution of traps E_m, $N_t(E)$, H_t, σ and ν could be obtained from Eqs.(1-4).

Results

1. α-CuPc evaporated film.

The IDC data of α-CuPc evaporated film have been determined at different temperatures and fields. The results of IDC data measured at fields 2.5, 3.75 and 5×10^4 V/cm with temperature 300K are shown in the Fig.3. The field dependence of IDC curves indicates a field assisted detrapping probably in the sense of Poole-Frenkel mechanism.

The rate of current decay depends on the temperature and applied field; the higher the temperature or the applied field, the higher the current and the faster it decays. The observed IDC data measured at temperatures 300, 313 and 324K with applied field 2.5×10^4 V/cm and their least square fitting curves are shown in Fig.4, with correlation coefficients 0.998, 0.995 and 0.993 respectively. The parameters are given in Table.1.

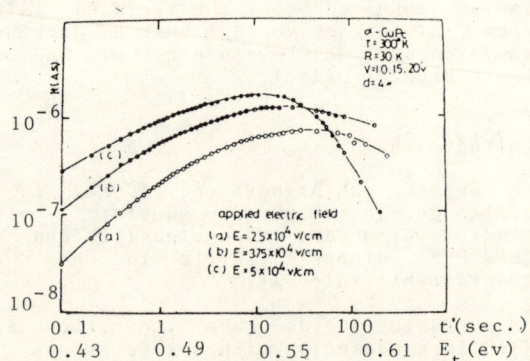

Fig.3 Field dependence of IDC for α-CuPc evaporated film

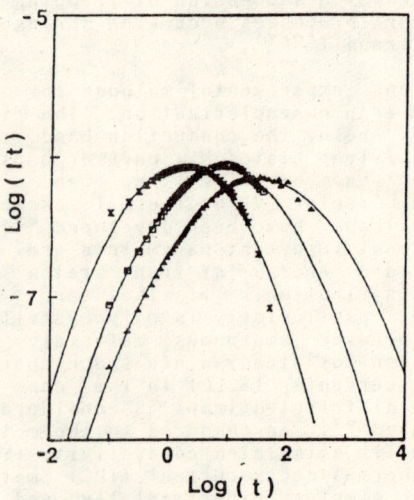

Fig.4 Temperature dependence of IDC for α-CuPc evaporated film at 300, 313 and 324K
Points: observed data
Curves: fitted Gaussian distribution

Table.1   α-CuPc evaporated film
----
trap energy level E_m = 0.47   eV
distribution width σ = 0.036   eV
total states density H_t = 2.4×10^{20} cm^{-3}
frequency factor ν = 1.8×10^7 s^{-1}

2. α-CuPc/PVK dispersed film.

The IDC data measured at temperatures 253, 263 and 283K are shown in Fig.5. Obviously, on the curves of Log(It) vs. Log(t), no sharp peak is apparent due to overlapping distribution. The experimental data (at 253 K, 10^4 V/cm) were fitted by the least square method for two Gaussian distributed traps (Fig.6). The correlation coefficients of the fitting are between 0.93 and 0.99.

The obtained parameters of two Gaussian distributions by the use of Eqs.(1-4) are given in Table.2.

Table.2		1	2
E_m (eV)	=	0.14	0.17
σ (eV)	=	0.021	0.019
H_t ($\times 10^{18}$cm^{-3})	=	3.03	5.9
ν ($\times 10^2$s^{-1})	=	2.4	2.1

The I-V characteristics of dark/photo conduction for α-CuPc/PVK film are given in Fig.7. Ohm's law was observed below ca. 5V; at higher voltage both of dark/photo conductions became non-Ohmic. When the values of Lg(I/V) were plotted vs. $V^{1/2}$ (the insert of Fig.7), two good straight lines was obtained. Thus the non-Ohmic conductivity can be represented by:

$$\sigma = \sigma_0 \exp[1.5 \times 10^{-4} \times E^{1/2}/KT] \quad (5)$$

For α-CuPc/PVk dispersed film, since both the IDC curves and the I-V characteristics were symmetric with respect to the change of polarity of the applied field, the observed field effects should be a bulk effect. The coefficient 1.5×10^{-4} eV/(V/cm)$^{1/2}$ in Eq.(5) close to the value required by Poole-Frenkel equation. Three kinds traps would be existed in α-CuPc/PVK, however, the trap between the interface of α-CuPc and PVK is dominant the carrier transport[3,14].

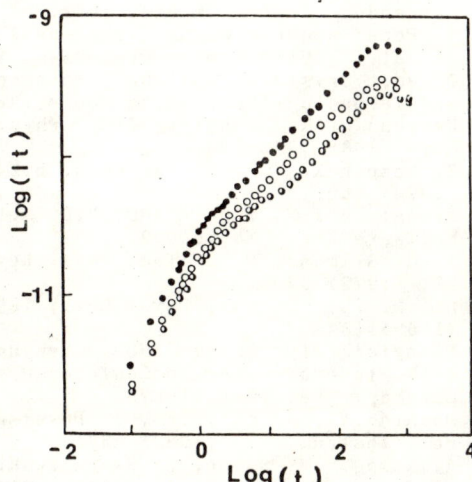

Fig.5 Temperature dependence of α-CuPc/PVK dispersed film at 253, 263 and 283K

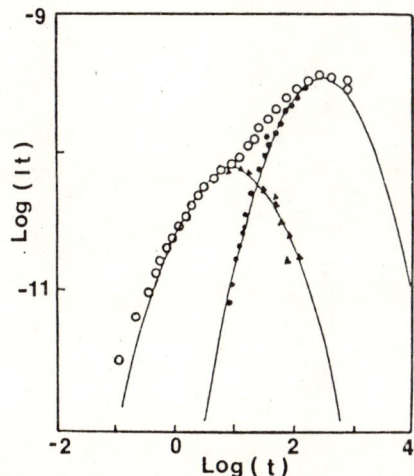

Fig.6 IDC curves of α-CuPc/PVK(253K)
Open circles: observed data
Curves: fitted Gaussian distributions

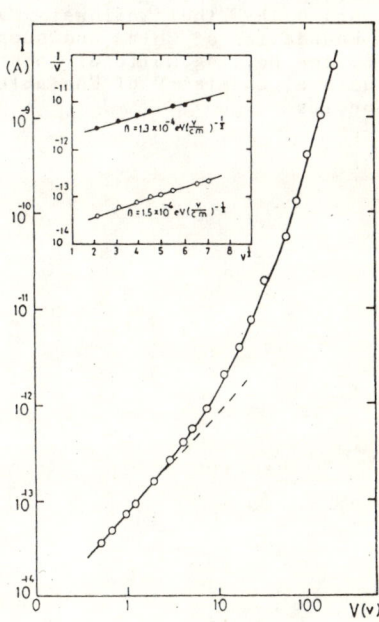

Fig.7 I-V characteristics of dark/photo conduction for α-CuPc/PVK film

Some evidence of Gaussian distributed traps have been obtained in the I-V experiments too. The I-V characteristics of α-CuPc/PVK film with a negatively biased metal electrode measured at 300K are shown in Fig.7 Ohmic conduction are observed below ca. 1×10^4 V/cm; at higher voltage the slope in the region of supper linear conductivity are changing rapidly with the applied voltage. According to the theory of SCLC[2], this observed I-V characteristics indicates the existence of Gaussian distributed traps.

Discussion

From the data of two organic solid films, it is apparent that the Gaussian distribution of trap levels is appropriate for the analysis of IDC curves for these materials. The IDC method offers a convenient and useful means for characterizing the traps in a simple manner. In IDC as well as in TSC, however, both carriers are involved and a discrimination of the sign of carrier is difficult unless a blocking contact to one of the carriers is employed during the trap filling by field injection[15].

Acknowledgement

The authors wish to thank Professor QIAN Renyuan for many clarifying discussions and Wang Zeng-qi who made the linear voltage amplifier. This research was supported by the National Natural Science Foundation of China and supported in part by the Beijing National Laboratory for Structural Chemistry of Unstable and Stable Species.

References

[1] Renyuan QIAN, Shuqin ZHOU, Xiangfeng JIN et. al, J. Appl. Sci. $\underline{5}$, 1, (1987) 37.
[2] Shuqin ZHOU, Xiangfeng JIN, Photographic Science and Photochemistry, No.2 (1988) 94-102.
[3] Y. Hoshino, J. Appl. Phys., $\underline{52}$, No.9 (1981) 5655.
[4] C. Hamann, Phys. Stat.Sol., 26(1968) 311.
[5] J. P. Fillard, Ann. Phys. 4, (1969) 617.
[6] M. Pope, C. E. Swenberg, Electronic Processes in Organic Crystal, Clarendon Press, Oxford, 1982.
[7] M. Pope, Renyuan QIAN, Xiangfeng JIN et al., Electronic Processes in Organic Crystal, Shanghai Sciences and Technology Publishing House,1987.
[8] Renyuan QIAN, Xiangfeng JIN, Phys., No.6 (1988) 325-334.
[9] S. Nespurek, Czech. J. Phys. B.$\underline{24}$, (1974) 660.
[10] Xiangfeng JIN, Shuqin ZHOU, "IDC-2.1" Software Reg. No.860082.
[11] J. G. Simmons, M. C. Tam, Phys. Rev. $\underline{B7}$, (1973) 3706.
[12] H. P. D. Lanyon, Phys. Rev., $\underline{130}$, (1963) 134.
[13] Xiangfeng JIN, Shuqin ZHOU, Renyuan QIAN, in Proceeding on Conference of Hardcopy '88, 1988, Tokyo.
[14] Hoshino,Y., Arishima,K., J. Phys. E: Sci. Instrum., $\underline{16}$ (1983) 427.
[15] A. Samoc, M. Samoc, J. Sworakowski, Phys. Stat. Sol., (a)36, (1976) 735.

The Residual Potential of Xerographic Materials As-Se

Min Szukwei, Zhang Gancheng, Chen Linrong and Zhang Fuzhen
Shanghai Institute of Ceramics, Academy of Science
Shanghai, P.R.China

ABSTRACT

This paper presented the influence of composition and halogen dopants in η, $\mu\tau$. these parameters can be used to estimate the residual potential of drum which is the source of dark background of xerographic copies.

Residual potential is the potential left on the surface of drum after exposure, which is the source of dark background of xerographic copies. Two kinds of dark background have been observed, one is the dark background on the first copy, that is due to deep trap of materials, the other is the background getting darker and darker during repeated copying, that is due to build up of positive space charges near the substrate. Normal xerographic discharge curve is shown in Fig 1., while dv/dt is the photoreponse.

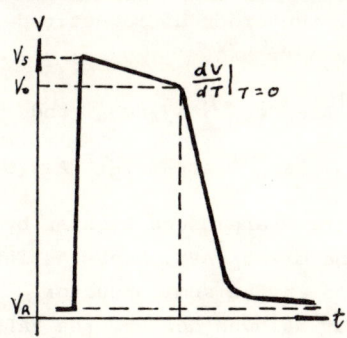

Fig 1. Xerographic discharge curve

The surface potential decreased rapodly during exposure until a nearly constant value V_r, which is the residual potential. But usually we use values of the decresed surface potential at a certain time t as V_r to charaterize drums, hence the tail part of discharge curve play an important role in residual potential. Photoresponse closely related to carrier transport, which is the result of $\eta\mu\tau$. This paper presents investigations of the influence of As and halogen dopants on $\eta\mu\tau$ of As-Se amorphous semiconductors.

Experiment

Samples were prepared by weighing proper amounts of 5N As and Se in to quartz amples which was sealed in vacuum and melted in rotating furnes at 800°C for 24 hrs. Amorphous films was formed by quenching the ample from 800°C to dry ice -alc. mixture, then vacuum deposited a 60μm thickness film on Al substrate.

Samples of the composition $As_{24}Se_{76}$, $As_{40}Se_{60}$, $As_{50}Se_{50}$ were used to investigate the influence of As in photoquantum efficiency, the composition As_5Se_{95} were used to investigate the effects of halogen dopant in residual potential.

Xerographic spectrum was determined with a home made apparatus. Samples were mounted on a movable sample stand. The change of surface potential during exposure was measured with a KZB-3302 surface potential meter and traced with a X-Y recorder.

I. Compositional dependency of η

In our previous paper we found that the residual potential will gradually decrease when As content incread from 10 at% to 40 at%.[1] and explained this phenomeno as the increased unit of As_2Se_3. In this paper we studied the influence of As in η with the As content higher than that in As_2Se_3.

For sample thickness less than Schubway ($\mu\tau E$), traps inside the body can be neglected. The rate of field strength change with η is expressed as

$$dE/dt = \eta Fe/\epsilon\epsilon_0$$
$$\eta = \epsilon\epsilon_0(dE/dt)/Fe$$

where F is the light intensity, and ϵ is the dielectric constant. Plot η vs. dE/dt and compared with the Onsger curve obtained the thermal stimulated radius r_o, the results are shown in table 1., Fig 2.

Table 1. Photo-quantum efficiency and thermal-stimulated radius of materials with three compositions.

	Composition		
	$As_{24}Se_{76}$	$As_{40}Se_{60}$	$As_{50}Se_{50}$
$\eta(E=0)$	$1.2 \; 10^{-2}$	$1.35 \; 10^{-1}$	$3.9 \; 10^{-2}$
$r_o (\text{Å})$	16	30	29

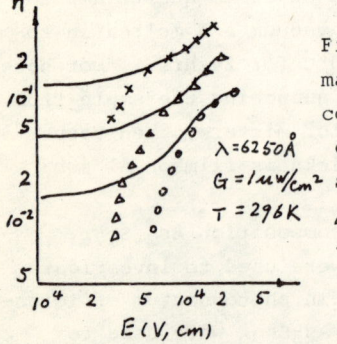

Fig. 2. η vs E of materials with three compositions
$\lambda = 6250 \text{Å}$
$G = 1 \mu w/cm^2$
$T = 296 K$
o: $As_{24}Se_{76}$
x: $As_{40}Se_{60}$
△: $As_{50}Se_{50}$
—: Onsager curve

It shows that both η and r_o are highest for $As_{40}Se_{60}$ among the three compositions.

II. Effects of halogen dopant on residual Potential.

Sample composition used for this determination was As_5Se_{95}, and 1000 ppm Cl, Br, I, were doped respectively.

Residual potential was measured with the same xerographic spectroscopy apparatus. According to Kanazawa[2], for weak trapped sample: $Vr/Vs = (0.5d^2/\mu\tau Vs)[-\ln(2V_R/Vs)]$ strong trapped sample: $Vr/Vs = 1 - (\mu\tau Vs/d^2)(\ln 2)$ where Vr, Vs are residual potential and surface potential respectively.

Enck[3] measured Vr of the first cycle and calculated the $\mu\tau$ value from the above equation, he obtained the same value as that measured with drift mobility. In this paper we use the residual potential of the first cycle to investigate the $\mu\tau$ value of halogen doped samples. The maximum absorbed light wave length 5200Å, and the light intensity $50 \mu w$ were used for determination. Results is shown in table 2.

Table 2. Vr, and Vs, $\mu\tau$ values of As_5Se_{95} and doped films.

Dopant	Electronegativity	Vs	Vr	Vr/Vs	$\mu\tau$
Undoped	--	1200	740	0.63	$1.4 \; 10^{-9}$
I	2.5	800	48	0.06	$4 \; 10^{-7}$
Br	2.8	700	21	0.05	$1 \; 10^{-6}$
Cl	3.0	540	1	0.002	$3.3 \; 10^{-5}$

Clearly the order of magnetitude of $\mu\tau$ values of these three samples are:
$$\mu\tau_{Cl} > \mu\tau_{Br} > \mu\tau_I$$

Discusion

Defects in chalcogenide glasses are Valence Alternation Pairs (VAP)[4]. The VAPs of pure Se glasses are (C_1^-, C_3^+), pure amorphous As are (P_4^+, P_2^-). stoichicmetric As_2Se_3 are (C_5^+, P_2^-)[5]. In case of off-stoichicmetric composition, thier VAPs may be distributed as table 3.

Table 3. The proportion of VAP defects in materials

Compositions	As_2Se_3 units	Proportion of VAP defects
$As_{24}Se_{76}$	$[As_{24}Se_{36}]Se_{40}$	$[C_3^+, C_1^-]_{20}[C_3^+, P_2^-]_{24}$
$As_{40}Se_{76}$	$As_{40}Se_{60}$	$[C_3^+, P_2^-]_{40}$
$As_{50}Se_{50}$	$[As_{35}Se_{50}]As_{17}$	$[P_4^+, P_2^-]_{8.5}[C_3^+, P_2^-]_{33}$

From table 1. the order of magnitude of values of η and r_o are

$$\eta_{As_{40}Se_{60}} > \eta_{As_{50}Se_{50}} > \eta_{As_{24}Se_{76}} \text{ and}$$
$$r^o_{As_{40}Se_{60}} > r^o_{As_{50}Se_{50}} > r^o_{As_{24}Se_{76}}$$

obviously carries are bound tighter by C_3^+, C_1^- pairs then the mixed VAPs, hense As introduce showller traps in the semiconductor.

In case of halogen doping, the VAPs of samples are the same, but the $\mu\tau$ values increased. In these samples the dominated defects are (C_3^+, C_1^-) which control the carriers transport. The eletronegtive halogen might combine with the terminal of Se chain or create a branch chain.

$$C_1^- + X_1^o = C_2^o + X_1^-$$
$$C_2^o + X_1^o = C_3^+ + X^-_o$$

In the first case (C_3^+, C_1^-) pairs will decrease because of $C_1^- + X_1^- = C_3^+$, in the second case, C_3^+

-256-

will trape e to form negative space charge
and compensate the positive space charge
therefore reduce the residual potential.
table 2. showed that the $\mu\tau$ value increased
with the increasing electronegtivity of halo-
gens, that is to say the average carrier drift
length in three doped samples is of the order
Cl>Br>I, which are same as the order of decreasing of residual potential.

References
1. Min, Szukwei et al, Xinxing Wuji Cailiao 1981. 45
2. Kanazawa, K. K.   J. Appl Phys 1972, 43 1845
3. Enck, R. C.   J. Non-cryst sol 1980 35, 831
4. Kastner, M.   Phys. Rev. Lett. 37, 1976, 1504
5. Adler, D.   Phys. Rev. Lett. 41, 1978, 1755

A MEASURING METHOD OF ELECTROPHOTOGRAPHIC TONER CHARGE

NOBORU KUTSUWADA and YOUICHI NAKAMURA
Nippon Institute of Technology
Miyashiro, Minamisaitama-Gun, Saitama-Ken, 345 JAPAN

I. Introduction

In order to investigate the behaviour of electrophotographic powder in the xerographic industry, several measurement methods have been proposed [1],[2]. The Blow-Off Tribo method [1] is well known as one of them for measuring methods of electric charge of toner.
By this method, the average charge on some amount of toner particles may be measured, but the magnitude and the sign of an electric charge on one particle of toner cannot be determined. In 1982, a charge spectrograph for xerographic toner reported by R.B.Lewis et al. [3].
This deals with a distribution of toner charge.
While, in 1983, relationships between the distribution of toner charge and its particle radius is simultaneously shown by Imamura and Tetsuya [4],[8]. The measuring device of the distributive tendency between charge and radius for toner particles is reported or saled by J.Bares [5] R.H.Epping [6] Y.Takahashi et al. [9] and M.K.Mazumder et al. [10](Mikrol Pul:E-Spart Analyzer). However, most of those can give no information of the electric charge on a single particle.

The direct measurement of the amount of electric charge on a particle can be made by applying the famous Millikan Oil Drop experiment. But few have tried to do such a measurement as this because there are difficulties in making accurate measurements on a rapidly moving particle whose position and velocity can be ralatively well controlled [2]. Nevertheless, it must be a valuable experiment for researchers in determining the charge on a particular single particle.

In this report, we describe an applied Millikan's method for measurement of radius and electric charge for a single particle [11],[12]. Furthermore, we describe an improved Millikan's method for measurement of radius and electric charge for a single toner particle in developer.
We propose an additional automatic separator and combine it with a Millikan's Oil Drop Apparatus.
The experimental results will be presented by this method, and the amount of the charge dependence on toner radius in developer will be also discussed.

II. Experiment

Charge Measurement Method

The principle of the Millikan's method is that three types of force act on a single moving particle.
Gravitational and aerodynamic forces always exist and the electrical field can be controlled artificially.
When Stokes law is applicable, the moving equations of a particle in free dropping are as follows;

$F_1 = -mg$ --[1] $-mg = 6\pi \eta r v'$ --[4]
$F_2 = -6\pi \eta r v'$ --[2]
$F_t = F_1 + F_2 = 0$ --[3] $m = \frac{4}{3}\pi r^3 \rho$ --[5]

where

F_1 gravitational force in the z direction
F_2 aerodynamic force
F_t total force on the particle
r effective particle radius
v' final free dropping velocity in the z direction
g acceleration due to gravity
m particle mass
η viscosity of air
ρ particle density of toner

The force F_3 in which the electric field acts on a toner particle, is given by the following equation;

$$F_3 = Q \cdot E \quad \text{or} \quad F_3 = Q \cdot \frac{V}{D} \quad --[6]$$

Where Q amount of the charge on a particle
E electric field
D distance between the two parallel electrodes
V applied voltage between the electrodes

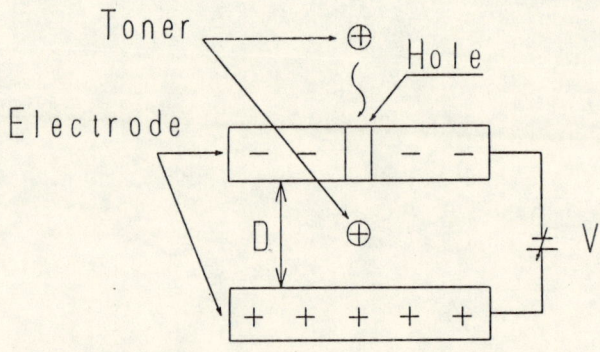

Fig.1 shows a cross-sectional view of electrodes.

Thereforce, a moving particle of toner with a velocity v in air by the electric field between the horizontal electrodes, and the equation of motion is as follows;

$$Q \cdot \frac{V}{D} = mg + 6\pi \eta r v \quad \text{or} \quad Q \cdot \frac{V}{D} = \frac{4}{3}\pi r^3 \rho + 6\pi \eta r v \quad --[7]$$

The geometry of the main electrodes is shown in Fig.1. Because the motion of a particle can be controlled by magnitude of the electric force, it may be stopped, rise upward or drop downward in a certain condition.
When these facts are taken into consideration, the following equations can be derived.

$$Q \cdot \frac{V_0}{D} = mg \quad \text{---- Stop motion}$$

$$Q \cdot \frac{V_1}{D} - mg = 6\pi\eta r v_1 \quad \text{---- Rise motion}$$

$$Q \cdot \frac{V_2}{D} - mg = -6\pi\eta r v_2 \quad \text{---- Drop motion}$$

Where V_0 applied voltage in a stop condition
V_1 applied voltage in a rise condition
V_2 applied voltage in a drop condition
v_1 rising velocity with the V_1
v_2 dropping velocity with the V_2

Using combination of these equations with the Eq. [5], we can solve r and Q as follows;

$$r = \left[\frac{6\eta(v_1+v_2)\frac{V_0}{V_1-V_2}}{\frac{4}{3}\rho g} \right]^{\frac{1}{2}} \quad \text{---- [8]}$$

$$Q = 6\pi\eta r \frac{\frac{v_1+v_2}{V_1-V_2}}{D} \quad \text{---- [9]}$$

Experimental apparatus I (Mono-component toner)

The schematic diagram of the experimental apparatus is shown in Fig. 2-a, b.

Dropping particles are supplied through a valve 1 by blowing off some amount of toner with compressed air.

In this apparatus, a free dropping motion in a steady state can be settled after a certain time period.
A telescope with a calibrated eyepiece is used to observe the motion of the toner particle between the two electrodes. A glass flask is used. The charging means of the experimental apparatus is similar to the principle as the powder cloud developing method(a).

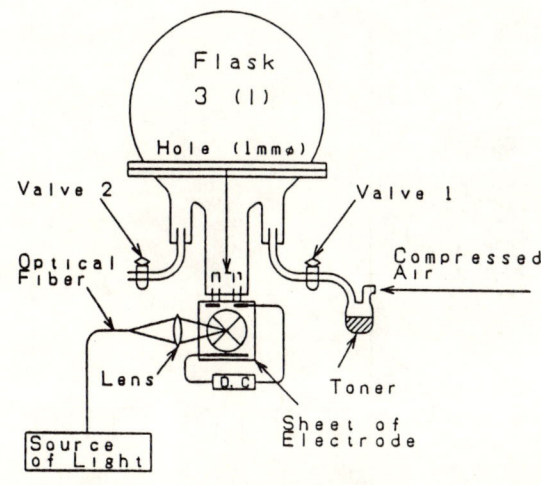

Fig. 2-a shows an experimental apparatus of the schematic disgram.

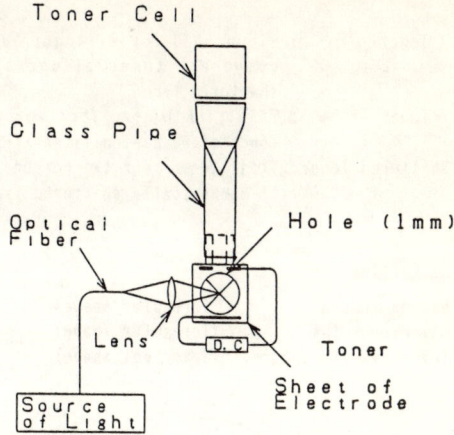

Fig. 2-b shows another experimental apparatus of the schematic diagram.

Another dropping particles are supplied through toner-dropping device with slits. In the case, we dropped the carrier after we agitated it in a beaker by a glass stick or non-agitated case. The charging means of the experimental apparatus is similar to the principle as the magnetic developing method.

Experimental Apparatus II (Two component developer)

The schematic diagram of new additional automatic separative apparatus is shown in Fig. 3. This apparatus is composed of aluminum case, magnetic plate, rubber plate, mesh, funnel and aluminum plate. Fig. 4 shows that the total device has an additional automatic separative apparatus on the Millikan's apparatus.

In this apparatus, a free dropping motion in a steady state can be settled on the upper part in an aluminum case for developer. In the case, the charging means of the experimenal apparatus is similar to the principle as the magnetic developing method.

III. Sample Materials

In this experiment, the five samples of toner and three samples of carrier were used.
All of them are of Japanese make, and these are classified into the two types according to their shapes.
One sample is a chemically spherical toner type and the other four are mechanically shattered ones.
The spherical one may be obtained with difficulty because of a trial product but it has the advantage of the ideal shape for electro-photographic toner. The sample names are as follows;

[I] Toner

(1) Spherical Toner: Trial AT-series toner is produced by chemical polymerization.
(2) C-Toner   : Canon NP-270 copier toner, for mono-component toner of mechanically shattered type

(3) R-Toner   : Ricoh 50 50 copier toner, for two-
                    component toner of mechanically
                    shattered type
(4) T-Toner   : Trial TDK toner, for two-component
                    toner of mechanically shattered type
(5) Shattered Toner: Trial toner, for two-component toner
                     of mechanically shattered type

[Ⅱ] Carrier

(1) Nippon Teppun   (Irregular shape)
(2) Dai-Nippon Ink   (Irregular shape)
(3) Tomoegawa A1～2, B1～2   (Spherical shape)

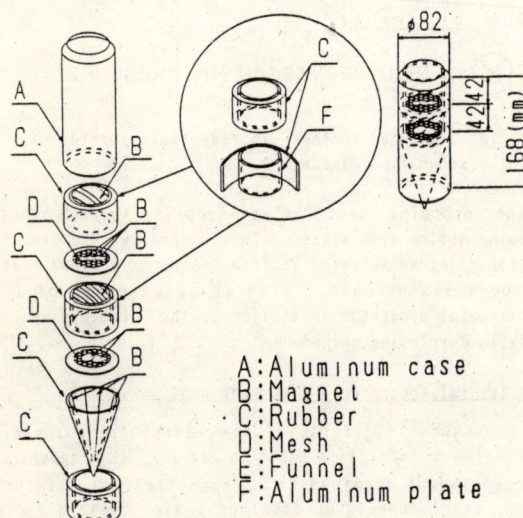

Fig. 3 separater apparatus.

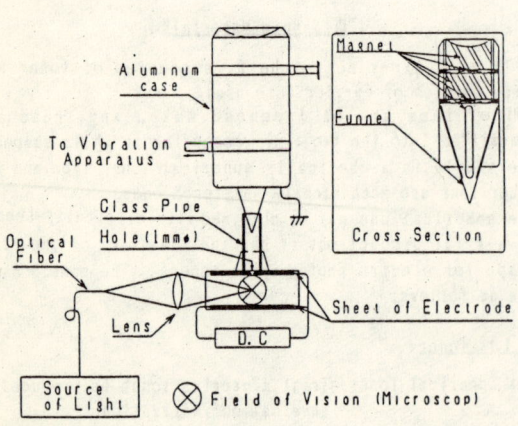

Fig. 4 all experimental apparatus.

Ⅳ. Results and Discussion

1) Mono component toner

In this expriment, a steady flow of free dropping particles was actualized after a period of more than one minute when toner in the reservoir was blown off by compressed air. Another dropping particles are set on the central position that apparatus has for a toner-dropping with slits. The single particle motion controlled by electric force was observed, and the rise time and the fall time to be required for a particle to move across the certain interval were measured. The applied voltages at the rise time, the fall time and stop time were individually recorded. Most of all particles observed in our experiment have the positive sign of the charge. This reason is as follows. Although the negatively poralized particles are generated by blow-off tribo, they will be captured on the inner surface of the glass flask which is known to be easily polarized to positive [7], and the positive charged particles drop down to the electrodes. This is an advantage of this apparatus in comparison with the other reported ones [4], [5]. The effective radius and charge quantity of a particle were deduced by Eqs. [8], [9]. The experimental results were presented as graphs of radius dependence of electric charge on a single particle in Fig. 5～9. A log-log plot for the spherical type of two-component toner is shown in Fig. 5. The particles of spherical toner are produced by a chemically polymerization method and the shapes are nearly spherical and the sizes are distributed over a narrow regin within five times of radius.

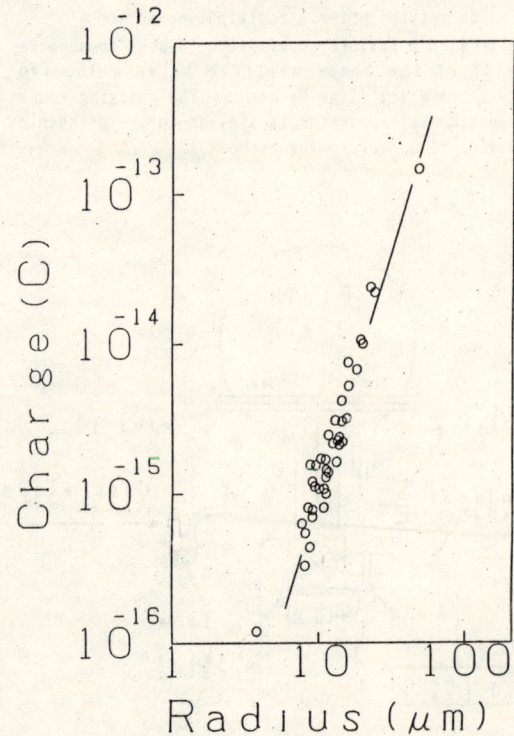

Fig. 5 radius dependence of electric charge for a spherical toner particle.

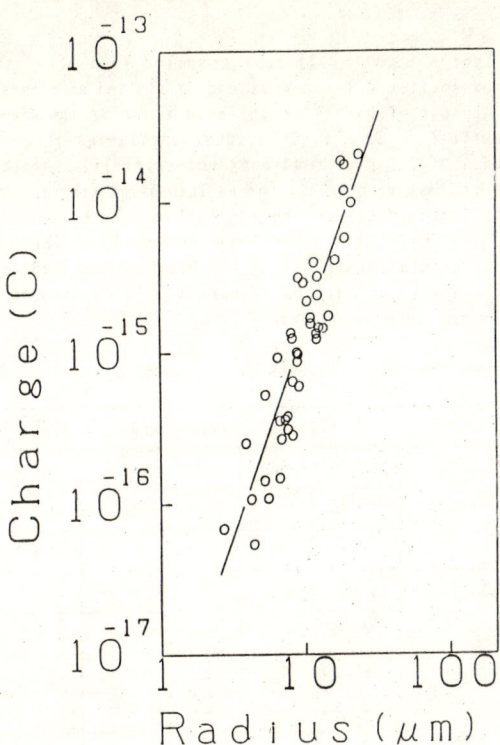

Fig.6 radius dependence of electric charge for a mechanical shattered toner(micro magnetic powder content) [CANON]

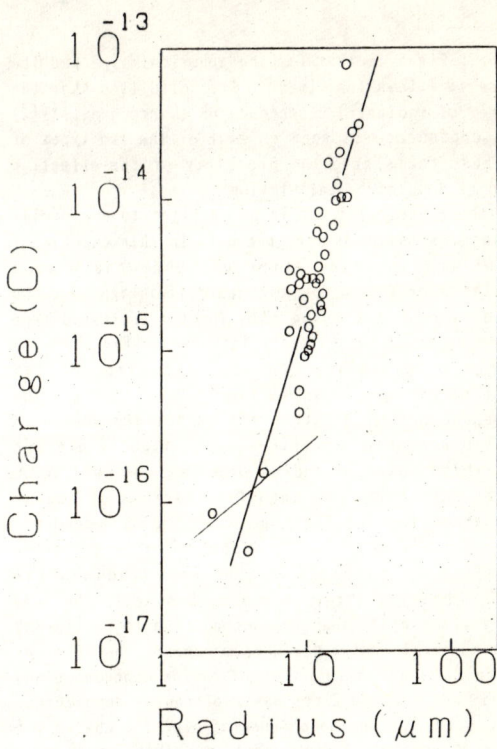

Fig.7 radius dependence of electric charge[RICOH]

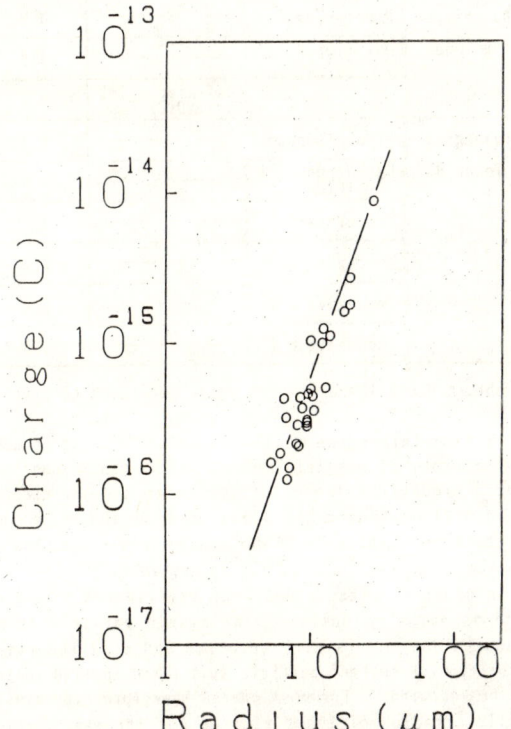

Fig.8 radius dependence of electric charge for a static tribocharging[Japan DIC].

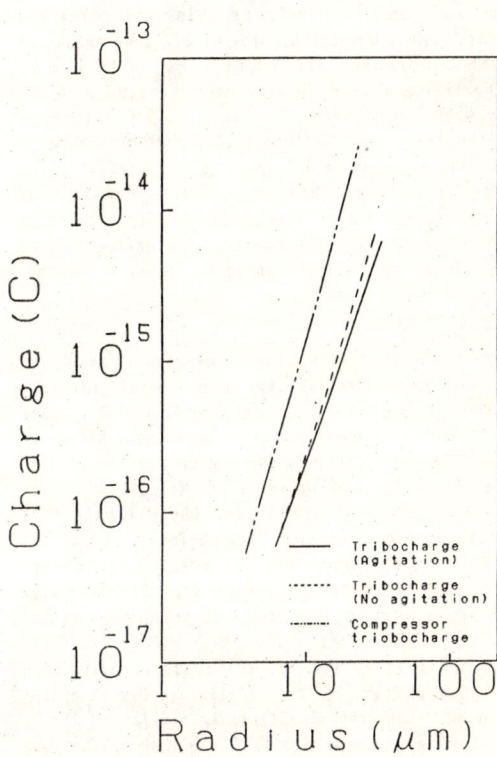

Fig.9 radius dependence of electric charge for three type of contact charging.

The graph in Fig. 5 is that for the spherical type and the difference is within ten times. The solid line show the third power of radius[r] dependence on charge quantity[Q] and this dependence is seen on both of the two types of toner. These facts give the propriety of the effective radius to be deduced by calculation.

Although the spherical toner is not applied to commercial plain paper machines as yet, the data in this experiment show that the spherical type of toner has a good electrostatic or electrophotographic advantage over the shattered type. A log-log plot for the shattered type of mono-component toner(C-Toner) is shown in Fig. 6.

Three log-log plots for the shattered type of two component toners are shown in Fig. 7~9. In Fig. 7 the radius dependence of electric charge for the shattered type of two component toner[R-Toner] slightly differs from the third power of radius dependence in the region of smaller particles less than five microns of radius and in that region the first power of radius dependence appears. The different radius dependence in the smaller particle region in Fig. 7 can be explained on the assumption that the first term will become to the same order or smaller order than the second term in Eq. [7] and the first power of radius dependence appears, because the first term has the third power of radius dependence and the second term has the first power of radius dependence.

In Fig. 8 the radius dependence of electric charge for the shattered type of two-component toner(DIC-Toner) have nearly 2.2~2.5 power of radius dependence of particles. The experimental results were presented as graphs of radius dependence of electric charge on a single particle in Fig. 5~9. In Fig. 5~7, dropping particles are supplied through a valve 1 by blowing off some amount of toner with compressed air as Fig. 2-a. In Fig. 8, dropping particles are set in the central position of the apparatus with a toner cell as Fig. 2-b, and a log-log plot for the two component toner(DIC-Toner) of shattered type is shown. In Fig. 9, dropping particles have three type of contact charging. The quantity of electric charge of toners are difference from charging mechanism of each type, but shows a similar tendency of radius dependence of electric charge for toner particles.

2) Two component toner

The experimental results were presented as graphs of radius dependence of electric charge on a single particle in Fig. 10~12. In this report, most of all toner particle observed in our experiment have the positive of the charge. The effective radius and charge quantity of a particle were deduced by previous Eqs. (8), (9).

The graph in Fig. 10 is that for the difference of tribocharging between toner-toner(Ricoh-Toner) and toner-iron powder carrier(Nippon DIC Carrier). The charge quantity of a toner particle consists of tribocharging toner and it is smaller than the one which consists of tribocharging toner-carrier.

The graph in Fig. 11 is similar to Fig. 10 except a used toner. The graph in Fig. 12 is similar to Fig. 11 except used carriers. The graph in Fig. 13 is similar to Fig. 10 except used toner and carrier. Table 1 shows that the index coefficient of toner radius depends on charge quantity for several kinds of the composed developer.

Terris & Jaffe[13] have compiled that the data of index coefficient are most easily plotted as curves Q/d vs. number of particles for each toner if the shape is spherical. They find various dependences of $Q \propto d^n$ with $n \simeq 2$ for a broad range of carriers. Where : Q is the charging quantity for an insulating sphere. d is the diameter for an insulating sphere. n is the index coefficient of sphere diameter. The index coefficients obtained from our Millikan's apparatus are most easily calculated as curves of Q vs. number of a toner particle for each radius.

Table 1 Index coefficients (n) in composed developers

Toner	Carrier	Tribocharging	Index co-efficient
Ricoh Type 5050	Nippon Teppun		3.6
	Dai-Nippon Ink (DIC)		2.8
		Ricoh 5050 Toner	3.0
TDK	Nippon Teppun		3.5
	Dai-Nippon Ink (DIC)		3.0
	Tomoegawa A-1		3.1
	Tomoegawa A-2		3.1
	Tomoegawa B-1		3.0
	Tomoegawa B-2		3.0
		TDK	3.0
Dai-Nippon Ink Toner	Nippon Teppun		3.8
	Dai-Nippon Ink (DIC)		3.0
		Dai-Nippon Ink (DIC)	2.1
Tomoegawa Toner B	Nippon Teppun		2.8
	Dai-Nippon Ink (DIC)		3.0
	Tomoegawa A-1		2.4
	Tomoegawa A-2		2.3
	Tomoegawa B-1		2.4
	Tomoegawa B-2		3.0

We obtain Min 2.1 and Max 3.8 value for Index coefficient.

By calculation the criterion point is laid on about five microns of particle radius. This can be concluded that a product of the electric charge and the electric field will depend on the first power or the third power of the toner particle, if our assumptions for Stokes law and the effective particle radius are valid.

In usual Millikan's Oil Drop experiments, the final free dropping velocity of the single particle is not measured due to its high velocity and only the rising velocity of a charged particle at a fixed applied voltage may be measured. For most cases, therefore, the similar particle radius of toner wich is not the same particle with the observed rising velocity must be previously determined by another methods.

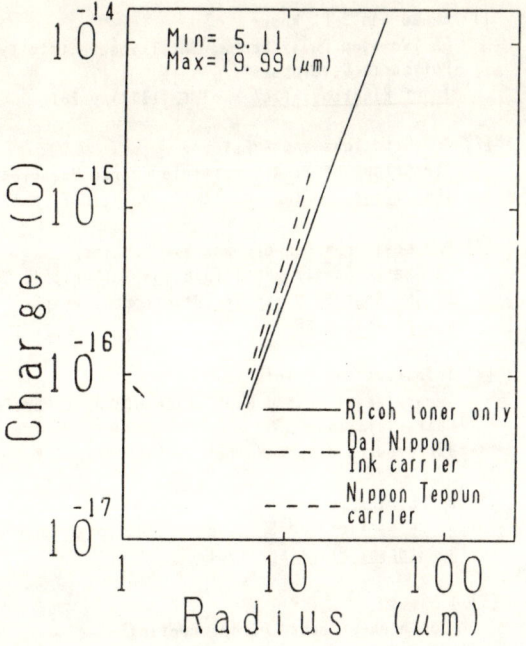

Fig. 10 radius dependence of electric charge[RICOH].

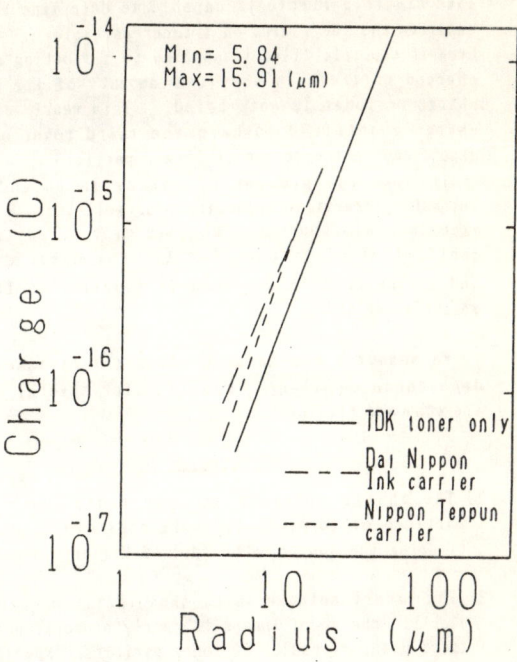

Fig. 11 radius dependence of electric charge[TDK-1].

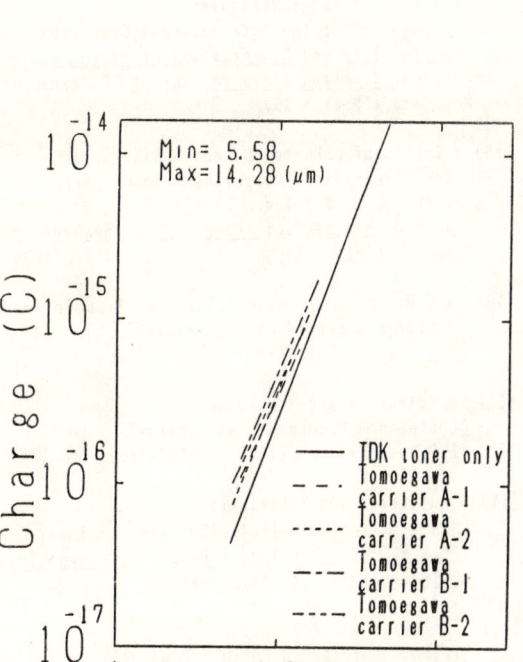

Fig. 12 radius dependence of electric charge[TDK-2].

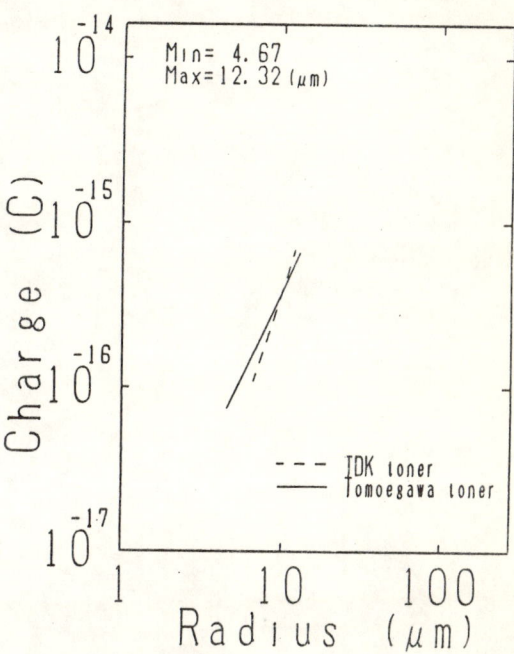

Fig. 13 radius dependence of electric charge [TOMOEGAWA Co. produced carrier A-1].

By our applied Millikan's method the five values of the same single particle is capable to determine the charge quantity and the radius on a toner particle.

Even if electric filed is applied to a floating or moving charged particle in air, the amount of the electric charge of toner is not varied. This nearly causes the charge quantity Q to have the third power of radius dependence in the region of larger particle.

Through the process of making in an additional automatic separative apparatus plus well known Millikan's experimental apparatus, we have discovered the index coefficient of a toner for two-component developer and the method; this method is superior to the other measuring method.

We measured the Index Coefficient of toner radius dependence on charge quantity for several kinds of the composed developer.

V. Conclusion

1) The charge quantity and the radius on a single particle of toner in air were shown to be determined by using the apparatus on applied Millikan's method.

2) The experimental results show that the quantity of electric charge of toners is nearly proportional to the third power of radius of toner paricles, especially for the chemically spherical toner.

3) In different types of tribocharging, the quantity of electric charge of toners are different from charging mechanisum, but shows a similar tendency of radius dependence of electric charge for toner particles.

References

[1] L. Lee and J. E. Weser,
"A Two-Step Decay Scheme for Triboelectricity of Polymeric Developers",
J. of Electrostatics. vol. 6, (1979), p. 281.

[2] C. D. Hendricks and K. F. Yeung,
"Technique of Single-Particle Charge Measurement",
IEEE Trans. on IA, vol. IA-12, No1, (1977), p. 56.

[3] R. B. Lewis, E. W. Connors and R. F. Koehler,
"A Charge Spectrophotograph for Xerographic Toner",
Denshi Shashin Gakkaishi [Electrophotography] vol. 22, 1983, p. 85.

[4] T. Imamura and T. Tetsuya,
"Measurement Method and Device of Toner Particls Characteristics",
Japanese Laid Open Number 116542/1983, J.P.O.

[5] R. H. Epping,
Guide Book of 2nd S.P.S.E International Congress on Non-Impact Printing, 1984.

[6] J. Bares,
"Conference Record of IAS Meeting",
IEEE, Tronto, Canada, (1985), p. 125.

[7] J. Henniker,
Nature vol. 196, (1962), p. 474.

[8] N. Kutsuwada and Y. Nakamura,
"Proceedings of the International Symposium"
The Stability and Conservation of Photographic Images Chemical, Electronic and Mechanical, Bangkok, Thailand Nov. 3-5, (1986), p. 133.

[9] Y. Takahashi, H. Horiguchi and T. Sakata,
"New Toner-charge Measuring Method Charge Measurement of A Toner Particle",
The 3rd International Congress on Advances in Non-Impact Printing Technologies. Aug. 24-28, (1986), p. 23.

[10] M. K. Mazumder, R. E. Ware, T. Yokoyama, B. Rubin and D. Kamp
"Conference Record of IAS Meeting"
IEEE Atlanta, USA, (1987), p. 1606-1614.

[11] N. Kutsuwada and Y. Nakamura
"Conference Record of IAS Meeting",
IEEE, Atlanta, USA, (1987), p. 1597-1601.

[12] N. Kutsuwada and Y. Nakamura
"A Measurement of Electrophotographic Toner Charge"
S.P.S.E Fourth Non Impact Printing International Conference. Mar., (1988), p. 88-93

[13] B. D. Terris and A. B. Jaffe,
"Tribo-charging in Polymer Powders",
SPSE's 40th ANNUAL CONFERENCE and SYMPOSIUM on HYBRID IMAGING SYSTEMS,
Rochester, New York, May 17-22, (1987), p. 219.

THE INFLUENCE OF As⁺ ION IMPLANTATION ON THE OPTICAL AND ELECTROPHOTOGRAPHIC PROPERTIES OF AMORPHOUS Se PHOTORECEPTOR

Wang Yong Yue

Physical Department of North-Western University
Xi'an, China

Abstract

In order to improve Xerographic performance and the manufacture technique, we present a new method to make the double structure; i.e. Evaporation-Implantation method.

In this paper, the new method and the influence of As⁺ ion implantation on the optical and electrophotographic properties of amorphous Se photoreceptor are presented.

Sample Preparation and Measurement

introduction to the new method

The process is firstly to evaporate amorphous αe with 50μm thick on Al substrate, subsequently implant high dose As⁺ ion in the Se surface, so that we obtain a photoreceptor with the double layer structure. As shown in Fig.1 in which the thin amorphous Se-As layer is a charge generating layer(CGL), the other thick layer is a charge transport layer(CTL).

Implanted plan

In order to understand the effect of As⁺ ion beam on the α-Se film surface optical and electrophotography properties, we selected two beam currents (1μA, 30μA) four doses (10^{13}, 10^{14}, 10^{15}, 10^{16} atom/cm²) for 150 kev energy.

Measurement instrument and condition

The optical properties was determined by ellipsmometry at the wave-lengh 6328 Å and incident angle 70°. The electrophotographic characteristics of the photoreceptor was obtained by the Sp-428 electrostatic instrument (made in Japan),
the experiment condition were: room temprature 26°C, relative humidity RH=70%, charge voltage 6 kv, charging time 0.4 second, light illuminance on surface of the sample 10 lux.

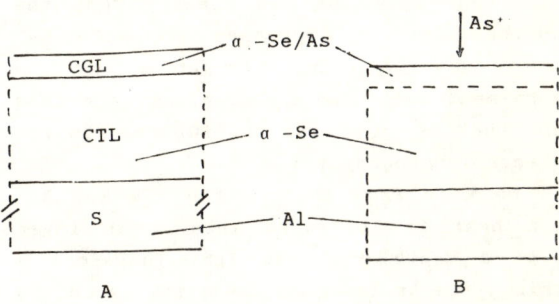

Fig.1 The double structure of photoreceptor for electrophotography.
A. All-evaporation method. B. evaporat-implantotion menthod

Results and Disscussion

The influence of As⁺ ion implantation on the optical properties of α-Se

As shown in table 1, we can see:
1. The refractivity n is reduced at the doses of 10^{13}--10^{14} atom/cm² range, but increased at 10^{15} atom/cm² dose.
2. The oxtinction index k and the absorption coefficient α general increases, as predicted. It may be contribute to the effect of implantation doping and damage, which formed some absorption centre on the surface layer.

The effect of As⁺ ion implantation on the α-Se electrophotographic characteristics

As shown in Fig.2 we can see:
1. As⁺ ion implanted α-Se may be modulate the electrophotographic properlies of α-Se layer.
2. The variational quantity of after implantation depends on the As-sample and the implantation condition. For this

Table 1 The variety of optical properties at before and after As⁺ ion implanted α-Se

specimen No.	refractivity η before	after	extinction index k before	after	absorption coefficient α before	after	reflectivity R before	after
A	2.89	2.79	0.065	0.074	1.3	4.1	23.6	23.5
B	2.83	2.79	0.05	0.07	1.0	3.8	23.1	22.8
C	2.78	2.82	0.038	0.079	2.1	4.9	22.4	23.8

reason we conder the variety ratio of before and after ion implanted, as show in Fig.3-4.

From Fig.3, we see clearly that the variety ratio of maximal surface potential reduces with the dose is increased.

From Fig.4, we see clearly that the photosensitivity increases with the As⁺ ion quantity. This tendencey is in agreement with the all-evaporation method and that is just our hopeful results for electro-photography.

3. We know from theory, the 150 kev As⁺ ion beam is implanted into α-Se layer with a depth less than 1μm. Therefore a thin α-Se-As layer on the thick (about 50 μm) α-Se layer is formed. This is just the double layer structure.

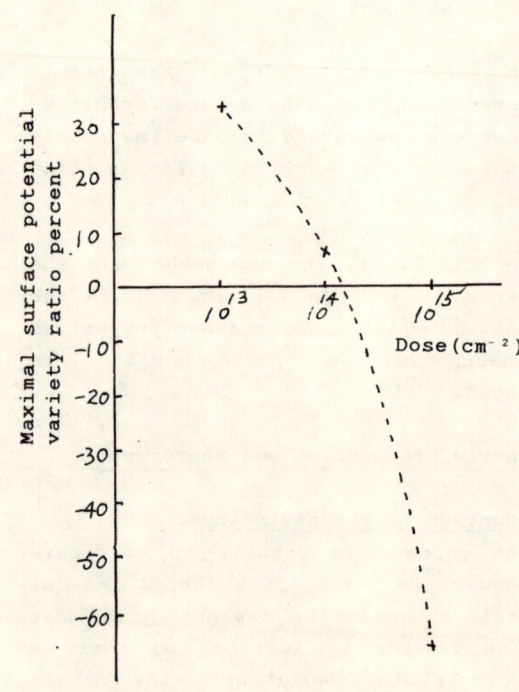

Fig.3 Maximal surface potential variety ratio percent of α-Se versus dose at As⁺ ion implantation before and after

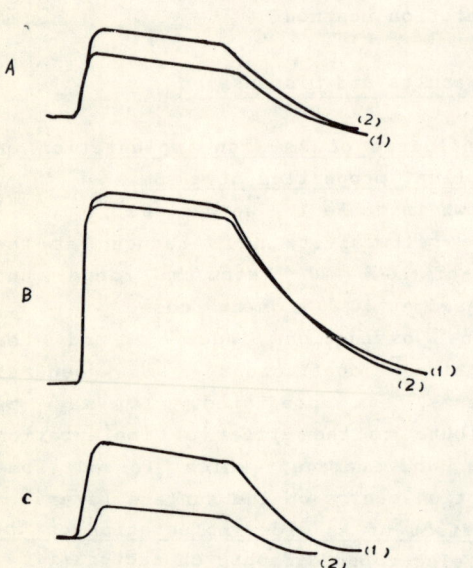

Fig.2 phot-induced discharge characteristics(PIDC) of α-Se at before¹ and after² implantation
  A. Dose 10^{13} As⁺/cm²;
  B. Dose 10^{14} As⁺/cm²;
  C. Dose 10^{15} As⁺/cm².

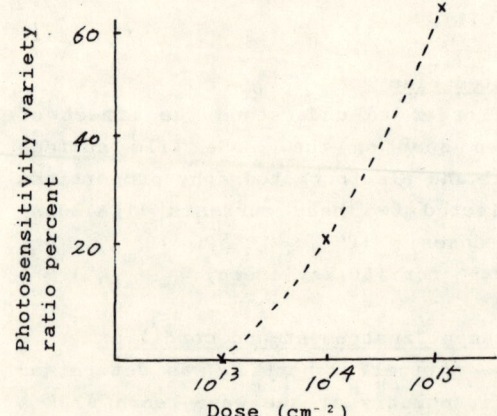

Fig.4 photosensitivity vriety ratio percent of α-Se versus dose at As⁺ ion implantation before and after

The advantage of the new method
1. The implanted ion concentration may be earily controlled by charging the dose .
2. The interface in the double layer structure by ion beam method gets better comparing with all eavaporation method.

Conclusion

Summarizing our experiment results and according with theory analyse, we can conclude that evaporate-implantation method has the follow merits:
1. It can modulate the properties of α-Se layer.
2. The photoreceptor with double-layer structure could be obtained.
3. Suitable selected implantation condition could be used to increase the electrophotoraphic velocity and photosensitivity certainly. As a new method, it is firstly presented and our work is only in begining.

Acknowledgement

The authors are greateful to Ju Xiu-yan for α-Se sample preparation, Zhong Shi-qian for ion implanted and Liu Shu-guo his assistance in the electrophotographic plate testing.

Reference

1. R.M. Schaffert, ELECTROPHOTOGAPHY, THE FOCAL PRESS, 1975
2. Wang Yon-yue, Journal of North-Western University, Natural Science Edition, Vol 16, No.1,pp34 (1986)

ADVANCED TECHNOLOGIES OF ELECTROPHOTOGRAPHY IN JAPAN

Kenichi Kōda
Konica Coperation, Tokyo, Japan

1. Modern Office Work

The typical office working consists of five different types of working such as 1) preparing documents, 2) calculating, 3) copying and printing, 4) storage and retrieval image information, 5) transmitting image information.

These different types of work is carried out by using the automated business machine, such as Japanese word processer, personal computer, copier and duplicator, electronic filling system and facsimile, respectively.

The caracteristics of Japanese user's needs and domestic market are shown as follows:

User's Needs
1. The documents written by hand.(using Kana Kanji character)
2. The paper size, A series and B series.
3. The office space per head is very narrow.
4. Price of machine should be cheap.

Market
1. The size of firm is smaller than that of America and Europe.

The all of OA machine should be required not only smaller and cheaper, but also better image quality.

2. Progresses of copying technologies

The output of copiers made in Japan is shown Figure 1.

The changes of compactness is shown Figure 2.

The ratio of compactness is expressed by the value of the quotent devided machine weight by copy speed.

2.1 The key technologies for compact low cost machine.
 1. Optical fiber lens array
 2. Organic photorecepter coated on the thin aluminum drum
 3. Small size developing device
 4. Blade cleaning device
 5. Precision molding for plastics
I will mainly talk about the progress of organic photoreceptor.

2.2 The new technologies for high quality imaging.
 1. Two component toner
    a) Microtoning development
    b) Panafine development
    c) Developer using resin coated carrier
 2. Mono component toner
    a) Bipolar magnetic toner process
    b) Jumping developing process
    c) Floating electrode effect development

3. The new PPC model in 1987
   Table 1 shows new model in 1987

4. The activities of The Society of Electrophotography of Japan.

This society was established by Prof. Doctor S.Kikuchi on 1985, so we had Mini-International Conference "Japan Hard Copy '88" in Tokyo and "Post Symposium Hard Copy '88 in Kansai" in Osaka for the memory of 30th anniversary.

CPM	Model	CPM	Wt/CPM	Year	Model	CPM	Wt/CPM	Year
80	Xerox 1075 DS*	70	7.57	'83	Canon NP 8580	82	2.31	'88
60	Xerox 3600R	60	9.33	'71	Sharp SF 9700	60	3.16	'87
40	Xerox 4000	42	8.33	'73	Toschiba LEO 8811	40	1.75	'84
25	Canon NP 5500	26	8.50	'77	Toschiba LEO 7610	30	2.00	'87
15	Xerox 3103	14	8.14	'76	Ricoh FT 1520 *	15	1.66	'86
10	Fuji Xerox 2200	5	15.00	'73	Sanyo SFT 600 *	10	1.70	'85

* = OPC

Table 1 New Model 1987

CPM	Konica	Fuji Xerox	Canon	Ricoh	Minolta	Toshiba	Matsushita	Sharp	Mita	Sanyo	Silver & Sony	Total
-10cpm			PC-9 6 OPC PC-7 8 OPC	Cuvax MC-50 4 Digital Cuvax MC-50 IM-A 4 Digital				Z-61 6 ZnO			MCP-100 6	-6
-15cpm	Konica 1012 12 OPC				EP-370 15 OPC EP-3702 15 OPC	LEO-3705 12 Se LEO-4810 15 Se LEO-5110 18 Se	LEO-4135 14 Se	FP-1540 15 Se FP-1510 15 Se	SF-7750 15 OPC SF-7700 15 OPC	DC-1205 12 Se DC-1225 12 Se DC-1785 18 Se	DC-1786 18 Se DC-2055 20 Se		-12
-20cpm													5
-25cpm	Konica 2022 22 Se	FX 3950 21 Se	NP-3225 25 OPC	FT-4630 21 Se	EP-4152 21 OPC EP-405RE 21 OPC					DC-1655 16 Se DC-2555 25 Se	DC-2585 25 Se	SFT-2120 21 Se SFT-2120F 21 Se	10
-30cpm		FX-5020 28 Se FX-5030 28 Se	NP-3725 27 OPC	FT-5630 30 Se	EP-4902 30 OPC	LEO-7550 28 Se	LEO-7610 30 Se	FP-3007 30 Se FP-3037 30 Se					9
-35cpm	Konica 3032 32 Se											SFT-2123 33 Se SFT-2123F 33 Se	3
-40cpm		FX-5270 37 Se FX-5041 40 Se		FT-5840 40 Se	EP-5702 40 Se					DC-4055 40 Se	DC-4085 40 Se		6
-45cpm	Konica 3042 42 Se	FX-5043 45 Se											2
-50cpm			NP-7050 50 a-Si NP-7650 50 a-Si	FT-6550 50 Se FT-6950 50 Se	FT-6850 50 Se			FP-4660 45 Se		DC-5055 50 Se			7
50⁺cpm	Konica-5070 70 As2Se3 Konica-4055 55 As2Se3				EP-850 55 Se	LEO-9110 55 Se			SF-9700 60 Se SF-9750 60 Se				6
Special Use		FX-100Printer 100 2nD	Color Copier 5 OPC	FW-550 14 Se FW-940 14 Se				FN-P300		DC-A22 14		FL-2200 7	8
Total	6	7	7	10	7	7	6	4	5	12		4	2	74

-269-

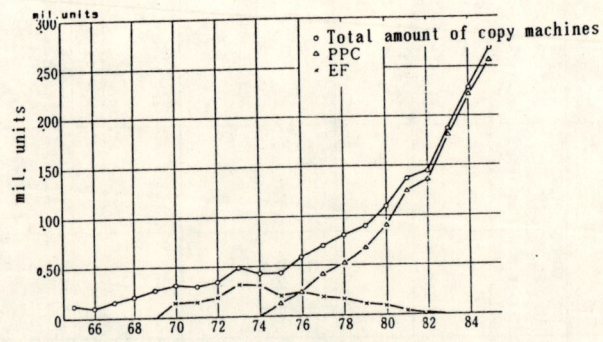

Fig.1 The output of Japanese Business copying machine(1965-1985)

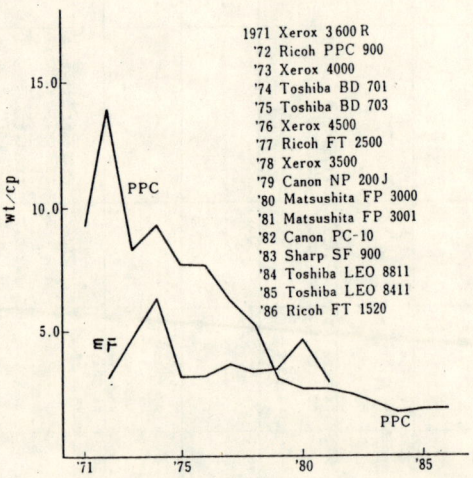

1971 Xerox 3600 R
'72 Ricoh PPC 900
'73 Xerox 4000
'74 Toshiba BD 701
'75 Toshiba BD 703
'76 Xerox 4500
'77 Ricoh FT 2500
'78 Xerox 3500
'79 Canon NP 200J
'80 Matsushita FP 3000
'81 Matsushita FP 3001
'82 Canon PC-10
'83 Sharp SF 900
'84 Toshiba LEO 8811
'85 Toshiba LEO 8411
'86 Ricoh FT 1520

Fig.2 The compactness changing(1971-1987)

THE SURFACE AND INTERFACE OF α-Se FILM STUDIED BY XPS

Li Hongyi, Deng Xiaohong, Zhang Guanming* and Zhong Guirong
The Center of Analy.and Measure. *The Department of Physics
Wuhan University, Wuhan, Hubei, China

ABSTRACT

The impurities, oxygen and carbon, deposited on the Al-substrate by evaporation are analysed by XPS in the light of the correlations between the valence band spectra and the structure form of Se. It has been shown that the structure form of the deposited Se film is amorphous but there are crystalline layers with different thickness at the interface between Se film and Al-substrate on condition that T_s is below 80°C. The existence of the crystalline layer has been comfirmed by TEM. A lot of reasons lead to crystallization at the interface. One of the important reasons is the existence of trace amount of oxygen.

The crystalline layer has an obvious effect on photoelectrical properties of α-Se film. The crystalline layers not only decrease the electrical thickness of α-Se film as a layer absorbing light, but also enhance the probability of the field injection of charge carriers from the crystalline layer to the α-Se film and, therefore, cause the dark decay rate of the α-Se film to increase.

INTRODUCTION

Selenium is an important material as xerographic photoreceptors. When Se was deposited onto Al-substrate by vacuum evaporation, its electrical and optical properties are related to the purity of the Se material, the structure form of the Se film, and the technological process. These interrelations are always taken notice by a number of research workers[1-4] Some previous research work[1,2,4], though the influences of various factor for the crystalline layer between the a-Se film and Al-substrate were investigated from various angles, the influence of surface state of a-Se film and interface crystalline layer for electrostatic properties of a-Se film are yet short of essentisl investigation. However the states at the surface and interface of a-Se film occupy an important status in the xerographic process. This paper made use of X-ray Photoelectron Spectroscopy(XPS), one of a strong measure, obtained direct the informations of electric state on the surface, interface of a-Se film, and observed corresponding that one affected the photoelectric properties of a-Se film. This provided valuable basis for the productive technique of photoreceptors.

EXPERIMENTAL METHODS AND RESULTS

1. The a-Se film were prepared by vapour deposition of Se onto an Al-substrate in a vacuum chamber with the base pressure of $2-5 \times 10^{-5}$ torr. Al-substrate temperature(T_s) was choose any constant value from room temperature to 90°C. The raw materials Se with the first-grade purity(99.992%) and high-grade(99.9992%) purity were to use respectively. The molybdenum boat was used as a crucible.

2. The XPS measurements were performed with an XSAM800 photoelectron spectrometer. To inspect the surface, interface states of Se film, the samples were prepared and selected as follows:

Sample 1#: The raw material Se is the first-grade pure one. when the indicative crucible temperature was held at 200°C, a fine film of Se was deposited. Its XPS is shown in Fig.1. It is clear from Fig.1 that peaks O1s and C1s are comparatively strong. Even if they were etched by Ar^+ for a long time, peaks O1s and C1s keep their shaps.

Sample 2#: The raw material Se is with the first-grade purity yet, but it was purified physically once and used as evaporating source. Its XPS spectrum of the

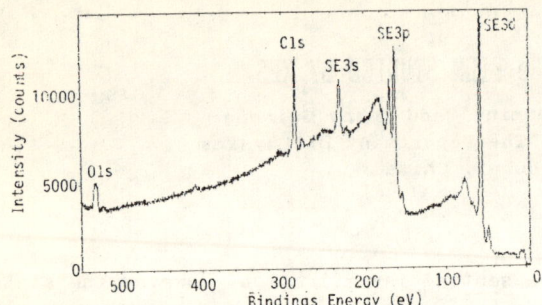

Fig.1 The core levels of initial evaporated Se film

surface of this sample is shown in Fig.2. As a consequence, the peak O1s may be seen faint from Fig.2 and the peak O1s disappeared after etching slighly by Ar^+.

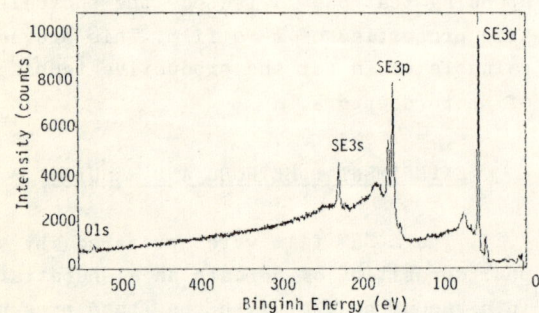

Fig.2 The core levels of the surface of sample $2^\#$

When the sample $2^\#$ passed the etching by Ar^+, the peak O1s disappeared. But if the sample was taken out from the spectrometer and was exposed to air for a period of time then reenter this sample into the XSAM800 and obtained a spectrum again, the peaks O1s and C1s can be seen obviously, which is show in Fig.3.

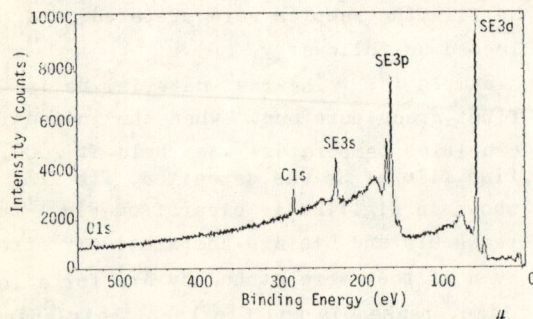

Fig.3 The core levels of sample $2^\#$ after exposing to air for 24hr.

The valence band electron state levels of sample $2^\#$, as shown in Fig.4.

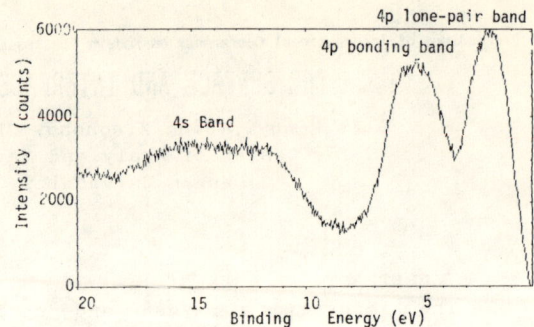

Fig.4 The valence band of the surface of sample $2^\#$

Fig.5 shows the valence band spectra obtained from the interface of sample $2^\#$, which had been peeled from the Al-substrate.

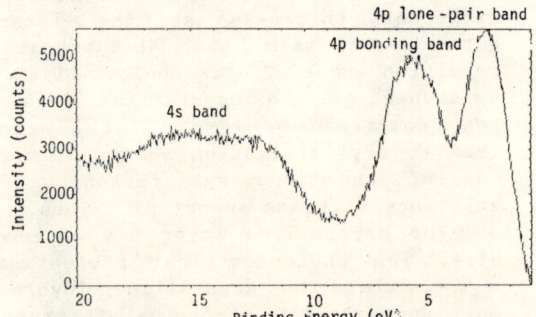

Fig.5 The valence band of the interface of the sample $2^\#$

Sample $3^\#$: The raw material Se with high-grade purity was used. The XPS spectra of the surface and interface of this sample are similar to the Fig2,3,4 and 5. Only the peak O1s is extremely weak.

3. In order to observe and check the crystalline situation, the cross-section of sample $2^\#$, $3^\#$ were measured by using TEM respectively. The test results presented that all were similar to each other except the thickness of crystalline layer. Fig.6 and Fig.7 show the TEM image of the cross section of sample $3^\#$ and the electron diffractions patterns of its a-Se film and interface crystalline layer, respectively.

Fig.6 A cross-sectional TEM micrograpg of sample $3^\#$

a-Se film   c-Se layer

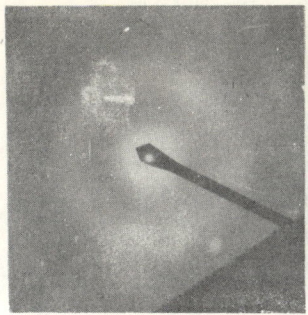 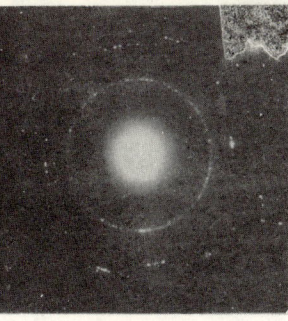

a-Se film   c-Se layer

Fig.7 The electron diffraction pattern of sample 3#

4. For observing the relationship between the electrostatic properties and thickness of crystalline layer in Se film, a set of samples, which have evident differences on electrostatic properties, were selected. Its electrostatic properties and the thickness of crystalline layer were observed, respectively. The results are shown in Table 1:

Sample	Dark Decay Rate	Thickness
4#	1.6 V/sec	5 μm
5#	3.2 V/sec	23 μm
6#	5.5 V/sec	30 μm
7#	8.9 V/sec	44 μm

Table 1 The dark decay rate and crystalline thickness of samples

5. To observe the process of the phase change, which is used as the purpose of investigating the inquiry into the origins of interface crystalline, a series of Se film were carried on the differential thermal analysis using DT-30B thermal analyzer. The results showed that all the exothermic regions and shaps of different Se film are not same. With the degradation of the disordered degree of the Se film, the exothermic region is reduced gradually, the heights of the exothermic peaks are increased. But the exothermic regions of all Se film lie in the region from $94\pm2°C$ to $132\pm2°C$. Their DSC curves are presented in Fig.8.

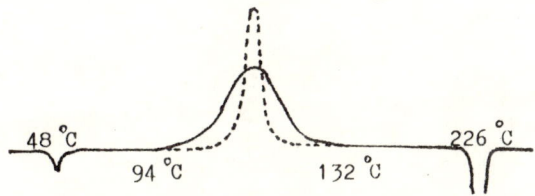

Fig.8 The DSC curves of Se film

DISCUSSION

Some previous research work[3,5] have proved the following: When an amorphous form was transformed into a crystalline form, the 4s band spectra of Se will be changed correspondingly. The centric part of 4s band of c-Se is slightly hollow downward and the 4p bonding band in point of 4.2eV(binding energy) is slightly protuberant compared with amorphous of Se(shown in Fig.4,5). It can be shown from the experimental results that.

1. In the Se films deposited by vacuum evaporation, even if the raw-materials with high-grade purity were selected to use, a very thin crystalline layer exists at the interface between the Se film and substrate.

2. So long as T_s was chosen below 80°C the structure form of the Se films which deposited using the first-grade purity Se, they are still amorphous.

3. The surface of a-Se film (include raw-material Se) very easily adsorbs oxyge and carbon in the air (as shown in Fig.3).

4. The dark decay rate on the electrostatic properties of a-Se film was increased along with the thickness of crystalline layer was increased.

5. In the Se film, which deposited by vacuum evaporation at the begining, the content of oxygen is higher than a-Se film of the later period (as shown in Fig.1)

Above experimental results provided the dependable basis for the analyses of the causes of producing crystalline layer in the interface of a-Se film. Previously Montrimas and Petreties regared that when T_s surpassed 60°C, the crystalline layer is mainly composed of trigonal Se with small amounts of monoclinic Se, and when T_s surpassed 75°C, the probability of dendritics crystalline growth was obviously increased. Frank Jansen's work reported that oxygen incoprated in Se at low ppm levels it will increase the interfacial crystalline rate of the Se film. The research results presented in this paper are completly agreement with the results of above authors. Nevertheless it is worth putting forward that under 92 C(phase change temperature)the major causes leading to the interfacial crystallization in a-Se film are not only the substrate temperature but introducing oxygen. From DSC's results, when Ts was far below the phase change temperature, the crystalline layer in pure Se film produces provided oxygen exists. And about the way introducing oxygen, we concluded the following major manners:

1. One of them was introduced in the course of purifying the raw materials Se. Because the raw materials Se were prepared by reducing SeO_2 using SO_2. If the reduction was infull or the excessive SO_2 was not completely cleared away at the later processing, oxygen could be introduced.

2. Leaving the raw materials Se in atmosphere, oxygen and carbon in the atmosphere were adsorbed, even SeO_2 was formed, and sneaked into the raw materials.

3. There is residual oxygen in a vacuum chamber during the period of deposition of Se film.

Therefore at the initial stage depositing a-Se film by vacuum evaporation. It isn't avoidable to contain a trace of oxygen in the Se film deposited onto an Al-substrate, and evaporation temperature of SeO_2 is lower than that of Se (shown in Fig.1). On the Al-substrate surface, there is a infinitely thin layer of Al_2O_3 too, all these lead to increase the oxygen content of the Se film at the interface. Of course with the rising of T_s, the crystalline action of oxygen was enhanced and the thickness of the crystalline layer was increased too, with the result that the dark decay rate of a-Se film was risen rapidly (as shown in Table 1).

CONCLUSION

1. Oxygen is one of the major causes which bring about crystalline of the interface between pure Se film and aluminium substrate.

2. With T_s rises the thickness of the crystalline layer at the interface of pure Se film will be increased.

3. With the thickness of the crystalline layer at the interface adds the dark decay rate of a-Se film rises.

4. The surface of Se adsorb oxygen and carbon in air very easily.

5. To prepare a-Se film with good photoelectric properties, the purity of raw materials Se isn't below 99.992%, the oxygen content of which must be lessened to the full. It isn't suitable that the raw materials Se in store are exposed in air over a long period of time.

REFERENCES

1. E. Montrimas and B. Petretics,
   Phys. Status Solidi(a) 15, 361 (1973)
2. Frank Jansen,
   J.Vac.Sci. Technol., 18(2) 215 (1981)
3. Takashi Takahashi and Takasi Sagawa,
   Phys. Rev. B 26 7039 (1982)
4. Zhong Guanming Li Hongyi and Zhong Ying
   Reprography Vol. 1 (5) (1987)
5. Li Hongyi Deng Xiaohong,
   "Japan Hardcopy'88" paper sum. P.309

ION FLOW CONTROL TECHNOLOGY AND ITS APPLICATIONS

Y. HOSHINO, M. OHTA, T. TANAKA, M. OMODANI AND S. SHIWA

Electrical Communications Laboratories, NTT, Yokosuka, JAPAN

abstract

Applications of corona discharge phenomena to printing and display devices are described. As a fundamentals of these applications, ion flow control characteristics by a pair of aperture electrodes are investigated. The ion flow control technologies are applied to the following two devices: ① a high quality continuous tone printer, and ② a novel electrophoretic display.

Introduction

Recent progress of non-impact printing technologies has been rapid, with the rapid progress of data processing and communication technologies. Non-impact printing technologies including such varied technologies as thermal, ink jet, electrophotographic, electrostatic, and so on, have found applications in many different fields. The most important considerations for printing technologies are, ①print quality, ②printing speed, ③volume of equipment, ④cost, and ⑤noise. Continuous-tone image reproduction, which is one aspect of print quality, has become an important theme in printing technology.

Ion flow technology that ion flow from corona discharge is controlled by aperture electrodes has the following advantages:
· Simple mechanism in electrostatic image formation,
· Precise controllability of ion flow.
The latter merit is most suitable for continuous-tone reproduction.

In this paper, first, the application of electrophotography to printers will be discussed and a comparison between electrophotography and ion flow technology will be described. Next, ion control characteristics by aperture electrodes and application to high quality continuous-tone reproduction are described. Finally, application to a novel electrophoretic display will be mentioned.

Application to printer

Electrophotography, which was first implemented in a copying machine,[1] has been applied to printers by adding light scanning unit. The first application to a printer produced a very high speed laser printer for peripheral use with large computers(for example, printing speed: 89cm/s, volume: 325cm(W) × 88cm(H) × 152cm(D)).[2] In subsequent years, efforts to develop compact size[3] and high-quality reproduction have continued. The expanded application of electrophotography to printing is shown as Fig.1. (An elliptical laser printer is described in Ref.[4]) Ion flow technology[5] is one of the most promising technologies for continuous-tone reproduction. The process is close to electrophotography, so we have included it in Fig.1.

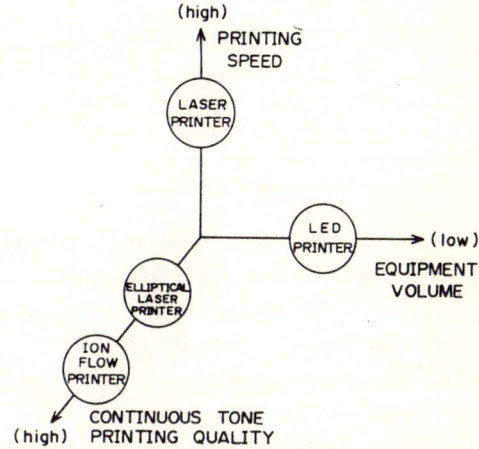

Fig.1. Expansion of electrophotography.

Electrophotography and ion flow technologies compared

The printing mechanisms of these two technologies are shown in Fig.2. In order to form electrostatic

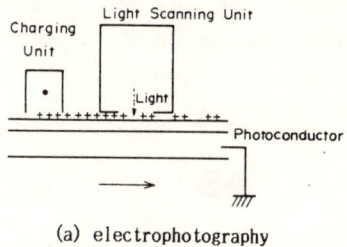

(a) electrophotography

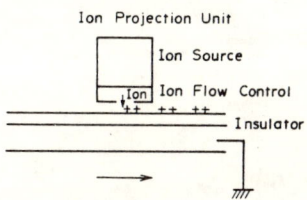

(b) ion flow technology

Fig.2. Comparison between electrophotography and ion flow technology.

latent images, electrophotography requires two processes--charging and light image irradiation, while ion flow technology requires only one process--charge image irradiation. Ion flow is controlled by the directions of electric field between a pair of apperture electrodes, as shown in Fig.3.

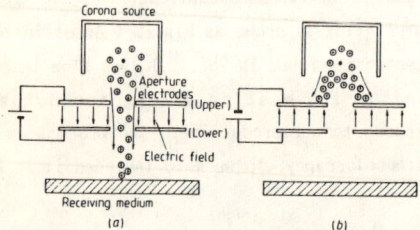

Fig.3. Modulation of ion flow using the electric field between the aperture electrodes. (a) Ions pass through the aperture electrodes when the electric field between the electrodes is in the ON state. (b) Ions are blocked by aperture electrodes when the electric field between the electrodes is in the OFF state.

Corona ion generation (positive case) has the properties of good uniformity and little fluctuation. The electric field between the electrodes and the duration of the ON state can be precisely controlled by electric circuits. Therefore, the amount of ions passing through the aperture electrodes can be precisely controlled and high-fidelity charge patterns can be formed on insulating film.

Control characteristics by aperture electrodes

Developing a through understanding of control characteristics by aperture electrodes is the first step toward actual applications. A simplified model for investigating these characteristics is shown in Fig. 4. The current I passing through the apertures is expressed,

$$I = J_1 S_1 = J_3 S_3.$$

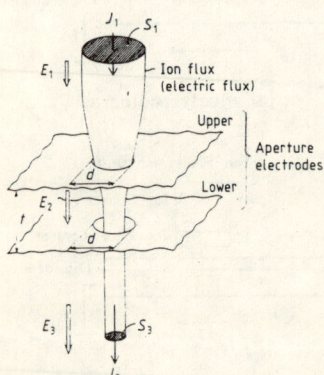

Fig.4. A schematic drawing of ion flux passing through the aperture electrodes.

We define factor f as,

$$f = S_1 / \pi / (d/2)^2.$$

A good approximation of factor f can be determined as follows,[6]

$$f = f(E_2/E_1, E_3/E_1, t/d).$$

Factor f is important, because the current which passes through the apertures is estimated by,

$$I = J_1 \pi (d/2)^2 f.$$

The value of factor f is measured using the apparatus shown in Fig.5. The measured results are shown in Fig.6.

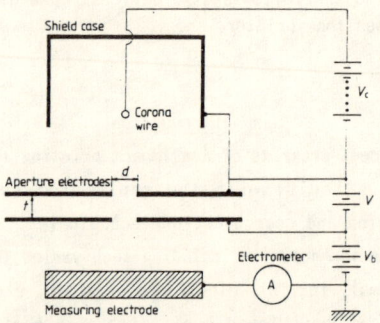

Fig.5. Measuring apparatus for ion flow control characteristics.

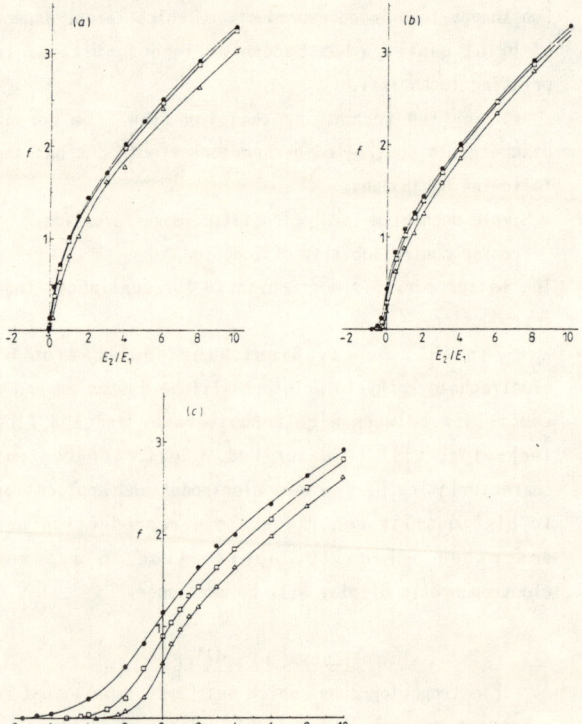

Fig.6. Measured ion transmission rate f vs. electric field strength between the aperture electrodes: (a) $t/d=2$, (b) $t/d=1$, (c) $t/d=1/4$. Symbols: ●, $E_3/E_1=8$; □, $E_3/E_1=4$; △ $E_3/E_1=2$.

It is found that factor f is increased from certain threthold, as the value E_2/E_1 is increased.

To obtain a more detailed understanding of the control characteristics, the trajectories of ions are calculated, as shown in Fig.7. The results agree well with the experimental results.

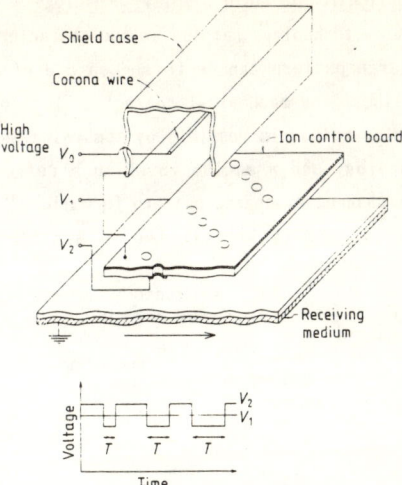

Fig.8. The exeperimental apparatus for forming charge images where T is the ion projection time.

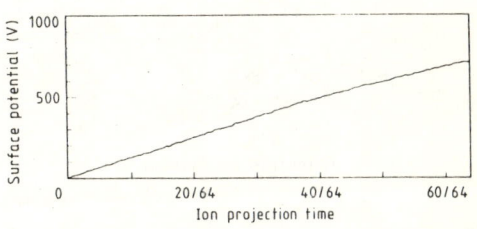

Fig.9. The surface potential of a gray-scale image.

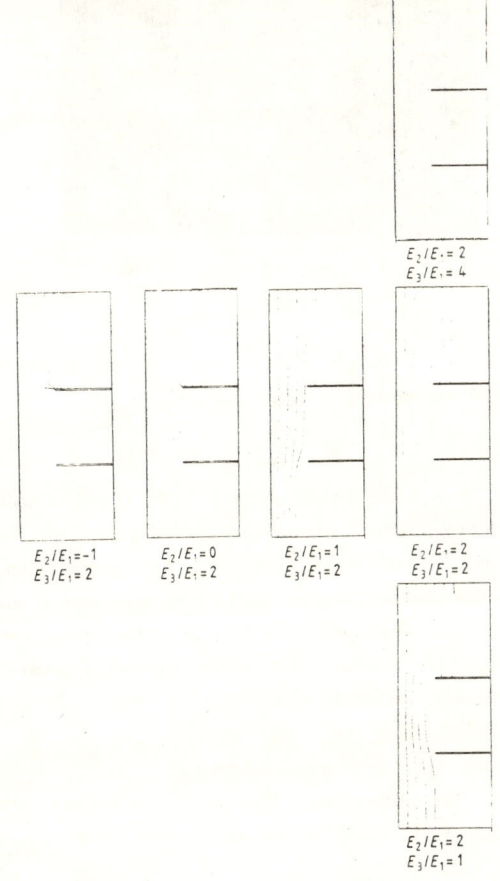

Fig.7. Typical ion trajectories obtained by numerical analysis where t/d=1.

Application to continuous-tone printing

Ion flow printing is a promising process for producing high-quality continuous-tone image printing, because the amount of ions through apertures can be precisely controlled. The feasibility of the process for continuous-tone printing was investigated by the apparatus shown in Fig.8.[7] The receiving medium, which consists of a thin dielectric layer formed on top of a conductive layer, is moved in accordance with ion projection. The amount of ions can be controlled by the amplitude and the pulse duration of applied voltage to aperture electrodes. In this experiment, pulse duration is modurated into 64 levels with voltage amplitude being constant. The surface voltage vs. pulse duration is shown in Fig.9. It can be seen that the surface voltage is well controlled by the modulation of pulse duration. The electrostatic charge images, observed by SEM (scanning electron microscopy), is shown in Fig.10.

This shows that the diameter of the dots increases as the pulse duration increases.

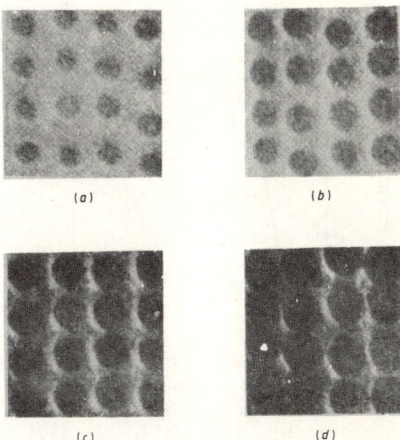

Fig.10. SEM images of electrostatic charges on a gray-scale sample. (a) T=3/64; (b) T=7/64; (C) T=12/64; (d) T=23/64, where T is the ion projection time.

Application to novel display technology

Ion flow technology has the unique characteristic that a charge pattern can be formed on a dielectric surface without any mechanical contact. Therefore, a novel display method is possible by combining the ion flow technology and a medium in which a refrection spectra is changed by charge pattern formation.[8] The display principle is illustrated in Fig.11. The

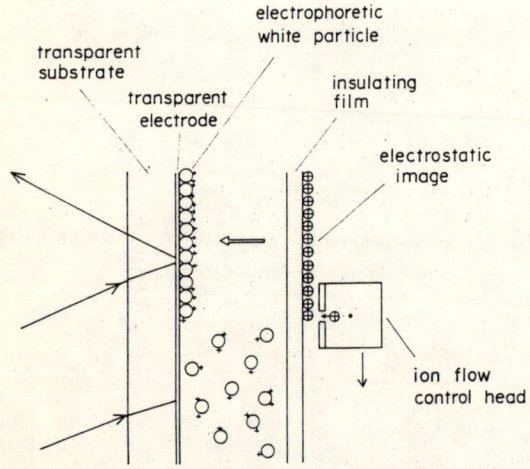

Fig.11. Principle of display method.

display medium is composed of electrophoretic white particles and colored insulating fluid. The electric field is generated in the region where the charge pattern is formed, and electrophoretic white particles move to the surface, causing the surface to turn white.

The measured contrast ratio vs. surface charge density results and a photograph of displayed image are shown in Figs.12 and 13, respectively. This display method is suitable for static large-screen image displays,

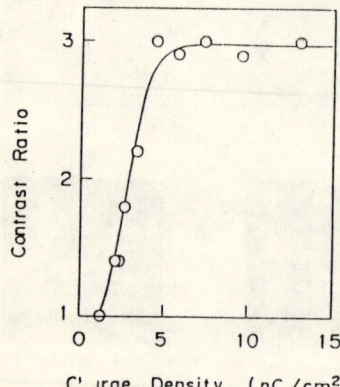

Fig.12. Contrast ratio as function of charge density generated from ion-flow control head.

Fig.13. Sample of displayed image.

because this type of display has intrinsic memory and can be implemented in large scale.

Summary

Ion flow technology offers the following advantages: simple mechanism, precice control of ions, and non-contact charge formation. As a application of this technology, continuous-tone printing and a novel display are described. Ion flow technology has a number of superior characteristics that make it a promising technology for future applications.

Acknowledgements

The authors would like to express sincere thanks to Drs. H. Yasuda and K. Komiya for encouragement during this work.

References

[1] R. M. Schaffert,"Electrophotography," (Focal Press, London, 1975).

[2] I. Fujimoto, K. Nishimura, H. Kamada and Y. Hoshino, Elec. Comm. Lab. Tech. J. 29(1980)1489(in Japanese).

[3] K. Tateishi and Y. Hoshino, IEEE Tr IA IA-17(1981)169.

[4] Y. Hoshino, T. Tanaka and M. Omodani, Rev. Elec. Comm. Lab. 35(1987)421.

[5] G. L. Pressman, 2nd Int. Conf. on Electrophotography, at Columbia by SPSE, (1974)37.

[6] M. Omodani, Y. Hoshino and T. Tanaka, J. Phys.D:Appl. Phys. 18(1985)153.

[7] M. Omodani, T. Tanaka and Y. Hoshino, J. Phys.D:Appl. Phys. 20(1987)1224.

[8] S. Shiwa and Y. Hoshino, SID 88 (Int. Conf. at Anaheim USA).

V. Measurement Techniques in Applied Electrostatics and Electrostatic Source

DEVELOPMENT OF HIGH TEMPERATURE FARADY CUPS FOR IN-SITU MEASUREMENTS OF POWDER SURFACE CHARGE UNDER THERMAL PLASMA-POWDER FLOW

J.S. Chang[1], T.G. Beuthe[1], N. Hayashi[1], M. Keila[1] and F.Y. Chu[2]

1. Department of Engineering Physics, McMaster University, Hamilton, Ontario, Canada L8S 4M1

2. Ontario Hydro Research, 800 Kipling Avenue, Toronto, Ontario, Canada M8Z 4M1

* On leave from Department of Electrical Engineering, The University of Tokushima, Tokushima, 770 Japan

Abstract

A special high temperature Faraday cup has been developed for in-situ measurement of powder surface charge under plasma flow conditions. Primary experimental results for powder surface charge to mass ratio measurements under thermal plasma flow are presented.

Introduction

Injection of powdered materials into thermal plasmas is becoming an important industrial process for various chemical and metallurgical industries (Akashi et al. 1986). However, the role played by an electrified powder during the injection process is unknown at this moment since most of the diagnostic techniques presently available are not designed for use in a high gas temperature (500 to 4000 [K]) and high gas velocity (few [m/sec]) environments. In this work, a high temperature Faraday cup was developed for in-situ measurements of powder surface charge under thermal plasma flow conditions, and initial tests to investigate the applicability of the device using silica powder injection into an 80 [kW] DC thermal plasma torch have already been conducted. The surface temperature of the powder particles and the Faraday cup were monitored by an infra-red (IR) image camera, and the mass of the collected powder was measured by an electronic scale. Two types of powder feeders -- a turntable and a fluidization type -- have been used, and the powder fraction in the injection system was measured by a capacitance transducer.

Experimental System

The present experimental investigation was conducted using an Acurex/Aerotherm 80 [kW] DC plasma torch with associated diagnostic instrumentation, a reaction chamber and powder feeders (Chang et al. 1987) as shown in Figure 1. The plasma gas is supplied via a system of four tangential injection ports at the anode of the torch. From here, the gas is excited and/or ionized in the arc column which is struck between the tungsten anode and the copper cathode. The arc column was stabilized under the floating potential environment provided by a series of segmented electrodes placed between the anode and the cathode. In this work, the plasma gas was argon in all cases.

The powder feeding system consisted of a carrier gas supply, a powder feeder and a feed line with a capacitance transducer (Irons and Chang, 1983). In this work, silica powder with a size distribution of 149-200 [μm] diam. was used. The carrier gas was argon in all cases. The carrier gas flow rate was controlled by a needle valve located upstream of the powder feeder. Two different types of powder feeders were used in the present investigation. The schematic of these feeders (turntable type and fluidization type) are shown in Figure 2. In the turntable type feeder, the powder flow rate is controlled independently of the carrier gas flow rate by varying the turntable speed on a graduated scale of 1 to 10. The solid particles are discharged at the bottom of the feeder and mixed with the carrier gas stream. The feed rate of the powder in the fluidization type feeder is controlled by the carrier gas flow rate which is used to fluidize the powder inside the tank. After fluidizing the powder, the carrier gas exits the top of the feeder, carrying some of the powder with it in the process. Therefore it can be expected that the electrification process between these two feeders is very different.

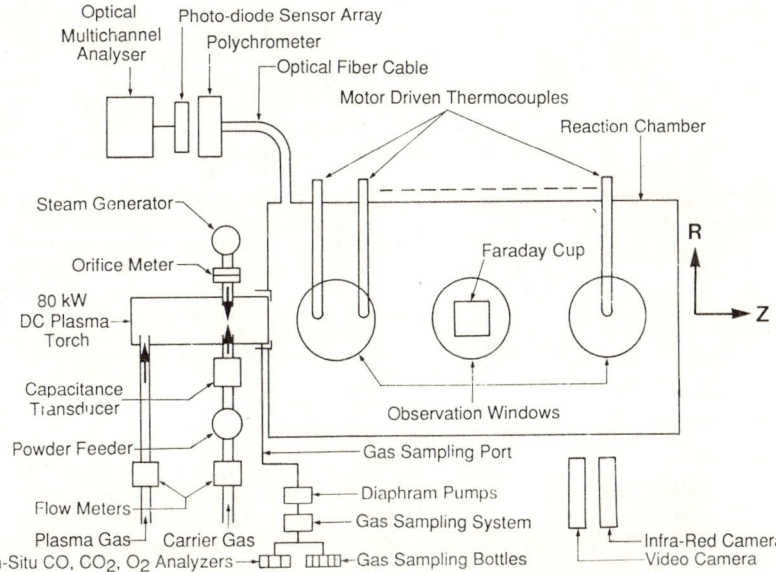

Fig. 1. Schematic of the diagnostic systems and overall experimental setup.

The two-phase powder/gas flow supplied by the feeder was conveyed pneumatically in a 6.4 [mm] i.d. feeding pipe through a capacitance transducer before being injected into the torch. The capacitance transducer consisted of two copper plates or a series of copper rings wrapped around the outside wall of a polycarbonate or alumina tube through which the powder/gas mixture flowed. The copper electrodes were electrically shielded. The capacitance between the copper electrodes was measured by a Boonton Model 72B capacitance meter (1 MHz). This meter can detect the difference between two capacitances applied to the input terminals. In our experiments, one of the two was set to a fixed capacitance equal to that of the capacitance transducer (Ce) without powder loading. The other was the capacitance transducer itself. This arrangement made it possible to measure a small change in capacitance of the probe (down to 10^{-4} [pF]) in comparison with Ce as powder flowed through the system.

At one window of the reaction chamber, an AGA IR camera (Thermovision model 782) was installed to measure the velocity and surface temperature of the particles ejected from the torch. The Faraday cup was situated downstream of the torch exit to determine the charge on this powder. The charge collected by the Faraday cup was measured using a high impedance Keithley 610C electrometer. The surface charge-to-mass ratio of the ejected powder was calculated by weighing the total mass of powder collected using an electronic scale. A schematic of the Faraday cup is shown in Figure 3. In order to operate this device in a high gas temperature environment, all insulating material directly exposed to high temperatures were constructed of boron nitride

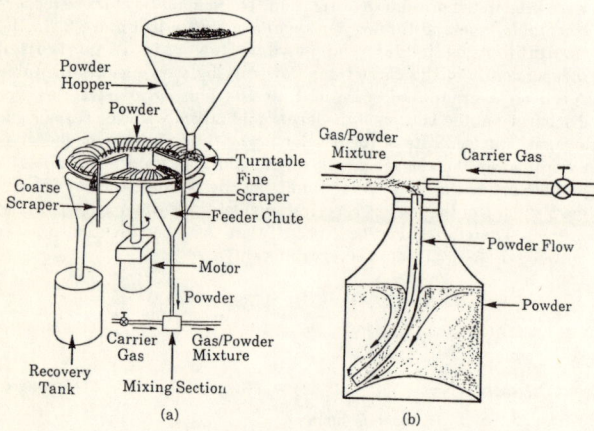

Fig. 2. Types of powder feeders used in this investigation (a) turntable type (b) fluidization type.

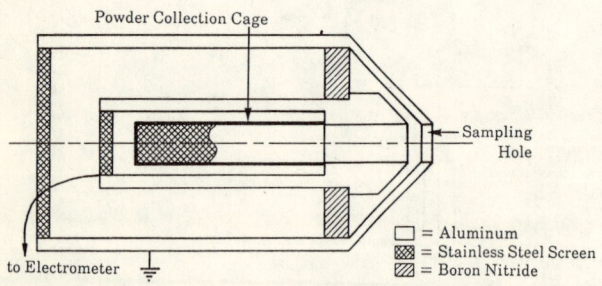

Fig. 3. Schematic of the Faraday cup.

Temperature Field

Typical gas temperature profiles measured by an array of thermocouples and the IR camera are shown in Figure 4. This figure shows that the temperature environment in which the Faraday cup resides is 500 to 1500 [K]. Typical IR camera images observed on the Faraday cup surface with and without powder injection are shown in Figure 5. This figure shows an abnormally high temperature spot on the surface of the Faraday cup. This high temperature spot may be caused by radiative heating from the plasma jet. Although it is difficult to determine the absolute temperature of the plasma with the IR camera, Figure 5 shows that the downstream gas temperature becomes even higher when powder is injected into the plasma. Using thermocouples in conjunction with the IR camera, the results indicate that the powder temperature is generally higher than the downstream gas temperature. Since the powder velocity is in the range of 20 to 60 [m/s] (Hayashi at al, 1985), it is conjectured that the gas and powder are not in local thermodynamic equilibrium. This may lead to the observed slower gas temperature decay when powder is injected into the system.

Particle Charging Characteristics

When the powder was injected into the gas stream without igniting the plasma torch, most of the powder was ejected radially at the exit of the torch. Typical waveforms of the powder fraction measured by the capacitance transducer, and the charge collected by the Faraday cup are shown in Figure 6. Since the powder flowrate and powder fraction does not fluctuate significantly, the charge to mass ratio of the powder can be obtained by weighing the collected powder. Faraday cup measurements show that the powder feeders and feed lines have a tendency to impart a negative charge to the powder as illustrated in Figure 7. This figure shows the time averaged charge-to-mass ratio as a function of powder flow rate for the two different types of powder feeders is always negative and less sensitive, within a factor of four, to both the powder and gas flow rates than has been observed by other investigators (Cole et al. 1969, Masuda et al. 1976).

When powder was injected into the plasma, the powder was ejected almost completely axially, and carried far downstream by the plasma gas. Typical waveforms of the powder fraction measured by the capacitance transducer, and the charge collected by the Faraday cup with and without powder flow under these conditions are shown in Figure 8. Figure 8a shows that the Faraday cup tends to be electrified to a negative bias when only a plasma without powder impinges on the cup. This phenomena is similar to that seen in an electrostatic probe in a plasma (Langmuir and Mott-Smith 1926), or spacecrafts charging in a space plasma environment (Knot 1972). In this case, the Faraday cup, and hence also the powder sampling hole is negatively biased. Thus, negative ion flow into the powder collection cage is limited by the sheath generated in front of the sampling hole (Chang et al. 1976, Chang and Laframboise 1977). Figure 8b shows that the negatively charged powder supplied by the powder feeders changes to positively charged powder after passing through the plasma. The charge-to-mass ratio of the powder in the plasma is shown as a function of powder flow rate in Figure 9 for silica powder in an Argon gas flow. The mechanism to explain the observed conversion from negative to positively charged powder can be found in the theory presented by Chang et al. (1977, 1978). This work indicates that the equilibrium charge on the powder particle surface is controlled by the ratio of the diffusion coefficients of the positive and negative species in the plasma. If the diffusion coefficient of the positive species is larger than that of the negative one, the particle surface will become negatively charged by bipolar ion or plasma environments. In this case however, an Argon plasma is present, which means that the powder surface will become positively charged (as shown in Figure 9), since the electron diffusion coefficient is much larger than the positive ions found in the plasma.

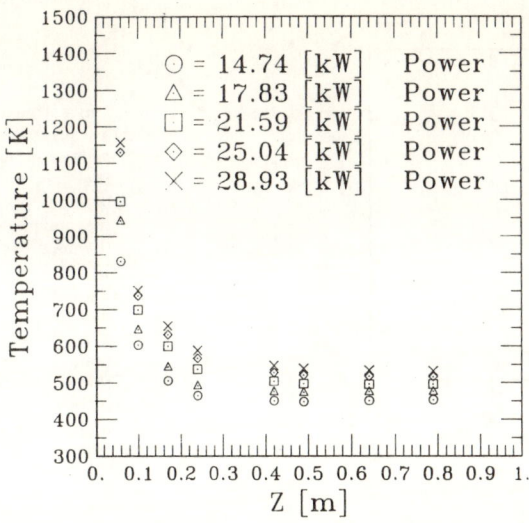

Fig. 4. Downstream temperature profiles at various torch power levels. (3.7 [g/sec] plasma gas flow, no powder injection)

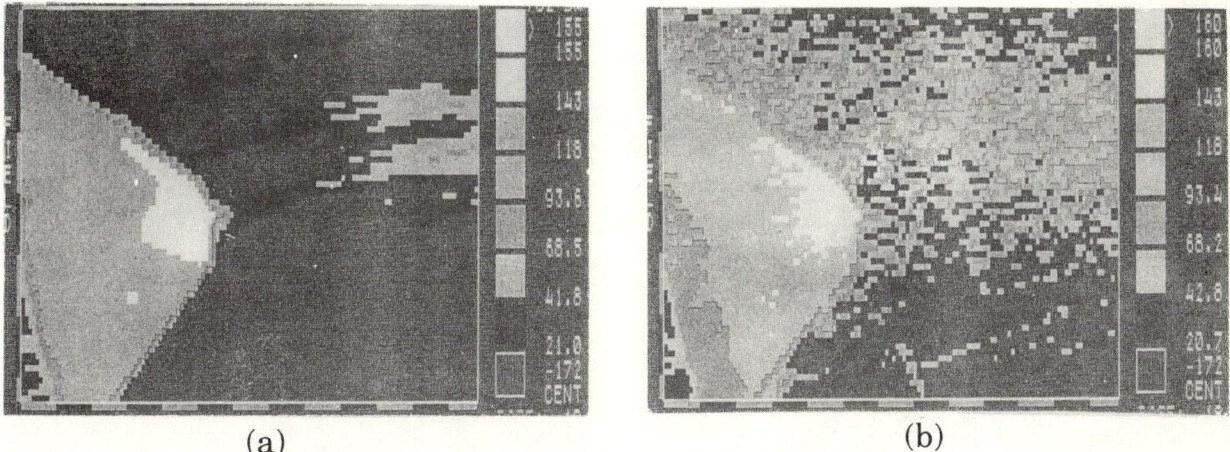

Fig. 5. Downstream IR camera images with the Faraday cup in place (a) without powder injection (b) with powder injection

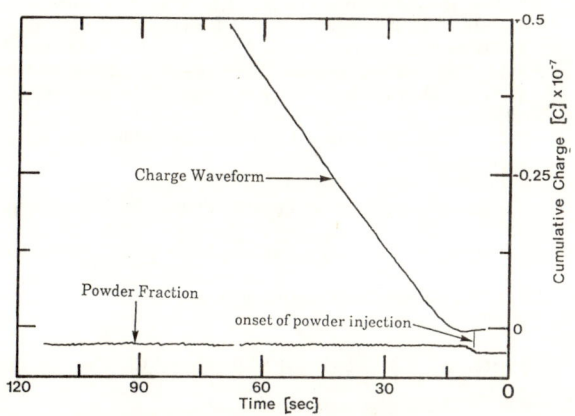

Fig. 6. Typical charge and powder fraction waveforms.

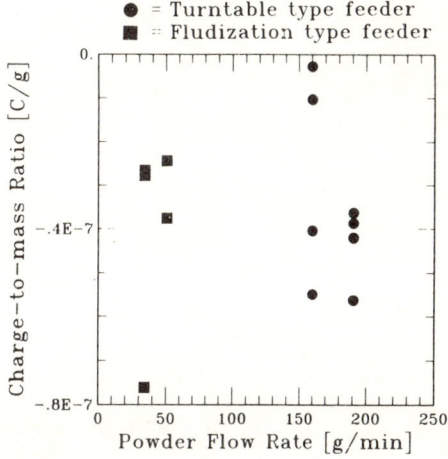

Fig. 7. Charge-to-mass ratio as a function of powder flowrate.

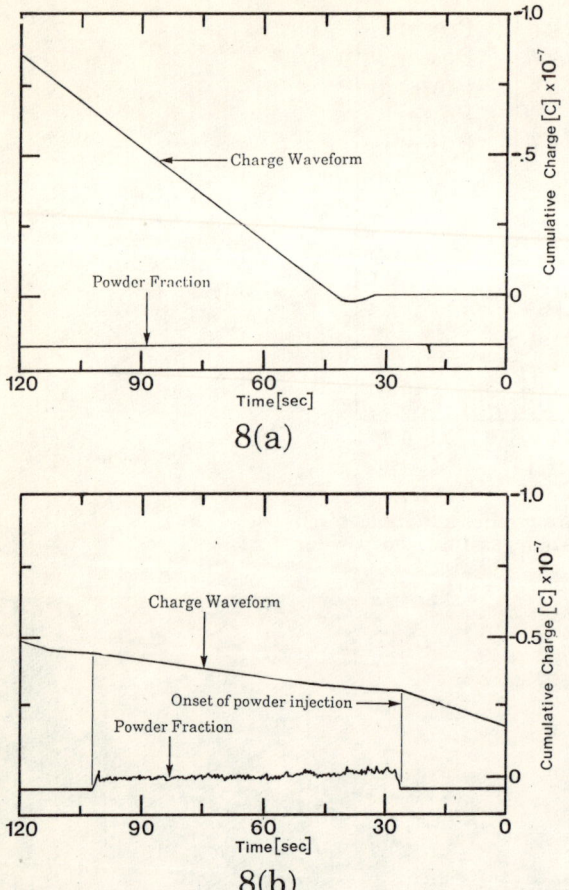

Fig. 8. Typical charge waveforms detected by the Faraday cup (a) with plasma gas only, and (b) with plasma gas and injected powder impinging on the cup.

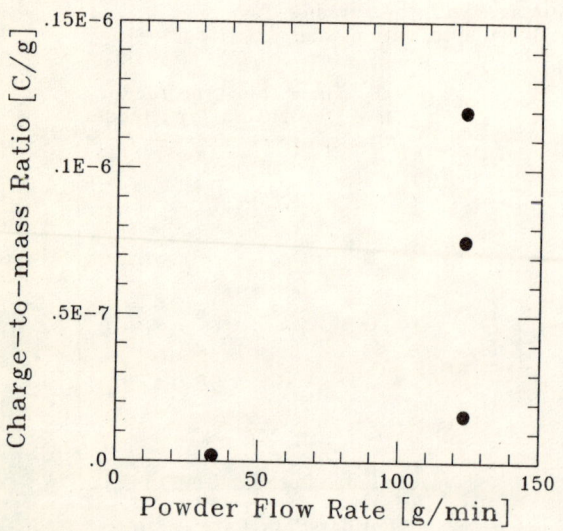

Fig. 9. Charge-to-mass ratio of the powder after injection into the plasma.

Concluding Remarks

Primary investigations have been conducted to study the charge-to-mass ratio in a thermal plasma/powder flow using a high temperature Faraday cup. The results show that:

(1) the originally negatively charged silica powder becomes positively charged after passing through a thermal plasma;
(2) the magnitude of the charge-to-mass ratio for silica is 10^{-8} [C/g];
(3) the Faraday cup is significantly electrified by thermal plasmas;
(4) a boron nitride insulator can be used for gas temperatures up to 1000 [K] for current applications;
(5) the downstream gas temperature rises on powder injection, the powder temperature is significantly higher than the gas temperature.

Acknowledgements

The authors wish to express their appreciation to G.A Irons, A.A. Berezin, R. Cilic, W.K. Lu and M. Sugaya for valuable comments and discussions. This work was supported partly by Natural Sciences and Engineering Research Council of Canada.

References

1. K. Akashi, S. Hattori and O. Matsumoto, ed. "Optical and Plasma Processing", Nikkan Kogyo Shinbun Press, Tokyo, 1986.

2. J.S. Chang, K. Kodera and T. Ogawa, "Computer Simulation of Electrostatic Charging of Aerosol Particles by Bipolar Ions. Conf. Record IEEE IAS 1978 meeting, pp. 38-43, 1978 (also Proc. Jpn. Atm. Elect. Soc., 20, 22-31 (1978)).

3. J.S. Chang, N. Hayashi, T.G. Beuthe, I. Ishii, A.A. Berezin, M. Sugaya, G.A. Irons, W.K. Lu, R.M. Cilic and F.Y. Chu, "Development of Plasma-Coal-Steam Injection Technology", Proc. 37th Canadian Chem. Eng. Conf., pp. 422-424 (1987).

4. J.S. Chang and J.G. Laframboise, Int. J. Mass Spect. & Ion Phys, 24, 225-235 (1977).

5. J.S. Chang, S. Matsumura and S.L. Chen, J.Phys. E.: Sci. Instr., 9, 894-895 (1976).

6. B.N. Cole, M.R. Baum and F.R. Mobbs, Proc. I. Mech. Eng., vol. 184, pt. 3C (1969).

7. N. Hayashi, J.S. Chang, T.A. Myint, W.K. Lu, G.A. Irons, A.A. Berezin, F.Y. Chu and L. Mannik, "Effect of Powder Loadings on Downstream of 80 kW DC Plasma Torches", IEE Japan Study Conf. Plasma and Gas Discharge, Vol. 1 EP-85-16/EP-85-92, pp. 93-102 (1985).

8. G.A. Irons and J.S Chang, Int. J. Multiphase Flow, Vol. 8, No. 3, pp. 289-297 (1983).

9. K. Knott, Planet Space Sci. 20, 1137-1146 (1972).

10. I. Langmuir and H. Mott-Smith, Phys. Rev. 28, 727-63 (1926).

11. J.G. Laframboise and J.S. Chang, J. Aerosol. Sci., 8, 331-338 (1977).

12. H. Masuda, T. Komatsu and K. Iinoya, A.I.Ch.E. J., vol. 22, 558 (1976).

RESISTIVITY MEASUREMENTS OF DIELECTRIC FILMS IN ION-RICH ATMOSPHERE

Tetsuji Oda and Tadashi Takahashi

Department of Electrical Engineering, the Faculty of Engineering, the University of Tokyo
3-1 Hongo-7chome, Bunkyo-ku, Tokyo 113, Japan

Abstract

To understand the surface and bulk charge behaviour of dielectric materials, an exact resistivity value is observed for a long time range in ion-rich and normal conditions by using the personal computer data processing. The former condition (ion-rich) is generated by the needle corona and the surface potential of the sample film is controlled by the mesh grid electrode which is applied the voltage, Vg. The current through the film is monitored by the electrometer and is averaged by the computer through an AD converter and interfaces. The resistivity of teflon is found to be very large, that is, more than 10^{18} ohm cm although the existing of ions. It was found that the evaporated electrodes on films are very effective to reduce the local discharge between the film and the electrode.

Introductions

The exact electrical resistivity measurement of dielectric materials is the very important technique to understand the surface charge behaviour on them, but is also very difficult because of a very small conductive current, such as pA range even though the applied voltage is in order of kV. Especially, the resistivity of powder is very difficult to be determined, that is, it depends on the circumstances, such as, pressure, humidity, temperature and so on. Existing of the ions on and in the powder influences greatly the resistivity of the powder[1]. When spatial ions enter into the powder, the local electrostatic field becomes strong and small partial discharge easily occurs at that part. Preliminary experiments to measure the thin film resistivity by the injection of ion carriers suggesting the very large resistivity compared with the conventional resistivity measurement[2]. On the other hand, the smaller resistivity in the ion-rich condition than in the normal condition was reported by other group[3]. The authors have newly developed the resistivity measuring system by using the personal computer to increase the SN ratio by the digital averaging function of the computer and the long time data stability etc. This data processing method is also applied to the conventional resistivity measuring tool. The comparing results of those measurements of resistivity and the effect of contactness of the electrode to the sample film are reported.

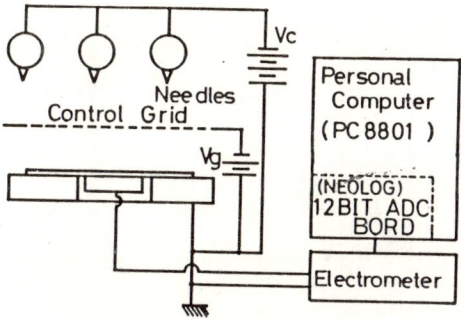

Fig.1 Experimental apparatus to measure the conduction current of the dielectric film under the ion-rich condition with a personal computer system.

Experimental

Resistivity Measuring Apparatus in Ion-rich Condition

A newly developed experimental setup to measure the film resistivity under the spatial ion-rich condition is shown in Fig.1. Fifteen needles (5 X 3) are applied by high voltage (Vc) of + or - 25 kV to supply corona-ions which are located 12 cm above the sample. Between the sample and needles, a mesh-like flat grid electrode is inserted to control the surface potential of the test film. When the grid voltage, Vg, is not so large such as + or -3. kV, the surface potential is mostly as same as the grid voltage. The back electrode of the film has the probe electrode whose area is 39.5 cm^2 surrounded by the grounded guard electrode. The bulk current just passed the film vertically is in order of pA and is amplified by the feedback-type electrometer of Keithley 616 to the normal impedance voltage signal. As the signal is influenced by the corona noise, that is digitalized by the commercial AD converter board and is transferred into the computer. The signal is digitally averaged to reduce the noise level. The practical measuring limit is about 10^{-13} A in order. As the applied voltage to the sample is mostly as same as the grid voltage, Vg, the resistivity of the dielectric material is easily calculated from the thickness of the film while the area is known already. In this system, applying the high voltage to the thin film is easy without large troubles for the maximum current is also limited by the very high output impedance of the voltage source. As the current is mostly limited by the ion-supplying capacity, the electric spark does not occur and the discharge damage is very small to realize the high voltage application to the sample film. At the same time, spatial-ion effect on the conduction of the film should be also added. Therefore, the standard resistivity measurement is also carried out by using the commercial high resistance meter (YHP4329A) which has the voltage supply of 1000V and the standard sheet resistance measuring adapter (YHP16008A) connected with analog recorder or the digital computer system. In this case, the stability for a long time measurement is also checked compared with the former one.

Samples

Experimental samples are PTFE, FEP and PFA teflons (Toyoflon formed by Toray) whose thickness is from 25 μm to 125 μm. Some samples are metalized by vacuum the evaporation of aluminium on one side or on both sides, to reduce the intermittent discharge and so on. The specific resistivity is more than 10^{17} ohm m in every case.

Experimental Results

Ion Currents versus Time

As the ion source for the resistivity measurement, only the negative corona is used in this experiment. The experimental procedure is as followings:1.The corona excitation voltage Vc is applied to the 15 needles where the typical value is -25 kV and the grid voltage Vg is zero. 2.The grid voltage is increased

-285-

to be the pre-determined value gradually. Just after the application of Vg, a very large current is observed, for that is a charging current of the capacitor. One example of the current change with time is in Fig.2 where the sample film is PTFE teflon of 100 um thickness without any metalized electrodes. The grid voltage, which is also assumed to be the surface potential of the film, is -250 V, that means the average electrostatic field strength of 25 kV/cm which is very small compared with the normal breakdown field of the teflon, MV/cm. In this case, the capacitance of the film is about 700 pF or a little bit smaller. If the total impedance of the voltage source and electrometer is assumed to be 1.3 gigaohm, the time constant of the system is about 1 second. The total charge of the capacitor may be 175 nC at Vg of 250 V. The initial peak current integration of about 10 or 20 seconds agrees the this charge in order. However, after the 200 s, still a large current of 10 pA is seen in Fig. 2. Those may be due to the dielectric relaxation effect, ionic charge motion including space-charge redistribution and so on. That relaxation time is surely more than 3600 sec. After 6 hours, the current still decreases a little bit. At sometimes, the noisy-like current can be observed in the Fig. which may be due to the internal local discharge in the film. The average current is very small and surely about 0.25 pA (after 10000 sec) which means the resistance of about 1×10^{15} ohm. In this case, the resistivity may be 4×10^{18} ohm cm. Details will be discussed later.

Figure 3 is another example of the current measurement through the film in ion-rich condition where the grid voltage, Vg, is -1 kV and the average current is surely large (more than 1 pA). In this case, some pin-holes exist and the small local discharge (back corona) occurred. In normal case, the average current is smaller than that in this case. The spike-like burst current is also large and frequent compared with the former. The origin of the pin-hole is not yet understood. The native pin-hole and the other type of pin-hole caused by the breakdown of the film are assumed to be the source of that spike current even though after 1.5 hours. This local back corona on the film is observed by the image intensifier at sometimes. The steady-state glow discharge and intermittent discharge are seen by chance. The very strong glow is photographed by the image intensifier which is shown in Fig. 4. When the grid voltage is small, such as 1 kV and so on, that glow is very weak and not so easy to be observed in general. When the glow point was identified, no pin-hole can be seen by eyes after the discharge. As the most extreme example, the film has the artificial pin-hole made by the pin in diameter of about 0.1 mm. The test results are recorded such as in Figs. 5 (a) and (b). The existing of the pin-hole surely causes very high current (more than 10 times large) compared with the normal film shown as the original in Fig.5(a). The current in Fig.5(a) is rather steady or the time constant for such case is small. Only after 15 min. application of the grid voltage, Vg, the back-corona current becomes constant.

The Effects of Electrode

Some spike-like current in Fig. 2 is also assumed to be due to the discharge through the microgap between the back surface of the film and the back ground electrode (probe in this case). When the film surface is charged, some ions (in this case, conducted ions and so on) is accumulated on the back side surface of the film and the electric field between the probe and the back surface increases to ignite the discharge through the microgap. To identify that effect, the back side of the film is metalized by aluminium just as the same configuration as the probe area and the guard electrode. Figure 6 is the example of a such sample. The spike-like current is surely smaller than that in Fig.2 and the steady-state current noise (width of the current) is also very small. In that case, the decrease of the current is still observed even after 10000 sec.

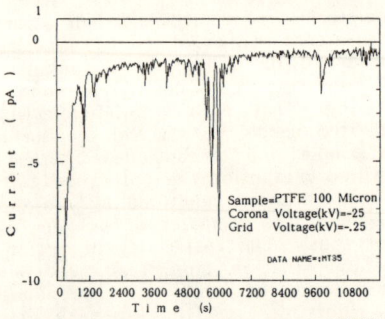

Fig.2 Time dependence of the current of PTFE at Vg=-250V.

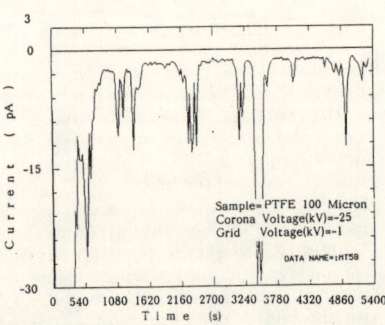

Fig.3 Time dependence of the current of PTFE at Vg=-1kV where the local discharge is assumed to occur.

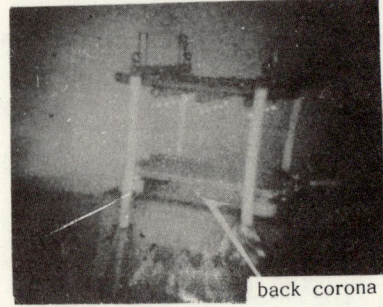

Fig.4 A photograph of back corona on the film by the Image Intensifier.

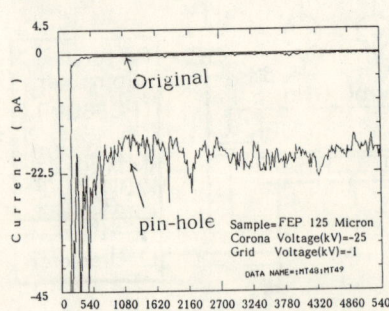

Fig.5 (a) Currents of FEP with and without pin-hole at Vg=-1kV.

Resistivity in Ion-Rich Conditions

The resistance and the resistivity is easily calculated from the current data such as Fig.2 and so on. As already mentioned, resistance of 100 μm PTFE is easily obtained by the equation: the surface voltage (= Vg) /steady-state current. In Fig.2, the resistance, R, after 10000 sec is $-250V/-0.25pA = 10^{15}$ ohm as already shown the same result. The resistivity is calculated from R*S/t where S is the probe area and t is the thickness of the film. In that case, the resistivity is $10^{15}*40/10^{-2} = 4 \times 10^{18}$ ohm cm which is very close to the specification value by the manufacturer. The upper mentioned equation is very simple and resistance or resistivity for each current value shown in such as Figs.2,3,5 and 6 is easily displayed in the similar figure Such examples are listed in Figs. 7 (a) - (g). All samples were metalized by aluminium on their back side to reduce the backside discharge to stabilize the resistivity value and so on. In Figs., two line are seen where the small resistivity value is directly calculated value from the current. In general, such long measurement (at some time, 6 hours) causes the drift error of the current measuring system (not only the electrometer but also including the stability of probe, cable temperature and so on) is very large. The large resistivity values with large noise (vibration) shown in Figs. are calculated from those drift-calibrated current value. The large vibrations are only due to the real small current value, that is, the same fluctuation value causes relatively large change when the current is small. In every case, the stabilized resistivity is surely more than 10^{18} ohm cm in the specifications and those values have the tendency of increasing even after 6 hours application of the voltage. The resistivity for the thinner film is larger than that of the thicker film in general, which is as usual matter.

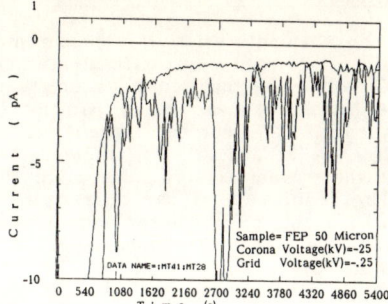

Fig.5 (b)
Currents of FEP with and without pin-hole at Vg=-250V.

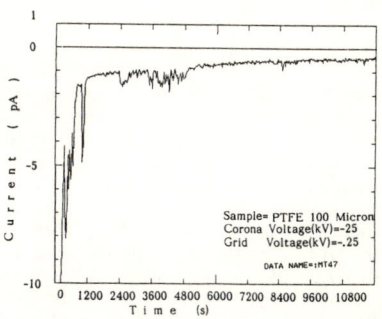

Fig.6 The currents where the film is coated by Al.

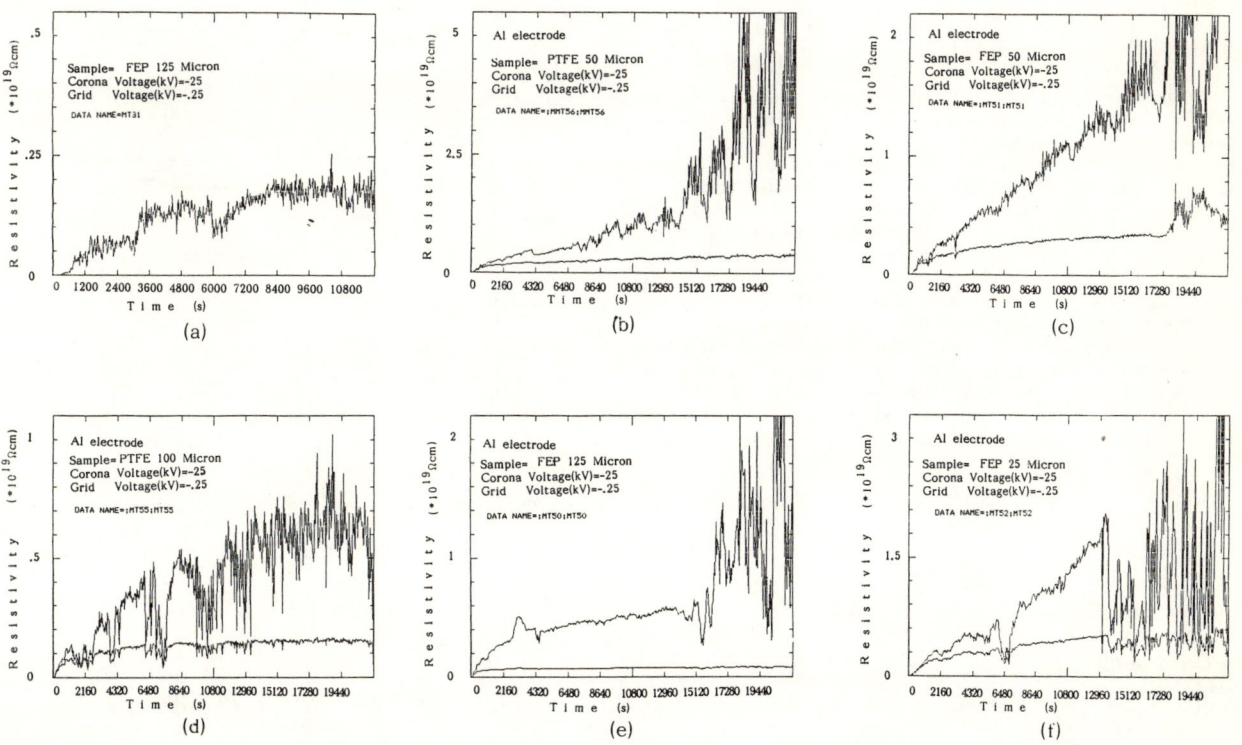

Fig.7 Resistivity of the film in ion-rich conditions.

Resistivity Measured by Conventional Method

In the conventional resistivity measurement, the sample film is sandwiched by two flat plate electrodes and the DC voltage is applied between their electrodes. The current is measured by parametric amplifier in this case. The current signals are recorded shown in Figs. 8 (a) and (b). In this case, relaxation time of the current is very large more than 1 or 2 hours. The resolution of the current is a little bit smaller than the electrometer, but the stability is very well. The effect of the metalization is very large shown in both Figs. The current value is very small and the fluctuation range of the current is also very stable in both metalized films. In that case, the metalization is made on both surfaces (front and back) opposite to the former sample for ion-rich condition. In every case, the resistivity is always more than 10^{19} ohm cm.

Discussions

Test results for many different samples are listed in Table 1 including the informations for differences of metalization, drift cancel, measuring method and so on. By the conventional method, every sample have high resistivity of more than 10^{19} ohm cm and the resistivity of the thinner sample is larger than that of the thicker one. Compared with that, the resistivity under the ion-rich condition is a little bit smaller than that measured by the conventional method. The thickness dependence of the resistivity is as same as the conventional data. The calibrated resistivity value is mostly the same order of the conventional results indicating that the ion-rich condition does not so affect the resistivity when the careful measurement is carried out. When the sample is porous, a very small electric field causes the back-ionization under the ion-rich condition and only a very weak field can be used compared with the conventional method[4].

Conclusions

The resistivity of the thin film is measured by two methods, that is, in the ion-rich condition and the conventional mode. Every teflon has very large resistivity of more than 10^{19} ohm cm after the relaxation time of more than one hour. Before the back corona, the resistivity under the ion-rich condition is also not so small as that by the conventional method if the measurement is carefully done. However, surely the back corona easily occurs in the ion-rich condition. At sometimes, a new method can measure the resistivity under such discharging mode.

The contactness of electrode is very effective to reduce the spurious current, that is, the resistivity of the sample, when that surface is metalized without any air gap, is very high and stable to be measured indicating that the microdischarge between the sample surface and the electrode is very effective to the resistivity measurement.

References

1) S.Masuda and Y.Nonogaki:"Bi-Ionized Structure of Back Discharge Field in an Electrostatic Precipitator", Rec.IEEE/IAS Annual Meeting pp.1111-1119(1981)
2) T.Oda:Report on Committee of IEJ (1984 Jan)
3) M.Ieda et al:"Electrical Condition and Chemical Structure of Insulating Polymers",IEEE Tran.Electr. Insul. EI-21, pp301-306(1986)
4) T.Oda and J.Ochiai:"Charging Characteristics of a Non-Woven Sheet Air Filter" Proc. ISE6 (1988)

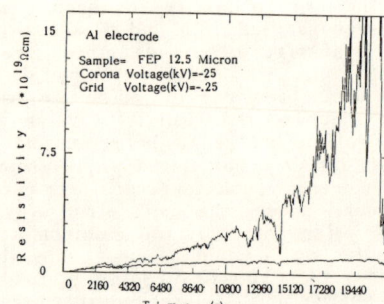

Fig.7 (g) Resistivity of the film in ion-rich conditions.

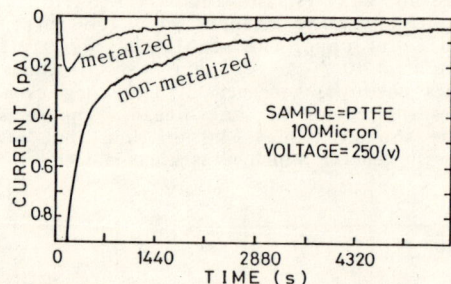

Fig.8(a) The current measured by the conventional method.

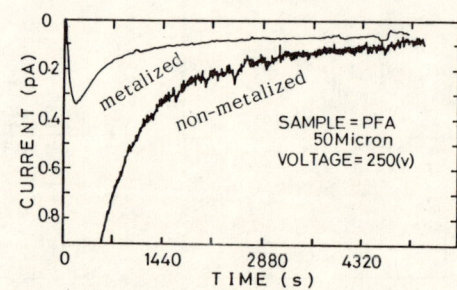

Fig.8(b) The current measured by the conventional method.

Table 1 Resistivities of various samples in two types measured by new and conventional methods.

Samples (thickness in parenthesis:μm)	by the high resistance meter (19 C, 56%) Va=250V, X 10^{19} ohm cm		by the new method(ion-rich condition)(19 C,56%) Vg=-250V, Vc = -25 kV, X 10^{19} ohm cm	
	without metalized(90min)	with metalized(90min)	without metalized(200min)	with metalized(360min)
PTFE (100)	1.225	2.72	0.2	0.15 (0.7)*
PTFE (50)	2.45	3.26	0.3	0.4 (4.0)
FEP (125)	1.12	1.12	0.15	0.1 (0.8)
FEP (50)	1.78	2.8	0.55	0.5 (2.0)
FEP (25)	2.45	3.92	0.65	0.6 (4.0)
FEP (12.5)	14.5		2.6	1.0 (15.0)
PFA (50)	1.96	3.06	0.6	

* Values in parenthesis are exact resistivities calibrated by the offset currents and so on.

ELECTROSTATIC MEASUREMENTS BY MEANS OF MICROWAVES

Chen Guorui   Guo Chenjiang

Dept. of Electronic Engineering
Northwestern Polytechnical University   Xian   P.R.China

Abstract

Under certain condition, measurements of charged particle media may be transformed into conventional microwave measurement of Q factor and polarization, etc. By doing so, some information related to electrostatic environment under test is obtained. An on line electrostatic measurement system configuration is given. Theoretical analysis and preliminary experiments show that integration between microwave and electrostatics makes it possible to develop a new type instrument for electrostatic measurement.

Introduction

As known, electrostatic measurement is essential and necessary for the study on electrostatic applications, hazards, etc. For electrostatic systems, charge is an important parameter. Others are charge density, electric field intensity and potentials. Besides, some parameters such as resistivity, permittivity and loss tangent are related with the intrinsic characteristics of dielectrics. In order to measure those parameters, scientists have been making great efforts. Various methods and instruments of electrostatic measurements have been developed [1]. For the time being, most available field meters consist of metallic probes, which are based on electrostatic induction. The original field distribution may be disturbed and some correction measures should be made during measurement. In some sensitive area, the probe may cause discharge as well.

Recent years the interdisciplinary research has greatly promoted the development electrostatic measurements. For example the electric quantities of the charged particles can be determined by laser speed meter [2]. Application of radioactive effect to static eliminator was also reported.

Modern electronics provides good opportunity for research and development of new type static instruments. Radio receiver can serve as a simple, sensitive sensing device at a distance. Experiments show that spectrum of spark dischare is rather wide. It may be up to 20GHz. The operating frequency of the available spark receiver is around 40MHz. It is believed that some information of the electrostatic environment can be obtained through the change of microwave transmission characteristics. As a new member, microwave has joined the club of static measurements.

Theoretical Basis

Usually microwave frequency is ranged from 3×10^8 to 3×10^{10} Hz. Millimeter and even light waves can be treated as microwaves, to which microwave people also show strong interest. The integration between microwave and electrostatics forms a new topic.

For propagation of plane waves in homogeneous and isotropic medium, electric field may be expressed as follows:

$$\bar{E} = \bar{E}_0 e^{-\gamma z} e^{j\omega t} \quad (1a)$$

where
$$\gamma = \alpha + j\beta \quad (1b)$$

α and β are attenuation and phase constant of wave respectively. Magnetic field has similar expression.

For low lossy medium

$$\alpha \approx \frac{\sigma}{2}\left(\frac{\mu}{\varepsilon}\right)^{1/2} \quad (2)$$

$$\beta \approx \omega(\mu\varepsilon)^{1/2}\left[1 + \frac{1}{8}\left(\frac{\sigma}{\omega\varepsilon}\right)^2\right] \quad (3)$$

where σ is conductivity.

Wave propagation within multilayer dielectric seems complicated, however, it can be dealt with equivalent transmission line. For three-layer dielectric (Fig.1),

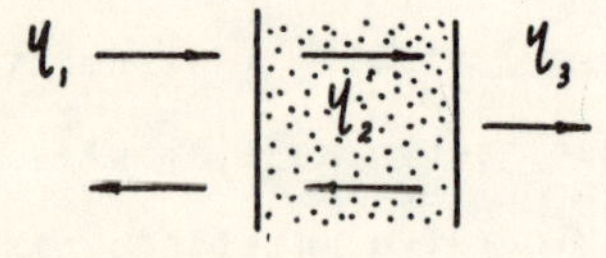

Fig.1 Sketch of three-layer dielectric
the equivalent load impedance of the first layer is:

$$Z_{L_1} = \gamma_2 \frac{\gamma_3 \cosh \gamma l + \gamma_2 \sinh \gamma l}{\gamma_2 \cosh \gamma l + \gamma_3 \sinh \gamma l} \quad (4)$$

where γ is the propagation constant of medium 2.

For lossless medium, expression (4) becomes:

$$Z_{L_1} = \gamma_2 \frac{\gamma_3 \cos \beta_2 l + j\gamma_2 \sin \beta_2 l}{\gamma_2 \cos \beta_2 l + j\gamma_3 \sin \beta_2 l} \quad (5)$$

Reflection coefficient caused by multi-layer dielectric is:

$$\Gamma_1 = (Z_{L_1} - \gamma_1)/(Z_{L_2} + \gamma_1) \quad (6)$$

Medium 1 and medium 3 may be taken as air, and medium 2, ionized gas or dust. Electrostatic information is implied in expressions (1) to (6), where media exert effect on wave magnitude and phase.

Besides, microwave resonant characteristics reacts sensitively to change of medium parameters. As known, Q factor of a cavity is:

$$Q_0 = Q_c/(1 + Q_c \, tg \, \delta) \quad (7)$$

where Qc is the quality factor due to conducting loss, $tg\delta$ is the loss tangent of the dielectric.

Let Qc=10,000, then Qo=10,000 for perfect dielectric. If $tg\delta$ =0.001 for some low lossy dielectric, then Q factor will drop to Qo=909. It indicates that slight change of $tg\delta$ may cause significant change of Qo.

Further investigation shows that the polarization of wave can find application in static measurements, too.

Pockels and Kerr cells are two kinds of electro-optical devices. Due to applied electric field, the permittivity of the device material will present anisotropic nature. The tensor will be a diagonal form so long as coordinate system is properly chosen:

$$[\varepsilon] = \begin{bmatrix} \varepsilon_{11} & & 0 \\ & \varepsilon_{22} & \\ 0 & & \varepsilon_{33} \end{bmatrix} \quad (8)$$

Since permittivity of the electro-optical crystal is a function of applied electric field, corresponding change of phase will take place when electromagnetic waves pass through the device. Here applied field may be electrostatic field itself.

For Kerr cell, if applied field is perpendicular to propagation direction of wave, phase shift is proportional to the square of electric field, i.e:

$$\Delta \beta \propto E^2 \qquad (9)$$

For Pockels cell, phase shift is directly proportional to electric field, i.e:

$$\Delta \beta \propto E \qquad (10)$$

Thus if a linearly polarized wave with E_y and E_x compontents propagates in Z direction, the phase shift will become:

$$\Delta \beta = \beta^{(y)} - \beta^{(x)} = \omega \sqrt{\mu_0}(\sqrt{\varepsilon_{22}} - \sqrt{\varepsilon_{11}}) \qquad (11)$$

Application

Pockels and Kerr devices can be used as modulators, switches and polarization deflector in optical communication(Fig.2). The modulating field is D.C. or A.C. field. The modulated microwave (or optical)signal is sent to a photo-electric detector.

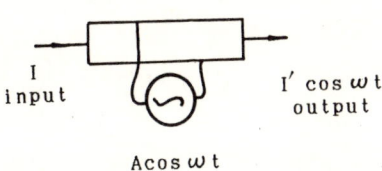

Fig.2 Electro-optical modulator

For electrostatic measurement, it is just a reverse problem. The applied field i.e, the electrostatic field under test, can be determined through phase or polarization deflection measurement. After calibration, the output of the photo cell represents the applied electric field.

The measurement of space charges by means of Pockels device has been reported [3]. The authors also described the piezo-electric application to the electrostatic measurements[4].

Because of the extremely small size of IC device, strong field intensity is produced by a low electric potential, which may cause degration or breakdown of the device. Severe loss up to billions of U.S. dollars is suffered in electronic industry yearly. It is meaningful to investigate and measure the charge dissipation for IC chips. A paper on measuring charge dissipation and surface resistance by using a rotary vane field meter was reported. [5] Based on the properties of Pockels crystal, the authors suggest an on line electrostatic system to measure the charge dissipation (Fig.4).In the system a rotary vane field meter can be replaced by Pockels device, which causes little disturbance to the original field. And r.f source is a laser with low level output. Work on this project is under way.

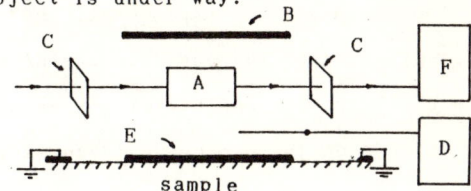

Fig.3 On line electrosatatic measurement
    A - Pockels cell B - electrode
    C - polarizer   D - HT supply
    E - surface electrode
    F - photo-detector

Preliminary experiment of microwave transmission and reflection for charged powder is made at 9370MHz (Fig.4). The sample is SbS powder. At beginning let applied voltage be zero, sample is kept in a box and remained in natural state. The reading of the meter M2 is 0.6 w, which represents the reflection power due to sample powder. It can be considered as inherent reflection. Then D.C. voltage is applied and SbS powder falls down along a slip trough into the box. Different electrification corresponds to different applied voltage, and corresponding readings of meter M2 as follows:

Applied Voltage (kv)	0	0.8	2.0	-1.0	-2.0
Meter M2 Readings (w)	40	24	8	2.4	30

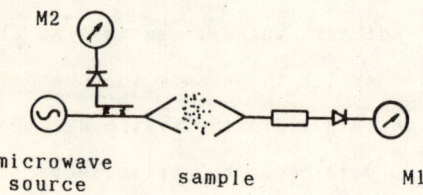

Fig.4 Waveguide method

Test indicates that different charged state causes quite different reflection. It is also observed by experiment that the sliding speed of the powder, charging history of the powder and humidity have obvious effects on the microwave transmission.

Conclusion

Theoretical analysis and preliminary experiments show that microwave link with electrostatics creates an opportunity to develop some new type static instruments. After proper calibration, the electric parameters such as charges, potential, electric field, etc, can be determined. Further research and development are needed to give a practical mathematic-physical model.

Acknowledge

The authors are grateful to Mr.Ke Lin Senior Engineer of Qing Hua Electric Apparatus Factory for his help.

Reference

[1]. P.E.Secker and J.N. Chubb, Instrumentation for Electrostatic Measurements, J. of Electrostatics 16 (1984) PP1-19

[2]. J.N.Chubb, W.D.Bamford and J.B.Higham Experimental Studies of Airborne Particle Behaviour in Corona Discharge Field, IEE Colloquium on Electrostatic Precipatation, 19 Feb, 1985

[3]. Kunihiko Hidaka and Teruya Kouno, A method for Measuring Electric Field in Space Charge by means of Pockels. Device, J. of Electrostatics 11(1982) PP195-211

[4]. Guo Chenjiang, Chen Guorui, Application of Piezoelectric Devices to Electrostatics Measurement, Proc. of 1987 Annual Conf. on Electrostatics, China PP321-326

[5]. D.M.Taylor, D.R.Owen, An Instrument for Measuring Static Dissipation from Materials, J. Of Electrostatics 19 (1987)PP53-64

IMPROVING THE RESOLUTION OF SURFACE CHARGE DENSITY MEASUREMENT

Zhang Yewen and Liu Yaonan

Department of Electrical Engineering,
Xi'an Jiaotong University, Xi'an, P.R.China

Abstract

In this paper, the concept of "probe characteristic function" which is used in case of two-dimensional plot is described, and an experimental method for getting the probe characteristic function is given. To improve the resolution for measurement of surface charge density distribution, a new method is suggested. It is based on eliminating the edge effect of the probe from the measured surface potential distribution by using the probe characteristic function. The results shows that this technique is useful for improving the resolution of surface charge density measurements.

I. Introduction

In many cases, it is necessary to measure the surface charge density and its distribution. The induction method using a capacitive probe that is currently used was suggested by the end of 60s[1,2]. In 70s, some authors studied and improved this method[3-5]. Recently, some scholars still advanced the research on this method[6-11]. In many situations, it is necessary to make the measurement of surface charge density distribution with high resolution[12-14]. For example, with a probe diameter of 50μm[11], a resolution of about 100μm is obtained.

The smaller the diameter of the probe, the more serious its edge effect, and the resolution is decreased. For probes with larger diameters, the edge effect can be neglected[15]. In this paper, we intend to increase the resolution of a probe by eliminating its edge effect.

II. Theoretical Analysis

1. The edge effect of electrodes and the probe characteristic function

On the basis of work by Haenen[11] and the author in reference [16], no obvious increase of resolution is obtained by decreasing the edge effect of a probe with improving screening of a probe. The main reason may be that the edge effect of an electrode can not be neglected thoroughly by screening. Hence, if we try to increase the resolution of a probe without decreasing the diameter of a probe, the effort must be made in another way.

Now, we define the probe characteristic function with a point charge as follows:

$$f(x,y) = U_s(x,y)/q, \quad (1)$$

where, $U_s(x,y)$ is the reading of electrometer (or surface potential meter), q is the quantity of the point charge, (x,y) is the coordinate of the probe (taking the position of the point charge as the coordinate origin). In SI units system, the unit of $f(x,y)$ is F^{-1}.

2. Method of eliminating the influence of the probe edge effect

According to principle of superposition, when a probe potential $U_s(x,y)$ derived from a surface charge $\sigma(x,y)$ is measured, we obtain

$$U_s(x,y) = \int_{-\infty}^{+\infty}\int_{-\infty}^{+\infty} \sigma(x',y') f(x-x', y-y') dx' dy', \quad (2)$$

where, (x,y) is the location of the probe in Cartesian coordinates, (x',y') is that of the charge.

But, when the $U_s(x,y)$ and $f(x,y)$ are known, it is difficult to solve $\sigma(x,y)$ from the above equation. The discrete solution method may be used to calculate the numerical value of $\sigma(x,y)$.

To make $U_s(x,y)$ and $\sigma(x,y)$ discrete, the x-axis and y-axis are subdivided into (n-1) parts of width Δx, and (m-1) parts of width Δy respectively. Then we have m×n measured values $U_s(x_i,y_j)$ and m×n values of surface charge $\sigma(x_k,y_l)$. Eq. 2 can be expressed in form of an algebraic sum as follows,

$$U_s(x_i,y_j) = \sum_{l=1}^{m}\sum_{k=1}^{n}[\sigma(x_k,y_l)f(x_i-x_k,y_j-y_l)]\Delta x \Delta y \quad (3)$$

$$\begin{cases} i=1,\ldots,n \\ j=1,\ldots,m \end{cases}$$

But, in contract with the case of a one-dimensional plot, Eq. 3 can't be expressed easily in the form of matrix. It is necessary to make a transformation of Eq. 3.

Because $U_s(x_i,y_j)$ has m×n values and $\sigma(x_k,y_l)$ also have the same values, they can be written in a matrix form of order of (m×n)×1. We determine that the subscript of x is varied prior to the replacement of the subscript of y. Thus,

$$[U_s]^T = [U_s(x_1,y_1), U_s(x_2,y_1), \ldots, U_s(x_n,y_1),$$
$$U_s(x_1,y_2), U_s(x_2,y_2), \ldots, U_s(x_n,y_2),$$
$$\ldots\ldots,$$
$$U_s(x_1,y_m), U_s(x_2,y_m), \ldots, U_s(x_n,y_m)] \quad (4)$$

$$[\sigma]^T = [\sigma(x_1,y_1), \sigma(x_2,y_1), \ldots, \sigma(x_n,y_1),$$
$$\sigma(x_1,y_2), \sigma(x_2,y_2), \ldots, \sigma(x_n,y_2),$$
$$\ldots\ldots,$$
$$\sigma(x_1,y_m), \sigma(x_2,y_m), \ldots, \sigma(x_n,y_m)] \quad (5)$$

Hence, Eq. 3 can be expressed in matrix form as

$$U_s = [F] \cdot [\sigma]. \quad (6)$$

The matrix [F] is a square matrix of which the order is (m×n)×(m×n). The elements of matrix [F] are follow,

$$\begin{cases} F_{rs} = f(x_i-x_k, y_j-y_l)\Delta x \Delta y \\ \quad = f[(i-k)\Delta x, (j-l)\Delta y]\Delta x \Delta y \\ i = r - n\,\text{Fix}[(r-1)/n] \\ j = \text{Fix}[(r-1)/n]+1 \\ k = s - n\,\text{Fix}[(s-1)/n] \\ l = \text{Fix}[(s-1)/n]+i \end{cases} \quad (7)$$

Knowing all elements of matrices [F] and [U_s], we can acquire the solution of the matrix [σ], i.e. the discrete solution of the surface charge density distribution σ(x,y), by solving Eq. 6 as linear equations. As a result of σ(x,y) in Eq. 2 being the real surface charge density distribution without the influence of edge effect of the probe, [σ] is identical to the discrete value with the same configuration.

3. The measurement of the probe characteristic function f(x,y)

-293-

In this paper the diameter of the probe is 1.0mm. For this kind of probe, the measurement of the probe characteristic function may be done by means of a very small charged film to simulate a point charge. As the size of the charged film is very small compared with that of the probe. We may consider that the charge density on it is uniform and $f(x,y)$ remains almost unchanged in this small region.

When $U_s(x,y)$ is measured with the said charged point, $U_s(x,y)$ depends on both the size of charged point and its charge. Assuming $\sigma(x',y')$ is the density of charge and $(\Delta x) \times (\Delta y)$ is the size of charged point, from Eq.2 and let $\sigma(x',y') \approx \sigma_s$, we have

$$U_s^*(x,y) \approx \sigma_s \Delta x \Delta y f(x,y) . \qquad (8)$$

So, $\quad f(x,y) = U_s^*(x,y)/(\sigma_s \Delta x \Delta y) = C U_s^*(x,y) , \qquad (9)$

where, $\quad C = 1/(\sigma_s \Delta x \Delta y) = \text{constant} .$

Substituting Eq.9 into Eq.2, we have

$$U_s(x,y) = \int_{-\infty}^{+\infty}\int_{-\infty}^{+\infty} [\sigma(x',y') C U_s^*(x-x',y-y')] dx'dy' . \qquad (10)$$

When the charged sample is an infinite plane with uniform charge density σ_s, it can be simulated by an infinite number of charged points spread continuously on an infinite plane. Thus,

$$U_s(x,y) = C\sigma_s \int_{-\infty}^{+\infty}\int_{-\infty}^{+\infty} U_s^*(x-x',y-y') dx'dy' \qquad (11)$$

Now, $U_s^*(x,y)$ is independent of the position of point (x,y), assuming $x=y=0$. The above equation can be written as follows,

$$U_s = C\sigma_s \int_{-\infty}^{+\infty}\int_{-\infty}^{+\infty} U_s^*(x,y) dxdy \qquad (12)$$

On the basis of the theory of measurement of surface charge density, in this case, we have

$$\sigma_s = \varepsilon_0 \varepsilon_r U_s/s \qquad (13)$$

where s and ε_r are the thickness and the relative permittivity of the sample respectively, ε_0 is the permittivity of vacuum, U_s is the surface potential of the sample. Substituting Eq.13 into Eq.12,

$$C = s/[\varepsilon_0 \varepsilon_r \int_{-\infty}^{+\infty}\int_{-\infty}^{+\infty} U_s^*(x,y) dxdy] \qquad (14)$$

Hence, by measuring $U_s^*(x,y)$, the probe characteristic function $f(x,y)$ is expressed as follows,

$$f(x,y) = s U_s^*(x,y)/[\varepsilon_0 \varepsilon_r \int_{-\infty}^{+\infty}\int_{-\infty}^{+\infty} U_s^*(x,y) dxdy] \qquad (15)$$

III. The Experimental Results and Analysis

1. Experimental determination of the probe characteristic function

A PP film of $0.2 \times 0.2 mm^2$ ($\varepsilon_r = 2.2$, $s = 12.0 \mu m$) is charged with the corona charging method. The film is placed on the lower electrode plated with gold in vacuum. Then the measurement of $U_s^*(x,y)$ can be carried out by scanning the device.

To study the influence of the size of charged sample, simulated as point charge on $f(x,y)$, we measured $U_s^*(x,y)$ for some small charged samples of different size and calculated the value of $f(x,y)$ in these conditions. If the standard deviation of the results of $f(x,y)$ is very small, we believe that the size of charged sample used is good enough to simulate a point charge.

As from Eq.15, it is obvious that if the size of the charged sample is too large, $f(x,y)$ will be too large. But if the size of the charged sample is too small, the $U_s^*(x,y)$ will be so small that the reading error on the instrument increases. By our experiments, the proper size of sample is $0.15 \times 0.15 \sim 0.25 \times 0.25 mm^2$. The probe characteristic functon $f(x,y)$ is obtained from the average value of measurement taken in three times as shown in Fig.1. Because of the manufacturing and mounting defect, the probe is not perfactly cylindrically symmetric. So the probe characteristic function is nonsymmetric.

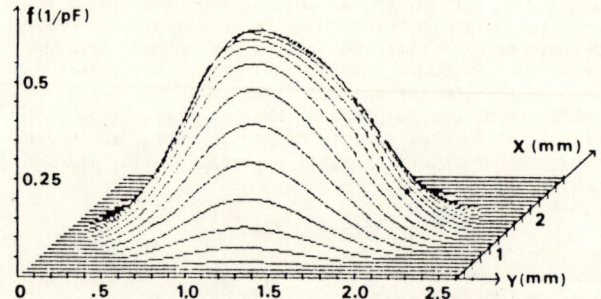

Fig.1 the probe characteristic function $f(x,y)$

2. The experimental method

(1) Evaluation of the probe characteristic function

It is necessary to determine the step length Δx, Δy and its corresponding $f(x_i-x_l, y_j-y_l)$ for solving the $\sigma(x,y)$ with the method of linear equations.

In general taking $\Delta x = \Delta y = h$, it is relatively convenient to make h equal to the diameter of the measuring electrode in the probe. This way, the error of calculation is also relatively small. In our experiments, $\Delta x = \Delta y = h = 1.0 mm$, where d is the diameter of measuring electrode.

Now, how to evaluate the probe characteristic function $f(x_i-x_l, y_j-y_l)$ should be considered carafully. The method recommended in this paper is as follows.

The non-zero domain of $f(x,y)$ can be divided into a number of square domains in which the length of the side is equal to h with the peak point at the center of the square domain. The average value of integral is calculated in the square domain of $x_i-h/2 \leq x \leq x_i+h/2$, $y_j-h/2 \leq y \leq y_j+h/2$, and take this average value as the value of the probe characteristic function at $f(x_i, y_j)$. Thus

$$f(x_i, y_j) = \frac{1}{h^2} \int_{y_j-h/2}^{y_j+h/2} \int_{x_i-h/2}^{x_i+h/2} f(x,y) dxdy \qquad (16)$$

Table 1: the value of $f(x_i, y_j)$ when $h=1.0mm$ (1/pF)

y \ x	-1.0	0.0	+1.0
-1.0	0.001593	0.022451	0.001202
0.0	0.045905	0.364489	0.057083
+1.0	0.009883	0.100322	0.013012

The evaluation method of the probe characteristic function has its clear and definite meaning in physics. It represents the average value of the responses of the probe at (0,0) to the surface charge on the domain $x_i-h/2 \leq x \leq x_i+h/2$, $y_j-h/2 \leq y \leq y_j+h/2$. Hence, the data can be processed to eliminate the edge effect of the probe by using the probe characteristic function. The value of surface charge density for different points will be considered as the average value of surface charge density on the domain with an area of h×h. When $\Delta x=\Delta y=h=1.0mm$, $f(x_i,y_j)$ of the probe shown in Fig.2 has nine non-zero values as shown in Table 1.

(2) The experimental verification of the elimination of edge effect

An experiment was done to verify this method for eliminating the influence of edge effect.

A polmer film (12μm) on which a layer of aluminium film is vaporized in vacuum was kept close the lower electrode of the surface potential meter. The aluminium film is conducted electrically to 100V D.C. source by means of a reed, as shown in Fig.2. Although the surface potential of Al film changes suddenly at edge Al film, the $U_s(x,y)$ measured falls gradually due to the edge effect of the probe, as shown in Fig.3.

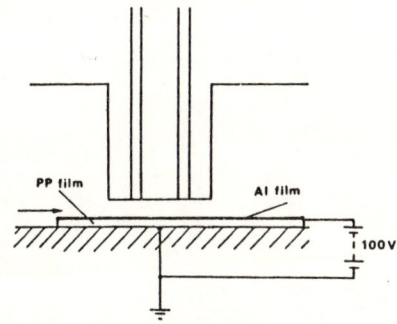

Fig.2 the diagram of experimental set, the arrow means the direction of scanning

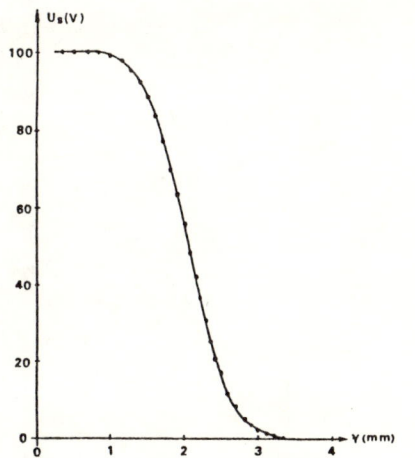

Fig.3 the relationship of $U_s \sim y$ on the center on Al film

In Fig.4, the calculated results $\sigma(x,y_0)$ and the value of $(\varepsilon_0\varepsilon_r/s)U_s(x,y_0)$ are plotted where point of $\sigma(x,y_0)$ represents the average value of surface charge density on the corresponding domain. By studying the data in Fig.4, it shows that the influence of edge effect of the probe on the measurement of the surface charge density is eliminated essentialy.

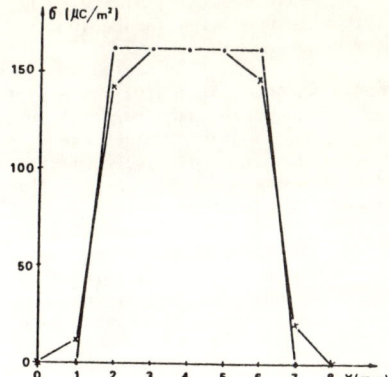

Fig.4 the surface charge density distribution $\sigma(x,y_0)$ on the aluminium film
×---measured directly
•---correct with $f(x,y)$

From the results in Fig.4, the resolution of the probe during the measurement is equal to the length of step, 1.0mm. However, the resolution of the probe before the data processing is approximately equal to 1.4mm[16]. Moreover, if the measured step is decreased once more, the resolution may be improved further.

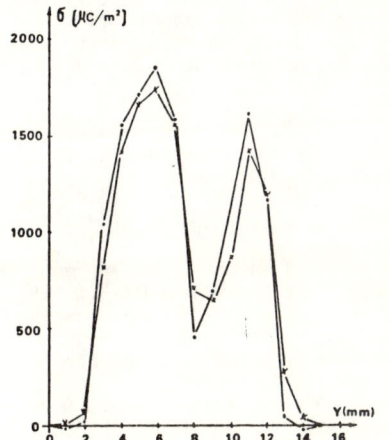

Fig.5 the surface charge density distribution $\sigma(x_0,y)$ on the charged PP film
×---measured directly
•---corrected with $f(x,y)$

3. The experimental results of surface charge density

A PP film (10×10mm², s=12.0μm) that had been charged with corona discharge was kept close to the lower electrode, and then the resolution of $U_s(x,y)$ was measured. Then, the data processing was carried out to eliminate the influence of edge effect. In Fig.5, $\sigma(x_0,y)$ and value calculated directly by $U_s(x_0,y)$ are plotted. Every point of $\sigma(x_0,y)$ represents the average value of surface charge dendity on th domain of 1.0×1.0mm².

IV. Discussion

1. The accuracy of this data processing method for eliminating the influence of edge effect of the probe strongly dopends on the accuracy of the probe characteristic function. If the test condition is promised,

it is better to measure $U_s(x,y)$ of the probe by an instrument with high precision to acquire $f(x,y)$ at higher accuracy.

2. In the above theoretical analysis and experiment, the definition of the probe characteristic function is the probe response to a unit point charge. And this function is related to the relative permittivity and the thickness of the sample. So, the above mentioned probe characteristic function is only good for the sample with $\varepsilon_r=2.2$, $s=12.0\mu m$. For other samples, the probe characteristic function can be obtained conveniently by the function stated multipied by a factor $(2.2/12.0)*(s_1/\varepsilon_{r1})$, where ε_{r1} and s_1 are the relative permittivity and the thickness of another sample respectively.

3. During the measuring procedure of the probe characteristic function $f(x,y)$, the lower electrode must have no oxide layer at all, because a little charge on it may affect the results of measurement[3]. This is the reason why the lower electrode has to be plated with gold.

V. Conclusion

1. The experimental method suggested in this paper to obtain the probe characteristic function for the probe with diameter larger than 1.0mm is feasible.

2. The data processing method used in this paper to eliminate the influence of edge effect has significant effect on the improvement of resolution of the probe.

3. The theorem stated in this paper is not only limited to the inductive probe, but also suited for other similar probe.

VI. Reference

[1] D.K.Davies, J.Sci.Instrum., vol.44, p.521-4(1967)
[2] T.R.Foord, J.Phys.E, vol.2, p.411-3(1969)
[3] G.M.Sessler and J.E.West, Rev.Sci.Instrum., vol.42, p.15-9(1971)
[4] N.Nordhage and G.Baekstrom, J. of Electrostatics, vol.2, p.91-5(1976)
[5] H.T.M.Haenen und L.M.Hosselet, ATM.Arch., v.942-17, p.23-6(1976), (in German)
[6] H.Kraemer und D.Meßner, etz.Archiv, Bd.2, p.43-8(1980), (in German)
[7] R.Gerhard-Multhaupt and W.Petry, J. Phys. E, vol.16, p.418-20(1983)
[8] K.Ohara, Jpn. J. Of Static Electricity, vol.7, p.50-4(1983), (in Japanese)
[9] M.Hennecke, Rev. Sci. Instrum., vol.51, p.803-5(1980)
[10] H.J.Wintle, J.Phys.E, vol.3, p.334-6(1970)
[11] H.T.M.Haenen, J.of Electrostatics, vol.2, p.203-22(1977)
[12] E.A.Baum, T.J.Lewis and R.Toomer, J.Phys.D, vol.20, p.487-97(1977)
[13] E.A.Baum, T.J.Lewis and R.Toomer, J.Phys.D, vol.11, p.963-77(1978)
[14] E.A.Baum, T.J.Lewis, IEE Conf.Pub., No.129, p.79-82(1975)
[15] M.Wolff et al, J.Phys.E, vol.2, p.921-24(1969)
[16] Zhang Yewen, M.Sc.thesis, Xi'an Jiaotong University, 1984(in Chinese)

FIELD CALCULATION IN RING-TYPE ELECTRODE CAPACITANCE TRANSDUCER FOR PARTICLE FRACTION MEASUREMENTS

N. Hayashi, M. Matsuuchi, Y. Yokoi

Department of Electrical and Electronics Engineering, Faculty of Engineering,
The University of Tokushima, Tokushima, 77 Japan.

Abstract

This paper presents numerical analysis of the electric field profiles in the capacitance transducers with the ring-type electrodes used for particle fraction measurements. Two phase gas/solid core flows were assumed in the calculation. Theoretical relationships between the normalized capacitance and solid fraction of powders carried in the sensor pipe were discussed.

Introductions

The capacitance transducer methods have been successfully used to measure the solid fraction of powders conveyed pneumatically in a duct pipe [1,2].

In general, prototype transducers are constructed and used to calibrate capacitance output. Some investigations on the development of analytical expressions to estimate the capacitances as a function of the solid fractions have been conducted[3,4], but they used extremely simplified assumption as to the electric field distributions in the powder flowing region, i.e. uniform field is assumed.

This paper examines the electric field profiles in the sensor pipe of the ring-type electrode capacitance transducer under various solid fraction conditions. Also analyzed are the capacitance characteristics under various geometric conditions of the sensing electrodes.

Calculation Model of Capacitance Transducer

Figure 1 diagrams schematics of the capacitance transducer employed for the calculation. Two ring-type conductors are attached on the outer surface of plastic pipe in which powders are carried as two phase flow. These conductors serves as the main electrodes #1 and #2 and are connected to the capacitance meter. That is, capacitance between the main electrodes are used to determine the solid fraction of powders carried in the sensor pipe. The shielding electrode of ground potential surrounds the whole transducer to avoid electrostatic induction from the ambient electric field. The main electrodes are insulated from the shielding electrode with a insulating tape.

In the diagram of Figure 1, a two phase air/powder core flow is assumed, that is, the powder flows coaxially in the pipe and the air fill the gap between the powder and inner surface of the pipe.

The solid fraction that is defined as the ratios of the volume of the powder to the total volume of the flow region is denoted by the notation "a_p" and is calculated as:

$$a_p = (d_s/d_p)^2 \quad \ldots\ldots\ldots\ldots\ldots\ldots\ldots(1)$$

where, d_s and d_p are the diameter of solid core and the inner diameter of the pipe, respectively.

Electric Field Distributions

Method   Surface charge method was used to calculate both the electric field profiles in the sensor pipe and the induced electric charges on the electrodes. The main and shielding electrodes and the dielectric boundaries were divided into ring-type segments because of the symmetrical geometry of the transducer, and the electric charges on each segment were represented by a electric charge of constant magnitude. The width of each segment was chosen so that the segments have almost the same order of electric charges each other.

Results   The electric field calculations were performed under the following conditions. The specific permittivities of the solid phase, the gas phase, and the transducer pipe and insulating tape were assumed to be 3.0, 1 and 4.6, respectively. The separation between the main electrodes and the shielding electrode that surrounded the main electrodes was chosen to be 1 mm. Positive unit voltage (+1 V) was applied to one electrode (main electrode #2 in Figure 1) and negative unit voltage (-1 V) was applied to another electrode (main electrode #1 in Figure 1).

In Figure 2, described are the equipotential lines in the positive region of the sensor pipe, i. e. in the space under the

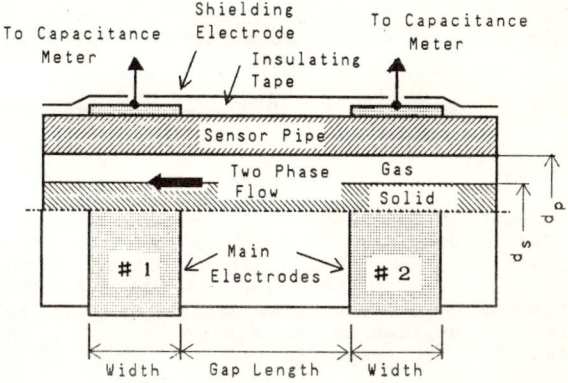

Fig. 1   Schematic Diagram of the Capacitance Transducer.

-297-

positively charged electrode #1, for the solid fractions of α_p=0, 0.5 and 1.0. The equipotential lines in the negative potential area are completely symmetrical to those in the positive potential area.

The lines of force are everywhere perpendicular to the equipotential lines and are tangent to the electric filed, and the magnitude is proportional to the density of lines. Therefore, it is found that the direction of the electric field in the area directly surrounded by the main electrode is nearly perpendicular to the probe axis. As well, it can be seen how the concentration of the equipotential lines, or the enhancement of the field strength, takes place in the gas phase region when the solid phase is exist.

Figure 3 represents the profiles of the axial and radial components of electric field in the sensor pipe at three radial positions.

It can be seen from Figures 2 and 3 that the axial components are predominant in the sensor pipe but the space directly surrounded by the main electrodes (i.e. directly beneath the main electrodes in Figure 2) where the radial components are predominant than the axial components. The electric field is almost parallel to the probe axis and their magnitudes are almost constant in the interelectrode region, i.e. the electric field is uniform. However, the field uniformity is disturbed as being apart from the probe axis.

The field distributions calculated under various geometry of electrode widths and gap lengths indicated that the field patterns in the sensor pipe had almost similar tendency as those in Figure 2 and 3, but the areas the field was uniform depend on these parameters, particularly on the electrode gap length.

Capacitance Characteristics

<u>Method</u>   The charges on the electrode #1 can be expressed in terms of both the potential differences between the conductors and of the direct capacitances between the conductors.

$$Q_1 = C_{10}V_1 + C_{12}(V_1 - V_2) \quad \ldots\ldots\ldots(2)$$

The direct capacitances between the conductors, including the shielding electrode, are indicated as C_{10} and C_{12}. Here, the subscripts 0, 1 and 2 is used to denote the shielding, #1 and #2 conductors, respectively.

If $V_1=0$ and $V_2=1$ are provided, then from (2)

$$Q_1 = -C_{12} \quad \ldots\ldots\ldots\ldots\ldots(3)$$

Equation (3) means that the direct capacitance between the electrode #1 and electrode #2 can be calculated as the negative of the total charges on the electrode #1 providing that the ground potential and the unit potential are applied to the electrode #1 and the electrode #2, respectively.

Surface charge method is well suit for this purpose since the total charges on a given electrode can be easily calculated by summing the charges on each segment which constitutes the electrode concerned.

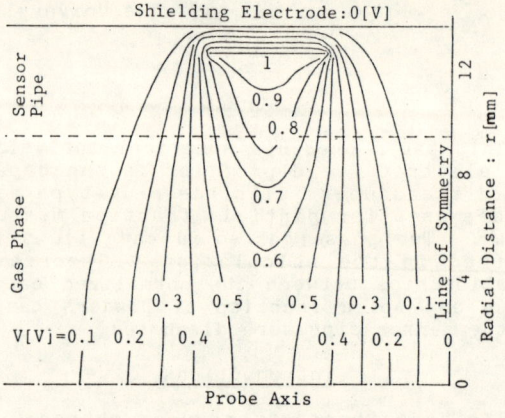

(a) $\alpha_p = 0$

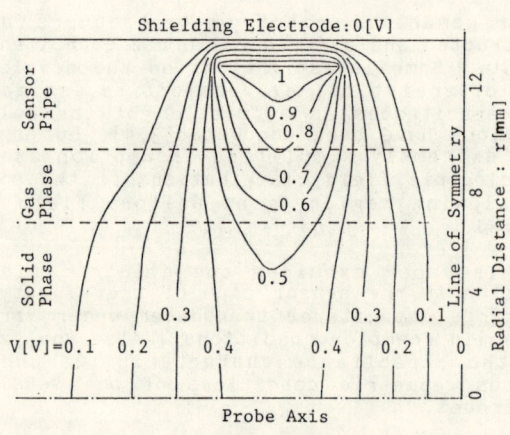

(b) $\alpha_p = 0.5$

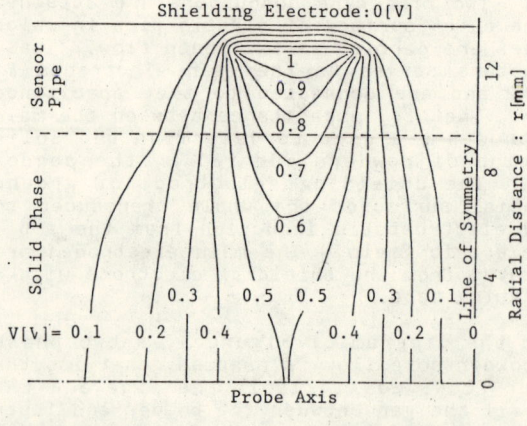

(c) $\alpha_p = 1.0$

Fig. 2   Equipotential Lines.

<u>Results</u> Figures and 5 show the normalized capacitance characteristics between the main electrodes as a function of the solid fraction. In these figures, C_e and C_f are the capacitances when the pipe is empty and when the pipe is completely filled with the powder, respectively. That is, C_e is a capacitance of the transducer itself.

Figure 4 illustrates how the electrode width affects the capacitance characteristics of the sensor. The electrode width were chosen to be 10, 20 and 30 mm for experimental reasons in future. The electrode interval was chosen to be 20 mm and kept constant. It can be seen from Figure 4 that the capacitance characteristics of the transducer are almost independent of the electrode width.

While, Figure 5 illustrates the effects of the electrode gap length on the transducer capacitance characteristics. The electrode gap lengths were chosen to be 10, 20 and 30 mm and the electrode width was chosen to be 9 mm for experimental reasons in future. Remarkable dependency of the normalized capacitances on the electrode gap length can be seen from Figure 5 unlike the case of electrode width.

Although it is difficult at present to

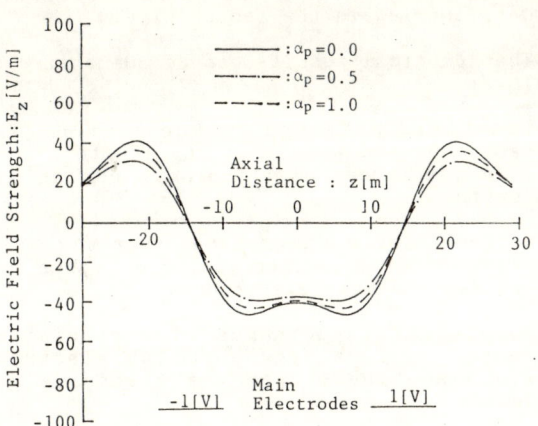

(a) On the Probe Axis, Axial Component.

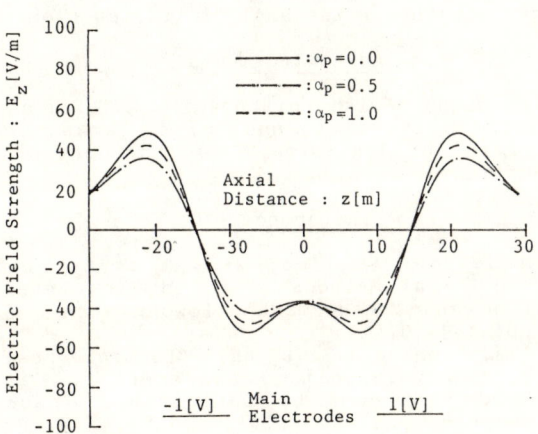

(b-1) r=4 mm, Axial Component.

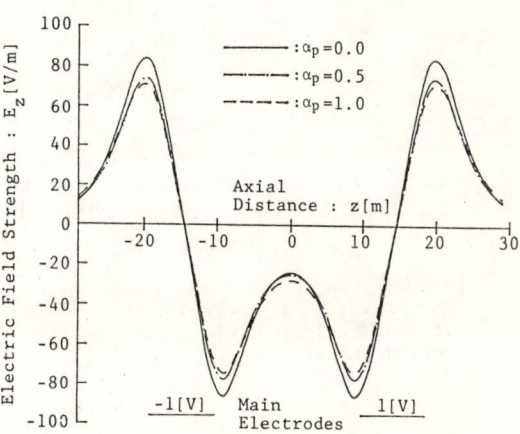

(c-1) r=8 mm, Axial Component.

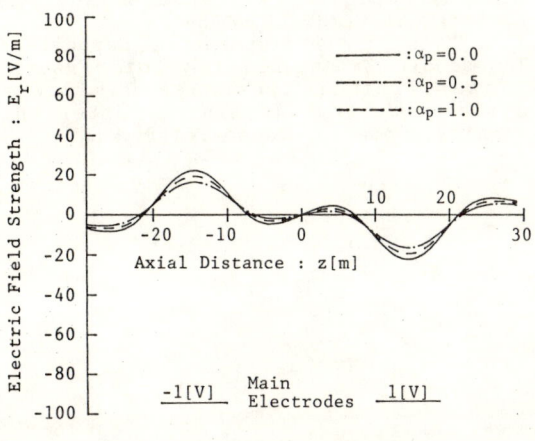

(b-2) r=4 mm, Radial Component.

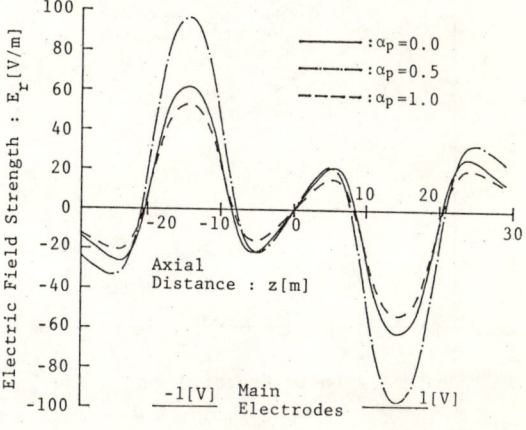

(c-2) r=8 mm, Radial Component.

Fig. 3 Electric Field Distributions.

measure quantitatively the dependency of these normalized capacitances on the geometrical parameters of the sensor electrodes due to the short of data as to the electric field profiles, these dependencies might be attributed to the changes in the uniform field areas in the sensor pipe as the electrode parameters are changed.

From the point of sensitivity of the transducer, the transducer with small gap length causes large change in the capacitance for the unit change in the solid fraction while the transducer with large gap length causes large change in the capacitance for the unit change in the solid fraction.

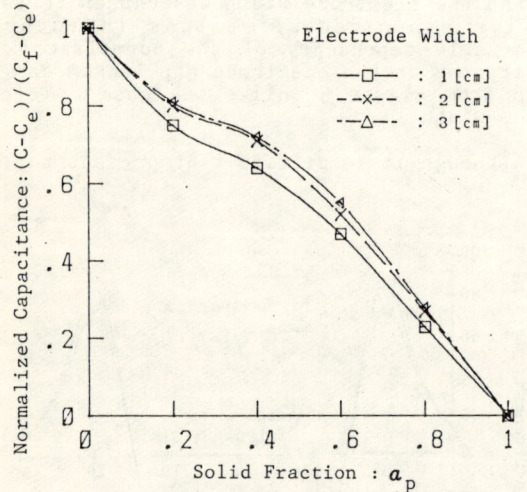

Fig. 4 Normalized Capacitance as a Function of Solid Fraction. Electrode gap length is 20 mm.

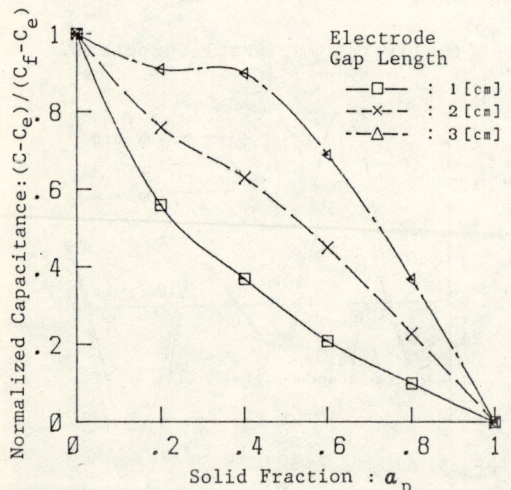

Fig. 5 Normalized Capacitance as a Function of Solid Fraction. Electrode gap width is 9 mm.

Conclusions

This paper presented the theoretical analysis of both the electric fields in the pipe of the capacitance transducer with the ring-type electrodes and of the resulting capacitance as a function of the solid fraction of powders carried in the sensor pipe.

The numerical results can be summarized as follows:

1) The field patterns in the flow region depend on the width and gap length of the main electrodes, particularly on the gap length.

2) Uniform fields are achieved near the probe axis and in the interelectrode space in-between the main electrodes.

3) Normalized capacitance characteristics depend in less degree on the electrode width and greatly on the electrode gap length.

4) High sensitivity to the solid fraction change is obtained by using the transducers with the small gap length for the large solid fraction and with the large gap length for the small solid fraction.

Acknowledgments

The authors wish to thank Dr. J.S. Chang, McMaster University, and Dr. K. Isaka, The University of Tokushima, for their helpful suggestions.

References

[1] M.S. Beck & N. Wainwright : "Current Industrial Methods of Solid Flow Detection and Measurement", Powder Tech., 2, pp. 189-197(1969).
[2] G.A. Irons & J.S. Chang : "Particle Fraction and Velocity Measurement in Gas-Powder Streams by Capacitance Transducers", Int. J. Multiphase Flow, 9, pp. 289-297(1983).
[3] H. Chun & C.K. Sung: "Flow Regime Characterization and Void Fraction Measurement by Capacitance Transducers", ASME Meeting, 84-WA/HT-56, 1984.
[4] T.A. Myint : "Development of Capacitance Transducer Techniques for Void Fraction Measurement in Two-Phase Gas/Powder System", M. Eng. Thesis, McMaster University, Ontario, Canada, 1986.

THE PRINCIPLE OF OPTICAL PHASE COMPENSATION A NEW METHOD FOR MEASURING ELECTROSTATIC FIELD BY MEANS OF POCKELS DEVICE

Zhang Ping   Ge Zuhuai   Lin Tai
Hebei University, Baoding, China

ABSTRACT

In this paper, a new method for measuring electrostatic field by using Pockels device is presented.

The principle of the measurement is based on optical phase compensation. The electro-optic crystal used in the measurement is crystal of LiNbO3.

The measurement result can be obtained by readings of a volt-meter.

1 INTRODUCTION

Since the linear electro-optic effect or Pockels effect was discovered in 1893, such effect has been widely used in many fields.

People have proposed some methods for measuring electric field by using Pockels device in recent years. One problem is how we can get a new and convenient method, while the measuring device is not complicated and the measuring result is easy to get. So, in this paper, we propose a method for measuring electrostatic field by using Pockels device. The principle of optical phase compensation is the basis of the new method.

2 THE PRINCIPLE AND APPLICATION OF OPTICAL PHASE COMPENSATION

Suppose two identical electrooptic crystal samples are placed in a path of ray, whose principal sections are perpendicular to each other.

Electric fields are applied to the two samples along a definite direction respectively. We can define ordinary rays and extraordinary rays according to the direction of the principal section of the first sample.

Let a beam of linearly polarized light be normal incident to and propagate in the two samples. If the electro-optic effect of the first one is positive, the effect of the second one will be negative, or vice versa.

If the intensity of the electric field applied to the first sample is equal to that applied to the second one, the total electro-optic effect of the two sample is zero, it can be expressed as follows:

$$\delta_E^{(1)} - \delta_E^{(2)} = 0 \tag{1}$$

Where $\delta_E^{(1)}$ and $\delta_E^{(2)}$ are the phase shifts of ordinary rays and extraordinary rays, the superscripts (1) and (2) are numbers of the samples respectively.

If the intensity changes of the applied electri fields are same, the variation of the total electro-optic effect of the two samples is zero. It can be shown in equation(2)

$$\Delta\delta_E^{(1)} - \Delta\delta_E^{(2)} = 0 \tag{2}$$

In accordance with the principle, we can develop a new method to measure an unknown electric field in comparison with a known one.

The phase differences of ordinary rays and extraordinary rays $\delta_E^{(1)}$ and $\delta_E^{(2)}$ are all the independent variables of the function of light intensity. The operation relation of $\delta_E^{(1)}$ and $\delta_E^{(2)}$ is subtraction. So, in the process of measurement, $\delta_E^{(1)}$ and $\delta_E^{(2)}$ can get any values within the range of π.

3 MEASURING ELECTROSTATIC FIELD BY MEANS OF POCKELS DEVICE

(1) Selection and processing of Pockels crystal samples

Pockels crystal samples are chosen and processed in accordance with the following principles:

(a) A crystal can give rise to a large Pockels effect in some directions. Therefore, the direction corresponding to the largest Pockels effect is required.

(b) The dielectric constant of a crystal should be as small as possible, in order that the electric field is disturbed slightly.

Thus the electric field should be applied along the direction coinciding with that of the small dielectric constant.

(c) Because the natural briefringence of an electrooptic crystal is very sensitive to the variation of the abient temperature, we must try to eliminate its effect.

So, in the experiment, we choose LiNbO3 crystal which is one of widely used electro-optic crystals.

The crystal possessing the largest Pockels effect is avaible from the market.

The sample of LiNbO3 crystal is shown in figure 1.

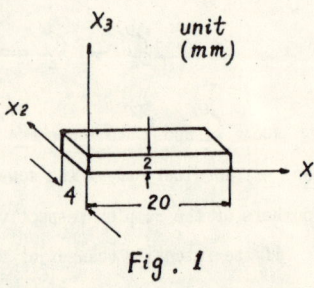

Fig. 1

In figure 1, the direction of X3 is the direction of the optical axis of the crystal.

In the experiment, the direction of the applied electric field is parallel to the direction of X3.

A LiNbO3 crytal sample shown in figure 1 is processed, which has got the largest Pockels effect and the smallest dielectric constant. The half wave voltage of the crystal sample is 294V, the dielectric constant of the crystal in the direction of X3 is 28.7.

Moreover, even if there exists a component of the applied electric field E2 in the direction of X2, the principal section of the crystal sample will not rotate, so that the accuracy of an optical system using the crystal sample is ensured.

(2) The new method for measuring electrostatic field

The arrangement of the electrostatic field measuring device is shown in figure 2.

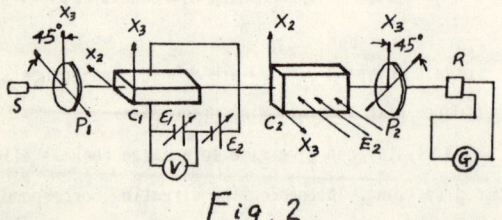

Fig. 2

Where S : a laser (The wave length is 6328Å); P1: a polarizer; C1: the first electro-optic crystal sample; C2: the second electro-optic crystal sample (the probe of the electric field sensor); P2: a polarization analyzer; $\varepsilon_1, \varepsilon_2$: two D.C Power supplies; v : a voltmeter (We get the measuring result by reading the meter); R: a photo detector; G : a sensitive galvanometer.

From figure 2, the intensity of light received by the photo detector can be expressed in the following equation:

$$I = I_0 \sin^2(\delta_E^{(1)} - \delta_E^{(2)} + \delta_0) \qquad (3)$$

Where δ_0 can be written as follows:

$$\delta_0 = \frac{2\pi}{\lambda}(n_o - n_e)(L_1 - L_2) \qquad (4)$$

Where n_o and n_e are refractive indices respectively for ordinary rays and extraordinary rays; L_1 and L_2 are the lengths of the crystal samples respectively in the direction of propagation. If the crystal samples are precisely machined, the lengths L_1 and L_2 are equal, δ_0 can be negligible.

When an electric field E2 is applied to the second crystal C2, a phase shift $\delta_E^{(2)}$ occurs. If another electric field E1 is applied to the first crystal C1, another phase shift is produced. If the intensity of the electric field E1 is proper, the total phase shift of ordinary rays and extra-ordinary is zero, the optical phase compensation is realized.

There are two electrodes mounted on both sides of the first crystal sample, The electric field E1 is applied to the crystal sample via them.

The separation of the two electrodes of the first crystal sample is known. We can read the potential difference between the two electrodes with the voltmeter, so that the intensity of the electric field E1 and its variation can be known.

But there are no electrodes mounted on the second crystal sample C2. It is only a piece of dielectrics. The existance of the crysal sample will disturb the electric field E2 to be measured.

So, we use an electric field between two parallel plates to determine the compensation relation of the two electro-optic crystal samples.

Because the dimensions, the dielectric constants and the Pockels coefficients of the crystal samples are known, the compensation relation of the two cyrstal samples will be known too.

Once the function relation is established, the intensity of an unknown electric field E2 can be determinted.

The separation of the two parallel plates is 30mm, The crystal sample C2 is placed in the middle of the parallel plates.

An electric field is applied to the crystal sample C1 by an adjustable D.C Power supply first, then an optical bias is set up in the path of ray.

So the measurement will be done in the sensitive range of the function of light intensity $I(\delta)$.

When potential difference is present between the parallel plates, an electric field E2 is applied to the crystal sample C2. In the path of ray, a phase difference occurs.

Due to the phase shift, the intensity of light received by the detector R is changed Such change is indicated by reading the shift of optical marker of a sensitive galvanometer G.

The adjustment of the output voltage of the D.C Power supply $\mathcal{E}_i$ is followed, so that another electric field is applied to the crystal sample C1 to make the optical marker of the sensitive galvanometer back to the original position.

On the basis of voltmeter readings and potential difference between the parallel plates, the compensation relation of the two crystal samples then can be determined.

4 COMPARISON BETWEEN EXPERIMENT RESULTS AND CALCULATION

The finite element method is used in our calculation. For simplification, the length of propagation of the crystal sample C2 is assumed infinite. Such a crystal sample is placed in the middle of the parallel plates.

The experimental results and the results of the theoretrical calculation are shown in table 1.

In accordance with the experimental result, the compensation relation of the crystal samples is shown in figure 3.

Table 1

potential difference between parallel plates	3KV	4KV	5KV	6KV	7KV
values of compensation voltage	11.79V	15.66V	19.16V	23.17V	28.20V
ratio of the two electric fields intensities	16.96	17.03	17.40	17.26	16.55
theoretrical ratio of the two electric fields intensities	18.08				

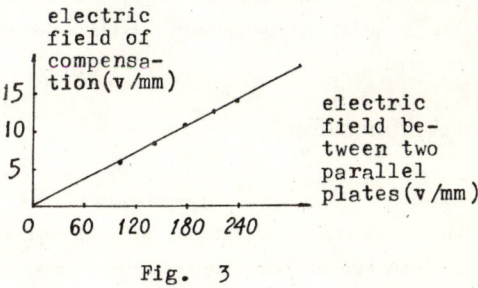

Fig. 3

If the scale of the voltmeter V is calibrated in accordance with the relation, we can get the result of measurement directly.

References:

1 Introduction to optical electronics, Holt, Rinehart and Winston, New York, 1976.
2 Kunihiko Hidaka and Teruya Kouno, Joural of Electrostatics, 11 (1982) 195-211
3 Koji Tada; Hirokuni Nanba; Yoshki Kuhara; Masayoshi Tatsumi, all of Osaka, Japan. United States Patent, Patent Number: 4,465,96, Data of Patent: Aug. 14, 1984

THE STUDY OF POWDER ELECTROSTATIC ACCUMULATION TEST DEVICE

Liu Yanyi Wang Changying

Machinery Industrial Committee of the State No.213 Research Institute

P.O.Box 99 Xi'an, Shaanxi, China

ABSTRACT

In this article the working principle and main functions of an instrument for the test of accumulation static charge of powders, "the Digital Electrostatic Accumulation Parameter Test Instrument (Model SJJ)" is developed. The instrument is one of the latest achievements in scientific research for powders and may be used in lab. The instrument may follow and measure the whole process of the accumulation and dissipation of the electrostatic charges of the powder sample, record and monitor the tested data and work on them at the same time. It may measure the ability of electrostatic accumulation of the same powder sample under different conditions and different powder samples under the same condition. It can also analyse quantitatively and study the sequence of electrification and the ability of electrostatic accumulation of various materials.

I INTRODUCTION

Extremely dangerous potentialities are usually encountered in menufacturing powders and explosives not only due to their characteristics of easy explosions and combustion, but also to their activity to electrification. Electrostatic accumulation usually appears in the handling and processing of explosives. Close attention has already been paid to such explosion and combustion accidents coming out of the discharge of accumulation charges.

The electrification in the processing of powders and explosives (such as pouring, sieving and dividing) generallly belongs to the triboelectrification also called the contact-separating electrification. The assessment and comparison of the electrification capacities were both carried out by means of the test pouring explosive along guiding channel either by German scientists in 1940s, by Japanese investigators in 1970s or by English scholars in 1980s.

In addition to the guiding-channel test, the Digital Eletrostatic Accumulation Parameter Testin instrument (Model SJJ) (See Fig. 1) is developed for the quantitative study of the electrification capacity and the electrostatic hazard of one dust material under various conditions as well as that of one dust materials under the same condition. The simulation of different experimental conditions. the continuous measurement of electrostatic accumulation and dissipation of dust explosives and the analysis of the experimental results can all be achieved by using this instrument. Experimental results of some dust powders and explosives indicate that this instrument can be used to measure the electrification characteristics of various dust materials. The reproducibility and the precision of the experimental result are excellent.

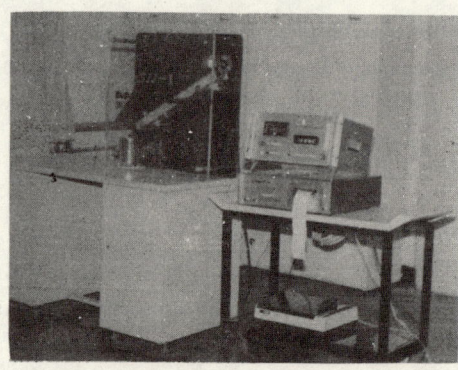

Fig. 1 Digital Electrification Accumulation Parameter Test Instrument (Model SJJ)

The characteristics of this instrument can be summarised as follows:

1. The electrification environment and testing condition can be artificially varied and the electrification capacity of a dust material under any environmental condition can be determined. The material of the guiding channel and the explosive container, the inclination of the channel and the explosive pouring rate are all adjustable to fulfill the requirements of the test.

2. The surface potential or the entire undergoings of the electrostatic accumulation and dissipation can be measured continuously and the data can be automatically processed. A four-digit indicator is adopted and the measuremant accuracy is excellent.

3. The explosion protection system in the instrument ensures safe operation in testing the explosible and combustible materials.

II MEASUREMENT

The realization of the measurement is as following:

The dust material is poured into the guiding channel at a constant velocity, then it slides down the inclined channel into a Faraday Cup. The static charge produced in the friction of the dust against the channel is accumulated the Faraday Cup or in some other dust container and measured automatically. The data are processed in time, and the electrostatic accumulation parameters under certain experimantal conditons of the dust are thus obtained.

The time constant of such measuring system and the Faraday Cup is $\tau = RC$, the measured potential is V

$$V = RI_0 (1 - e^{-\frac{t}{\tau}}) \quad (1)$$

where I_0 is the charge current due to the dust triboelectrification. And the accumulated charge

$$Q = CV_m = CRI_0(1 - e^{-\frac{t}{\tau}}) \quad (2)$$

Granted that actual charge carried by the dust is Q_0, then

$$Q_0 = I_0 t_m \quad (3)$$

where t is the time for the dust to slide down the channel. If $t_m \ll \tau$ the relationship between the actual charge Q_0 and the measured charge Q may be as follows according to Taylor Series Expansion

$$Q_0 = Q(1 + \frac{t_m}{2\tau}) \quad (4)$$

Suppose the potential is V_e at the end of the sliding t_m, then

$$V_e = V_m e^{-\frac{t_m}{\tau}} \quad (5)$$

For the same reason of approximation in obtaining Eq.(4), we have

$$V_e = (1 - \frac{t_m}{\tau}) V_m$$

$$\frac{t_m}{\tau} = \frac{V_m - V_e}{V_m} \quad (6)$$

Therefore, the actual charge carried by the dust is

$$Q_0 = Q(1 + \frac{V_m - V_e}{V_m}) = CV_m(1 + \frac{V_m - V_e}{V_m}) \quad (7)$$

The insulating resistance of the Faraday Cup is greater than 10^{16} ohms, the input resistance of the measuring system is greater tha 10^{14} ohms and the dissipation current is very small, so the value of $(V_m - V_e)/2V_m$ in Equ. (7) is less than 10^{-3} and can be omitted. In this way the actual charge carried by the dust Q_0 is approximately equal to the product of the capacitance of the system C and the measured potential at the end of the sliding V_m

$$Q_0 \doteq Q = CV_m \quad (8)$$

The electrification capacity q_0 is usually represented by means of the accumulated charge per unit volume or per unit mass

$$q_0 = CV_m / M \quad (\mu C/kg) \quad (9)$$

Where M is the mass of the dust material.

III SPECIFICATION OF THE INSTRUMENT

1. Kinds of samples
   (1) Chemicals,
   (2) Metal powders,
   (3) Non-Metal powders.

2. Mass of the sample
   5~25 grams (the mass of the primary explosive should be less then 10 grams).

3. Sample pouring time
   1~10 seconds (adjustable in 6 steps).

4. Material of the guiding channel
   alumunium, copper, iron, stainless steel, coated conductive rubber, the paperboard used to brush shellac varnish.

5. Length of the guiding channel
   500 mm.

6. Measuring range
   (1) 0 ~ 1,000 volts,
   (2) 0 ~ 10,000 volts.

7. Measuring accuracy
   (1) -1,000 ~ +1,000V   2.5%
   (2) -10,000 ~ +10,000V   5%

8. Display and data record
   (1) digital dispaly of the accumulated potential.
   (2) automatical display of the polarities of the potential.
   (3) print-out of the results of the accumulated potential and the accumulated charge.
   (4) data sampling rate: 2.5 data/second.

9. Power source and the operating environment of the instrument
   (1) power source
       220V±10%   50±2Hz
   (2) environment temperature: -10—40°C
   (3) environment relative humidity 40—80%

IV BLOCK DIAGRAM OF THE INSTRUMENT

The Digital Electrostatic Accumulation Parameter Testing Instrument(Model SJJ) consists of four parts, the block diagram of the instrument is presented in Fig. 2. The part of the guiding channel is composd of the dust pouring device, the channel inclination setting device. The measuring probe, the signal control circuit and the digital voltmeter compose the potencial testing part. The controling part is composed of the stepmotor driving circuit, the logical control circuit, the print-out circuit and the power source circuit. So the automatical pouring of the dust at different rates, the sliding-down of the dust along the channel, the display of signal conversion and the print-out of the data of the acumulation parameters are all achieved by the testing system; and the accumulation parameters of one dust material under various conditions and of different materials under the same condition are obtained.

The CMOS and TTL digital integrated circuit together with the linear circuit form the test-control network in the controlling part of the instrument. The sample in the container can be poured into the guiding channel at certain rate due to the adjusting of the sample pouring time by means of the logical circuit. Static charge is produced while the dust slides down the channal and the friction condition of the dust against the channel can be varied by changing the material of the channel. A hand-lever adjusts the inclination of the channel between 45—75°. The electric field intensity of the dust-carrying static charge accumulated in the Faraday Cup is sensed by the probe and the voltage signal is converted into four-digit result and is displayed by fluorescence indicator. The testing environment conditions such as the mass of the dust, the capacitance of the Faraday Cup as well as the accumulated parameters calculated from Equ. (8) and (9) are printed out by the data output system.

The channel inclination, the dust sliding rate and distance as well as the environment temperature and relative humidity will all affect the amount of the dust-carring static charge and the testing result. In testing, one of these factors is varied while the other factors remain unchanged in order to determine quantitatively the effect of this factor on the electrification capacity of the dust.

V TESTING METHOD

A dust sample of a certain weight contained in special papercup under environment temperature of 20±10°C is put in a desiccator for 24 hours. Then the sample is divided into five portions of equal amount. After setting the pouring rate by the pouring-set switch, adjusting the channel inclination and choosing the channel material, adjust the height of the operation panel so that the distance between the top of the Faraday Cup and the bottom of the channel is suitable to the requirement. Insert the probe into the Faraday Cup. Adjust the indicator so the potentiometer displays zero. When all the above procedures have been performed, put the sample in the paper-cup onto the pouring plate, press the pouring-start button on the operation panel. The indicatior will show the result of the accumulated potential and its polarity as the dust sample is poured along the channel into the Faraday Cup. After all of the sample has been poured, the potential value ceases to increase and might have a slight decrease. The entire variation of the charge accumulation and release will be printed out and the results of the accumulated charge and the charge density per unit mass are both obtained. The above experiment ought to be performed for five times. The mean value of the five maximum values in the five repeated experiments will be considered as accumulation parameter. The authors have studied the electrification of powers and explosives as PETN, LTNR, antimony sulfide, pistol composition. The result was presented in "Studies on the Electrostatic Accumulation of Dust Powder and Explosives".

*****

The Digital Electrostatic Accumulation Parameter Tsting Instrument (Model SJJ) is a special instrument used to study the electrification of dust material and to prevent any electrostatic hazard. Its development has provided the combination of the conventional guiding channel test with the advanced data processing system and realized the quantitative study of the electrifiction of dust material. It is believed that this instrument will play an important role in anti-static activities, in protection of human body from electrostatic hazards, in safe opperation as well as in electrostatic investigations.

Finally the authors wish to acknowledge their debt to Engineers Wang Xiufeng, Qian Chong and Xu Renshan for their assistance in the development of the intrument and also to Senior Engineers Ke Lin and Li Shengyue and Professor Chen Jiaxing for their instructions.

REFERENCE

1. D.P. Dongan, "Trends in Electorstatic Precautions in filling Factories", AD-A036015, P 363-383.

2. Wang Changying and Liu Yanyi, "Studies on the Electrostatic Accumulation of Dust Powders and Explosives", 1986.

3. J.F. Sumner and R.M.H. Wyptt, "The Effect of Temperature and Relative Humidity of the Accumulation of Electrostatic Changes and Primary Explosives", ERDE offPrint 1972.

4. J.N. Chubb, "Electrostatic instruments and measurements", in International Workshop on Electrostatics, 28/9,1983. Landadno, North Wales.

5. 菅義夫: 静電気ハンドブック、地人書館(株)、1972

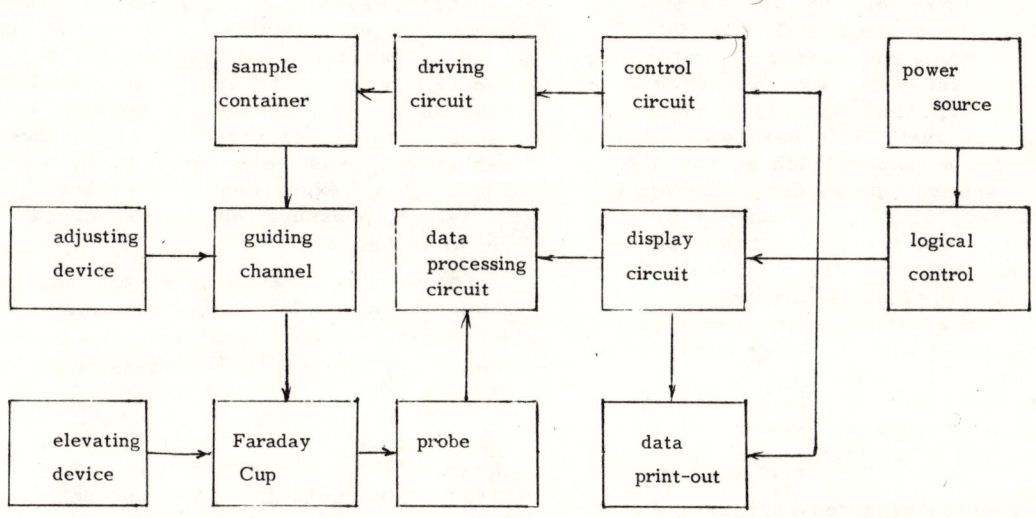

Fig. 2 Block Diagram of the Digital Electrostatic Accumulation Parameter Testing Instrument (Model SJJ)

ON RESISTIVITY MEASUREMENT OF CONDUCTIVE SPONGE

Xu Lianghua
Xiang Fan Rubber Factory

Xiang Fan, China

Wei Doutan, Zou Dongdong
Li Yumei, Wan Aimin
Wuhan University of Water
Transportation Engineering.
Wuhan, China

Abstract

Based on resistivity measurement of conductive sponge with different kinds of methods, preliminary evaluations and analysis on electrode systems and contact resistance are presented in this paper.

Introduction

Conductive sponge is a new kind of antistatic material. It is important and meaningful to measure the conductivity of materials in investigation of its electrification, charge dissipation and the charge transfer processes. As the conductive sponge is a kind of soft solid, it is rather difficult to test accurately with current methods. Some different methods of measurement have been tried to evaluate and analyse the resistivity of the conductive sponge. The purpose of this investigation is to select the electrode systems, optimize the geometric configuration of electrodes and find out the factors affecting the contant resistance.

Resistivety measurements

Contrast experiments are conducted under ambient temperature of 20°C and relative humidity of 70%. The sample is prepared in accordance with ISO 1653[1], and sharp cutter and lubricant water are used in processing the sample. It must be careful to keep the touching pressure on the sponge as small as possible to avoide defectes of the sample, such as the deformation of the sponge surface. In accordance with ISO 1923[2], the sample size is determined by the probable minimum reading of the measurement. Electrostatic voltmeter of type Q_2-V and a galvanometer with graduation of 2.5×10^{-9} A/div are used for quadri-electrode system, while a megameter of type ZC-36 is used for bi- and tri-electrodes systems. The average results of a number of repeated measurements are given in Table 1.

Table1 Average measured resistivity

electrode system	quadri	tri	bi
resistivity (Ωcm)	4.03×10^4	6.45×10^6	7.25×10^6

From the table, it can be seen that the value of resistivity measured with quadri-electrode system is much smaller than that of resistivity measured with the other two electrode systems. There may be various factors to affect the results, however, the key factor is due to the contact resistance and property of the material itself. In fact, during the all test procedures contact resistance always exists. When two smooth bodies contactes each other, it is point contactes instead of planar contact. Such contact resistance is rather larger than that of anti-static material. For bi- and tri- electrode systems, the contact resistance is in series with the resistance of the sample, therefore the total measured resistivity become one or two orders larger than that measured with quadri-electrode system. In other words, with quadri-electrode system the effect of contact resistance could be avoided and reliable results could be obtained. Because of the porosity of the sponge, it is impossible to have planar contact between the electrode and the sponge, and there is a certain contact resistance. As known, the resistivity is a constant value for a specified material under a given condition. Thus the variation of measured resistivity along with the reduction of porosity can only be explained by the change of contact resistance. Effect of contact resistance on resistivity has been verified by experiment. The experimental results under different pressure on the electrode are shown in Table 2.

As a rule, the porosity of conductive sponge would be reduced significantly under sufficient pressure. The experimental results could be explained briefly as follows:

1. The resistivity of material changes very little with additional weight on quadri-electrode system, it shows the effects of contact resistance in quadri-electrode system on measured resistivity is small.
2. For bi- and tri- electrode systems, the measured resistivity would be reduced with the increase of the pressure on the elec-

Table 2

W1+W2 (kg) electrode system		W1+0	W1+0.5	W1+1.0	W1+1.5	W1+2.0	W1+2.5
quadri-	resistivity (Ω-cm)	4.03×10^4	3.50×10^4	2.80×10^4	2.50×10^4	2.30×10^4	2.50×10^4
tri-		6.45×10^6	1.72×10^6	1.23×10^6	7.50×10^5	6.45×10^5	5.05×10^5
bi-		2.96×10^6	1.85×10^6	1.18×10^6	6.45×10^5	4.73×10^5	3.33×10^5

W_1: weight of electrode itself. W_2: additional weight. The weight of electrode W_1 is as follows: 0.92 kg for quadri-electrode system, 0.46 kg for tri-electrode system, 2.00 kg for two-electrode system

trode. It shows that the contact resistance is reduced too.

3. Larger pressure causes larger deformation and less porosity of the material. Under such condition, the contact state is improved and the contact resistance is reduced. Therefore the measured resistivity of is decreased accordingly.

4. From Table 2, the measured resistivity with tri-electrode system is a bit larger than that with bi-electrode system. It seem to be abnormal. The crux of the problem is that the measuring condition is some thing different for bi- and tri-electrode systems. The electrodes of megameter ZC-36 are used as tri-electrode. However, these electrode are much lighter than that used in bi-electrode system. Then the contact resistance gets larger, which causes a bit larger measured resistivity.

Further discussion on tri-electrode system

The effect on resistivity measurement of some factors, such as geometrical configuration of electrodes, insulation condition, etc, are examined. The layout of the test arrangement is shown in Fig.1 Three cases different tri-electrode systems are as follows:
1. Megameter ZC-36 associated with its own tri-electrode system.
2. Megameter ZC-36 associated with a specially designed tri-electrode system. The diameter and weight of the electrode are 5 mm and 2 kg respectively.
3. Megameter ZC-36 associated with another designed tre-electrode system. The diameter of the electrode is 40 mm, while its weight is still 2 kg and a polyster thin film cylinder is used as the insulator between electrodes. The measured resistivity are listed in Table 3.

Table 3

electrode No.	1	2	3
volume resitivity (Ω cm)	6.45×10^6	2.27×10^6	4.6×10^6

Table 3 shows that the discrepancy of the measured resistivity is small. Usually the condition of $r > 2d$ is required to obtain a reliable result, where r is the

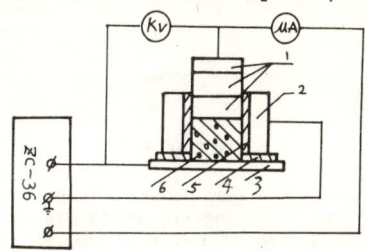

Fig.1 The measurement system. 1. the principal electrode and the additional weight; 2. the protect electrode; 3. another princpal electrode; 4. polyester ring 5. polyester cylindrical tube; 6. conductive sponge sample of $\phi = 50$ mm and 18 mm in height

radius of the testing electrode and d is the height of the sponge. Besides the gap between the protect electrode and test electrode should be small and uniform as possible so as to reduce the electric leakage and to weaken the fringe effect. Thus, the second configuration of electrodes is better. As pointed before, the original electrode of megameter is not heavy enough, so case 1 is not in favor of resistivity measurement of soft material like the conductive sponge.

Conclusion

The accuracy of resistivity measurement for soft material is mainly subject to contact resistance. Conductive sponge is a kind of soft solid material which could be compressed as ten times smaller as its original volume. In view of porosity, quadri-electrode system is reliable. But there are still some disadvantages in measurement. For example, due to the softness of the sponge and contact state with cutter edge, it is difficult to keep the electrode stable. Tri-electrode system with a specially designed electrode is also suitable. It is noticed that during experiment the external electrostatic field may interfere the measurement and cause errors. So some necessary shield measures should be taken to protect the electrode system and signal wire from the interference.

Reference

[1]. International Standard ISO 1653.
[2]. International Standard ISO 1923.

A NEW METHOD FOR MEASURING CHARGE DENSITY IN AN OIL-PIPING SYSTEM

Hao Wenyi, Beijing Kang Hua ES Research Center
Chen Juyin, Guang Zhou Research Institute of Elec. Appliances
Tan Fenggui, Tianjing No. 59165 Army Unit   China

ABSTRACT

This paper presents a new apparatus and its principle of operation for measuring charge density in a liquid medium. The feature of this new apparatus is using a fixed sensor thrust inside a pipe. It simplifies the structure and facilitates to maintain. It can be widely and beneficially used in many aspects.

1. Introduction

It is necessary to find a simple but accurate method of measuring the ES charge density generated within a dielectric liquid medium. At present, the measuring method used in most countries has the same complicated structure with a rotating sensor in the piping line. Is it possible to fabricate a new charge-density measuring apparatus that has the sensor statically fixed in the pipe which makes its design simpler and its maintenance easier? This paper is just to study and find a solution for this problem.

2. Principle of measurement

According to Gauss Theorem in electro-static field, the potential inside the unifromely charged infinite cylinder can be expressed

$$\phi = \frac{\rho r_0^2}{4\varepsilon}\left(1 - \frac{r^2}{r_0^2}\right) \quad (1)$$

where:
ϕ-potential at any point r inside the cylinder;
ρ-charge density of the medium inside the cylinder;
r_0-radius of the cylinder;
r -distance between point inside the cylinder and its axis;
ε-dielectric constant of the medium inside the cylinder.

It can be seen that under this assumption, the potential inside the cylinder is directly proportion to the charge density, in other words, the charge density could be measured with the measurement of the potential.

3. Method of measuring the potential at a given position inside the cylinder

It is suggested to use a rod-ball shaped probe which is extended into the axis of the cylinder and completely insulated from the ground. (see Fig. 1) When the probe is introduced into the medium the evenly charged liquid mediium starts to charge the probe..The potential of the probe will be gradually increased, and approches the equilibrium state with a stable value.

If the medium, the radius of the cylinder, the dimension and position of the rod-ball probe are definite, the charge density of the medium inside the cylinder will direct proportion to the potential of the ball . The proportional constant can calibrated by certain medium with known charge density. Then the ES potential meter can be calibrated as a charge density meter.

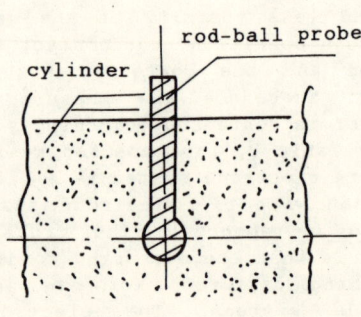

Fig.1

4. Calibration method

To calibrate the charge density of a dielectric liquid, one may use "the insulated tank method". A micro-ampere-meter is connected in series between the insulated metal tank and the earth. The impedance of the insulatio of the tank should be at least 2 orders larger than the imput impedance of the micro-amperemeter. The charged medium to be measured is introduced into the insulated tank at a constant flow rate. The dynamic current I_S and the volume flow rate U of the medium are recorded, the charge density of the electrified medium is the result of the value I_S divided by U, expressed as follows:

$$\rho = \frac{I_S}{U} = \frac{dq/dt}{dv/dt} = \frac{dq}{dv} \quad (2)$$

where:
q-electric charge;
v-volume of the medium;
t-time.

5. Experimental results

Experiments are carried out in a earthed metal pipe of 2.5 inches in diameter. The liquid medium used is aeronautic kerosene under flow rates of 250-1000 litres per minute. When there is a turbulent flow the electric charge could be approximately regarded to be uniformly distributed inside the pipe. Thus, the relation between the potential at any point within the pipe and the charge density could be apparoximately calculated by the expression (1).

A rod-ball shaped probe with dimensions as shown in Fig. 2,was thrust into the centre of the pipe, and is insulated from the pipe and fixed onto the wall by using a polytertrafluorethlene bolt. In orde to meet the requirement that the insulation impedance of the rod-ball probe to the ground must be at least 2 orders greater than that of the liquid medium, it was necessary not only for the polytetrafluorethlene bolt but also for the impedance input of the secondary apparatus to meet the same requirement. The specific resistance of the aeronautic kerosene used in the experiment was about 10^{12} Ω/m, and that of polytetrafluorethlene bolt is $10^{14} - 10^{16}$ Ω/m, while the imput impedance of the secondary apparatus was greater than 10^{15} Ω/m. Under these conditions, the rod-ball probe could be approximately taken as being in an ES field practically without any electric leakage to the ground. Thus, the measured value ϕ_b the stable potential at probe was also in direct proportion to the charge density of the electrified liquid.

Experiments have been repeated manything with different conductivity kerosenes under conditions of differnt flow rates, charge polarities, charge densities, temperatures and humidities, as shown in Fig. 3. The dots represent experimental values, while the normalized straight line represents the average value.

From Fig. 3, it can be concluded that the change of charge density under varied conditions is almost in direct proportion to the corresponding change of the potential of the probe.

6. Analysis of the range of errors
The average valur of factor defined by in root-mean-square could be obtained by the follwing expression:

$$\beta = \frac{\sqrt{\frac{\varepsilon(\phi_i - \phi)^2}{n}}}{\sqrt{\frac{\varepsilon(\rho_{0i} - \rho_0)^2}{n}}} = 10.18 \quad (3)$$

Where:
ϕ-potential of the rod-ball probe, in volt
ρ_0-charge density of the medium, in $\mu C/m^3$

To facilitate calibration of the apparatus, the value of β is taken as 10, and a calibrated straight line representing an average value can be obtained (see Fig. 3). By use of this calibrated line, it is possible to recalibrate the secondary ES potential meter and use it as a charge-density measuring apparatus. If the calibrated valu of charge density is represented by ρ, the absolute error Δ in measurement may be obtained as follows:

$$\Delta = \rho - \rho_0 \quad (4)$$

This expresion is used to calculate the absolute errors for 93 groups of experimental data, each of which being less than $\pm 10 \mu C/m^3$. This is similar to the error range when the American A.O. Smity type of charge-density measuring apparatus is used.

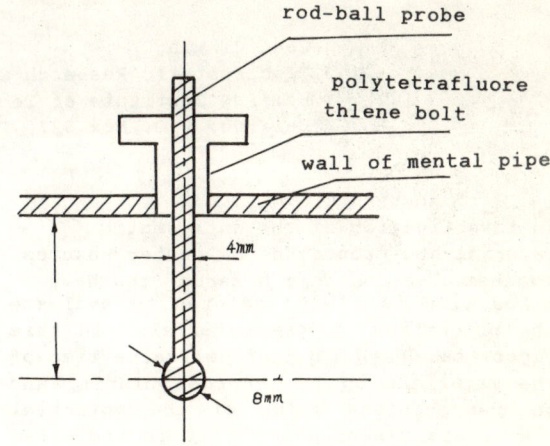

Fig.2

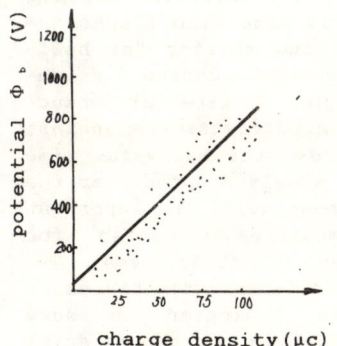

Fig.3

STUDY ON DECAY CHARACTERISTICS FOR STATIC ELECTRICITY ON BOTH CONDUCTIVE AND ANTI-STATIC MATERIALS

Chen Jiaxing
Electrostatic Research Center
Beijing Institute of Technology
Beijing, P.O. Box 327, P.R. China

Abstract

In investigation of the anti-static material and proposing the coutermeasures against electrostatic hazards, the half-value time is always adopted to evaluate the properties of the materials. In this paper, the dependence of halfvalue time of the materials on the charge polarity and on the absolute value of the potential itself are discussed, and it is indicated that the surface potential on the material decay according to the formula:

$$V = V_0 - \beta \ln(1+\alpha t)$$

Introduction

The property of charge dissipation of materials could be described by its half-value time, which is the requried period for charge or potential on the material to decay to its original halfvalue[1]. The longer the halfvalue time, the slower the charge disspation; the shorter the half-value time, the faster the charge dissipation. In investigation of material conductivity and the countermeasures against electrostatic hazards, the halfvalue time parameter has been widely adopted over the world[2,3]. It is meaningful and important to evaluate the measurement system for halfvalue time and to study decay behaviors on the materials for charge or potential. It was indicated in some data[4] that the potential on the material follows the exponential law of decrease. The different conclution is given in this paper based on our experiments.

Experiment

In this paper, the halfvalue time is measured with corona charging method as shown in Fig. 1.

The sample is charged when circulating it under the point electrode. The surface potential on the sample is measured and recorded when charged sample passing through the inductive probe. After charging the sample to a certain potential, the charging by corona discharge is turned off and the natural charge leakage process is recorded and analysed. The typical charging and discharge whole

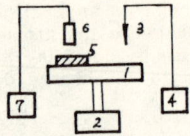

Fig.1 Schematic measurement system, in which 1, refers to the earthed metal plate; 2,the motor to spin the plate, 3 and 4 indicate the point electrode and the high-voltage supply respectively; 5,the sample while 6 and 7 refer to the inductive probe of the volt-meter and the recorder.

processes are shown in Fig.2, in which curve ab indicates the charging process with the increase of the applied voltage on the point electrode, bc part refers to the stationary state under a given applied corona voltage, while cd part illustrates the natural leakage process of charge on the sample.

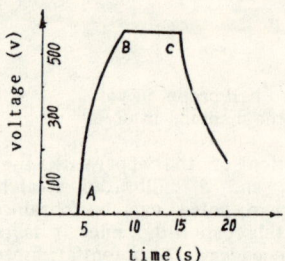

Fig.2 Charging and discharging plot

Results

In this paper, a series of experiments have been done with different samples and various charging processes. The typical measured results are given in Table 1 and in the Fig.3, in which the stationary potential on the sample is measured with TREK Model 344 electrostatic voltage meter.

In our experiment, results concerning the dependence of the corona current with the voltage applied to the needle, for different samples under a given condition, are shown in Fig.4, while the dependence of the halfvalue time on the voltage applied to the point electrode is shown in Fig.5.

Table 1 Measured halfvalue time and surface potential with different applied voltage on the corona point.

Sample	Voltage on the needle (kv)	Measured surface potential of the sample (v)	halfvalue time (s)
Anti-static silk	-6.0	-270	5.0
	+6.0	+220	12.0
	+7.2	+260	6.0
Surface layer of conductive floor	-6.0	-430	14.3
	+6.0	+330	16.7
	+6.5	+400	15.0
Anti-static rubber	-6.0	-150	1.7
	+6.0	+100	4.0
	+7.2	+140	1.9
Anti-static cardcloth	-6.0	-240	3.3
	+6.0	+160	5.0
	+6.5	+230	3.7

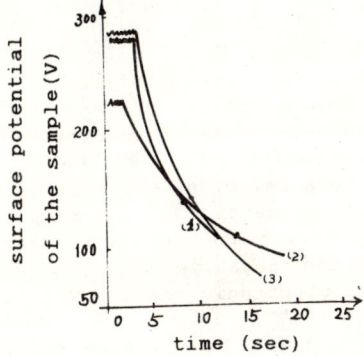

Fig.3 The discharging plot of anti-static silk sample in which curves 1,2 and 3 correspond to the applied voltage on the corona point electrode -6.0kv, +6.0kv and +7.2kv respectively.

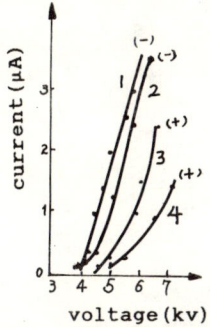

Fig.4 Dependence of the corona current on the voltage applied to the needle when different sample is put on the earthed plate. Curves 1 and 3 correspond to the sample of anti-static silk, while curves 2 and 4 to anti-static rubber.

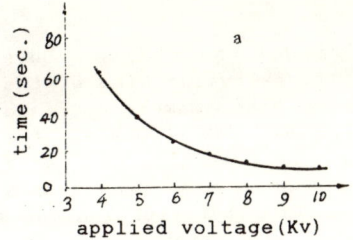

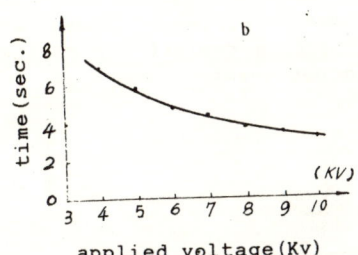

Fig.5 Dependence of halfvalue time on the applied voltage on the point electrode, (a) for polyester fiber and (b) for polyamide fibre.

Discussion and Conclusion

1. It can be obviously seen from Table 1 and from Fig.3 and 5 that the halfvalue time decreases with increase of the value of the voltage applied to the needle i.e. the stationary state surface potential on the sample. This kind of phenomenon could be explained with the non-linearizing resistance of the sample which increases with decrease of the voltage on the sample[5]. Generally speaking, the larger the resistance of the sample, the longer the halfvalue time.

2. It can be seen both from Table 1 and from Fig.3 that the decay process of the charge on the sample is dependent on the charge polarity. Of all the experiments indicate it is always true that the halfvalue time for the negative charged sample is smaller than that for the positive charged sample when equal but oppositive voltages are applied on the corona needle. This reason could be illustrated by the different characteristics of corona discharge. Becaues the threshold voltage for positive corona is always larger than that for negative corona[6,7] (see Fig.4), the negative corona current is larger than the positive corona current under the equal applied voltage (see Fig.4). Then many charges (or

higher surface potential)on the sample could be obtained in the negative corona case, and halfvalue time is shorter in this case.

3. It is clearly seen from above discussion that the halfvalue time of the sample depends on the sample surface potential itself. In other words, the decay behavior of the potential or charge on the sample relates to the sample potential itself. This behavior indicates that the charge or the potential on the sample is not exponentially decresed with the time,which can be easily seen in lnV - time phase diagram as shown in Fig.6. This is because resistivity of the sample is field dependent.

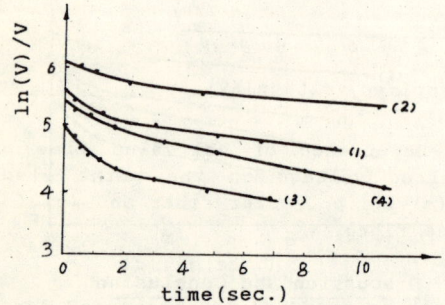

Fig.6 Dependence of the surface potential on the sample on time in the coordinate system of lnV - t in which curves 1,2,3 and 4 correspond to the samples of the silk,the layer, the rubber and the card cloth respectively

4 It is found from our experimental results that the surface potential or the charge on the sample follows the following law of decrease

$$V = V_0 - \beta \ln(1+\alpha t)$$

where V_0 is the initial value of surface potential on the sample at t=0.
α and β are the constants which are dependent on the sample, and can be determined with the experiment.

The correctness of this relation can be easily verified by results of the experiment. The correlation data of the measured value and calculated value according to the above relation are given in Table 2 - 5 with the proposed α and β parameters. It can be clearly seen from the data that the calculated values correspond to the measured values within the limits of the error.

Reference
1 Hwang Chaochien, Electrostatic Safety terminology 1987
2 DIN 53486
3 JISL 1049
4 菅义夫, 静電気ハンドブック 1972
5 Chen Jianxing, Liu Jiyuan, Discussion on assessment of conductive properties of conductive and anti-static products,1987 Pro.of Electrostatics Con. 1987(China)
6 G. F. Leal Fereira, O. N. Oliveira Jr. and J. A. Giacometti, point-to-point corona: current-voltage characteristics for positive and negative polarity with evidence of an electronic component J. Appl. Phys. 59(9),1 1983
7 Zhu Deheng and Yan Zhang, High-Tension Insulation 1980

Table 2 The measured and calculated data for the silk $\beta=50$, $\alpha=2.6$

	Time (s)	0	0.3	1.2	3	5	6.5	9
Measured value	Surface potential (v)	270.0	231.0	192.3	156.2	135.0	123.4	108.0
Calculated value	Surface potential (v)	265.0	236.1	194.1	156.3	133.0	120.8	105.3
Relative error(%)		1.9	2.2	0.9	0	1.5	2.1	2.5

Table 3 The measured and calculated data for the surface layer of conductive floor $\beta=60$, $\alpha=2.0$

	Time(s)	0	0.3	1.0	2.2	4.5	10.5	14.3	20.3	28.2
Measured value	Surface potential (v)	430.0	411.8	366.0	320.3	274.5	228.8	215.0	201.3	183.0
Calculated value	Surface potential (v)	414.5	386.3	348.6	313.3	276.3	229.0	211.2	190.8	171.5
Relative error(%)		3.6	6.2	4.8	2.2	0.7	0.1	1.8	5.2	6.3

Table 4 The measured and calculated data for the sample of the anti-static rubber $\beta=26$, $\alpha=8.7$

	Time (s)	0	0.2	0.5	0.8	1.2	1.7	3.0	4.2	4.7	6.0
Measured value	Surface potential (v)	150.0	130.4	107.6	97.8	87.8	75.0	65.2	58.7	55.4	48.9
Calculated value	Surface potential (v)	151.0	125.0	107.4	97.2	88.4	79.3	65.4	56.9	54.0	47.8
Relative error (%)		0.7	4.1	0.2	0.6	0.7	5.7	0.2	3.1	2.5	2.2

Table 5 The measured and calculated data for the sample of the anti-static card cloth $\beta=52$, $\alpha=2.8$

	Time (s)	0	0.2	0.3	0.8	1.7	3.2	3.3	5.0	8.0	10.7
Measured value	Surface potential (v)	240.0	229.8	204.2	178.7	153.2	127.7	120.0	102.1	76.6	61.3
Calculated value	Surface potential (v)	239.5	216.4	207.8	178.3	148.4	120.0	118.5	98.7	75.5	61.0
Relative error (%)		0.2	5.8	1.8	0.2	3.1	6.1	1.3	3.3	1.4	0.5

THE REALIZATION OF SUPERHIGH VOLTAGE PULSE WITH MULTIPLYING CIRCUIT

Wang Qiang

(Institute of Electrostatic Technology, Noreast Normal University, Changchun, China)

Wang Rongyi

(Dept. of Phys., Datian University of Technology, Datian, China)

Abstract

The realization of superhigh voltage pulse with multiplying circuit has been discussed, in which a bipolar pulsed inverter is utilized as the power supplier for the transformer and the superhigh voltage pulse is obtained by innovating the output terminal of the multiplying circuit. The magnetic saturation of the transformer is avoided by meams of series connecting a coupling capacitor in the primary side.

Introduction

The demands for pulse energization, especially the high voltage pulse, is currently getting increased. The high voltage pulse energization shows its superiorities to conventional power in the new fields such as electrostatic precipitator, electro-chemistry and electro-biology effect etc.

The pulsed power in present use is mainly to series install the D.C voltage and the pulsed voltage which are generated separately. One of the shortcomings of this method is that one end of the pulsed power must be installed onto the high voltage terminal of the D. C source, which makes the peak voltage of the pulse not very high. Besides, its large bulk and high cost also limit the wide spread application.

Having no such demerits, the superhigh pulsed power researched and manufactured by us is a self-combination of pulsed voltage and D.C voltage. Its outstanding advantages are high pulse peak, easy production, low cost and small volume.

Fundamental

Removing a capacitor near the output terminal will not change the maximum value of the output voltage in the multiplying circuit, but will diminish the minimum value. The D-value between the maximum and the minimum is twice as much as the output peak voltage of the transformer.[1]

Its essence lies in that the secondary coil of the transformer is connected in series to the capacitors circuit as an output. As shown in Fig. 1, traditional output of the multiplying circuit is from point 2 and point 3, which draws out a steady D.C voltage. But if we choose point 1 and point 3 as the output terminals, then the output voltage is a superimposition of the D.C voltage and the voltage on the secondary coil of the transformer.

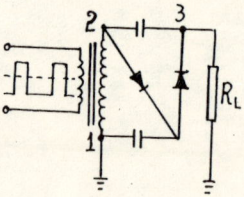

Fig. 1. Diagram showing the acquirement of superhigh pulsed voltage in the multiplying circuit.

If the traditional sine voltage wave is changed into the bipolar pulsed wave on the primary side, the superimposition of D.C voltage and pulse voltage can be obtained from the terminals 1 and 3 (see Fig. 2).

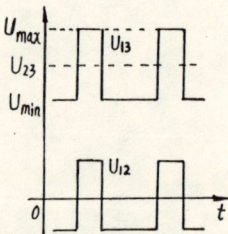

Fig.2. Superimposed ideal waveform

On ideal condition, if the peak voltage of bipolar pulse is Um, then the maximum and the minimum values of the output voltage are:

$$U_{max} = U23 + U_m \quad (1)$$
$$U_{min} = U23 - U_m \quad (2)$$

With the pulse amplitute being:

$$\delta U = 2U_m \quad (3)$$

The output voltage Uo is divided into D.C component Ud and pulse component Up,

$$U_o = U_d + U_p \quad (4)$$

where $U_d = U_{min}$, $U_p = 2U_m$

This kind of power pack is equivalent to installing a D.C source with voltage Ud and a pulse source with voltage Up. For example, to a one-stage multiplying circuit, if the output pulse peak voltage of the transformer is 50 kv, a total voltage of 150 kv can be obtained with 50 kv being the D.C voltage and 100 kv the pulsed, whose effect is the same as that of the superimposition of a 50 kv D.C source and a 100 kv monopolar pulse source. But the manufacture of a 100 kv monopolar pulse transformer is much more difficult than the making of a 50 kv bipolar one.

However, this approach needs a bipolar power source in primary side. It is realized by using a inverter which is made of power semiconductor apparatus such as GTR, SCR or GTO etc.[2] In order to obtain a wider range of frequency and pulse width modulation, the GTR and GTO are preferred.

THe Output Of Bipolar Pulse From Capacitor Coupling

When the width of positive pulse is equal to that of negative pulse, if the frequency is high enough the magnetic saturation of the transformer will not be produced. In order to get a superhigh voltage pulse in which the frequency and duty factor can be adjusted, the frequency and duty factor of the low voltage bipolar pulse must be adjustable. But if the positive pulse width in the bipolar pulse is not equal to the the negative pulse width, especially while having great width difference at low frequency, the transformer can easily get to magnetic saturation as soon as the wide pulse emerges, and this can lead to overcurrent of the whole system or to the direction changing failure of the current in the inverter.

Because of the great D.C component in the transformer, a coupling capacitor is series connected to the primary side between the inverter and the primary coil, as shown in Fig. 3. The effect is that it not only avoids the magnetic saturation but also keeps the the peak-to-peak voltage of the bipolar pulse unchanged. To ensure the waveform and power supply, the coupling capacitance must be large enough.

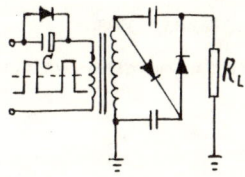

Fig.3. Connection of the coupling capacitor

The Power System For Superhigh Pulse Voltage

A power system has been designed following the analysis mentioned above, the measurements to it show that amplitude of the superhigh pulsed voltage coincides with the theoretical value, the waveform of the high pulse voltage is different from that of the low pulse voltage and has some oscillation at the top and bottom parts of the pulse, as demonstrated in Fig. 4. This phenomenon differs from narrow pulse to wide pulse, the rising time of the pulse is obviously increased.

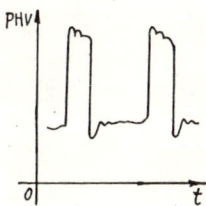

Fig.4. Measured waveform of high pulse voltage

Conclusion

A simple and economic method is utilized to obtain superhigh pulse voltage. Tests show that this method is applicable and practicable.

On the basis of the conventional multiplying circuit, we put a inverter at the primary side to generate bipolar pulse. By using a coupling capacitor, the problem of the magnetic saturation has been solved at the expense of not reducing the amplitude of the high pulse voltage.

Reference

[1] Wang Qiang and Wang Rongyi, Research on the High Voltage Multiplying Circuit With Abitrary Capacitance. Proceedings of 1988 International Moden Electrostatic Technology Symposium, Beijing, 1988

[2] Handbook for Silicon Controlled Rectifier Designers, 1970

Proceedings of International Conference on Modern Electrostatics

RESEARCH ON THE HIGH VOLTAGE MULTIPLYING CIRCUIT WITH ARBITRARY CAPACITANCE

Wang Qiang
(The Institute of Electrostatic Technology, Northeast Normal University, Changchun, China)
Wang Rongyi
(Dept. of Phys., Dalian University of Technology, Dalian, China)

Abstract

This paper makes a thorough theoretical study of the integer-stage and the half-integer-stage high voltage multiplying circuit with arbitrary capacitance. The universal calculating formulas are deduced. The calculating problem of the circuit is solved when the capacitance used in it is not equal. The method to generate the superhigh pulse voltage with the multiplying circuit is given.

I. Introduction

The important historical effect of the high voltage multiplying circuit (simply called HVMC in the following) is that in 1932 Cockcroft and Walton for the first time discovered the nuclear reaction of the artificial accelerating particles when they used it as a multiplying voltage accelerator. Afterwards, its application has played an important part in nuclear science and other sciences.

In recent years in scientific research and on technology market, there has been an urgent need for the high voltage power supply with miniature and low cost but advanced performance. Therefore, it is imperative to have a new and further study on the HVMC. The studies in the past were made under the condition of all capacitors used being equal or changing with their position orders in the form of kC in the HVMC, [1-4] and of the capacitors used having a little change in the half-integer-stage HVMC.[5] This paper will study the HVMC under the condition of all capacitors used being arbitrary to deduce the universal calculating formulas.

II. The calculation of the integer-stage HVMC with arbitrary capacitance

Fig.1 shows the integer-stage HVMC. C_k and C_k' ($k=1,2,\ldots,n$) respectively stand for the main output capacitors' capacitance and the auxiliary one. If the power supply transformer output sine wave voltage, the output voltage waveform with a load on the HVMC is shown in Fig.2. τ_0 which is the time of C_k' charging C_k is very small as compared with T, the period of AC power. So the charge lost due to the load during τ_0 can be ignored.

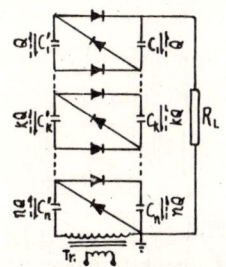
Fig.1. The integer-stage voltage multiplying circuit

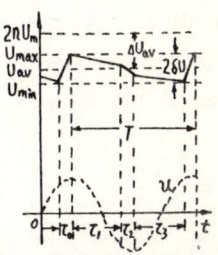
Fig.2. The output voltage waveform loaded

Suppose the charge supplied for the load is Q during T, the peak voltage of the power transformer output U_m and the average load current I_{av}, the voltage fluctuation on the output capacitor C_k will be:

$$\delta U_k = \frac{kQ}{2C_k} \qquad (1)$$

Therfore, the total voltage fluctuation across the load is

$$\delta U = \sum_{k=1}^{n} \delta U_k = \sum_{k=1}^{n} \frac{k}{2C_k} \frac{I_{av}}{f} \qquad (2)$$

where, $f=1/T$. From (2), the total voltage fluctuation value of the integer-stage HVMC with arbitrary capacitance can be obtained.

According to the relation between charge and voltage of capacitor, the maximum voltage on the output capacitor C_k is given as follows:

$$U_k = 2U_m - \frac{nQ}{C_n} - \sum_{i=k}^{n-1}\left(\frac{iQ}{C_{i+1}} + \frac{I_{av}(\tau_1+\tau_2)}{C_{i+1}} + \frac{iQ}{C_i'}\right) \qquad (3)$$

Since the maximum output voltage is the sum of the voltage on every capacitor, we obtain

-319-

$$U_{max} = \sum_{k=1}^{n} U_k = 2nU_m - \left[\frac{n^2}{C_n} + \sum_{k=1}^{n-1}\left(\frac{2k^2+k}{2C_{k+1}} + \frac{k^2}{C'_k}\right)\right]\frac{I_{av}}{f} \quad (4)$$

Where we have used (3), $I_{av} = Q \cdot f$, and $I_{av}(\tau 1 + \tau 2) = I_{av}/2f$. Fig. 2 shows the average value of the output voltage is $U_{av} = U_{max} - \delta U$. So by using (2) and (4), we obtain

$$U_{av} = 2nU_m - \left[\frac{n^2}{C_n} + \frac{1}{2C_1} + \sum_{k=1}^{n-1}\left(\frac{2k^2+2k+1}{2C_{k+1}} + \frac{k^2}{C'_k}\right)\right]\frac{I_{av}}{f} \quad (5)$$

From (5), the average output voltage of the integer-stage HVMC with arbitrary capacitance can be obtained.

The above-mentioned formulas have universality, while the others are the special case of them.[1-4]

III. The Calculation of the Half-Integer-Stage HVMC With Arbitrary Capacitance

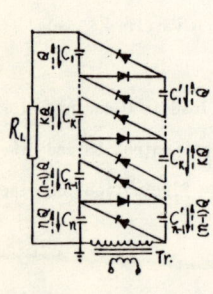

Fig. 3 shows the half-integer-stage HVMC, in which the output voltage unloaded has a value of odd number times U_m. The capacitor C_n is called a half-stage one because its maximum voltage is U_m instead of $2U_m$. C_k ($k=1,2,\cdots,n$) are the output capacitors, C'_k ($k=1,2,\cdots,n-1$) the auxilary capacitors.

Fig.3. The half-integer- stage HVMC

With the ignorance of the charge lost due to the load during τ_o, we can find the total voltage fluctuation value of the half-integer-stage HVMC,

$$\delta U = \sum_{k=1}^{n} \frac{k}{2C_k} \frac{I_{av}}{f} \quad (6)$$

By the similar method used in the calculation of the integer-stage HVMC, we obtain the maximum voltage on the output capacitor C_k

$$U_k = 2U_m - \sum_{i=k}^{n-1}\left(\frac{2i+1}{2C_{i+1}} + \frac{i}{C'_i}\right)\frac{I_{av}}{f} \quad (7)$$

With $k = 1, 2, \cdots, n-1$.
Since the maximum output voltage is as follows:

$$U_{max} = \sum_{k=1}^{n} U_k = U_n + \sum_{k=1}^{n-1} U_k ,$$

using $U_n = U_m$ and (7) we obtain

$$U_{max} = (2n-1)U_m - \sum_{k=1}^{n-1}\left(\frac{2k^2+k}{2C_{k+1}} + \frac{k^2}{C'_k}\right)\frac{I_{av}}{f} \quad (8)$$

According to $U_{av} = U_{max} - \delta U$ and from (8) and (6), the average output voltage is given by

$$U_{av} = (2n-1)U_m - \left[\frac{1}{2C_1} + \sum_{k=1}^{n-1}\left(\frac{2k^2+2k+1}{2C_{k+1}} + \frac{k^2}{C'_k}\right)\right]\frac{I_{av}}{f} \quad (9)$$

In order to unify the calculation formulas of the integer-stage and the half-integer-stage HVMC, the n (n is integer) representing the output capacitors' number in the above formulas has to be changed into the stage number n (n is half-integer). As the stage number n plus 1/2 equals the output capacitors' number, (6), (8) and (9) may be rewritten as

$$\delta U = \sum_{k=1}^{n+\frac{1}{2}} \frac{k}{2C_k} \frac{I_{av}}{f} \quad (10)$$

$$U_{max} = 2nU_m - \sum_{k=1}^{n-\frac{1}{2}}\left(\frac{2k^2+k}{2C_{k+1}} + \frac{k^2}{C'_k}\right)\frac{I_{av}}{f} \quad (11)$$

$$U_{av} = 2nU_m - \left[\frac{1}{2C_1} + \sum_{k=1}^{n-\frac{1}{2}}\left(\frac{2k^2+2k+1}{2C_{k+1}} + \frac{k^2}{C'_k}\right)\right]\frac{I_{av}}{f} \quad (12)$$

The above-given formulas have universality, while the others are the special case of them.[5]

IV. The Universal Formulas

Suppose n is the stage number of the HVMC, which is an integer for the integer-stage HVMC and a half-integer for the half-integer-stage one. With the use of the integer function $J = INT(n-1/2)$ and the absolute value function $|\cos n\pi|$, (2) and (10), (4) and (11), (5) and (12) may be respectively written as

$$\delta U = \sum_{k=1}^{J+1} \frac{k}{2C_k} \frac{I_{av}}{f} \quad (13)$$

$$U_{max} = 2nU_m - \left[\frac{n^2}{C_n}|\cos n\pi| + \sum_{k=1}^{J}\left(\frac{2k^2+k}{2C_{k+1}} + \frac{k^2}{C'_k}\right)\right]\frac{I_{av}}{f} \quad (14)$$

$$U_{av} = 2nU_m - \left[\frac{n^2}{C_n}|\cos n\pi| + \frac{1}{2C_1} + \sum_{k=1}^{J}\left(\frac{2k^2+2k+1}{2C_{k+1}} + \frac{k^2}{C'_k}\right)\right]\frac{I_{av}}{f} \quad (15)$$

Since the average interior voltage drop on the HVMC $\delta U_{av} = 2nU_m - U_{av}$, using (15), we obtain

$$\delta U_m = \frac{n^2}{C_n}|\cos n\pi| + \frac{1}{2C_1} + \sum_{k=1}^{J}\left(\frac{2k^2+2k+1}{2C_{k+1}} + \frac{k^2}{C'_k}\right)\frac{I_{av}}{f} \quad (16)$$

The above-given formulas have universality can be applied to the calculation of either the integer-stage or the half-integer-stage HVMC.

V. The Super-high Pulse Voltage Generated by the HVMC

From (13) and (14) we know that the output voltage fluctuation value is related to C_1 while the peak voltage has no connection with it. For this reason, we take C_1 away, so that the output voltage becomes DC voltage adding the sine voltage of the transformer secondary. The whole voltage waveform is thus similar to pulse one. The peak value of the pulse is the same as the one when C_1 is not taken away. The pure pulse amplitude is two times the peak voltage of the transformer secondary. Generally speaking the peak voltage of the pulse can be very high, because the DC voltage component can be modulated by changing the stage number n.

If a bipolar pulse power adjustable for frequence and width is used as the power of the transformer, we can obtain the super-high pulse voltage of which the frequence and pulse width are adjustable. Compared with other similar products, this power supply has the features of very high pulse voltage, small volume, low cost and easy making.

VI. Conclusion

This paper has deduced the calculation formulas of the high voltage multiplying circuit with arbitrary capacitance. By the analysis of them, we have got a better understanding of the electric characteristics and the new applications of the circuit. This will provide a theoretical guidance for the optimizing design of the high voltage power supply. The method given in the paper by which the superhigh pulse voltage can be realized with the high voltage multiplying circuit is simple and practical for developing a new high pulse voltage power supply.

References

[1] J.D. Cockroft. and E.T.S. Walton. Pro. Roy. Soc. (Landon), A136,619(1932)

[2] P. Lorrain, etc., Can. J. Phys., 35, 229(1957).

[3] 静电気学会言志, 6(4), 205(1982)

[4] A.A.BoPoBbEB, СВЕРХЫСОКИЕ ЭЛЕ-КТРИЧЕСКИЕ НАЛРЯЖЕНИЯ, 1955

[5] 王荣毅，吉林师大学报，2, 45(1979)。

HIGH VOLTAGE MODULATING PULSE ELECTROSTATIC POWER AND ITS APPLICATION IN ELECTROSTATIC DEMULSIFICATION

Feng Zhaolin, Wang Xiangde, Zhang Xiujuan
Environmental Science Institute, South China University of Technology, Guangzhou, P.R. China

Abstract

A new type of high voltage modulating pulse electrostatic power is developed. The influences of frequency, waveform and the average voltage and peak voltage of the modulating pulse wave on the demulsifying rate are discussed. Finally two kinds of high voltage electrostatic coalescer using modulating power are made and used in liquid surfactant membrane separation and solvent extraction for demulsification in an industrial scale.

Introduction

High voltage electrostatic coalescer (HVEC) is widely used in many processes for demulsification such as crude oil dedsalting in petrolem imdustry. Liquid surfactant membrane separation (LSMS) is a new technique invented in 1968[1]. Demulsification is a key and extremly difficult step in this technique and HVEC is regarded in the would as the only method can be used effectively. HVEC used for crude oil desalting can't be used in LSMS, because the emulsion much more stable due to its content of potent surfactant and large amount of water, and prepared by appointed emulsifying technique. Serval papers have reported on the HVEC of liquid membrane demulsification[2-5]. The author reported a type of HVEC named EC-1[6] which has been successfully used in the industrial application of phenolic wastewater treatment by LSMS. In this paper, on the base of discussing the influences of frequency, waveform and the average vlotage (Va) and peak voltage (Vp) of modulating pulse (MP) wave on demulsifying rate(Rd), a much more effective power-high voltage modulating pulse electrostatic power (HVMPEP) is developed. By using this power two kinds of HVEC are made and successfully applied in LSMS and solvent extration (SE).

Equipment and Procedure

(1) High voltage power
The block diagram and main parameters of HVMPEP are:

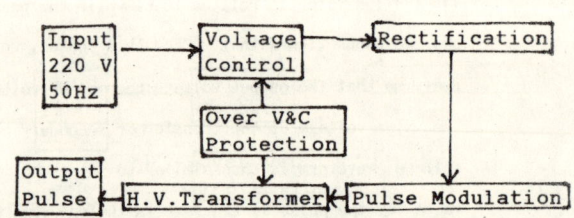

Power rating	1.2 kw
Maximun pulse voltage	45 kv
Output voltage rating	15 kv
Output current rating	60 mA
Pulse frequency	50-5000 Hz
Pulse duration	5-500 μs
Repetition frequency	5-500 Hz

(2) Demulsifing equipment:

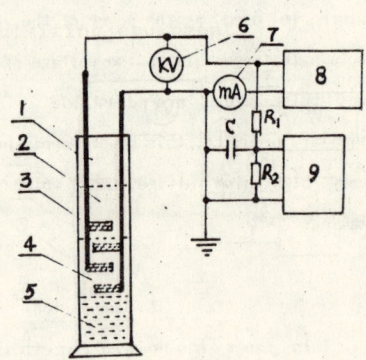

Fig.1. Sketch of the equipment for measuring Rd. 1.Demulsifing cylinder; 2. Oil phase; 3. Insulated electrode; 4. Emulsion phase; 5. Aqueous phase; 6. Electrostatic voltage meter; 7. High frequency current meter; 8. HV power supply; 9. Oscillograph.

(3) Formula of emulsion:
2%(wt) LMS-2(emulsifier) in kerosene and 5%(wt) NaOH aqueous solution with volume ratio 1:1.

(4) Emulsion preparation:
mechanical agitation 7500 rpm 15 min.
colloidal mill δ=2.0 3 min.

(5) Test procedure:
Pour 200 ml emulsion into the cylinder, turn on the power and measure meter, record the volume of water coalesced at definite time. Rd is expressed as

the percentage of water coalesced in unit time or the volume of broken emulsion in unit time.

Results and Discussion

It is reputed that Rd is directly proportional to electric-field strength. It means to set up an effective electric field is the key for demulsification.
In [6] we found that applying high voltage of 50 Hz is not a valuable way to increase the effective electric-field strength, because the voltage drop at the insulated layer of electrode is very high. For this reason, in this paper the influences of frequency and waveform on Rd are studied. In our system, when the applied voltage is the same, if we increase frequency, the voltage drop at the insulated layer of electrode will decrease and the voltage drop at the emulsion will increase comparatively. Thus the effective electric filed strength exerted on the emulsion will increase so that the Rd will increase consequently. The test results as shown in Fig. 2 coincide with this deduction, when the frequency increases from 500 Hz to 4000 Hz, Rd increases approximately by 2.3 times. Fig.4 shows the order of Rd for defferent waveforms i.e. MP wave>sawtooth wave >square wave >sine wave

Accordng to the characteristics of the four kinds of waveform under same of Va, the order of Vp is:
MP wave >sawtooth wave >square wave >sine wave

We consider that Vp may be an important factor in deciding Rd. Fig.5 shows the influence of Vp and Va of MP wave on Rd. Obviously in curve 1 Va is the lowest, Vp is the highest and Rd is the fastest; in curve 2 and 4 Va is the same but Vp is higher in curve 2 than in curve 4, while Rd follows the same rule. Therefore Vp is really an important factor for Rd but Va is not.

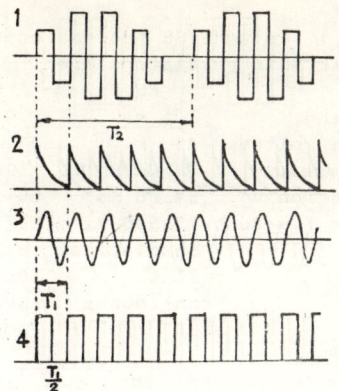

Fig.3 Four types of waveform. 1. MP wave; 2. sawtooth wave; 3. sine wave; 4. square wave. T_1-pulse cycle; T_2-repetition cycle.

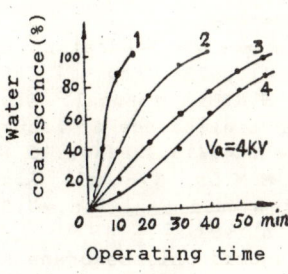

Fig.4 The influence of waveform on Rd. 1 MP wave; 2. sawtooth wave; 3. square wave; 4. sine wave.

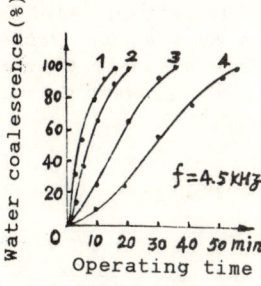

Fig.5 The influence of Vp and Va(gived by MP wave) on Rd.
1. Va=4.6 kv   Vp=47 kv
2. Va=8.7 kv   Vp=38 kv
3. Va=6.8 kv   Vp=27 kv
4. Va=8.6 kv   Vp=20 kv

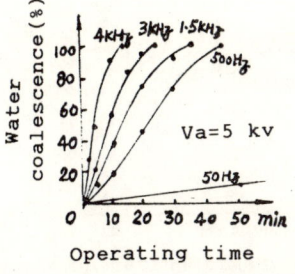

Fig.2 The influence of frequency on Rd

Application

On the basis of the studies stated above, two kinds of HVEC using MPEP are made for industrial application, one is used in LSMS named EC-2 and the other is

used in a 110 stages SE process named
EX-1. Their main parameters are:

	EC-2	EX-1
Power rating	1.5 kw	2.0 kw
Max. pulse voltage	45 kv	35 kv
Pulse frequencty	50-5000 HZ	300-1500 Hz
Pulse duration	5-500 μs	60-250 μs
Repetition frequency	5-500 Hz	30-150 Hz
Electrode	insulation	insulation
Volume	20 l.	600 l.
Operation	continuous	discontinuou
Rd	300 l/h	200 l/h

Rd in both of the two industrial applications are very satisfactory. MPEP is truely an ideal power supply and will be widely used in other electrostatic applications.

Literature

[1]. N.N.Li, Membrane Separation Process, U.S. Patent 3,410,794 (Nov. 12,1968)

[2] E.C.Hus, N.N.Li, T.Hucal, Electrodes for Electrical coalescence of Liquid Emulsion, U.S.Patent 4,415,426(Nov.15,1983)

[3] E.C.Hus, M.M.Li, T.Hucal, Electrical coalesence of Liquid Emulsion, U.S.Patent 4,419,200(Dec.6,1983)

[4] E.C.Hus, and N.N.Li, Membrane Recovery in Liquid Membrane Separation Process,Sep. Sci. Technol., 20 (2&3) 115 (1985)

[5] R.Marr, and M.Proetsch, Zinc Recvery from Wastewater, GER,offen DE,3,318,109 (08 Dec. 1983)

[6] Feng Zhaolin, Wang Xiangde, et. al., A New High Voltage Electrostatic Coalescer EC-1 USED for Industrial Application in Liquid Membrane Separation, Proceeding of the Second Interational Conference on Separation Technology, West Germany, 1987.

VI. Fundamental Research of Static Electrification and Discharge

Proceedings of International Conference on Modern Electrostatics

CONTACT ELECTRIFICATION BETWEEN METALS AND POLYMERS: EFFECT OF SURFACE OXIDATION

Dan A. Hays

Xerox Corporation, Webster Research Center, Rochester, NY 14644

Abstract

Contact electrification measurements between preformed metals and as-cast and ozonized polyethylene films are described. The ozonization increases the negative contact charging of the polymer, especially for aluminum with a low work function. A second set of measurements was obtained for contacts between vacuum evaporated metals and as-cast and ozonized co-polymer films. For contacts with aluminum and nickel, the charge exchange decreased by a factor of two when the metals were exposed to the ambient for ~1/2 hr. The decrease in contact chargng is attributed to growth in the metal oxide thickness. The measurements are interpreted in terms of a polymer surface states model for metal-polymer contact electrification.

Introduction

The phenomenon of contact electrification between materials is not well understood in spite of its ever present nature and exploitation in several major industries including electrophotography. The contact electrification properties of insulators such as polymers are of particular importance since the charge is retained for long times. In most situations the charge retention presents a nuisance or safety problem. In some industrial applications where the polymeric material is in the form of powder or fiber, the charge retention is beneficial since it enables control of the material by an electrostatic force.

In studies of polymer contact electrification[1], different metals are oftentimes used as the contacting material since the electronic properties of metals are thought to be well understood and characterized by an electronic work function. For some polymers the contact charge exchange tends to vary linearly with the work function of the metal and is consistent with electron transfer to or from the polymer.[2] For other polymers the charge exchange is low and independent of the work function.[3] Since both types of contact electrification have been reported by different observers for nominally the same polymer, it seems that the metal-polymer contact electrification characteristic must also depend on the purity and surface condition of the materials. It is the purpose of this paper to discuss the effect of both polymer and metal oxidation on contact electrification. For common polymers such as polyethylene and polystyrene, oxidation of the polymer by exposure to the ambient and laboratory produced ozone can have large effects on contact electrification by mercury. Measurements on mercury contact charging of the polymers used in this study are also described and found to be consistent with the earlier observations.[4]

A survey of the literature on solid metal-polymer contact electrification reveals that all measurements to date have been obtained with preformed metals for which the surface is usually finished through machining, polishing and cleaning operations. Many of the more recent measurements have been obtained under moderate vacuum conditions (~10^{-6} Torr), primarily for the purpose of circumventing air breakdown which can limit the charge exchange under ambient conditions. It is well known that metal oxide layers are always present on preformed metals. In spite of this awareness, there have been no studies explicitly examining the role of metal oxides in contact electrification. It has been assumed that the metal work function value which can be measured for the actual surface is the primary physical parameter for controlling contact electrification by metals. Dervos and Truscott[5] have described ultra high vacuum (UHV) measurements on contacts between in-situ vacuum evaporated metals and semiconductors. The tools of modern surface science characterization have yet to be employed in contact electrification studies under UHV conditions.[6]

This paper describes two sets of contact electrification measurements obtained under ambient conditions with two different polymers and metal surface preparations. For the first set of measurements, the contact charge exchange between preformed metal sheets and as-cast and ozonized polyethylene films is described. The ozonization increases the negative contact charging of the polymer, especially for contacts with low work function metals. The second set of measurements was obtained for contacts between vacuum evaporated metals and as-cast and ozonized co-polymer films. For contacts with evaporated aluminum and nickel, the highest charge exchange was observed for contacts immediately following vacuum deposition. The charge exchange for both metals decreased by a factor of two when the metals were exposed to the laboratory ambient for ~1/2 hr. The decrease in contact charging is attributed to growth in the metal oxide thickness which increases the spacing between the metal and electron accepting oxidative groups on the polymer. These observations suggest that the contact electrification of polymers by metals depends not only on the metal work function but another parameter which we presume to be the metal oxide thickness. A dependence of metal-polymer contact electrification on the oxide layer thickness could explain why Fabish, et al.[7] and Lowell[3] did not observe a monotonic dependence of the contact charge exchange on the measured work functions of different metals.

Contact Charging Measurements

Polymer Contacts with Preformed Metal Figure 1 shows the dependence of the contact charge exchange density on the work function of preformed metal sheets pressed against as-cast and ozonized films of high-density polyethylene. The as-cast polyethylene films were cast on a heated aluminum substrate (100°C) from a hot solution of xylene (130°C) and subsequently vacuum dried. The ozonized films were exposed to a 1% concentration of ozone for 1 min. The charge exchange density on the polyethylene was calculated by measuring the surface potential of 5 µm thick films (dielectric constant of 2.3) assuming the charge is trapped at or near the surface. The metal sheets were polished with steel wool and pressed against the polyethylene in a vise. The work function values of 3.6, 4.0, 4.6 and 4.8 eV for aluminum, nickel, copper and platinum, respectively, were determined by measuring the contact potential (Monroe Isoprobe Model 146) relative to a gold reference, assumed to have a work function of 4.8 eV. For the as-cast polyethylene films, the contact charge exchange density is -1.4 nC/cm^2 and independent of the metal work function. For the ozonized polyethylene, the negative contact charging increases and depends on the metal work function. The contact charging is zero for a metal work function value of ~5.3 eV. For contacts to the liquid metal mercury (work function of ~4.5 eV), the charge exchange density increased from -6 nC/cm² for the as-cast film to -30 nC/cm² for the ozonized film. The higher charge exchange densities obtained for mercury contacts are attributed to a higher area of actual contact. In summary, we observe both a work function independent and dependent contact electrification characteristic for as-cast and ozonized polyethylene, respectively. Since a low level of ozone is continually present in the atmosphere, one can expect the surface of unprotected polymers such as polyethylene to be oxidized upon exposure to the ambient over a period of days.[4]

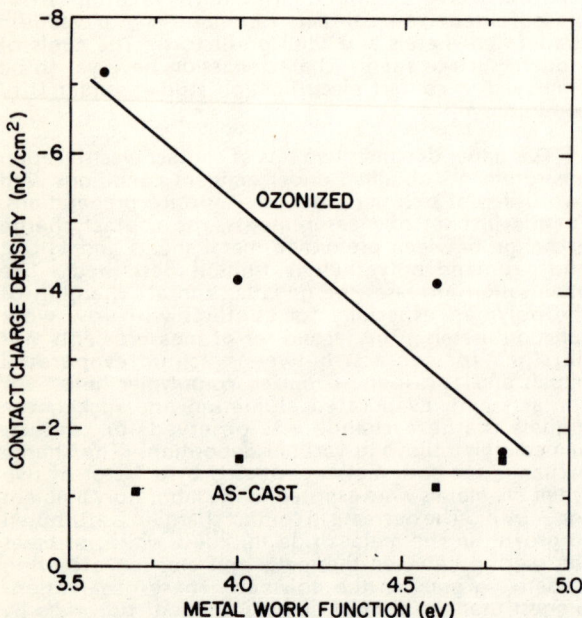

Figure 1. Dependence of the polymer contact charge density on the work function of preformed metal sheets pressed against as-cast (■) and ozonized (●) films of high-density polyethylene.

<u>Polymer Contacts with Vacuum Evaporated Metals</u> Since one expects a higher quality surface with vacuum evaporated metals, we obtained a second set of contact electrification measurements with different metals vacuum evaporated (~10^{-6} Torr) onto convex aluminum substrates and repeatedly contacted under laboratory ambient conditions to co-polymer films of styrene and n-butylmethacrylate. The radius of curvature for the metals was 3.8 cm and the contact force was held constant at 35 N. As-cast co-polymer films of thickness ~250 μm were solution cast from xylene onto aluminum substrates and vacuum dried. The charge exchange measured by an electrometer (Keithley 610C) connected to the metal increased upon repeated contacts, a characteristic commonly observed.[7,8] The evaporated metals of aluminum, indium, nickel and gold as well as preformed platinum provided a work function range (determined by contact potential measurements) of 1.2 eV. For each metal evaporation, a set of six samples was prepared. Figure 2 shows the dependence of the polymer contact charging (after 20 contacts) on the work function of the metal for the as-cast and ozonized co-polymer films. The data for the aluminum and nickel contacts are numbered to indicate the sequence of contacts during the time the metals were exposed to the laboratory ambient, from ~5 min following evaporation for the first sample to ~1/2 hr for the last sample. The lettered data are for a repeat sequence of contacts with aluminum for which the work functions were not measured (they were less than 3.8 eV, however). The arrows between the solid and dashed lines illustrate the decrease in contact charging upon exposure of the metals to the ambient for ~1/2 hr. For freshly evaporated metals, the contact charging of both the as-cast and ozonized co-polymer films tends to vary linearly with the metal work function and converges to zero charging for a work function value of ~5.0 eV. The slope of the curve for the ozonized film is approximately 5 times as large as the slope for the as-cast film. The charge exchange density for contacts between indium and the ozonized co-polymer was estimated to be -11 nC/cm². A geometric contact area of 3 X 10^{-3} cm² was calculated by noting the size of a highly reflecting circular spot (surrounded by dull appearing indium) formed by compressive yield of the relatively soft indium. The contact charge exchange density for mercury contacts was -2 and -20 nC/cm² for the as-cast and ozonized films, respectively.

In reference to Fig. 2, the contact charging for aluminum and nickel contacts to both the as-cast and ozonized co-polymer films decreased by a factor of two upon exposure to the laboratory ambient for ~1/2 hr; during this period the work function of the metals tended to increase by 0.1 to 0.2 eV. Lower levels of charge exchange were obtained for contacts between the air exposed metals and the as-cast films; for this case the contact charging was essentially independent of the metal work function (as also observed for the first set of measurements for contacts between preformed (oxidized) metals and as-cast films). The decrease in contact charging obtained for contacts between the air-exposed metals and ozonized films represents a larger change than expected from the work function shift. We attribute this decrease in contact charging to growth in the metal oxide thickness; this oxide layer increases the spacing between the metal and the electron accepting oxidative groups on the polymer. Support for the validity of this interpretation requires a discussion of metal oxide growth characteristics at room temperature.

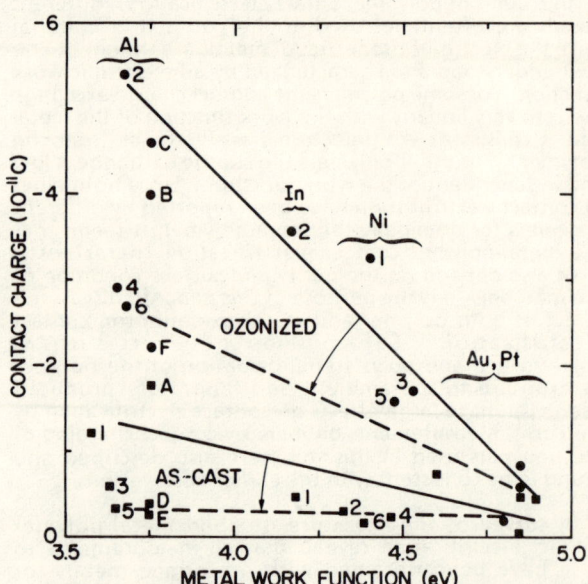

Figure 2. Dependence of the polymer contact charging (after 20 contacts) on the work function of vacuum evaporated metals for as-cast (■) and ozonized (●) for 11 min co-polymer films. The contact charging of the as-cast and ozonized films by the freshly evaporated metals is represented by the solid lines. The contact charging by metals exposed to the air for ~1/2 hr is represented by the dashed lines.

Discussion

Metal Oxide Growth The oxidation of metals has been studied for many years in view of its technological importance.[9] The discussion here is limited to the oxidation of aluminum at room temperature under atmospheric conditions. It is well known that aluminum rapidly forms an amorphous Al_2O_3 oxide layer which is ~20 Å thick at room temperature.[10] As discussed by Hart[11], the oxidation of evaporated aluminum films is at first very rapid and then comes virtually to a standstill after about 40 min. This time scale for oxide growth is consistent with the decrease in contact charging we observe when evaporated aluminum is exposed to air. Boggio and Plumb[12] have reported elliptically polarized light measurements of the oxide thickness. They measured a thickness of 13, 15 and 17 Å for exposure times of 2, 20 and 200 min, respectively. After 11.6 days, the thickness was 21 Å. Although the oxide thickness is somewhat less than that reported by Cabrera and Mott[10] and Hart[11], they observe a thickness increase over a time interval from minutes to hours. Boggio and Plumb also describe parallel measurements of the contact potential which indicate that the work function is nearly constant during the first few hours of exposure and then increases by 0.35 eV for 11.6 days. During our contact charging measurements obtained within ~1/2 hr, the aluminum work function increased by 0.1 to 0.2 eV. Our contact charging and contact potential measurements would be consistent with the data of Boggio and Plumb provided our oxidation rate happened to be higher, for some unknown reason.

Cabrera and Mott[10] described a theory for very thin (< 50 Å) metal oxide growth which provides the basis of many theoretical descriptions of metal oxidation.[9] For oxidation near room temperature, it is assumed that electrons from the metal tunnel through the oxide to acceptor energy levels provided by oxygen to produce O^- or O_2^- ions on the surface. The high electric field in the oxide layer due to the oxygen ions reduces the activation energy for the transport of (aluminum) cations from the metal and through the oxide layer to the surface where they combine with the oxygen ions to form the oxide. The rate limiting step which controls the kinetics of oxidation is the release of cations across the metal-oxide interface. In the Mott-Cabrera model, it is assumed that the electrostatic potential established across the oxide is independent of the thickness and set by the difference in energy between the sum of the electron affinity of oxygen and adsorption energy of an oxygen ion and the work function of the metal. As the film thickness increases, the growth rate rapidly decreases since the electric field for transporting the metal cations across the metal-oxide interface diminishes. Since this model cannot account for the observed changes in the contact potential with oxidation time, Boggio and Plumb[12] modified the Mott-Cabrera model by assuming that the electric field rather than the electrostatic potential across the oxide layer is constant during film growth.

Electronic Model for Metal-Polymer Contact Electrification A polymer surface states model proposed by Bauser, et al.[13] and discussed by Krupp[14] represents a useful theoretical framework for describing many metal-polymer contact electrification measurements.[1] We have previously used this model to describe the effects of polymer ozonization on mercury-polymer contact charging.[4] The polymer ozonization forms extrinsic surface states that are assumed to be uniformly distributed in energy as illustrated in Fig. 3a. The polymer surface states are filled to an energy ϕ_p below the vacuum level. The work function for an oxidized metal ϕ_m is also illustrated in Fig. 3a where we show the potential energy increase across the oxide layer due to the adsorption of oxygen ions. When the materials are brought into contact as shown in Fig. 3b, equilibrium is rapidly established by electron tunneling from the metal into the surface states near ϕ_p, assuming $\phi_p > \phi_m$. There are two contributions to a change in the electrostatic potential energy between the metal and polymer when contact is established. One contribution is due to the electric field associated with the charge transfer. The other contribution is due to an interfacial dipole layer formed at the interface by a polarization interaction over atomic dimensions. For mercury-polystyrene contacts, Hays[15] has reported an interfacial dipole energy shift ΔE of 0.25 eV which lowers the energy levels for electron transfer to the polystyrene. For polyethylene in contact with oxidized aluminum, Pong, et al.[16] have described a photoemission experiment to measure the interfacial dipole energy and they observe a lowering of 0.4 eV.

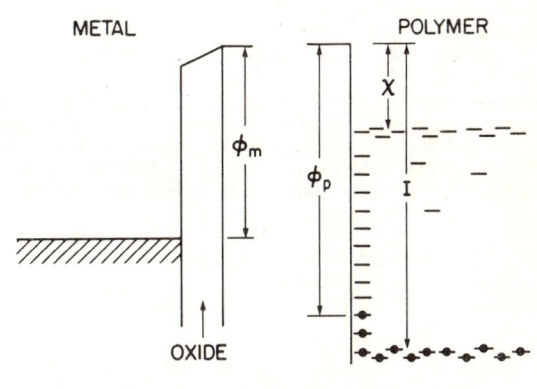

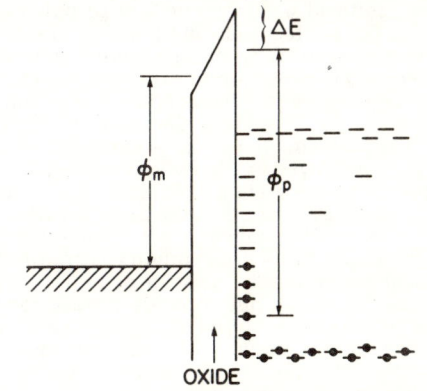

Figure 3. (a) An electronic model for an oxidized metal of work function ϕ_m and polymer with surface states before contact. The lower empty molecular orbitals, the highest filled molecular orbitals and the highest filled surface states lie at energy levels of X, I and ϕ_p below the vacuum level, respectively. (b) After contact, the energy levels between ϕ_p and ϕ_m are filled, taking into account a potential energy shift due the charge transfer and an interfacial dipole energy ΔE.

The contact charge exchange density on the polymer for the metal-polymer interface illustrated in Fig. 3b is given by the expression (SI units)[15,16]

$$\sigma = eN_s[\phi_p + \Delta E - \phi_m][1 + (e^2N_s d/\varepsilon)]^{-1} \quad (1)$$

where N_s is the uniform density of surface states per unit area per unit energy, e is equal to -1.6×10^{-19} C, d is the distance of separation between the metal and surface states which is set by the oxide thickness (in previous model discussions, d was assumed to be of atomic dimensions) and ε is the permittivity of the oxide layer. For a low surface states density ($e^2N_s d/\varepsilon \ll 1$), Eq. (1) becomes

$$\sigma = feN_s[\phi_p + \Delta E - \phi_m] \quad (2)$$

where we have included a factor f which denotes the fraction of the geometric area which is available for intimate contact. Furthermore, since the charge density on the polymer will be reduced by electron tunneling as the metal and polymer are separated, the charge density will be less than that given by Eq. (1). The low density of states expression of Eq. (2) is equivalent to that described by Pong, et al. For the contact charging data shown in Figs. 1 and 2 for the as-cast polyethylene and co-polymer films, we ascribe both the low charging and metal-work-function-independent charging to a low surface states density.

In the limit of a high surface states density ($e^2N_s d/\varepsilon \gg 1$), Eq. (1) becomes

$$\sigma = f\varepsilon[\phi_p + \Delta E - \phi_m]/ed. \quad (3)$$

The contact charging data shown in Figs. 1 and 2 for ozonized films of polyethylene and co-polymer show a nearly linear dependence on the metal work function. Assuming that ΔE is 0.4 eV, then ϕ_p is equal to 4.9 and 4.6 eV for the ozonized polyethylene and co-polymer films, respectively. For contact charging between the ozonized co-polymer and evaporated metals, we presume that the oxide thickness d increases upon exposure to the ambient during the time between ~5 min and 1/2 hr. From Eq. (3), an increase in d will cause a decrease in the contact charging as observed. If Eq. (3) is correct, the contact charging data displayed in Fig. 2 suggests that d increases by a factor of two for aluminum and nickel contacts. Although there is evidence that the oxide growth rate for aluminum and nickel is rapid at room temperature, we do not have the quantitative information to confirm that the oxide thickness increases by a factor of two. We cannot rule out possible effects due to oxide surface roughness, moisture uptake or hydrocarbon adsorption. Confirmation of the model will require measurements under well controlled conditions for forming and characterizing metal oxides.

Summary

Data obtained from the two sets of measurements have demonstrated that surface oxidation on both metals and polymers can have an important effect on metal-polymer contact electrification. It seems that the contact charging of polymers by metals depends not only on the metal work function but also the oxide layer thickness. It seems likely that a cause of some of the variability reported in the literature on metal-polymer contact electrification can be due to differences in the degree of metal and polymer surface oxidation.

References

1. Lowell, J. and Rose-Innes, A.C., Adv. Phys. 29, 947 (1980).
2. Davies, D.K., Adv. Static Electrication 1, 10 (1970).
3. Akande, A.R. and Lowell, J., J. Electrostatics 16, 147 (1985).
4. Hays, D.A., J. Chem. Phys. 61, 1455 (1974).
5. Dervos, C. and Truscott, W.S., J. Electrostatics 16, 137 (1985).
6. Briggs, D., *Electrostatics 1979* (The Institute of Physics, London and Bristol, 1979), Conf. Ser. No. 48, p.201.
7. Fabish, T.J., Saltsburg, H.M. and Hair, M.L., J. Appl. Phys. 47, 940 (1976).
8. Lowell, J., J. Phys. D: Appl. Phys. 9, 1571 (1976).
9. Fromhold, A.T. Jr., *Theory of Metal Oxidation*, Vol. I, (North-Holland, Amsterdam, 1976).
10. Cabrera, N. and Mott, N.F., Rep. Prog. Phys. 12, 163 (1949).
11. Hart, R.K., Proc. Roy. Soc. A 236, 68 (1956).
12. Boggio, J.E. and Plumb, R.C., J. Chem. Phys. 44, 1081 (1966).
13. Bauser, H., Klopffer, W. and Rabenhorst, H., *Advances in Static Electricity* (Auxilia, Brussels, 1970) Vol. 1, p.2. (1971).
14. Krupp, H., *Static Electrification*, 1971 (The Institute of Physics, London and Bristol, 1971), Conf. Ser. No. 11, p.1.
15. Hays, D.A., *Electrostatics 1979* (The Institute of Physics, London and Bristol, 1979), Conf. Ser. No. 48, p.265.
16. Pong, W., Brandt, D. and He, Z.X., J. Appl. Phys. 58, 896 (1985).

OBSERVATION OF LOCAL DISCHARGE PHENOMENA ON CHARGED SURFACE OF DIELECTRIC FILMS

Tetsuji Oda and Masatoshi Yanagida

Department of Electrical Engineering, the Faculty of Engineering, the University of Tokyo
3-1 Hongo-7chome, Bunkyo-ku, Tokyo 113, Japan

Abstract

Local electrostatic surface discharge phenomena on charged surfaces of dielectric films such as FEP teflon sheets were observed related with their pulse-like discharge current form, residual charge-distribution and so on. When the sphere electrode approaches to the uniformly charged dielectric film surface, the discharge occurs and, under the good measuring conditions, a very sharp pulse-current with a very fast rise-time, such as 1 ns, and narrow pulse width of 2 ns is observed. After the discharge, crater-like potential wells are observed on the surface potential profile. In the case of negatively charged surface, the trace of the surface discharge travelling horizontally was recognized by the dust figures and photos by the image intensifier.

Introductions

The surface of the dielectrics is easily charged, and the unexpected discharge following that charging happens to cause large electrostatic hazard, trouble and so on at some times. When the corona-charging creates the uniform charge distribution on the film, such an electrostatic discharge disturbs the uniformity of the surface charge distribution. That discharge is localized and the discharge area is very small in general because the surface resistivity is very high and the travelling of the surface discharge is not so fast as the discharge itself. The some trouble of radio-frequency noise generated by the electrostatic noise have been studied in related to the error of the electronic device and so on. However, the approach to the discharge mechanism is not yet done enough. Kobayashi and his group measured only the amount of transferred charge by such discharge where the potential was very high[1]. The authors have observed that localized discharge occurred by approaching the grounded sphere electrode to the uniformly corona-charged thin electret film. Test results of discharging current-wave forms and the change of surface potential distributions by that discharge are reported here.

Experimental

Samples   Test sample films are mainly FEP teflon films (Toyoflon by Toray of 100 μm thickness). The square film (10 cm X 10 cm) is cleaned by the alcohol and is charged by the needle corona with the mesh-like grid electrode. The basic diagram of the charging system is shown in Fig. 1 where 15 needle points of + or - 22.5 kV are located about 11 cm above the sample. The grid voltage V_g was usually from + or - 2 kV to 6 kV. After the charging, the series experiments were carried out at the fixed time schedule. Charging time is 30 min. at room temperature in most cases.

Observing Equipment for Local discharge   A new test apparatus has been designed and fabricated to observe the local discharge between the earthed sphere metal electrode and the charged surface of the dielectric film with very small disturbance. That is, the approaching speed and the position of the electrode is controlled by the personal computer (Fujitsu FM77AV). The whole experimental setup is shown in Fig. 2. The discharge phenomenon is sensed by the trigger unit of the digitizer (Tektronix 7212AD with 7A19 and 7B92A). This triggering signal was also sent to the stepping motor controller to identify the discharge. The motor stops for preventing the contact of the sphere electrode to the sample surface. The discharge position (distance from the sample surface when the discharge occurs) is also displayed on the CRT of the computer system through the interface of the controller. The discharge electrode is a stainless steel sphere whose diameter is 5mm and is mounted to the z-axis pulse-motor stage with a flexible joint to realize a soft contact to the film which is exhibited in Fig.3. As the apparent discharge current is not observed to trigger the detectors at sometimes, the electrode hits the surface. Figure4 is a basic procedure of the one measuring loop.

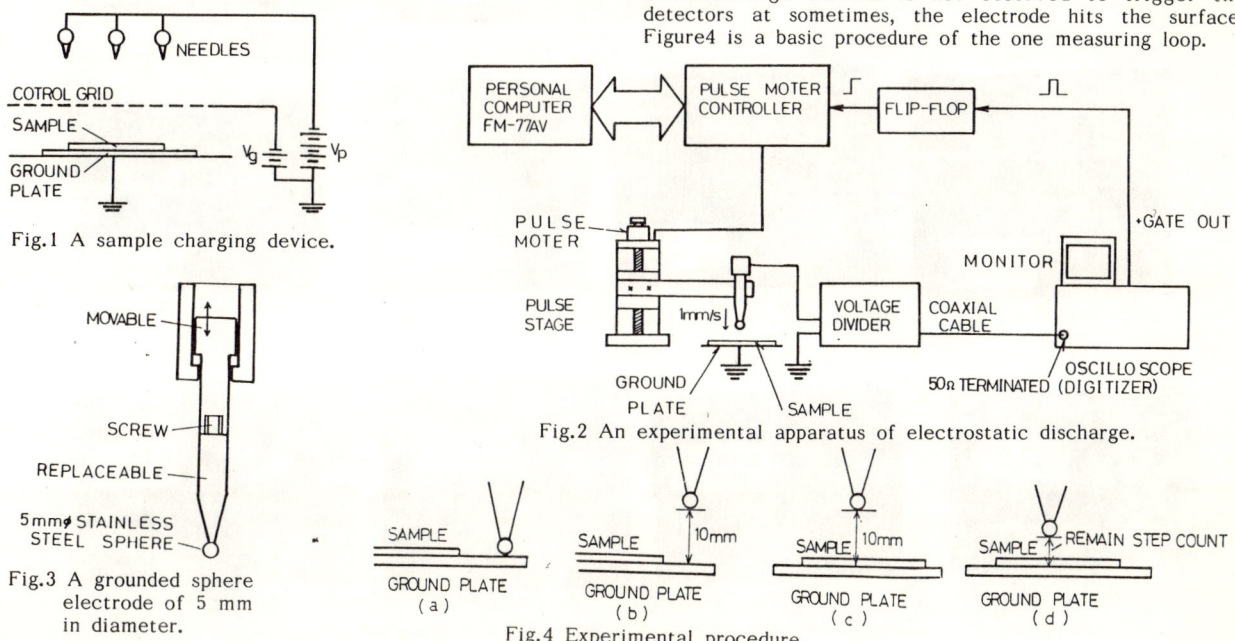

Fig.1 A sample charging device.

Fig.2 An experimental apparatus of electrostatic discharge.

Fig.3 A grounded sphere electrode of 5 mm in diameter.

Fig.4 Experimental procedure.

The first step is to identify the back electrode (ground plate which supports the dielectric film) position. Then the discharge electrode rises 10 mm. The film is moved to the discharging place with a back electrode and the sphere electrode begin to fall down with a constant velocity, typically 1 mm/sec. When the discharge is detected, the residual step number is checked and the the real distance is calculated by the computer.

One of the most important informations about the electrostatic discharge is the discharging current wave form. It may be strongly dependent on the circumstances, including characteristics of the detector itself. The upper sphere electrode is connected with a hand-made voltage divider shown in Fig.5. As the digitizer input impedance is also 50 ohm, the voltage dividing ratio is 1:100 in total. In this case, the discharge occurs through 50 ohm and other spurious impedance. The signal means the current which flows at that condition.

Surface Potential Profile of the Film after Discharge
The surface potential profile of the charged dielectric film is monitored by using the electrostatic surface potential probe (Monroe 244) and a digital X-Y stage whose resolution is 10 μm. All system is also controlled by the similar personal computer (Fujitsu: FM-77AV) and other devices. Data are also processed by that. The details of the system will be described in near future. The dust figure of the dielectric film after the electrostatic discharge is also examined by using the special powder.

Experimental Results

Discharging Current-Wave Forms

Negatively Charged Film Figure 6 is three different types of discharging current wave forms where the grid voltage, V_g, is -3 kV. In every case, rising time of the current is about 1 ns order which is very fast compared with previously reported value. That is, the discharge phenomena are surely very quick, less than nanosecond indeed. Figure 6 (a) shows only a single large negative (homo-) pulse current and other small oscillations. On the other hand, the double peaks where peak values are mostly same order and smaller than that of the single pulse, is also observed such as in Fig.6 (b). At special case, the opposite polarity current (hetero-) is also recorded such as in Fig. 6 (c). The discharge occurs statistically, even though the film is the same. The wave form is assumed to be related to the distance when the discharge occurs. The pulse width in Fig. 6 is surely less than 2 ns meaning the discharging phenomena are very speedy which is limited by the system response time. Figures 7 (a) - (c) are typical current forms for the grid voltage, V_g, - 5 kV, -4 kV and -2 kV, respectively. When V_g is large, such as - 4kV or -5 kV, the value of discharge current increases nonlinear. That is, the current at V_g = - 5 kV is about twice as large as that at V_g = - 4 kV. The rising time (2.5 ns) and pulse width of the current at V_g = - 5 kV are larger than those at V_g of - 4 kV for the discharge distance is larger at high surface charge density. When V_g is - 2 kV, the current value is small and the discharge is not detected easily although the rising time is very small as about 1 ns. When V_g is - 1 kV, the trigger is not so constant because of the small discharge current.

Positively Charged Films Figures 8 (a) - (d) are the typical discharging current wave forms for positively corona-charged film surfaces. In this case, the peak values are not so much different except that at V_g = 2 kV with large contrast to the case of negatively charged film. The rising time is a little bit large at high V_g in good agreement with the negative case. It may be dependent on the discharge distance.

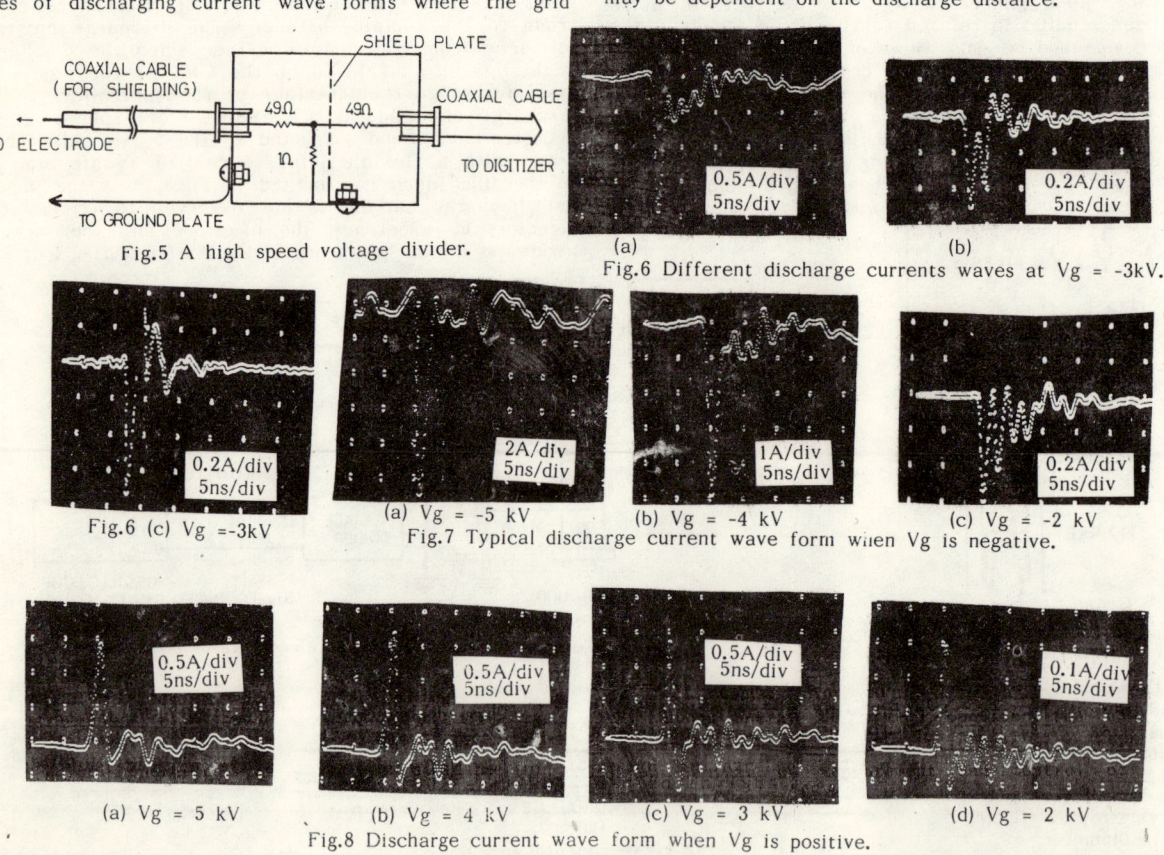

Fig.5 A high speed voltage divider.

Fig.6 Different discharge currents waves at V_g = -3kV.

Fig.6 (c) V_g = -3kV

Fig.7 Typical discharge current wave form when V_g is negative.
(a) V_g = -5 kV (b) V_g = -4 kV (c) V_g = -2 kV

Fig.8 Discharge current wave form when V_g is positive.
(a) V_g = 5 kV (b) V_g = 4 kV (c) V_g = 3 kV (d) V_g = 2 kV

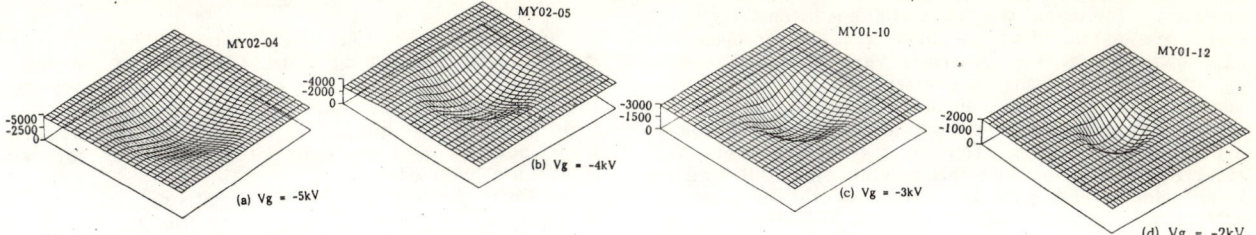

Fig.9 Surface potential profiles of negatively charged film after the electrostatic discharge.

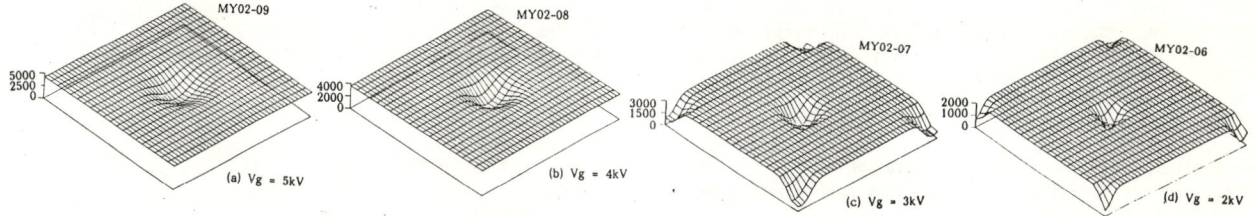

Fig.10 Surface potential profiles of positively charged film after the electrostatic discharge

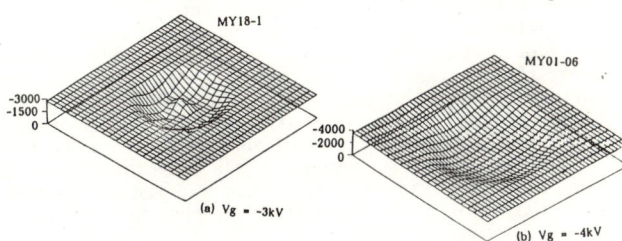

Fig.11 Surface potential profile of the film where the electrode contacted the film after the discharge.

The long distance causes a large inductance and degrade of the rising time. The peak values are fluctuated at each experiment, but only the single pulse and small following oscillation are observed at every case in general.

Surface Potential Profiles

Negatively Charged Surface   The surface potential profile just after the charging (before the test of electrostatic discharge) is very uniform and mostly as same as the grid voltage Vg when Vg is not so high as -3 kV. This boundary value is dependent on the conditions, especially on the humidity. Potential profiles after the discharge are shown in Figs. 9 (a) - (d) where the shown area is 20 X 20 mm square. When Vg is large, such as - 5 kV, the crater-like hole of the potential can be seen as large as about 15 mm in diameter where the needles are negative. On the other hand, the diameter of the hole is about 10 mm when Vg is - 3 kV. By the way, the surface potential of 1 kV_2 agrees with the surface-charge density of about 18 nC/cm^2.

Positively Charged Surface   Figures 10 (a) - (d) are surface potential profiles of positively charged teflon film just after the discharge experiment. There are two large differences compared with the formers. That is, the size of the crater hole, charge eliminated area, is very small of only 6 mm in diameter even though the Vg is as large as +5 kV. The crater diameter at Vg = +2 kV is about 4 mm, which is a little bit smaller than the former one at Vg = 5 kV. This relation is similar to that in negative charging case. The boundary wall of the potential well is very abrupt compared with that in the case of negative charging. The initial stage of this experiment, the stepping-motor stopping circuit triggered by the discharge, was not installed and the sphere electrode contacted to the film surface with some small pressure. At the center of the hole, a small hill of the potential was observed at sometimes like in Fig. 11. In other case, the level of the center hole was not so below as that when the electrode does not contact with the film. Those potential patterns (hill and so on in Fig.11) are not observed when the sphere electrode does not contact to the film surface. Therefore that center hill or some charge at the center part is surely due to the contact effects after the electrostatic discharge. As those phenomena occur occasionally, the reproducibility of the discharge is very wrong.

Dust Figures

The example photos of dust figures of the electrostatic discharge are exhibited in Figs. 12 and 13. In the case of negative charging in Fig.12, every dust figure is circle-like and the radial trace of the surface discharge is apparently observed. The diameter of that circle is monotonously with the grid voltage, Vg. On the other hand, the circle dot of the dust figure is small and no trace of the surface discharge is found when the film surface is charged positively. This surface discharge trace is effectively distinguished in the dust figure technology.

Photographs of The Electrostatic Discharge

The light emission of the electrostatic discharge is

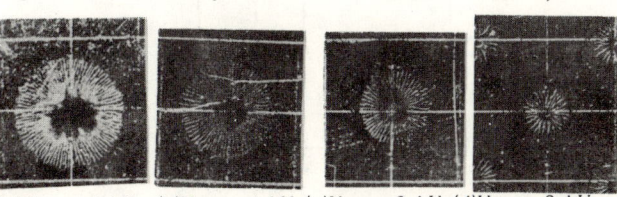

(a)Vg = -5 kV (b)Vg = -4 kV (c)Vg = -3 kV (d)Vg = -2 kV

Fig.12 Dust figures of the discharge (negative film).

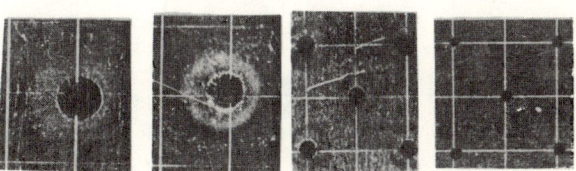

(a)Vg = 5 kV (b)Vg = 4 kV (c)Vg = 3 kV (d)Vg = 2 kV

Fig.13 Dust figures of the discharge (positive film).

very weak. It is impossible to take photographs of that directly. By using the handy-image intensifier tube, that trace of the discharge light was photographed such as shown in ig. 14. As it is very difficult to take pictures with the image converter camera, the time sequence of the discharge cannot be checked. In the case of negative charging, some surface discharge phenomena are taken in pictures. That is, a relatively weak vertical light bars between the moving electrode and the film surface which may be a trace of the electrode itself is seen and the bright surface discharge trace is observed to be divergent to the horizontal direction. On the other hand, a very bright but small point-like light trace is pictured when the surface is charged positively indicating the surface discharge (maybe surface glow corona[2]) occurs only when the dielectric film surface is charged negatively. This tendency is in good agreement with the results of dust figures, surface potential profiles and so on. As the polarity described here is that of the surface charge or the potential, that is quite opposite when the discharge term of the negative surface corona and so on (now it is positive here).

Discharge Distance

The electrostatic discharge distance between the sphere electrode and the film surface is not so constant and a little bit dependent on the surface potential of the film. Their relation is shown in Fig.16 where the distance range is plotted for each grid voltage which determines the surface potential of the charged film. When the grid voltage is only - 2 kV, it is not so easy to identify the spatial discharge, for the checking system shown in Fig. 4 is poor resolution of more than several ten microns. The authors have also examined at Vg = - 1 kV, but the result is very unstable and difficult to be listed in this paper.

The relation between the electrostatic discharge distance and the peak value of the current is also shown in Fig. 15 when surface potential of the film is about - 4700 V at Vg of - 5000 V. The inverse proportionality of the distance and the current is observed very well which will be explained by the following mechanism, that is, the large reactance at long discharge distance causes a little bit large rising time and smaller peak current values.

Conclusions

One example of the electrostatic discharge between the uniformly charged dielectric film (FEP of 100 μm in thickness) surface and the grounded sphere electrode (the ground means the same potential of the back electrode of the film here) is studied with relation of their polarity and surface potential. The following results are obtained:

1. When the film is charged negatively, the weak electrostatic discharge occurs from the sphere to the film just below the electrode at first and then stronger surface glow corona is observed by image intensified photographs and dust figures. The absolute peak value of the discharge current increases with the large absolute surface potentials. Especially, the change of the peak is very large (about twice) when the grid voltage increases from -4 kV to - 5 kV.

2. When that is charged positively, only the straight electrostatic discharge from the sphere electrode to the center surface of the charged film is recognized. The total discharge area in this case is smaller than that in the case of negative surface charging. The peak value of the discharge current is mostly independent on the surface potential.

3. The discharge distance is widely distributed and the electrostatic discharge occurs statistically. However, that distance is also strongly dependent on the surface potential in both case. The rising time of the discharge current is also affected by that distance. When that is small, the small rising time of only 1 ns is observed which is limited by the experimental setup. The discharge distance in the case of positive charging is a little bit larger than that in the case of negatively charging.

References

1) S.Kobayashi et al:"Discharge on Electrified Polymer Surface" Proc.2nd Int.Conf.ESP pp.1039-1044(1985)
2) Ed.by S.Masuda et al "Handbook of Electrostatics" p.224(1981) in Japanese, Ohm.

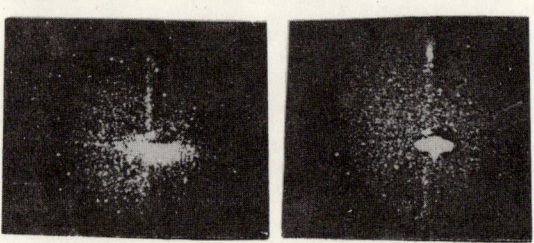

(a) Vg = -5 kV   (b) Vg = 5 kV
Fig.14 Photographs of the electrostatic discharge.

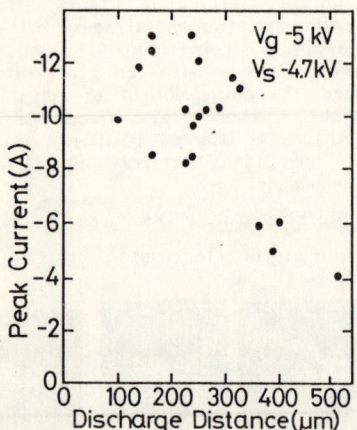

Fig.15 The relation between the electrostatic peak current and the discharge distance at Vg = -5 kV.

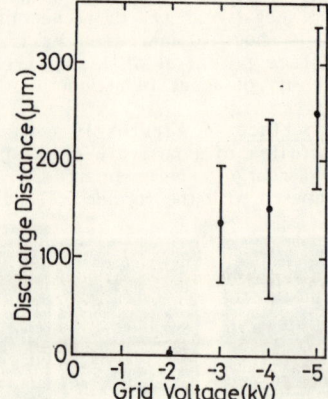

Fig.16 The distribution of the electrostatic discharge distances for different negative Vg.

Proceedings of International Conference on Modern Electrostatics

EXPERIMENTAL SEARCH FOR FREE QUARKS IN MATTER

C. D. Hendricks and R. S. Hornady
Lawrence Livermore National Laboratory
University of California
Livermore, California, U.S.A. 94550

Abstract

An apparatus has been constructed for the measurement of very small electric charges on macroscopic particles with a charge resolution better than 0.05e. Together with the charge measurement system, a generator of small liquid particles has been assembled and used to test the system. To achieve the charge resolution desired, some of the problems which were solved included particle interactions resulting from induced dipoles, particle position and timing measurements to accuracies of a few microns and nanoseconds, operation in high vacuum to avoid aerodynamic effects, and spurious charging of the particles by external effects. The technique, which depends on simple, straight-forward deflection of particles in an electric field, is described and some particle charge measurements are presented.

Introduction

The detection of free stable quarks in matter depends on the electrical charge carried by the quarks. Quarks are considered to carry $\pm 1/3$ or $\pm 2/3$ of an integer electron charge. Thus, a fragment of matter containing a single, free, stable quark should have a net charge equal to some integer number (positive, negative, or zero) of electron charges plus a fractional charge contributed by the quark. A system has been designed and constructed by which the charge on small particles of matter can be measured with respect to integer electron values to an accuracy of about 0.05e. Small particles of the material to be studied are generated and passed through an electric field deflection system. The trajectories of the particles depend on the q/m of the particles, the electric field intensity, and the time the particles are in the deflection field. The deflection field is produced by applying an electric potential difference to two parallel plates, between which the particles pass. The particles are all made with the same masses and the same velocities so the individual particle trajectories depend only on the net charge carried by the particle. Thus, the presence or absence of a free quark on the particle can be determined. If, for example, 1 gram of a material (6.023×10^{23} nucleons) is separated into small particles and analyzed with the charge spectrometer and one particle is found to carry a fractional charge, the conclusion is that there is 1 free quark per gram of that material. To obtain an adequate confidence level to make such a conclusion would, of course, require many grams of material be analyzed.

Experimental System   The system used to detect the presence of free quarks takes the form of a charge spectrometer and consists of several major parts:

(1) A particle generation system which produces a stream of uniform particles which are equally spaced, and in the absence of external fields or other disturbances, travel at the same velocity along the same trajectory. The particles are produced at rates up to a few hundred thousand per second and thus provide a mass throughput of a few grams per hour.

(2) A deflection system through which the beam of particles pass so the particles can be directed along different trajectories according to the net charge on each particle.

(3) A system for detecting the particles after they have been deflected and separated. The detection system is capable of a lateral resolution of 1-5 micrometers and can detect every particle which traverses the deflection system.

(4) A computer controlled data acquisition, analysis and storage system, which presents on a screen and on hard copy a histogram of the lateral separation of the particles which have traversed the deflection system.

The various parts of the apparatus are described in more detail in the following paragraphs.

Particle Generator   For the initial experiments, a low vapor pressure diffusion pump fluid is used for convenience. The fluid is non-toxic, does not evaporate in a high vacuum environment, has a relatively low viscosity, and has a moderate value for its surface tension. The values of the fluid parameters permit the formation of small mass particles by utilizing the controlled disintegration of a cylindrical liquid jet by the application of a capillary wave on the jet. The particles are liquid drops of the fluid and can be made quite easily with diameters from about 8 micrometers to more than 3 millimeters. For the research described in this paper, the lower range of sizes is most useful. There are several reasons for using smaller particles. It is expedient to have particles with the least possible net charge carried by the particles—a condition which is ensured by having small particles. It is also advantageous to have the particle mass be very small, which will be the case if the particles have small diameters. With small net charge values and small masses, adequate physical separation (~1 mm/electron) of the particles in the charge spectrometer can be obtained without having a particle spread which becomes larger than the dimensions of the apparatus. Further advantages accrued by using small particles will become apparent in the following paragraphs.

The liquid jet from which the particles are formed is produced by forcing the liquid under a pressure of 50 to 2000 kPa through a small orifice. The pressure is regulated to about 0.5 Pa so that the velocity of the liquid jet is maintained at a constant value. A capillary wave is launched on the jet by modulating the velocity of the jet with a periodic signal. The amplitude of the capillary wave grows as the wave and the jet travel away from the orifice (NOTE: This phenomenon was described by Lord Rayleigh for the small perturbation, inviscid case in 1878.) and produces from the jet, one liquid drop per wavelength of the capillary wave. It is possible to set the initial amplitude of the wave and the wavelength so that a satellite drop appears between each of the major drops produced, but for the best operation of this system, only the major drops are used and satellite drops are avoided. There is generally a shortest wavelength at which controlled drops are produced without difficulty and at such a wavelength the drops are formed approximately three diameters apart along the trajectory of the jet.

Deflection System

The particles (liquid drops in this case) are formed in a high vacuum and introduced vertically downward along the mid-line between deflection two plates (5 meters long, 15 cm wide and spaced 1.4 cm apart). The deflection plates are carefully aligned to be vertical in order that the drop stream with no deflection potential applied will exit the deflection zone along the same mid-line of the plates along which it entered at the top. Alignment is accomplished by the relatively simple technique of using a plumb line stretched between the plates. When the plumb line is along the mid-line at both the top and bottom of the plates, the plates are vertical. It is necessary, of course, that the plates be flat, parallel and rigidly supported before the vertical alignment can proceed.

Particle Detection System Particles traveling through the deflection zone will be deflected according to the net charge on the particles. A deflection of 1 mm/electron provides sufficient resolution that a particle containing a charge of 1/3 or 2/3 electron, in addition to an integer number of charges would be adequately separated and observable. This is true only if the detection system itself can resolve and measure the spacing of two particles which are spaced 0.33 mm apart laterally and may be between 0.06 and 1.2 mm apart vertically. The particles travel through the deflection zone sequentially, having started in a single file from a point source. The particles are illuminated after they leave the deflection zone by a moving light spot. The light spot moves laterally along a 1.5 cm long horizontal line immediately below the deflection plates. The line of the moving light spot is in the plane defined by the trajectories of all the particles. Reflected light from the particles is detected by a photomultiplier. Signals from the photomultiplier corresponding to a deflected particle will have a time relationship to the initiation time of the sweep of the light spot. By initiating a time counter at the beginning of the light spot sweep and stopping the time counter when a particle is detected, the lateral position of the particle is determined relative to the end of the line where the light spot sweep started. The moving light spot is generated by the motion of the spot on the face of a cathode-ray-tube monitor. The phosphor has an extremely short decay time and the spot size is small. The CRT sweep is imaged optically into a line at the bottom of the deflection zone where the particles are to be illuminated. The output of the master-oscillator is divided down in frequency and adjusted in phase to provide a sweep ramp to the CRT monitor. A reference gate, produced at the start of the ramp, is used to start a TDC (time-to-digital converter) and the PMT output pulse is used to stop the TDC. The interval times are stored in a histogram memory directly. After a suitable accumulation time, the memory is read by the control computer and the histogram presented on the computer monitor screen. The screen parameters are the number of PMT counts as a function of the horizontal endpoints of the particle trajectories. The trajectory endpoints correspond the the net charge carried by the particles which have been detected by the optical system, therefore, the histograms display the numbers of particles as a function of the particle net charge. The charge resolution is about $0.05e$ and the particle deflection per electron is nominally 1 mm so that with only integer electron charges present, a set of narrow spectral lines are displayed corresponding to net particle charges which differ by 1 integer electron charge. The presence of a free quark would be evidenced by a point on the histogram either at 1/3 or 2/3 of the distance between two integer electron peaks.

Computerized Data Acquisition and Analysis System

The experiment is monitored and controlled by a small computer (HP 310). The instrument interface is through CAMAC. The essential controls may be operated either locally by hand or remotely by computer. When switching to remote control, the software first determines the present adjustments and increments from that starting point. Thus, the control shift is "seamless" to the experiment.

All essential parameters are read at the time the histogram is captured to guarantee that all values are current. All parameters are archived along with the histogram to guarantee the data reconstruction.

The computer can run the experiment completely automatically acquiring, plotting and archiving the data; or the user may, under manual control, sequence through each operation. The user interface is by means of screen-labled-soft-keys. The software is written as a layered structure, three levels deep.

Experimental Difficulties and Possible Error Sources

A number of difficulties have made what appears to be a simple student-physics-laboratory experiment into a very difficult and relatively long project. As has been pointed out, the desired resolution in charge measurement is about $0.05e$. It was also decided to try to achieve a 1 mm separation between particles whose charge differs by one electron. Thus, all accumulated external random vibrations, velocity variations, particle mass variations, residual gas in the vacuum system, particle-particle interactions, and other effects on the particle's trajectories should not cause more deflection than is equivalent to an additional charge of $0.05e$. It is appropriate to enumerate and discuss the sources of variation and the means used to reduce the variations to the lowest possible values.

Liquid reservoir pressure variations The velocity of the liquid jet from which the particles are made, and hence the velocity and size of the particles are directly dependent on the pressure used to force the liquid through an orifice to form the jet. To avoid velocity fluctuations detrimental to the experiment, the pressure is controlled and monitored to an accuracy of about one part in 10^6. The viscosity of the fluid is about 20 cps, the orifice diameter is between 5 and 20 micrometers, the fluid density is 910 kg/m^3, and the pressure is 20 to 300 psig depending on the particular orifice in use. For these parameters, the velocity is very nearly a linear function of the pressure, so controlling the pressure to one part in a million maintains the velocity to about the same minimal variation.

External mechanical vibrations and shocks   There are constant vibrations and random movements associated with any occupied building, particularly one in which there are mechanical pumps, air conditioners, and other mechanical devices which move. Mechanical disturbances which couple to the charge measurement apparatus give rise to spurious movements of the particles and either broaden the lines in the charge spectrum or make it virtually impossible to acquire meaningful data. In order to mechanically decouple the apparatus from the surrounding environment, the entire system; including the diffusion pumps, 5 meter tall vacuum enclosure, optical detector system (including the light source), and other attached parts; is supported on low pressure elastomer air-bags. The pressurizing mechanism for the bags contains valves which are actuated by sensors which maintain the tower so that the deflection plates are vertical. That is, the orientation is such that if a particle is dropped along the center line at the top of the plates with no electric field present between the plates, the particle will emerge from the bottom of the deflection plates along the center line of the plates.

Induced dipole interaction effects   A material particle in an electric field becomes an induced dipole and if two such particles are close enough together, they will interact through the dipole fields and suffer deflections which may entirely mask the charge spectrum of the particles. As generated, the liquid drops used in this research are about three diameters apart, center-to-center. With this spacing, the dipole forces totally overshadow the net charge-field interaction. If, on the other hand, the drops are 60 diameters apart, the dipole interaction forces are at least 2 orders of magnitude less for a given electric field intensity than the force on one electron in the same field. To provide the wide spacing of the drops while permitting the generation of the drops under optimum conditions, a rotating, slotted, chopper disc is used to allow every twentieth drop to pass into the deflection zone and to remove from the system the other nineteen undesirable drops. The rotational speed of the disc must be well synchronized with the drop frequency to allow one and only one drop out of twenty, to pass the disc. The parameters which determine the chopper process are the width, radius and number of the slots in the disc; the disc thickness; the rotational speed of the disc; the drop velocity, frequency and initial spacing; and the thickness of the disc. The nominal chopper parameters around which adjustments can be made are: Motor speed = 60 rps; slot radius = 0.1 meter; slot width = 450 micrometers; disc thickness = 12.5 micrometers; and number of slots = 75. These parameters are chosen to permit removal of 19 of 20 drops and to allow every 20th drop to pass through the system.

Several other phenomena must be considered before the charge spectra can consist of narrow lines with totally clean field between the lines.

Deflection plate potentials   The electric potentials applied to the deflection plates must be constant DC with no fluctuations or ripple. Modern power supply technology easily provides up to 60 kv with only a few millivolts of ripple or fluctuation. This is particularly the case when the load on the power supply is virtually zero as it is in this experiment.

Residual gas   The residual gas in the vacuum must be reduced to a value that leads to negligible deflections of particles as the result of particle-gas molecule interactions. A pressure of $10^{-6} - 10^{-7}$ torr ensures that deflections from this source will not be significant.

Particle charge changes in flight   If a particle changes charge in flight, the end-point deflection measured will not provide an accurate measure of the particle charge. If a particle changes charge from N to N+1 electrons at a point 43% of the way along its trajectory, it will reach the same position at the end of its flight as would a particle having N+1/3 electrons from the beginning of its flight. Charge changes can occur because of several phenomena. These include collisions with ions or electrons from photoionization, field ionization, cosmic ray debris and radioactive decay. The first two can be avoided by appropriate choice of operating conditions. The use of coincidence counters and background measurements can detect events which may cause spurious data or provide information to assist in the analysis of data and elimination of unlikely events from serious consideration.

Conclusions

An apparatus has been constructed and techniques have been developed which will permit the assay of matter for the presence of free, stable quarks. Preliminary results and tests of the system indicate a resolution of one free quark in 10^{23} nucleons. This resolution is almost three orders of magnitude improvement as compared to past results. Additionally, the throughput of material in this apparatus is grams per day. Published results indicate the total mass sampled to data by all other systems is less than 50 milligrams.

A schematic diagram of the major components of the experiment and of the control system is shown.

This work was performed under the auspices of the U.S. Department of Energy by the Lawrence Livermore National Laboratory under contract number W-7405-ENG-48.

1499A-5/23/88

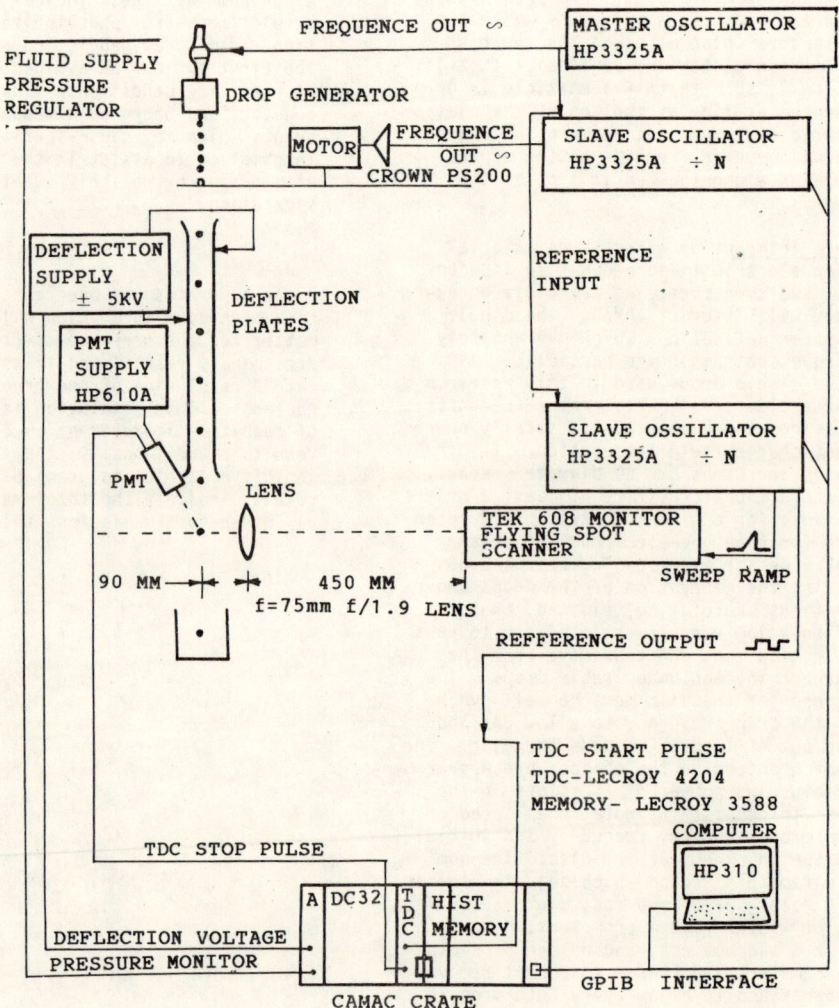

SEPARATING DISCHARGE IN FILM/ROLLER SYSTEM

Xianggang Ji*, Yutaka Komai, Yuzo Takahashi and Shigeo Kobayashi
Department of Electrical Engineering
Tokyo University of Agriculture and Technology
Koganei
Tokyo 184, Japan

* on leave from
Beijing Institute of Technology
Beijing, China

Abstract

Separating discharges occur in film/roller system, where an electrified PET sheet is separated from a grounded metal cylinder. Pictures of discharge light are taken with image intensifiers aided camera, and powder figures are made by sprinkling toners on the sheet. Tree-like discharges occur on negatively electrified sheet, and oval-shaped discharges appear on positively electrified sheet. For negative electrification at surface charge density about $-0.4 \times 10^{-4} C/m^2$, there occurs large tree-like discharge whose branches may be longer than 70 mm, and charge transfer can be as much as 5000pC.

Introduction

Electrostatic discharges occur in industrial, domestic and even space environments, and lead to failures and hazards. Such discharges often result in degradation of products in manufacturing of plastic and paper materials, photographic materials and in printing processes. The discharge may trigger explosion of inflammable liquids, gases and powders. The discharges will cause EMC interference in electronic equipments for communication, control and computer use.

This contribution deals with the discharge in film/roller system---it takes place when an electrified polymeric sheet is separated from an earthed metal cylinder, i. e. separating discharge. There are a lot of publications on EOS (Electrical Overstress) and ESD. Studies on characteristics of separating discharge, however, have rarely been reported before. Separating discharges on PMMA sheet have been investigated by the authors [1] [2]. The authors measured also the light and charge transfer of separating discharges on a PET loop specimen [3]. In the present study, silicone coated PET sheet is used for investigating the influence of surface condition on the discharges. Light of separating discharges is measured with image intensifiers aided camera, and powder figures of the discharges are made by sprinkling toners on the sheet. Charge transfer of the discharges is estimated through the measurement of current pulses. The development of surface discharge is explained with charge neutralization by preceding discharges.

Method of Experiment

Figure 1 is a schematic diagram of the apparatus for studying separating discharges. The sheet which was wound on to the metal cylinder is electrified by the corotron(corona wire), the surface potential of it is measured with field-mill, and it is finally separated from the cylinder as the cylinder rotates. One end of the sheet is fixed to the cylinder and a weight of 1 kg is hung at the other end. The diameter of the cylinder is 230 mm, and peripheral velocity of the cylinder was 70 mm/s.

The sheet used is PET sheet with silicone coating on one side, 38μm in thickness, 100 mm in width and 240 mm in length. A new sheet was used for each trial.

The discharges occur on plain PET surface when the sheet is electrified on silicone coated surface and vice versa. The discharge bridges the air gap and spreads on the sheet. Discharge light was photographed with image intensifiers aided camera. Intensifiers used are VARO type 3603 and HAMAMATSU type V1366P. Powder sprinkled to get discharge figures on the sheet is mixture of red and black developers for facsimile use. Red powder sticks to negatively charged sites, and black powder sticks to positively charged sites. The experiments were carried out in a room at 21-23°C and RH 37-45%.

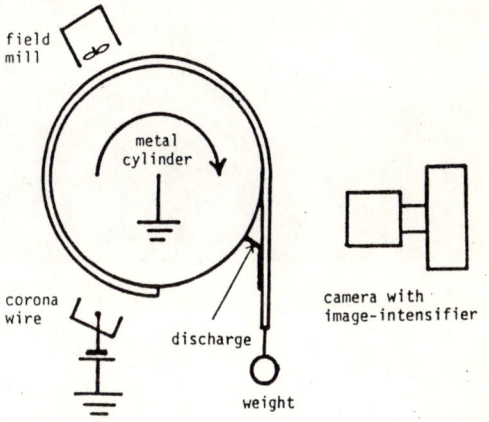

Fig.1 Schematic diagram of the separating apparatus

Discharge Light

There are no big differences between light of discharge on PET surface and that on silicone coated surface. Photograghic pictures of light of discharges on PET surface are shown in Fig.2 for negative electrification and in Fig.3 for positive electrification.

For negative electrification, the figures of the discharges are tree-like, and the branches are sharp. These show typical features of positive discharge figures. The positive discharge figures appear because the metal cylinder electrode or the discharge column bridging the air gap is positive in comparison with the insulating surface. The figures of discharges are of extreme asymmetry, showing difference to normal radial discharge figures of partial discharge in uniform field. The asymmetry in discharge figures is probably ascribed to the wedge-shaped air gap.

In contrast to the discharge figures for negative electrification, figures of discharge light for positive electrification are round or oval-shaped. The edges of the discharge figures are fuzzy, showing features of negative discharge figures.

The figures of discharge light differ also for different surface charge density on the sheet. For negative electrification at $-0.38 \times 10^{-4} C/m^2$, that is near the lowest level of electrification at which the separating discharge occurs, there appear large tree-

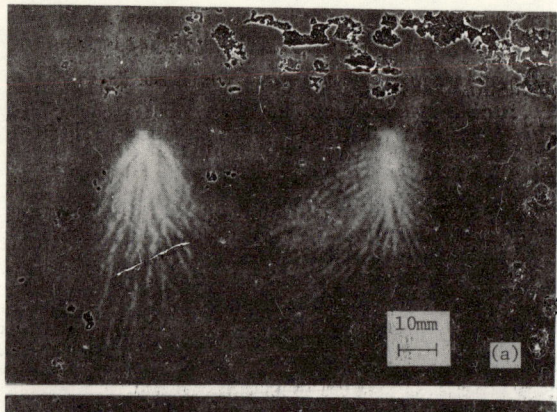
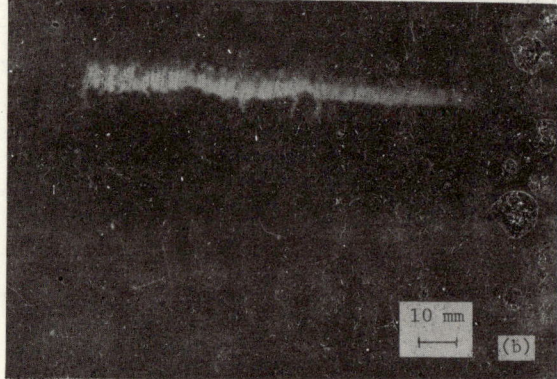

Fig.2 Photographic pictures of discharge light for negative electrification
a) $\sigma=-0.38\times10^{-4}C/m^2$ b) $\sigma=-1.10\times10^{-4}C/m^2$

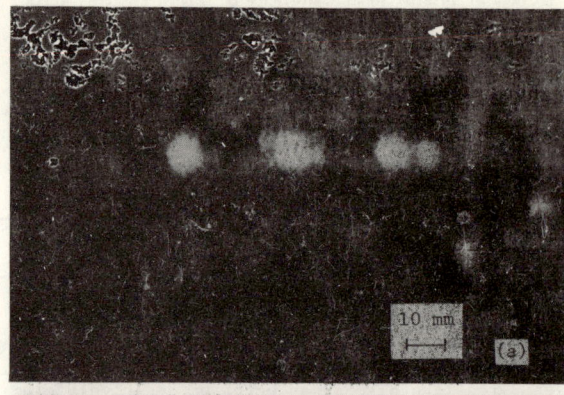
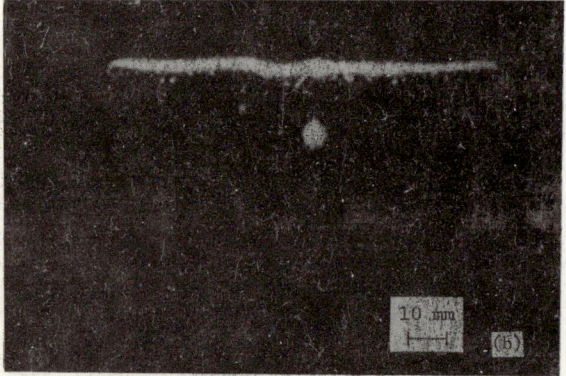

Fig.3 Photographic pictures of discharge light for positive electrification
a) $\sigma=0.45\times10^{-4}C/m^2$ b) $\sigma=1.81\times10^{-4}C/m^2$

like discharge figures which have a lot of branches. Changes in discharge figures will be shown in more detail later in powder figures of the discharges.

In Fig.3(b) two types of discharges are observed: many small discharges which look like a bright line and larger discharges that we call secondary discharges. The secondary discharges tend to occur at large surface charge density, and for negative electrification they are tree-like and can be large [2].

Powder Figures of the Discharges

Powder figures of discharges on silicone coated surface are almost same as those on PET surface. Examples of powder figures of the discharges on PET surface are shown in Fig.4; (a)-(c) for negative electrification and (d)-(f) for positive electrification. Characteristic features of the powder figures are very similar to the figures of discharge light in Figs. 2 and 3.

At surface charge density $-0.38\times10^{-4}C/m^2$, that is near the lowest level of electrification at which the separating discharge occurs, there appear large tree-like discharge figures which have a lot of branches. At $-0.64\times10^{-4}C/m^2$ many asymmetrical tree-like figures are observed, the branches of which are shorter than those in Fig.4(a). At $-1.74\times10^{-4}C/m^2$ discharge figures become lateral stripes. At $-2.4\times10^{-4}C/m^2$ large tree-like figures are observed, which are the traces of "secondary discharges".

Relation between the length of the discharge figures and the surface charge density for negative electrification is plotted in Fig.5. The length of the branches of the tree-like figures is more than 70 mm at $-0.38\times10^{-4}C/m^2$.

For positive electrification, at $0.45\times10^{-4}C/m^2$ the discharge figures are round or oval-shaped, the diameter of which is 4-10 mm. At $0.91\times10^{-4}C/m^2$, lateral stripes are observed, and at $1.81\times10^{-4}C/m^2$ there appear some narrow lateral stripes and a few round figures. The round discharge figures in this case, are probably the traces of the "secondary discharges" shown in Fig.3(b).

Charge Transfer of the Discharges

Current pulses of separating discharges were detected with RC impedance connected to the cylinder, where R=2.5kΩ and C=4000pF (τ=10μs) were chosen. The pulses were observed with oscilloscope, and peak values of them were recorded with an electromagnetic oscilloscope through a diode-detector for estimating the apparent charge transfer of the separating discharges.

Figure 6 shows the maximum charge transfer of the discharges at various levels of electrification. For negative electrification, at surface charge density about $-0.35\times10^{-4}C/m^2$ the charge transfer of one pulse was more than 5000pC. The charge transfer decreases as the surface charge density increases up to about $-2.1\times10^{-4}C/m^2$. At larger surface charge density, charge transfer of the discharges becomes the order of 10^3pC because of the occurrence of the tree-like

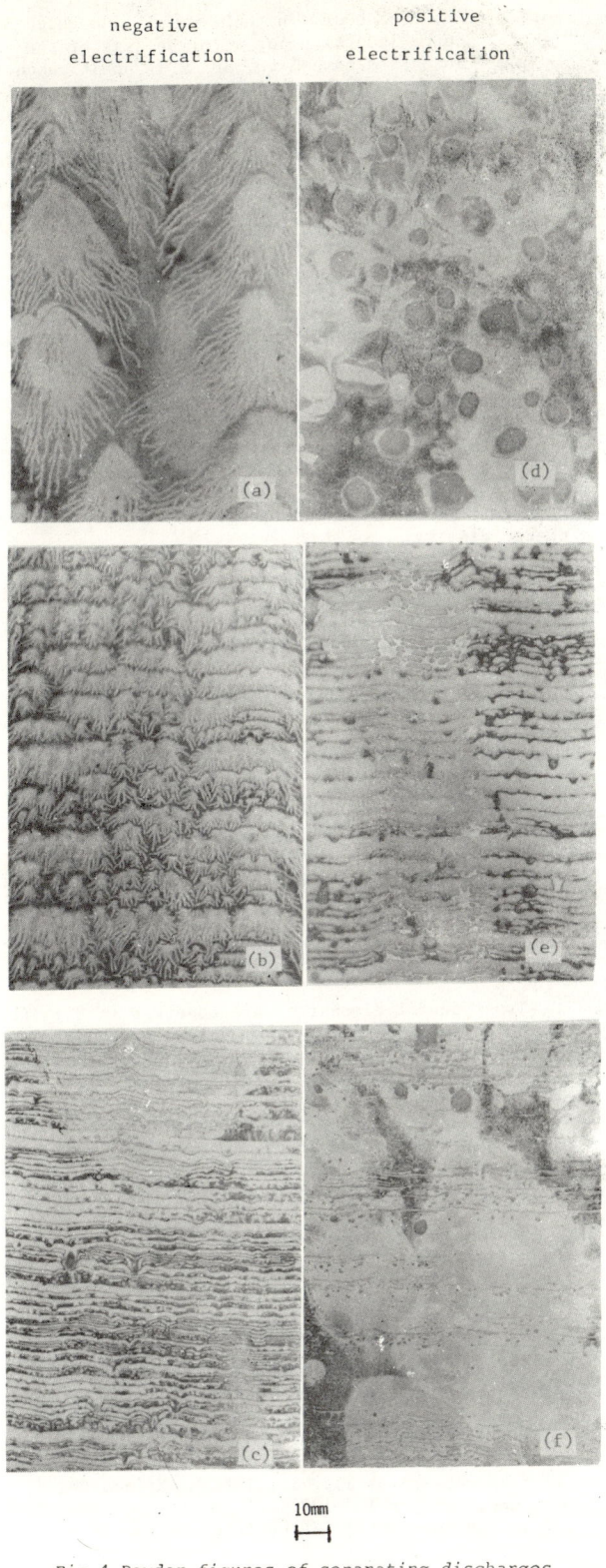

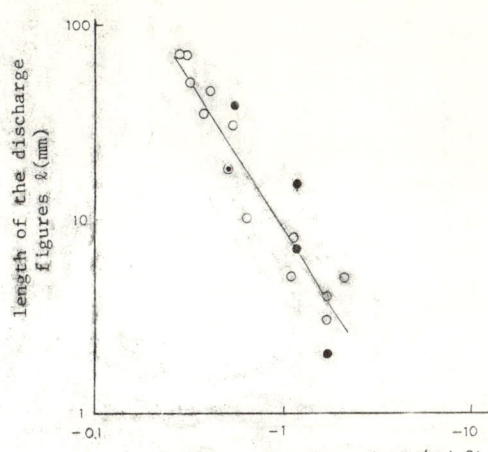

Fig.5 Relation between the length of the discharge figures and the surface charge density with negative electrification

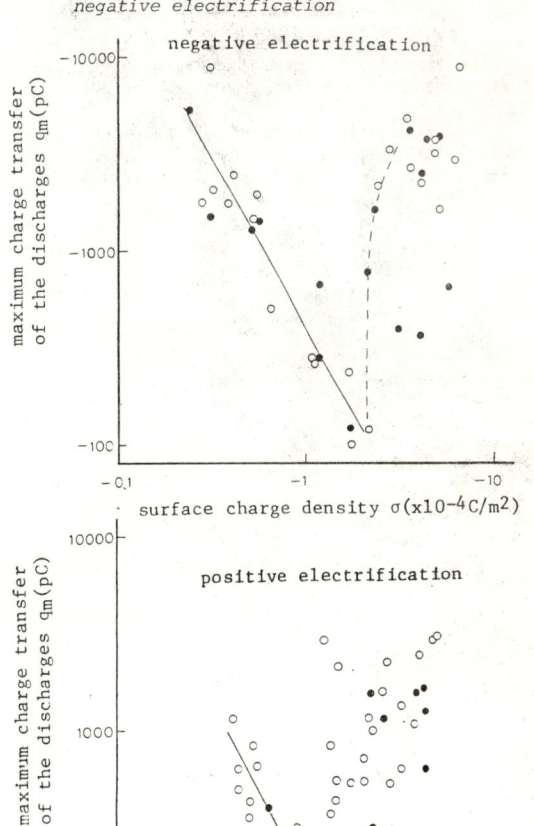

Fig.4 Powder figures of separating discharges on PET surface of the sheet
a) $\sigma=-0.38 \times 10^{-4} C/m^2$ d) $\sigma=0.45 \times 10^{-4} C/m^2$
b) $\sigma=-0.64 \times 10^{-4} C/m^2$ e) $\sigma=0.91 \times 10^{-4} C/m^2$
c) $\sigma=-1.74 \times 10^{-4} C/m^2$ f) $\sigma=1.81 \times 10^{-4} C/m^2$

Fig.6 Maximum charge transfer vs. level of electrification

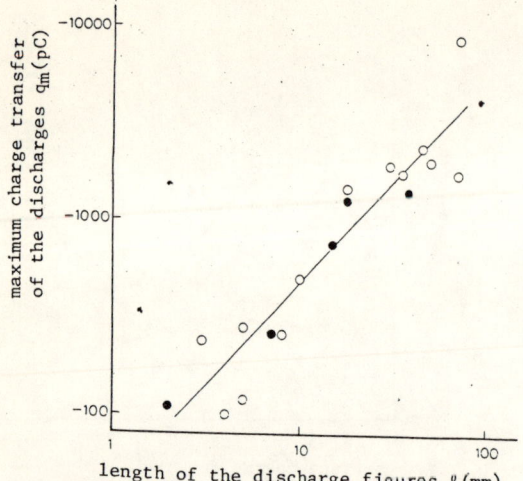

- ○ discharges on PET surface
- ● discharges on silicone coated surface

Fig.7 Maximum charge transfer vs. the length of the discharges with negative electrification

secondary discharges. One can not find the difference in charge transfer between discharges at PET surface and at surface with silicone coating.

For positive electrification, changes in charge transfer with the change of surface charge density are similar to those for negative electrification. However, the charge transfer of the discharges is less than that for negative electrification on the whole: for example, 1000pC at $0.4 \times 10^{-4} C/m^2$. Furthermore, charge transfer of discharges on silicone coated surface seems to be less than that on PET surface.

The charge transfer characteristics are considered to be related to the features of discharge figures. In Fig.7, relation between the charge transfer and the length of discharge figures is depicted for discharges with negative electrification. The relation is found to be $q_m \propto \ell$ approximately.

Estimation of the Surface Potential

A partial capacitance model which is illustrated in Fig.8, is developed to estimate the inner surface potential of the sheet. The horizontal line which is in the same horizontal plane with the axis of the cylinder is called base line, and x is position on the sheet relative to the base line. Assuming the electric field at x is determined by partial capacitances, and dissipation or movement of the surface charge is permissibly small, one can calculate voltage V_d across the air gap, or the inner surface potential at x for a given σ, the surface charge density. Here c_t represents the capacitance per unit area of the sheet, and c_d that of the air gap. We get

$$c_d = \varepsilon_o/d, \quad (1)$$

and

$$V_d = \sigma/c_d = \sigma d/\varepsilon_o, \quad (2)$$

where d is the air gap length and ε_o permittivity of free space.

Comparing the equation (2) to breakdown voltage of air gap (Paschen voltage)[4], one can estimate that the minimum electrification level for separating discharge is about $0.35 \times 10^{-4} C/m^2$ and that the discharge air gap length is about 2 mm. The separating discharges will occur at shorter gap for larger surface charge density.

Discussions and Conclusions

Figures of separating discharges differ for different surface charge density on the sheet. This may be accounted for by effect of charge neutralization due to the preceding discharges. Powder figures of discharges with negative electrification indicate that the tree-like surface discharge tends to stretch on electrified sheet until the tips of it meet the neutralized zone produced by the preceding discharges. Consequently, at low electrification level the harmful large discharge occurs easily, since the temporal and spatial interval of the discharges is quite long. At higher level of electrification, the branches of surface channel are short or even invisible: the discharges occur so frequently that they meet the neutralized sites very near to them. We see the lateral stripes in this case.

Two types of discharges are observed in Fig.3(b). The occurrence of them is considered in the following process: at small gap many small discharges occurred according to Paschen's law and they neutralized part of the charge on the sheet. The charge neutralization is not enough, and secondary discharges occur on the sites where small discharges occurred, at large gap. In this way, at very high level of electrification, large discharges occur again [1] [2].

No remarkable difference was found between the separating discharges on PET surface and those on silicone coated surface.

Acknowledgement

Thanks are due to the support by Yazaki Memorial Foundation for Science and Technology.

References

[1] Y. Takahashi et al., 1st ICPADM, Xi'an, China, June, 1985, p.135-138.
[2] Y. Takahashi et al., submitted to the IEEE Trans. EI.
[3] X. Ji et al., presented at the 2nd ICPADM, Beijing, China, 1988.
[4] H. Lau, "4432 Durchbruchspannungen in Gasen- 44321 Homogenes Feld", in Landolt-Börnstein Zahlenwerte und Funktion, 7. Aufl., IV Band, 3. Teil, Berlin, Springer, 1957, p.107.

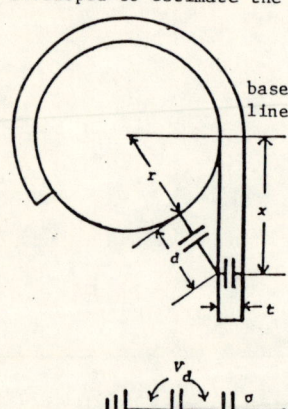

Fig.8 Equivalent representation consisting of partial capacitances.

CHARGE TRANSPORT PROCESSES AT CORONA-CHARGED POLYMER FILM SURFACE

Yang Baitun   Liu Yaonan

Department of Electrical Engineering
Xi'an Jiaotong University, Xi'an, China

Abstract

In this paper surface potential of the corona-charged polymeric film was detected by using a non-contact induction probe 1mm in diameter. It was found that some charges could move along the surface of Polyethylene terephthalate(PET) in the initial stage after charging, in case of decay of the average surface potential of the sample. But the phenomenon was not observed at the surface of Polyethylene(PE) film. Based on experiment, the surface process and bulk process for charge transfer were analysed further. It suggests there may be three reasons for the lateral motion of charge on the thin film. An mathematical treatment of the standard transport equation leads to an expression by which the bulk process is described.

Introduction

It is important to investigate the mechanism of charge transport in polymer for eliminating electrostatic hazards and developing the applications. In recent years, several authors have carried out much research work by using the technique of surface-potential decay[1-12]. Various theories have been developed for surface potential decay from Ohm's law, to partial instantaneous injection of charge, completed injection, continuous injection, field dependent mobilities and surface states as well as dynamics of trapping[1-15]. However, the mechanism of charge transport is not known well up to now due to the complicated phenomena and the limitations of theoretic consideration.

In this paper, a non-contact induction probe 1mm in diameter was used to detect the surface-potential of corona charged Polyethylene terephthalate(PET) and Polyethylene(PE) film at various times after charging. The transfer of charge along the surface on the charged PET film at room temperature was observed. But it did not occur on that of charged PE. The surface process and bulk process for charge transfer were further analysed and some useful results were obtained.

Experiment procedure

The thicknesses of PET and PE films are 25 and 50 μm respectively. The samples, 4cm^2 in area were cut from a large sheet of film. The surface of the samples were thoroughly cleaned with ethanol and deionized water and then dried. They were mounted on a grounded metal turntable. The corona charging device consists of many needle electrodes situated above the surface of the turntable. The samples were exposed to the charging source for a short time, say 30s, and the corona ions deposited on the free surface. The surface of films could be scanned and the distribution of surface potential Vs is determined at various times after charging at temperature 25°C and relative humidity 50% approximately.

Surface processes

Suppose an insulating film of thickness $d(0 \leq x \leq d)$ on one surface($x=d$) there is equipped with a grounded electrode, while the free surface($x=0$) is corona charged. At the atmosphere the most predominant ion produced by a negative corona would be CO_3^- and by a positive corona it would be the proton with various degree of hydration $H^+(H_2O)_n$ and also hydrated versions of NO^+ and NO_2^+. Of course, the large quantities activated neutral species M*, metastable states and phonons are likely to arrive at the surface together. The ion could remain as a stable entity adsorbed on or within the surface layers of polymer and be attracted by the induced field between it and the grounded plane. One of three possible processes will then occur as shown in Fig.1. In (a) the occupied state E_I of negative ion is lower than surface state E_A' of the polymer, the ion remains on the surface, if the activation energy provided by the species M* is below the potential barrier between E_I and E_A'. In (b) the occupied state E_I is within the distribution of E_A' or M* activation makes it so, and the charge transfers into the polymer surface but remains there since E_A' is below E_A of adjacent bulk states. In (c) the condition are same as in (b), expect that now $E_A' > E_A$ and the charge may now migrate into the bulk and thence to the grounded plane. Charge transfer in (b) and (c) may be acheived by activation hopping or quantum mechanical tunnelling. The similiar analysis can be carried out for positive ion as shown in Fig.2, where the charge transfer is realized by hole.

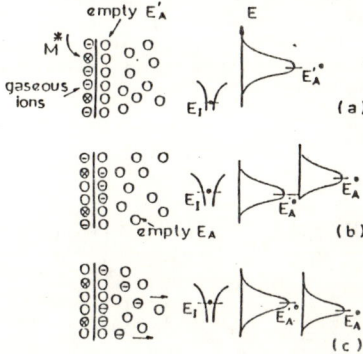

Fig.1 Electron transfer from gaseous negative ions on a polymer surface in presence of activated neutral species M*

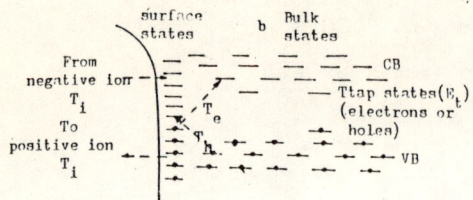

Fig.2 Electron(T_e) and hole(T_h) transfers from surface states to conduction (CB) or valence band (VB) bulk states. Also shown is the possibility of direct transfers (T_i) from incident ions to bulk band states.

When the material and the way of charging were given, one could assume that a discrete set of such transfers exists so that a proportion a_n of the initial charge deposited transfer to bulk states at a specific rate α_n, then if $V_s(0)$ is the surface potential immediately after charging, i.e. at t=0, the field at the surface is given by[15].

$$E(0,t) = \frac{V_s(0)}{d} \sum_{n=1}^{\infty} a_n \exp(-\alpha_n t) \quad (1)$$

where the summation is over all the surface states and $\sum a_n = 1$, we note that charge injected "instantaneously" corresponds to $\alpha_n = \infty$ and deeply trapped surface charge corresponds to small α_n and the limiting value zero.

Bulk processes

Charge injected from the surface transfers throughout the bulk, and then reaches the rear electrode and leaves the sample. It is assumed that the material has unipolar conduction, the charge of the carriers is q, the mobility of carrier is μ, and the dielectric constant of the material is ε. It is also assumed the material contains deep traps with density Nt, capture cross section Sc, and with a trapping time τ, liberation from traps will be neglected, and we will assume that the traps are far from fully occupied.

Neglecting the diffusion component of the current, the basic equations expressing the variation of surface potential with time are, respectively, Poission's equation, current continuity equation, open-circuit condition, and dynamics of trapping, as written below:

$$\varepsilon \frac{\partial E(x,t)}{\partial x} = p_1(x,t) + p_T(x,t) \quad (2)$$

$$\mu \frac{\partial [p_1(x,t) E(x,t)]}{\partial x} = -\frac{\partial [p_1(x,t) + p_T(x,t)]}{\partial t} \quad (3)$$

$$\mu p_1(x,t) E(x,t) + \varepsilon \frac{\partial E(x,t)}{\partial t} = 0 \quad (4)$$

$$\frac{\partial p_T(x,t)}{\partial t} = \frac{p_1(x,t)}{\tau} \quad (5)$$

where $E(x,t)$ is electric field, $P_1(x,t)$ and $P_T(x,t)$ are the desity of mobile and trapped charge, respectively.

Using the method of characteristics[29] form Eqs. (3), (4) and (5), we have

$$dV_s(t)/dt = -(\mu/2)[E^2(d,t) - E^2(0,t)] + V_s(t)/\tau \quad (6)$$

Equation (6) gives the relation between surface potential and time. If $E(d,t)$ and $E(0,t)$ are given, we can get the result by solving the differential equation. Neglecting the trapping process, equation (6) can be reduced into the following equation

$$dV_s(t)/dt = -(\mu/2)[E^2(d,t) - E^2(0,t)]$$

it is the Sonnonstine's linear decay process[3].

The charges injected from the surface may spread in the direction of transfer before they reach the rear electrode. Assume the time required for the front of charges reaching the other electrode is t_d. For the time interval, $0 \leq t \leq t_d$, there is the relation:

$$E(d,t) = \frac{V_s(0)}{d} \quad (7)$$

substituting Eq.(1) and (7) into Eq.(6), we have

$$\frac{dV_s(t)}{dt} = -\frac{\mu}{2} \left(\frac{V_s(0)}{d}\right)^2 \left[1 - \left(\sum_{n=1}^{\infty} a_n \exp(-\alpha_n t)\right)^2\right] + \frac{V_s(t)}{\tau} \quad (8)$$

For partial instantaneous injection, corresponding to $\alpha_1 = 0$, by solving the equation (8), we get

$$V_s(t) = -\left[\frac{\mu\tau}{2}\left(\frac{V_s(0)}{d}\right)^2(1-a_1^2) - V_s(0)\right]e^{t/2} + \frac{\mu\tau}{2}\left(\frac{V_s(0)}{d}\right)^2(1-a_1^2) \quad (9)$$

The equation (8) gives typical decay form. The transfer time t_d can be determined following

$$t_d = \frac{d^2}{\mu V_s(0)} \quad (10)$$

For $t \geq t_d$, the charge reaches the rear electrode and begins to leave the sample, the equation (1) is still valid, but $E(d,t)$ has a different form determinable by the method of characteristics.

Results and Disccussion

A set of curves showing the surface potential distribution of PET measured at different time after charging is described in Fig.3 (a), by comparing the curve b with curve a, we see that some charges move along the surface of film. This phenomenon was also observed by E.A.Baum[16]. He proposed that the physical reason for the lateral motion behaviour is due to charge diffusion. In Fig.3 (b), we can see that the average potential of the sample is descent with time.

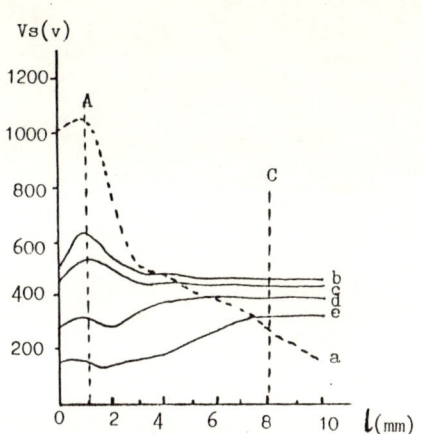

Fig. 3(a) Decay of Vs(t) for negative corona-charged PET. Elapsed time after the charging are, a: 10min, b,c,d, e: 4,10,80, and 180 hours, respactively.

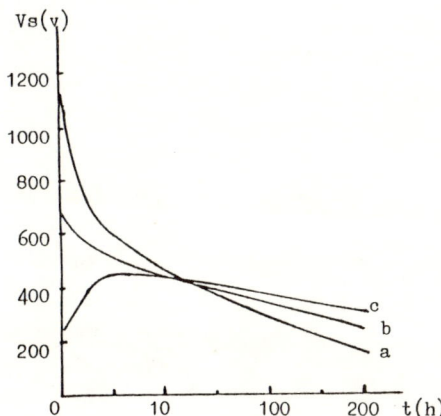

Fig.3(b) Surface potential of PET depends on time, curve a,c corresponds to A,C shown in Fig.3(a), and curve b is the average potential of the sample.

The investigation of the mechanism for the lateral motion is necessary since the behaviour does not occur in other polymer such as PE. It may be caused by the interaction between the charge ions and the surface of polymer. The possible reasons are: (1). The time for mobile charges remaining on the surface of PET is longer. This is a necessary condition, since the charge, once moving into surface state or bulk state, are impossible to move again along the surface, unless they were activated. Takada etal[22] drew attention to the fact that, in comparison with PE, both positive and negative charge would remain on the surface of PET for a long time even in the field of 2×10^{-8} v/m. Our experiment also shows that the surface potential of PET almost does not decay in initial ten minutes after charging. The cause is not due to the deeper traps existing in the surface states of PET. As the open circuit TSC experiments, carried out by Heine Von Segger[23], show that the surface state traps in PET are lower than that of FEP-A and PE. Based on results of experiment and the analysis described above, we conjecture that ion charges do not complate the direct transfer into the traps in the surface of PET when they are deposited on the surface. The charge transfer from ion state into surface state, likely needs some activation energy which could be supplied by the activated neutral species M^* as shown in Fig.1. So it requires a time process. (2). There is a proper surface resistivity. The surface resistivity of polymer is generally very high. However, the electrical properties of polymer will be changed by charging[24,25]. Some authors[26] indicated that the ion implantation results in the formation of conductive grains or islands in polymers and causes the increase in electric conductivity by some even more than ten orders of magnitude. We have enough evidence to say that the surface resistivity of corona charged polymer is so low that lateral motion of charge is possible. (3). There is a tangential electric field. The field decreases the energy barrier height over which the charge transfer along the surface.

The typical decaying characteristic of surface potential of PE is shown in Fig.4. There are some difference between positive(curve a) and negative(curve b,c) corona charged. The experimental result(curve a) and theoretical value calculated by equation (9) are tabulated in table 1. for comparison. The parameter, a1=0.3, is used in calculation, and we obtain the trapping time τ=265s and the trap-modulated mobility $\mu=1.16 \times 10^{-10}$ cm^2/v.s at room temperature.

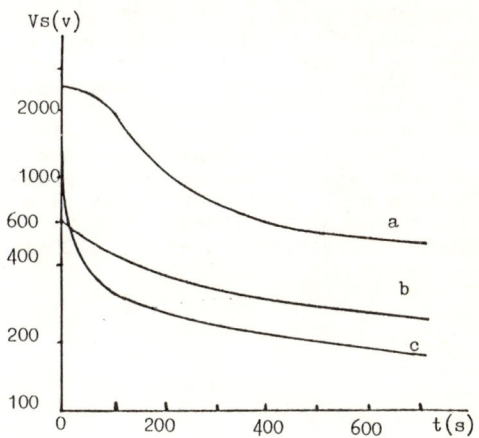

Fig. 4 Decay of Vs(t) for positive(curve a) and negative(curves b,c) corona-charged PE

The equations (8) and (9), can also be applied to curve b and c in Fig.4 by choosing proper parameter, we can obtain satisfactory results,too. The equations are particularly effect for the decaying curve of surface potential of which initial stage is of arched form. The curve, was also observed by provious authors[11,27], in contrast to typical form, its quadratic derivative in initial stage is smaller than

-345-

zero. They can be interpreted by the theroetical analysis stated in section 3.

Table 1. Comparing of the experimental data and theoretical value

time(s) Vs(v)	0	25	50	75	100
experimental	2500	2403	2300	2166	2040
theoretical	2500	2401	2292	2173	2043

Reference
[1] B.Gross,etal,J. Appl. Phys.45,2841(1974)
[2] G.M.Sessler,etal,J.Appl.Phys.43,922(1972)
[3] T.J.Sonnonstine,etal,J.Appl.Phys.46,3975(1975)
[4] M.Iede,etal,Jpn. J.Appl.Phys.6,793(1967)
[5] M.Iede,etal,Electr.Eng. Jpn 88,67(1958)
[6] I.P.Batra,etal,J.Appl.Phys. 41,3416(1970)
[7] I.P.Batra,etal,J.Appl.Phys. 42,1124(1971)
[8] H.Seki,etal, J.Appl.Phys. 42,2407(1971)
[9] H.J.Wintle, J.Appl. Phys. 41,4004(1970)
[10] M.Campos,etal,Appl.Phys.Lett.32,794(1978)
[11] E.A.Baum,etal,J.Phys.D.Appl.Phys.10,487(1977)
[12] E.A.Baum,etal,J.Phys.D.Appl.Phys.10,2525(1977)
[13] M.Campos,etal,J.Appl.Phys.52,4546(1981)
[14] Heinz Von Seggern,J.Appl.Phys.50 7039(1979)
[15] R.Toomer,etal,J.Phys.D.Appl.Phys.13,1343(1980)
[16] E.A.Baum,etal,J.Appl.D.appl.Phys.11,963(1978)
[17] Shain,etal,J.Chem.Phys.42,2600(1966)
[18] R.S.Segmod,etal, Inst.Phys.Conf.Ser.66,81(1983)
[19] T.John Lewis, IEEE Tran.Elec.Insu.EI-21, 3,289(1986)
[20] Chudleigh PW,J.Appl.Phys.48,4591(1977)
[21] G.M.Sessler,"Electrets" New York ,1980
[22] Takada,etal,Elect.Eng.Japan,92,28(1972)
[23] Heinz Von Seggern, J.Appl.Phys.52,4086(1972)
[24] Rapos,etal,Proc.IEE Vol.152,No.2,162(1978)
[25] D.T.Clark & W.J.Feast,"Polymer Surface" New York 1978
[26] M.S.Dresselhaus,etal, Mat.Res.Soc.Symp. Proc.Vol.27, 413,(1984)
[27] H.T.M.Haenen, J.Electrostatics,1, 173(1975)
[28] O.N.Oliveiva,etal,Appl. Phys. A 42, 213(1987)
[29] T.Kanra,etal,J.Electostatics 19,45,(1987)
[30] R.A.Moreno, etal,IEEE Trans.Elec.Insu. EI-21,No. 3, 319(1986)
[31].Zahn,etal,J.Electrostatics 1,235(1975)

CALCULATION OF THE ONSET VOLTAGE OF STREAMER
From A Grounded Hemispherically Capped Cylinder To Negative Well-charged Dielectric Plate

Ni ZhiQuan

Department Of Electronics, Hebei University, China

Abstract

For a grounded cylinder with a hemispherical end to a negatively well-charged dielectric plate, the onset voltage of streamer is calculated. The results show some similarities to that of Nasser et al's[1] whose model is a hemispherically capped metallic cylinder to a grounded plate.

In order to meet the demands in electrostatic measurement, the grounded cylinder with a hemispherical end to a negatively well-charged dielectric plate model has been developed and calculated, and a comparison is given between this model and Nasser et al's. The similarities and differences are found in the streamer onset processes between the two models.

Fig.1 illustrates the dimension and arrangement of the model, with $h=0.4$m, $r=0.01$m, $R=0.1$m, the gap distance being d and the surfacial charge density being σ.

The charge simulation method is adopted in calculation[2], in which N matching points on the hemispherically capped metallic cylinder are assumed and the same number N of simulating charges are virtually placed in the metallic cylinder.

The potentials at the matching points generated by the charged dielectric plate and the simulating charges are calculated. By satisfying the boundary condition of the grounded cylinder, we must let the algebraic sum of the potentials at every matching point to be zero, N linear coupled equations can be derived. Solving these equations, the N simulating charge magnitudes can be found and the potential at any given point in space calculated.

The calculation results are listed in Table 1. From the Table it can be seen that the onset voltage Vo increases with the increase of the gap distance d. The relation between the variables Vo and d is shown in Fig.2.

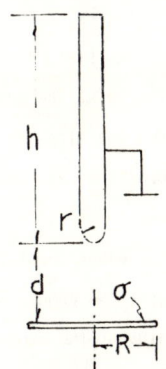

Fig.1 Arrangememt of the model

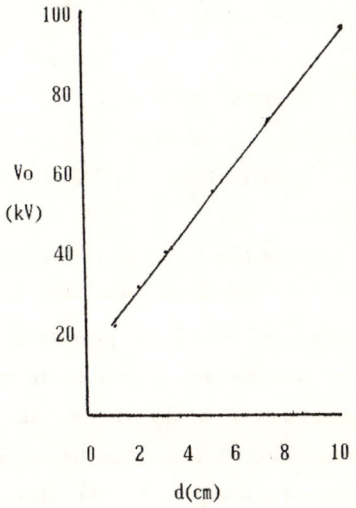

Fig.2 Onset voltage vs. gap distance

Table 1 Parameters at streamer onset

d (cm)	Vo (kV)	Eao (kV/cm)	Eco (kV/cm)	e^Q	t (μs)	Lmin (cm)	r (cm)	E'av (kV/cm^2)	σ (C/m^2)
1.0	21.94	40.85	13.36	2.26×10^4	0.0264	0.33	0.00584	50.18	-7.55×10^{-6}
2.0	32.34	40.90	8.82	2.49×10^4	0.0264	0.33	0.00584	50.21	-8.40×10^{-6}
3.0	40.40	40.88	7.65	2.41×10^4	0.0264	0.33	0.00584	50.21	-9.33×10^{-6}
5.0	55.38	40.94	7.65	2.62×10^4	0.0264	0.33	0.00584	50.33	-11.49×10^{-6}
7.0	70.73	40.86	8.59	2.28×10^4	0.0265	0.33	0.00585	50.33	-13.95×10^{-6}
10.0	95.89	40.94	10.72	2.53×10^4	0.0254	0.32	0.00572	51.16	-18.23×10^{-6}

Table 2

d (cm)	Vo (kV)	Eao (kV/cm)	Eco (kV/cm)	e^a	t (μs)	Lmin (cm)	r (cm)	E'av (kV/cm^2)
1.0	25.67	44.9	18.89	2.77×10^4	0.03369	0.5	0.009322	43.84
2.0	35.53	45.25	10.50	2.13×10^4	0.02593	0.4	0.00818	52.89
3.0	41.00	45.48	6.965	2.42×10^4	0.02592	0.4	0.00818	53.88
4.0	44.44	45.44	5.083	2.23×10^4	0.02597	0.4	0.00819	54.11
5.0	47.00	45.45	3.96	2.21×10^4	0.02598	0.4	0.00819	54.28
10.0	54.06	45.39	1.788	2.00×10^4	0.02604	0.4	0.008197	54.43

Parameters at streamer onset, Nasser et al's[1]

d gap distance;	Vo onset voltage;
Eao anode field;	Eco cathode field;
e^a exponential o, i.e., total no. of electrons in the initial avalanche;	
t electron transit time;	*Lmin minimum avalanche length;
r space-charge radius;	E'av average field gradient;
σ surfacial charge density on dielectric plate.	

Table 2 is abridged from Nasser et al's results as a comparison.

From Table 1 and Table 2, the parameters in both models have the same tendency. This is due to the fact that their mechanism at the streamer onset is the same. Owing to the stronger electric field near the anode, when the voltage between electrodes reaches the value of the onset voltage, an initial avalanche will be created near the anode. In the avalanche, all electrons concentrate at the head, while the positive ions leave behind. When electrons flow into the anode the positive ions remain near the anode. The electric field in front of the anode will be enhanced. New avalanche can be successively created to produce the streamer, leading to breakdown of the gap.

Changing the distances between electrodes, the average field gradient E'av is calculated at various gap distances within the minimum length of avalanche Lmin.

$$E'_{av} = \frac{1}{n} \sum_{i=1}^{n} \frac{\Delta E(x)}{\Delta x}$$

where, $n = L_{min} / \Delta x$

It can be seen from Table 1 and Table 2 that at various gap distances, the anode field at the onset of the streamer, the minimum avalanche length and the average field gradient within this length remain constant. This illustrates that even at different gap distances and voltages, the field distribution within the minimum avalanche length at the onset of streamer is essentially the same. Since the streamer onsets occur under the same physical condition, all the significant parameters remain reasonably constant.

The streamer onset is determined by the field distribution near the anode, which is a function of the curvature radius of the anode. The curvature radius of the anode in Table 1 and Table 2 both are 0.01m, and under this condition, the electric field distribution between electrodes at the onset of streamer is computed for the model presented in this paper with the electrode distances varied. When the distance from anode exceeds the minimum length of avalanche, the electric field distributions are quite different with gap distances varied. The difference of Eco values shown in Table 1 identifies this. That is why changed electrode distances make the onset voltage of streamer different. Moreover, it can be apparently seen that even though the electrode distances are the same, the two models in Table 1 and Table 2 have inconsistent field distributions and different onset voltage of streamers.

The potential at every point on the charged dielectric plate at the onset of streamer can be computed, the surfacial charge density on the plate is known, as shown in Table 1. So its potential energy can be computed. For example, at d=0.05m, the potential energy is about 1.7×10^{-2} Joule and it is independent of the process of the development of streamer and breakdown. The computed potential energy should be the total energy released duing the gap breakdown. This data should be of some value for references.

In this research, the influence of the dielectric coefficient of the plate has not been taken into consideration.

*Note: The trigger electron is considered to originate from a point on the axis of the gap where the field-to-pressure ratio E/P is approximately 30 $V\ cm^{-1} Torr^{-1}$. The streamer onset voltage will be a minimum at this situation. The distance from this point to the anode is the minimum length of avalanche Lmin[1].

REFERENCES

[1] Nasser, E. and Heiszler, M. " Mathematical--physical model of the streamer in nonuniform fields " Journal Applied Physics, 45(1974) 3396

[2] Shen, J.N. et al " Numerical analysis of electro-magnetic field "(in Chinese), China Science Publisher, 1984, 1st edition

STANDING IONIZATION WAVE IN THE POSITIVE COLUMN OF GLOW DISCHARGE OF OXYGEN

Chen Jianlin
Beijing Institute of Labor Protection, China

Li Ruinian
Beijing Institute of Technology, China

Abstract

In this paper, a series of reactive diffusion equations concerning six species are built, as a model of ionization waves observed in the positive column of glow discharge of oxygen. A dispersion relation of ionization waves is derived by using the approximation of linear perturbation. Then the wavelengths of standing striations are obtained. The results agree well with the experiment.

Introduction

Ionization wave is one kind of oscillations with low frequencies, which appears to be a group of striations in the positive column of glow discharge, standing or moving in the direction from cathode to anode or from anode to cathode. In 1951, Wichman found the two types of glow discharges of oxygen (H and T type discharges). In the positive column of H type discharge, the field strength was very high, up to 30 V/cm. While in the positive column of T type discharge, the field strength was much lower, approximately 5 V/cm, and moving striations were observed. In 1960, Pekarek [1] also studied the two types of discharges. He obtained that when the radius of discharge tube was 1.8 cm, pressure of oxygen was 2 mmHg and discharge current was 5 mA, the ionization wave would move from cathode to anode with a speed of 2.7×10^5 cm/s and a spatial period of 8.5 cm. If the current is increased to 8 mA under the same conditions, the moving speed of striation would become 3.6×10^5 cm/s and the spatial period would be 10 cm.

Since the first model of ionization wave was proposed by Pekarek [2] in 1962, many studies have been done on the rare gases. But, advances in the understanding of ionization wave in the molocular gases such as oxygen have been much slower, bacause the phenomena were nearly always studied under nonlinear conditions. Also the discharges were not quite clear generally so that there was no control on the species of ions or complex reactions that can occur with mixture of dissociated products. However, the present availability of reliable electronic dynamic data for elementary reactions permits fairly detailed theoretical analysis of ionization wave in the positive column of glow discharge of oxygen. This paper discusses the ionization wave with the reactive diffusion equations of concentrations of six kinds of particles [3]—the electron, negative ions O^-, positive ions O_2^+, metastable O_2^* ($^1\Delta g$), atoms O and molecules O_2. All the rate coefficients of considered reactions have been calculated accurately by Maska et al. [4] in 1978 and by other people, as functions of E/N. It is shown that the theoretical and experimental wavelengths of standing striations are in good agreement.

Experiment and Theory

The experimental apparatus are illustrated in Fig. 1.

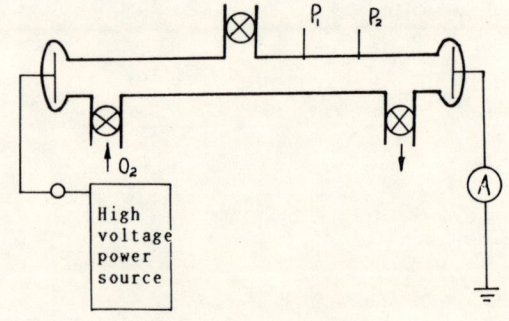

Fig. 1 The experimental apparatus

Where the radius and length of discharge tube are 1.5 cm and 40 cm respectively. On the wall of the tube there are two probes with a distance of 10 cm, which are used to measure the average field strength. Two discharge electrodes with radius of 2 cm are installed at the ends of discharge tube.

Before the high-purity oxygen is filled, the discharge tube must be vacuumized in order to remove the impurity sufficiently. Close tightly the discharge tube at a given pressure of oxygen and exert a high voltage on electrodes. The discharge current is changed by adjusting the applied voltage until the clear standing striations occur. Write down the discharge current of this time and count the number of striations between two given places. The wavelengths of standing ionization waves are then obtained.

In order to explain the standing ionization waves in the positive column of glow discharge, the reactive diffusion equations of the six species are set up. All the elementary reactions which are taken into account are listed as follows [3]:

P1 $e + O_2^*(^1\Delta g) \longrightarrow O + e$
P2 $e + O_2^*(^1\Delta g) \longrightarrow O_2^*(^1\Sigma_g^+) + e$
P3 $e + O_2 \longrightarrow O_2^*(^1\Delta_g) + e$
P4 $e + O^- \longrightarrow O + 2e$
P5 $e + O_2^*(^1\Delta g) \longrightarrow O + O^-$
P6 $e + O_2 \longrightarrow O + O^-$
P7 $e + O_2 \longrightarrow O + O + e$
P8 $e + O_2^*(^1\Delta_g) \longrightarrow O_2^+ + 2e$
P9 $e + O_2 \longrightarrow O_2^+ + 2e$
P10 $O^- + O \longrightarrow O_2 + e$
P11 $O^- + O_2 \longrightarrow O + O_2 + e$
P12 $O^- + O_2^*(^1\Delta g) \longrightarrow O_3 + e$
P13 $O^- + O_2 \longrightarrow O_2^- + O$
P14 $O^- + O_2^+ \longrightarrow O_2 + O$

The reactive diffusion equations of the six species mutually bound by the above reactions are:

$$\frac{\partial X_E}{\partial t} = D_E \frac{\partial^2 X_E}{\partial Z^2} + B_E X_E \frac{\partial E}{\partial Z} + B_E E \frac{\partial X_E}{\partial Z} + X_E X_2 (k_9 - k_6)$$
$$+ Q X_E X_M (k_8 - k_5) + Q X_E X_{N1} k_4 + X_{N1} X_1 k_{10}$$
$$+ X_{N1} X_2 k_{11} + Q X_{N1} X_m k_{12} - \frac{\lambda^2}{R^2} D_{eff}^E X_E$$
$$+ X_E X_2 [(\frac{\partial k_9}{\partial E})_{E_o} - (\frac{\partial k_6}{\partial E})_{E_o}] \Delta E \quad (1)$$
$$+ Q X_E X_M [(\frac{\partial k_8}{\partial E})_{E_o} - (\frac{\partial k_5}{\partial E})_{E_o}] \Delta E$$
$$+ Q X_E X_{N1} (\frac{\partial k_4}{\partial E})_{E_o} \Delta E + X_{N1} X_1 (\frac{\partial k_{10}}{\partial E})_{E_o} \Delta E$$
$$+ X_{N1} X_2 (\frac{\partial k_{11}}{\partial E})_{E_o} \Delta E + Q X_{N1} X_M (\frac{\partial k_{12}}{\partial E})_{E_o} \Delta E$$

$$\frac{\partial X_{N1}}{\partial t} = D_{N1} \frac{\partial^2 X_{N1}}{\partial Z^2} + B_{N1} X_{N1} \frac{\partial E}{\partial Z} + B_{N1} E \frac{\partial X_{N1}}{\partial E}$$
$$+ X_E X_2 k_6 + Q X_E X_M k_5 - Q X_E X_{N1} k_4$$
$$- X_{N1} X_1 k_{10} - Q X_{N1} X_{P2} k_{14} - Q X_{N1} X_M k_{12}$$
$$- X_{N1} X_2 (k_{11} + k_{13}) - \frac{\lambda^2}{R^2} D_{eff}^{N1} X_{N1} \quad (2)$$
$$+ X_E X_2 (\frac{\partial k_6}{\partial E})_{E_o} \Delta E + Q X_E X_M (\frac{\partial k_5}{\partial E})_{E_o} \Delta E$$
$$- Q X_E X_{N1} (\frac{\partial k_4}{\partial E})_{E_o} \Delta E - X_{N1} X_1 (\frac{\partial k_{10}}{\partial E})_{E_o} \Delta E$$
$$- Q X_{N1} X_{P2} (\frac{\partial k_{14}}{\partial E})_{E_o} \Delta E - Q X_{N1} X_M (\frac{\partial k_{12}}{\partial E})_{E_o} \Delta E$$
$$- X_{N1} X_2 [(\frac{\partial k_{11}}{\partial E})_{E_o} + (\frac{\partial k_{13}}{\partial E})_{E_o}] \Delta E$$

$$\frac{\partial X_{P2}}{\partial t} = D_{P2} \frac{\partial^2 X_{P2}}{\partial Z^2} - B_{P2} X_{P2} \frac{\partial E}{\partial Z} - B_{P2} E \frac{\partial X_{P2}}{\partial Z}$$
$$+ Q X_E X_M k_8 + X_E X_2 k_9 - Q X_{N1} X_{P2} k_{14} \quad (3)$$
$$- \frac{\lambda^2}{R^2} D_{eff}^{P2} X_{P2} + Q X_E X_M (\frac{\partial k_8}{\partial E})_{E_o} \Delta E$$
$$+ X_E X_2 (\frac{\partial k_9}{\partial E})_{E_o} \Delta E - Q X_{N1} X_{P2} (\frac{\partial k_{14}}{\partial E})_{E_o} \Delta E$$

$$\frac{\partial X_M}{\partial t} = D_M \frac{\partial^2 X_M}{\partial Z^2} + X_E X_2 k_3 - Q X_E X_M (k_1 + k_2 + k_5 + k_8)$$
$$- Q X_{N1} X_M k_{12} + X_E X_2 (\frac{\partial k_3}{\partial E})_{E_o} \Delta E - Q X_E X_M [(\frac{\partial k_1}{\partial E})_{E_o}$$
$$+ (\frac{\partial k_2}{\partial E})_{E_o} + (\frac{\partial k_5}{\partial E})_{E_o} + (\frac{\partial k_8}{\partial E})_{E_o}] \Delta E - Q X_{N1} X_M (\frac{\partial k_{12}}{\partial E})_{E_o} \Delta E \quad (4)$$

$$\frac{\partial X_1}{\partial t} = D_1 \frac{\partial^2 X_1}{\partial Z^2} + X_E X_2 (k_6 + 2k_7) + Q X_E X_{N1} k_4 + Q X_E X_M k_5$$
$$+ Q X_{N1} X_{P2} k_{14} + X_{N1} X_2 (k_{11} + k_{13}) - X_1 X_{N1} k_{10}$$
$$- \frac{1}{2} X_1 K_R \frac{\bar{V}}{R} + X_E X_2 [(\frac{\partial k_6}{\partial E})_{E_o} + 2(\frac{\partial k_7}{\partial E})_{E_o}] \Delta E$$
$$+ Q X_E X_M (\frac{\partial k_5}{\partial E})_{E_o} \Delta E + Q X_E X_{N1} (\frac{\partial k_4}{\partial E})_{E_o} \Delta E$$
$$+ Q X_{N1} X_{P2} (\frac{\partial k_{14}}{\partial E})_{E_o} \Delta E + X_{N1} X_2 [(\frac{\partial k_{11}}{\partial E})_{E_o}$$
$$+ (\frac{\partial k_{13}}{\partial E})_{E_o}] \Delta E - X_1 X_{N1} (\frac{\partial k_{10}}{\partial E})_{E_o} \Delta E \quad (5)$$

$$\frac{\partial E}{\partial Z} = \frac{q}{4\pi\epsilon_o}(X_{P2} - X_E - X_{N1}) \qquad (6)$$

$$X_2 = N - X_M - \frac{1}{2}X_1 \qquad (7)$$

In these equations N is the initial concentration of molecules. X_E, X_{N1}, X_M, X_{P2}, X_1 and X_2 are the average concentrations of electrons, O^- ions, metastables $O_2^*(^1g)$, positive ions O_2^+, atoms and molecules. k_1-k_{14} are the rate coefficients. Q = 1.45 is a constant which is introduced in the quadratic terms because of the radial dependence of the densities. λ = 2.405 is the first root of Bessel function. K_R is the recombination coefficients of atoms on the wall. R is the radius of discharge tube. $\bar{V}$ is the average speed of atoms. B_i and D_i (i = E, N_1, P_2) are the mobility and diffusion coefficient of electrons, negative and positive ions and $\alpha = X_{N1}/X_E$. D_{eff}^E, D_{eff}^{N1} and D_{eff}^{P2} are the effective diffusion coefficients of electrons, negative ions and positive ions.

$$D_{eff}^E = \frac{(1+\alpha)(D_E B_{P2} + B_E D_{P2}) + \alpha(D_E B_{N1} - B_E D_{N1})}{B_E + \alpha B_{N1} + (1+\alpha)B_{P2}} \qquad (8)$$

$$D_{eff}^{N1} = \frac{(1+\alpha)(B_{P2} D_{N1} + B_{N1} D_{P2}) + (B_E D_{n1} - B_{N1} D_E)}{B_E + \alpha B_{N1} + (1+\alpha)B_{P2}} \qquad (9)$$

$$D_{eff}^{P2} = \frac{(B_{P2} D_E + B_E D_{P2}) + \alpha(B_{N1} D_{P2} + B_{P2} D_{N1})}{B_E + \alpha B_{N1} + (1+\alpha)B_{P2}} \qquad (10)$$

It is assumed that
$$X_E = X_E^o + X_E^o \exp i(\omega t - kZ) \qquad (11)$$

$$|X_E^o| \ll X_E$$

and similarly for X_{N1}, X_M, X_{P2}, X_1 and E. Here X_E^o is a known constant which is taken as the steady-state solution. X_E^o is a constant, yet unknown, perturbation amplitude.

Under this assumption, Equations (1-7) are reduced to the following form

$$(A) \times \begin{pmatrix} X_E^o \\ X_{N1}^o \\ X_{P2}^o \\ X_M^o \\ X_1^o \\ e^o \end{pmatrix} = 0 \qquad (12)$$

From Equation (12), we learn that if the X_i^o (i = E, N1, P2, M, 1) and e^o are all not equal to zero, we must have

$$|A| = 0 \qquad (13)$$

Equation (13) is the dispersion relation derived under the conditions of linear perturbation approximation, in which the high degree of and K are contained. Suppose that ω and K are all real variables and use Equation (13) to do iterative calculation, then we get the values of K corresponding to every ω.

Results and Disussion

The standing striations are easily observed visually. The two curves in Fig. 2 illustrate the change of wavelength with discharge current at given pressure of oxygen.

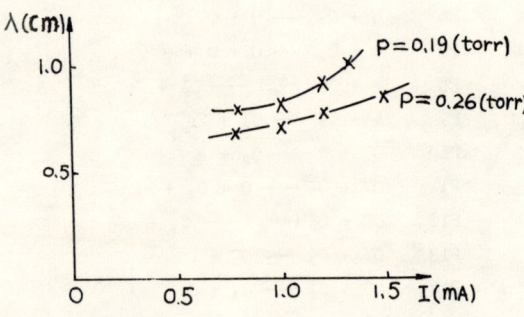

Fig. 2 Wavelength in dependence on discharge current

It is found that at the same pressure of oxygen, the wavelength increases with discharge current. For the same discharge current, the wavelength at lower pressure is longer than that at higher pressure.

After calculation, the theoretical wavelengths of standing striations at given discharge currents and oxygen pressures are obtained, which are shown in Fig. 3.

For comparison, the calculated and measured curves when the pressure of oxygen is 0.26 torr are drawn on Fig. 4, from which we learn that the theoretical results correspond well to the experimental results.

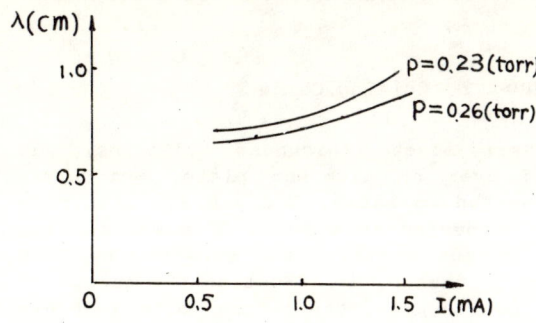

Fig. 3 Wavelength in dependence on discharge current

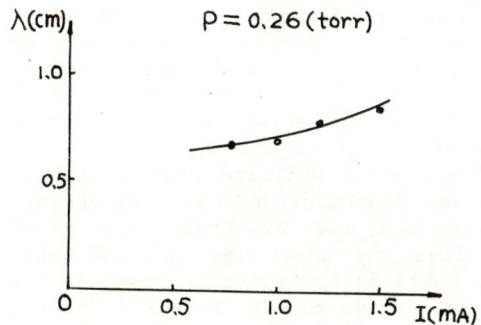

Fig. 4 Wavelength in dependence on discharge current
Open points—experimental values
Full line—calculated results

Compared with the first model of Pekarek, our system of equations does not include the continuous equation of electron engery. This is because in the plasma of oxygen discharge, there are large collisional frequencies for both elastic and inelastic collisions. The electronic energy distribution function is assumed to follow any electric field changes instantanously. So it is reasonable to omit the equation of electron energy from the system and the influence of electric field on rate coefficients is represented directly by the terms $(\frac{\partial k_i}{\partial E})_{E_o} \Delta E$ in equations. The values of them are modified properly in our calculations.

In this paper, we have assumed that both K and ω are real variables. However the calculation can be done in the same way to find $\hat{\omega}$ and $\hat{K}$ denoting the complex forms of $\hat{\omega}$ and $\hat{K}$. The imaginary parts of $\hat{\omega}$ and $\hat{K}$ indicate the changes of ionization wave with time and spatial place respectively. So the decay characteristics of ionization wave can be discussed by means of $\hat{\omega}$ and $\hat{K}$ and the critical conditions on which the self-excited ionization waves occur can be determined then.

Conclusion

In this paper, the standing ionization waves in the positive column of glow discharge of oxygen have been discussed with a series of reactive diffusion equations. Through the calculation, we obtained the theoretical wavelengths of standing striations, which correspond well to the experimental results. So it is concluded that the model of this paper is able to describe rather satisfactorily the self-excited ionization waves of oxygen discharge. Also this model might be used to deal with the other characteristics such as the decay of ionization wave and the critical condition on which the self-excited ionization waves occur. Then the transform from H type discharge to T type discharge can be predicted theoretically.

References

1. L. Pekarek and M. Sicha (1960), Czech. J. Phys., B10, 749.
2. L. Pekarek and V. Krejci (1962), Czech. J. Phys., B12, 296.
3. L. Laska, K. Mašek, T. Růžicka (1979), Czech. J. Phys., B29, 498.
4. K. Mesek, L. Laska, T. Růžicka (1978), Czech. J. Phys., B28, 1321.

THE EFFECTS OF A SMALL ISOLATED CONDUCTOR ON PROPAGATING BRUSH DISCHARGE

Li Guoxiang
Safety Technology Research Institute
National Committee of Mechanical Industry, Beijing, China

Abstract

When there is a small isolated conductor on a charged insulating surface, the development and incendivity of propagating brush discharge(PBD) will be changed a lot. Experiments have been carried out by means of a small steel plate or water drops on a square polymer sheet. The results show that the propagating area of PBD was extended apparently and the incendivity of PBD increased considerably when PBD occured just above the steel plate or in the vicinity of it. These facts demonstrate that electrostatic hazards due to PBD increase greatly under the conditions of presence of small isolated conductors.

1. Introduction

Up to now many investigations have dealt with the phenomena of brush discharge, which takes place above the charged surface of insulator[1][2][3]. Discharge of this type are usually confined to a small area just oppsite the electrode, and the equivalent energy released may relatively low, in the order of 1 mJ[4][5]. But if the insulator takes the form of sheet (the thickness<8 mm) lying over an earthed metal plate, at definite high charge density($\sigma > 2.5 \times 10^{-4} c/cm^2$), the discharge initiated by an earthed electrode approching to the sheet will be very violent, and the equivalent energy may rise to the range of several Joules[1][6]. The discharge of this type is called propagating brush discharge PBD.

The energy liberated from PBD is so large that it may ignite almost all flammable gas/vapors and some combustible dusts/powders. In order to assess and control this type of risky discharge, a large number of experimental work has been carried out, including the studies on characteristics and incendivity of PBD and factors that effect them, such as the thickness of insulating sheet, surface charge density, the polarity of charge, diameter of sphere electrode and resistivety of sheet etc. On the base of these studies, we recently found that a small isolated conductor (a small steel plate or water drops, which may present on the insulating surface practically) can also affect the characteristics and incendivity of PBD apparently.

2. Experimental method

The experimental arrangement we used is shown in Fig.1. An area 420×420 mm² of polymer sheet (thickness 0.125 mm) was laid over an aluminum plate, which was connected to earth. A tip brass electrode was connected to a 50 kV DC.power supply, the output of which was measured by Q3-V high voltmeter. By adjusting the output of DC.power supply, the corona voltage of tip electrode could be changed in the range -30— +30 kV. Another spheric brass electrode (diameter 20 mm) was connected to earth. When test was being conducted, a steel plate of diameter 40 mm or water drops was first placed on the polymer sheet. Then the tip electrode was moved downward approaching to the polymer sheet and swept over the sheet for a while. The surface of the sheet was then charged homogeneously to a definite charge density by corona discharge of tip electrode. Afterward the tip electrode was moved away. After the sheet was charged well, the PBD could be initiated by lowering the spheric electrode toward the center of the steel plate or moving it near to the sheet. The pattern of the discharge of PBD was traced by powdering the surface of the sheet with lycopodium powder. If necessary, the picture of the pattern could be taken by camera.

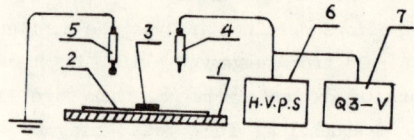

Fig.1 Experimental arrangment for PBD
1.the earthed aluminium plate 2.polymer sheet 3.isolated conductor 4.tip electrode 5.spheric electrode 6.high voltage power supply 7.high voltmeter

3. Experimental results

a.The effects on discharge characteristics

The development and propagating area of PBD can easily be determined from the discharge pattern. Fig.2a shows the pattern of PBD with a steel plate at the center of the sheet while Fig.2b shows the pattern of PBD without small conductors. Both patterns were taken at the same corona voltage -25 kV. It is very clear from Fig.2 that the propagating area of PBD with a small conductor is apparently more extensive than the one without small conductor. On the other hand, under the conditions of presence of small conductors, the shape of discharge pattern changes considerably. In Fig.2a there is a wide discharge channel which connects the steel plate to the point where the discharge was initiated, and in addition to the initiative point, there is another discharge propagating center formed by the steel plate.

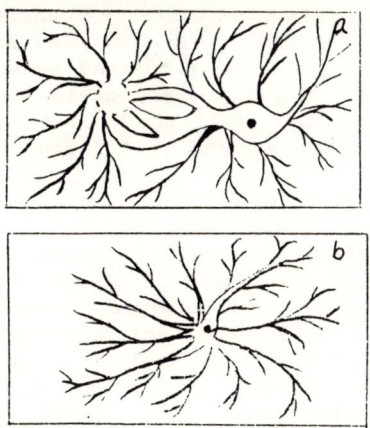

Fig.2 The pattern of PBD. (a)with a small metal plate (b)without isolated conductor. ○ steel plate, • initiative point.

Fig.3 PBD patten with two steel plates. ○ steel plates, • initiative point.

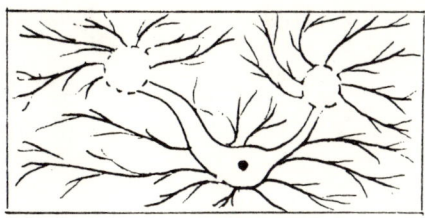

Fig.4 Another PBD pattern with two steel plate. ○ steel plates, • initiative point.

Fig.3 and Fig.4 show discharge patterns of PBD with two steel plates on the sheet. It can be seen that similar discharge characteristics appear there. The differences from Fig.2 are that there are two wide discharge channels which connect the initiative point to two steel plates and two additional discharge propagating centers.

Fig.5 shows the discharge pattern of PBD with four drops of water on the sheet. As Fig.2-4, similar discharge characteristics can also be seen from the pattern.

Fig 5 PBD pattern with water drops. ○ water drop, • initiative point.

b. Incendivity

Instead of steel plates, a coin of five cents was used to test the incendivity of PBD for powders. When ignition test was conducted, the coin was first placed at the center of the sheet, then the test powder was spread on the sheet and coin. Subsequently the discharge of PBD was initiated above the coin or in the vicinity of it. The powder on the sheet and coin might be ignited by the spark of PBD.

Two kinds of powder, Xeroy Coner and lycopodium powder, were used to test the incendivity of PBD. The experimental results show that, in the case when a coin was present, both powders were easily ignited by PBD at the corona voltage of -20 kV (the igintion probability was 100%). But if the coin was taken away, both powders were hardly ignited by PBD at the same corona voltage (the ignition probability < 20%).

By pouring some flammable liquids onto the steel plate, we tested the incendivity of PBD for liquids. Two kinds of liquid, alcohol and acetone, were tested. The experimental results show that, in the case of presence of steel plate, both liquids were very easily ignited by PBD at the corona voltage of 12 kV (the ignition probability was 100%), but if the plate was taken away, the ignition for both of them was very difficult (the probability decreased to 0%) at the same corona voltage.

4. Discussion

a. Fig.2-5 show clearly that, in the presence of the small isolated conductors, PBD propagated radially not only at the center around initiative point but also at the center around small conductors. Thus an additional discharge propagating center was produced and the propagating area will be certainly extended apparently. The equivalent energy liberated from PBD will certainly increase.

b. The experimental results show that, in the presence of small conductors, the incendivity of PBD increased a great deal either for flammable liquids or for combustible power. There are two reasons which may explain the phenomena mentioned above: One is due to the extension of propagating area of PBD or the increase of equivalent energy of PBD spark, which connect the initiative point to small conductors, with extra-high temperature and high level of energy.

c. As is well known, the equivalent energy of PBD may be high up to several Jouls. Almost all flammable gases/vapors and some conbustible powders/dusts, of which the minimum ignition energy is less than 1 J, can ignited by PBD. Under the condition of presence of small conductors, the incendivity of PBD will be increased futher. Thus, in the practical processes, in order to prevent the ignition risk due to PBD, we should not use metallic pipes or containers internally coated with a nonconducting layer. In case such pipes or containers have to be used, we should do our best to eliminate the electrostatic charges on processed materials (e.g. inserting an earthed metal rod into processed materials) as well as to avoid every small isolated conductor, which may attach to the inner coating layer, especially for small water drops.

References

[1] E.Heidelberg, Static electrification 1967,Inst,Phys,Conf.SER. No.4ppl 47-55

[2] E.Heidelberg,Advances in static electrity.Proc.1st Int.Conf.Static.Electr. Vienna May 4-6 1970 PP351-59

[3] K.G.Lö Trund "The ignition power of brush discharge-experimental work on the critical charge density"
Journal of Electrostatics 10(1981)162-168

[4] E.Heidelberh, PTB-MITT 80(1970)440

[5] M.Glor, "Ignition of gas/air mixture by discharges between electrostatically charged plastic surfaces and metallic electrodes"
J.Electrostatics 10(1981) 327-332

[6] M.Glor, "Hazards due to electrostatic charging of powders"
J.Electrostatics 16(1985)175-191

ON THE PROBLEM OF CHARGE DISTRIBUTION ON A CONDUCTOR AND THE CALCULATION OF GEOMETRIC FACTOR OF AN EMITTING CATHODE

Peilian Lee Prof. of the Basic Science Dept.
Anshan Institute of Iron and Steel Technology,
Anshan, Liaoning, China

Introduction

The general relation between the surface charge density on a charged conductor in electrostatic equilibrium and its surface curvature k is a problem long left open. Up to now this problem for only a few cases of isolated condutor with regular shape have been solved. People doubted if there would be any such generally valid relation. Such doubt is natural because the charge density σ at a point on condutor surface is not only related to the local form of the surface at that point but it is also closely related to the global form of the condutor suvface. However it was claimed recently[1] that this very problem was solved and it attracted much attention[2]. We give here an analysis of this problem, so as to show that this problem may not have a solution in general. We have also described the correct statement of this problem and pointed out the errors in [1]. The suggested formula for calculating the geometric factor of emitting cathode in [1] is another result. As an approximation formula it is a useful complement to the existing formula of the kind. We give here an improvement on it.

I. The problem of charge distribution on a charged conductor in electrostatic equilibrium

According to the electromagnetic theory[3], the charge distribution on a charged conductor in electrostatic equilibrium is uniquely determined. But if we deform locally the conductor surface at any point, the charge will redistributed altogether.

Thus we see, the surface charge density σ can not be determined only by the local mean curvature k at the point. But it does not mean we can not have a relation between σ and k. For we can still expect that, together with other parameters α_i, β_i, describing the local and global properties of the surface shape, there may be a functional relation between σ and these parameters, i.e.

$$\sigma = \sigma(k, \alpha_i, \beta_i, Q) \qquad (1)$$

where Q is the total charge on the conductor. Clearly, σ is not determined only by k in (1), but if we can reach such a function, we may say that we have solved the problem of the relation between σ and k. It is easy to see, even under this meaning, the relation between σ and k does not exist in general. Let us see what does the above relation really mean. According to the electromagnetic theory, the σ at point (x,y,z) on a conductor surface is determined as soon as the charges, shapes, dimensions and the relative positions of the charged condutors are known. Let α_i, β_i denote these parameters and Q_j denote the charges on these conductors, then we have the following functional relation in general:

$$\sigma = (x, y, z, \alpha_i, \beta_i, Q_j) \qquad (2)$$

But this is not equivament to (1). Only as the dependence of σ on (x,y,z) can be represented through k(x,y,z) as a compound function of (x,y,z), where k(x,y,z) is the mean curvature at (x,y,z), we may say that we have obtained the real relation between σ and k. If the relation obtained contains k and (x,y,z) as well, then we may not say that we have obtained such a real relation. For if (x,y,z) is allowed to appear in the expression of σ, then the relation between σ and k can not be determined uniquely. In fact, we can adjust the relation between σ and (x,y,z) so as to make σ to be an arbitrary funtion of k. Clearly the dependence of σ on (x,y,z) can not be expressed as acompound function through k(x,y,z) in general. But we really have some cases where this expression is posssible. So the pro

-blem of reltion between σ and k is of meaning and interest.

Let us give the correct statement of this problem. It is well known, as the quantities used to describe the local properties of surfaces, besides mean curvature k, we also have Gaussian curvature K. These independent. From the theory of differential geometry[4], we know the local properties of surfaces are uniquely determined by the first and second fundamental form coefficients g_{ij} and Ω_{ij} (i, j=1,2). g_{ij} and Ω_{ij} are symmetric with i,j, so there are only six independent parameters, from which k and K can be calculated. So the correct statement of the problem of the relation between surface charge density and the geometric form of conductor may stated as follows:

"Under what conditions the surface charge density σ may be expressed as a function of g_{ij}, Ω_{ij} and other parameters β_i describing global geometric properties of the surfaces? And how is this function expressed in general when the conditions were fulfilled?"

Let us now examine the formula suggested by [1] to be the general expression we expected, which is as follows:

$$\sigma = \frac{2\varepsilon k \Delta V}{\exp(-2k\Delta n) - 1} \quad (3)$$

In fact, the above formula is not a real relation between σ and k, for the Δn in it is also a function of (x,y,z), as is indicated in Fig.1. Of course, in some special cases $\Delta n(x,y,z)$ may be expressed as a function of k, but it is not possible in general. So we can not insist that formula (3) solves our problem.

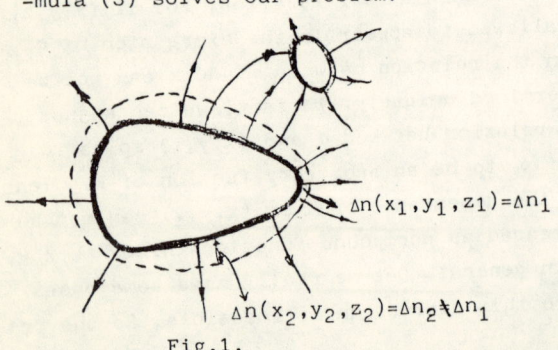

Fig.1.

II. An improved approximate formula for calculating the geometric factor of emitting cathode

From the classical formula in electrostatics[3]

$$\frac{dE}{dn} + 2kE = 0, \quad (4)$$

An approximate formula for calculating the geometric factors of emitting cathode is derived in [1] as follows

$$\beta \cong 2k/(\exp(-2k\Delta n) - 1) \quad (5)$$

The derivation of above formula is based on the following equation which determines the variation of electric field strength along a force line

$$E = E_0 \exp(-2kn) \quad (6)$$

This is obtained by solving the differential equation (4), regarding k as constant. It is evident that k is variable with n. An improved formula will be obtaoned by considering this variation. The solution to differential equation (4) will thus as follows

$$E = E_0 \exp\left(-\int_0^n 2k(n)dn\right) \quad (7)$$

Expanding k(n) as a Taylor series in n

$$k(n) = k_0 + \left(\frac{dk}{dn}\right)_0 n + \cdots \quad (8)$$

and substituting it into (7) gives

$$E(n) = E_0 \exp\left(-2k_0 n - \left(\frac{dk}{dn}\right)_0 - \cdots\right) \quad (9)$$

This in turn gives the following result

$$V = \int_n^\infty E dn = -\frac{E\Delta}{2K}\left\{\exp(-2K_0\Delta n)[1-(\frac{dK}{dn})_0(-\frac{1}{2K_0^2}+\frac{\Delta n}{K_0}+\Delta n^2)] = [1-(\frac{dK}{dn})_0\frac{1}{2K_0^2}]\right\} \quad (10)$$

The improved approximate formula for geometric factor of emitting cathode is thus obtaineed as

$$\beta \cong \frac{E}{\Delta V} = 2K/\left\{\exp(-2K_0\Delta n)[1-(\frac{dK}{dn})_0(-\frac{1}{2K_0^2}+\frac{\Delta n}{K_0}+\Delta n^2)]-[1-(\frac{dK}{dn})_0\frac{1}{2K_0^2}]\right\} \quad (11)$$

When the variation of k with n is neglected, that is, let $(\frac{dk}{dn})=0$, $(\frac{d^2k}{dn^2})=0$,...,formula (11) is reduced to (5).

Formula (12) as compared to (5), not only its approximation is improved, but it also has another merit, that it can reflect the influence of the curvature of anode on β with the term involving $(\frac{dk}{dn})$ in it.

1. Lou Enze, J.Phys. D Appl. Phys. $\underline{19}$ (1986) 1-6
2. Lung Vie,"Science News" March 20,1987
   Desen Fan, J.Phys. D Appl.Phys. (to be published)
3. J.D.Jackson, Classical Electrodynamics, (Wiely, New York, 1975)
4. Chuan-Chih Hsiung, A First Course in Differential Geometry, (Wiley, New York, 1981)

RESEARCH AND SIMULATED TEST ON SPACE ELECTROSTATIC FIELD

Luo Hongchang, Zhong Xiaolian
Xiong Jigu, Zhou Ziyun, Wang Liqun
Shanghai Ship & Shipping Research Institute
Ministry of Communications

Abstract

This paper presents the data of electrostatic field under high tention power line and its distribution according to the theoretical calculation and actual measurement of the electrostatic field under a simulated three-phase high tension power line. The presentation also proves that the use of both E-1 space field intensity gauge and VC-4 electrostatic meter is feasible.

Studies on static-electricity such as application of electrostatic technique and prevention of static-electricity, as well as study on the intensity and distribution of electrostatic field are of great importance to industrial production. In order to learn the influence of the static-electricity upon the ships passing the high tension power line, we carried out a simulated test. In the test, the alternative field under the three-phase high tension power line was calculated and measured, the field intensity and its distribution were studied and E-1 space field gauge and VC-4 electrostatic meter were proven to be successful in the measurement of the electric field.

I. Simulated Electrostatic Field

According to the existing conditions of the laboratory, a simulated space electrostatic field is designed with a space of 3.72m in length, 3.20m in width and 2.80m in height. (See Fig.1.)

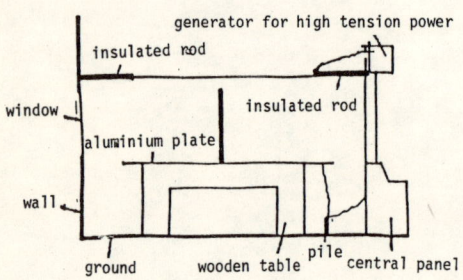

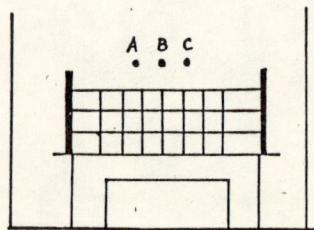

Fig.1 Diagram of the simulated space field

The three wires of the three-phase power line are arranged with a distance of 0.28m between each other and in a height of about 1m to the floor made of aluminium plates of 2 X 2m. The length of each wire is 2.07m and the diameter 0.01m. The voltage used in the test is 10kv.

A cross section is taken in the middle of the line, where the three dimension space is considered as a two-dimension space. The intersection between the cross section and the floor is taken as the X axis on a plane, while the Y axis goes through the neutral line in order to simplify the theoretical calculations of the field. Besides, a symmetric mesh is designed under the plane coordinates. The cross points which are arranged 3 in line and 7 in row on the mesh are the points for measurement and theoretical calculation. The arrangement of the points are shown in Fig.2.

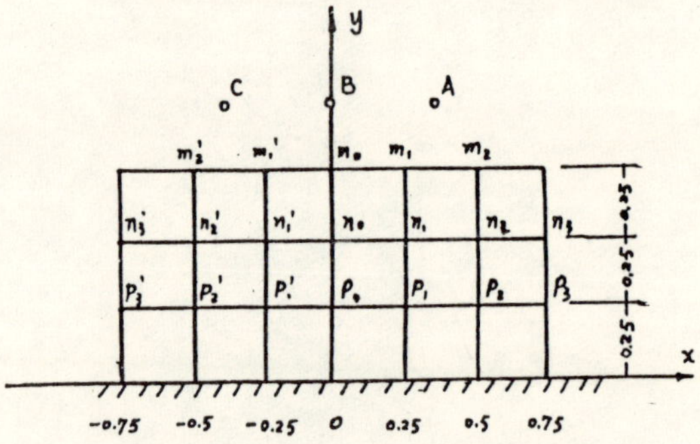

Fig.2 Mesh for measurement

II. Instrumentation Used in the Test

An adjustable power supply of 10kv, 3-phase, 50Hz together with a digital indicator was used in the test.

The field intensity gauge used in the test is E-1 Space Field Intensity Gauge developed by Shanghai Ship & Shipping Research Institute (SSSRI). The gauge, with a measure capacity of 0-1000 kv/m, can measure both DC and AC electrostatic fields. Besides, the gauge has a digital indicator and a printer which prints the reading of the gauge automatically. The detector, though without any need for power, is able to measure the field 10 meters away with an accuracy of less than 5%. It can work continuously for a long time when with DC 12v regulated power. While with battery, it can work continuously for two hours at the designed accuracy. (See Fig.3.)

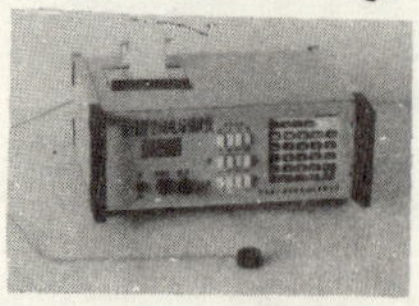

Fig.3 E-1 Space Field Intensity Gauge

The space electrostatic potential is measured by the VC-4 Electrostatic Meter developed also by SSSRI.

III. Theoretical Calculation of the Field Under High-Tension Power Line

Method of equivalent charge was used in the calculation of the field intensity and potential on each space point under the high tension line because the field and potential of any point out of the electrified bodies in the unbounded field can be calculated with this accurate method. Based on the mirror image principle of the electrostatic field, the method is composed of both the calculation of the equivalent charge on a conductor on unit length and the calculation of the field caused by the electric charge.

The formular for the calculation of the field intensity and the potential of an arbitrary space point is as follows:

1. Potential coefficient of conductor m at the point for calculation

$$Pom = \frac{1}{2\pi\varepsilon} \ln \frac{\sqrt{(Ym + Yo)^2 + (Xm - Xo)^2}}{\sqrt{(Ym - Yo)^2 + (Xm - Xo)^2}}$$

Where
Xo, Yo ---- coordinates of the point
Xm, Ym ---- the coordinate position of Conductor m

2. The potential to ground of the point

$$VRo = \sum_{m=1}^{N} Pom \cdot QRm$$

$$VIo = \sum_{m=1}^{N} Pom \cdot QIm$$

$$Vo = \sqrt{VRo^2 + VIo^2}$$

Where
VR, VI ---- real part and imaginary part of the voltage to ground of the conductor
QR, QI ---- real part and imaginary part of the charge on the conductor

3. Vertical component of the field intensity at the point

$$ERv = \sum_{m=1}^{N} \frac{QRm}{2\pi\varepsilon} \cdot \left[\frac{Ym - Yo}{(Ym-Yo)^2 + (Xm-Xo)^2} + \frac{Ym + Yo}{(Ym+Yo)^2 + (Xm-Xo)^2} \right]$$

$$EIv = \sum_{m=1}^{N} \frac{QIm}{2\pi\varepsilon} \cdot \left[\frac{Ym - Yo}{(Ym-Yo)^2 + (Xm-Xo)^2} + \frac{Ym + Yo}{(Ym+Yo)^2 + (Xm-Xo)^2} \right]$$

$$Ev = \sqrt{ERv^2 + EIv^2}$$

Where
ERv, EIv ---- vertical component caused by the charge in the real part and the charge in the imaginary part
ε ---- the dielectric constant of air

IV. A Comparison between Theoretical Calculation and Actual Measurement

The theoretical calculation and actual measurement of the simulated electrostatic field under the 10kv three-phase power line are showing the following results:

1. The result from both the calculation and the measurement, either for potential or for field intensity, are almost the same and their charge patterns are similar. (See Table 1 and Table 2.)

Table 1. Potential of the points in the space under the power line (kv)

Point	Value from Calculation	Reading of the Electro-Meter
Mo	1.08	0.82
M_1	2.25	1.416
M_1'	2.25	2.621
M_2	2.34	1.785
M_2'	2.34	2.834
No	0.254	0.298
N_1	0.89	0.355
N_1'	0.89	1.295
N_2	1.30	0.787
N_2'	1.30	1.638
N_3	1.64	0.656
N_3'	1.64	1.499
Po	0.06	0.104
P_1	0.688	0.572
P_1'	0.688	0.872
P_2	0.496	0.270
P_2'	0.496	0.958
P_3	0.780	0.177
P_3'	0.780	0.979

Table 2. Field intensity of the points in the space under the power line (kv/m)

Point	Value from Calculation (vertical component)	Reading of the Field Gauge
M_0	7.24	5.40
M_1	9.29	10.60
M_1'	9.29	10.80
M_2	5.86	11.20
M_2'	5.86	12.00
N_0	1.34	1.20
N_1	2.14	4.20
N_1'	2.14	4.60
N_2	2.73	5.70
N_2'	2.73	6.90
N_3	1.83	5.00
N_3'	1.83	5.20
P_0	0.36	0.30
P_1	1.58	1.62
P_1'	1.58	1.82
P_2	2.57	2.30
P_2'	2.57	2.70
P_3	3.35	2.30
P_3'	3.35	2.30

From the tables we can see:

1) The higher the point locates, the higher the potential, the stronger the field intensity and the larger the fluctuation. Otherwise, the closer the point locates to the ground, the gentler the fluctuation is.

2) The field intensity value and the potential value from the actual measurement are higher than that from the theoretical calculation. One reason for this is that the detector (10mm X ⌀20mm) and a piece of thin wire led to, more or less, influence upon the concentration of the electric force line and the distortion. The closer to the power line, the larger the effect will be. The other reason is that the horizontal component of the field intensity had not been taken into the theoretical calculation.

3) The potential value and the field intensity value of the points with " ' " are higher than that of the points without " ' " according to the actual measurement. The higher the point locates, the clearer the phenomenon. This is because the points without " ' " are near the wall which decentralizes the electric force line as the earth.

4) When the theoretical calculation of the distribution of the field intensity was carried out, peaks appeared at phase A and phase C. This is about the same as actual measurement. The higher the point locates the higher the peak will be.

2. A ship model of 0.18m in height (0.35m if including the mast), 0.2m in width and 0.72m in length was put upon the floor of the simulated space to cause distortion to the space field under the power line. But the change of the field intensity and the potential measured were not so obviously as shown in Table 3 and Table 4. The reason for this might be the simulated space is not large enough. From the tables we can see when the ship model is put in, the distribution of the electric force line in the space under the power line is concentrated to the ship and the right wall, thus causes the potential and the field intensity of the points with " ' " such as N_1', N_2', N_3', P_2', P_3' lower than that before distortion.

V. Conclusion

The identical results of the theoretical calculation and actual measurement show that both VC-4 electrostatic meter and E-1 space field gauge of simple operation and good stability are feasible to be used in the measurement.

Besides, a trial was carried out on a ship which is sailing under the 500kv power line across over a river in Pingwu. Considering the safety factors of the ship, the points at bow and at the mast top where the field intensity exists concentratively were selected as the measure points. There is a distance of 5 meters between the bow and the water surface, and 10 meters between the mast top and the water surface. During the trial, the ship started from a point at the symmetric centre of the water, 40 meters in the lower reaches of the river, and headed to a point 10 meters in the upper reaches at a course perpendicular to the power line. Every point was measured at an interval of 10 meters. The potential measured was similar to that of the theoretical calculation, while the field intensity was 20-30 times as large as the calculation. This means the space electrostatic field will be distorted by the ship. But the pattern of the field is similar to the calculation. So we can say the trial is a positive one. However, this is only the first trial onboard ship. There are still a lot of questions to be solved in the further tests and researches.

Table 3. Comparison of potentials between normal field and distorted field (kv)

Point	Normal Field	Distorted Field (with ship model)
M_0	0.820	0.627
M_1	1.416	1.419
M_1'	2.621	1.962
M_2	1.785	1.808
M_2'	2.834	1.994
N_0	0.288	0.073
N_1	0.355	0.473
N_1'	1.295	0.636
N_2	0.787	0.741
N_2'	1.638	0.873
N_3	0.656	0.661
N_3'	1.499	0.756
P_0	0.104	—
P_1	0.572	0.099
P_1'	0.872	0.125
P_2	0.270	0.205
P_2'	0.958	0.232
P_3	0.177	0.168
P_3'	0.979	0.233

Table 4. Comparison of the space field intensity between the normal field and the distorted field (kv/m)

Point	Normal Field	Distorted Field (with ship model)
M_0	5.40	5.80
M_1	10.60	10.20
M_1'	10.80	11.30
M_2	11.20	11.50
M_2'	12.00	11.90
N_0	1.20	1.20
N_1	4.20	4.20
N_1'	4.60	4.20
N_2	5.70	5.60
N_2'	6.00	5.70
N_3	5.00	4.70
N_3'	5.20	4.90
P_0	0.30	0.20
P_1	1.63	1.20
P_1'	1.85	1.90
P_2	2.30	2.20
P_2'	2.70	2.20
P_3	2.30	2.10
P_3'	2.30	2.10

A METHOD OF NUMERICAL ANALYSIS AND CALCULATION OF STREAMING CURRENT BY A HOPPING MODEL

ZHOU FENG* WU ZONG-HAN* G.TOUCHARD**
*Dept. of Phys. and Chem., Nanjing Inst. of Tech .
** University of Poitiers, France, Laboratory of Physics & Fluid Mechanics

Abstract

The electric current generated by the laminar flow of an insualting liquuid through a cylindrcal pipe has been calculatted with the equilibrium boundary conditions by virtue of Hoppong Model. The result is presented in dimensionless from and it is shown that the streaming current depends on two parameters which are dimensionless representation of interface potential and the radius of pipe. The radial charge density distribution is given. The influence of the conductivity and interface potential on the charge density are discussed. The result is consistent with others that have more adjustable parameters than ours .

The generation of electic currents by the flow of dielectric liquids through pipes is a well-known phenomenon which has been extensively studied both theoretically and experimentally. There are many models to describe the charge generation process. As we don't know enough about the boundary condition at the liquid/pipe interface and the electrochemistry of hydrocarbon solvents, so all models are made to rely on some adjustable parameters .

Two typical models were put forward by Gibbings and Touchard.Gibbings et.(1966) assumed that all ions to reach the wall were discharged. They gave the conservation equation with the non-equilibrium boundary condition, but it couldn't be solved for there are some difficulties to determine the boundary conditions.Touchard(1978) suggested that ions are sorbed and desorbed at equal rate. He gave streaming currents in terms of the axis charge density and other parameters which were used as adjustable ones .

We analyse the experimental result and above models, and suggest that the non-equilibrium boundary condition fit for the region far from the pipe exit. We describe a new model for current generation which incorporates quilibrium bouundary conditions at the liquid/pipe interface. The model equation is derived by using Hopping Model. The result shows that the number of adjustable paramenters in our medel is fewer than others .

Hopping model

In this paper, we use a Hopping Model for calculating the radial change distribution and streaming current. This model explains the ionic conduction as a result of hopping of ions between potential wells. The model enables us to analyze the transient state in the space-charge formation without diffusion constant and mobility taken into account. In the continuous model,the equation of current continutity is a second-order nonlinear equation which is difficult to solve under the bounndary condition,while in the Hopping Model,the current-continuity condition is a first-order nonlinear equation which is easy in dealing with the transient phenomena.In the Hopping Model, potential wells of height $W_i(j)$ are arranged at equal interval within the dielectric (Fig.1). We assume that there are $U(v)$ kinds of positive (negative) in the well. The transition probabilities of ions from the Kth well to the (K+1)th well and from the Kth well to the (K-1)th well are :

$$(1)\begin{cases} \text{Positive}: W^i_{k,k+1} = \gamma_i \cdot exp[-\frac{q_i}{kT}(\phi_{k+\frac{1}{2}}-\phi_k)] \cdot exp[-W_i/kT] \\ \qquad\qquad W^i_{k,k-1} = \gamma_i \cdot exp[-\frac{q_i}{kT}(\phi_{k-\frac{1}{2}}-\phi_k)] \cdot exp[-W_i/kT] \\ \text{Negative}: W^j_{k,k-1} = \gamma_j \cdot exp[-\frac{q_j}{kT}(\phi_k-\phi_{k+\frac{1}{2}})] \cdot exp[-W_j/kT] \\ \qquad\qquad W^j_{k,k+1} = \gamma_j \cdot exp[-\frac{q_j}{kT}(\phi_k-\phi_{k-\frac{1}{2}})] \cdot exp[-W_j/kT] \end{cases}$$

Where $\gamma_{i(j)}$ is the oscillation frequency of positive(negative) ions in the well, q is the ionic charge,k is the Boltzman constant, T is the absolute temperature,ϕ_k is potential at the Kth well .

The current that ions travel from the Kth well to the (K+1)th well is :

$$(2)\begin{cases} \text{Positive}: j_P = \sum_{i=1}^{U}(W^i_{k,k+1} \cdot q_i \cdot P^i_k - W^i_{k+1,k} \cdot q_i \cdot P^i_{k+1}) \\ \text{Negative}: j_N = \sum_{j=1}^{V}(W^j_{k,k+1} \cdot q_j \cdot N^j_k - W^j_{k+1,k} \cdot q_j \cdot N^j_{k+1}) \end{cases}$$

Where $P_k (N_k)$ is the ion number in the Kth well .

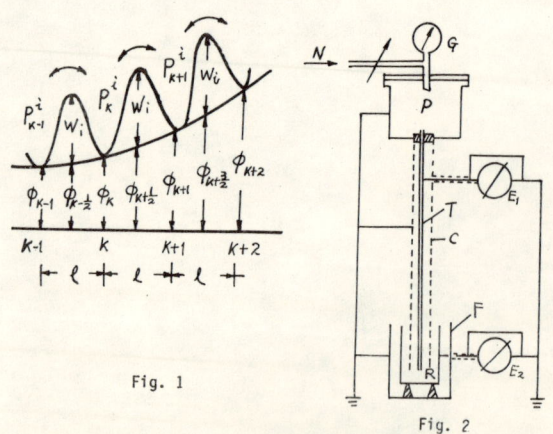

Fig. 1

Fig. 2

We will use this model to deal with the electric current generated by the laminar flow as follows.

Experiment

In order to analyse the boundary condition. Experiments performed to look for the relation between the wall current and pipe length.

A schematic diagram of the experimental pipe line system is shown in Fig.2. the liquid such as kerosene can be blown by nitrogen from the pressure vessel(P) (stainlesss steel) through the pipe line (t) into a stainless steel receiver(r). The pressure vessel is fitted with a nitrogen control system (ṅ) and pressure gauge (G). the pipe line (0.8mm internal dia., 20cm to 100cm long.) are made of stainless steel. The pressure vessel and pipe line were insulated one another by PTFE. the insulation resistance was maintained greater than 1×10^{15} Ω. The receiver was enclosed by earthed metal box by a 8cm dia. of the flow pipe earthed steel tube (c) to prevent electrical interference. The pressure vessel was earthed too. The currents were measured by electrometers (E, E) (KEITHLEY 642).

The unstable length due to pipe exit in the laminar flow is: $l_x = 0.0575 d \cdot Re$ where d is diameter of pipe, Re is Reynold number. In this experiment, $l_x \leq 0.11m \leq 0.2m$. So we have not considered the effect of the unstable region of the flow.

The result was shown on Fig.3, where I_w is wall current, l_r is tube length. We have received from these data the value of currnet denstiy j_w and plotted them in terms of length (Fig.4). it shows that the wall current density is always decreasing with the length. Therefore we assume that the fluid can be divided into two regions. In region I, there is ion flux density at the pipe/liquid interface. In region II, equilibrum boundary contion may be applied

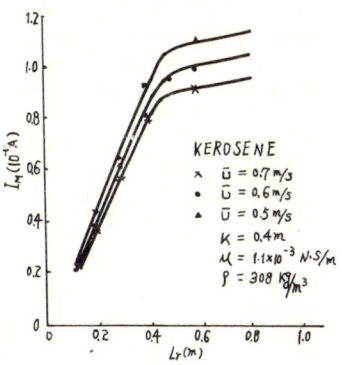

Fig. 3

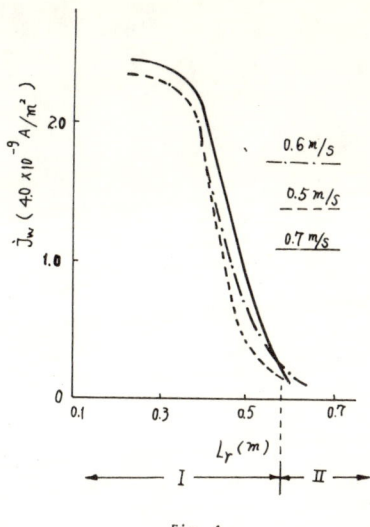

Fig. 4

The Model for Charge Generation
* Derivation of the model Equation

Firstly we calculate the radial charge distribution in liquid with equilibrium boundary condition.

We assume that there are uniform potential well in fluid dielectric (Fig.5). Ion travel from potential well 1 to 0 due to sorb and desorb. Using eqilibrium boundary condition:

$$(3) \quad P_0 \gamma e^{-Q/kT} \cdot e^{q \Delta \phi / kT} = P_1 \gamma e^{-q \Delta \phi / kT}$$

Where P_0 is ion surface density on the pipe wall, in the 0th well, So is P_1 in the 1st well. Q sorb heat, Q/q is defined as the interface potential. It describes the influence of the pipe on the steaming current. $\Delta \phi$ is potential difference between the 0th well and 1th well.

Ions travel from the 1st well to the 2nd well, from the 2nd well to the 3rd, due to drift and diffusion. From the equilibrium of ions between wells:

$$(4) \begin{cases} W^P_{k,k+1} \cdot q \cdot P_k - W^P_{k+1} \cdot q \cdot P_{k+1} = 0; \\ W^N_{k,k+1} \cdot q \cdot N_k - W^N_{k+1} \cdot q \cdot N_{k+1} = 0, \end{cases} (k=1,\cdots,N-1)$$

Using Gauss's Law between the Kth well and (K+1)th well, the electric field is given by:

$$(5) \quad E_{k+1} = E_k \cdot \frac{R-kl}{R-(k+1)l} - q \cdot (N_k - P_k) \cdot \frac{[R-(k+\frac{1}{2})\cdot l]}{[R-(k+1)\cdot l] \cdot \epsilon}$$

Where ϵ is the permittivity.

We assume that the liquid is neutral before flowing in pipe and the ion concentration (positive or negative) is denoted as D. We have

$$(6) \quad \sum_{k=1}^{N} N_k \cdot [R-(k-\frac{1}{2}) \cdot l] = R^2 \cdot D/2$$

The potential is obtained :

$$\phi_{k+1} = \phi_k - l E_{k+1} \quad (7)$$

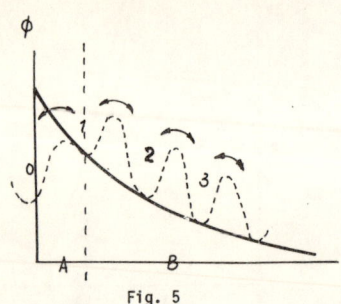

Fig. 5

Equations (3), (4), (5), (6), (7) can be solved for P_k, N_k, E_k, ϕ_k with the help of a computer program.

Having calculating R_k, N_k, the streaming current in the pipe is obtained by intergrating the axial flux over the tube diameter using the parabolic velocity distribution $V = V_0(1 - r^2/R^2)$. That is :

$$I_L = 2\pi V_0 q \cdot \sum_{k=1}^{N} \left\{ \frac{N_k - P_k}{l} \cdot [1 - \frac{(R-kl)^2}{R^2}] \cdot (R-kl) \cdot l \right\} \quad (8)$$

* **The Model Equation in Dimensionless Form**

The equations are normalised using the following characteristic values, the width of the diffusion layer $l_0 = (\frac{\epsilon kT}{2q^2 D})^{1/2}$, the equilibrium charge density, the thermal energy $\phi_0 = kT/q$.

Axial velocity $U_0 = V_0$ we can first derive a characteristic surface density $\delta_0 = (D\epsilon kT)^{1/2}$, a characteristic current $I_0 = \pi R^2 V_0 q D$. Substituting above characteristic values into equations (1) - (5), we get dimensionless form of (8):

$$I_L^+ = I_L/I_0 = \sum_{k=1}^{N} \left\{ 4 \cdot \frac{(R^+ - kl^+)}{R^{+2}} \cdot \frac{(N_k^+ - P_k^+)}{l^+} \cdot [1 - (\frac{R^+ - kl^+}{R^+})^2] l^+ \right\}$$

$$= f(R^+, W^+)$$

Where only W.D are adjustable parameters. The number of adjustable parameters is less than Walmsley's [1]

The Calculation Results

---

* **The raidal charge distribution**

In Fig.6, we have given example of the radial charge density and radial electric field. Where $\rho_{(N)}$ is positive (negative) charge density. $\rho = \rho_N - \rho_p$. It is consistent with others [4]

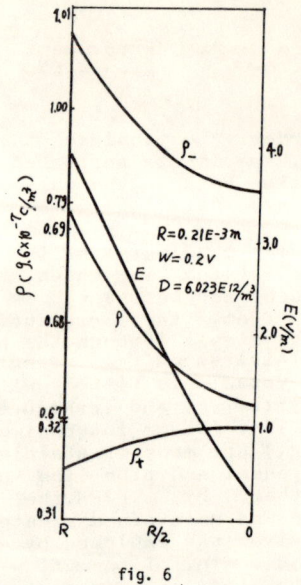

fig. 6

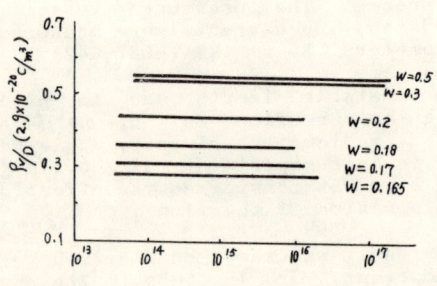

Fig. 7

* **The charge density influence of conductivity**

Definite : $\rho_V = \int_0^R \rho(r) 2\pi r \, dr / \pi R^2$

In Fig.7, we have presented the variation of ρ_V with the ion concentration D. As the conductivity of liquid is proportional to D. It shows that ρ_V is proportional to the conductivity of dielectric liquids, which conform with experimental results[2]. When the conductivity is large enough, this conclusion is untenable. Because there is no equilibrium boundary condition at the wall then.

* **Streaming Current**

Fig.8, Fig.9 give some examples of the variation of the I_L^+

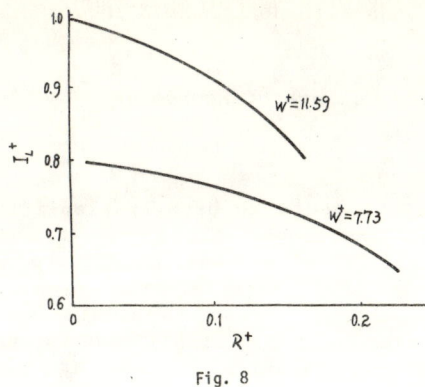

Fig. 8

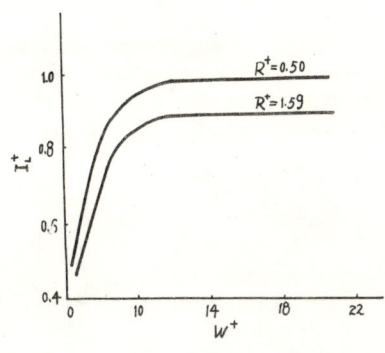

Fig. 9

With W^+, R^+. The systematic study on there parameters has allowed a comparsion between predictions and experimental data.

Conclusion

The electric current generated by laminar flow of an insulating liquid has been calculated numerically using a Hopping Model with equilibrium boundary conditions. The result shows that the dimensionless streaming current depends on two parameters which are dimensionless radius R^+ and dimensionless interface potential W^+. The adjustable parameter number is less than preceding result.

Acknowledgements

We wish to thank Prof.Ou-Yang Yi, Prof.Xiao Busheng and Mr.Xu Fudong, for their helpful suggestions and great support. We are particularly grateful to Miss Zhu Yafei for her assistance.

References:

#1. Hl Walmsley & G.Woodford, J.Phys.D: Appl.Phys.14 (1981)1761-82

#2. J.C.Gibbings & G.S.Saluja, Journal of Electrostatic. 3(1977)333-370

#3. Jc. gibbings, Journal of Electrostatic. 19 (1987) 115-119

#4. G.Touchard and P.Dunargue, Electrochim, Acta. 20 (1975) 125-135

#5. G. Touchard P.Dunargue, Journal of Electrostatic. 14 (1983) 209-223

#6. Mitsumasa Iwamoto & Taro Hino, Electrical Engineering in Japan, 100 (1980) NO.5.

RESEARCH ON ELECTRIC FIELD IN STORAGE TANK WITH NUMERICAL METHOD

Yang Youqi   Zhao Lianqing
Department of Safety Engineering, Beijing College of Economics
Beijing, China

Abstract

It is well-known that the explosion of storage tank caused by electrostatics is due to the energy stored in the electric field. This article discusses the distribution of electric field in storage tank by means of numerical method. By calculating, we get a few groups of charactristic curve to indicate electrostatic potential distribution, draw field diagram, point out the quantity and distribution of field intensity and energy density in storage tank.

1. Introduction

The source of electric fields in the storage tank is the volume charge on the insulators and the field is very complicated. To find out the perilous point and evaluate dangerous extent, the distribution of electric field strength in both liquids and upper space of the oil in the tank should be studied.

As we know, the potential on liquid surface inside the tank varies with surface levels. As electrostatic discharge in the tank usually happens on the liquid surface, its regularity must be studied.

Capacitance is one of inherent parameters of conductors and it depends on factors such as the configuration of conductors and the dielectric around. The magnitude of capacitance restricts the relation between the charge and the potential on the conductor. However, with our calculation the similiar relationship between the volume charge and the potential is existed.

Based on the classical electromagnetic theory and finite difference analysis with the computer, one method for calculating the electric fields in the tank has been proposed. It can also be applied to other equipments such as tubes, containers and grooves through proper replacements and exchanges.

2. Numerical Computation of the Electric Field in Storage Tank

Suppose that the liquid inside the earthed tank is electrified uniformly and no free space charge exists upon the liquid surface. The electric field can be considered as a two-dimensional field in uneven medium after proper simplification. Although most of the fields are in three-dimension, it could be usually simplified as a two-dimensional field due to its weaker dependence on one coordinate in engineering. For example, the electric field at central part of a highly round storage tank can be simplified as the field between parallel planes. If both sides of the tank are spherical, the field near them can be simplified as an axisymmetric one.

In rectangular coordinate system as shown in figure 1, the potential functions in both oil and air inside the tank could be expressed as following:

$$\begin{cases} \frac{\partial^2 \Phi_1}{\partial x^2} + \frac{\partial^2 \Phi_1}{\partial y^2} = -\frac{\rho}{\varepsilon_1} \\ \frac{\partial^2 \Phi_2}{\partial x^2} + \frac{\partial^2 \Phi_2}{\partial y^2} = 0 \end{cases}$$

with boundary conditions of
(1) The potential on tank is zero, that is $\Phi_1 = \Phi_2 = 0$
when having $(Y-R)^2 + X^2 = R^2$
(2) On the interface between air and liquid they satisfy:

$$\begin{cases} E_{1t} = E_{2t} \\ D_{1n} = D_{2n} \end{cases} \quad \begin{cases} \Phi_1 = \Phi_2 \\ \varepsilon_1 \partial \Phi_1/\partial n = \varepsilon_2 \partial \Phi_2/\partial n \end{cases}$$

To solve these partial differential equations are the basis of solving the static field. Those equations must all be put into distive algebra equations for the calcultions. The cross point number of network and the pitch are determined by the precision and calculation time. Because of the symmetry, it can be seen that it is enough to get the solution of the field in first quadrant.

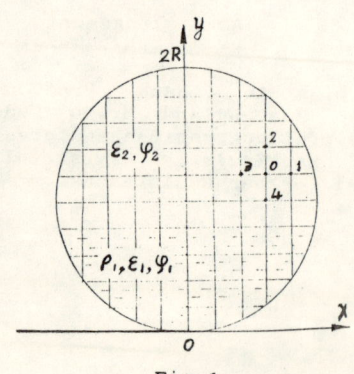

Fig.1

A linear algebra equation on every dispersive point by using Taylor series can be derived which corresponding to the Poisson's or Laplace's Equation. With the iteration between these algebra equations, the field could be calculated, with the proper first order approximative value of the field. To reduce the computation time, Overrelaxation Iterative Method is used in this paper.

For the Nth approximation of the potential could be expressed as:

$$\phi^{(n)}_{1(i,j)} = \phi^{(n-1)}_{1(i,j)} + \frac{\alpha}{4}\left[\frac{h_2 h_3 h_4}{(h_1+h_3)(h_1 h_3 + h_2 h_4)}\phi^{(n-1)}_{1(i+h_1,j)} + \frac{h_1 h_3 h_4}{(h_2+h_4)(h_1 h_3 + h_2 h_4)}\phi^{(n-1)}_{1(i,j+h_2)}\right.$$

$$+ \frac{h_1 h_2 h_4}{(h_1+h_3)(h_1 h_3 + h_2 h_4)}\phi^{(n)}_{1(i-h_3,j)} + \frac{h_1 h_2 h_3}{(h_2+h_4)(h_1 h_3 + h_2 h)}\phi^{(n)}_{1(i,j-h_4)}$$

$$\left.+ \frac{h_1 h_2 h_3 h_4}{2(h_1 h_3 + h_2 h_4)}\cdot\frac{\rho_1}{\epsilon_1} - 4\phi^{(n-1)}_{1(i,j)}\right]$$

$$\phi^{(n)}_{2(i,j)} = \phi^{(n-1)}_{2(i,j)} + \frac{\alpha}{4}\left[\frac{h_2 h_3 h_4}{(h_1+h_3)(h_1 h_3 + h_2 h_4)}\phi^{(n-1)}_{2(i+h_1,j)} + \frac{h_1 h_3 h_4}{(h_2+h_4)(h_1 h_3 + h_2 h_4)}\phi^{(n-1)}_{2(i,j+h_2)}\right.$$

$$\left.+ \frac{h_1 h_2 h_4}{(h_1+h_3)(h_1 h_3 + h_2 h_4)}\phi^{(n)}_{2(i-h_3,j)} + \frac{h_1 h_2 h_3}{(h_2+h_4)(h_1 h_3 + h_2 h_4)}\phi^{(n)}_{2(i,j-h_4)} - 4\phi^{(n-1)}_{2(i,j)}\right]$$

To the dispersive points on the interface, the difference eqution is expressed as:

$$\phi^{(n)}_{(i,j)} = \phi^{(n-1)}_{(i,j)} + \frac{\alpha}{4}\left[\frac{K_1}{K_0}\phi^{(n-1)}_{(i+h_1,j)} + \frac{\epsilon_2 K_2(K_2+K_4)}{K_0(\epsilon_1 K_4 + \epsilon_2 K_2)}\phi^{(n-1)}_{(i,j+h_2)} + \frac{K_3}{K_0}\phi^{(n)}_{(i-h_3,j)}\right.$$

$$\left.+ \frac{\epsilon_1 K_4(K_2+K_4)}{K_0(\epsilon_1 K_4 + \epsilon_2 K_2)}\phi^{(n)}_{(i,j-h_4)} + \frac{K_5 \rho}{K_0(\epsilon_1 K_4 + \epsilon_2 K_2)}\right]$$

where i and j stand for the coordinate value in X and Y axes respectively; h_1, h_2, h_3 and h_4 refer to the space lengths from point 0 to 1, 2, 3 or 4 in figure 1; and α is the convergence constant, which controls the computation speed; while for K_1 there are

$$K_1 = \frac{2}{h_1(h_1+h_3)}; \quad K_2 = \frac{2}{h_2(h_2+h_4)}; \quad K_3 = \frac{2}{h_3(h_1+h_3)};$$

$$K_4 = \frac{2}{h_4(h_2+h_4)}; \quad K_5 = \frac{2(h_1 h_3 + h_2 h_4)}{h_1 h_2 h_3 h_4}$$

To the symmetric star shape network, that is to say there is the relation of $h_1 = h_2 = h_3 = h_4$, the equation above could be simplified greatly. And to the points closely near the inside surface of the earthed tank, one or more items in the equation will disappear.

The number of discrete points inside the field region and on the interface will be computed by the computer.

In order to analyse the electric field easily, one comprehensive computation program has been programmed with the functions of drawing the field lines and printing the potential values, when given a set of parameters such as diameter of the tank, height of the liquid surface, dielectric constant, pitch and the distribution density of volume charge.

3. Computation Results

(1) Potential Distribution

The result indicates that the surface potential of liquid varies with its height, as shown in figure 2, where Φ represents the potential on center of liquid surface and Φ_{max} is its maximum; while H and D represent the height of tank and the inside diameter of the tank respectively. The result agrees with the practical measurements very well.

It had been considered that the maximum potential always keeps at the center of liquid surface. But according to our computing, it is illusion. The maximum potential point moves down gradually with the rising of liquid surface as H/R-H/D curve shown in figure 2 (the dotted line). Here R is the radius of tank while H is the difference between the height of liquid's surface and the height of maximum potential point. The maximum point approaches to the center when the tank is filled fully, which is identical with the conclusion by Gauss's Law.

As Φ/Φ_m - S curve shown in figure 3, the potentials from the center of liquid surface to the edge of tank is reduced gradually. Here Φ_m is the central potential of liquid surface and S is the distance to the center.

(2) Field's Graph

The field's graph is not only audio-visual but also can be adopted to calculate many parameters. At the beginning, the equipotential lines are drawn according to the principle of curve squaring. The typical field inside the tank is shown in figure 4. From which it is known that there are the maximum electric fields near the inner wall of the tank in liquid, near the interface above the liquid or near the center parts of the liquid on the surface.

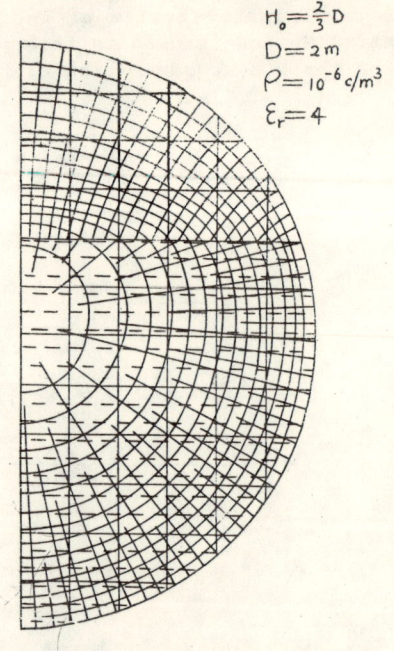

Fig.4

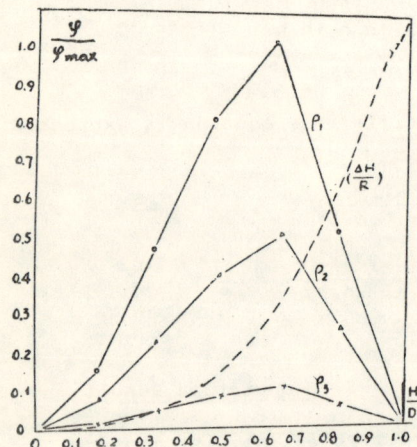

Fig.2

Bibliography

1. 冯慈璋，电磁场，1984
2. 盛剑霓，电磁场数值分析，1984
3. J.A.Stratton, Electromagnetic Theory, 1941
4. K.J.Binns and P.J.Lawrenson, Analysis and Computation of Electric and Magentic Field Problems, 1973
5. W.M.Bustin and W.G.Dukek, Electrostatic Hazards in the Petroleum Industry, 1983

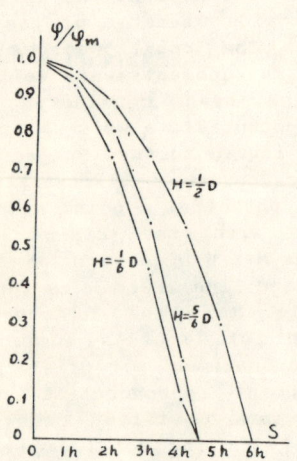

Fig.3

Electronic States and Contact Charging of Polymers

Yuji Murata
Department of Electrical
Engineering, Faculty of Science and Technology
Science University of Tokyo, Noda, Chiba, JAPAN 278

ABSTRACT

Photoelectric emission from some polymers are measured. The number of electrons emitted from polymer surface is found to be changed with illumination time. The tendency of the change depends on the photon energy of the illumination and the material of the sample. The minimum photon energy of photoemission measured is around 4 eV which is close to the WF of polymers measured by contact charging.

INTRODUCTION

Contact charging between metals and high polymers has been considered to be caused by electron transfer because of the dependence of polymer charge on the work function of contacting metals [1,2]. So the distribution of electronic states of polymers should be the most important factor which determines the charging tendency of polymers.

There are some experimetal results which make us doubt the mechanism of contact charging above mentioned. The charge might be carried by ions. In order to clarify the charging mechanism, we must know about the electronic structure of the polymer surface. Though some researchers have reported about the distribution of the electronic states of polymers [3,4], they estimated it from the experimental results of contact charging. We need some direct methods for determining the electronic structure of polymer surface.

The authors estimated the electronic structure (both filled and empty states) from the measurement of photoemission current using an electrometer[5,6]. In the present paper, some new results of photoemission measurement will be shown, which are measured by electron counting method.

APPARATUS

Fig.1 shows the schematic view of the apparatus. A sheet of film sample is mounted on the curved surface of a cylindrical sample holder. The emitted photoelectrons from the sample surface are captured by an electron multiplier and are converted to charge pulses. The output pulses from the photoelectron multiplier are amplified and the number of electrons per unit time is counted with an electronic counter.

After the each measurement by certain wavelength light is finished, the sample holder turns slightly to prevent the charge up on the sample surface caused by electron emission. The light source used is a xenon lamp. The light from the lamp is introduced to a monochrometer and the output monochromatic light from it is focused onto the sample surface with ellipse mirrors. The measuring system is mounted in a vacuum chamber. The operating vacuum pressusre is 1×10^{-6} Torr.

Samples measured are high density polyethylene, polyester and nylon 6. In order to release the trapped charge in/on the sample, the films of each polymers are heat treated in Argon gas before the measurement.

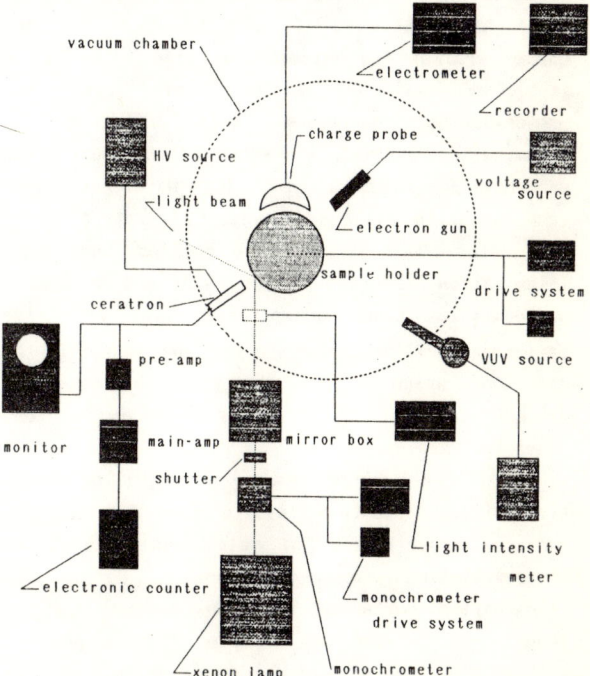

Fig.1 Schematic view of apparatus.

RESULTS AND DISCUSSION

1. Photoemission from sample surface in the dark

It is found that some noise pulses are observed from charged polymers without photo-illumination. Since this phenomenon was observed just when the sample was mounted on the sample holder, this was not from the measuring system. Electrons are emitted from the sample surface even in the dark. This exo-electron emission is remarkable when the sample surface is previously charged. After heat treatment, it becomes very samll.

2. Dependence of electron emission on illumination time

In the case of highly insulative materials, the electron emission is considered to decrease with the illumination time because of positive charge accumulation on the surface of the sample resulted from the decrease of surface electrons. In addition to this, another interesting effect was revealed in the present experiment. Fig.2 shows the dependence of the rate of photoemission on illumination time. For low energy photons, the number of electrons emitted per unit time decreases with illumination time. On the contrary, for high energy photons, the nunber of electrons per unit time increases with time. In the case of polyester, photoemission for wavelength of 200 nm light (6.18 eV) increases up to 130% to the original value at 40 s after illumination is started. By the illumination of 230 nm light (5.37 eV), the photoemission does not change against the illumination time. For 270 nm (4.57 eV) and 300 nm (4.12 eV) light, the photoemission decreases with duration of illumination. Polyethylene shows a little different tendency. For illumination of short wavelength light (200 nm, 6.18 eV), photoemission increases up to 160% after 100 s illumination. For long wavelength illumination (279 and 300 nm, 4.57 and 4.12 eV), photoemission tends to increase slightly at the beginning of illumination and then decreases. In the casse of nylon 6, the change of photoemission against illumination time seems to be smaller than other two samples.

The increase of photoemission with illumination time is considered to be resulted from the change in surface distribution of electrons caused by photoexcitation during measurement. During the illumination, surface electrons are emitted from the sample surface, and, at the same time, a part of excited electrons on/near the sample surface must be trapped in electron traps. The difference in

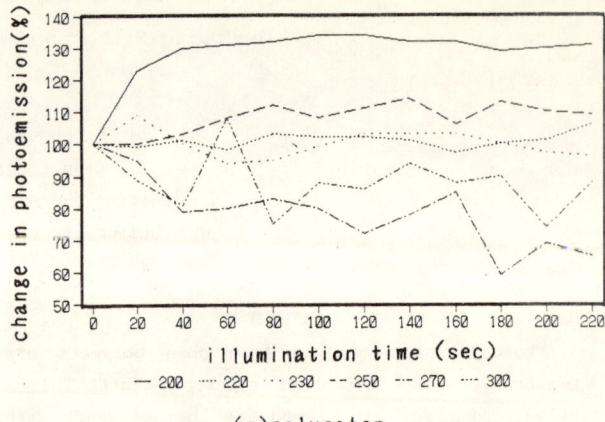

(a) polyester

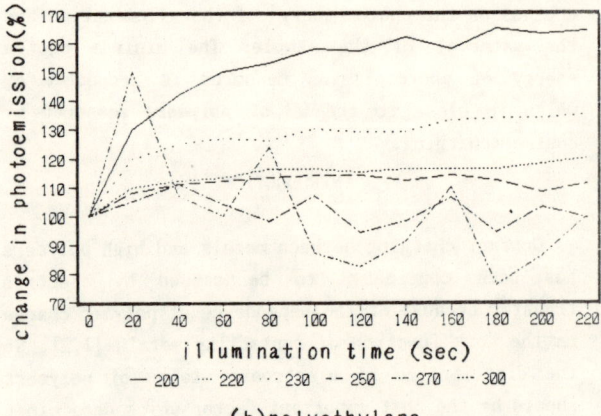

(b) polyethylene

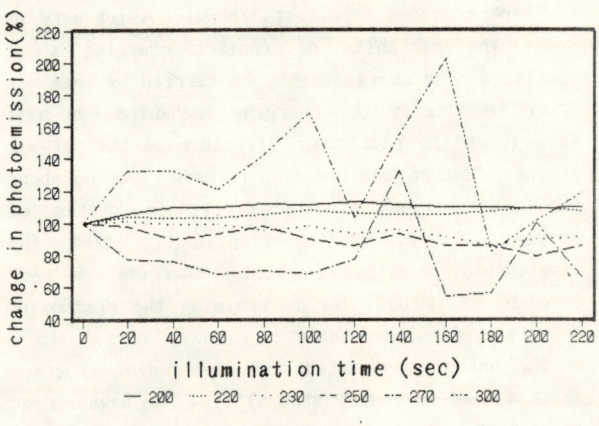

(c) nylon 6

Fig.2 Change in photoemission with illumination time.

tendency of photoemission with illumination would depend on the difference in the distribution of surface electronic states of the sample. The number of surface electron traps of polyethylene was estimated to be larger than that of nylon 6. This results is considered to interpret the difference in the characteristics of time-dependent photoemission. In order to avoid the time effect of photoemission, once illuminated area on the sample is was never used for the following measurement by rotating the sample holder by a small angle after the area is illuminated, and photoemission from the new area is measured.

3. Characteristics of photoemission

Fig.3 shows the relation of the photon energy of incident light and the quantum yield of electron emission from polymer samples. Photoemission is observed from illumination of about 4 eV.

In the case of polyethylene, threshold energy of photoemission was experimentally estimated as 8~9 eV [7]. This value almost coincedes with the band gap energy of polymers, so that the threshold energy of the electron emmision estimated in the past is thought to show the upper edge of the valence band of polyethylen. In the present measurement, it is difficult to decide the threshold energy because the curves in Fig.3 do not go down straight at the region of low energy illumination. The minimum energy values of photoemission obtained here is 4.31 eV for nylon 6 and polyethylene, and 4.0 eV for polyester. They are too smaller than the threshold energy obtained in the past. The difference in the threshold values between the two cases arises from the difference in sensitivity of the measurement.

From the present results, the following conclusions are obtained:
(1) There are electronic states in the bandgap of the polymer measured.
(2) The depth of the states are nearly 4 eV from the vacuum level.
(3) This values are close to the value of the work function estimated from the measurement of contact charginig.
(4) The energy value of the measuring limit of photoemission of each sample is not coincide with the tendency of contact charging.

Nylon 6 which has the tendency of most positively charging among the three samples is thouught to have the shallowest states. But the photon energy of the measuring limit is not the smallest. This is interpreted by considering that the characteristics of contact charging is determined by the distribution of both the filled and empty states. Nylon 6 does not have large number of empty states (electron traps) in comparison with polyethylene. We do not have data for polyester.

4. Effect of corona exposure

It is known that the charging tendency of the polymer is changed when its surface is exposed to

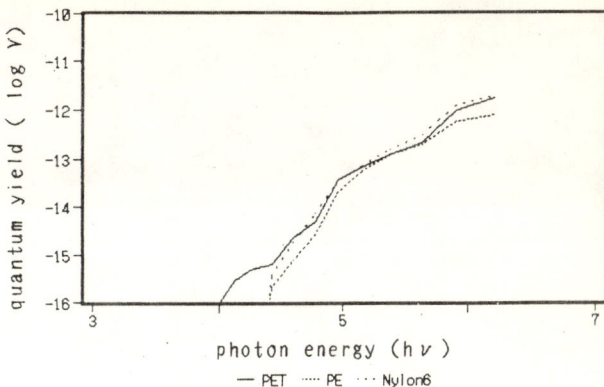

Fig.3 Photoemission yield vs. photon energy of some polymers.

corona discharge. In the present measurement, the photoemission characteristics is also changed by corona exposure before the measurement. The photon energy of the measuring limit of photoemission tends to be smaller value after corona exposure. The characteristics of contact charging after corona exposure will be interpreted by the change in electronic states.

REFERENCES

1. R.G.C.Arridge: Br.J.Appl. Phys., 18 (1967) 1311
2. D.K.Davies:Br.J.Appl.Phys.,Ser.2 2 (1969) 1533
3. H.Krupp: Proc. 2nd Conf. Static Electrif., London, 1971 (Inst. Phys. and Phys. Soc., Conf. Ser., 11, 1971) p.1
4. T.J.Fabish and C.B.Duke: J.Appl. Phys., 84 (1977) 4256
5. Y.Murata: Jpn.J.Appl.Phys., 18 (1979) 1
6. S.Kittaka and Y.Murata: Jpn.J.Appl.Phys.,18 (1979) 515
7. J.Delhalle, J.M.Amdre and S.Delhalle:J. Chem. Phys., 60 (1974) 595

Proceedings of International Conference on Modern Electrostatics

VII. Electrostatic Pollution in Electronic Industries

ELECTROSTATIC POLLUTION AND CONTROL TECHNOLOGY IN SUPER
CLEAN ROOMS (Extended Abstract)

Dumitru Cucu, Prof. Dr. Eng.

Filtronics GmbH, Kurfürstendamm 178, D-1000 West Berlin 15

Abstract

The paper presents in an extended form the problem of "electrostatic pollution" in the microelectronics industry.

It describes the causes and sources that generate electrostatic charges in the manufacturing process of microelectronic elements and the destructive effects of these charges on the microelectronic product (ESD-electrostatic discharge).

It also presents the most recent technologies for eliminating electrostatic charges in the manufacturing places.

Introduction

The so-called "electrostatic pollution", a new notion, has been introduced for the first time in parallel in Japan by S.Masuda and in Europe by the author.

The notion of electrostatic pollution has been elaborated by similitude to the particle pollution. The necessity to introduce this new terminology derived - on one hand - from the actuality of the problematic in the microelectronics industry, and on the other hand from the necessity to define and control the electrostatic charges (ESC) and electrostatic discharges (ESD). This control technology of electrostatic charges became more actual, as the manufacturing of IC chips took a dramatic increase of density of microelectronic elements per volume unit.

If about 20 years ago a destructive charge voltage was of about 1000 V, nowadays the destructive values are of a few tens of volts.

Another reason for the development of this new domain (ESC and ESD) was the elimination of destructive effects on communication means in the civil and military field, determined by a so-called electromagnetic pulse that appears at atomic explosions or in other situations.

Since 1986 when the notion of electrostatic pollution was elaborated by the two authors until today, no standards and official regulations have been settled yet, due to the complexity of this problem. Many studies, researches and recommendations have been made especially in Japan and the USA. In contrast with these two countries, the control of electrostatic pollution has been critically neglected in the field of European microelectronic industry.

The Effects of Electrostatic Pollution
on IC

As the density of electronic elements per volume unit increases and electrostatic charges become more dangerous, damages of the IC appear, caused by phenomena that represented no danger in the past.

The so-called destructive electrostatic discharges (ESD) - as demonstrated by studies made with an electronic microscope - can determine short circuits between certain microbridges, interruptions of some conductive elements, erosion of certain metallic microparts, or the destruction of semiconductor microelements, a.s.o.

Besides the destructive charge in the form of pulse, there are also so-called latent destructions caused not within the manufacturing process, but afterwards, namely during the IC deposit and transport, or later, after fitting and when in function. In the ESD - during the manufacturing process - it can be so noxious as to lead to wastes production in a proportion of 20-30% of the whole production.

Electrostatic Pollution Sources in the
Microelectronic Industry

To know the electrostatic pollution sources and to interpret their effects is absolutely necessary for conceiving the "model" for how to control and eliminate the electrostatic charge and discharge. Basically, each working place has its own configuration and characteristics. This leads to the necessity of making measurements and identifying the ESC sources. Each working place has different process configurations.

ESC sources can be: the electric potential of a working human being (operator), the air condition and air filter plants, walls and floors, lighting devices, frictions between different materials, a.s.o.

Other ESC sources can derive from the production process itself, such as wafer cutting, wafer or IC transport, frictions of any nature.

Control Technologies of ESC and ESD

These control technologies are of different kinds and chosing the optimal one depends on very many factors.

One of the classical methods to reduce ESC and ESD is to increase the relative humidity to such a level where ESC is no longer possible.
But it isn't always possible to apply this solution, and not always economic either.

Another control technology is to use bipolar ionizers. But this technology is expensive and - in many cases - either inconvenient or insufficient. Moreover, the ionizers - as certified by researches made in the USA and Japan - are generators of metallic microparticles, the so-called killer particles, which pollute the microelectronic product.

There are also radioactive neutralizers which present well-known disadvantages.

A device generating bipolar ions of a new conception has been recently presented within several international conferences and it belongs to MASUDA RESEARCH Inc. - Tokyo. The integrated system (hightech static - HTS) made known by FILTRONICS GmbH - West Germany, is also new. This HTS can realize both air filtration with an efficiency of 99.9995% and bipolar ions generation in super clean rooms.

In general, "treating" a working place against ESC and ESD is done not only with the above mentioned means, but also with a number of other complementary measures, such as using conductive garments, conductive walls and floors, grounding the operators and tools, using devices to indicate ESC, transportation of IC wafers in conductive containers, a.s.o.

This complexity of measures is called "anti ESC and ESD arsenal".

Conclusion

Taking into consideration that the IC elements density per volume unit is permanently increasing, it is more and more necessary to elaborate new methods and means of controlling ESC and ESD, as well as an urgent elaboration of standards and official regulations for electrostatic pollution in the microelectronic industry.

FUNDAMENTALS OF STATIC CONTROL FOR THE ELECTRONICS INDUSTRY

George R. Berbeco
Charleswater Products, Inc., W. Newton, Mass. USA

Soon after the first production quantities of MOS and CMOS devices, a major and unique reliability problem was identified. The problem in the use of these devices manifested itself in the devices becoming short circuited after assembly into a circuit. That is, the devices that were shipped by the manufacturer appeared to be in good order, but after assembly into a printed circuit board, had experienced a failure. It was found that these failures were due to the extremely high input impedance of the MOS device, whereby static electricity caused electrical breakdown of the gate oxide. A large portion of the failures were due to over-voltages or to defects in the oxide, resulting in gate shorts and excessive leakage. Since most of the MOS devices used thermally grown silicon dioxide about 1000 angstroms thick as a gate dielectric, such oxides result in a gate voltage for breakdown of as little as 1000 volts. Since normal static charges generated by people in the environment very often exceed this voltage level, static electricity became a particularly important problem.

Static electricity is generated in the work environment by people and by machinery operating at high speeds. The problem becomes a unique one to evaluate and deal with since the amount of voltage necessary to destroy the device is far less than is required to cause a visible spark to jump from one object to another. That is, the device itself can experience failure without any visible change or event taking place.

It is important to note that electrostatic discharge (ESD) can result in device failure at any point in the manufacturing process from receiving to field use. The relative cost of failure increased exponentially with the process stage. A modest repair or replacement cost is incurred in receiving, and a much higher cost of repair results from failure in the field.

Generation of Static Charges

Static electrical charges are brought about by the movement of dissimilar materials against one another. Although numerous theories have been presented as to the technical description of the generation of static charges, most often these charges are attributed to some form of contact between two dissimilar surfaces. A common observation of static electrical charge generation is summarized in the partial tribo-electric series shown in Table 1. The tribo-electric series is a list of materials in order of static charge generation. Materials which are further apart when rubbed together generate the greatest static charge, with the materials nearest the top retaining the positive charge. The further apart the materials are in the table, the greater the charge generated. Thus, in rubbing polyethylene and human hair the static charge generated will be greater than that generated by rubbing nylon and cotton.

Table 1. Tribo-Electric Series

+ Asbestos	Cotton
Glass	Steel
Human Hair	Acetate Rayon
Nylon	Orlon
Lead	Saran
Silk	Polyethylene
Paper	− Teflon

One possible explanation is that ion transfer between one surface and another contacting surface is a major cause of the static charge buildup. Other possible causes of electron transfer between contacting surfaces include differences in

tacting surfaces include differences in dielectric constant, thermal effects, and piezoelectric or pressure effects. Also, it has been suggested that electrons can "tunnel" back through a potential barrier as two surfaces are being separated. Overall, inconclusive evidence of the behavior of electrostatic charges has been presented, although results indicating ion transfer between surfaces are rather strong. It is suggested that this mode of static charge generation fits the category for most plastics used in the electronics industry, and therefore most charge generation which would be of concern.

Static charges are generated in numerous ways by the contact of dissimilar materials and actual daily encounters. Examples of static charges evidenced in day-to-day activities are shown in Table 2. For example, a man walking on a dry carpet(at a low relative humidity) can generate up to 5000V of static charge. Similarly, automobiles traveling on dry pavement or conveyor belts running over pulleys generate substantial static charges. Many microelectronic circuit facilities include clean room conditions that have been designed to maintain the absence of dust and other particles in the manufacturing environment. While materials used in this environment, for example finger cots, eliminate contamination during handling of devices, their use in conjunction with other materials often results in a very high static charge generation. Specifically, synthetic materials are utilized frequently since they are nonporous, flexible, and easily fabricated. However, these materials do have a tendency to generate static charges when in contact with other dissimilar materials. Table 3 illustrates the static charges that can be generated from common materials found in clean room environments or within the industrial manufacturing work place, in general. For example, latex finger cots rubbed against a plastic box can generate 6000V of charge, and yet a bare finger can generate only 200V of charge. Even 200V of static charge can degrade or destroy some static-sensitive

Table 2. Common Electrostatic Charges

- Man walking on dry carpet, up to 5,000 V
- Car on dry pavement, up to 10,000 V
- Belts running over pulleys, up to 25,000 V

Table 3. Static Charges From Rubbing Workplace Materials (volts)

	Bare Finger	Latex Finger Cots
Plastic Box	200	6,000
Polyethylene Bag	3,000	8,000
Glass Petri Dish	200	4,000
Metal Forceps	0	1,000
Antistatic (nonconductive) Polyethylene Bag	0	2,000

Latex finger cots can generate substantial static charges when rubbed against most common laboratory items, and even against "antistatic" polyethylene bags. Such "antistatic" bags are simply those with a lower surface resistivity (10^{12} ohms versus 10^{15} ohms of conventional polyethylene), but with a slightly greater ability to conduct static charges to ground at 50% relative humidity. That is, these products are dependent on relative humidity to provide an equilibrium moisture layer on the surface which provids the path to ground - not anything indigenous in the bag itself. Likewise, a bare finger can generate substantial static charges when rubbed on glass or plastic items. The evidence with respect to the antistatic polyethylene bag varies considerably in technical reports, depending upon the relative humidity, how much the bag was handled, and the length and duration of the task.

During the manufacturing of hybrid microcircuits, most static problems that are generated lie in the choice of handling and packaging materials, and also in the ambient humidity level. Latex finger cots, as well as rubber or plastic gloves, are often used extensively and these materials are common sources for static charge generation. Also, common plastic tote boxes and trays are frequent problem areas. Other areas of concern may result from the manufacturing process itself.

For example, thick film resistors may be trimmed to value by air-abrasive techniques. This results in the use of high velocity air-alumina powder mixture that physically removes portions of the resistor. A dry gas is frequently used as the vehicle to avoid clogging of the lines. Such a large volume of ambient air drawn across a surface may lower the ambient humidity level on the surface, which in combination with the other materials in the environment allow static charges to build up on the device itself. The absence of the equilibrium water moisture in the air at the surface reduces the "conductivity of the surface", thereby decreasing the conduction to ground and increasing static charge maintenance at the surface.

While the decision to change or upgrade a room floor protection system is not an easy one, protective floor surfaces can be compared systematically and the results related to the required application. Because protection from degradation is as important as outright destruction of the IC, the level of floor protection should be sufficient for the sensitivity of the devices being handled.

Where highly sensitive devices are fabricated, surface resistivity should be 10^9 ohms/sq. or lower, static decay time should be less than 0.1 sec. (5000V to zero) and, most importantly, static charge generation should be as near zero as possible. By dissipating triboelectric charging as it occurs, when personnel or carts are moving across the clean room floor, many static control problems can be prevented. Although ESD will always be a factor, it is a problem that can and is being overcome.

Unlike other types of ESD control, floor surfaces affect the entire area and, in turn, the entire operation. Floor protection is the largest, and many times, the most expensive type of ESD protection a facility purchases, and it is the first line of defense for dissipating static charge from personnel. A commitment to a particular type of floor protection system is usually one that has ramifications for years to come. To ensure that the optimum decision or upgrade is made then, the future direction of the facility should be considered as well as the more technical aspects of protection.

Given the parameters that exist in a clean room environment, an effective ESD floor protection system can be defined as one that has the following characteristics:
(1) Ability to discharge static charge rapidly from clean room personnel;
(2) A low propensity to generate static charge on personnel and carts that travel across the floor;
(3) Clean, high-gloss, hard surface with excellent durability for foor traffic;
(4) Cleanable, non-curling surface to prevent tripping of workers.

There are a variety of factors to consider in comparing floor protection systems. However, with the aforementioned goals in mind, electrical characteristics should be considered first. Previous investigators have addressed the issue of static control of surfaces primarily in terms of decay of static charge, but not static charge generation,[1],[2],[3] which is the one factor that strikes most directly at the source of the problem. Testing procedures have included measuring charge disspation from a seated operator through shoes onto floor surfaces, as well as charge decay of a highly charged person in the room. Surface resistivity has also been commonly used as an indication of static charge decay time.

An evaluation of static charge generation involves the measurement of charge generated by a person walking on a floor surface with chrome leather or neolite soles and heels. Using the AATC (American Association of Textile Chemists and Colorists) Test Method 134-1979, a grounded metal plate is placed under a 27"x36" tile on a plywood floor and precondtioned to 15 to 20 percent relative humidity. Voltage generated is measured with an electrometer and chart recorder. Ideally, a surface should not generate any static charge. The average peak values of the static charges are shown in the bar graph (Figure 1).

The results of these tests are summarized in Table 4. Antistatic surfaces and conductive multilayer mats are shown for comparison, although it should be clear that these are unsuitable for clean room floors. Conventional vinyl asbestos tile, with or without surface treatment, provides reasonable static charge decay characteristics. Both conductive polyethylene tile and conductive multilayer mats provide static discharge, but less than perfect static charge suppression. The colloidal film, however, was able to achieve both rapid static discharge and extremely low static charge generation simultaneously. (In this test procedure, zero static charge was generated when walking on the colloidal film with either neolite or chrome leather soles.)

To prevent static generation, one must notice the types of materials present in the environment; and to eliminate a static charge once it is present, one must provide a rapid path to ground potential. Therefore, we wish to maintain all things (tables, floors, components, people) always at the same potential. Should an aberration occur so that a charge is present, we would want to return to our static safe position as rapidly as possible. Our general system of static protection is then summarized as follows:
▲ Remove/reduce the static generating materials in the system.
▲ Provide a means of rapid static removal via a conductive path to ground potential.
▲ Maintain all parts of the system at the same ground potential.

The most frequently used technique involves the use of conductive materials which permit static dissipation independent of the relative humidity in the environment and which provide a permanent route of the static charge to ground potential. As a practical matter in the electronics workplace, the types of products which are utilized are bags, foam, table tops and floor mats, personnel wrist straps, parts trays, boardshorts, and stool covers, all of which are electrically conductive.

In the use of these conductive materials, it is vitally important to remember the safeguards to people working in this environment. It is desirable to keep the current less than one milliamp in order to maintain a safe environment for workers. Grounding straps which attach conductive materials to the ground potential should have a one megohm resistor in series as part of its material construction.

A summary of the limitations of the available products is shown in Table 5. Electrically conductive bags are preferable in use to antistatic polyethylene bags, since the antistatic bags may have problms of non-permanency, as well as migration during shipment prior to use. Also, such antistatic bags do not provide a Faraday cage for shielding the contents.

High density conductive foam is recommended to package IC's in transportation and storage. Such foam should be noncorrosive, so that moisture pick-up from the air does not result in acid formation and subsequent corrosion on the IC leads. High density conductive foam is available in sheets and in perforated foam. Perforated foam is used for ease of removing small pieces for individual IC's. Low density conductive foam provides a soft cushion for wrapping of PC boards or packaging of components. Again, corrosion resistance is important.

Electrically conductive table tops and floor mats provide good surface conductivity and volume conductivity. However, a limitation in the use of electrically conductive polyethylene in the application is that frictional wear often increases the barrier potential of the material, and conductivity decreases to a point where the material is no longer useful. Periodic checks of the electrical resistance are worthwhile as a common factory practice.

Personnel wrist straps should be utilized on the electrically conductive material, in series with a resistor to reduce the chance of injury to the personnel involved. Also, grounding straps which are electrically conductive should be utilized with a large resistor to prevent any personnel injuries due to very large shock currents.

The general guidelines for static protection include keeping people and devices at ground potential with the use of conductive wrist straps, bags, foam, containers, and table and floor coverings. Also, it is desirable to shunt PC boards with conductive boardshunts during transportation within the factory to prevent any potential static discharge problem. Likewise, it may be desirable in some environments to provide ionized air in the general work environment to promote corona discharge of static charges from the surfaces. All materials which have a potential for generating or transmitting a static charge should be properly grounded.

A static-safe environment necessitates careful planning, a complete static protection system, and personnel education. A safe program should include initial and continuing personnel awareness, as well as regular monitoring of work areas.

References
(1) Jowett, C.E. <u>Electrostatics in the Electronics Environment</u>, John Wiley and Sons, 1976
(2) IBID, page 72
(3) Cusak, et. al., Journal of Electrostatics, 1972
(4) Berbeco, G.R., <u>Charaterization of ESD Safe Requirements for Floor Surfaces</u>, EOS/ESD Symposium, 1982

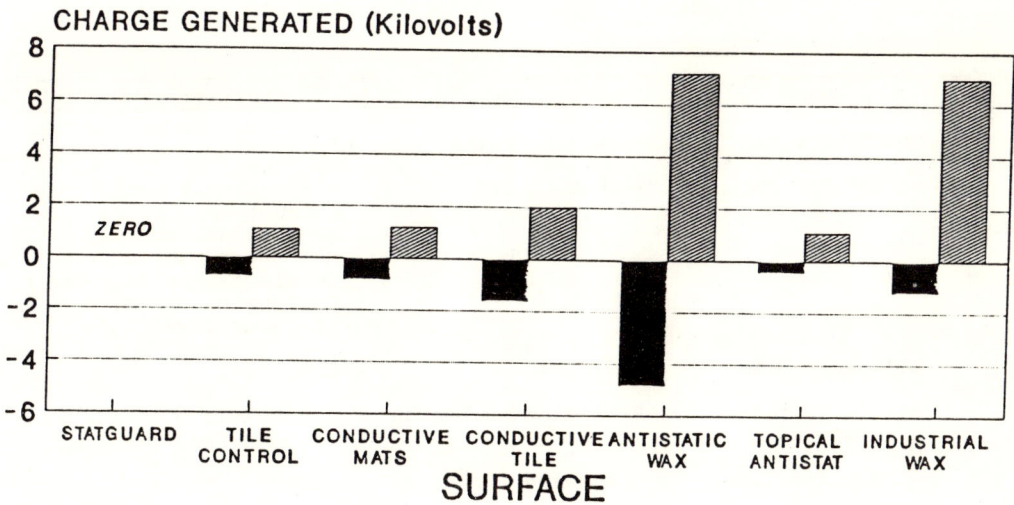

AATCC Test Method 134-1979; 15 - 20% RH

Fig.1 CHARGE GENERATION By AATCC STEP Test

TABLE 4 — COMPARISON OF FLOOR SURFACES

	Static decay time*	Charge generation*	Resistivity +	Total discharge of personnel
Ideal characteristics	0.1 sec.	Zero	10^9/sq.	1 sec.
Colloidal film on tile	0.05 sec.	Zero	10^9/sq.	1 sec.
Vinyl asbestos tile				
w/wax	0.36 sec.	2 to 3.5 kV	10^{13}/sq.	10 sec.
w/topical	0.90 sec.	4 to 7 kV	10^{13}/sq.	10 sec.
anti-stat	0.36 sec.	0.5 to 1.2 kV	10^{11}/sq.	10 sec.
Conductive P.E., tile, or multi-layer mats	0.01 to 0.1 sec.	2.0 to 4.0 kV	10^6 to 10^9/sq.	1 sec.
Anti-static polyethylene	10.0 sec.	1.0 to 3.0 kV	10^{13}/sq.	10 sec.

* Federal Test Method Standard 101B, Method 4046.
* AATCC Test Method 134-1979 (electrostatic propensity of carpets).
+ Measured using an HP4341 high resistance meter at 100 V input, and an ETS concentric circle "Rupe" fixture (ASTM D257).

TABLE 5 - STATIC PROTECTION PRODUCTS

PRODUCT	APPLICATION	LIMITATION
Conductive bag	Packaging PCB's	None
Antistatic bag	Packaging PCB's	Nonpermanent antistatic character; shouldn't be handled excessively; use fresh bags only; should employ incoming QC; nonshielding, danger of induced voltages.
Colloidal Film	Wax floor	None
Static shielding bag	Packaging PC boards	Antistatic property is nonpermanent.
Conductive foam	Packaging PCB's, IC's	None
Conductive table top and floor mats	Table and floor coverings	Conductivity decreases with frictional wear.
Wrist straps	Personnel grounding	Use 1 megohm resistor
Grounding straps	Path to ground	Use 1 megohm resistor

Reliability Study on MOS Electrostatic Protection from Charge on Human Body

Xia Hong, Liu Jiyun
Applied Physics Department
Beijing Institute of Technology,
P.O.Box 327, Beijing, P.R.China

ABSTRACT

In handling of MOS devices, static charge on human body may result in damage or failure of the devices. To prevent static generation, systematical counter-measures are taken in working environment, including operator's wearing, the floor, tabletops and so on[1,2]. It is important to select which parameters used to evaluate the reliability of the measures. Besides the resistance of hyman body to the earth and one's charge decay time, one's static potential should be one of the parameters for reliability study on the measures. In this paper, special consideration on operator's potential is emphasized based on our experimental results.

Introduction

MOS devices are of static sensitivity. Electrostatic discharge (ESD) can cause direct, indirect, and latent failures in MOS devices[3-6]. Various models of the failure mechanisms have been proposed[7,8] The human body model is one of them. In terms of this model, it is proposed that the staff working at the stations where devices are handled should not be charged to a high static potential. Static charge on human body are usually specified as the discharge source. Table 1 lists the ESD susceptibility range i.e. critical voltage of some devices[9]. A charged operator with the potential over the critical voltage of a device may lead the device to failures while handling of it, especially handling with hands.

Basically, whenever two different materials are rubbed together, static charges are generated. While one's working, walking, rubbing his arms across a work table surface, and so on, many thousands of volts of static potential could be generated on himself. Thus the charge on human body is dangerous to MOS devices, in the case of without electro-static protection. The amount of one's charge buildup is affected by the dissipation of the charge and the body capacitance to the ground. Anti-static materials are usually applied as one of the measures to provide a charge dissipative path with small earthed resistance and to retard the tribo-electrification as well.

Table 1 ESD susceptibility of some devices[9]

Device type	Range of ESD susceptibility(v)
VMOS	30--1800
MOSFET	100--200
CMOS	250--3000

*(courtesy 3M static control systems)

Current main measures against the static charge accumulation on human body are as follows[2]:

(1) Installing a conductive floor mat and a conductive tabletop;

(2) Covering the seats with conductive materials;

(3) Wearing either conductive shoe or boot strap to ground himself to the conductive floor mat when standing, or wearing a conductive wrist strap.

Whatever, if, with the anti-static measures, the operator's potential can be kept below the critical voltage of ESD failure of the devices, it is said that the measures are reliable for handling of MOS devices safely. Our experiment results demenstrate that vairations of both one's equivalent resistance to the earth and the the capacitance between human body and the earth are rose because of the spatial non-uniformity of the electrostatic properties of the anti-static floor mat. As a result, the potential value of a charged operator would be changed, while his standing at different positions on the mat, even the charge buildup on the body is same. In this case, besides the equivalent resistance and the charge decay time, the operator's static potential must be taken as an important parameter to evaluate the reliability of the anti-static measures against charge on human body.

Experiment and Result

1. Measures against static charge accumulation on human body

In order to eliminate the charge buildup on human body, the measures against the static charge are taken. There are different ways for the installation of the conductive or anti-static floor mat[1,2]. The installation of the anti-static floor mat used in the experiment is schemically shown in Fig.1. The anti-static rubber mat is on the normal floor in the work station, and between them there is a inset of earthed Z-shape copper strap with the width of about 5.5 cm. The rate between the floor mat aera touched on the copper straps and the total mat aera is about 20%. Due to the way of the installation, the anti-static floor mat should be devided into three typical kinds of parts according with one's standing positions as referred by the position I, II and III in Fig.1.

In addition, in our experiment an operator standing on the mat wears anti-static clothing and rubber shoes.

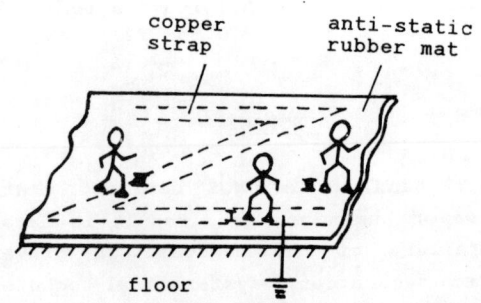

Fig.1 Schematic diagram of the construction of the anti-static floor mat (2m × 2m) and three kinds of typical positions.
1--normal floor; 2--anti-static rubber mat; 3--earthed Z-shape copper strap under the rubber mat, with width of about 5.5 cm.
Position I: both feet are on the place where copper strap are underneath the rubber mat;
Position II: both feet are on the place where there are no copper straps under the mat;
Position III: only one foot is on the place where the copper straps are under the mat.

2. Results

With four kinds of the anti-static floors, the experiments are carried out under the condition mentioned above. The

Table 2 Effect of operator's positions and states on the capacitance C and equivalent resistance

Anti-static mats & normal floors	Blue rubber mat				Red rubber mat			
	cement		terrazzo		cement		terrazzo	
R and C Position and state	R (Ω)	C(pF)	R(Ω)	C(pF)	R(Ω)	C(pF)	R(Ω)	C(pF)
I Standing	6.7×10⁷	414	5.0×10⁷	304	1.0×10⁸	293	4.5×10⁷	500
Sitting		287		242		233		444
II Standing	2.8×10⁸	240	7.7×10⁷	284	1.3×10⁸	177	8.0×10⁷	406
Sitting		206		232		164		355
III Standing		328		316		242		458
Sitting		287		242		188		397

Note: Experimental environment condition: t--15~22°C; RH--20%~30%.

capacitance C between the tip of the human body and the earth and the equivalent resistance R of human body to the earth, which is the resistance of the charge-transfer-path i,e. the path from the operator's feet via the soles and the floor mat to the earth, are tested. The results are listed in Table 2.

As Table 2 indicated, at different kinds of positions on the floor mat, the capacitances are quite different, and the equivalent resistances are also not in same. The equivalent resistance R of an operator at position I is greater than that at the position II. And the capacitance C of the operator at position I is smaller than that at position II. These variations are probably caused by the changed efficient distance by the grounded copper straps. The efficient distance is the interelectrode distance of the capacitor formed by the human body and the earth.

In addition, the capacitances are different while the operator is in different habit states. In our cases, when one's standing, the capacitance is greater than that when one's sitting by amount of about 20--130 pF, which can also be concluded from Table 2.

Discussion

The potential of one charged human body can be derived from the formula:

$$V = \frac{Q}{C} \quad (1)$$

where V is the potential of the body, C is the capacitance to the earth, while Q is the charge buildup on human body.

Supposing an operator is charged to 30 volts (the minimum critical voltage of ESD failure of VMOS), the capacitance to the earth is 500 pF, while his standing on position I, the charge buildup on his body is equal to 1.5×10^{-2} μc, derived from the equation (1). Furthermore, if the same amount charges are accumulated on the operator standing at the position II, the potential will be greater than 37.5v due to the capacitance decreased by about 100 pF. This means that even with the same amount of static charges on the operator, it is possible to cause the ESD failure of MOS device due to the operator's static potential may be greater than its critical voltage value caused by the variation of the capacitance.

In fact, since the charge decay time calculated by τ =RC from Table 2 is changed slighly for the different typical position, the same quantity of charge on human body by tribo-electrification of a certain moving of the body could be always obtained, while one's standing at the different positions on the floor mat. Therefore, if, at the position I, the potential of a human body reaches to the critical voltage of VMOS in ESD failure, the potential, while one's standing at position II and III, is certainly over the critical voltage. In this case, the ESD failures of the devices may appear since the devices are very sensitive to the static potential. It is means that the measures against static charge on human body taken nowadays are not always in success, except considering the human body's potential.

Conclusion

In general, the non-uniformity of the electrostatic properties of the anti-static floor mat or the spatial difference of the floor mat installations can lead to the variation of an operator's static potential even though the charge buildup on him is in same. As handling of the MOS devices which is very much sensitive to static or ESD, special consideration should be taken on the design, installation and construction of the anti-static floor mat. In brief, whenever and wherever, the potential of the human body must be controlled below the critical voltage of ESD failure of MOS devices, in spite of the equivalent resistance R is in the range of 10^4--10^8 Ω required by anti-static measures. The increase of the potential resulted from the decrease of the capacitance may be great enough to lead one's potential to reach up the critical potential of MOS device. Therefore, not only the equivalent resistance but also the potential should be considered as the important parameters to evaluate the reliability of the anti-static measures against static charge on human body.

Reference

[1] George R.Berbeco
Electrical Overstress/Electrostatic Discharge Symposium Proceeding, 1982,pp124-130
[2] C.J.Strand, A.Tweet, and M.E.Weight, EOS/EDS Sym. Proc.,1982, pp145-156
[3] W.D.Greason and G.S.P.Castle, IEEE Trans. Ind. Appl.,vol.20, no.2, pp247-252 Mar./Apr. 1984
[4] B.A.Unger, IEEE Int. Reliability Physics Conf. 1981,pp193-199
[5] D.K.Davies, J. Electrostatic,vol.16, pp329-342, 1985
[6] P.G.Pierce and D.L.Durgin, EOS/ESD Sym. Proc. EOS-3, Reliability Analysis Center, 1981,pp120-131
[7] Reliability Analysis Center, Electrostatic Discharge (ESD) Susceptibility of Electronic Devices, VZAP-1, RADC/RAC, Griffiss AFB, NY,1983
[8] S.Y.Youn, N.Hartdegen, and M.Sharp EOS/ESD Sym. Proc. 1982, pp157-163
[9] Protection ICs From Electrostatic Discharge, Electronic Products Magazine, June, 1980

VIII. Electrostatic Hazards in Petro-Chemical Industries and the Prevention Measures

Proceedings of International Conference on Modern Electrostatics

RELAXATION TIME OF ELECTRIFIED PETROLEUM SURFACE POTENTIAL IN STORAGE TANKS WITH DIFFERENT CAPACITY

Naotake KITAMURA
Dep. of Electrical Eng., Faculty of Eng., Nagoya University
Furo-cho, Chikusa-ku, Nagoya, 464-01 Japan

Ichiro FUSHIMI
Mitubishi Oil Co. Ltd.
1-2-4 Toranomon, Minato-ku, Tokyo, 103 Japan

Minoru UEDA
The ex-professer Emeritus of Nagoya University and the ex-president
Emeritus of Institut of Electrostatics of Japan, died Feb. 15, 1987

Abstract

A tendency of the relaxation time for the three kinds of tank capacity can be classified into three patterns by each conductivity of the five kinds of petroleum used in our experiment. The measered relaxation time is larger one or two orders than a relaxation time obtained from time constant which is calculated by the conductivity and the dielectric constant of each petroleum measured through our experiment. Therefore, a measuring or sampling operation at the manhole on a storage tank is never begun on the rest time that is estimated by the time constant of the sample petroleum.

Introduction

A. Klinkenberg and J. L. van der Minne have described in their work that the relaxation time is dependent only on the properties of the petroleum and not on the dimentions and shape of a container|1|. W. M. Bustin and W. G. Dukek describe in their book as follows|2|. There are two charge relaxation formulas. The first applies to ordinary relaxation assuming the conductivity measured in a test cell adequately characterizes the liquid. This formula is an exponential. The second non-exponential formula applies to very low conductivity liquid where charge relaxation differs substantially from the rate predicted by the conductivity. Bustin describes also that charge relaxes exponentially at a rate determined by the conductivity measured in a cell when the conductivity is one picosiemen/meter (pS/m). Charged distillates having conductivity less than one pS/m can be assumed to relax according to the hyperbolic formula with a charge mobility of approximately 10^{-8} meter^2/volt sec.|3|. H. Krämer and G. Schön have investigated a relaxation correlation in which the conductivity effective during charge relaxation equals a constant plus a term linearly proportional to the absolute value of the charge density|4|. Others have developed more complex formulas|5|-|10|.

Experimental Method

The experimental systems are shown in Fig. 1(a) and (b). For the present, Fig.1(a) and (b) are named "probe method" and "direct method" respectivly. The reson to use the two methods is able to give more reliable data. Authers have used "probe method" to measurre a potential in electrified petroleum. However, there is one ploblem. A value of the potential varies with a resistance supporting the probe. The true value of the potential must be given by the infinite resistance supporting the probe. Therefore, authers read till now the true potential on a graph when a leak current at the probe suport is zero. This time we measure with a constant suppoting resistance shown Fig.2.

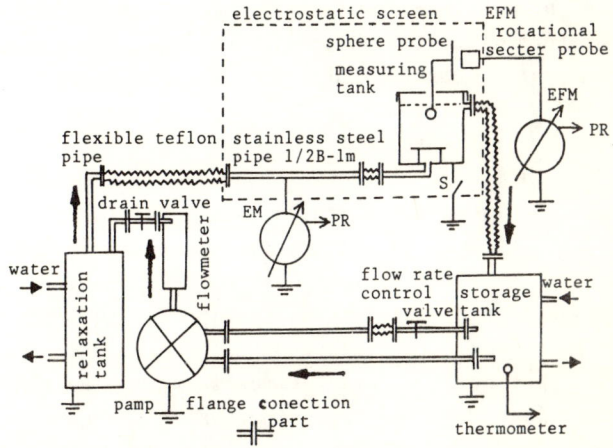

Fig.1(a) Experimental apparatus of "probe method"

The "direct method" measures a electrical force from the petroleum surface by a rotational secter probe of electrical feald meter(EFM).

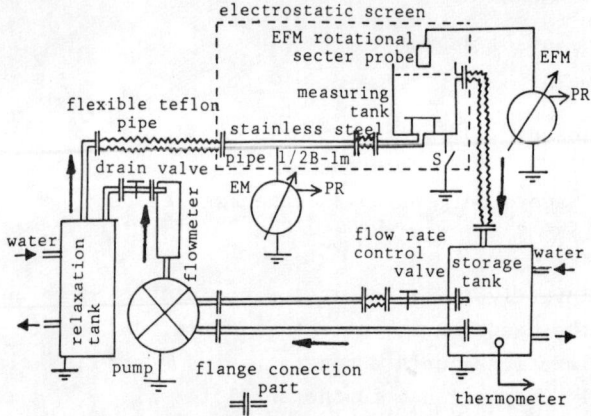

Fig.1(b) Experimental apparatus of "direct method"

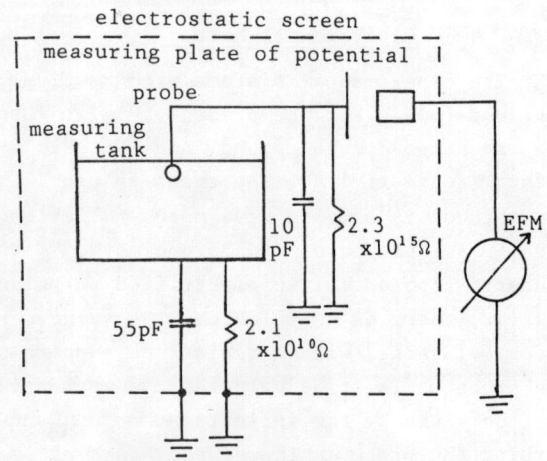

Fig.2 Resistance and electrostatic capacity of apparatus

The flow system of petroleum is made of stainless steel except the flexible teflon pipe and the flow meter. Then a petroleum is flowing through the flow system, the petroleum temperature rises for viscousloss. Therefore, its temperature can be controled the range of 19±0.5°C at the relaxation tank and the storage tank by city water.

The static charge in measuring tank is generated only in the stainless steel pipe(1/2B-im). The pump is driving till the potential saturat. After the pump was stopped and the measuring tank was earthed, the potential decay is recorded. Initial potential is named by V_0. and a potential on every times is V_t. The geoometrical sizes of measuring tank are shown in Table 1. The tank height is all 0.3m. The used probe is made of a steel ball(15mm diameter). The properties of each petroleum are given in Table 2. SS-1500 is the solvent name sold by Mitsubishi Oil Co. Lid. It is made of aromatic hydrocarbon composed of C_9. $\tau_{1/2}$ is culculated by Equation(1), here $\tau = \frac{\varepsilon_0 \varepsilon_r}{\sigma}$.

$$\tau_{1/2} = \tau \ln 2 = 0.693 \tau \text{ (sec.)} \quad (1)$$

tank type	capacity(m^3)	radius(m)
#1 tank	0.94×10	0.1
#2 tank	2.1 ×10	0.15
#3 tank	4.8 ×10	0.23

Table 1 Geometrical sizes of each tank

sample of petroleum	conductivity σ (pS/m)	dielectric constant ε_r	$\tau_{1/2}$ (sec.)
toluene	39.0	2.7	0.19
SS-1500	18.0	2.7	0.92
JP-4	4.2	2.0	2.9
gas oil	0.77	2.3	18.0
kerosene	0.29	2.25	48.0

Table 2 Properties of each petroleum

Expelimental Results and Discussion

Two cases of kerosene and toluene are shown in Fig.3(a),(b). This is the reason which they have the extremely different conductivity shown Table 2.

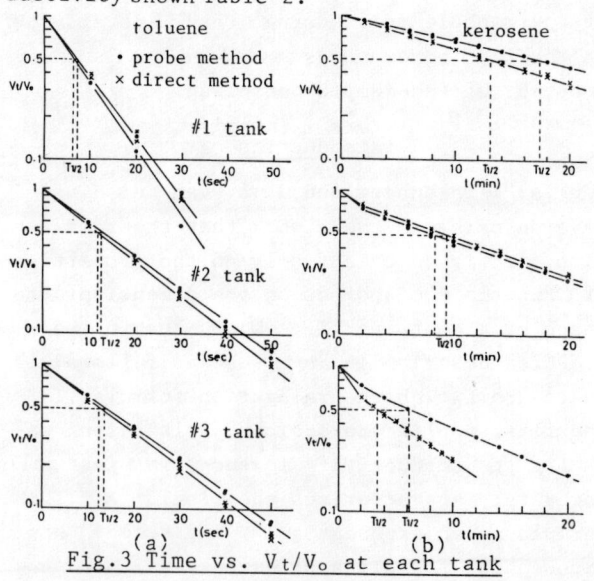

Fig.3 Time vs. V_t/V_0 at each tank

Vertical axes are weitten with the numelical value V_t/V_0 on logarismic scale. Every result except the results of kerosene #2 and #3 tank is satisfied next equation with the probe and the direct method.

$$Q_t = Q_0 \exp(1 - t/\tau') \quad (2)$$

Here, Q_t is charge quantity after t secound, Q_0 is initial charge quantity, τ' is effective time constant decided by the petroleum and the rate of electrification. τ' has to need great care, because high V_0 decease fast than low V_0

through each 3 or 5 runs of every petroleum. The values of V_o are measuered from 701 to 2328V, most case of V_o is 1200V order through every measuerment.

Tank Capacity and Half Value Time of Relaxation

Fig.4(a)-(e) are shown the half value time $T_{1/2}$ versus the tank capacity.

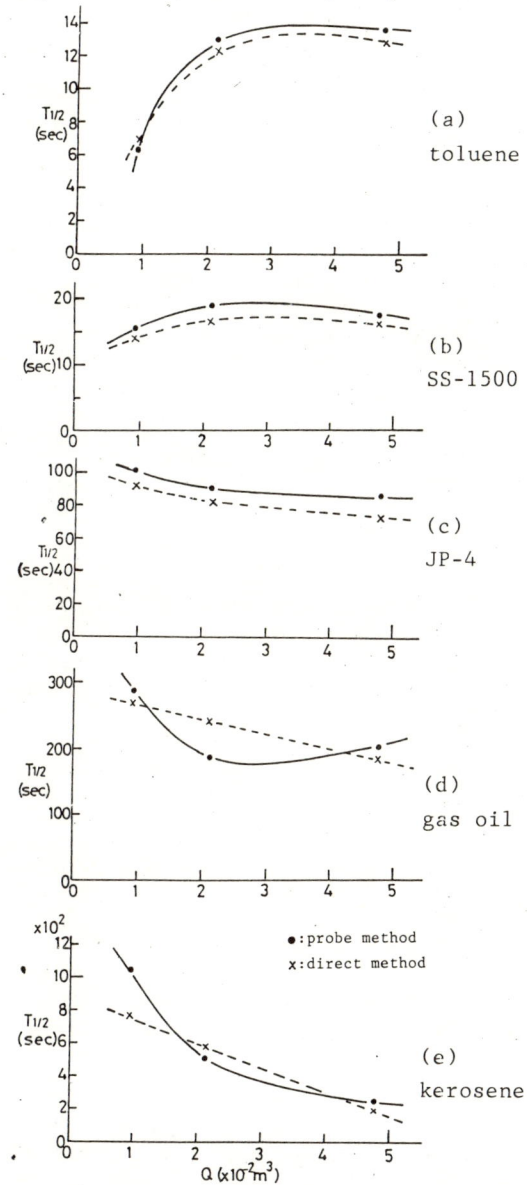

Fig.4 Half value time $T_{1/2}$ vs. tank capacity

The tendencies of tolene(Fig.4(a)) and SS-1500 (Fig.4(b)) are different from other samples. The tendencies of the probe method(solid line) and the direct method(dotted line) are same respectively, even if each value has a little difference. This reason means that the time constant carcurated R and C of probe in Fig.2 is larger one or more orders than measuring relaxation time constant.

There may be three relaxation modes for each petroleum attending increase of the tank capacity. SS-1500 and toluene have the increasing $T_{1/2}$, this is named the first mode. Second mode has the constant $T_{1/2}$ as JP-4. Third mode shown kerosene and gas oil decreses $T_{1/2}$. It knows that each mode has different rest conductivity σ respectively. The first mode petroleums have higher conductivity than pico-siemen(pS/m). The conductivity of the second mode is about pS/m. The third mode has lower conductivity than pS/m.

One reason mey be thought by a difference of potential distribution in the measuring tank. The potential in the higher conductivity petroleum should distribute more uniformly from the center to the wall of the tank. A current dipendenced by next equation recages from the tank wall to the earth along the surface of teflon insurated the tank while drive the pump.

$$J = \Sigma (q_+ \mu_+ + q_- \mu_-) \nabla \psi + D \nabla (q_+ - q_-) - v (q_+ - q_-) \quad (3)$$

Here, q_+ and q_- are plus and minus charge density, μ_+ and μ_- are pulus and minus charge mobility, ψ is potential determind by Poisson's equation, D is ionic diffusibity, v is mean velocity of petroleum. On the stage of charge relaxation, each term of equation(3) decreases with time. The third term of light hand in the equation may be related to a kinematic viscosity of each petroleum. Becuse charge carriers in lower kinematic viscosity petroleum may be moved with the petroleum for a long time in a larger capacity tank after stoped pump. Table 3 shows the kinematic viscosity of each petroleum at the temperatur 10 and 50°C.

Toluene has the lowest kinemati viscosity. Therefore, Fig.3(a) and Fig.4(a) show a more large half time $T_{1/2}$ according as increase the tank capacity. This result should be explained also by the kinemati viscosity.

sample of petroleum	temperature of sample (cSt)	
	10°C	50°C
toluene	0.73	0.45
SS-1500	1.0	0.62
JP-4	1.15	0.7
gas oil	4.7	2.1
kerosene	1.9	1.05

Table 3 Kinematic viscosity of sample

Fig.5 shows the relation between the product of rest conductivity σ and tank capacity Q and $T_{1/2}$. The mean $T_{1/2}$ of each petroleum decreases in propotion to $(\sigma \cdot Q)^{-0.78}$. As Klinkenberg described, this result points out that the relaxation time depends on the material property of petroleum all over.

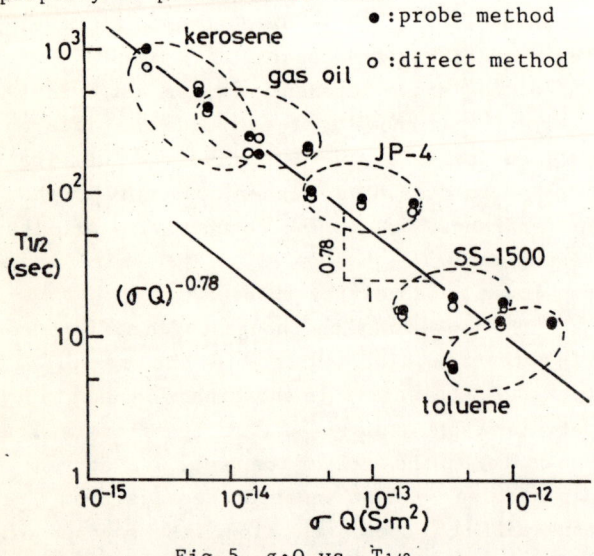

Fig.5 $\sigma \cdot Q$ vs. $T_{1/2}$

Fig.6 shows that is compared the measured values with the hypabolic theory(o) and the ohmic theory(Δ) on kerosene. Parameter is the tank capacity. The relaxation characteristic forms the hypabolic theory when τ is given by the rest conductivity according as increase the tank capacity.

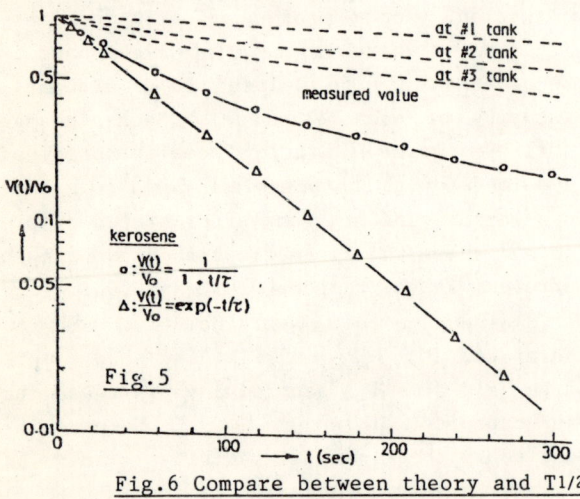

Fig.6 Compare between theory and $T_{1/2}$

CONCLUSION

The results made clear from this study are described as following.

1. The relaxation time decreases accordind as a saturated petroleum surface potential make up high.
2. The relation of the relaxation time varsus the tank capacity has three pattern with the petroleum conductivity.
3. The scale of the tank capacity and the kinematic viscosity of petroleum may be connected with the relaxation time.
4. Kerosene had lower conductivity than pS/m approaches the hyperbolic relaxation form according as the tank capacity is large.
5. The mean half value time of each petroleum decreases in proportion to the -0.78th power of each petroleum conductivity.

References

[1] A.Klinkenberg and J.L.van der Minn,"Electrostatics in the Petroleum Industry", 1958, Elsevier Publishing Co., Amsterdam

[2] W.M.Bustin and W.G.Dukek,"Electrostatic Hazards in the Petroleum Industry", 1983, Reseach Studies Press LTD., England

[3] W.M.Bustin, I.Koszman and I.T.Tobye,"mew theory for static relaxation from high-resistivity fuel", Vol.44(III), pp.548-561, American Petroleum Institute, Dev. of Refinery Proceedings

[4] H.Krämer and G.Schön,"Ladungsrelaxation stark aufgeladener Kraftstoffe", Vol.72, pp.361-272, 1974, 2Int. Tagung über Elektrostatische Aufladung, Dechema-monographien No.1370-1409

[5] M.D.Foster,"The dissipation of electrostatic charges in purified petroleum product", pp.78-88, 1967, Inst. Phys. Conf. Sries No.4

[6] J.Gavis,"Transport of electric charge in low dielectric constant fluids", Vol.19, pp.237-252, 1964, Chem. Eng. Sci.

[7] J.Gavis,"Non-linear equation for transport of electric charge in lowdielectric constant fluids", Vol.22, pp.359-364, 1967, Chem. Eng. sci.

[8] J.Gavis,"Relaxation electrically charged hydrocarbon liquids", Vol.22, pp.633-635, 1967, Chem. Eng. Sci.

[9] J.Gavis,"The effect of ionic dissociation and recombination on the relaxation of charge in low dielectric constant fluids", Vol.24, pp.451-460, 1969, Cem. Eng. Sci.

[10] S.J.Vellenga and A.Klinkenberg,"On the rate of discharge of electrically charged hydrocarbon liquids", Vol.20, pp.923-930, 1965, Chem. Eng. Sci.

TECHNIQUES OF ANTI-ELECTROSTATIC HAZARD OF THE PETROCHEMICAL INDUSTRY IN CHINA AND OUTLOOK

Wang Shili, Dong Changui, Yang Jin
Production Management Division, SINOPEC,
Anwai Xiaoguan, Beijing, China

Song Guangcheng, Bi Zaijun
Fuel Dept. RIPP. 18 Xueyuan Road
Beijing, China

Abstract

This paper presents the recent developments of some anti-electrostatic techniques and supervision that have been launched with great effort in the petrochemical industry of China, and these result from the combination of the successful efforts of foreign electrostatic specialists with the practical conditions of China. The prospective plan in this field from now on is also stated.

Introduction

Along with the rapid development of modern industry, larger in scale, faster in speed, and more diversify in variety, the industrial hazards caused by static electricity repeated day by day. Expecially to the combustible and explosive hazardous crude oil and petrochemical products they make people in great anxiety. Therefore, many writings, handbooks and safety codes about the prevention of electrostatic hazard are published by many organizations and experts of developed countries all over the world, and in these papers provided are many effective technical measures and reference data [1-6] for control and minimization of the electrostatic hazards in different extents.

The petrochemical enterprises in China have made great effort to increase their production, at the same time consult the advanced experiences of foreign electrostatic workers linked with the practical conditions of China, and took various means and supervision systems of the prevention of electrostatic hazard to ensure the overall completion of their production plans safely.

1. General description

In the process of production, storage and transportation of petrochemical products, pressing, stripping, filtering and spraying processes are often used. Consequently, friction, impact, collision etc., a series of contact and separation phenomena occur. These actions induce and accumulate static electricity inevitably and tend to induce ignition of the combustible explosive petrochemical product and cause serious consequences. In the past twenty years, more than twenty relatively serious electrostatic disasters happened in Chinese petrochemical enterprises and caused a loss of million yuan.

The fire and/or explosion hazardds ever happened in Chinese petrochemical industry may be classified into three kinds:

1) Fire and explosion hazards induced by the discharge of static electricity of floating bodies in the storage tank. For example, the accident of oil storage tank in Shanghai Gaoqiao Petrochemical Company, the accident of acid storage tank in Shandong Qilu Petrochemical Company and the accident of dilute ammonia tank in Fushun Petrochemical Company, they are all caused by the discharge of floating ball or floating objects with the tank wall.

2) Electrostatic fire and explosion hazards induced by careless operation of sampling or check of temperature. The accident of Shanghai Gaoqiao Petrochemical Company was caused by lifting the temperature measuring element too quickly in a toluene storage tank; the accident of Fushun Petrochemical Company was caused by sampling in a jet fuel storage tank which had received fuel just a moment ago; and the accident of Hubei Jingmen Refinery was caused by a plastic bucket used for receiving oil etc. All of them created electrostatic fire and explosion. A plastic tube of gravimetric measuring system was swaying by the wind, static electricity was induced, causing an explosion of that gasoline tank. This accident happened in Shandong Qilu Petrochemical Company.

3) Fire and explosion created by the static electricity produced by spraying or friction of the media at the loading of the tank car of storage tank. These accidents happened in the petrochemical companies of Daqing, Beijing and Dalian etc. relatively generally in the past. In a loading operation of train tanker in Qilu Refinery, Qilu

Petrochemical Company, the reason was the velocity of the petroleum product in the crane pipe was too high. Massive static electricity accumulated at the outlet of the pipe. When it was contacted to the well grounded tanker, sparks occurred immediately, then the train tanker exploded. After the accident happened, traces of electric sparks were found obviously at the aluminum bosh tube of the crane pipe outlet and the upper manhole of the train tanker.

2. Works launched

In order to ensure the safe operation of the petrochemical enterprises and minimize the electrostatic hazards, China Petrochemical Corporation (SINOPEC) has undergone the following works:

1) Extablished a static electricity special committee of petrochemical industry.

In 1982, under the guidance of electrostatic committee of the Society of Physics of China, a static electricity special committee of petrochemical industry was established with the members of warm-hearted technical staffs. Under the strong backing of the petrochemical enterprises, research works, establishment of standards, academic exchanges and training works were launched. In 1983 a static electricity technique center was also established in Dalian.

2) Launched a special subject study project.

In the past few years, we have finished the studies of: "Anti-electrostatic additives of petroleum products", "Test instruments of the conductivity of petroleum products", "The technique of under-liquid loading with small crane pipe", "Anti-electrostatic ground surface of the bottling house of propylene oxide", "The prevention measures of the electrostatic hazard of pneumatic drying of powdered materials", etc. We have got many good results from these studies.

3) Formulated a number of safety technical regulations of static electricity, such as: National standards: "The test method of the conductivity of light petroleum products", "Safety static conductivity of light petroleum products", and "Safety potential of the oil surface at the loading of light petroleum products". Special codes: "Safety regulation of the static electricity of combustible and explosive liquids", "10 regulations of the prevention of the electrostatic hazards", "Safety technical codes of the static electricity of synthetic fabrics in the petrochemical enterprises", etc.

4) Conducted academic exchanges and personnel training at regular time.

3. Main technical measures adopted

Through the study of the prevention technology of the electrostatic hazards in the past few years, SINOPEC has taken a series of technical measures one after another, which are mainly:

1) Determine the permissible maximum flow velocity of the light combustible petroleum products.

When the light petroleum product flows in the pipeline with high velocity, it will induce massive quantity of electrostatic charge. Experiments have shown that there is a certain relationship between the average flow velocity (V) and the maximum electrostatic potential (V_{max}) liquid product flows along the pipeline (Table 1).

Table 1. The relationship between the flow velocity of petroleum product and the maximum electrostatic potential.

Product	d(mm)	V(m/s)	V_{max} (v)
gasoline	100	7.4	500
gasoline	100	2.9	250
jet fuel	80	12.8	13400
jet fuel	80	5.24	9600
diesel oil	100	14.1	-11000
diesel oil	100	8.7	-1700

Experiments also show that the filter in the pipeline has notable effect on the induced maximum electrostatic potential (V_{max}) (Tables 2 and 3).

Table 2.

Product	V (m/s)	Filter	V_{max} (v)
gasoline	4.67	no	2,400
gasoline	5.20	yes	9,300
jet fuel	4.46	no	8,500
jet fuel	4.43	yes	11,000

In order to restrict the production of static electricity in the pipeline, lowering the flow velocity or friction is suitable. When the

Table 3. The effect of the material in the filter on the max. electrostatic potential.

Type of filter	Material in filter	V_{max} (v) before filtering	V_{max} (v) after filtering	Surface of tank
I	silk fabric	350	8100	22500
II	paper	140	15000	28000
III	glass fibre	130	10000	24000

combustible and explosive gas may occur in the storage tank, the initial flow rate of petroleum product should be limited strictly below 1 m/s. When the loading pipe submerged in the petroleum product, the flow rate may be increased properly. For loading operation of the tank car or train tanker, the following equations for the relationslhip between the loading pipe diameter d (in meter) and the permissible flow rate V(m/s) may be applied:

$V_d \leq 0.5$ (for tank car loading)

$V_d \leq 0.8$ (for train tanker loading)

2) Control the mode of injecting and mixing of the oil to avoid occurrence of impact or spray.

Which mode is used to inject the oil is also an important measure to control the production of static electricity. It is proved in practice that to fill oil into a large storage tank from the top is not permitted, and this restriction may reduce the generation and accumulation of static electricity obviously. Because, if oil is filled from the tank top, impact and form foam may occur, then space charges are produced and the charge quantity of the petroleum product increases. In addition, this can promote the formation of oil vapor and increase its concentration.

When the oil is filled in a storage tank, the filling pipe should be inserted in from the bottom and horizontally to the center of the tank. Be careful not to stir up the water in the tank bottom since if the oil is mixed with water may intensify the production of static charges.

The research work has shown that stirring, spraying, sedimentation, splashing, foaming and flowing can produce massive amount of static electricity in oil products. It is the potential ignition source of gasoline, kerosine etc. Practical test has indicated that when the air as small bubbles is mixed with petroleum product, the initial charge density can be 100 times more than that of the petroleum product normally flowing in the pipeline. Therefore, it is better to use mechanical stirring with a pump or on-line blending to blend petroleum products. In the operation of circulation by pump, use single- or multi-nozzle to circulate via the inlet and outlet of the tank bottom is advisable, and the blending time will be reduced by 50% with the multi-nozzle process. In case of the total circulation of oil by sucking out from the tank bottom and pump back through the top of the tank, since the oil spraying or splashing down from the tank top, it is inevitable that the oil will contacts with air and massive static electricity is produced. So, this mode of pumping circulation is not advisable.

3) Popularize the floating roof storage tank.

Floating roof storage tank and arched top tank with inner-floating roof not only can reduce the vaporization loss of oil, ensure the quality of product, minimize air pollution, but can also prevent the accumulation of static electricity. Recently, the success of soft seal development has opened a wide route to the floating roof tank.

The floating roof storage tank is so designed that a floating boat is on the oil surface and it can move up and down with the oil surface. Two bus bars are tied to the floating boat and along the stairs contact with the tank wall. The oil is highly insulating material, the conductivity is very low, and the leakage of electrostatic charge is also rather low. As metal floating boat is mounted, a great deal of electrostatic charge on the oil surface will flow to the boat, then it will grounded along the wire, and the electrostatic charge can also be grounded through the circular rubber seal.

4) The use of closed loading facility.

The major disadvantages of loading cars with splashing mode are: first, the difference of height between the car and oil is too high, so the amount of static electricity accumulated on the oil surface is high. Second, a lot of oil vapor is vented out and will produce explosive gas and contaminate the air, safety problem arises seriously.

In the closed loading process, a good conductive metal crane pipe is inserted into the bottom of a tank car and a distributoris mounted, so the oil stream will not impact the bottom of the tank car directly. Most oil gets out under the oil surface, and the oil surface increases steadily, and the friction between the oil stream, tank wall and air is fairly low, and there is no serious seethe of the liquid surface, and so less static electricity will be produced. The most important thing is that the metal crane pipe is grounded, large conductors allow efficient charge leakage, so this can solve the problem of static electricity accumulation by loading oil properly. Since the loading crane pipe is inserted into the bottom of the tank car, the vaporization and loss of oil will be reduced greatly. The result of on-site test has shown that the static electricity potential of conventional splashing mode loading can be up to 10,000-30,000 volts, and the potential of closed loading may be lower than 10,000 volts, and reaches the safety potential of oil surface.

In addition, in the closed loading process, since the pressure in the tank car is higher than 0.2 kg/cm² (G), the outside air can not enter the tank car, so there is lack of oxygen inside the tank car, no combustion will happen, and safe loading is achieved basically.

5) The use of static electricity eliminator.

The static electricity eliminator is made according to the principle of static electricity induction. It is a special nipple mounted in the pipeline. It contains: a steel pipe of 1 m in length lined with 5 mm thick organic glass tube, 5 rows of tungsten needles (3-5 needles/row) mounted in the cavity of the pipe through the lining glass tube, and the tungsten needles fixed on the steel pipe and grounded. When the oil with electrostatic charge flow into the eliminator, the ground capacity decreases and the inside potential increases, the high field strength occurs, then the opposite charge appears along the nearaby space of the needle tips, and so neutralizes the charges in the oil, and the static electricity is partially eliminated.

For powdery media, the flange-typed static electricity eliminator can be used. Its electrodes for the elimination of static electricity are either round or square flange in feature. It injects compressed air on one side and sprays ions on the other side. Because it is used at the place of pipe flange, it is effective to eliminate the static electricity carried by the flowing powder in the pipelines.

6) Addition of anti-electrostatic additive.

The conductivity of light petroleum products, such as gasoline, jet fuel, diesel oil etc., are lower than 10 ps/m. In case of adding 1 ppm of anti-electrostatic additive, the conductivity of petroleum products may increase to higher that 100 ps/m (Table 4).

Table 4. The effect to conductivity of petroleum product with anti-electrostatic additive.

Name of petroleum product	Amount of anti-electrostatic additive added	Conductivity (ps/m)
No. 2 jet fuel	0 1ppm T1501 1ppm ASA-3	0 520 130
No. 1 jet fuel	0 1ppm T1501 1ppm ASA-3	0 660 250
No. 3 jet fuel	0 1ppm T1501 1ppm ASA-3	5 620 110

The research work has shown that anti-electrostatic additive T1501 has obvious effect of eliminating static electricity from gasoline, jet fuel and diesel oil (Table 5).

4. Preliminary prospective plan

Although there are other various measures to prevent or reduce the electrostatic hazard in petrochemical processing, such as: grounding, protective measures for personnel safety etc., these measures do not mean no risk at all. Because the factors that cause electrostatic hazard are very complex, and there still exist some unexpected factors.

In order to control and eliminate the electrostatic hazard, two aspects of work should be undertaken from now on. First, strengthen the organization and management, popularize the safety training of the anti-electrostatic hazard technique. And second, continue the development and research work of the new anti-electrostatic hazard techniques.

In the past years, both foreign and domestic specialists on static electricity have tried to

Table 5. The effect of the elimination of static electricity of anti-electrostatic additive T1501.

Humidity (%)	Ambient temp.(°C)	Name of petroleum product and the amount of additive added	Conductivity (ps/m)	Potential on oil surface (V)
33	33.5	jet fuel	7	2900
33	34	jet fuel+0.5ppm T1501	230	-20
40	25	gasoline	2	-20000
38	9	gasoline+1ppm T1501	520	-500
68	1	diesel oil	5	>10000
33	4	diesel oil+1ppm T1501	90	500

find generally acknowledged precautions against static electricity, but the mechanism is so complicated, especially there exist various sudden interfering factors, that the static electricity is often looked upon as a kind of natural or probability phenomenon with very poor reproducibility. That causes the research work of the prevention of the static electricity to slow down. In fact, a great deal of experimental and research work needs to be carried on.

The preliminary prospective plan of anti-electrostatic technique the subordinate enterprises of SINOPEC will carry on is to develop and explore the following fields:

1) Conduct the study on safety measures for preventing static electricity for the vapor phase and powdery materials in petrochemical enterprises and draft safety operation code.

2) Conduct the study on safety measures for preventing static electricity for the synthetic fabrics in petrochemical enterprises and draft safety operation manual.

3) Conduct the study of the filtering materials for the filtration of petroleum products in order to reduce the static electricity in the filtration process.

4) Conduct the study of the conductive seal rubber of floating roof storage tank and conductive foamed plastics.

5) Develop and manufacture the new type anti-electrostatic monitoring instruments and connect them to the computer for synchronized monitoring.

Along with the modernization of anti-electrostatic technique, the use of computer has significant meaning such as statistic analysis of a large number of electrostatic hazards, policy dicision study, computer simulation of the detonation in the electrostatic accident and the centralized monitoring of the large scope environment etc. All of these topics need further study.

References

1. American Petroleum Institute API RP2003 4th edition, March 1982.
2. British Standard BS 5958: 1982, part II, British Standards Institution, London. (Code of practice for control of undesirable static electricity)
3. Handbook of Electrostatics, 1981, Japan.
4. Code of static electricity, AS 1020-84, Australia.
5. Справочник по охране труда и технике безопасности в нефтеперерабатывающей и нефтехимической промышленности, стр279-320
6. A. Klinkenberg, J.L. Vander Minne, Electrostatics in the Petroleum Industry, 1958.

STATIC ELECTRIFICATION DUE TO LIQUEFIED NATURAL GAS FLOWS

G. TOUCHARD*, P. HUMEAU*, L. MARCANO*, WU ZONG-HAN**, J. BORZEIX*, J.P. NOSSENT*, S. WATANABE***

* Labo. de Physique et Mécanique des Fluides, U.A. 191 C.N.R.S., Univ. POITIERS, FRANCE
** Dept. of Phys. and Chem., Nanjing Institute of Technology, CHINA
*** A.I.C.H.I., Institute of Technology, TOYOTA, JAPAN.

Abstract

It is a well know fact that flows of dielectric liquids like hydrocarbon liquids generate static electrification (1), (2), (3), (4), (5). The Liquefied Natural Gas is also a dielectric liquid with a resistivity even greater than hydrocarbons, more it is very easily inflammable, so it can be a real worrying phenomenon.

In the first part of the paper we expose the fundamental and applied purpose of our research according to the preoccupations of the French Gas board Company (Gaz De France) who has sponsored a part of this work.

Then, in the second part, we analyse the experimental results obtained on a small size equipment. The space charge density convected in a capillary tube is obtained in terms of the Reynolds number of the flow (which is always very high), the diameter of the pipe, the composition of the L.N.G. and the rate of gas in the case of two phases flow. These results are compared with the theoretical predictions.

The third part of the study is probably the most important on account of its applications. Experiments made with a full size equipment are related. This equipment corresponds to a system of decanting in the case of an incident in a big storage tank. The liquid is move very quickly to another vessel in the open air or covered by a plastic sheet.

The electrical potential and the charge generated are measured in different part of the equipment and analysed in order to point out the principal safety measures which must be taken for such operation.

In the conclusion, we summarized the different results of as well fundamental order and applied one, and we expose the future research we expect to make soon in this domain.

Introduction

Flows of dielectric liquids such hydrocarbons through pipes or filters generate electric charges. It could be predicted that flows of liquefied gas as they are also very insulated liquids might give the same kind of phenomena. However because of their very high resistivity and very low coefficient of permittivity most authors estimated that the charge generated with such liquids might be very small. In an other hand, if experiments on streaming currents with hydrocarbons are indeed very difficult, with liquefied gas it seems to be even more difficult due to the very low temperatures, and one must have a special motivation given by a real preoccupation to enter upon such experiments.

Thus, as liquid hydrogen has been early used in aerospacial and also as an energy vector, some studies have already been made for this liquid, CASSUTT (6), SAUNIERE (7). Recently GAZ DE FRANCE has built very important installations to receive, store, and gasify liquefied natural gas, and in order to prevent electrostatic hazards, they have asked us to make different studies on these phenomena. Hence, we have first make a theoretical analysis and perfected one equipment in order to settle the fundamental aspect of the evolution of the space charge generated in flows of L.N.G. through metallic pipes in terms of different parameters and then, made several experiments on a real size installation.

Theoretical analysis

We can make the same analysis than for hydrocarbons (8) and as the viscosity of L.N.G. is very low (≈ 0.12 cp) we have very important Reynolds number and so the non dimensional space charge density is given by the following equation :

$$Qt_+ = - \frac{64\, Ump_+}{R_+^2} \left(\frac{1}{8} + \frac{1}{\rho_{+0} R_+^2 - 8} \right)$$

Where R_+ is the non dimensional radius of the pipe, Ump_+ the non dimensional mean velocity, and ρ_{+0} the non dimensional space charge density on the pipe axis. This means that the space charge density convected must be independant of the Reynolds number. We have first try to verify this assumption on a small size equipment and to analyse the experimental results.

Equipment for the fundamental study

Measurements of streaming currents are always very difficult because of the very small values of such currents and the necessity to keep the liquid in the same state of purity in order to have repeatable results. So even for hydrocarbon liquids the equipment is generally complex (9). With L.N.G. we have two other additionnal constraints resulting from the very low boiling point of such liquids, and the risks of ignition.

Thus all the equipment is located in a building far from experimenters and controlled by a pneumatic keyboard. Moreover all the containers and the pipes must be very well thermally insulated.

A general diagram of this equipment is given figure (1). The L.N.G. stored in a large vessel (1) moves to another vessel (2) and then we can work in closed circuit. The liquid is stored in (2) for a sufficient long time so that it is electrically neutral. This vessel which is superinsulated with high vacuum has also a nitrogen liquid guard so the L.N.G. can be stored in it a very long time.

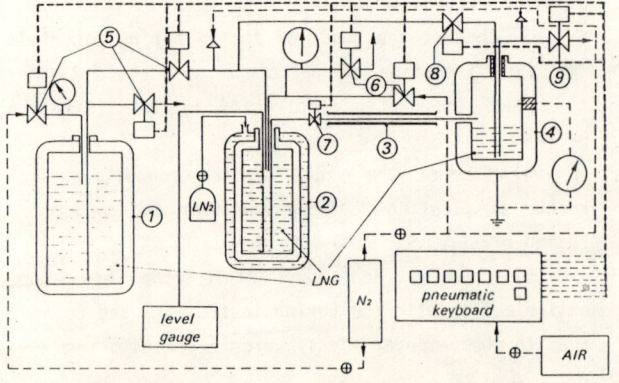

Figure 1

-400-

To transfer the liquid we apply the pressure given by a nitrogen pressure bottle. Nitrogen is used because it is a neutral gas and also it has a boiling point lower than L.N.G.

The pressure applied is controlled by a pressure gauge and the level by a capacity probe in the liquid. Then the liquid is flowing through the capillary tube (3), where the electric charge is generated. At the exit of the pipe the liquid falls down into a vessel (4) thermally superinsulated with high vacuum and also electrically insulated.

When an experiment is finished the liquid is returned in the vessel (2). All the gates (5) (6) (7) (8) (9) are pneumatic.

All the installation is earthed except the collecting vessel in which the electrical current is measured with a KEITHLEY Picoammetter 610C.

Experiments on the small size equipment

We have tested several kind of L.N.G. with differents compositions and we have made experiments for an applied pressure between 0 to 2 bars. Three pipes have been tested of one meter long and 1.16 mm, 2.0 mm, and 2.5 mm of diameter. Even if the storage vessel (2) and the pipe are very well thermally insulated we have noticed that at the exit of the pipe the fluid is strongly diphasic, and thus the flow may be different of those predicted in terms of the applied pressure. So we made experiments for four kinds of L.N.G. which gave different streaming currents.

In order to settle the proportion of liquid we make the further analysis (11) :

$$\rho_m = X \rho_L + (1 - X)\rho_g$$
$$\mu_m = X \mu_L + (1 - X)\mu_g$$

X being the volume of gas in a unit volume of fluid. ρ_m, ρ_L and ρ_g being the fluid, liquid and gas specific mass respectively.
μ_m, μ_L and μ_g being the fluid, liquid and gas dynamic viscosity respectively.

The value of X is obtained for each set of experiments by fitting the evolution of the experimental flowrate in terms of the pressure drop with the Blasius equation (figure 2). So we determine the ratio $\mu_m^2 \rho_m$, and then X is computed numerically.

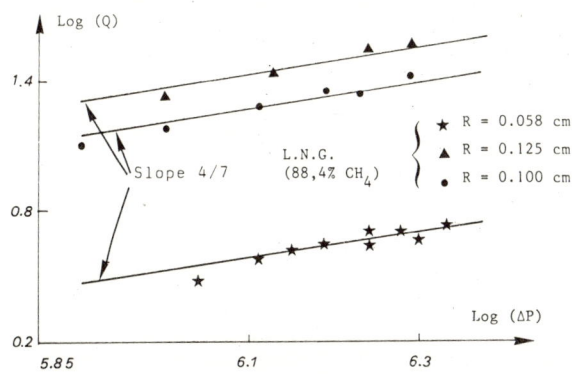

Figure 2

Knowing X we can compute Y the massic proportion of liquid and we get the following table :

R(mm)	0.58	1.00	1.25
Y			
L.N.G. with 74 % CH_4	0.594	0.853	0.819
L.N.G. with 88.4 % CH_4	0.997	0.998	0.999
L.N.G. with 88.5 % CH_4	-----	0.996	0.995
L.N.G. with 88.7 % CH_4	0.793	0.936	0.948

The second set of experiment were more repeatable and we can see that the ratio of the liquid increases with the diameter of the pipe as it could be predicted by a thermodynamic analysis. The values of the charge convected are given figure 3 for this second set of experiment, the first one being already published (11). We can see that in spite of the dispersion the space charge convected seems to be independent of the Reynolds number as it was predicted more the evolution of this charge with the radius of the tube is decreasing as it was predicted by a previous theory (10).

Comparing this set of experiments with other ones (11) we see a difference in the charge convected. We think that theses differences between the various set of experiments are partly due to the composition of L.N.G., but there is specially an important influence of the ratio liquid/gas on the charge generation.

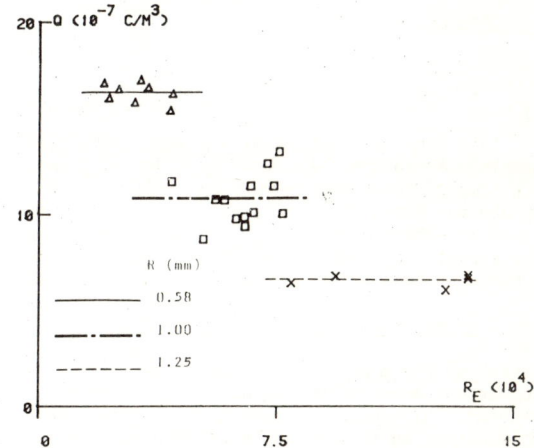

Figure 3 : L.N.G. with 88.4 % of methane.

Installation for real size measurements

The diagram is given figure 4. We have made experiments of charge generation during a decanting of a tank truck of L.N.G. (1). At the exit of the truck is connected a pipe of 8 cm of diameter (2), the liquid flowing through the pipe, flows then through a by pass (3) on one branch of which is placed a flowmeter (4). The pipe is about sixty meter long and made of ten pieces. Each piece of the pipe is thermally insulated with a surrounding sponge of polyurethane of 10 cm thick. One piece of the pipe (5) is electrically insulated with two big disks of P.T.F.E. figure 5. So we measure (9) the current generated on this part of the pipe. At the exit of the pipe the L.N.G. falls down in a big rectangular tank (6) of twenty meter long, one meter large and twenty centimeters height.

-401-

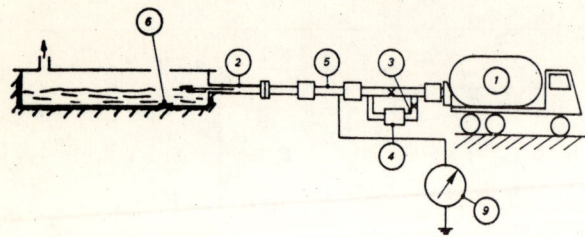

Figure 4 : Real size equipment.

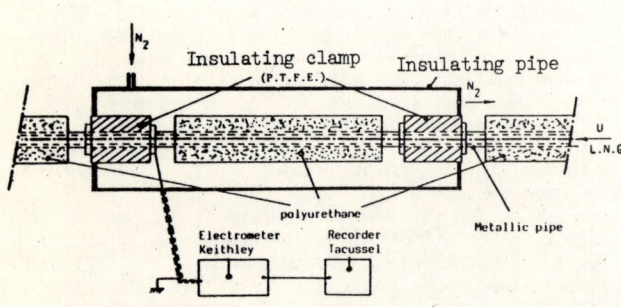

Figure 5 : Piece of pipe on which the current is taken.

This tank was covered by a plastic sheet. Such experiments was specially made in order to improve this type of tanks which could be placed around the big storage tanks of L.N.G., and would retain the liquid in case of an incident on the storage tank. Four electrodes placed at different part of the rectangular tank were connected to four picoammeters.

Experiments on the real size equipment

Current on the pipe   On figure 6 we have plotted the current measured on the insulated piece of the pipe. This current is about 10^{-8}A. The first ten minutes, as the pipe is cooling down, the condensation of the humidity of the air reduce the current, then with an important flow of nitrogen gas around the pipe the humidity disapears and so the current is growing to 10^{-8}A. Then when the nitrogen stop, the current decreases again.

Currents in the tank   Inside the tank the electrode 2 is in the jet, the electrodes 1, 3 are very closed to the ground of the tank and the electrode 4 is on the sheet near the hole figure 7.

At the beginning, the flow is mainly diphasic and pulsative, the pressure in the tank is oscillating and the sheet is unstable. So the currents are not really significants. Then the L.N.G. is more and more composed of liquid, and like the pressure, the sheet is more stable. Only the electrode 2 is in the jet, the electrode 4 is in a region of gas and electrodes 1 and 3 near the bottom are not much affected by the flow. It is probably the reason why the current on electrode 2 reaches an important constant value when on electrodes 4, 3 and 1 it is about ten times smaller (figure 8).

In fact it seems that the current measured on electrode 2 (10^{-7}A) is more or less equal to the current generated on ten pieces of pipe (10.10^{-8}A).

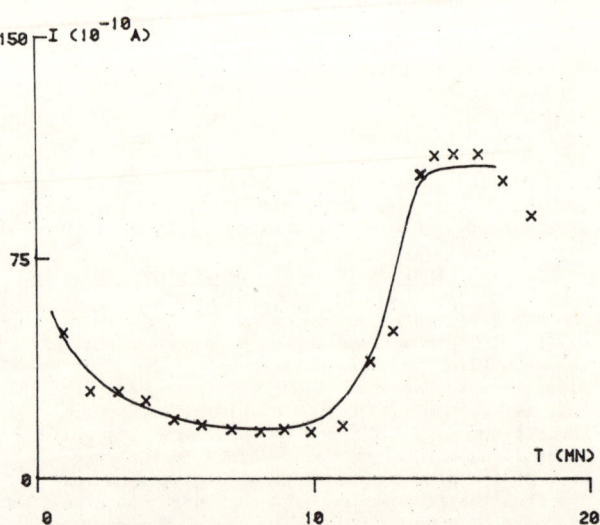

Figure 6 : Current on the piece of pipe.

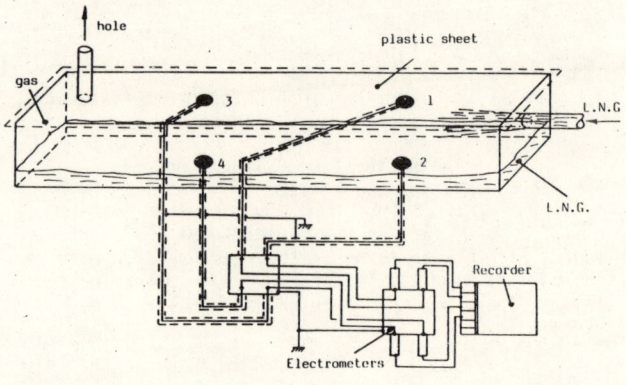

Figure 7 : Rectangular tank with a plastic sheet.

Development of electrical potentials inside a big storage tank   The greater storage tanks of Gaz De France are about 50m high and 55m of diameter (125 000 m^3). The relaxation time is given by : $\tau = \varepsilon/\sigma_1$ where ε is the permittivity and σ_1 is the conductivity. For L.N.G. $\sigma_1 \approx 5.10^{-14}$ Ω^{-1} m^{-1} and $\varepsilon \approx 2.10^{-11}$ so $\tau \approx 400$ s, thus for this big storage tanks an accumulation of the charge may remain several hours. If we suppose that the space charge density generated by the flow inside the filling pipe is about 10^{-8} C/m^3, then with a very simple approximation assuming that this space charge density is constant on a radius of 20 m, taking into account the Poisson equation $\Delta \psi = -\rho/\varepsilon$ the difference of potential is about 10^5 V. Even if it is only a gross approximation this means that very high potentials are generated inside such tanks and explain the large sparks which have been already observed.

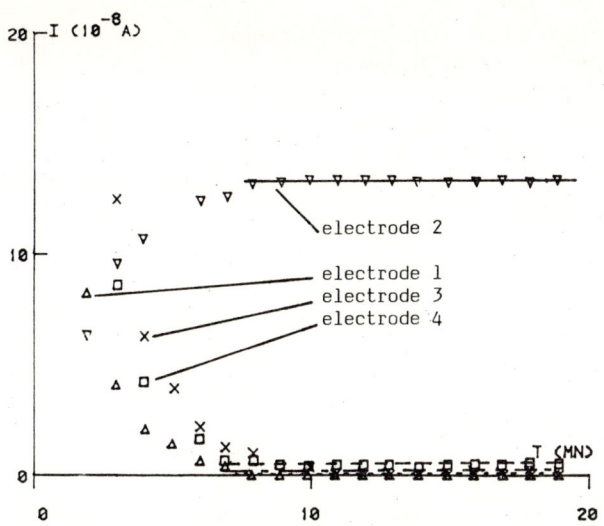

Figure 8 : Current on the electrodes in the rectangular tank.

Conclusion

The streaming currents generated by flows of L.N.G. are nearly of the same order of magnitude than those measured in hydrocarbons, in spite of their resistivity much larger.

Such streaming currents are a risk for industrial plants of L.N.G. and the magnitude of the potential during the filling of a big storage tank is an example of these risks.

Happily as there is no air inside the tank an ignition is not possible. However this risks increase with the new international recommendation to use an insulating piece of pipe between the methane boat and the metallic pipe. It is the reason why we are now making different experiments on such device to prove the inconsistency of this recommendation.

References

(1) A.A. BOUMANS,
"Streaming current in turbulent flows and metal capillaries",
Stichting voor fundamenteel onderzoek der materie, Utrecht, 1007-1055, 1957.

(2) I. KOSZMAN and J. GAVIS,
"Development of charges in low conductivity liquids flowing past surfaces",
Chem. Eng. Sci., 17, 1013-1023, 1962.

(3) E.T. HIGNETT and J.C. GIBBINGS,
"Electrostatic streaming current developped in the turbulent flow through a pipe",
J. Electroanal. Chem., 16, 239-249, 1968.

(4) A. KLINKENBERG,
"On the electric streaming current",
Electrochim. Acta, 12, 104-105, 1967.
"Static electricity in liquids",
Static. Elec. Proc. Conf. London, 63-68, 1967.

(5) N.J. FELICI, J.P. GOSSE and A. SOLOFOMBOAHANGY,
"Liquid flow electrification and zeta-potential in hydrocarbons",
Proc. Conf. 7 ICDL, 284-288, 1981.

(6) L. CASSUTT, D. BIRON and B. VONNEGUT,
"Electrostatic hazards associated with the transfer and storage of liquid hydrogen",
Arthur D. LITTLE, Inc. Cambridge Mass, report., 1968.

(7) J. SAUNIERE, C. TELLIER and G. TOUCHARD,
"Contribution to the study of electrical charge transport in liquid and gaseous hydrogen flow",
Proceeding of the 4th W.H.E.C., T. 4,
T.N. Veziroglu Editor Perg. Press., 1737-1748, 1982.

(8) G. TOUCHARD and P. DUMARGUE,
"Streaming current in stainless steel and nickel pipes for heptane and hexane flows",
J. of Electrostatics, 14, 209-223, 1983.

(9) G. TOUCHARD,
"Streaming currents developped in laminar and turbulent flows through a pipe",
J. of Electrostatics, vol. 5, 463-476, 1978.

(10) L. MARCANO et G. TOUCHARD,
"Courants d'écoulement dans les liquides cryogéniques",
J. of Electrostatics, vol. 15, n° 3, 321-328, 1984.

(11) L. MARCANO,
"ELectrisation par écoulement turbulent de liquides cryogéniques",
Thèse de Doctorat de l'Université de Poitiers, n° 46, septembre 1986.

RELATIONSHIP BETWEEN ADDITIVE CONCENTRATION AND STREAMING CURRENT

SHIGEO WATANABE* MOTOMU FUJII** KOUICHI TANABE** ASAO OHASHI*
WU ZONG HAN*** MOHAMED BENYAMINA**** GERARD TOUCHARD****

* AICHI INSTITUTE OF TECHNOLOGY, TOYOTA JAPAN.
** FUKUI INSTITUTE OF TECHNOLOGY, FUKUI JAPAN.
*** NANJIN INSTITUTE OF TECHNOLOGY, NANJIN CHINA.
****UNIVERSITE DE POITIERS, POITIERS FRANCE.

<ABSTRACT>

As is well known, when a liquid with good insulating properties flows through a pipe an electrical charge is generated. It is supposed that the size of this charge is determined on the one hand by circumstances such as liquid flow velocity, pipe diameter and length, and on the other hand by the physical properties of the liquid such as its conductivity and relative permitivity.

In the authors' previous investigations of charge volume in such cases, the charge was found to depend on the product of ① the quantity of dissociated ions in the liquid and ② the quantity of energy required to transfer these ions to the interface and there to adsorb them.

This dependence can be expressed in the equation:
$I = A \exp(-BX) / \exp[\exp CX]$
where A, B and C are constants relating to flow velocity, Gibbs' energy and degree of dissociation respectively.

In this paper we shall use Klinkenberg's experimental results and discuss the relation of values A and B in the above equation with pipe length and flow velocity.

<INTRODUCTION>

As is well known, an insulating liquid in flow will produce an electric charge. Among the best known research on this subject is that published by Klinkenberg. Klinkenberg introduced an interface activation agent into gasoline and measured the electric charge generated while varying the liquid's conductivity. His results indicate a maximum charge value in the area of conductivity 10^{-12} S/cm, with a decrease in the observed charge if conductivity is increased further. Klinkenberg's concluding explanation for this is that the increase in conductivity leads to a relaxation in the generation of charges [1].

The present authors have likewise conducted experiments on streaming current, but even when using liquids with a conductivity several powers higher than that used by Klinkenberg, a peak phenomenon could still be observed in the charges generated [2]. From this, the authors concluded that the charge peak phenomenon must be related to further factors than fluid conductivity alone.

Seeking a cause for the phenomenon in the field of colloid chemistry, the authors suggested that streaming electrification might be expressed by the following equation [3]:

$$I = \underbrace{A \exp(-Bx)}_{1} \cdot \underbrace{\frac{x_1}{\exp(\exp(Cx))}}_{2} \qquad (1)$$

where A, B and C are constants relating to flow velocity, Gibbs energy and dissociation respectively.

The first exponent of Eq. 1 (underlined term 1) corresponds to the term j_a proposed by Klinkenberg (the density of the charge generated from the pipe wall). The second exponent (underlined 2) corresponds to j_w (the density of the charge escaping to the pipe wall).

Another way of regarding these two terms is to suppose that 1 represents the probability of ions present in the liquid being adsorbed at the wall surface, while 2 represents the probability of them existing as dissociated ions within the liquid.

The two terms together are expressed as the product of the quantity of ions found from Langmuir's adsorption theory equation [4] and of the exponential dissociation [5].

Fig. 1 shows the quantity of ions adsorbed at the interface. If we substitute 1 for A and allow changes in B only, we see that as the number of ions in the liquid increases the probability of a given ion being adsorbed at the interface will diminish. Next, if we assign 1 mol as the quantity x_1 of ions present in the liquid, the changes produced in C, i.e. the dissociation rate, will be as shown in Fig. 2. From this figure we see that as the concentration of interface activation

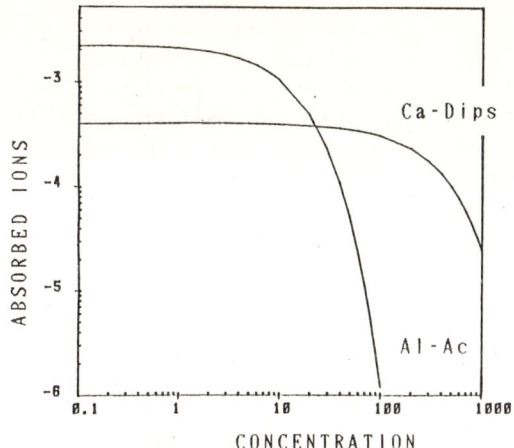

Fig.1 Quantity of ions adsorbed at the interface

agent (additive) increases, the dissociation rate rises to a maximum value at a certain concentration and then, with a further increase in concentration, begins to decline.

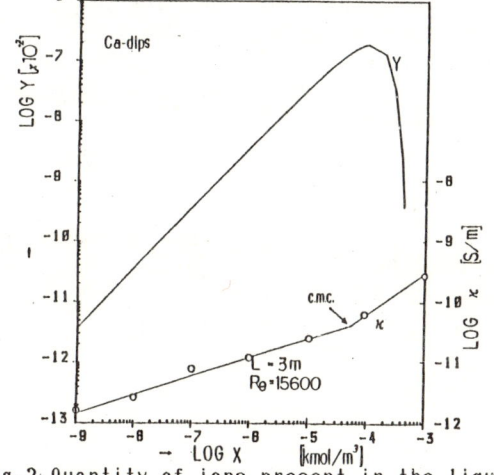

Fig.2 Quantity of ions present in the liquid

In colloid chemistry, the appearance of this peak is explained in terms of C.M.C.[6]. Moreover, collating these values with Klinkenberg's experimental results, we find that C.M.C. coincides with the maximum charge generation. This correspondence appears to be unrelated to the charge relaxation phenomenon, which depends on an increase in the electrical conductivity of the fluid.

From the above considerations, the authors have made use of C.M.C. in order to explain why the electric charge should show a peak value at a certain conductivity rate.

<FROM KLINKENBERG'S EXPERIMENT>

In this report we shall carry the discussion further, using Equation 1 and Klinkenberg's experiment on the relation between the properties of additives and the generation of charges. The apparatus used in Klinkenberg's experiment comprised an upper tank, a pipe and a lower tank. The pipe was earthed, and charge readings were taken by means of a very high precision ammeter connected to the lower tank.

The experimental liquid was gasoline, into which were introduced the anionic interface activation agents aluminium alkylsalicylic acid (valency 3) and calcium alkylsalicylic acid (valency 2). Charge readings were taken for pipe lengths of between 0.5 m and 3 m.

PIPE LENGTH	0.2m	0.5m	1.0m	2.0m	3.0m
VALUES OF A	0.65	1.2	1.6	2.2	4.0
VALUES OF B	0.13	0.18	0.20	0.25	0.28
VALUES OF C	5.0	5.0	5.0	5.0	5.0

※ $A \times 10^{-4}$, $B \times 10^{-2}$, $C \times 10^{-5}$,
Sample: Ca-Dips, Re=15600

PIPE LENGTH	0.2m	0.5m	1.0m	2.0m	3.0m
VALUES OF A	0.27	0.9	2.0	11.0	22.0
VALUES OF B	3.5	5.0	6.5	7.5	7.5
VALUES OF C	5.0	5.0	5.0	5.0	5.0

※ $A \times 10^{-4}$, $B \times 10^{-2}$, $C \times 10^{-5}$,
Sample: Al-Ac, Re=15600

Table 1 Values of A, B and C

The experimental results are shown in Fig. 3. If we apply Eq. 1 to these results, the closest corresponding values for A, B and C are found to be those given in Table 1

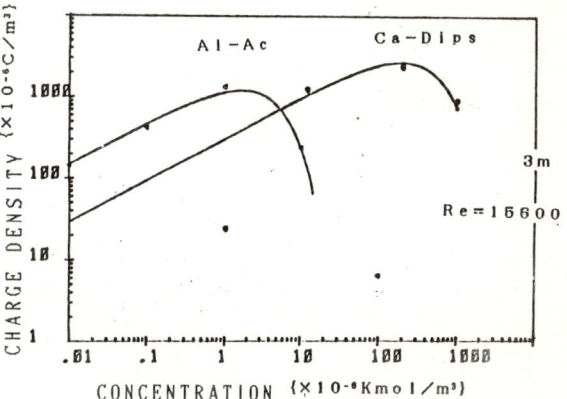

Fig.3 Experimental and calculated results

<CONCLUSION>

From Table 1 we find that the value of C, the constant relating to the dissociation of the additive, shows no relation to the particular type of additive used. It is conceivable that both additives produce the same values because both belong to the series of substances sharing the same phenol chain structure. The dependence of A on pipe length is shown in Fig.4. As can be seen, the shorter the pipe, the more apparent it becomes that the calcium alkylsalicylic acid produces the lesser charge.

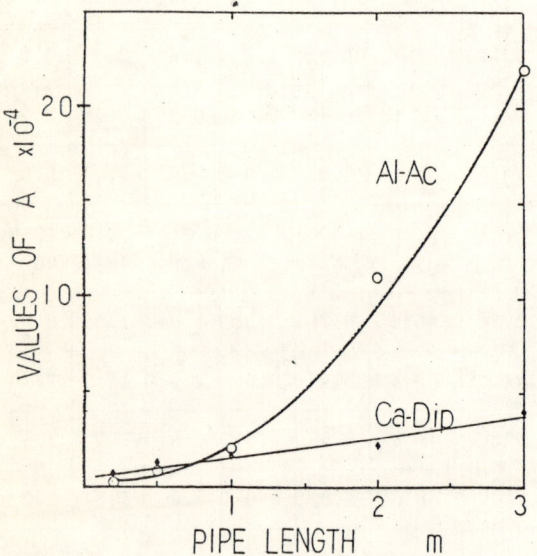

Fig.4 Dependence of A on pipe length

While interface activation additives are generally used mostly in long pipelines, calcium alkylsalicylic acid would still appear to be the more effective means of preventing charge generation. It may be surmised that these results bear some relation to valency.

As for the measured value of B, bearing in mind the known relationship $-Bx = \Delta G/(RT)$, by substituting the additive concentration, the gas constant and the absolute temperature we are able to obtain a value for the conduction activation energy. In general, we can say that:

$$\Delta G = \Delta H - \Delta T \cdot S, \quad [7]$$

where ΔH is enthalpy and S is entropy.

If we ensure constant temperature and no heat transfer so that $\Delta G = \Delta H$, ΔH can be taken as identical to the activation energy.

Thus, the activation energy is found to be 0.087 eV (for pipe length 1 m and additive concentration 10^{-6} Kmol/m^3).

Gasoline has a viscosity of 0.5 cP, which is close to that of n-heptane (0.4 cP at 30°C).

The positive and negative charge activation energies for n-heptane are, respectively, 0.117 eV and 0.087 eV [8]. Comparing these two measured values, we find that, as expected, the major influence comes from the negative charges. This accords with what past research has indicated.

<REFERENCE>

1) A.Klinkenberg, J.L.van der Minne, Electrostatics in the Petroleum Industry. Elsevier Publishing Co. 1958.
2) S.Watanabe, A.Ohashi, M.Ito, "The Effect on Streaming Electrification of Additives introduced into a Liquid", Trans. IEE of Japan, Vol. 105-A, No.3, Mar., 1985
3) S.Watanabe, M.Fujii, K.Tanabe, A.Ohashi, M.Benyamina, G.Touchard, "Factors Determining Streaming Electrification", Trans. IEE of Japan, Vol. 107-A, No.3, Mar., 1987
4) A.Kitahara, M.Watanabe, Electrokinetic Phenomena at the Liquid-Solution Interface. Kyoritsu Shuppan, 1972, p35.
5) The Chemical Society of Japan, Handbook of Chemistry. Maruzen, 1981, p1000.
6) A.Kitahara, K.Aoki, Colloid Chemistry and Liquid-Solution Interface. Hirokawashoten 1983, p74.
7) The Chemical Society of Japan, Handbook of Chemistry. Maruzen, 1981, p1010.
8) Industrial Electrical Engineers of Japan, Phenomena in Insulating Materials. Denkigakkai, 1973, p248.

A STUDY ON THE SAFE POTENTIAL OF CHARGED HYDROCARBON OILS

Qu Jianbang and Lu Mingfang
Hebei University, Baoding, China

ABSTRACT

Studies have been made using both photo-electric conversion (PEC) method and charge transfer method on the spark energies from the surface of charged hydrocarbon oils on simulation-scale system. The potential distribution over the oil surface has been calculated by computer. Results for the measurement of the discharge energy of the surface of kerosine with a rest conductivity of 3.4 pS/m reveal that the surface potentials corresponding to 0.2 mJ energy and 0.1 µC charge transfer are 14 kV and 32 kV, respectively.

INTRODUCTION

Some experiments both at home and abroad [1]-[11] have been made to investigate the rules of the discharge from the surface of hydrocarbon liquids and to determine the safe criteria of surface potential of the liquids. It have been proved by J.K.Johnson [1977, [3]] and H.R.Kramer and K.Asano [1979, [5]] that the amount of charge transfer in the discharge from the surface of hydrocarbon liquids sufficient to ignite hydrocarbon vapor/air mixture was at least the same level of the order of 0.1 µC. The ignition or safe potential of the liquid surface were ignition potential 60 kV, 1 pS/m, [3]; safe potential 40 kV 0.5 pS/m, [5]; ignition potential 25 kV, 18 pS/m,[7]; safe potential 35 kV, 3 pS/m, [9] and safe potential 13 kV 12 pS/m, [10], et al. This wide variation is primarily due to the differences in the test conditions (especially in the conductivity of the liquid) and instrumentation employed. In this paper, the spark energies from the liquid surface are measured by using both PEC method and charge transfer method.

TEST PROCEDURE AND INSTRUMENTATION

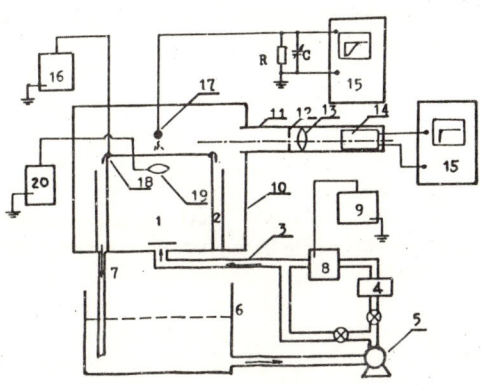

Fig. 1 Test equipment

The experiment equipment as shown in Fig. 1.

1. test vessel, measured ϕ0.70 × 0.59 m high, PVC; 2, 3, 5, 6, 7. oil circulation pipes; 4. microfilter; 8. charge injector; 9. high voltage power supply; 10. test chamber (black room); 11, 12, 13. light path; 14. photomultiplier; 15. storage oscilloscope (type DSS6521, KIKUSUI); 16, 20. Q4-V electrostatic voltmeter; 17. electrode; 18, 19. surface potential probe.

In the test process, kerosine (rest conductivity is 3.4 pS/m) is charged by passing through the microfilter and charge injector. The liquid can be either positively or negatively charged by changing the filter material and the polarity of power supply of the charge injector. The charge density of the liquid can be varied by changing the flow rates or sideway proportions, and then the charged liquid enters into the test vessel forming a charged liquid surface.

Surface potential

The surface potential of the liquid is measured with a probe mounted on the upper edge of the test vessel. It was previously calibrated against a floating ball probe at the centre of the liquid surface when the grounded electrode did not exist. The relation between the centre potential Vc and the edge potential Ve is measured and is shown in Fig. 2, it can be written in the following equation

$$V_c = 1.17 V_e \qquad (1)$$

and the computed potential equation is

$$V_c = 1.22 V_e \qquad (2)$$

When grounded electrode exists, the surface potential will decrease. Say, for 10 mm in diameter sphere-shaped electrode, gap 1.6 cm, the computed potential is

$$V_c = 1.04 V_e \qquad (3)$$

The constant in the equation varies with the different diameter and the gap. The real potential during discharging is measured with a probe mounted just below the centre of the liquid surface.

Spark energy (measured by the charge transfer method)

We have the formula

-407-

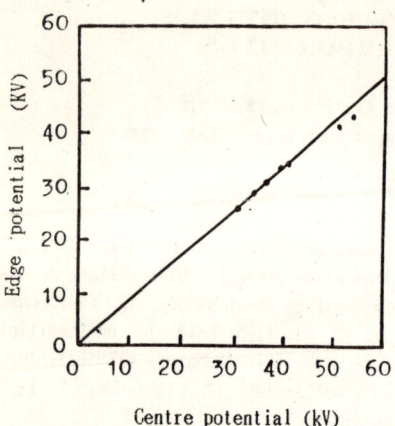

Fig. 2 The potential relation over the oil surface

$$E = UQ = UCV_o \qquad (4)$$

where, E - spark energy, U - surface potential of the liquid, Q - charge transfer in the spark, C - the distributive capcitance of the discharge loop and Vo - peak value of the charge transfer pulse. Vo and U are measured with storage oscilloscope and electrostatic voltameter, respectively. The spark energy E is obtained by the formula (4).

Spark energy (measured by PEC method)

The light output of the spark is measured directly by photomultiplier (PM, type 6097B, E.M.I, the maximum response wavelength 4300 ± 500 Å). The PM is previously calibrated against metal-to-metal spark which energy is known. As shown in Fig. 3. The breakdown process of metal-to-metal discharge is different from that of metal-to-liquid discharge. Complete breakdown readily occurs in the metal electrodes and forming concentrated channel and bright region; the charge transfer in the spark is relatively larger and, at certain discharge potential, the spark energy is larger. However pre-breakdown and partial breakdown spark are always happened in metal-to-liquid discharge, and the largest breakdown distance never exceeds the half of the gap between the electrode and the liquid surface. Because the work function of electron from the liquid surface (for negatively charged liquid) is very large, it can not emitte electrons from the surface. Electrons also can not move along the liquid surface because the resistivity of the liquid is very high. So the discharge from the liquid surface is mainly the process of the space gas. The release of charges from the discharge layer of the liquid is mainly caused by the bombardment of positive space ions on the liquid surface. The charge transfer in the discharge is relatively smaller and, at certain discharge potential, the spark energy is smaller. The spark energy is mainly distributed in a number of faint channels which is converged at the electrode and forming one or more bright "roots". For decreasing the differences between the two kinds of discharges, in the calibration process, the space between the metal electrodes is adjusted continuously that complete breakdown of the gap must avoid. Under this condition, it is assumed that the same light output is given by the two kinds of discharges if their energies are same. According to [11] and our results, comparison has been made to show that about 35% less light output is given by partial breakdown than by complete breakdown of the gap between the metal electrodes. Substituting conducting rubber or other suitable materials for the metal plane electrode in the calibration will make the condition more similar to that of metal-to-liquid discharge.

RESULTS AND CONCLUSION

Experiment conditions: ambient temperature 10 - 15 °C; oil temperature about 12 °C; RH 25 - 30%; rest conductivity of the oil 3.4 pS/m; flow rate 83 - 194 l/min.

Measurements for the spark energies from liquid surface have been carried out by cone electrode 60°, and sphere-shaped electrodes with diameters of 3, 5, 10, 20 and 44 mm. Results are shown in Fig. 4. The gaps between the electrode and the liquid surface are in the range from 0.5 cm to 10 cm.

Fig. 4 shows the maximum spark energies given by various electrodes. The outer envelope curve represents that no discharge would occur with energy exceeding that.

As shown in Fig. 4, the lowest potential corresponding to the minimum ignition energy (M. I. E, 0.2 mJ) is 14 kV, and the corresponding charge transfer is about 0.02 μC (for electrode with φ 3 mm). Energies all exceeding 0.2 mJ are obtained with electrodes larger than φ 10 mm.

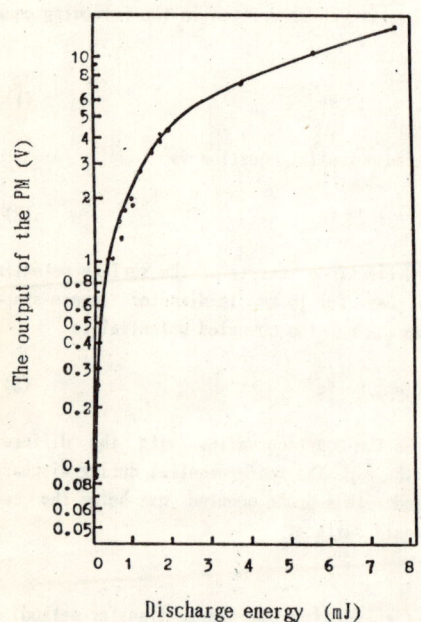

Fig. 3 The calibration curve of the PM

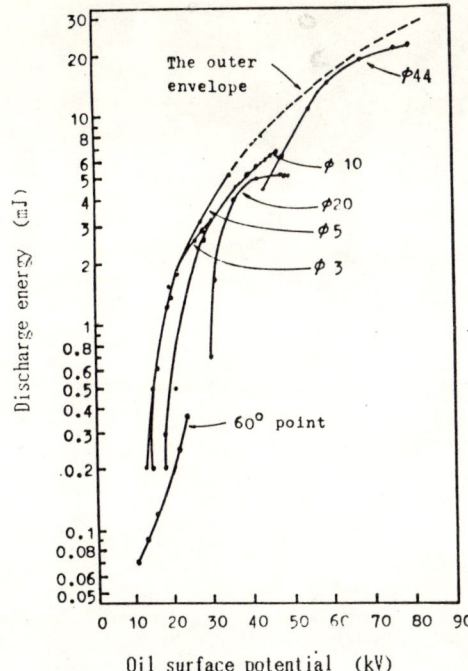

Fig. 4 The maximum discharge energies from the oil surface by various electrodes and their outer envelope curve measured by PEC method

The safe surface potential of hydrocarbon oils

Some experiments on the ignition of hydrocarbon vapor/air mixture on liquid surface had been carried out by J. K. Johnson [3], H.R. Kramer and K. Asano [5], etc. Their results showed that the charge transfer in the discharge necessary to ignite hydrocarbon vapor/air mixture was at least 0.1 μC. According to our results carried out with kerosine with a rest conductivity of 3.4 pS/m, that the surface potential corresponding to 0.1 μC is 32 kV. For consideration of safety, the safe potential of this kind of liquid is suggested to be 25 kV.

The minimum ignition energy (M. I. E) for hydrocarbon vapor/air mixture (0.2 mJ) was carried out with spark discharges between metal electrodes, and of not desirable for liquid surface discharges.

It have been found in the present and previous works that the charge transfer in the discharge primarily depends on the liquid conductivity, as shown in Fig. 5. For the given surface potential of the liquid, the charge transfer and the spark energy increase with the increasing conductivity, on the other hand, as most of the practical equipment is grounded, the quantity of charge release in liquid increases with the increasing of the liquid conductivity. It is necessary to put these two factors together into consideration for the more accurate determination of the safe potential on the oil surface.

The "oil bridge" formed between the liquid surface and the electrode affects the breakdown of the discharge. The liquid Taylor Cone on the electrode increases the field strength for breakdown, and more breakdown channels exist in the spark. The spark energy is also some greater than that without the liquid Taylor Cone.

The discharges from liquid surface are all overvoltage breakdown. For electrodes with diameters smaller than 10 mm, the discharges are of mainly corona type and pre-breakdown type; and pre-breakdown and partial breakdown spark for electrodes with diameters larger than 10 mm.

The widths of the spark pulses are in the range from 0.5 μs to 2.0 μs.

It has been found in the experiment carried out with the sphere-shaped electrode of 44 mm in diameter that a number of spark pulses are contained in one charge transfer pulse, the time intervals between every two adjacent spark pulses are in the range from several μs to about 200 μs. This is caused by multi-channel breakdown at the different parts of the electrode. Meanwhile the front of the waveform of the charge transfer shows a complex structure.

Under the model assuming the charge density in the test vessel is uniform, potential distribution is given by numerical calculation with computer, and the following equations hold:

$$V_c = 6.08 \rho \quad (kV) \qquad (5)$$

(without grounded electrode)

$$V_c = 4.66 \rho \quad (kV) \qquad (6)$$

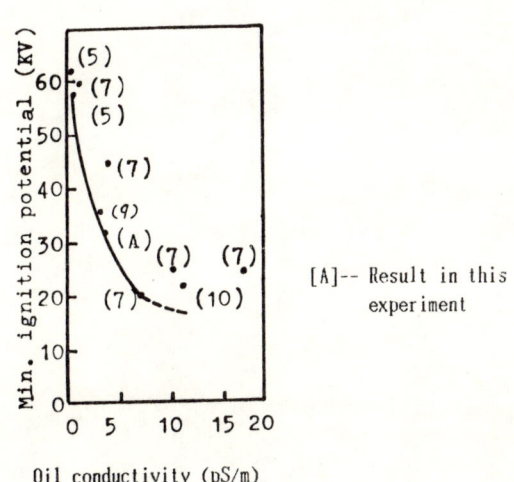

[A] -- Result in this experiment

Fig. 5 The realtion between minimum ignition surface potential and oil conductivity

(for existance of electrode with 10 mm in diameter, gap 1.6 cm). ρ is the charge density in the liquid of the test vessel in $\mu C/m^3$. For sphere-shaped electrodes from 10 mm to 50 mm in diameters, the surface potential will be lowered from 1 kV to 3 kV by single discharge, the recovery time for the potential are from 0.5 s to 5 s.

For the consideration of safety, it is always necessary to measure the energy of discharges from liquid surface in applicable systems and to evaluate for the safety of the systems. But accurate measurements of liquid potential and charge transfer and the released energy in the discharges in practical systems may be very difficult or even impossible. Under these conditions, the PEC method can be used as a distinct method because the influence of the factors mentioned above can be avoided (say, to determine the discharge energy as a function of charge density of the liquid). In this paper, the light output properties of both metal-to-metal discharge and metal-to-liquid discharge are studied by PEC method and the results show a resonable relation between the light output and the discharge energy under certain discharge conditions, and the discharge energy from the liquid surface is measured by this method. Studies on using the PM calibrated against metal-to-liquid discharge on simulation-scale system to measure the discharge energy from liquid surface on practical systems are being undertaken.

REFERENCES

[1] J.T.Leonard, H.W.Carhart: Static Electrification Conference Series No. 4, Inst. Phys. Soci. (1967) p 100

[2] H.Strawson, A.R.Lyle: Static Electrification Conference Series No. 27, Inst. Phys. Soci. London (1975) p 276

[3] J.K.Johnson: J. Electrostatics, 4 (1977) p 53

[4] I.G.Haig, A.N.Bright: 3rd Int. Congress on Static Electricity, Grenoble (1977) 29-a

[5] H.R.Kramer, K.Asano: J. Electrostatics, 6 (1979) p 361

[6] W.D.Rees: J. Electrostatics, 11 (1981) p 13

[7] L.G.Britton, T.J.Williams: J. Electrostatics, 13 (1982) p 185

[8] J.T.Leonard: J. Electrostatics, 10 (1981) p 17

[9] G.J.Butterworth, K.P.Brown: J. Electrostatics, 13 (1982) p 9

[10] Shi Chong-Yue, Pan Gang, Tao Tao, Liang Wen-ji: Electrostatics, 1 (1984) p 26 (in Chinese)

[11] A.R.Lyle, H.Strawson: Static Electrification Conference Series No. 11, (1971) p 234

[12] Li Bai-gua, Ge Zu-huai, Qu Hao, Mi Hong-sheng: Proc. of Conf. on Electrostatics' 87 (1987) p 343 (in Chinese)

[13] The Institute of Electrostatics of Japan: Hand-book of Electrostatics (1981) (in Japanese)

[14] Yang Jin-ji: Discharge in Gases, Chinese Publisher of Science, (1981) p 146, 193 (in Chinese)

[15] E.Nasser, M.Heiszler: J. Appl. Phys., Vol. 45, No. 8 (1974) p 3396

SAFETY SURFACE POTENTIAL FOR HYDROCARBON PRODUCTS

Shi Chongyue, Liang Wenji, Pan Gang, Tao Tao
Shenyang Fire Research Institute of Ministry of Public Security
Shenyang, China

Abstract

This paper discusses the discharge between earthed electrodes and the charged oil surface, gives the outer envelopes of discharge energy and spark charge transfer. Based on the obtained results and consulting other results from abroad, suggest the safety surface potential to be 13 KV for the hydrocarbon products, and point out that the basic criterion for judging the danger of the charged oil is the minimum ignition energy.

Experimental Apparatus

The experimental apparatus consist of oil circulating system, oil surface potential measurement system and the measurement system for charge transfer as shown in Fig.1.

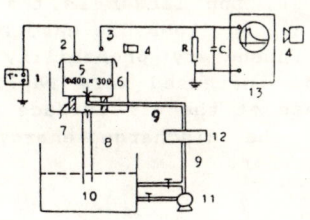

Fig.1 Apparatus for the surface discharge experiment. 1:potential meter, 2:floating ball, 3:electrode, 4:camera, 5:test tank, 6:collecting tank, 7:insulated support, 8:oil return back pipe, 9:oil pipe, 10:oil storage tank, 11:pump, 12:filter, 13:oscilloscope

Kerosine (conductivity 12 ps/m) is charged by pumping it through pipes and filter into test tank. When the oil surfarce potential is rised to an approprite value, it will cause a discharge between the oil surface and the earthed electrode. The discharge can be recorded by an oscillosocope. The oil surface potential is measured by the floating ball method (with ball diameter of 30 mm). The oil surface potential can be varied by changing the materials of the filter, the oil flow speed. During the oil circulation, the oil surface in the test tank is mainly smooth. The eletrodes are conical copper bares with 30° and 60° apex angle and steel balls with diameters of 3.0, 5.0, 10.3, 20.6, 30.0, 40.0 and 50.0 mm.

Method of Measurement

Supposing that the oil surface potential during discharge is V (it is assumed that the local fluctuation of the oil surface potential during the discharge could be neglected), the charge transfer is Q, then the discharge energy W can be expressed by following equation:

$$W = VQ \qquad (1)$$

Supposing the current, voltage on the leak resistance R in the discharge measurement system are i(t), U(t) (displayed on the oscilloscope) respectively, then

$$Q = \int_0^\infty i(t)dt = \frac{1}{R}\int_0^\infty U(t)dt = \frac{S}{R} \qquad (2)$$

where, $s = \int_0^\infty U(t)dt$, which can be called the voltage area and supposing the rise time of the voltage pulse is very short then the voltage pulse attenuates exponentially, i.e.

$$U(t) = U_0 e^{-\frac{t}{RC}}$$

where U_0: peak voltage of the pulse,
C: stray capacitance of the charge transfer measurement system.
then equation (2) can be expressed as following:

$$Q = \frac{1}{R}\int_0^\infty U_0 e^{-\frac{t}{RC}} dt = CU_0 \qquad (3)$$

hence, $W = VCU_0$

It can be seen that the charge transfer and the discharge energy can be calculated with the peak voltage of the pulse U_0. The observation of the oil surface potential indicates that the intrduction of the electrode causes the potential decrease considerably for almost parts of the surface. The smaller the electrode is, the voltage reduction effect is more obvious. With large diameter spherical electrode the potential could be caused decreasing by 10 KV or more from 100 KV, while the smaller electrode could cause a potential reduction of more than half of its original value. This is attributed to the corona discharge, in addition to the serious interference of the earthed electrode to the original field. When a brush discharge appeares, the surface potential would be further disturbed. The discharge has less interference to the surface potential when the discharge frequency is low enough. (The time interval between two successive discharge pulse is more than few seconds). However, the oil surface potential directly underneath the electrode decreased considerably, when the discharge frequency was high (more than 5 times per second). The potential will reduced along with the increasing of the discharge frequency and the discharge electricity. Generally, the discharge frequency for electrodes of larger diameter is low, and its interference to the oil surface potential is small. However, for the small electrodes there is a greater interference to the potential when the discharge frequency is high. For small electrode or for high frequency discharge (which occurs when the surface potential is higher or the discharge gap is small) which has greater interference, the potential should be measured with the floating ball located adequately near to the discharge electrode when calculating the discharge energy. For voltage pulses with longer rise time, it is needed to integrate the waveform when calculating the discharge energy and charge transfer. The stray capacitance C of the discharge measurement system is approximately equal to 260 pf, the leakage resistance R is about 16 MΩ, under room temperature of 20-22°C and the relative humidity in air of 30-40%.

Results and analysis of the experimentes

The brush discharge which is very dangerous is emphatically observed in this experiment, and there is no spark channel between electrode and oil surface been observed, all the term "discharge" which appears in the paper stands for brush discharge. This experiment also proved that the discharge energy from the negatively charged oil surface is much greater than that from the postively charged oil surface. This means that the negatively charged oil surface is much more dangerous than the positively charged oil surface. Therefore, the experiment is conducted in particularly with respect to the discharge resulted from the negatively charged oil surface. Except the polarity of the potential is pointed out, the surface potential quoted in this paper is negative.

1. discharge with a definite discharge gap

The spherical electrode of 10.3 mm diameter is fixed at the point above oil surface in a height of 16 mm, the charge transfer and discharge energy is measured with respect to different oil surface potential. Results with greater discharge energy (in respect to different surface potentials) are selected and shown in Fig.2a. When potential ranges from 20 to 44 kv, the curve in the Fig. is the outer envelope of the all experiment points, which means that any discharge with a larger value of the energy is impossible. Therefore this envelope can also be called the curve of the maximum discharge energy for the 10.3 mm spherical electrode. In Fig.2b is shown an outer envelope of the charge transfer (This is also called the maximum charge transfer curve for 10.3 mm spherical electrode). It is indicated in Fig.2 that the charge transfer and discharge energy increase rapidly with the potential, while the electrode diameter and the discharge gap remain unchanged. For lower surface potential, the discharge frequency is small and the interference of the discharge to the potential will not influence the successive discharge. The discharge frequency increases with the potential. For a potential of 40 kV, 7-8 times discharge will take place each second. For a given discharge energy, the higher the discharge frequency is, the larger is the possibility of igniting flammable gas. That is to say, the incendiary probability of discharging is increased more rapidly with the increase of the oil surface potential than with the discharge energy or the charge transfer.

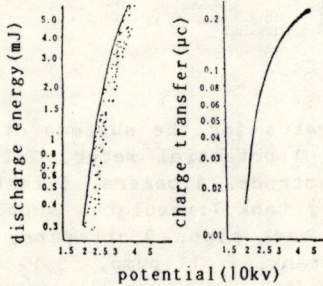

Fig.2 Discharge of 10.3 mm electrode (a) discharge energy verse the potential,(b) envelope of charge transfer verse the potential

2. Discharges with different electrodes and the minimum discharge potential

The charge transfer and the energy were measured under different surface potential for different kinds of electrodes without any limit on discharge gap. The result is shown in Fig.3. The curves in Fig.3 are the maximum discharge energy and maximum charge transfer for different electrodes. In Fig.3, the lowest points of the envelopes indicates the minimum discharge potential and the maximum discharge energy and charge transfer at this potential for that electrode.

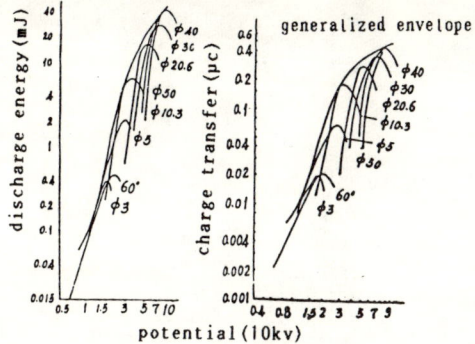

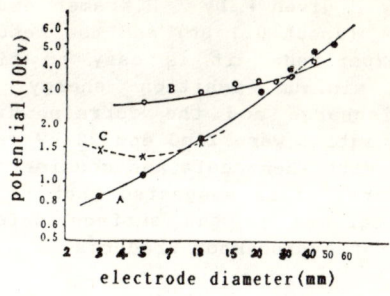

Fig.3 Envelops of discharge energy and charge transfer of different kinds of electrodes. (a) envelopes of discharge energy. (b) envelopes of chaege transfer

For evaluating the danger of different electrodes, the minimum potential of different spherical electrodes and the minimum potential corresponding to the discharge energy of 0.2 mJ (which corresponding to the minimum ignition energy of hydrocarbon oil vapour and air mixture) are dotted in Fig.4. It is indicated that the discharge energy for all spherical electrodes with the diameter greater than 10 mm exceed 0.2 mJ. As the discharge frequency for smaller electrodes is comparatively large and the surface potential of 10 kv—30 kv is often occured in petrochemical enterprises, the electrodes of 10 mm diameter is much dangerous. Minimum discharge potential curve given in reference[1] is also shown in the Fig. It could be seen that the difference between both curves are small when electrodes diameter is larger. However, the difference becomes greater when the diameter gets smaller. This might be attributed to the interference of small electrode to the surface potential which is not considered in reference[1].

3. Maximum charge transfer and discharge energy

For convenience of comparison and observation, the discharge energy envelope for different electrodes is shown in Fig. 3a, by which an overall outer envelope is drawn. It menas that the discharge energy for all fixed electrodes (irrespective of their diameter) can not exceed the value on this envelope. Therefore, it can also be called the maximum discharge energy curve. Fig.3b show the outer envelopes of charge transfer for different kinds of electrodes and overall outer envelope-maximum charge transfer curve.

Fig 4 Minimum discharge potential of the electrodes verse electrode diameters
A: Minimum discharge potential curve presented by this paper
B: Minimum discharge potential curve given in reference [2]
C: Potential corresponding to discharge energy of 0.2 mJ

The existance of the overall envelopes of both the charge transfer and discharge energy indicates that the maximum charge transfer and discharge energy are impossible to be obtained only by means of limited experiments. The significance of the maximum charge transfer and energy curve lies in that the maximum values could be obtained without repeated experiments with different electrodes.

4. Safety surface potential for hydrocarbon products

The minimum ignition energy is one of the important parameters for the mixture of air and conbustible gases. The minimum igntion energy is the minimum value of discharge energy between metal electrodes to ignite combustible materiales. For the discharge between oil surface and conductor, however, the minimum ignition energy mentioned above has lost its original meansing. It is necessary to redefine the minimum ignition energy under this particular condition and to be used as the safety measure. The experiment [2] made by J.K.Jahson, H.Kramer and K.Asano is belong to that of the exploration of the minimum ignition energy of oil surface discharge. Although research work in this region is just in beginning, results have demonstrated that it need at least a charge transfer of 0.1 μc to ignite mixture the hydrocarbon vapour and air.

Based on the experimental results, it is determined that the oil surface charge transfer and the surface potential corresponding to discharge energy of 0.2 mJ are

0.016 μc and 13 kV respectively. Based on the minimum oil surface discharge ignition electricity given by H.Kramer and K.Asano [1] (about 0.1 μc) and the result of this experiment, it is easy to find that the minimum ignition energy of surface discharge and the corresponding surface potential were 2 mJ and 23 kV respectively. With these data and considering other factors, it is suggested that 13 kV could be defined as the surface safety potential for hydrocarbon products.

5. The effect of surface charge polarity on discharge

In order to assess the effect of different surface charge polarity on discharge, the discharge experiment of a positively charged surface under the same condition as given in Fig.2 is conducted. The similar experiment results and the envelopes as shown in Fig.3 are drawn in Fig.5. It can be seen from this Fig. that the discharge energy from the negative charged oil surface is one order greater than that from the oil surface with positive charging. Moreover, the discharge energy with positive charging is distributed dispersively in space. Therefore, the negatively charged oil surface is much more dangerous than the positively charged oil surface.

The difference of the discharge energy and its distribution in space between the positive and negative charged oil surface is attributed to the discharge mechanism.

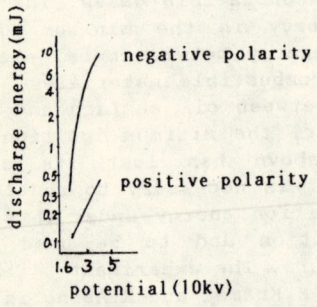

Fig.5 Comparison of the discharge energy for different charging polarities of oil surface with electrode diameter of 10.3 mm, discharge gap of 16 mm

References
[1] H.Kramer, K.Asano
    J.Electrostatics,6(1967) 361
[2] J.K.Jonhson
    J.Electrostatics,4(1977) 53

EFFECT OF MATERIAL SURFACE CHARACTER ON ELECTRIFICATION OF FUEL OIL

Tan Fenggui, Dong Changgui
Safety Technique Research Institute, China General Petroleum and Chemical Industry Corporation
P. R. China

Abstract
This paper mainly discusses the chemical interaction on the surface of a fuel oil treatment system. Through an analysis of the water film on the sueface, the effect of chemical properties of the surface on electrification and charge polarity in fuel oil and the nature of acceleration of electrification by water are explained.

1. Background and Scope of the Research

People are able to estimate the triboelectrification between solids with the aid of the electrifying sequence table or Coehn formula. In fact, it could be similiarly explained by the charge transfer process due to different work fuctions. Electrification of liquid flow is another type of electrification. The formation of an interface double layer and the mode are comparatively complicated and there are many factors that affect the electrification of liquids. Therefore, it seems that it is not realistic to put forward a similar electrifying sequence table for liquid/solid electrification. Some even relieve that the electrifying tendency is unpredictable[1]. However, within the limited scope of engineering, what is the major factor that controls the electrification of fuel oil? Is there a law governing the effect of surface material on the electrification of fuel oil ? Is it in any way connected with the chemical property of the surface material? Problems as such have been paid attention to for a long time. As to the effect of the surface of metal pipeline on electrification. Felice et al (1982) believe that it follows the same mode as the charge injection from the electrode and it is restrained by the image force (Schottky potential barrier)[2]. However, electrification mechanism for microporous filter materials and the filter that is still not clear yet. As Leonard pointed out before the basic surface chemistry is well understood, the electrification of hydrocarbon liquid and fuel oil is still a mystery[3]. That is not only a theorectical problem, it also directly affects the precautions against electrostatic hazards and the basic countermeasures to be taken by people. For example, in recent decades the basic contermeasures taken usually emphasize the elimination or dissipation of charges after the electrification but seldom touch upon the reduction or elimination of the electrification of the material used.

In view of the fact that there are quite many factors affecting the electrification of oil products[4], it is necessary to limit the research condition to a certain scope. The writers have carried out analyses for more than 40 different kinds of materials in three oil refineries These materials cover nearly all the engineering materials generally used in the past and some latest ones. Experiments show that under the condition that the fuel oil is neither added with antistatic agent nor contaminated, there exists a comparable arranging order in the electri- fication. Even when contaminated there arises a great variation in the magnitude of electrification, but the order is basically unchanged. This indicates that the nature of material surface is the most essential factor that affects the electrification of fuel oil. Both the liquid and solid phases are is essential factors that determine the effective force on the interface while the dispersion phase (absorbent) only affects the magnitude.

What has been mentioned is to show why we believe that within the limited range of engineering, it is possible to make clear the effect of surface material on the electrification of fuel oil and understand certain surface phenomenon which has been puzzling over a long period of time. For expample, people have wondred why there are such big differences in the tendency of electrification for different materials, The profound mystery of polarity variation in the electrification of fuel oil and the effect of water on the electrification of fuel oil etc have also been puzzling.

2. Experimental Statistics of the Effect of the Surface Material on Electrification

Phenolic resin is the surface binder generally used in the manufacturing of industrial filter paper, but there is a big difference in electrification between different phenolic resins. Fig.1 gives a the comparision of electrification of the original-paper (II_1) after having been treated with different surface binders. When the industrial flow speed is about 0.14 cm/s, the electrification of the surface (II_3) treated with common phenolic aldehyde is approximately twice as much as the untreated; after the surface having again been treated with 5% polysiloxane (II_5), electrification is increased by 9.6 times; if it is treated with pure polysiloxane(II_4) electrification is increased by approximately 19 times. On the contrary, if the surface is treated with water soluble phenolic resin (II_2), electrification is decreased by 9 times. among the materials mentioned above, the polarizability on the surface of II_4 filter paper is the strongest. Polarizability of II_5 is second-strogest. II_3 and II_2 rank third and fourth. This sequence is consistent with their electrifying tendency. The greater the difference in ploarizability, the greater the difference in electrification. A similar phenomenon appears in the treatment of other fibres. When different surface agents are used for treating glass fibre filter paper, the amount of charge due to electrification at rated flow speed (about 0.9 cm/s) may differ by 35 times as shown in Fig.2 . Of all these samples, that treated with acryl acid ester(WG) has the lowest surface polarizability, but its charge due to electrification is the largest. After mixing it with 5-6% polar group (including carboxyl, amino group and hydroxyl), electrification descends (JU). If it is treated with polyvinyl alcohol (B), the surface polarizability becomes stronger and electrification will further descend. There are surfaces having not been treated, but the electrification is still the same as those mentioned above. For polytetrafluoroethylene, polyethylene and pure polyester fibre etc, the surface polarizability is small but their electrifying ability is all rather high. However, as to wool, silk cloth and certain water absorpting materials, such as sodium CMC (Aquacon) etc. containing mang polar groups, their electrifying ability is all fairly small. In short, from experiments we made, the electrifying ability of over 40 kinds of materials has some relations to the chemical properties of their surface material. And those with stronger surface polarizability electrify the oil more.

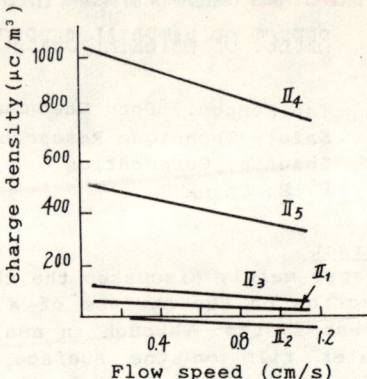

Fig.1 Charging character of different filter papers

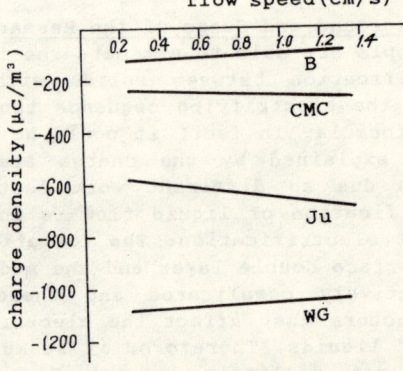

Fig.2 Charging character of different glass filter paper

3. Analysis of Effect of Surface Chemical Nature on Electrifkication of Fuel Oil

The major factor that dominates the dissociation of liquid/solid interface charge is the selective absorption of the interface. However, if the experiment mentioned above is explained according to the ordinary principle of chemical absorption, that is with phase system absorption, it will cause great confusion and even lead to a completely opposite conclusion. The mistake here is that the effect of the water film on the surface is overlooked. This layer of water film first comes from the air and after it is soaked with oil, it is just like a part of the solid and moreover, as the fuel oil continues to flow, the free water will gradually thicken the water film. As water has a strong polarizability, this layer of water film will apparently change the nature of the liquid/solid interface, thus making the absorption on the interface similar to the typical polar surface absorption of a polar dispersed substance (absorbent) in a non-polar liquid. This analysis will not only make the dissociation of polar ion compounds in fuel oil easily understood but also explain the difference in electrifying ability of different filtering materials. For example, non-polar or weakly polar surface with low hygroscopicity repels free water in the oil and has a "water film" with typical properties resulted from its interception or accumulation of much water over it. There, the interface is easy to absorb polar ion compounds in the fuel oil and the electrification is high. When the hygroscopicity of polar surface is high, both oil and water penetrate through the microporous surface, and the accumulation of water on the surface is less and the capacity of the surface to absorb the ion compounds in the fuel oil is low, consequently electrification is rather ssmall. This analysis is also in agreement with the expression of filters of different types.

Works done in the past has pointed out that there is an apparent effect of water on the electrification of fuel oil, but the mechanism remains a riddle[3]. Some of them guess that there possibly exists a certain indirect effect[3] while others consider that it possibl includes the dissociation effect of free water on the polar matter. Experiments show that the latter is not the essential factor. The essential factor is still the accumulation of free water on the surface. Table 1 shows the charge density data (in $\mu c/m^3$) of aviation kerosine with free water electrified by different filtering materials (at the same flow speed as the above). It is discovered that not all materials are sensitive to water, glass fibre filter paper (B214) treated with 214# phenolic aldehyde, polyester fibre (LPVA) treated with polyvinyl alcohol (PVA) and filter paper (I_2) treated with water soluble phenolic resin etc. are, but some of the examples The effect of "electric acceleration" of water is very small and only those hydrophobic materials, such as I_5 filter paper treated with polysiloxane etc. will bring about variations in electrification. The difference in elecrification between these two kinds of materials lies apparently in the amount of water on their surfaces. As to the first kind of hydrophilic material, the increment of free water will not apparently change the amount of accumulation of water on the surface, therefore the variation in electrification is not great, but as the accumulation of water on the hydrophobic surface of the second kind apparently increases, the variation in electrification is great.

This shows that the traditional argument that free water can apparently increase the electrification of fuel oil is not correct and only those materials with water dissociation effect are sensible. Before 70's, the filter paper widely used belongs to this category (such as I_5 filter paper etc.).

Table 1 Charge density ($\mu c/m^3$) of aviation kerosine with free water electrified by different filtering material

Amount of water added	Material				
	I_2 Filter paper	I_3 Filter paper	I_5 Filter paper	L(PVA)	B214
~0%	50	220	430	-40	-40
1-3%	85	320	1630	-55	-50

In the past, reports regarding electrifying polarity of fuel oil were also confusing and even mystifying. Experiment show that these reports were based on experiments done under different conditions Table 2 gives the electrifying polarity of certain materials under different oil product conditions and with a wide flow speed range. For some materials, the variation in polarity arises only when there is a change in flow speed. When carefully observed, it will be discovered that only in those materials with low electrify ability, such as I_2 filter paper, woolen felt and glass fibre filter paper treated with PVA or 214# resin etc., a change in polarity can easily occur. As to high electrifying materials, such as filter paper I_4 and I_5 treated with polysiloxane, glass fibre (Ju) treated with acrylate and polyester fibre etc. no reversal of polarity has been discovered during experimenting. This phenomenon is related to their surface structure and the absorption pappened. Carefully observing the transient process of electrification, one will discover that the liquid/solid interface absorption is not a single polar type, but there simultaneously exists on absorption of ions of two polarities and only the tendency toward absorption and the speed of absorption are different. The "water film" on the low electrifying surface is

Table 2 (*: conductivity less than 50 C.U.)

Filter material	Basic aviation Kerosine		Adding antiablation agent		Adding antiwear agent		Adding antistatic(*) agent		Adding water	
	Variation of electrification	Variation of polarity	Variation of electrification	Variation of polarity	Variation of electrification	Variation of polarity	Variation of electrification	Variation of polarity	Variation of electrification	Variation of polarity
Filter paper I_5	200~500	+	←	+	←	+	←	+	←	+
Filter paper I_3	90~200	+	unchanged		←	+	←	+	←	+
Filter paper I_2	-6~100	-/+	unchanged		→	-	→	-/+	not great	-/+
Polyester fibre	-20~-90	-	unchanged		→	-	←	+	←	-
Woolen felt	10~-80	+/-	unchanged		←	-	→	-/+	←	-
B(PVA)	25~-100	-/+	unchanged		unchanged	-	→	-/+	←	-
B214	-30~120	-/+	unchanged			-	←	-/+	←	-

not typical and usually there is no big difference in the absorption tendency of positive and negative ions. When the effective force of the interface is interfered externally, the proportion of the absorption of positive and negative ions changes easily and a variation in polarity occurs. On the contrary, most of the arrangements of fuctional groups in high electrifying materials is rather regular. The orientation of water molecules is comparatively apparent. The difference in absorption tendency toward positive and negative ions is a great and it is not only high in electrification but also comparatively stable in polarity, filter paper coated with methy polysiloxane is an example and is shown in Fig.3. The surface functional group $-CH_3$ is taken as the major group. The positive ends of water molecules are mostly so arranged that they face to the outside. " Water films " easily absorb negative ion compounds and fuel oil readily carry positive electricity. The surface structure of glass fibre filter paper (HB) coated with a mixture of acrylonitrile and acrylate is in the form shown in Fig.4. The negative ends of water molecules are mostly so arranged that than face outside, the outer surface easily absorbs positive ion compounds, therefore the fuel oil carries negative electricity. Experiments show that for surface with an alkyl functional group as the major group, fuel oil easily carries positive electricity and for surfaces containing a great amount of hydroxy (-OH), carboxy(-COOH), nitrile group(-CN) and easter bond (-COOR), fuel oil easily carries negative electricity, which are basically in compliance with the analysis mentioned above.

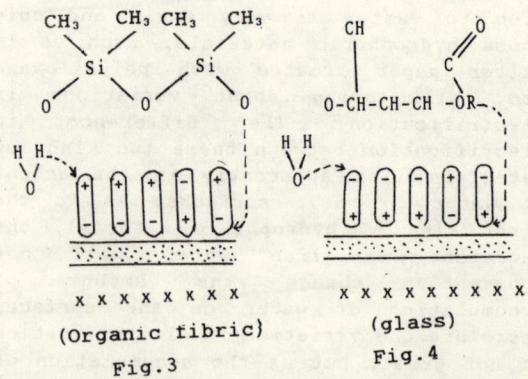

Fig.3 (Organic fibric)   Fig.4 (glass)

4. Conclusions

In this text, emphasis is put on the discussion of the liquid/solid interface phenomenon within the range of engineering, the electrifying tendency and polarity of fuel oil and the effect of water etc. connected with the nature of the surface material. The arguments given in the text can help us not only to understand the electrification mechanism of micro-porous filter materials but also to select correctly filtering materials or surface treatment to effectively restrain electrification of the filter. For surface treatment, either a chemical method or a physical method can be adopted. In these two respects, the writers have made an attempt and have obtained quite satisfactory results. Fig.5 is the actually measured result of electrification on two kinds of new type filter cores put forward by the writers, which not only satisfy the requirement of MIL-F-8901E filtering but also are in compliance with the antistatic requirement of API. America, England and USSR etc. there are also antistatic special filter cores and people basically adopt the scheme of mutual offset with materials of different electrifying polarities, with which there generally brings about a common defficiency, that is the selected materials are mostly the ordinary high electrifying materials. Fluctuation amplitude of electrifying tendency varies with the variation of in the oil product, flow speed and temperature is great and it is hard to ensure a reliable compensation for a wide range. All the schemes of the matches of the two kinds of new filter cores the writers propose are made up of low electrifying materials. The variation amplitude of electrifying tendency is small and the performance is more reliable. Since the basic research "Study on the Effect of Surface Nature on Dissociation Effect of Production of Charge "[3] was put forward by Mr.J.T.Leonard, it has attracted great attention and notice in many respects. It is hoped that by putting forward this text there will be more discussions regarding this problem.

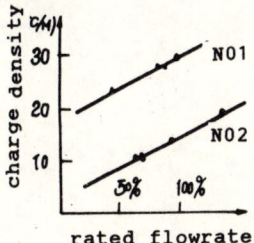

Fig.5 Charge character of new type filter core

References

1. Kenneth C.Bachman,J.C. Munday "Evaluation of the hazard of static electricty in nonmetallic pot system-static effects in handling jet fuel in fiberglass reinforced plastic pipe" AD 764558, p98
2. N.J.Felici,J.P.Gosse, A Solofomboahangy "Liquid flow electrification and Zeta-potential in Hydrocarbons" Journal of Electrostatics, 12 (1982) p369-376
3. J. T. Leonard "Static Electricity in Hydrocarbon and Fuel" Journal of Electrostatics, 10 (1981) p29
4. Tan Fenggui, Dong Changgui, "Analysis on Material Gausing Electrification of Hydrogen Liquid" Symposium of Static Science Public Lecture, 87, p381-385

IX. Electrostatic Hazards
and
the Prevention Measures

ELECTROSTATIC HAZARDS OF EXPLOSIVE ARTICLES AND THE CONTROL MEASURES

Zhang Guoshun

National Commission of Machinery

P. R. China

Abstract

Several accidents caused by the static electrification of gunpowders, explosives and igniters are cited in this paper, thus alerting people to the serious menaces to safety in the production of explosive. The criterion for electrostatic hazards in the production of dangerous articles and the munimum ignition energies of several powders, explosives and electrical detonators are also expounded. In the later part of this paper the author points out main measures for eliminating static hazards, such as grounding, humidification, air-ionizing, dissipating charges on human bodies and the floor, setting up suitable monitoring equipments and improvement of the structures of the ignitors themselves.

Electrostatic Hazards of Explosive Articles

Primers, black powder, gunpowder, explosives, detonator, caps and other ammunition are all potentially hazardous in bring about combustion and explosion. That is not only related to their own inflammability but also the external conditions for ignition. Of all the external factors causing hazards, electrostatic discharge seems to be one of the most dangerous. There are many accidents due to the static electrification of explosive articles in our history. A few of the examples are listed below:

1. Four explosive accidents in granulating black powder between 1961 to 1980 took place in one single factory of Shanxi with 13 victims. All these were considered to be closely related to static electrification. However, no such accidents ever occurred since adoption of comprehensive measures against static electrification were taken.

2. Many accidents have ocurred in sifting bursters and emptying containers with bursters in a factory in Shanxi. For example, Once when lead styphnate was been poured into a paper container a grounded conductor touched the latter accidently, an explosion suddenly occurred.

3. In 1981, two explosive accidents with two men killed occurred while handling electrode wires of electro-explosive detonators for seismic prospecting. They were also considered to be due to the electrification static on a worker's clothing.

4. When nitro-cotton was transmitted through a metallic tube pueumatically to a cloth bag, a spark suddenly flew off. There was no igniting source around whatsoever except for static electricity in the gunpowder. This accident happened in a factory in Liaoning in 1973.

5. Artillery shells were let to roll over foamed plastics soaked with anti-rust grease, when plastic surface suddenly burst into flames, Investigation showed that it was because an electrostatic spark had ignited the gasoline vapor.

6. An explosion occurred, when a worker was carrying electric fuses with the electrode wires in open circuit. The accident was attributed to static electrification of the human body.

From the cases mentioned above one can conclude that safety measures against static electrification are absolutely necessary in manufacturing, using and transporting explosive articles.

Electrostatic Spark Sensitivity of Some Explosive Articles

In China electrostatic spark sensitivity in terms of minimum ignition energy is usually measured with JGY-50 Electrostatic Sensitivity Tester, whose schematic circuit diagram is shown in Fig.1. Test procedure is in accordance with standard AH0003-79. Both the voltage and energy for 50% firing probabability (V_{50} and E_{50}) and the standard deviations are determined by using Bruceton method with $E_{50}=1/2\ CV_{50}^2$. If the standard deviation is adequately small, E_{50} can be defined as the minimum ignition energy E_{min}. At E_{min}, the firing probability is no more than one millionth. Some of the electrostatic spark sensitivity of some primary explosives are given in Table 1.

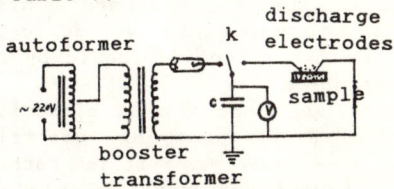

Fig. 1 Circuit diagram for measurement of minimum ignition energy of explosives

Table 1 Electrostatic spark sensitivity of some primary explosives (pin-plate electrode geometry)

name	polarity of pin electrode	C (pf)	gap (mm)	V_{50} (kv)	E_{50} (mJ)	$E_{min}=E_{50}/20$ (mJ)	remark
lead styphnate	−	485	0.03	0.58	0.0802	0.00401	
	−	546	0.12	0.92	0.23	0.0116	
	+	487	0.12	1.18	0.339	0.016	
lead azide (fine crystal)	−	487	0.12	0.90–1.16	0.197–0.328	0.0099–0.00164	
	+	546	0.12	1.56	0.664	0.0332	
CMC lead azide	+	546	0.12	3.52	3.38	0.169	
mercuric fulminate	+	9700	0.12	1.30	8.2	0.41	
	−	10000	0.12	1.37	9.83	0.492	
DDNP	+	9900	0.12	2.83	39.6	1.98	10kΩ in series

By using this method the 50% ignition energy of agglomerated black powder was found to be 4.06 mJ. This agglomerate has a minimum spark ignition energy of 0.20mJ or E_{min}= 0.20mJ. It is up to the same grade as the primers.

Problems arise in measuring spark ignition energy of high explosives, because of its dependence on the state of sample, the structure of electrode system, test procedure, and the judgement of the firing of the sample as well. Recently, a fairly reasonable and reproducible result was obtained by measuring the pressure rise in sample chamber after been ignited by a spark. High explosives and gun-powders fall into three different grades of danger: grade I (sensitive explosives), E_{min} < 1 mJ, grade II (common explosives) 1mJ < E_{min} < 450mJ and grade III (inactive explosives).

Electro-explosive devices (EED) measure the electrostatic spark sensitivity by simulating a discharge from the human body. The capacitance and resistance of a human body are 487pf and 5 kΩ in series respectively. Table 2 shows some of the results.

Main measures against electrostatic dangers

The fuction of the most anti-static measures consists in prompt dissipation of static electricity generated. Sometimes neutralization techniques are also adopted. For electro-explosive devices, improvement of the structure should by no means be overlooked.

1. Grounding is one of most effective measures for dissipation of electric charge, but sometimes it is rather complicated. Moving equipments should be earthed through conductive brush or coating layers. Non-conductive equipments should be coated with a layer of conductive paint and then earthed. The earth resistance of less than 10^6 Ω is considered adequate. Working tables and the floor are best covered with conductive rubber so as to keep the comprehensive earth resistance near 10^6–10^8 Ω. For some cases the grounding point for dissipation of static charges can be the same as that for the power supply and lightning rod.

2. Humidification is effective for preventing static hazards, which can be realized from the statistics shown in Table 3. Of all the 23 accidents in black powder processing, 22 happened in dry areas of North China. Of the 14 accidents happened in one single factory in North China, 12 happened in dry seasons. Moisture in air not only lowers spark sensitivity of black powder, but also lessens the electrified charge. Table 4 shows the relationship.

With reference to the data given above, conditions in black powder workshop is kept as follows: room temperature 18°C, RH 65%, absotute humidity > milibar and moisture content in the air > 8 g/kg (air).

For other explosives, things are quite similar, but there are some difference for explosives with different hydrophilicities

3. Air Ionizing can be adopted under some conditions with the aid of radio-isotopes. Ionic wind is blown out of an appropriate equipment and injected to the electrified area, such as the charged primer.

4. Elimination of static charge on human bodies is vital importance in the production of electro-explosive devices. The main measures are adoption of conductive clothing and conductive floor. Five kinds of conductive floor have been popular in China. They are made of conductive rubber, unsaturated polyester, polyaminoresin, non-ignitable tar and cement concrete.

Table 2 Electrostatic spark sensitivity of some typical electrical detonators

type	electrode wire--tube case		electrode wire--electrode wire		first filling
	V_{50} (kv)	E_{50} (mJ)	V_{50} (kv)	E_{50} (mJ)	
spark	1.36	0.45	-	-	lead azide
bridge filament	7.1	12.27	6.36	9.85	lead azide
	4.5	4.93	12.5	38.05	lead styphnate
	5.6	7.64	13.9	47.05	lead azide
	7.3	12.98	endurable to 20 kv		lead azide
heatproof detonator (180°C), endurable to 20 kv					$CMX-pbN_6$

Table 3 Correlation between probability of accidents in black powder and climate

Influential factors	Condition for more probability	Probability in general	Conditions for less probability
Areas	Northern dry areas (22/23)*		Southern humid areas (1/23)
Seasons (North China)	Dry seasons(Jan.-May, Nov. Dec.) (12/14)		Humid seasons(Jun.-Oct.) (2/14)
Time	10-15 o'clock(10/14)		Before 10 & after 15(4/14)
Weather temperature	<18°C (13/14)	<0°C(5/14), 0-10°C (5/14), 10-18°C(3/14)	≥18°C (1/14)
Relative humidity	$\rho \leqslant 65\%$ (22/23)	$\rho <35\%$ (14/23), 35-50%(5/23), 51-65%(3/23)	$\rho >65\%$ (1/23)
Absolute humidity	B < 6 milibar (10/14)	6-11 milibar (2/14)	B > 11 milibar (2/14)
Moisture in atmosphere	<4g(H_2O)/kg(air) (10/14)	4-8g(H_2O)/kg(air) (2/14)	>8g/kg(air) (2/14)

* (probability)

Table 4 Relationship between properties of black powder and atmospheric humidity*

Relative humidity in air (%)	18	30	50	70	90
Properties of black powder					
Moisture content(%) in black powder	0.51	0.62	0.85	0.92	1.10
Bulk resistivity ($\Omega \cdot cm$)	$>10^{13}$	10^{11}	10^9	10^8	10^7
Mass charge density ($\mu c/kg$)	0.58	0.46	0.27	0.14	0.018

* 4# granules of black powder is reserved at 25±1°C in air with different humidity and electrified in the same process.

According to the standard WJJ-1-82, their ground resistance Rg should be:

$5 \times 10^4 \Omega < R_g < 10^6 \Omega$ or $10^5 \Omega < R_g < 10^7 \Omega$

There are three kinds of materials for clothing available in China, i,e. cloth with fibre coated with anti-static agents, cloth with fibre internally mixed with anti-static additives and cotton-stainless steel fibre blend fabric.

Resistance of conductive shoes is limited within the range of $10^7 \Omega$ - $5 \times 10^4 \Omega$ so as to prevent from charge accumulation and avoid shock from electric mains.

5. Comprehensive Safety Measures against Static Electrification

Table 5 shows the comprehensive anti-static measures for the workshops, where explosive articles of different dangerous grade are manufactured. These measures are aimed at the above mentioned electrostatic accidents, which occurred before these measures were taken.

6. Improvement of the Anti-static Property of Ignitors and Electro-explosive Devices

In order to avoid discharge between the bridge wire and the case of bridge detonators, following measures are suggested:

(1) Setting up a protective discharge channel to avoid discharge through an unexpected dangerous channel. The proctective channel is located in the plug and has a breakdown voltage (3 kv in

Table 5 Comprehensive anti-static measures for explosive articles' workshop

Dangerous grade of the place	I	II	III
Electrostatic sensitivity of the explosives	Sensitive explosives ($E_{min} < 1$ mJ)	Common explosives 1 mJ $\leq E_{min} < 450$ mJ	Inactive explosives $E_{min} > 450$ mJ
R.H. of air	$\geq 65\%$	$\geq 65\%$	No requirements
Grounding	All the fixed and movable equipments should be grounded. $Rg \leq 10\Omega$	Ground fixed equipments. $Rg \leq 10\Omega$	Ground large conductors
Floor	Conductive, $Rg=10^5-10^7 \Omega$	Lie between I and III	Common
Staff workers	Wearing conductive workclothes and conductive shoes. Putting on and taking off clothes and wearing chemical fibre clothes are prehibited.		No allowance of wearing chemical fibre clothes
Monitors	Monitor of overall resistance of human body		No requirements
Others	No allowance to use insulating material		No requirements

general) of less than a quarter of that of the dangerous channel.

Fig.2 shows an example of the structure of a detonator for a perforating bomb. On the electrode wires in the plug there are points. Arround the wires is set a metallic ring. Discharge between the points and the ring happens at a lower voltage than the breakdown voltage across the unexpected gap.

(2) Insert a leakage resistor (100Ω) between every electrode wire and the case.

(3) Coat the surface of the plug with conductive paint ("2512" conductive lacquar) with a thichness of 0.05 mm. The resistivity of the layer is 25- 290 $\Omega \cdot$ cm.

A detonator with a coated plug can torterate a discharge through a resistance of 5 kv from a 500 pf capacitor charged to 25 kv. Without the coating V_{50} is only 16.8 kv with other conditions unchanged.

It may thus be concluded that static electrification seriously threatens the safe production of explosive articles. A series of anti-static measures have been adopted and the probability of accidents has been remarkablely lowered. We are still trying our best to develop more technical measures to ansure safety in the manufacturing of explosive articles.

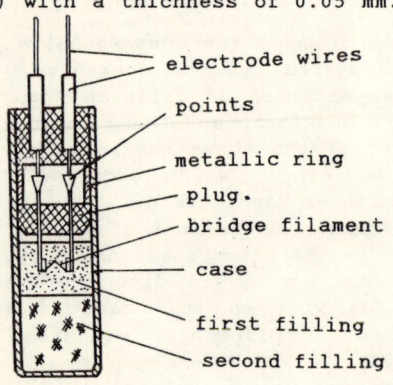

Fig.2 Structure of the protlective discharge of a channel

References

[1] N.Gibson "Electrostatic Hazards-- a Review of Modern Trends" "Electrostatics 1983" Institute of Physics, Oxford pp 1-11

[2] J.Nagy et al "Explosibility of Niscellaneous Dusts" VSBMRI 7208(1968)

[3] B.S.5958:1981 Part 1, 1982 Part 2

[4] Li Ruinian et al "The Optimum Point of Electric Spark Ignition of Agglomerated Black Powder" Proc.Conf of Electrostatic Society, Japan (1986) 357-358.

APPLICATIONS OF ELECTRICALLY CONDUCTIVE FIBERS TO ELECTROSTATIC SAFETY

Yasuyuki TABATA* and Yoshiteru MATSUO**

*Industrial Safety Institute
Ministry of Labour
5-35-1, Shiba, Minato-ku, Tokyo, Japan 108

**Central Research Laboratories
Kuraray Co., Ltd.
1621, Sakazu, Kurashiki, Okayama, Japan 710

Abstract

Two kinds of electrically conductive fibers which differ basically in structure as well as properties have been developed to reduce safely a static electricity on a synthetic fabric-ware. The purpose of this study is to investigate an electrostatic elimination quality of electrically conductive fibers required for electrostatic safety. An anti-static effect of the electrically conductive fibers woven partly into the fabric-ware was examined experimentally, and their weaving methods were specified for various kinds of fabric-ware. The anti-static effects of a few kinds of fabric-ware into which they are woven partly are considered in this paper.

Introduction

An electrically conductive fiber "ECF" made of a metal or carbon can produce a strong non-uniform electrostatic field when it is spaced near charged materials, and raises faint corona discharges around itself; as a result, electrostatic charges on the materials are neutralized by positive or negative ions from the corona discharge (1,2). Similarly, the electrostatic charge on a fabric-ware into which ECFs are woven partly is also neutralized by the corona discharges generating around ECFs (3). Therefore, ECFs are applied to reduce a static electricity on various charged materials as well as fabric-ware (4).

However, the former ECFs, especially one made of a metal, are not completely suitable to eliminate safely a static electricity. Newly, two kinds of ECFs, one is mainly made of a carbon and the other nickel, have been developed by KURARAY Co., Ltd. (5). As described later, some improvements have been made on them so that they can be applied to various kinds of charged materials and appropriated effectively for eliminating the static electricity on the charged materials.

The purpose of this research is to investigate a practically electrostatic elimination effect and safety about the ECFs developed. Experiments were made on the anti-static effect of the ECFs developed to specify the application method of ECFs to various kinds of synthetic fabric-ware.

Properties of Electrically Conductive Fiber

Electrically conductive fibers developed newly, ECF-A and ECF-B, have an entirely different structure and electrical properties, respectively. ECF-A consists of 4 electrically conductive cores, carbonaceous materials, covered over with a polyethylene sheath as shown in Fig. 1 (a). Therefore, an apparent resistance of a single fiber of ECF-A was about $10^7 \Omega$/cm and fairly large as shown in Table 1. And also a corona on-set voltage was high, about 6 kV, since corona discharges from ECF-A occurred due to an electrical break-down of the polyethylene sheath. However, hazardous spark discharges hardly occurred in electrostatic elimination processes, and further, the properties of ECF-A were similar to those of a general polyethylene fiber.

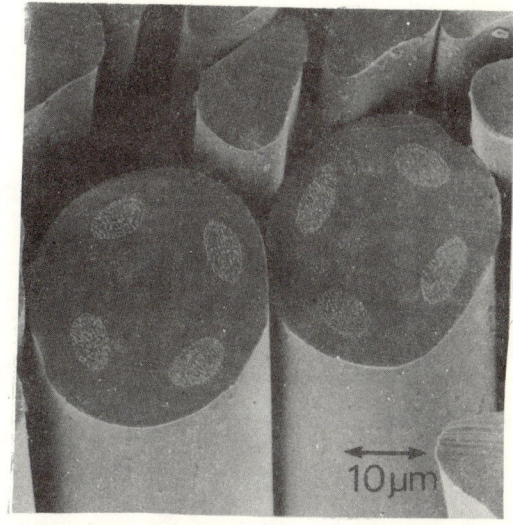

(a) Cross Section of ECF-A.

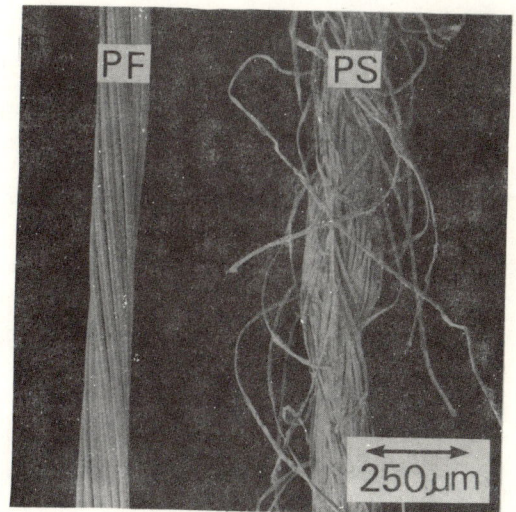

(b) Threads Made of ECF-B.

Fig. 1 Microscopic Photos of ECFs.

ECF-B is a filament or staple made of a polyethylene. Threads, "PF" and "PS", twisted together with filaments and staples of ECF-B are shown in Fig. 1 (b), respectively. A surface of ECF-B is plated with nickel in thickness of about 0.2 μm. A fiber size is about 10 μm in diameter as shown in Table 1. ECF-B raised faint corona discharges in low potential, about 1.2 kV, since the diameter was very small and there were many needle-like projections on the surface of thread, especially on that of one made of staples of ECF-B.

Table 1 Properties of ECFs Developed.

Items	Samples	
	ECF-A	ECF-B
Apparent Resistance (Ω/cm)	$10^2 - 10^3$	10^7
Resistivity* (Ω cm)	3×10^{-3}	7×10^{-6}
Diameter (μm)	30	10
Tensile Strength (g/d)	3.4 - 3.8	4.0 - 5.0
Extension Rate (%)	42 - 53	20 - 26
Young Modulus (kg/mm^2)	600 - 900	800 - 1,300
Specific Gravity	1.35	1.72

* Value is resistivity of conductive material in/on ECF.

Anti-Static Effect of Electrically Conductive Fiber

An anti-static effect of fabric-ware woven partly with ECFs was examined by experiments to specify the effetive and safe application method of ECFs to fabric-ware. All experiments were carried out in the controlled atmosphere, temperature of 25 °C and relative humidity of 30 %. Results obtained from experiments are as follows:

<u>Working Garment</u> Samples used in tests were sewn with pieces of cotton, polyethylene 65 % / cotton 35 % "PET/Cotton" and polyethylene woven partly with ECFs "ECF/PET" cloths. ECF/PET is a cloth that ECF-A developed is woven into PET cloth in the checkered pattern with the pitch of 10 mm.

An anti-static effect of each sample was investigated by the testing method, "Take-off Test", recommended by Industrial Safety Institute, Ministry of Labour (6). Measurements were made on the peak value of electrostatic potential of human body on the metal plate grounded when the sample was taken off, and also made on charge of the samlple for reference. The results, mean value obtained throughout 20 trials, are shown in Table 2.

<u>Carpet</u> Two samples, one was made of polyethylene "PET" and the other polyethylene 98 % / ECF-B 2 % "ECF/PET" threads, were used in experiments to examine the anti-static effect of ECF-B. Measurements were made on the peak values of electrostatic potential of human body putting on conductive shoes under walking for 60 secs and also the potential decay of human body after walking-stopped on the sample, respectively, where the sample was set on the metal plate grounded or non-grounded. Results, mean value obtained throughout 20 trials, are shown in Table 3.

Table 2 Anti-Static Effect of Working Garment Made of ECFs.

Samples		Potential of Human Body (kV) and Charge of Working-wear (μC)*		
		Under-wear		
		Cotton	Wool	Acrylic
Cotton	g**	-0.1(+0.03)	-1.2(+0.46)	+1.8(-0.70)
	ng**	-0.8(+0.05)	-8.4(+0.53)	+12 (-0.78)
PET/Cotton	g	+2.4(-0.88)	+2.4(-0.90)	-1.3(+0.58)
	ng	+14 (-0.82)	+15 (-0.84)	-9.3(+0.58)
ECF/PET	g	+0.1(-0.12)	+0.2(-0.12)	-0.3(+0.12)
	ng	+0.5(-0.22)	+0.6(-0.31)	-0.5(+0.22)

* Value in parenthesis is charge of working-garments.
** g and ng show that human body is grounded/non-grounded through conductive/non-conductive shoes, respectively.

Table 3 Anti-Static Effect of Carpet Made of ECFs.

Samples		Potential of Human Body (kV)*		
			Base	
		Non	Polymer	Hemp
PET	g**	+3.9(2.9)	+4.8(2.9)	+4.6(2.5)
	ng**	+4.4(3.2)	+4.8(2.2)	+3.4(2.9)
ECF/PET	g	+0.2(0.1)	+1.2(0.7)	+1.2(0.5)
	ng	0.2(0)	+1.0(0.8)	+0.9(0.6)

* Value in parenthesis is potential of human body in 60 secs after walking-stopped.
** g and ng show that metal plate under sample is grounded and non-grounded, respectively.

<u>Gloves</u> Samples used in experiments were cotton articles on the market "Cotton" and gloves knit out of cotton and ECF-B "ECF/Cotton" threads. Experiments were made on an electric shock caused by electrostatic discharges occurring between charged materials and workers putting on samples, since it was expected to relaxate or suppress the electric shock by use of the ECF/Cotton gloves. The suppression effect of electric shock was examined in practical processes of industry, handling of moulded acrylic plates and dried nylon cloths. Electrostatic potentials of charged plate and cloth were about 60 and 70 kV, respectively. Results, the occurrence rate of electric shock throughout trials of 100 - 200 times x persons, are shown in Table 4.

On the other hand, model experiments were made to examine discharge characteristics occurring between the charged materials and human body putting on gloves using Van de Graaff generator, since the electric shock was concerned with the characteristics of electrostatic discharges. Measurements were made on on-set voltages and characteristics of discharges occurring at the distance of 15 cm between the generator and gloves, and results, mean value obtained throughout 20 trials, were shown in Table 5.

Table 4 Anti-Static Effect of Gloves Made of ECFs.

Samples		Rate of Electrick Shock (%)*	
		Handling Precess	
		Acrylic Plate	Nylon Cloth
Cotton	New	100(100)	67.5(80.2)
	Used	100(100)	66.7(93.3)
ECF/Cotton	New	0(45.2)	0(15.8)
	Used	0(60.6)	0(45.3)

* Value in parenthesis is rate obtained under condition that human body is not grounded.

Table 5 Discharge Characteristics Occurring between Generator and Gloves.

Samples		Discharge Characteristic	On-set Voltage (kV)
Cotton	g*	Streamer	22.0
	p*	Spark	26.8
ECF/Cotton	g	Corona	14.1
	p	Corona	16.3

* g and p show only gloves and gloves were put on hands, respectively.

<u>Bag Filter</u> Samples were made of pieces of polyethylene "PET" and polyethylene mixed partly with ECF-B "ECF/PET" unwoven cloths. ECF/PET is mixed partly ECF-B developed into PET unwoven cloth in the checkered pattern with the pitch of 25 mm x 50 mm. A performance of the sample, PET and ECF/PET bag filters, was examined by experiments in handling processes of ferro-silicon particles of about 40 μm in diameter. Results obtained after operation for 6 months are shown in Table 6.

Table 6 Anti-Static Effect of Bag Filter Made of ECFs.

Samples	Potential of Filter (kV)	Draft Rate of Air (cc/cm s)	Attraction of Powder (g/m^2)
PET	-42 – 50	6.0	85.1
ECF/PET	-0 – 3	14.6	47.1

Discussion

There are some differences between 2 electrically conductive fibers developed newly, ECF-A composed mainly by 4 carbonaceous cores and ECF-B plated with nickel, in corona discharge characteristics as well as fiber properties. They influence directly a suitability of the application to fabric-ware and anti-static effect. Considering these difference and characteristics, ECF-A is suitable to reduce gradually electrostatic charges on the fabric-ware which has a small capacitance and requires various fitness between fibers, since the surface of ECF-A is the same as that of general fibers as shown in Table 1 and the corona currents are suppressed not to shift to hazardous discharges, such as spark and streamer.

On the other hand, ECF-B is widely applicable to various charged materials as well as fabric-ware and suitable to reduce safely electrostatic charges on them, since both the corona on-set voltage and current are low, about 1.2 kV, and small, about 10 μA/cm at 3 kV, respectively.

In the applications of ECF-A or ECF-B to the fabric-ware for the purpose of eliminating a static electricity on them, it is sufficient to mix ECF-A or ECF-B below about 1.0 % into fibers, and the high elimination quality is practically realized by means of this simple mixing method as shown in Tables 2 and 3. In case that the high anti-static effect, such as electrostatic potential below a few hundreds V or charge density below 10^{-6} C/m^2, is required, such the anti-static criteria are also realized by means that ECF-A or ECF-B of a few percent is mixed into fibers or that ECF-A or ECF-B is woven partly into the fabric-ware in the checkered pattern with the pitch of a few centimeters.

Next, a working garment, carpet and gloves are made of cloths woven partly with ECFs developed newly, respectively; as a result, both electrostatic charges on them are eliminated and also a potential of human body putting on them is suppressed in low level as shown in Tables 2 and 3. Furthermore, electrostatic discharges between charged materials and human body are also relaxated as shown in Table 5; as a result, an electric shock is nearly solved as shown in table 4.

Consequently, ECFs are fairly available to prevent electrostatic hazards caused by a human body working in electrostatic environments, but it is impossible to realize perfectly an elecrtrostatic safety on human body only putting the fabric-ware woven partly with ECFs. Besides putting the fabric-wares of ECFs, conductive shoes must be put for the electrostatic safety of human body.

As described above, ECFs developed hardly brings out dangerous discharges and neutralized safely electrostatic charges by corona discharges occurring around themselves. However, there is a limit to accompanish an electrostatic safety in industry using only ECFs. Careful considerations are required in the applications of ECFs and an evaluations of their anti-static effect.

Furthermore, although no other quality of ECFs developed newly is reported in this paper, it has been confirmed by practical tests that they are applicable as shielding materials to avoid industrial problems introduced by electromagnetic fields and waves. Noise radiated with discharges of high voltage as well as electrostatic charged materials and induction currents from commercial power lines of high voltage have been shielded by cloths made of ECFs.

Conclusions

Electrically conductive fibers developed newly were applied to a few kinds of fabric-ware required for the electrostatic safety in industry, and their anti-static effects were investigated experimentally.

Furthermore, in view point of the practical circumstances, the results obtained were considered, and the following conclusions were reached.

(1) Electrically conductive fiber composed mainly 4 carbonaceous cores, ECF-A, is useful to solve electric shock and ignition hazards caused by electrostatic discharges.
(2) Electrically conductive fiber plated with nickel, ECF-B, is effective to reduce safely a static electricity accumulated already or generating in great quantity on materials.
(3) By means that ECFs, both ECF-A and ECF-B, of a few percent in weight are mixed into the fabric-ware, a static electricity on the fabric-ware is eliminated and potential is also suppressed on a safe level.
(4) ECF-A is suitable to be applied to outer clothing, such as garment for working and protecting against cold, and ECF-B to fabric-ware in industry, such as carpet, bag filter and conveyer.
(5) ECFs are available to neutralize a static electricity on charged materials, but careful considerations are required to evaluate the electrostatic safety of systems which ECFs are applied.

References

(1) Y. Tabata: Research Report of Industrial Safety Institute, RIIS-RR-18-5 (1970)
(2) Y. Tabata: Proceedings of 2nd International Colloquium on Safety Work, 57, Berufsgenossenschftliche Institut fuer Arbeitssicherheit, Bonn and Hennef (1985)
(3) Y. Tabata and S. Masuda: IEEE Transactions on Industry Applications, IA-20, 1206 (1984)
(4) Y. Tabata: Technocrat, 18, 10 (1985)
(5) Y. Matsuo and Y. Tabata: Proceedings of Annual Conference on Electrostatics, Japan, 235 (1987)
(6) Industrial Safety Institute, Ministry of Labour: Recommended Practice for Protection against Hazards Arising out of Static Electricity in General Industries, 121 (1978)

Proceedings of International Conference on Modern Electrostatics

CONTROL OF INCENDIARY DISCHARGES OCCURRING FROM ELECTROSTATIC ELIMINATOR

Yasuyuki TABATA*, Teruo SUZUKI** and Shohnosuke KAMACHI***

*Industrial Safety Institute
Ministry of Labour
5-35-1, Shiba, Minato-ku
Tokyo, Japan 108

**Kasuga Denki Inc.
2-16-18, Higashi-Kamata
Ohtaku, Tokyo, Japan 144

***Technical Institution
of Industrial Safety
1-4-6, Umezono, Kiyose
Tokyo, Japan 204

Abstract

An active electrostatic eliminator of corona type is available to reduce static electricities on non-conductive materials. It, however, causes rarely incendiary discharges in explosive atmosphere, which happens to go on fires and explosions in industry. The purpose of this research is to investigate incendiary discharges from the eliminator and to control safely them. Experiments were made on corona discharges from eliminators drived by high voltage power sources of AC and HF to specify their safe operations, respectively. Experimental results obtained are presented in this paper.

Introduction

An active eliminator of corona type generally consists of a corona electrode and high voltage power source. It is classified into two kinds of eliminators by the high voltage power source: one consists of AC and the other HF power sources (1). The former has generally a high elimination quality and is applicable conveniently to various charged materials, such as films, papers and cloths. The latter operates in low voltage and is suitable to reduce uniformly electrostatic charges on materials running at high speed over about 100 m/min (2).

On the other hand, the corona electrode is generally made of needles, and each needle is connected to the high voltage power source directly or via a capacitor. In the eliminator of the capacitive coupled electrode, corona currents are limited to suppress damages to a minimum even though a needle is accidentally grounded (3).

From backgrounds described above, a capacitive coupled electrode has been widely used in both corona eliminators drived by AC and HF power sources in hazardous environments. However, the capacitive coupled eliminators don't always operate safely in explosive atmosphere, and there have been not a few accidents caused by their unusual operations (4,5).

The purpose of this study is to determine specifications for developing an active eliminator operating safely in explosive atmosphere. Experiments have been made on the corona discharges occurring from both eliminators drived by AC and HF power sources to control the incendiary discharges caused by their abnormal operations.

Experimental

Apparatus and Procedure The apparatus used in experiments consists of a rod-like eliminator "E" of about 30 cm in available length and ignition chamber as shown in Fig. 1. The eliminator was drived by AC (50 Hz) or HF(20 kHz) high voltage power source "PS", where a control system shown in Fig. 1 was attached to detect and suppress abnormal discharge currents transitting to incendiary ones, and besides, a grounded metal plate "G", length of about 50 cm, width of 20 cm and thickness of 0.01 cm, was set at the distance of 2 cm from the eliminator as a dummy load.

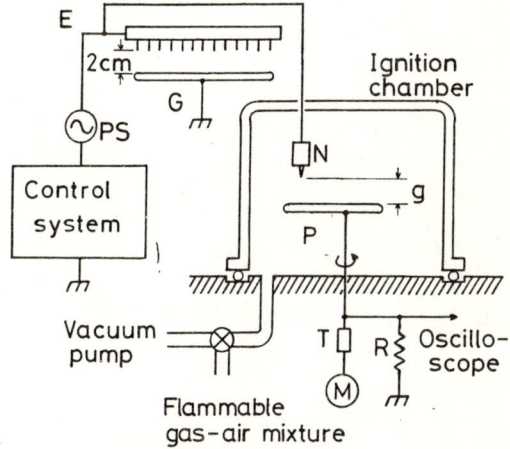

Fig. 1 Experimental set-up.

The chamber of about 260 cm^3 in volume was filled with explosive gas-air mixture or dried air of 1 atm, and in the inside, a piece of needle "N" involving a coupling capacitor and metal plate without edges "P", length of 30 mm, width of 20 mm and thickness of 3 mm, were spaced out length g apart, respectively. This needle with coupling capacitor was appropriated for one of needles of the test eliminator as shown in Fig. 1.

On the other hand, the rectangular metal plate with an axis at the center-off was set up to trigger artificially hazardous discharges from a needle inside the chamber, and was rotating at 60 rpm so that the relation between discharge characteristics and incendivities could be examined under various conditions, because the eliminator operates safely and neutralizes electrostatic charges under normal conditions but it occasionally takes place hazardous discharges under abnormal conditions such as grounded materials approach needles or the voltage of the power source rises pulsively up.

Results First, preliminary experiments were made

on corona discharges occurring from a neele of the eliminator without safety control system to investigate dangerous discharges under conditions that the metal plate was spaced at the constant gap from a needle and not turnning in the chamber filled with air. It became clear that corona discharge currents "I_d" depended primarily on a voltage "v", frequency of the power source "f", coupling capacity "C" and gap length "g", and larger pulsive discharges with light and sound appeared at random under positive half cycles and gap length below about 5 mm in 50 Hz and about 15 mm in 20 kHz, respectively. Figures 2, 3 and 4 are one of results obtained from experiments performed using a needle with coupling capacity of 4.2 pF set in gap length of 0.5 mm, and show that large pulsive discharges appear even at lower voltage in high frequency.

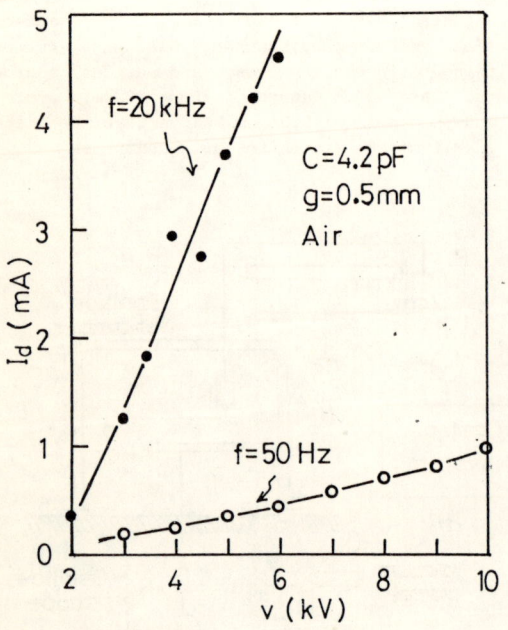

Fig. 2 Current-voltage curves of corona discharges.

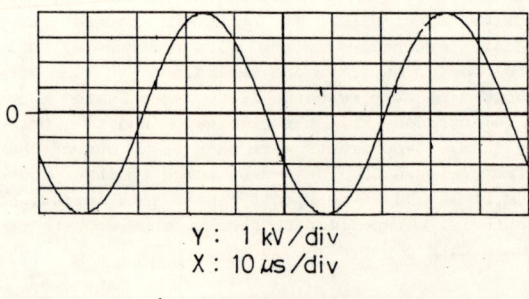

(a) Voltage wave form

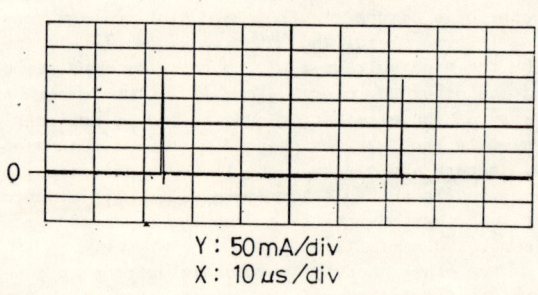

(b) Current wave form

Fig. 4 Voltage and current wave forms in 20 kHz.

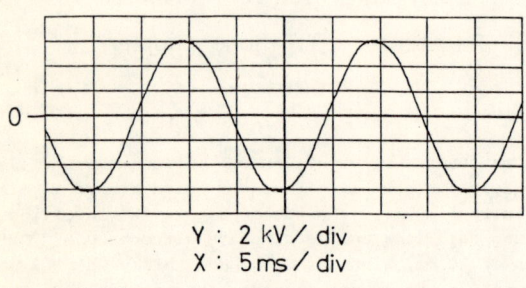

(a) Voltage wave form

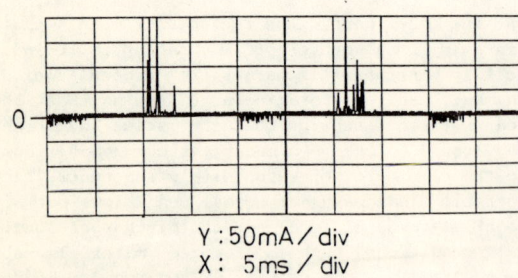

(b) Current wave form

Fig. 3 Voltage and current wave forms in 50 Hz.

Furthermore, it was observed by turning the metal plate that large pulsive discharges were triggered and longer discharge channels with light were formed between a rotating metal plate and needle in the eliminator driven by high voltage of 20 kHz when the metal plate approached or left a needle and in the eliminator of 50 Hz when the metal plate approached, respectively. Peak values of pulsive discharge currents are shown in Table 1, and these large pulsive discharges seemed to be such dangerous discharge as incendiary.

Next, based on the research results outlined above, large pulsive discharges should be noted to design a safe eliminator. Similar experiments were made in the chamber filled with explosive gas-air mixture, hydrogen and ethylene-air, to examine whether large pulsive discharges were ignitable or not, respectively; as a result, it was confirmed that ignitions of explosive gas-air mixtures were caused by large pulsive discharges from a needle and concerned closely with a coupling capacity, applied voltage and frequency of the eliminator. Furthermore, the minimum ignition voltages against the coupling capacities were

Table 1 Peak values of pulsive discharge currents.

Frequency of PS (Hz)	Coupling Capacity (pF)	Voltage of PS (kV)	Peak value of Pulsive Current* (A)
50	4.2	6.0	0.23 - 0.26
		8.5	0.23 - 0.26
		10.0	0.25 - 0.30
	11.0	6.0	0.60 - 1.0
		8.5	0.70 - 1.0
		10.0	0.70 - 1.2
20,000	3.3	2.4	0.76 - 0.92
		4.0	0.23 - 0.32
		6.0	0.25 - 0.40
	4.2	2.4	0.80 - 1.0
		4.0	0.35 - 0.45
		6.0	0.30 - 0.50

* Value was obtained from experiments carried out by turning metal plate spaced at 0.5 mm from needle.

identified only for the eliminator driven by high voltage of 50 Hz from these ignition experimrents. Results obtained were shown in Table 2.

However, in the eliminator driven by high voltage of 20 kHz, even though the voltage applied to a needle was low such as 3 - 4 kV, incendiary discharges of the high ignitability were come from a needle when the rotating metal plate approached or left a needle, and then it was also made clear that the peak value of pulsive discharge currents in the eliminator of 20 kHz was generally times larger than that in 50 Hz.

Table 2 Minimum ignition voltage in eliminator driven by high voltage of 50 Hz.

Coupling Capacity (pF)	Minimum Voltage for Igniting (kV)					
	Hydrogen-Air (vol%)					Ethylene-Air 6.5 (vol%)
	20	30	40	50	60	
8.5	2.1	2.2	2.8	3.7	4.9	6.3
7.4	3.0	2.7	3.1	3.8	5.1	6.7
5.8	3.5	3.8	4.5	5.4	7.1	9.2
3.9	6.2	6.0	5.7	7.4	>10	>10
1.7	>10	>10	>10	>10	>10	>10

Control of Incendiary Discharge

Since a capacitive coupled eliminator without safety control system was made clear to cause ignition in explosive atmosphere under abnormal operations such as a grounded conductor approached a needle, a trial, as one of safety control systems for eliminators, was carried out to control incendiary discharges occurring from the corona electrode and to avoid ignitions of explosive gas-air mixtures. This trial was a control system consisting of a corona discharge current detector and breaker, and the point was the operation that the breaker cut off the high voltage applied to a needle if a peak value of corona current exceeds the threshold adjusted already.

Experiments were made on incendivities of explosive gas-air mixtures caused by pulsive discharges to examine the safety performance of the eliminator with a tried control system in the chamber filled with hydrogen and ethylene-air mixture, respectively, and then a peak value of pulsive discharge current causing ignitions of gas-air mixture was introduced to be different in eliminators driven by high voltage of 50 Hz and 20 kHz each even though gas-air mixtures were same.

Consequently, the thresholds, peak values of pulsive discharge current that the beaker operates, were adjusted according to the eliminators of 50 Hz and 20 kHz, respectively, and the performance of the tried control system was considered experimentally and evaluated by the ignition probability obtained throughout 100 trials. Table 3 shows results obtained from whole trials, where a mark "n" represents non-ignition throughout whole trials and "i" 1 time of ignition at least, respectively.

Table 3 Suppression of ignitions by tried control system for eliminator.*

Frequency of PS (Hz)	Applied Voltage (kV)	Hydrogen-Air (30 vol%)			Ethylene-Air (6.5 vol%)		
		Coupling Capacity (pF)					
		5.8	7.4	8.5	7.4	8.5	11
50**	6.0	n	n	i	n	n	n
	7.0	n	i	i	n	n	n
	8.0	i			n	n	n
20,000**	3.0	n	i	i	n	n	n
	4.0	i	i		n	n	i
	5.0				n	i	i

* Suppression effect of tried control system was examined in case ignitions if no control system attached.
** Thresholds were set on 0.2(0.5) A for 50 Hz and 0.4(1.0) A for 20 kHz in peak value of pulsive currents, respectively. Values in parenthesis was for ethylene-air mixture.

Experimental results shown in Table 3 suggested that the tried control system was effective to control hazardous discharges from the eliminator caused by the abnormal operations. Especially, the tried control system seemed to be indispensable for suppressing incendiary discharges from the eliminator driven by high voltage of 20 kHz, since large pulsive discharges from it were mostly incendiary and it impossible to be controlled by the voltage applied to a needle and coupling capacity.

Conclusions

Corona discharges of the capacitive coupled eliminator were investigated experimentally in air to control dangerous ones occurring under abnormal operations such as a grounded conductor approached a needle, and a trial was also made to suppress them using the safety control system. Furthermore, the safety performances of the eliminator with safety control system were examined in explosive gas-air mixtures. Conclusions obtained from these experiments are as follows:

(1) Hazardous discharges, especially incendiary ones in explosive atmosphere, from the capacitive coupled eliminator are liable to occur when a grounded conductor approaches and/or leaves a needle, one of the corona electrodes.
(2) In the eliminator drived by high voltage of 50 Hz, incendiary discharges are primarily controlled by both voltage applied to needles and coupling capacity.
(3) In the eliminator drived by high voltage of 20 kHz, available corona discharge currents for neutralizing electrostatic charges are gained in lower voltage applied to needles, such as 3 - 4 kV, but the safety control system attached will be indispensable to avoid ignition hazards since incendiary discharges are hardly controlled by the voltage and coupling capacity.
(4) As a safety control system of the eliminator, a current breaker is available to reduce safely electrostatic charges in explosive atmosphere by means that it is adjusted in the threshold level of 0.2 - 1.0 A in a peak value of corona currents.

Acknowledgements

The authors would like to thank Mr. Tsutomu Kodama, Senior researcher of Industrial Safety Institute, and Mr. Shinji Yagi, Director of Kasuga Denki Inc., for their guidance, and Mr. Yukio Mikami of our colleague for his assistances given to this research.

References

(1) Institute of Electrostatics, Japan: Handbook of Electrostaics, 821, Ohm-sha, Japan (1981)
(2) Y. Tabata, T. Kodama, S. Kamachi and Y. Mikami: Proceedings of 1987 Annual Conference on Electrostatics, 251 (1987)
(3) Y. Tabata and S. Yagi: J. Japan Society for Safety Engineering, 18, 102 (1975)
(4) Y. Tabata and T. Kodama: Technical Note of Industrial Safety Institute, RIIS-TN-74-2 (1974)
(5) T. Horveath and I. Berta: Static Elimination, 89 Research Studies Press, Chichester (1982)

A STUDY ON INCENDIVITY OF DISCHARGE FROM CHARGED INSULATING PLATE WITH DIFFERENT ELECTRIC POLARITIES

Chen Jianlin and Zhao Luzhen
Beijing Municipal Institute of Labor Protection
Beijing, China

Abstract

In this paper a study is given in considerable detail on discharge between a grounded spheric electrode, the radius of which is 10 mm, and a PVC insulating plate charged with different electric polarities. Through experiments and calculations it can be found that under the conditions of the same potential and electrode, the energy of the single main discharge from the insulating plate with positive charge is less than that of the discharge from the insulating plate with negative charge by one order of magnitude. The energy of the single main discharge of the insulating plate with positive charge should be obtained from the discharge wave forms on the oscilloscope. Generally, it is not equal to the value obtained by means of Faraday cage.

Introduction

In the past decades, a series of experimental researches have been made on discharge between a conductor and an insulator. In 1967, after studying the effect of the thickness of insulation layer covering metallic body on the discharge ignition, E. Heidelberg [1] pointed out that the discharge ignition probability of the insulator with positive electric charge was lower than that with negative electric charge. In 1972, after a study on the spark figure of the discharge of insulator, O. Fredholm [2] pointed out that there was a big difference between the incendivity of discharges of insulators charged with positive and negative polarities. In 1981, after studying the ignition of brush-shape discharge between an insulator and a grounded conductor, K.G. Lovstrand [3] pointed out that with negative charge there were various ignition rates, while with positive charge no ignition occurred in several hundreds of tests. In the same year, A.R. Blythe and G.E. Carr [4] conducted an experimental study on the discharge between a grounded pole and an insulation film lined with a grounded conductor at the back. They pointed out that when the film carried positive charge, there were pre-discharges before the main discharge; when the film carried negative charge, there was no pre-discharge. In 1983, Shi Chung Yue et al. [5] conducted experiments on the static discharge energy of oil surface with different polarities. These experiments showed that the discharge energy of oil surface with negative polarity was more than that with positive polarity by one order of magnitude under the conditions of same surface potential. We have not found any reports of semi-quantitative study on the discharge of solid insulator with positive and negative polarities. In this paper the discharging energy of PVC round plate with charges of different polarities is given through different measuring methods, thus the ignition ability of different discharge is determined, and the phenomenon of pre-discharge of insulating plate with positive charge is studied.

Experiments and Calculations

There are many factors that affect the discharging energy and energy distribution. Besides the envionmental factors of humidity, temperature, atmospheric pressure and air flow velocity, they are mainly charged area S, surface charge density δ, radius of spheric pole a, resistivity of the insulator ρ, dielectric constant ε, charging polarity of the insulator and so on. The effects of polarities on discharge is the subject for discussion of this paper. We select a metallic spheric pole whose radius is 10 mm as the discharging pole (Experimental results show that when radius of the discharging spheric pole a=10 mm, the electric quantity of discharge is the maximum), and use PVC plate ($\varepsilon=6, \rho=10^{15}\,\Omega m$) as the insulator. To reduce the influence of polarization, we choose a round plate with a radius r=10 cm and thickness b=0.3 cm. The experimental device is shown in Fig. 1.

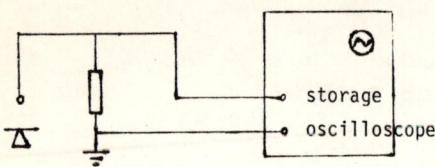

Fig. 1 Schematic illustration of measurement device

A. Calculation of the discharge energy for the PVC round plate charged in different polarities

Aim the grounded metallic sphere at the centre of the electrified PVC round plate and approach it gradually until discharge occurs. The charge energy is shown with the following equation:

$$\Delta W = W - W' \qquad (1)$$

Here W and W' are the static energies of the system before and after discharge respectively. The static energy of the system in MKS units before discharge is:

$$W = W_p + W_s \qquad (2)$$

$$\text{here, } W_p = 2\rho^2 r^3 / (3\varepsilon) \qquad (3)$$

$$W_s = \frac{6Q_s}{4\varepsilon} \{[(h-\delta)^2 + r^2]^{1/2} - (h-\delta)\} \qquad (4)$$

$$Q_s = -2\pi\alpha\delta[(h^2 + r^2)^{1/2} - h] \qquad (5)$$

In the equations $\delta = \alpha^2/(2h)$, r is the radius of the PVC round plate, δ is the density of surface charge on the round plate, Q_s is the image charge on the metallic sphere, ε_0 is the dielectric constant in vacuum, h is the distance between the centre of the metallic sphere and the centre of the round plate, α is the radius of the metallic sphere, W_p is the static energy of the isolated round plate (The round plate is very thin, effect of polarization can be ignored), W_s is the correction of static energy of the system after introducing the metallic sphere.

The discharge area can be measured by means of Lichtenger figure. From symmetry we can know this area should be a round surface whose centre is the centre of the round plate. Suppose that the density of charge outside the discharge circle holds unchanged, and the charge inside the circle is evenly distributed. In this case the static energy of the system after discharge can be given from the following equations:

$$W' = W_p' + W_s' \qquad (6)$$

$$\text{here, } W_p' = \frac{2}{3\varepsilon_0}\delta^2 r^3 + \frac{2}{3\varepsilon_0}(\delta' - \delta)^2 r'^3$$

$$+ \frac{2}{3\varepsilon_0}\delta(\delta'-\delta)r^3\{[1+(r'/r)^2]E(r'/r)$$

$$- [1 - (r'/r)^2]K(r'/r)\} \qquad (7)$$

$$W_s' = \frac{Q_s'}{4\varepsilon_0}\{\delta[\sqrt{(h-\delta)^2 + r'^2} - (h-\delta)]$$

$$+ \delta[\sqrt{(h-\delta)^2 + r^2} - \sqrt{(h-\delta)^2 + r'^2}]\} \qquad (8)$$

$$Q_s' = -2\pi\alpha\{\delta'[\sqrt{h^2+r'^2} - h] + \delta[\sqrt{h^2+r^2} - \sqrt{h^2-r'^2}]\} \qquad (9)$$

Here δ' is the residual density of surface charge in the discharge area, r' is the radius of the discharge round area, Q_s' is the image charge on the metallic sphere after discharge, K is the first kind complete ellipse integral function, E is the second kind complete ellipse integral function, W_p' is the residual static energy on the isolated round plate after discharge (polarization is ignored), W_s' is the correction to static energy of the system due to existence of the grounded metallic sphere after discharge. Some of the physical quantities included in formula (7) and formula (8) can be directly measured during discharge. Others can be measured before or after discharge.

B. Measurement of the discharge energy of positively charged PVC round plate by the discharge wave forms

The discharge of the PVC round plate with positive electricity is multi-pulse discharge. Many discharge pulses may appear in the movement of the sphere toward the plate until it is stopped. The discharge electric quantity measured with the Faraday cage will not be the result of a single discharge. In the experiment we recorded the wave forms of each discharge with a storage oscilloscope and analyzed the wave forms so as to obtain the electric quantity of any single discharge.

If the resistance to the earth in the discharge measurement is R_s, the voltage function of time is V(t), then the discharge electric quantity showed by the discharge wave forms is:

$$\Delta Q_1 = \int_0^\infty I(t)dt = \frac{1}{R_s}\int_0^\infty V(t)dt \qquad (11)$$

In ΔQ_1 the change of image charge on the sphere is not taken into account. So ΔQ_1 is not the actual electric quantity of single discharge. Before the image charge on the sphere is:

$$Q_S = -2\pi\alpha\delta[(h^2+r^2)^{\frac{1}{2}} - h]$$

After discharge the image charge on the sphere is:

$$Q_S' = 2\pi\alpha\{\delta[(h^2+r^2)^{\frac{1}{2}} - h] + [(h^2+r^2)^{\frac{1}{2}} - (h^2+r'^2)^{\frac{1}{2}}]\}$$

In order to strictly find out the electric quantity of single discharge, calculation should be made by iteration between ΔQ_1, Q_S and Q_S' with a computer. But, because the electric quantity of single discharge is very small, when the PVC round plate possesses positive charge, the change of image charge is also very small. So we use ΔQ_1 instead of the actual electric quantity of discharge ΔQ during the calculation of discharge energy.

Experimental Results and Discussion

We measured with two methods and recorded the discharge between a grounded metallic sphere whose radius is 10 mm and an electrified PVC round plate whose radius is 10 cm. The results of measurement by means of Faraday cage are given in Tables 1 and 2, in which h is the distance between the centre of the sphere and the centre of the round plate.

V_0(Kv)	-46.0	-50.0	-56.0	-60.0	-66.0	-72.0
h(cm)	3.60	4.80	5.00	5.50	6.20	6.50
Q(10^{-8}C)	-36.8	-40.0	-44.8	-48.0	-52.8	-57.6
r'(cm)	4.6	4.6	5.6	6.0	6.2	6.2
Q'(10^{-8}C)	-27.8	-26.5	-33.2	-30.7	-31.4	-33.1
ΔQ(10^{-8}C)	-9.0	-13.5	-11.6	-17.3	-21.4	-24.5
ΔW(mJ)	3.756	5.656	6.378	9.383	12.61	15.282

Table 1: T=31°C, rh=34%, R=2KΩ

V_0(KV)	27.6	35.3	48.5	55.9	68.5	72.3
h(cm)	2.4	3.2	3.4	3.6	4.4	4.8
Q(10^{-8})	22.1	28.2	38.8	44.7	54.8	57.8
r'(cm)	1.70	2.00	2.30	2.50	3.00	3.10
Q'(10^{-8})	14.9	18.5	21.1	29.8	36.2	40.0
ΔQ(10^{-8})	7.2	9.7	17.7	14.9	18.6	17.8

Table 2: T=23°C, rh=30%, R=2KΩ

If we ignore the change of image charge on the metallic sphere after discharge, then the results of discharge from the discharge wave-forms corresponding to Table 2 are shown in Table 3.

V_0(Kv)	27.6	35.3	48.5	55.9	68.5	72.3
h(cm)	2.4	3.2	3.4	3.6	4.4	4.8
Q(10^{-8}C)	21.1	28.2	38.8	44.7	54.8	57.8
ΔQ(10^{-8}C)	0.75	1.50	1.00	1.75	1.85	2.00
ΔW(mJ)	0.14	0.35	0.33	0.65	0.85	0.97

Table 3: T=23°C, rh=30%, R=2KΩ

If we increase the resistance value of the grounded resistor R, then we can monitor the pre-discharge phenomenon of the insulating plate with positive charge with the oscilloscope. When R=100 K, the results of measurement are given in Table 4.

V_0(Kv)	49.0	61.2	68.5	70.0	75.0	78.5
h(cm)	8.5	10.5	11.0	12.0	12.5	13.0
Q(10^{-8}C)	39.2	49.0	54.8	56.0	60.0	62.8
Q(10^{-8}C)	0.0050	0.0045	0.0050	0.0035	0.0055	0.0015

Table 4: T=17°C, rh=42%, R=100KΩ

From Tables 1 and 2 we can see that when the PVC round plate carries charges of different polarity, the discharge electric quantities measured with a Faraday cage are of the same order of magnitude. Therefore, the discharge energies are also of the same order of magnitude. Comparing Table 2 with Table 3, we can see that for discharge of round insulating plate with positive charge, the discharge electric quantity measured with Faraday cage is about 10 times of that measured with discharge wave forms. This is because the ΔQ in Table 3 measured and calculated from the discharge wave forms is the electric quantity of the single main discharge pulse, while that measured with Faraday cage is the total effect of the whole pulse discharges. Therefore, when calculating the single discharge energy of the insulating plate with positive charge, we should use the results measured from discharge wave forms. Table 4 shows that for PVC round plate with positive charge, when the gap is big enough, there exists pre-discharge, the electric quantity of the single pre-discharge pulse is very small.

From the above results we can see that for the insulator plate with negative charge, the discharge electric quantity can be measured with Faraday cylinder. Then the discharge energy can be obtained according to formulas (1) to (9). For the insulator plate with positive charge, the single main discharge energy can be simply found with the following formula:

$$\Delta W = V_0' \cdot \Delta Q \tag{12}$$

Here, ΔQ is the single main discharge electric quantity calculated according to the discharge wave forms. V_0' is the potential of the centre of the round plate before this main discharge. From Tables 2 and 3 we can see that the first main discharge emerges when the charge quantity of the insulator round plate has become about 2/3 of the initial electric quantity Q, so the formula (12) can be rewritten as:

$$\Delta W = \frac{2}{3} V_0 \cdot \Delta Q \tag{13}$$

Here, V_0 is the potential at the centre of the round plate measured with a static potentiometer after the round plate is charged. The results of the calculation of the discharge energy are given in Tables 1 and Table 3.

We can see that for the same potential at the centre of the insulator plate, the discharge energy for the negatively charged plate is larger than the single main discharge energy of the plate with positive charge by one order of magnitude. So, we come to the conclusion that the discharge of a PVC round plate with negative charge has stronger incendivity than that with positive charge of same density.

Conclusions

The results of our experiments show that the difference of the discharges of a PVC round plate charged in different polarities moving toward the metallic sphere with radius a=1.0 cm are as follows:

1) The gap of initial discharge of the round plate with positive charge is larger than that of the round plate with negative charge, the former is about two times of the latter.
2) The discharge gap of the round plate with negative charge is larger than that of the main discharge gap of the round plate with positive charge.
3) When the round plate is negatively charged, the discharge is a single discharge; when the round plate possesses positive charge, the discharge is multi-discharge.
4) The discharge energy of the round plate with negative charge is larger than the single main discharge energy of the round plate with positive charge by one order of magnitude.

References

1. Heidelberg, E., Static Electrification, Inst. Phys. Conf., Ser No.4 (1967) 147-53.
2. Fredholm, O. and Lovstrand, K.G., J. Sci. Instr., 5(1972) 1058-62.
3. Lovstrand, K.G., J. Electrostat., 10(1981) 161-68.
4. Blythe, A.R. and Carr, G.E., J. Electrostat., 10(1981) 321-26.
5. Shi Chong Yue et al., Safety potential of the oil surface of hydrocarbon oil.

SIMULATED ELECTROSTATIC TEST FOR THE FUEL TANK OF Q-6 PURSUIT

Luo Hongchang, Wang Liqun
Shanghai Ship & Shipping Research Institute
Ministry of Communications
Electrostatic Test Group
Aeronautical Engineering Division
Naval Aviation, China

Abstract

This paper describes the electrostatic situation inside the fuel tank and analyses some of the phenomena according to the simulated electrostatic test which was carried out to study the explosion reasons of the fuel tank of Q-6 pursuit. It also demonstrates that the explosion was caused by the fuel flowmeter. When the fuel tank rocks with the pursuit, the float impact the anti-sloshing cloth so that bringing about electrostatic discharge at the moment the float leaves the anti-sloshing cloth after touching it. Then the spark caused by the discharge ignites the mixture of the volatile matter and air, thus leading to explosion.

Besides, the paper gives a simple description on the test carried out to improve the fibres without anti-electrostatic features and its results.

Fuel tank explosion of Q-6 pursuit on many occasions occurs at the time of braking after landing. Study on these accidents shows that when the cover of the fuel tank is immediatly opened after landing to reduce the temperature, it actually brings some air into the tank and dilutes the over-concentrated gas inside the tank. This results in a mixture of gas which is within the explosion limit. Then the most posibility for fire is the spark caused by the electrostatic discharge.

How is the electrostatic situation inside the tank? Where is the electrostatic discharge? How much is the energy discharged? In a word, the simulated electrostatic test for the fuel tank of Q-6 pursuit was carried out to see whether the explosion is caused by the spark from electrostatic discharge.

I. Equipments and methods for the simulated test

The fuel tank of Q-6 pursuit was used in the test. It is fixed on a testbed made of steel as it is fixed on the aircraft. There were six wheels in the lower part of the testbed specially used to rock the tank, three windows with plexiglass in the upper part for observation and three holes in each side of the tank for operation. The complete testbed was installed in a specially sealed automobile. The humidity of this automobile was controlled by a hygroscopic system. When the test was carried out, approximately 750-800 litres of kerosene was inputed to the tank to simulate the actual fuel condition before explosion. Then the fuel tank was rocked manually.

The electrostatic potential and the electrostatic capacity of those measuring points such as fuel surface, float of the flowmeter and the anti-sloshing cloth, as well as the distributed capacity of the wire, were continuously measured by a VC electrostatic meter developed by Shanghai Ship & Shipping Research Institute (SSSRI). It can automatically record the curve of the electrostatic potential when working with a photo-oscilloscope.

VC Electrometer is specially designed to measure the electrostatic potential and the electrostatic capacity directly, and then calculate the electrocharge and energy according to the measurement. The error of the measurement is less than 5%. The electrometer operates with either battery or DC regulator, which is easy to use.

Since the wire used in the test was quite long and the distributed capacity was quite large, the reading for the electrostatic potential on the electrometer was quite low and needed to be appropriately corrected.

Suppose the actual electrostatic potential of a measuring point is Vo, electrostatic capacity is Co, the distributed capacity of the wire measured is C', and the electrostatic potential displayed on the electrometer is V, then

$$V_o = \frac{C_o + C'}{C_o} \times V \quad (V)$$

The time constant for the electrostatic discharge can be taken from the time scale which shows the attenuation period of the curve recorded by the photo-osciloscope. The attenuation period is the period from the initial Vo to 0.37 Vo.

The electrostatic discharge energy W can be obtained with the following equation:

$$W = 1/2 \, C_o \, (V_1^2 - V_2^2)$$

where
- C_o ---- electrostatic capacity at the measuring points, (F);
- V_1 ---- electrostatic potential before the electrostatic discharge at the measuring points, V;
- V_2 ---- residual electrostatic potential after electrostatic discharging at the measuring points, V.

II. Electrostatic Situation Inside the Fuel Tank and Its Analysis

When the fuel tank is rocking, the fuel inside will impact and splash the tank wall, the anti-sloshing cloth and the float. Besides, the fuel itself will stir. All these impact, splashing and stiring will easily cause static electricity.

Test result showed that not only there was an electrostatic charge on the fuel inside the tank, the anti-sloshing cloth and the plastic float but also the electrostatic potential was quite high. Fig.1 shows the arrangement inside the tank, the installation of metal float and the serial number of the measuring points on the anti-sloshing cloth. Whereas the electrostatic capacity and the maximum electrostatic potential are shown in Table 1.

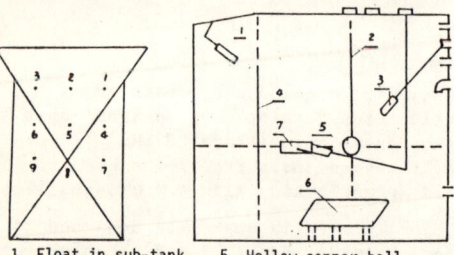

1. Float in sub-tank
2. polyemide fibre
3. Float in fuel tank
4. Anti-sloshing cloth
5. Hollow copper ball
6. Tank for inverted flight
7. Flowmeter

Fig.1. Serial number of the measure points in the tank, float ball and anti-sloshing cloth.

During the simulated test, the ambient temperature was controlled under 30°C, humidity over 40%, for the reason of safety. It is easy to image that the creation of static electricity is influenced by the factor such as the time and the difficulty of the high-altitude acrobatic flight, rolling, the sudden impact of fuel tank, humidity and temperature, etc. The more these factors influence, the higher the static-electricity occurs on each part, and the more possible the electrostatic discharge happens.

The addition of anti-electrostatic additives to the kerosene will not have distinct influence on the magnitude of the static electricity of the fuel, the anti-sloshing cloth, etc. The only effect of the additives is to change the electro-pole of the anti-sloshing cloth and shorten enormously the dissipating period of the static electricity. (See Table 2.) When a certain amount of anti-electrostatic additive is added into kerosene, it will do some help to prevent the accumulation of static electricity. But it will not prevent the creation of static electricity. What it does is simply increase the electric conductivity of kerosene to dissipate quickly the electro-charge caused by friction so that to prevent the accumulation of the electrostatic charge. However, the electric conductivity of the kerosene with additives is not immutable. It weakens in the course of storing and transporting of the kerosene. For example, it was told that the electric conductivity of kerosene was less than 200 cu. While what we meausred was 147cu, and later 88cu.

Besides, there were other things discovered in the test:

1) The static electricity on the surface of the fuel inside the tank will not change, no matter there is a piece of anti-sloshing cloth or not. This means the static electricity on the fuel surface is caused mainly by the impact and the friction between fuel and the tank wall, not between the fuel and the anti-sloshing cloth.

2) Those flowmeter with foamed plastic float or metal float will cause high static-electricity (See Table 3). Comparatively, the foamed plastic float will bring higher static electricity and stronger electro-discharging spark to the fuel surface. So it is easier to cause explosion. If a float with little electrification is required, it is better to use a float of subconduct.

III. The Position and the Magnitude of Electrostatic Discharge

It is proven in the test that static electricity is existing in the various parts of the fuel tank. The electrostatic potential of some parts is quite high. Additionally, electro-discharge was found from time to time.

1. The position where electrostatic discharge exists

When the electrostatic potential and its change were read and recorded during the test, a sudden drop of the electrostatic potential often occurs (not to zero) at the measuring points of the fuel surface and the anti-sloshing cloth. This kind of drop is caused by the nutral release of the electro-charge during the electro-discharge. But as the fuel and the anti-sloshing cloth are insulators it is impossible for static-electricity to be completely discharged. There always is some residuel, so it will not drop to zero. (See Fig.2)

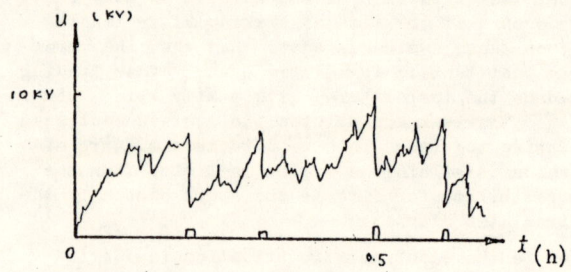

Fig.2. Curve of the electro-discharge on the anti-sloshing cloth

The electro-discharge spark was also observed during the test at night. However, the electro-discharge, according to both the record and the observation, neither happened between the float and the rod nor between the float of sub-tank and the anti-sloshing cloth. It always happened at the moment when the float left the anti-sloshing cloth after touching it. This is identical to the test result that the highest electrostatic potential always exists in this part of the anti-sloshing cloth and there was a sudden drop from time to time.

To validate the measurment and the observation, two more measurements are taken for verification: (1) A piece of anti-sloshing cloth was cut off from the part where the float touches; (2) The anti-sloshing cloth was kept intact, while the float was moved to a place far from the anti-sloshing cloth. Both of the measures aimed at keeping the float not to touch the anti-sloshing cloth while rocking. A number of tests showed that not only the electrostatic potential on the anti-sloshing cloth and the float was low, but also no electro-discharge was found in either the record or the observation. This showed clearly that the electro-discharge in the fuel tank occurs only between the float and the anti-sloshing cloth. It is because when the float impacts the anti-sloshing cloth during rocking, i.e. touching-friction-leaving, electro-discharge at high electrostatic potential will instantaneously happen, which leads to ignition and explosion.

2. Energy of electro-discharge

Normally, when a pursuit is rolling, the electrostatic potential on the various parts of the fuel tank goes up gradually. Even when it reaches a certain value which leads to electro-discharge, the electro-potential will not have a large drop. The drop normally is about 1/2 - 3/5. Therefore, the energy discharge is not large either. The discharge spark observed is small blue circle. But if a sudden rock which happens after the former one causes an impact, the electrostatic discharging observed on the anti-sloshing cloth will suddenly rise to about 10,000v. The discharging spark appears as a bright straight light of about 1cm. The energy of this kind of spark is high. It will ignite the mixture of kerosene vapour and air.

For example, the maximum electrostatic potential measured on an electro-discharging point of the anti-sloshing cloth is 9,300v. The electrostatic capacity of that point is 4 pF. The residual voltage after discharging is taken as 1/10. Then the energy of the discharging spark is about

$$W = 1/2 \times 4 \times 10^{-12} \times (9300^2 - 930^2)$$

$$= 0.17 \text{ mJ}$$

Normally, the minimum igniting energy of the explosive mixture of kerosene gas and air is considered as 0.2 mJ. 0.17 mJ seems less than the minimum igniting energy. Whereas the following factors should be taken into account:

(1) When an explosion happens in the fuel tank of Q-6 pursuit, the temperature inside the fuel tank is still 40-50°C and the ambient temperature is about 39%. While the ambient temperature is less than 30°C, and the humidity is about 50% in the simulated test. Actually, when a pursuit brakes in landing, its fuel tank will have a larger impact than it had in the simulated test.

(2) The minmum igniting energy of the mixture has a close relation with the temperature. When the temperature rises, not only the range of the explosive limit will increase, but also the minimum igniting energy will decrease relatively. That is to say, it is more possible to cause explosion.

(3) It is said, little anti-corrosion additive should be added to the kerosene to prevent the engine from corrosion. The main content of the additive is CS_2. However, the minimum energy of the CS_2 volatile gas is only 0.009 mJ.

In a word, the electrostatic spark has enough energy to ignite the explosive mixture of the kerosene gas and the air inside the fuel tank at the time.

IV. Improvement of Anti-Sloshing Cloth

The simulated test proves that the insulated anti-sloshing cloth inside the fuel tank is the key factor which leads to the explosion. To keep the functions originally designed for the cloth, (1) It is imposible to remove the cloth from the tank nor to cut off the cloth to prevent impact and friction with the float, either from the view point of strength or from the view point of safety; (2) The cloth can not be instead by metal plate; (3) The rod of the float can not be shortened or installed far away. The most realistic and effective measure is to improve the conduct function of the anti-sloshing cloth.

Then, three kinds of anti-sloshing cloth --- the original cloth without anti-static-electricity feature, capron fibre without anti-static-electricity feature and the capron fibre with anti-static-electricity feature and conducting rubber were tested and compared time and again. The results of these tests are shown in Table 4.

From the results, we can see clearly that the insulated function of the original cloth without anti-static-electricity feature is good. Its surface resistance is $2.5\text{-}16 \times 10^{11} \Omega$ and the volume resistance is $3\text{-}11 \times 10^{10} \Omega$. When it is with the same plastic float and the same way of rocking, the electrostatic potential and the discharging time constant are 3 to 5 times as large as that of the capron fibre with anti-electrostatic feature. Oppositely, the electrostatic potential of the capron fibre with anti-electrostatic feature and conducting rubber is as little as zero. According to the test carried out by Shanghai NO.2 Rubber Product Factory, the capron fibre with anti-electrostatic feature and conducting rubber is oil-proof and its strength meets the requirements concerned. After a continuous test of over one month, the fibre is proven to be stable and reliable in its performance. Moreover, it is not expensive.

* * * * * * * *

The naval Aviation took a number of measures according to the results of the simulated test. After that, no explosion of fuel tank happened in the last three or four years.

Table 1. Electrostatic capacity and potential on each measuring point

Item	Surface	Float			Anti-sloshing Cloth									Tank Wall	
		Flow meter	Sub-tank	Tank No.2 3, 4	I	II	III	IV	V	VI	VII	VIII		Front	Back
Electrostatic potential (V)	- 4200	200	8000	1600	1800	840	500	570	2508	9167	9300	4340		67	27
Electrostatic capacity (pF)	8	5	3	3	5	5	10	5	5	3	4	4		10	10
Environment	21°C 51%	28°C 63%	22°C 48%	28°C 63%	26°C 50%	26°C 50%	26°C 50%	21°C 67%	21°C 67%	28°C 62%	23°C 52%	23°C 52%		22°C 48%	22°C 48%

Table 2 Comparison between kerosenes with and without additives

Item	Oil Surface		anti-sloshing Cloth	
	Maximum potential (v)	Discharge time constant (s)	Maximum potential (v)	Discharge time constant (s)
With additive	-4200	4	+7400	28.5
Without additive	+6825	36	-9100	656

Table 3. Static electrification of different floats

Item	Max. potential on oil surface (v)		Max. potential on the cloth (v)	
	unanti-static cloth	anti-static cloth	unanti-static cloth	anti-static cloth
Metal Float	-2040	-3022	+6200VII	+1594 IIX
Plastic Float	-4200	-3928	+9166IIV	+4253 IIX

Table 4. Potential and time constant for different cloth

Item	unanti-static capron		anti-static capron		anti-static rubber capron	
	I	II	I	II	I	II
potential (v)	+744o	+6665	-2015	-2015	+6	+10
time constant for discharge (s)	28.5		6			

THE RESEARCH ON ELECTROSTATIC SPARK SENSITIVITY IN SUSPENDING BLACK POWDER DUST

Qian Chong, Wu Guirong, Ren Xiaoling
(Shaanxi Applied Phys.-Chem. Res. Inst. P.O. Box 99 Xi'an, Shaanxi, China)

ABSTRACT

The powder explosion hazards caused by static discharges have been generally recognized and paid great attention.

Black powder is a more sensitive mixture its dust content in production process is higher than any other one. And, the accident and hazard caused by electrostatic discharge or other causes would be more than others. In order to find out the ignition state of suspending dust of black powder caused by electrostatic discharge spark, and to furnish valuable data and theoretical foundation for guarding against accident, a research on electrostatic spark sensitivity in suspending black powder dust became an important subject

This article mainly presents the ignition of suspending powder dust by the energy of static discharge spark in explosive device, and lays particular emphasis on expressing the principle of test circuit and test method of suspending dust, as well as the factors affecting dust explosion or the optimum test conditions, test results, analysis and conclusions, thereby, to gain the minimum ignition energy and the minimum explosive concentration of suspending black powder dust with the test device.

1. INTRODUCTION

The report about people's recognition to the hazard of dust explosion can be traced back to the end of eighteenth century[1]. While the research on dust explosion is the thing happened within recent several ten years. Electrostatic discharge as a igniting source is giving rise to dust explosion or causing more accidents along with the development of industrial technology. Before the seventies, a vast amount of test data have been obtained by the Mineral Bureau of the United State of America after going through an explosive research on non-explosive material by simulated electrostatic discharge. Test criterion have been set up based on Hartman's Apparatus[2][3][4].

Economic losses and personnel injuries and deaths brought about by powder dust explosion caused by electrostatic discharge are more than those who are non-explosive materials. Research on it takes a slow pace. The Mineral Bureau of U.S.A. took an early time undergoing the research on black powder dust containing wax[5]. The sensitivity data of TNT have been reported by Romania Electrostatic Safety Standard[6]. In June 1979, Danger Corporation of Denville, New Jersey State of U.S.A. published a researching report, which introduced the relationship between detonation sensitivity of some explosives and influence factors. The sensitivity order has been given as well[7]. Explosive dust explosion brought about great losses under production. Accident caused by electrostatic discharge can not be underestimated.

Black powder is a more sensitive mixture. A large amount of dust can be produced during production. Therefore, the probability of accident happening is more than others. In order to solve the problems in relation to safety, to lay down the preventive measures so as to furnish theoretical basis, it is, thereby, urgent and necessary to investigate the electrostatic spark sensitivities of black powder suspending dust.

2. PRINCIPLE OF TEST CIRCUIT

The detonating method of test research on dust explosion is to simulate electrostatic discharge spark. To evaluate the safety of electrostatics, the method requires explosive devices to be both ensured the homogeneity of suspending dust state and suited the test conditions of electrostatic discharge spark. Energy accumulation capacitor may be selected arbitrarily; Discharge circuit is asked to be with high energy transfer rate, less loss, simple assembly, convenient to handle, to be used popularly and easy to operate. It is also asked to be with certain antidetonating quality, and convenient to observe. Its test principle is shown as Fig. 1.

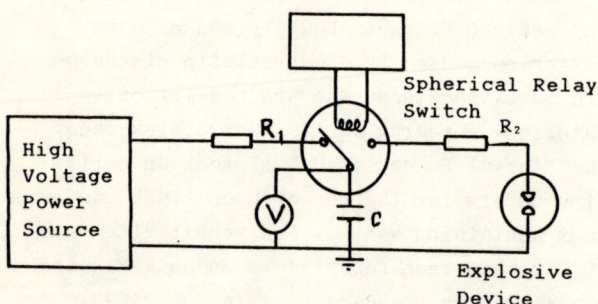

Fig. 1. Test Circuit Principle Drawing
Where in: R_1___charging resistor
R_2___discharge resistor
V ___electrostatic voltmeter
C ___charge, discharge capacitor

3. TEST METHODS

In the electrostatic spark sensitive test, different tester or electrostatic discharge circuit has distinct effect on ignite sensitivity value of black powder; and with certain tester or discharge circuit, different test conditions have great effect on ignite sensitivity. It was found in the test that the minimum ignite energy of powder dust is the related functions of its quality, state or test conditions. It possesses the character of polytropic function. Thus, ignite sensitivity data can be used as the basis to evaluate the safety of black powder after the most sensitive ignition conditions being found through optimized test.

3.1. Limitation of Flame Propagation

Two phenomenons can be found after dust explosion, one has flame propagation, the other has not. With the former, detonation forming process can be observed from slight reaction to its paper film split when black powder dust ignited. The standard to determine flame propagation is set as this: there is obvious burning flame around electrode, its length being 3omm and above.

3.2. Test Method of Minimum Ignition Energy

Dust weight in explosive devices is not changed. At this moment, the concentration is 3-5 times that of the minimum explosive concentration: set the spark gap (optimized gap being selected); air jar pressure is certain (to make optimum homogeneity for suspending dust). In this case, undergoing four times of successive test, when flame propagation occurs, ignite energy must be reduced, an other four times of successive test is done once more. In this way, do the test continuously until flame propagation will not occur. Yet, when flame propagation did occur after the above series of test done, whose energy being the smallest is the minimum ignition energy under this concentration.

3.3. Test Method of the Minimum Explosive Concentration

With the set spark gap, certain air pressure, unchanged ignition energy, let the suspending dust weight in the explosive device be changed. If four times of successive test is done with flame propagation, 5mg of suspending dust weight must be reduced. Then, do the other four times of test successively until flame propagation will not occur. Yet, when flame propagation did occur during the above series of test, whose weight being the lightest correponding to its concentration is the minimum explosive concentration under this ignition energy.

4. SELECTION OF OPTIMIZED TEST CONDITION OF BLACK POWDER SUSPENDING DUST

It is stated clearly by a host of electro static spark sensitivity test that there are more than 20 factors which have the effect on mixture (pile up) sensitivity. As for suspending dust, because of its dynamical property, there are so many influence factors in order to select the optimized test conditions, major or minor, independent or correlative influence factors must, first of all, be determined. Optimized selective ordering is to be used to constitute the optimized test conditions with many factors. Finally, correlative factor has found the relations between major factor and the optimized test conditions.

4.1. Optimized Test Condition for Air Pressure

Black powder dust is suspending in explosive devices. Suspending time is asked to be within 0.5-1.0 sec.. Dust distributes evenly and has no disturbance. Therefore, the optimized air pressure is most important. The pressure value of the gas-pressure meter is preferably 1.0-1.4 kg/cm^2 during the test.

4.2. The Effect of Electrode Gap

Electrode gap has many effects on ignition energy of black powder suspending dust. Test result is shown on Fig. 2. The optimized spacing is 6mm.

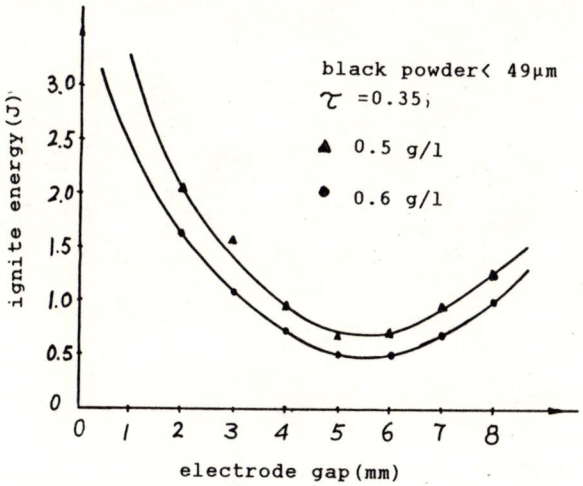

Fig.2. The Relationship Between Black Powder Ignition Energy and Electrode Spacing

4.3. The Effect of Discharging Time Constant

The discharging time constant of discharge circuit has effect on black powder ignition energy. The constant τ during discharging is $\geq$ 0.3ms as preferable.

4.4. The Effect of Series Resistance

The relationship between series resistance and ignition energy. The series resistance value is $10^4 \Omega$ or so as preferable.

4.5. The Effect of Capacitance of Energy Accumulation Capacitor

The capacitance of energy accumulation capacitor has a relation to ignition energy. The capacitance value of energy storage capacitor for black powder dust suspending state is preferably 0.01 μf.

4.6. The Selection of Radius of Curvature and the Height of Electrode Taper

It shows clearly from the test that the height of taper is 12mm, the radius of curvature is 1mm.

5. TEST RESULT AND ANALYSIS

5.1. The effect of black powder dust particle on minimum explosive concentration is shown on Fig.3.

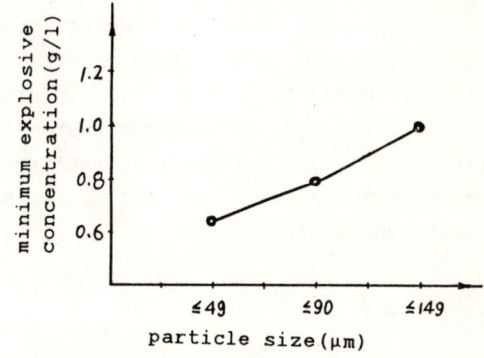

Fig.3. The relationship between particle size and minimum explosive concentration

5.2. The effect of black powder dust on minimum ignition energy is shown on Fig.4.

5.3. The minimum ignition energy of black powder dust (particle size $\leq$ 49μm) is 550mJ. The minimum explosive concentration is 665mg/l

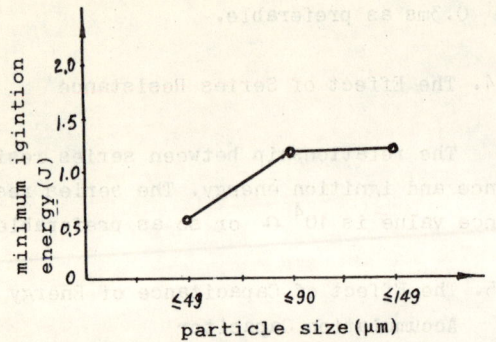

Fig.4. The relationship between particle size and minimum ignition energy

5.4. Test result analysis.

5.4.1. As the dust particle grows big, the minimum ignition energy value increases. This is because that the particle grows, the weight of each particle increases. So, gravity causes suspending time decreased. Moreover, as the particle size grows big, its thermal capacity increases, the ignition energy applied for its requirement is increased.

5.4.2. As the dust concentration decreases, the ignition energy is evidently increased. This is because the concentration decreases, the particle number applied for this fixed explosive space is decreased. The ignition probability caused by a certain energy decreased. In addition, even when a certain particle is ignited, it will not propagates easily because of a few particle scattering. Energy must be increased so that flame can propagate successfully.

6. CONCLUSION

Preliminary test apparatus has been defined through test research. The factor effects the electrostatic spark sensitivity of black powder suspending dust has been thoroughly studied. The beneficial interrelated data for evaluating the hazard of black powder dust have also been applied. Yet, as for practical use, the internal relations between suspending dust sensitivity and pile-up sensitivity, the spot data and laboratory data are still waiting for approaching.

7. REFERANCE

(1) Hazard of Dust Electrostatic Discharge M. Glor 1986
(2) RI Bureau of Mines Report of Investigations 5971
(3) RI Bureau of Mines Report of Investigations 7132
(4) RI Bureau of Mines Report of Investigations 7208
(5) RI Bureau of Mines Report of Investigations 5624
(6) Specifications of Electrostatic Safety. Chemical Industry of Romania.
(7) AD - A095353
(8) Research Scheme of Electrostatic Spark Sensitivity to Suspending Black Powder Dust. Technical Report. Shaanxi Applied Phys. - Chem. Res. Inst. 1986
(9) 安全工学 Vol. 21, No.5 (1982)
(10) 安全工学 Vol. 22, No.1 (1983)

Proceedings of International Conference on Modern Electrostatics

THE PROBLEM CONCERNING HOW TO ANALYSE AND IDENTIFY THE ELECTROSTATIC EXPLOSION ACCIDENT

Zhong Donglian
Air Force Research Institute
No 1 of PLA, Beijing, China

Zhao Luzhen
Beijing Institute of Labour
Protection Beijing, China

Abstract

Due to the traditional method of analysis and research about electrostatic explosion accident can not be used to get reliable evidences for the identification of the cause of explosion event, it is always difficult to identify the direct cause of the explosion by using the traditional method. The author of this article, suggests a new scientific method of analysis, based on the micro-morphological features and characteristics of the electrostatic discharge. In fact, it had been proven that this method of analysis can be successfully used to identify the source of electrostatic discharge and to discriminate the direct cause of explosion event. The detail of the above mentioned new method and also the way of how to use this method for practical application are recommeded in this paper.

Introduction

It is necessary to build up an effective procedure and method of analysis for the identification of the direct cause of explosion accident due to electrostatic discharge, in comparesion with the traditional method and procedure. According to our several years investigation and research, we suggest a new way to study the electrostatic discharge phenomena which causes explosion events. It is based on the electrostatic discharge micro-morphological features and characteristics to look for the source of explosion, and make sure the evidence of explosion. The analysis of the electrostatic explosion accident must include the following aspects:
(1) accident investigation;
(2) accident remains and wracks analysis;
(3) simulating test;
(4) synthetic analysis;
(5) making accident conclusion.

In the later part of this paper, the principle together with the applications are presented.

Accident investigation

The investigation of the accident scene is very important. The first important task of the accident scene investigation is to get a preliminary nature and regulation of the event. This is based on the record and data of the accident scene. For instance, an explosion accident occured at a petrochemical plant[3] during the process of liquidized gas refilling. According to the accident scene investigation, records and data, we got the first conclusion that the preliminary possibility of the explosion igniting source was the electrostatic spark discharge, because there are no any other kind of explosion igniting source able to be found, i,e, no open fire, high temperature body, electric circuit short cut sparking, and lightning.

The second important task of the accident scene investigation is to obtain accident remains and wracks which can provide the evidence of explosion and can help to find the direct cause, localization of ignition and the correct conclusion of the accident. Identification of the igniting of explosion is according to:
1. The witness' impression;
2. The explosion environment;
3. Experience and historical explosion events.

Remains and wracks analysis

The accident's remains and wracks really recorded the cause of explosion and its evidence, especially for the electrostatic discharge explosion. To analyse the accident remains and wracks include their micro- and macro- morpholohical analysis. The purpose of remains and wracks analysis is to fix the local position of electrostatic discharge or to point out igniting place. By this way, the reliable evidence of the cause of accident could be obtained.

With regard to the feature and characteristic of the electrostatic discharge explosion accident, according to several years research and investigation, we obatained the following conclusions:
1. Of all the electrostatic discharge, the energy of spark discharge is the highest, the most dangerous one and most of the electrostatic discharge explosion are caused by spark discharge.
2. The main macro-morphological characteristic of the electrostatic discharge explosion remains and wracks is the surface colour affected by the discharge, being yellow, grey or black. The spark

-447-

point is melted due to high temperature effect, being deep blue colour.

3. Electrostatic spark discharge is a kind of discharge with a high voltage but low current. If such kind of discharge has taken place, some special micro-pits called "vocanic craters" must appear. It is this major micro-morphological features found on remains and wracks of the electrostatic discharge explosion, to be used to identify the electrostatic explosion accident.

For instance, after the explosion of the liquidezed crude oil gas, it had been found a lot of "spark discharge micro-pit" existed on the local surface of the refilling nozzle as Fig.1. It can give a correct evidence of the explosion igniting source.

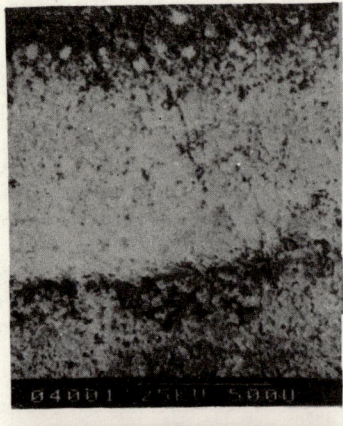

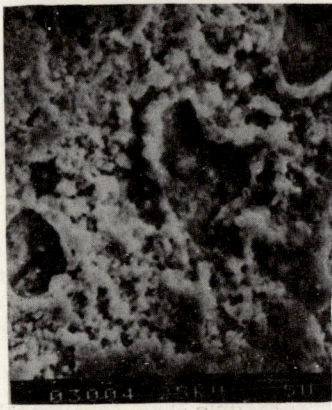

Fig.1 The electrostatic sparking discharge micro-pit found on remains and wracks of the liquidized crude oil gas explosion accident.

In the micro analysis, the most important problem is how to strictly discriminate different sorts of micro-pit, such as the "spark discharge micro-pit", "electric circuit short cut sparking micro-pit", "explosion flame attack micro-pit", "burn erosion pit", "corrosion pit", and the "mechanical pit". The detail of how to strictly discriminate different sorts of micro-pit has been pointed out[1].

Simulating test

The purpose of simulating test is to produce a sufficiently correct evidence of the accident with the repeated test. For instance, in the crude oil liquized gas explosion accident analysis, we also have done some of the simulating test, such as measuring the refilling nozzle electrostatic current. The value of static current is within 6.6×10^{-7}--8×10^{-7} A, the electric capacity of the refilling nozzle to ground is about 10 pF, the resistance to ground is $3 \times 10^{10} \Omega$. If the nozzle is not connected to ground for 0.5 sec., according to calculation, it will be charged to 33--44 kV with the stored energy of 5.45--8 mJ. These values satisfy the condition to ignite the liquidized gas.

A kind of air force flighter[4] used to get the ground fire accident during the aircraft landing. In the accident investigation procedure, we have found the electrostatic discharge localization with the micro-morphological analysis. With our simulating test results, the direct cause of the accident has been pointed out and after redesigning the fuel tank, such accidents have never been occured till now.

Synthetic analysis

In order to get precisely reliable conclusion of the accident, we must synthesize all the analysis. The accident always is due to several factors. For instance, in some times several igniting sources have been found in the accident scene, such as electric circuit short cut sparking, flamable material burning detonation sparking, moving frictional sparking. Except the moving frictional sparking, the other two kinds of sparking can cause different form of micro-pit on the igniting point. The problem to distinguish clearly the difference between these micro-features and charateristics is the base of key point to distinguish the ignition sources.

The electric circuit short cut sparking is the low voltage but high current discharge. The difference between the "sparking discharge micro-pit" and the "short cut sparking pit" is that, (1) The form of the short cut sparking pit is irregular and the size is greater, being seen with naked eye or with low maginification microscope; (2) The "short cut sparking pit" is not in the form of "vocanic crater", but is apparently in the form of "cell pattern", "splashing pattern"; (3) The "short cut sparking pit" always has the metal stuck tracing characteristic and between electrodes a lot of metal transfered and deposition phenomena could be found. The feature of the " short cut sparking pit" is shown in Fig.2.

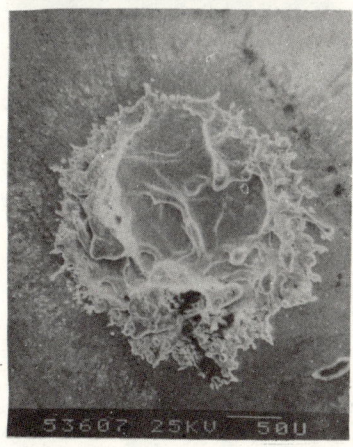

Fig.2 The micro-pit and characteristics of the electric circuit short cut sparking.

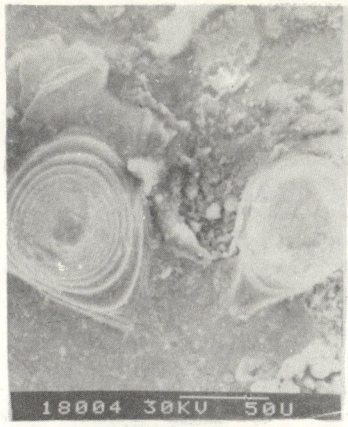

Fig.3 The electric circuit cut sparking micro-pit feature and characteristic found on remains and wracks of a fuel tank carrier's fire accident.

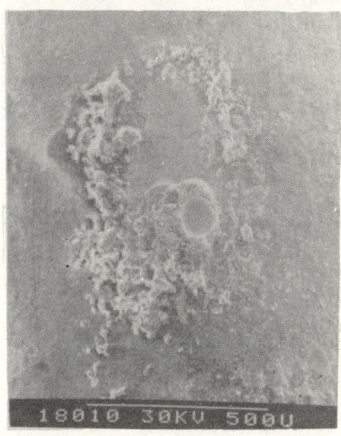

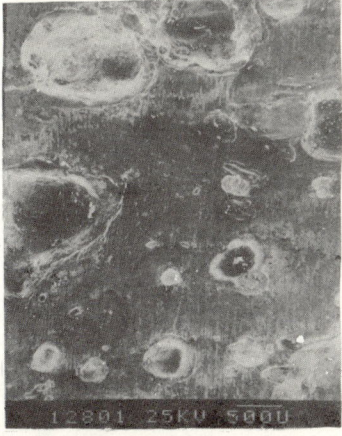

Fig.4 The explosion attack micro-pit found on the aluminum plate after TNT explosion.

Oil tank fire accidents caused by both electrostatic sparking discharge and short cut sparking have been identified with their micro-pit features on the remains and wracks as shown in Fig.3.

After explosion of detonator, dynamite and the flamable material, the micro-pit would also be formed on the surface of remains and wracks. Fig.4 is the tracing found on the aluminum plate after TNT explosion, which is apparently different from the electrostatic discharge micro-pit. According to those patterns, we are able to make a discrimination to the characteristics of different igniting source which can directly lead to the correct conclusion of explosion accident.

Conclusion

The method and procedure recommended in this paper have been successfully used for the analysis and identification of the electrostatic discharge accident. With micro-features and characteristics, the discharging point could be found to suggest the contermeasures against the electrostatic discharge hazard.

Reference

[1] Zhong Dongliang, "Research on the micro-characteristics and features of electrostatic discharge", Beijing Aircraft Maintenance Engineering Developing Corporation, Research Report 1987

[2] "The Shanghai Oil Factory benzene tank explosion accident and its actual cause", Research Report of Shanghai Oil Factory, China, 1982.

[3] Zhong Dongliang,"The analysis of explosion accident of the gas station in the south suburb of Beijing, Research Report,1983.

[4] Zhong Dongliang,and Lu Phong Lin, "The analysis of F-6 electrostatic discharge explosion accident and the proposal of a preventative measure."
"AVIATION MAINTENANCE" Beijing,198.

ELECTROSTATIC SHIELDING EFFECT OF BOX WAGON FOR AMMUNITION

Yu Yongfang, Li Huinan, Li Chunguang
Research Institute of Safety Technique,
State Commission of Machinery Industry, Beijing, China

Abstract

There is an electric field under the 25 kv 50 Hz power mains of electrical railway. Some test results in the laboratory and work site are presented in this paper. It is shown that wagon has good effect of electrostatic shielding. Field strength is almost zero inside the wagon when all windows and doors are closed. Maximum field strength is approximately 500 v/m when windows and doors are open. There is no risk of explosion caused by ESD for explosive goods transported in the wagon on electrical railway.

Introduction

There is an electric field around the power main of electrical railway. The induction charging of a conductor in this field will arise. Is it safe when the wagon with explosive goods is running on the electrical railway?

The field strength inside and outside the wagon can be measured and calculated. Comparing all results obtained from experiments and calculations, we will be able to answer whether or not a wagon loaded with ordinary ammuntions is electrostatically safe.

Basic principle

At present in China the frequency and voltage of the electric train are 50 Hz and 25 kv respectively. There is an alternating electromagnetic field around the power mains. This is a quasi-static field. If we neglect slowly varying magnetic field, the problem simply becomes electrostatic. An isolated conductor present in the alternating electric field will be charged and possessed of certain potential or inducing current depending on the shape and size of the conductor and the distance from the earth. It is sure that they are mainly decided by the distortionless field. So looking for the distortionless field is significant in the measurements and calculations. The induced charge of a conductor in the AC field is:

$$\dot{Q} = \dot{E}hc \quad \text{where } \dot{E} = E_0 e^{j(\omega t + \phi)}$$

The induced current is:
$$\dot{I} = d\dot{Q}/dt = j\omega \dot{E}hc$$

The induced potential is:
$$\dot{U} = \dot{I}/j\omega c = \dot{E}h$$

where
ω is the frequency of electric field
c is the capacitance of a conductor against the earth
h is the avarage distance between the conductor and the earth
$\dot{E}$ is the distortionless field strength

Therefore measuring of the induced current or the charge fluctuation of a proper conductor, even measuring of the current flowing from the electrode present in the field to the earth, distortionless field strength can be indirectly obatined. Two kinds of field-strength-meters recommended by the CIGRE wordshop group No. 3601 have been made on the basis of this principle.

Theoretical calculation of the field around the power mains is complicated. In order to simplify the situation, we just take the central profile of infinite parallel power mains as a model. (see Fig.1)

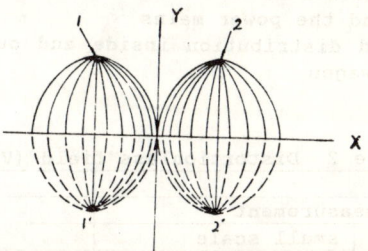

Fig.1 Profile of distortionless field under the power mains

The charge per unit length $[Q] = [C][V]$
where $[V]$ is the potential of the power mains; $[C]$ is the capacitance of the power mains; while 1,2 are the power mains infinately parallel with each other; and 1',2' are the images of the power mains.

The potential coefficients of the power mains are:
$$P_{11} = P_{22} = 1/2\pi\varepsilon_0 \ln(2h/r)$$
$$P_{12} = P_{21} = 1/2\pi\varepsilon_0 \ln(D'_{12}/D_{12})$$

where r is the radius of power mains; D_{12} is the distance from the first main to the second one; D'_{12} is the disatance from the second main to the image of the first main; h is the height of power mains against the earth

Because of $[C] = [P]^{-1}$
Hence $[Q] = [P]^{-1}[V]$
Electric field $\vec{E} = \sum_{i=1}^{2}(Q_i/2\pi\varepsilon_0)\vec{R}_i/R_i^2$
where R_i is the distance from a point in question to the power main $[i]$

Simulating and site test

The simulating test in the lab is

-451-

divided into two sorts: small scale and large one. The definitions of essential parameters in the experiments are:

voltage scale coefficient $K_u = U_s/U_p$
dimension scale coefficient $K_l = D_s/D_p$
frequency scale coefficient $K_f = f_s/f_p$
field strength scale coefficient $K_E = K_u/K_l$

where
  U_s: simuulating voltage
  U_p: practical voltage
  D_s: simulating dimension
  D_p: practical dimension
  f_s: simulating frequency
  f_p: practical frequency

The values of all these parameters see Table 1.

Table 1 Scale coefficients of main parameters

	K_u	K_l	K_f	K_E
small scale	1/250	1/15	400	0.06
large scale	4 (10)	1 (1)	1 (1)	4 (10)

Experimental tests in the lab and work site include following measurements:
(1) distortionless field distribution around the power mains
(2) field distribution inside and outside the wagon
(3) field distribution inside the wagon when the train running
(4) field distribution inside the wagon when the wagon loaded with explosive goods
(5) field distribution and variation inside the wagon when the voltage on the power mains increased

The field-strength-meter used in the experiments was calibrated in the homogeneous field of a large parallel plate calibration system in advance.
The work site see Fig.2.

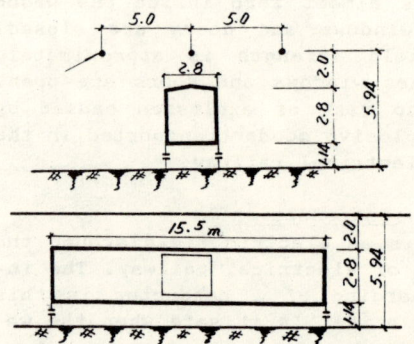

Fig.2 Profile of the work site

Table 2 Distortionless field (V/m)

		0.0 m	1.2 m	2.5 m
site measurement		2.4×10^3	3.5×10^3	4.4×10^3
simulating	small scale	1.9×10^3	/	/
	large scale	1.8×10^3	1.6×10^3	2.1×10^3
calculation		1.6×10^3	1.6×10^3	1.7×10^3

Table 3 Maximum field strength inside the wagon (V/m)

	material of the wagon	field strength E	location of measurements
small scale simulating	steel	593	at doors, on the floor
	wood (coating inside)	/	/
large scale simulating	steel	950	at doors, 1.55m high
	wood (coating inside)	740	at doors, 1.55m high
site measurement	steel	470	at doors, 0.90m high
	wood (coating inside)	403	at doors, 1.36m high

Table 4 Comparison of the results for the same point under conditions at rest and movement of the wagon

	material of the wagon	maximum field strength E_{max} (V/m)	location of measurements
at rest	steel	490	at doors (1) (doors open)
	wood (coating inside)	383	at doors (2) (doors open)
under movement	steel	197	at doors (doors half-open)
	wood (coating inside)	11	at doors (one door half-open)

Remark:
(1): The width of doors for all steel wagon is 3.05 m.
(2): The width of doors for the wagon coated with wood inside is 1.75 m.

Results

All results obtained from experiments can be summarized in Table 2,3,4.

All the experimental results can be summarized as following:
1. The results of measurements for a wagon of all steel and wood coating inside were almost the same.
2. The results of measurements for both rest and movement situation of a wagon were almost the same.
3. The field strength was basically zero when all the windows and doors of the wagon were closed, i.e, the wagon has a good effect of electric shielding.
4. The wagon under power mains made the electric field change a lot. The observed local surface field strength on the roof corner outside the wagon was 14.5 kV/m which is two orders more than the distortionless field and three orders more than the maximum field strength inside the wagon. It indicated that the field distribution inside and outside the wagon was quite different.
5. There was a maximum field strength 470 V/m at the doors when all windows and doors were open. It decreased rapidly to zero when the test point went away from the door.
6. No influence on the field distribution appeared inside the wagon when the wagon was loaded with ordinary ammunitions. The test results were similar to that of empty wagon.

Conclusion and discussion

1. All the wagons used on the electric railway have good effect of electrostatic shielding when the windows and doors are closed. It will not be possible that the electric field around the power mains with high voltage goes into the wagon through the steel shell so that the goods inside wagon can not be influenced by the outside electric field.

2. A little portion of the electric field may go into the wagon when the windows and doors are open. Assume the field inside wagons is homogeneous and the field strength is 500 v/m. Simple calculation gives 7.9 uA of our sample due to the electrostatic inducing. Compared to the data published until now, it is less than four orders of magnitude in relation to the safe current 0.05A for ordinary bridge type ignitor. Therefore no risk of ignition from static induction is present.

3. The wagon under the power mains of electric railway can make the field change a lot. Local field strength may reach a high value. It is preferable to pay attention on shotting the person climbing on the roof of the wagon.

Reference

[1] T.D.Bracken, Field measurements and calculations of electrostatic effects of overhead transmission lines, IEEE.vol.PAS-95 No.2 1976
[2] H.Singer et,al, A charge simulation method for the calculation of high voltage fields, IEEE PES Winter meeting.N.Y.1974
[3] Tadasu Takuma et,al, Analysis of calibration arrangements for AC field strength meters, IEEE vol.PAS-104 No.2 1985
[4] Shao Fangyin et,al, Electric and magnetic field induced by the power transmission systems.(in Chinese) 1984.2
[5] Shao Fangyin "The Electrostatic Field under 500 kV Transmission Lines" International Conference on Large High Voltage Electric Systems,Symposium 22-81, Stockholm 1981

STATIC ELECTRICITY OF HUMAN BODY

Ke Lin
State-owned Qin Hua Electrical Apparatus Works
East Suburb Xi'an, China

ABSTRACT

Very occasionally, fires and explosions occur during the handling of powders and flammable liquids in industrial processes. Some hazards have been attributed to ignition due to electrostatically charged human body.

The personal electrial parameters in workshops have been observed. Some electrostatic models, physical and mathematic, are to be created, which simulate discharge between charged person and a discharging object. In order to minimize the possibility of sparking from the body, safety can be achieved by using some conductive materials. Their antistatic behaviors are introduced.

Introduction

Anti-static protection has been developed into a specialized science. It is not only referred to electrostatic engineering, but also widely to combustion chemistry enginnering, enviromental engineering, materials engineering and system engineering. The stress and orientation of electrostatic hazards change with the development of modern science and high technology.

Static electricity of human body is recognized as ignition source of hazadous areas, contamination source of clean rooms and E.M.I. source of electronic devices.

The operator in workshop can be charged by contact and separation, or by induction. The charging potential may be very high. Under certain circumstances, that would allow an arc or spark to form. The whole discharging process may last for tens of milliseconds; the instantaneous power delivered may approach tens of kilowatts. M.I.E. of some substances is conducted as follows: primary explosives $10^{-3} - 10$ mJ, EED 10^{-3} mJ, flammable mixtures of gas and vapour less than 1mJ, dust cloud 100 - 150mJ [1]. Static potential above 10V may damage some electornic components and initiate high static sensitivity E.E.D. The probability of such an accident may be 10% around munition plants [2].

Investigation

Electrical parameters of an individual vary in charging and discharging process. The values of human resistance measured varies from 10Ω to $10m\Omega$, changing with applied voltage and variation in ambient conditions. Body to ground resistance depends upon the materials of shoes and flooring. The value of which has been experienced as high as $10^{13}\Omega$ for insulating system, less than $10^9\Omega$ for antistatic systems.

Body capacitance, including capacitance to conductive floor and that to conductive objects of surrounding, as obtained from 50 up to 700 pf in insulating situation; up to 2000pf in antistatic situation; up to 6000 pf in conductive situation [3].

Body inductance has a limiting effect on the rise in time of wave form, its value ranges from 4×10^{-6}H to 5×10^{-9}H, depending on the discharging circuit of human body.

The model of human body, can be simulated as a sphere, a cylinder, or a cross plate, under defferent considerations [4].

Charging of Human Body

Static generation of human body may be presented by equivalent electrical circuit shown in the Figure 1.

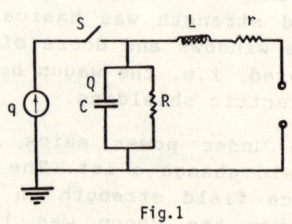

Fig.1

R—body to ground resistance
C—body capacitance
q—charging rate
Q—charge of the body at instant t
V—potential of the body at instant t
S—switch, on for charging, off for discharging.

suppose q=constant, i.e. uniformly charging, and if t=0, Q=Q$_0$, or V=V$_0$, then
$$Q = qCR + (Q_0 - qCR)e^{-t/CR}$$
or $V = qR + (V_0 - qR)e^{-t/CR}$
when $t \rightarrow \infty$ $Q \rightarrow qCR$ $V \rightarrow qR$
the schematic diagram is shown in Fig.2.

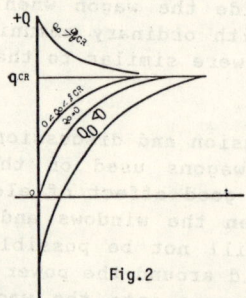

Fig.2

It is obvious that no matter what polarity and magnitude an individual has initially been charged with, the charging voltage he held tends asymptotically to a steady value qCR; this means that the charging rate and draining rate come into equilibrium. In reality, static electrification of a person is discrete. A typical charging voltage variations monitored on a person with different movements are shown in figure. 3.

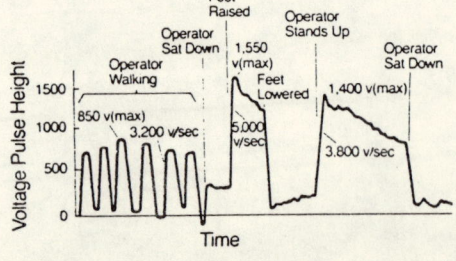

Fig.3

The abouve evidence leads to a conclusion that the body-to-ground resistance is one of the main factors controlling the magnitude of the charge on an individual.

Discharging of Human Body

The equivalent circuit of discharging of a human body may be simplified as follows:

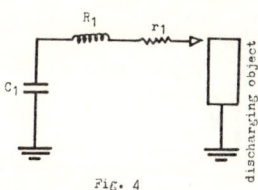

Fig. 4

C_1—body capacitance
r_1—discharge resistance of the body
L_1—discharge inductance of the body

The discharging object may be a grounded, or insulating conductive device, or a powder heap.

If R.L.C. are total resistancae, inductance and capacitance of a discharging circuit, at instant t=0, $u=u_0$, the max. current pulse can be calculated. The following shows the results.

R=1000   R=500   R=300
$L=4\times10^{-6}H$ $L=3.25\times10^{-6}H$ $L=4\times10^{-6}H$
C=200pf   C=50pf   C=50pf
$I_m=0.9\times10^{-4}u_0$ $I_m=1.5\times10^{-3}u_0$ $I_m=1.8\times10^{-3}u_0$

for u_0 higher than 10KV, the instantaneous discharging current may be as high as tens of amps. The importance of the values of body parameters lies on the fact that they affect on the waveform of current, on confining the possibility of current ringing and limiting decay time, and as a result, on the energy distribution in the discharge channel. Experiments have shown that, an ESD event may contain several ESD pulses from a single approach of th charged individual to the discharging point. The behavior of which is considered to due to quenching of the spark at the point, absorbtion of charge energy by the body resistance and formation of break-down field across the spark gap. Generally, the whole discharging period, including multiple sparking, continues for tens of milliseconds, while each single spark is passed at interval of tens of microseconds.

Static Protection

The most effective method for eliminating the charges accumulated on body is to ground the individual in such a way that the charging voltage of a person never rises to the safely controlled value around a specified area. This value is confined to that given by the following equation.

$$V_c = \sqrt{\frac{KE_{min}}{0.5C}}$$

$$R = \frac{V_c}{q_{max}}$$

K—safety coefficient
E_{min}—minimum ignition energy
V_c—safely controlled voltage
R—body to ground resistance
q_{max}—max. charging rate

First of all, the determining factor of body grounding depends on the conductivity of floor mats. In general, cement mortar, marble, terrazzo, magnesite floors are conductive. In China, there are many kinds of conductive and antistatic floors in use, such as misfire asphalt, unsaturated resin, polyurethane, rubber, plastics, lead and the like. Furthermore, a conductive flooring is effective only when a conductive foot wear is worn. Others conductive materials are also widely used in working stations. In some hazardous areas, or high clean rooms handling high static sensitivity components, wrist straps are necessary to reduce the human body potential to a safe value. The functions of conductive (antistatic) floorings, clothings, foot wears and wrist straps are preliminarily investigated in the following tables.

Table 1 Charge created while walking across flooring

No. of grouping	1	2	3	4	5	6
Clothing	#	*	#	*	#	*
foot wear	#	#	*	*	#	*
flooring	#	#	#	#	*	*
Body potential(KV)	3,2	3,0	1,8	1,6	0	0

* conductive   # insulant

Table 2 Charge created while wearing a wrist strap

Transient Peak Voltage (V)	Body to Ground Resistance (mΩ)
< 2	0,27
< 2	1
11	10
30	50
80	100
1550	∞

Enviromental factors should also born in mind, such as humidification of some occupancies where static hazard may exist. Relative humidity influences the antistaticity and resistivity of materials. To achieve the best static control, it is necessary to have a R.H. of 45-70%. The worker, to the full extent, keeps his operating process steady and normal, forbidding to cause abrupt activities in motion. Therefore, some regulations and academic training for static protection should be set up.

CONCLUSION

After many years of current research in static protection, we have not yet reached a full understanding of the static phenomena on the human body. In the industrial process, resistance and capacitance vary considerably; C. of one or two orders of magnitude, r. of several orders, depending on operating stuations. In general, the wave form of discharging current of a human body is non-oscillatory and discrete. In case of simulating a charged human body, the above situation should be taken into consideration.

References

1. N Gibson J. of Electrostatics
   11 (1981) 27-41

2. 柯林　人体静电问题

   1984, 8

3. 北京工业学院物理组, 火化工作业防静电危害

   1974 96-103

4. Минеев А. Н. Журавлес,
   Электричество NO.10. 1977, 88-90

5. James R. Huntsman EOS/ESD Symposium
   1982, 105

EVALUATION OF THE SAFETY PROPERTY ON TANKERS DURING THE DRAINAGE OF BALLAST WATER

Sun Keping

Research Section of Static Electricity, Shanghai Maritime Institute, 1550 Pudong Dadao Road, Shanghai, China

Abstract

This paper, from an explosive accident of a tanker, evaluates the safety property of staticelectricity on tankers during the drainage of ballast water. Based on the inverstigation and theoretical analysis about this accident, the paper indicates that the cause of the tanker explosion may be that the steam pipe fixed in the tank for heating leaked, a lot of steam sprayed and a very high static field was formed in the tank, then an insulated conductor (a copper rod or flashlight) happened to fall into the tank. Some discharging sparks spread and caused the explosion. This paper proposes some suggestions in order to avoid such accident.

1. Introduction

Explosive accidents on tankers or OBO ships caused by ballast water take place from time to time in the world. On Oct. 18,1986, an ampty tanker with a loading capacity of 24000 tons exploded. There has been a discussion about whether it is caused by the ballast water. The tanker was then berthed at a wharf with the ballast water being discharged through pump pipes. When the drainage of the ballast water in No.2 centre cargo was close to an end, an explosion suddenly took place. The whole ship sloped to left and then sank into the bottom of the sea. The distribution of the ballast water in every cargo before explosion is shown in Fig. 1. There is some ballast water in No.2 center cargo(about 50 tons), No. 5 center cargo(about 1000 tons) and in No. 2 left cargo, No.3 right cargo.

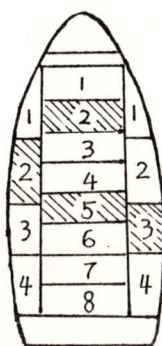

Fig.1 Scheme of distribution ballast water. (the volume of every centre cargo is 2000 tons)

As there was some remaining crude oil, some of which were blocks on the wall of the cargo and on the surface of the water, the steam through a row of pipes was used to heat the remaining oil. The saturated steam presure was 5 kg/cm^2. It had been heated for an hour and 20 minitues before the explosion took place.

At that time, the drainage work was in operation with port hole opened. Two sailors on duty were working on the deck (They both died in the accident).

Seven people were killed in this accident.

On that day it was fine. Wind of Beaufort force 5 blew from the stern to the stem with a wind speed 0f about 10 m/s.

2. Analysis of the accident

There was no operation with fire on the ship when the explosion took place.

Sailors were not allowed to smoke while working. All of the sailors usually observe this discipline. therefor, it was sure that this accident was not the result from fire.

The analysis of the sunken ship salvaged afterwards showed that the walls around No.2 centre cargo were obviosly deformed towards outside. Therefore it was clear to see that the explosion was due to the No.2 centre cargo.

(1) condition of the explosion

There was some inflammable vapor in the centre cargo because there was remaining oil in it. There was also sufficient oxygen in it since the port hole was open. Therefore explosive gas mixture was formed.

(2) Assessment of the formation and the safety property of the static field in the tanker during the drainage of the ballast water

During the period of 1967 to 1972 several large OBO ships exploded because of the strong static field formed by the shaky ballast water. When a ship is sailing in stormy waves.

The waves set up by the shake of the ballast water are similar to those when the cargos are being washed and can engender a high electrostatic field in the cargo. For example, when the ballast water containing oil is at an angle of $4°$, a charged mist can be easily formed in the cargo, with a charge density of about 50 nc/m^3 or more. The maximum space potential in the cargo can reach about -50 kv. Such a strong field can surely cause an explosion. The explosion of this tanker, however, could not be caused by the shake of the ballast water because the tanker was berthed at the wharf on that day. The wind force was only 5. The ballast water could not engender a strong static field as the tanker was in a smooth position.

The analysis, both experimental and theoretical indicates that the electrostatic field will be rather lower during the drainage than during the shake of the ballast water.

Fig.2 illustrates that although the ballast water does not shake, there is a layer of oil on the surface of the ballast water being discharged since there's some oil left. With the continuous discharge of the ballast water, liquid surface comes down constantly and is slightly stirred. There has indeed formed a static field in the cargo.

In this case, two situations may take place.

The first situation:

A copper rod might be dropped into the cargo by a sailor so as to measure the water level of the ballast water (see Fig.2). The copper rod obviously bacomes an insulated conductor.

W.M.Bustin and W.G.Dukek indicates that [1] the capacitance in the free space of long and thin isolated metal conductive object, can be caculated from the following formula:

$$c = 2\pi \xi_0 a \frac{(1 - \frac{b^2}{a^2})^{\frac{1}{2}}}{Ln[\frac{a}{b} + (\frac{a^2}{b^2} - 1)^{\frac{1}{2}}]}, a > b \quad (1)$$

where, c——capacitance, farad
ξ_0——permittivity of free space
a——length, meter
b——width, meter

Suppose the length of the copper rod is 30 cm, diameter is 3 cm, we can get its capacitance from the above formula as follow:
$$c = 5.57 \text{ pf}$$

The rod must be inducted when it enters the field full of charged mist. Suppose the inducted charge of the rod is Q (coulomb). When the discharge of the rod is taking place on the oil surface, the electrostatic energy E of the discharge will be:

$$E = \frac{1}{2} c v^2 = \frac{Q^2}{2c} \quad (2)$$

where, E——energy, joule
c——capacitance, farad
v——potential, volt
Q——charge, coulomb

W.M.Bustin indicates that [1] for the purpose of safety 0.25 millijoules

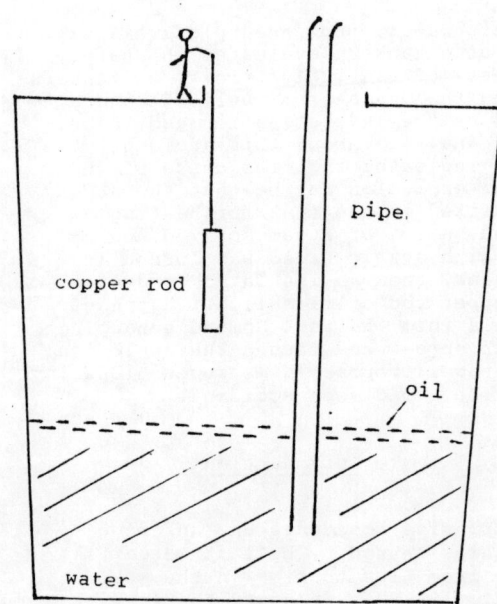

Fig.2 Scheme of the No.2 centre cargo

can be taken as the minimum ignition energy for most petroleum products and fuels.

Suppose that this accident is caused by the discharge between the copper rod and the surface of oil, the energy released during the discharge is 0.3 mj, then the potential of the oil surface shall be at least

$$v = (\frac{2E}{c})^{\frac{1}{2}} = 10.5 \text{ kv} \quad (3)$$

The megnitude of the static filed formed during the drainage of the ballast water, however, is far less than the one above.

The simulated test in a small vessel by us indicates that the space potential of the cargo is only hundreds of volts while the surface of the water coming down. This value is far less than 10kv.

Therefore, we think that discharge between the rod and oil surface can not cause the ignition and explosion.

The second situation:

The discharge between the flashlight and the surface of oil.

Through investigation, it is found that the dead sailors carried two flashlights with them during the work. One of the slashlights is explosive-proof, the other is a commom one. But both are missing after the explosion. It was possible that while one sailor was surveying the ballast water from the port hole with a flashlight, but the flashlight suddenly dropped. Suppose that there are no striking sparks during the drop and that the length of the flashlight is 30 cm, its diameter is 4 cm, from (1) we can know its

c = 6.10 pf.

Using equation (1), (2), (3), we can get the potential required for a dangerous discharge :

v = 9.92 kv

As mentioned above, the potential in the cargo during the drainage of the ballast water can not be so high.

Thus it can be seen that it is almost impossible for a dangerous discharge to take place between the flashlight and the oil surface.

(3) Assessment of the formation and the safety property of the static field when the steam pipe leaks

As everyone knows, if steam pipe leaks, high pressure steam will spray out through the small hole. This process will cause strong electrification. The charged mist will quickly spread all over the cargo and a strong electrostatic field appeares.

The problem is that whether or not a dangerous discharge can take place.

Van de Weerd made an experiment in 1971 on the discharge of a probe from the charged mist [2]. He put the end of a probe with the diameter of 5—35 mm into a 12000 m³ shore tank. The maximum space potential in it was raised to 40 kv by steam injection. The probe was hollow and fixed on a perforated hemispherical end. Propane gas flowed through out of the probe and formed a flammable gas mixture. Though corona discharge occurred at the end of the probe in the mist, no ignition happened in these experiments. The reason was that the charge could not be collected together from many seperated droplets and fed steadily to form instantaneously a spark discharge under a certain voltage level and physical scale of a tanker compartment.

Though the corona discharge between the mist and tanker structure was found, there was still no ignition. Since the discharge between the mist and the metal structure can not ignite, we'll pay attention to the second kind of discharge, that is, the one between the body of the tanker and an insulated conductor under the influence of the mist.

In 1975, Van de Weerd made an investigation of the discharge between the dropping conductor and the earthed conductor in the cargo [6]. The scheme of the experiments is shown in Fig.3. The dropping conductor, of course, is an insulated conductor. The earthed

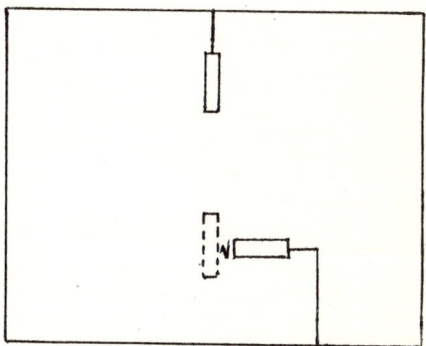

Fig.3 Scheme of static discharge between the falling conductor and an earthed condutor in the cargo

conductor indicates the body of the tanker or other metal structures in the cargo such as pipes. As there exists a strong electrostatic field formed in the charged mist, one end of the dropping conductor is inducted with a positive charge and the other end is negative. When the conductor comes down near the earthed conductor, a spark occurred between them. This spark discharge can ignite flammable gas mixture. It is this kind of spark that caused the explosines (VLCC's) in 1969.

American Petrolum Institute(API) concluded (1974c) that [7] if the maximum potentiel in the tank was below 10 kv (estimated to occur 90% of the time during conventional tank washing according to industry guidlines), any dangerous spark discharge could not occur with any objects like a less than 90 cm long rod or a less than 30 cm diameter sphere while these objects were dropping into the cargo. At 14 kv space potentiel the limits mentioned above would be a 60 cm long rod or a 22 cm diameter sphere; At 20 kv, a 30 cm long rod or a 10 cm diameter sphere.

In the explosive accident which we are discussing, a strong static field has been formed in tank by the leaking steam pipes. The simulated test by us indicates that the maximum potentiel in tank is more than 20 kv after the steam has injected for 10 minutes with 5 kg/cm^2 presure. But both the copper rod and the flashlight are about 30 cm long rods. When they suddenly drop, a dangerous spark can occur between the rod and the earthed conductor if the distance between them meets the need of the optimum spark gap one. It is quite possible this kind of spark causes the explosion of this tanker.

Of course, if a striking spark between the dropping rod or flashlight and the metal structure occur, this is perhaps the cause of the explosion that can't be ruled out.

3. Suggestions

The explosions of large tankers have happened several times in China. In order to draw a lesson and avoid such explosive accidents, besides the improvement of administrations and the further safety education to sailors, we put forward some technical suggestions as follow:

(1). Put an end to washing tank with steam.

(2). Examine carefully the pipes before using steam pipes to heat remaining oil. Take immedite measures if any leaks are found.

(3). Take static electricity-proof steps during the measurements of the level of water and oil.

(4). The flashlight used must be explosion-proof. Common flashlights are not allowed to be used during the work.

Of course these are only some emergency measures, the most important thing is that we should formulated as quickly as possible the rules and regulations about tank washing and ballst water drainage., which are suitable to the practical situations in China. All sailors must remenber the safety quidlines.

References

[1]. W.M.Bustin and W.G.Dukek(1981), Electrostatic Hazards in the Petroleum Industry.

[2]. J.M.Van de Weerd(1971), Electrostatic Charge Generation during the washing of tanks with water spray.

[3]. J.F.Huges, A.W.Bright and B.Makin (1974), Incendive spark originating from charged water mist.

[4]. J.N.Chubb(1975), Practice and computer assessment of ignition hazards during tank washing and during wave action in part ballsted OBO tanks.

[5]. A.W.Bright and J.F.Huges(1975), Reaserch on electrostatic hazards associated with tank washing in VLCC's.

[6]. Van de Weerd(1975), Electrostatic charge generation during tank washing: Spark mechanism in tanks filled with charged mist, J.Electrostatic,1,295-309

[7]. American Petroleum Institute (1974c), Final report, Tanker accident study committee, November 27,1974.

THE RELATION BETWEEN AIRCRAFT PRECIPITATION STATIC ELECTRICITY AND FLIGHT ENVIRONMENT

Zhang Hongzhen

China Flight Test Research Centre

Xian, P.R. China

Abstract

This paper describes the laboratory simulated tests and the flight tests on aircraft precipitation static electricity, and discusses the relation between aircraft precipitation static electricity and flight environment upon the test results.

Introduction

An aircraft in flight often charges electrostatically. The aircraft potential in relation to the atmosphere around the aircraft may reach several ten thousand volts and even a hundred thousand volts or more occasionally. The high aircraft potential may compromise the operation of the aircraft; especially, radio frequence noise caused by aircraft electrostatic discharge interferes severely with the operation of radio equipments onboard.

Aircraft electrostatic charging may result from flight through precipitation, electric cross fields, or from engine-produced ionization. The friction charging caused by impact of precipitation on aircraft skin, also defined as precipitation charging, is a principal cause for aircraft electrostatic charging and a cause affecting for a long time. It will take place on various types of aircraft.

Generation of Precipitation Static Electricity

An aircraft in flight is often impacted by precipitation as ice crystals, water drops, dust etc. As uncharged precipitation particles strike the aircraft skin, they acquire the charge of one polarity, leaving an equal charge of the opposite polarity on the aircraft. This is aircraft precipitation static electricity. The precipitation static electricity is created not only on the metal skin of aircraft, but also on such nonmetal material as windshield and radome.

Aircraft precipitation static electricity is affected by many factors. Flight velocity, aircraft size and flight environment are the principal factors. When an aircraft flies in a specific environment, the flight velocity and the aircraft size determine the amount of precipitation intercepted by the aircraft, as well as the magnitude of charging current. The effect of flight environment to precipitation charging is related to weather. In fair weather, there is very small precipitation in the aircraft. Only at the lower altitude the aircraft will encounter impact of dust. But there are a lot of water drops, supercooled water drops and/or ice crystals in cloudy, rainy or snowy weather. The striking to the aircraft by different property precipitations will cause charges different in amount and polarity.

Several kinds of applicable materials have been used to conduct triboelectrification test in laboratory. The test results are shown in Table 1.

Table 1. Triboelectrification of Several Kinds of Materials

Material	Snow to Al Plate		Aqueous Vapour to Al Plate	Sand to Al Plate	
Charge Polarity	+	−	−	−	+
Charge Quantity (Relative)	1		0.3	7	

Other tests also indicate that the friction of snow to organic glass causes positive charge on the organic glass, but the friction of snow to epoxy fiber glass causes negative charge on the epoxy fiber glass. Both charge quantities are considerable [1].

Flight Test to Measure Static Electricity on an Aircraft

In order to study the relation between aircraft precipitation static electricity and flight environment, flight tests were conducted with a medium-size propeller transport to measure the precipitation static electricity on the aircraft.

The static charging current may be regarded as a current source while an aircraft is charged. Its value only depends on the quantity and property of precipitation intercepted by the aircraft as well as their impact energy and is not related to the aircraft own potential. It is difficult to measure charging current because it is distributed everywhere over the aircraft. However, the charging state can be reflect by measuring other parameter. The charge amount accumulated on aircraft is the integral of charging current over time so the aircraft potential is

$$V = \frac{Q}{C} = \frac{\int i \, dt}{C}$$

An aircraft in flight may be regarded as an insulated conductor with a capacitance of hundreds PF to one thousand PF. The capacitance is about 300 PF for the medium-size common aircraft under test. If charging current is $10 \mu A$, the aircraft potential will be more than 30 kv in one second. The aircraft potential therefore rises rapidly where charging current exist.

However, the aircraft potential can not rise unlimitedly. When the potential rises up to a critical value, the corona discharges start at some sharp points on the aircraft. The corona current increases along with rise of the potential. The potential rises until a running balance is reached such that charging current equals the discharge current. The velocity to the running balance is related to the aircraft own capacitance and effective discharge resistance[1], that is

$$\tau = RC$$

A medium-size transport with dischargers has an effective discharge resistance of $1-10 \times 10^9$ ohms. When R is 10^9 ohms, C is 300 PF, the time constant τ will be 0.3 second. This shows that the process reaching the state of balance is very rapidly. Thus the aircraft charging may be reflected by measuring the discharge current.

For the aircraft with enough dischargers the discharge currents usually concentrate at installed dischargers, discharge will not occur at other parts. Thus, measured discharger current can show up all the discharge current of the aircraft. Five dischargers were located at wing tips, elevator tips and rudder tip of the aircraft under test respectively. Only the current of the discharger at the left wing tip was measured in this test. The five dischargers were distributed so for apart that their discharge could not interfere one another. The fields near discharger tips under the same potential were similar so that the discharge currents of dischargers would be equal essentially. Therefore, all the discharge currents of the aircraft could be judged from the current of the discharger at the left wing tip.

Flight tests were conducted in clear skies, rain, snow and clouds, as well as at high and low altitude in the performance range of the aircraft under test. The data maxima acquired in various flight conditions are presented in Table 2.

Table 2. Measured Discharge Current Maxima

Subject	Altitude (m)	Maximal Discharge Current (μA)	
		Left Wing Tip Discharger	Total
Under clouds	5290	− 52.4	−262
In clouds	5020	− 176.8	−884
In snow	740	− 33.4	−167
In rain	730	− 11.4	− 57
In clear skies at high altitude	6000	0	0
In clear skies at medium altitude	3600	0	0
In clear skies at low altitude	900	− 2.1	− 10.5

The Relation between Precipitation Static and Environment

Data acquired in flight indicated significant effects of environment on aircraft precipitation static. In clear skies, no discharge current could be measured out and at low altitude only very small current could be measured out. This showed that precipitation was so small in clear skies that the aircraft was hardly struck by them. Theoretically, an aircraft flying at low altitude should be struck by many particles of dust resulting a lot of precipitation static. But this is only applicable for helicopters. The downwashs from helicopter pivoting wings stir up dirt and dust from ground and distribute them around the pivoting wings causing quite a lot of precipitation static. In these flight tests, the aircraft could not fly that low and there was no environment typical for low altitude precipitation charging so no obvious discharge current could be measured out.

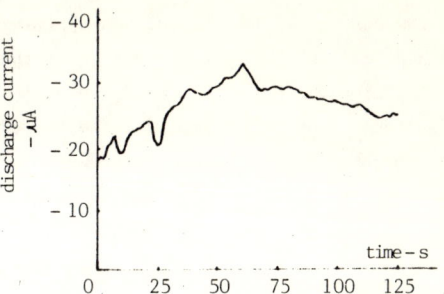

Fig 1. Typical discharge current in snow

The aircraft flying in rain, snow and cloud was charged with a lot of static charge and caused large discharge currents because there were large quantities of precipitation of waterdrops and ice crystals in the environment. But the property of precipitation has also an effect on the quantity of static. Ice crystals cause more static charge than do water drops[2], so that higher discharge currents were measured in clouds, especially in the clouds containing ice crystals, than in rain. If air temperature is very low and snow is dry, the precipitation static will be severe. But the flight test was conducted in the snow mixed with rain, so the discharge current measured out was lower than had been anticipated. All discharge currents measured in these flight conditions were negative. It indicate that the precipitation static on the aircraft is mainly negative. This is in accordance with the test results acquired in laboratory that the friction of an aluminum plate to snow or aqueous vapor causes negative charge on the aluminum plate. Because the skin of the aircraft under test is all made of aluminum plates, even if the windshield carries some positive charge due to triboelectric contact with precipitation, the electrostatic polarity of integral aircraft can not vary.

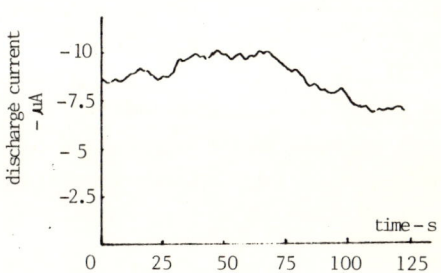

Fig 2. Typical discharge current in rain

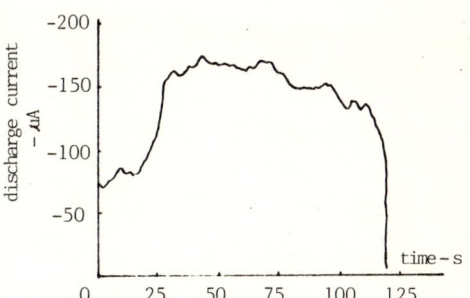

Fig 3. Typical discharge current in clouds

The curves of discharge currents measured in various flight conditions present different varying characteristic (Fig 1-4). In snow or rain, the current had less variation and speed of variation was slow (see Fig 1 and 2) owing to the equal charging condition of aircraft in snow or rain. When the aircraft was flying in clouds, the current varied a lot (see Fig 3) because aircraft charging conditions were different everywhere owing to the unequal density of the clouds on the way of the aircraft. But cloud density varied in a large

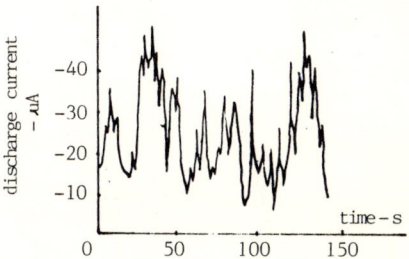

Fig 4. Typical discharge current under clouds

range so the current seldom varied. When flying under and near clouds, the aircraft was under the critical condition that it was in clouds sometimes and out of clouds some othertimes; and the cloud density was high somewhere, small elsewhere, so the discharge current varied a lot and frequently (see Fig 4).

Conclusion

Effect of meteorological condition is very obvious to aircraft precipitation static. When aircraft is flying through clouds, its precipitation static is the most severe; next is flying in snow or rain. When aircraft is flying in clear skies, there is little precipitation static on it. The conclusion is also demonstrated by the fact that communication equipment may be subjected to interference, radio-compass can not indicate direction when the aircraft is flying through clouds.

References

1. Li Ruinian. The Static Electrification of an Aircraft in Flight, Aviation Technique Information. 1977. NO. 4.

2. Nanevicz, J. E. Flight - Test Studies of Static Electrification on a Supersonic Aircraft. 1975 Conference on Lightning and Static Electricity. 14 - 17 April 1975 at Culham Laboratory. England. P 82055.

EXPERIMENTAL STUDY OF THE ELECTROSTATIC ACCUMULATION OF EXPLOSIVE POWDERS

Wang Changying, Chi Yunpong, Liu Yanyi

Xi'an Armd Police Technical College, Xi'an China

Abstract

The authors have carried out experimental study of electrostatic accumulation charges of five sorts of powder explosices with the Digital Electrostatic Accumulation Parameter Instrument and obtained main factors that affect the charge accumulation.

The analysis of the testing results can give the conceptions of "the unsafe explosives" and "the surface safe potential", which will be advantageous for the safe operation of explosives.

Introduction

Most of the powders and explosives are prone to electrification, thus high electrostatic potential appears either within or on the surface of the material of powders in industrial operations, handling, transportation and storage due to the friction of the powders with its containers. Surface discharge occurs when the electric field intensity has exceeded the breakdown intensity of the atmosphere. If those discharges produce sufficient energy, they can ignite sensitive powders or explosives resulting combustion or explosion. Attentions have already been paid to such terrible accidents(1).

The possibility of explosion coming out of the discharge of the explosive itself has been known since World War I to 1944. German scientists (2) investigated the electrostatic accumulation during pouring explosives. Their conclusion was that the electrostatic energy could ignited the sensitive primary explosives. In 1954, the electrifition experiment of the friction of powders and explosives with smooth channels of different materials was carried out (3). The result bore out that most of the powders and explosives are prone to electrifiction. In 1972, the relationship between the electrification and temperature, humidity, etc. was obtained by Waytt (4) in the extensive studies of the charge accumulation during explosive pouring process.

With digital electrostatic accumulation testing instrument (5) the experimental results of five powders and explosives and the related factors that affect electrification are obtained. The results approximately agree with those of Waytt's.

1. The accumulation fundamentals of static charge

The contact and seperation of two different materials are involved in the slidings of explosive powders along guiding channles, so it is a process that generates static charge. The electrostatic accumulation is the combination of both the generation and the leakage of static charges. Its equivalent circuit is shown in Fig.1.

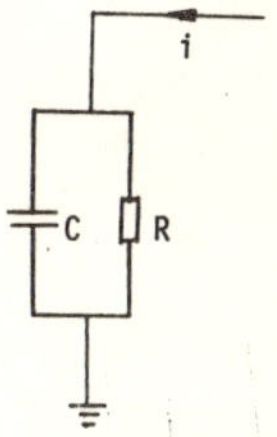

Fig. 1 Equivalent circuit for static accumulation

Where i—increasing rate of charge at the same moment;
C—capacitance for energy storage;
R—leakage resistance of the receiver box

An equation can be written out for electrostatic accumualtion from Fig. 1.

$$\frac{dQ}{dt} + \frac{Q}{CR} = i \qquad (1)$$

whose solution is

$$Q = e^{-\int \frac{1}{RC} dt} \left(\int i \times e^{-\frac{t}{RC}} dt - K \right) \qquad (2)$$

Generally speaking, the electrifiction rate i(t) is a function of time. In continuous electrification at a constant rate, $i = i_0 =$ const. and $Q=0$, $V=0$, at $t=0$, we have

$$Q = i_0 CR (1 - e^{-\frac{t}{RC}}) \quad (3)$$

$$V = i_0 R (1 - e^{-\frac{t}{RC}}) \quad (4)$$

It can be concluded from eq.(3) that the saturated amount of electrification charge is $i_0 CR$ and the saturated potential is $i_0 R$ in the explosive pouring process. In other words the electrification capacity and the amount of electrificatioon charge can be deduced granted that the electrification rate is known.

2. Experiment results

1) Electrification capacity of various powders and explosives.

Extensive experimental studies of electrification capacity of PETN, LTNR, antimony sulfide, pistol powder and percussioin composition are carried out under the conditions of certain thermperature and humidity fixed for the experiment, the results are shown in Table 1.

According to Table 1, the electrification and the value of charge accumulation of the five powders and explosives above can be divided into three kinds:

(1) The kind of high elecrification such as PETN and LTNR. Their saturated potentials are over 5KV and their saturated charges exceeds $60 \times 10^{-9}(C)$;

(2) The kind of medium electrification, such as antimony sulfide and percussion composition whose saturated potentials are $500 \sim 1000V$ and saturated charges are $17-34 \times 10^{-9}(C)$;

(3) The kind of low electrification, such as pistol powder whose saturated potential is about 100V and never exceeds 500V and saturated charge is $1 \times 10^{-9}(C)$;

2) Main factors affecting the electricfication of the powders and explosives.

(1) Variation of the accumulated charge with respect to time. Under the documentally stipulated experimental conditions, 5 grams of LTNR is allowed to slide along a guiding channel, and the variation of the accumulated charge with respect to time is shown in Fig.2 which indicates that the charge accumulation tends to be saturated as time elapses. The corresponding equation of the curve in Fig.2 is

$$V = i_0 R (1 - e^{-\frac{t}{\tau}}) \quad (5)$$

The characteristics of the curve can be summaried as follows. $dV/dt = $ const. within the first 1.5 seconds, then the slope (dV/dt) begins to diminish within the period of 1.5-2 seconds and later approaches to zero.

(2) Variation of the accumulated charge with respect to the weight of the explosive. Under stipulated experimental conditions, experiment was carried out, in which the weights of explosives were increased gradually from 3 grams to 25 grams while the other experimental factors remained unchaged. The results are illlustrated in Fig. 3. The accumulated potentials also have the limit values as the weights of the explosives increase.

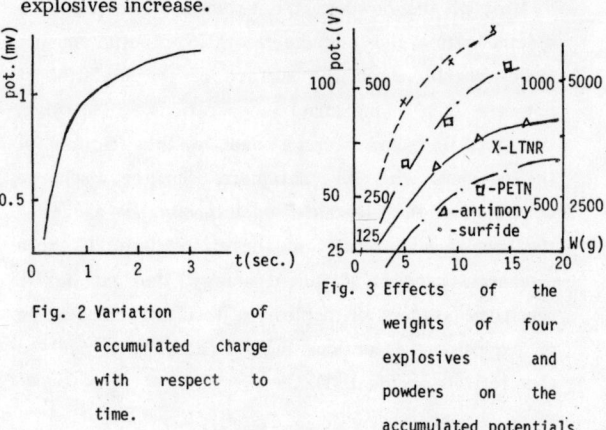

Fig. 2 Variation of accumulated charge with respect to time.

Fig. 3 Effects of the weights of four explosives and powders on the accumulated potentials.

Table 1

pot. and charge \ weight of expl.	Antimony sulfide		LTNR		PETN		Pistok powder		Percussion composition	
	pot. (V)	charge (10^{-9}C)	pot. (V)	charge (10^{-9}C)	pot. (V)	charge (10^{-9}C)	pot. (V)	charge (10^{-9}C)	pot. (V)	charge (10^{-9}C)
5	-156	4.6	-847	25.40	-1717	51.3	-35	1.0	-391	11.7
7	-204	6.12			-3677	110.3				
10	-225	6.7	-1480	44.40	-4680	140.4	-62	1.86	-584	17.4
15	-298	8.7	-1520	45.6	-5602	168.4				
20	-316	9.5					-84	2.50		
25	-338	10.1					-93	2.7		

The potentials become saturated when the weights of explosives are about 15 grams. Within 10 grams of explosives, the potential is proportional to the explosive weight.

(3) Relative humidity has great effect on charge accumulation. For hygroscopic materials, the accumulated potential is inversely proportional to the relative humididty, and with greater effect. For example, the accumulated potential at the humidity of 42% is 1.5 times that at 70%.

3. Analysis of the experimental results

(1) Derivation of the maximum saturated accumulated potential. On the basis of the above discusion the maximum electrostatic accumulation charge $Q_{max}=i_0CR$ and the maximum electrostatic accumulation potential $V_{max}=i_0R$ can be obtained after $dQ/dt=CdV/dt=i_0$ has been experimentally determined.

(2) Assessment of the electrostatic safety of explosives. According to Ref.(4) and the experimental curves in Fig.3, analysis result indicates that the maximum ignition energy of LTNR is less then 1mJ. In terms of "safe explosive" we mean that explosives which would not ignite due to its electrostatic accumulation at pouring 100 grams of it, and these explosives are called "safe explosives"; but explosives that ignite are called "unsafe explosives".

(3) The concept of "safe surface potential".
The experimental result of an institute in Japan (6) indicated that the surface potential accumulated in the discharge of the nonconductor is as high as 30 KV while the discharge energy is only hundreds of microjoules. It implies that the equivalent capacitance of discharge is 0.22_pF. the minimum ignition energy of LTNR E=0.3 J. From the following equation

$$V = \sqrt{2E/C} \qquad (6)$$

We have the safe surface potential of LTNR
$V < 1732$ Volts.

(4) The correlation of temperature and relative humidity. Experiment reveals the following aproximate correlation, which in any cases would be most appropriate for electrostatic release. So it is best to control the relative humudity at 65-70% in the processing rooms of powers and explosives.

4. Conclusion

1) Our experiment arrangment can be used to assess the electrification capacity as well as to study the progress of accumulation and divulgation of dust material.

2) The maximum saturated charges, the saturated potential as well as the maximum volume and plane charge density can be derived due to the experiment curves.

3) It can be seen from the factors affecting the electrostatic accumulation of dust material that the appropriade relative humidity in the processing room has great effect on the electrostatic release.

4) The development of the conception of "safe explosive" and "safe surface potential" has practical significance in the electrostatic safety and the safe operation of powers and explosives.

References

(1) "Survey of the electrostatic accidents happened in an explosive plant" May. 1987.

(2) D.P. Dongan "Trends in electrostatic precautions in filling factors" A.D.-A 036015 P. 363-383, 1976.

(3) 電気試験所:木脇久智"電試彙報". 1954. 5. 18 〔日〕

(4) R.M.H. Wyett et al. "The effect of temperature and relative humidity on the accumulation of electrostatic charges and primary explosives", ERDE Offprit 1972.

(5) liu Yanyi and Wang Changying "The study of powder electrostatic accumlation test devies" 1987.

(6) 静電気安全指針. 1978. 10. P. 47 〔日〕

DANGERS FROM STATIC ELECTRICITY
IN BLACK POWDER PRODUCTION AND ITS COUNTERMEASURES

Wang Wanlu

Jindong Chemical Plant, Yangquan, Shanxi, China

Abstract

There are serious dangers from the stattic electricity in the production and application of Black Powder (BP). The accumulation of static electricity can be reached 11 KV, the electrostatic voltage on the operators can be reached 15 KV, the stored energy will be up to 146 mJ, which is much larger than 0.203, i.e. E_0 of BP. Thanks to the adoption of the corresponding countermeasures against the dangers and calamities, remarkable progress has been obtained. The present paper mainly involves in discussion of the dangers from the static electricity, and analysis has been made about its countermeasures, both theoretically and practically.

I. Introduction

The BP plants such as Austin and Dupont in USA, Empire Chemical Co. Ltd. in Scotland were troubled with accidents of explosion, causing stoppage of production or temporary close down. The explosion accident in the BP plant of Dyno Company in Norway in 1976 caused injuries and deaths, the entire plant was distroyed, and it is not rebuilt up to now[1]. There were many accidents of explosion in the BP plants and the users in China. The accident of explosion in 1985 in a firework plant, which was caused by static electricity, caused great injuries and deaths, destroyed workshops, and the plant close down[2].

Due to the dried hot-pressing process which is adopted in the BP production, the materials are treated with low-moisture contents, the resistivity of the raw material, semi-finished product & finished product are extremely high (10^{11}—10^{16} $\Omega \cdot$cm). The metal parts of the equipment become isolated because of the dried enviornment, further more, the mechanical process of milling & frict separating, e.g. mixing of the three ingredients, beating, granulating can cause accumulation of static electricity fairly easy. The voltage of the static electricity of the three-ingredient-mixer during output can be reached 11 KV, the voltage will be up to several KV during the process of beating as the cloth covering was taken off the powder cake. Electrification experiment confirmed the charge density can be reached 12.830 μC/kg between the rubbing of the fine grains of BP and the cloth-trough, the charge between the powdered BP and Cu-trough can be reached 2.076 μC/kg.[3]

The static charge on the bodies of the operators during production is also very high. The electric charge on the operator, who worked in the process of moistened mixing and wrapping of the powder, had been measured. The operator was standing on the non-conductive rubber plate on the floor, and wearing a normal rubber overshoes. When the operator took up the wrapping cloth quickly from the platform, the electric charge on the operator produced a high voltage, more than 10 KV and reached 15 KV as maximum. When the capacity of the body to earth is 140 pF, under the same conditions, the stored energy can be reached 146 mJ. This was the situation before taking any countermeasures and precautions. According to the statistic figure from Janpan, the endangerment caused by electric charge on the operator takes 12%.[4] There have been many accidents of fire & explosion by the static electricity of the operators in China too, it is hard to mention them individually.

The three preconditions that cause calamities of static electricity are 1) the charge on the body can be up to a certain extent that would cause discharge, 2) the energy of discharge is large enough to ignite the inflammable materials, and 3) the inflammable materials are right in the channel of discharge. When W_{max} (or $W_{operator}$) is eequal or more then E_O, it will be possible to induce the endangerment of the static electricity.

Due to the different test condition and different state of explosion, the difference of value E_O of inflammable materials is very large. Taking the black powder for example, E_O was measured as 6mJ—800mJ by certain plants of other countries, and 4.06mJ—45mJ[5] by certain plants and institutions in China. If based on the minimum value E_O of 2.203mJ for black powder as the minimum ignition energy, the stored energy (max.) on the charged operator's body during processing is 720 times of the min. value. This is the reason why the static electricity is extremely dangerous in BP prosuction with the process of the dried-hot-pressing.

II. Study on Countermeasures

The accident can be effectively avoided if one of the three preconditions is eliminated. In practice, such following countermeasures are effective to reduce the accidents. The charge energy is not exceeding the value of E_O; during the entire process of BP production, there should be no equipment acting as an isolated conductor or an insulator; there should be a proper control for the flow-rate of the materials; there should be a certain stand-still after milling, granulating, sifting, lot-forming, particularly there should be a strict control to the creation and accumulation of the static charge to the operators; proper increase the relative humidity in production rooms and the surroundings.

Earthing methods both directly and indirectly are basic way to eliminate the danger of the static electricity. Direct earthing means that the metal equipment is directly connected to the ground, and the potential of the conductors will be under control. It is measured that when the relative humidity in the production line is not less than 65%, the resistance for the iso- lated conductor is below 10^{10}, and the resistivity for the dielectric material being processed is below $10^{10} \Omega \cdot cm$. Under such conditions, the result of earthing is obvious and effective in eliminating the electro-static charge. Generally speaking, the earthing resistance should be not more than 10^3 ohm for the individual earthing system.

In order to avoid the danger of discharging caused by the stronger electrical field between the charged powder materials and earthing interior surfaces of the equipment, the electro-static-proof materials (such as conductive rubber plate, liquid rubber, etc) are adopted, substituting for the conductive metals, connecting with the earthing items. This kind of method is called indirect earthing.[6] The motorized screen for the three-ingredient-mixer and the mixing drum has been safely treated by this kind of method. The indirect earthing could be also applied to the isolated conductor with small capacitance to avoid discharging with other objects. The indirect earthing resistance is expected within 10^4—10^6 ohms in general cases.

Another one of the main safety measures against the electro-static hazards in BP production is to increase the relative humidity in the production rooms.

TABLE I shows the dependence of moisture content, volume resistivity and charge density of the fine grain BP on relative humidity after keeping 12 hours at the temperature of $25\pm1°C$.[7]

TABLE I

Relative Humidity(%)	18	30	50	70	90
Moisture(%)	0.51	0.62	0.85	0.92	1.10
Volume Resistivity ($\Omega \cdot cm$)	10^{13}	10^{11}	10^9	10^8	10^7
Charge Density ($\mu c/kg$)	0.58	0.46	0.27	0.14	0.018

The results indicate that the volume resistivity of the black powder is decreased exponentially with the increase of the moisture content, there exists a linear relation between the charge density and the moisture content. As shown in Fig.1 & 2, the charge electrified is rather small when the moisture content occupies 1.9%. When the ambient relative humidity is more than 60%, the moisture content of the black powder being processed will be more than 0.90%, and the particle volume resistivity will be dropped down below 10^9 ohms.cm, and the corresponding specific charge density will be decreased by one or several orders. Under these conditions, the earthing resistances of the equipment will be greatly decreased. The resistance between the surface of oxhide and the mixing drum is 10^{10} ohms, and the wooden surface resistivity of the drum is 10^9 ohms. On contrary, as the ambient relative humidity is 18%, the moisture content of the black powder is less than 0.5%, and the particle volume resistivity is more than 10^{13} ohms.cm, therefore there will be unfavourable conditions for leaking of the electrostatic charge. The resistance over the surface of the oxhide to the mixing drum is up to 10^{13} ohms and the wooden surface resistivity of the drum is up to 10^{12} ohms, the serious accumulation of the electrostatic charge will be built up during the production.

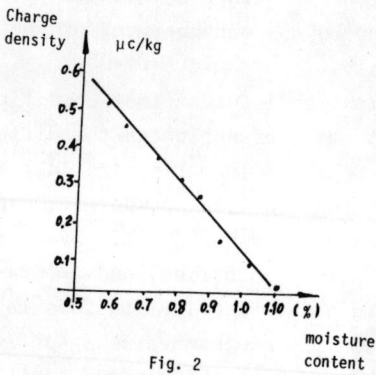

Fig. 2

According to the statistic analysis of the accidents of explosion that have happened in certain plant, of all the accidents, 85.7% took place during the dry season (Nov.—next May), 92.9% took place under low temperature conditions, 71.4% took place as the moisture content of the ambient air is lower than 4g/kg. This shows that the explosion accidents due to the static electricity are extremely serious as the production is performed under the conditions of dryness, low temperature and with low moisture content. Besides, in the intermittent production, the pure cotton bags of the powder are always charged with thousands of volts during the courses of filling, emptying, shaking and knocking under low temperature. However when the relative humidity is up to 65%, the surface resistance of cotton articles is below 10^9 ohms, the voltage will only reach 100volts even with shaking.

The author has made many experiments about the studies of the dependence of ignition sensitivity on moisture content. E_{50} for the black powder decreases firstly as the moisture content increases, and it will be increased when the moisture content gets larger as shown in Fig. 3 and Fig.4. When the moisture content is more than 7%, the powder is hard to be ignited even with a strong spark. Fig. 3 shows the sensitivity curve of the fine grain BP at the different moisture content with a temperature of $25\pm1°C$. Fig.4 shows the relationship between the moisture content of black powder No. 1 & 2 and E_{50}.

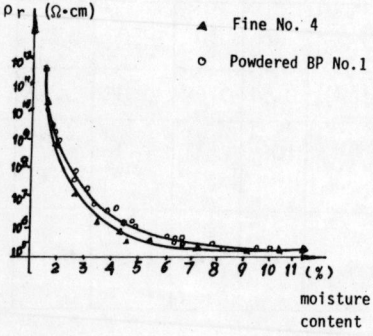

Fig. 1

It can be seen that there exists a general function relation, and the moisture contents corresponding to the minimum E_{50} varies with different kinds of black powder.

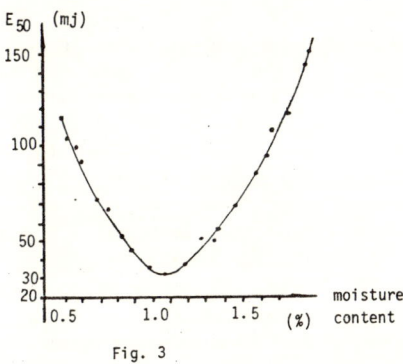

Fig. 3

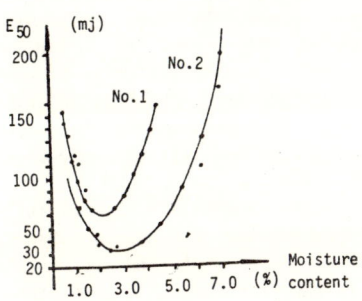

Fig. 4

In view of our experiments, the ambient relative humidity in BP production should be properly chosen as 65% for the dry & hot-pressing process. Under this situation, oprators could be out of the incendiary sensitive region of black powder due to electrostatic discharge. Wet (moistent) cold-pressing process has also been adopted in some other countries with the moisture content of 4% before pressing. There is no doubt that this will confirm the safety and get rid of the dangers caused by the electro-static discharge, i.e. out of the critical range of sensitivity.

Reference

1. James. E. Rose, "Black Powder — Contemporary Comments", Alexander P. Hardt 1979, (in English)

2. Wang Sigong, "Lesson from the Accident of Explosion in a Firework Plant", "Labour Protection" No. 4, 1985, p12, (in Chinese)

3. Wang Wanlu, "Study and Practice on Technology of Measures against Electrostatic Charge in BP Production", "Safety Technology of Explosives", No.3, 1984, p25—26 (in Chinese)

4. "On Measures against Static Electricity——Answer to Technical Staff on Spot", Research Institute on Industrial Safety (in Japanese)

5. Ren Baoming, Wang Wanlu, He Longwen, Li Ruinian, "Discussion on Dangers of Static Electricity in BP Production", "Thesis of Public Lecture on Static Electricity by China Physical Society", No. 87, p60 (in Chinese)

6. Wang Wanlu "Rubber Articles of Electro-static-conductor and Measures against Dangers of Static Electricity", "Labour Protection", No. 2, 1986, p18—19, (in Chinese)

7. Wang Wanlu "Function of Humidification in guard against of Dangers of Static Electricity in BP Production", "Technology on Static Electricity", No. 1, 1985, p37—38, (in Chinese)

THE SAFETY ELECTROSTATIC POTENTIAL OF NON-INCENDIARY DISCHARGES FROM CHARGED INSULATOR

Huang Jiu-sheng and Zhao Lu-zhen
Beijing Municipal Institute of Labour Protection

Abstract

This paper aims at the determination of electrostatic potential of charged insulator from which incendiary discharges will not occur basing on the charge transferred in a discharge as the ignition threshold. Many factors that have influence on charge transfers have been investigated. The relationship between maximum charge transfer and the potential of charged insulator has been founded. The results show that the minimum potential of charged insulator from which discharges will ignite the flammable gases with minimum ignition energy of being greater or equal to 0.2 mJ is about 35 kV. Considering safety margin, 15 kV is suggested to be the safety electrostatic potential of charged insulator for gases mentioned above.

Introduction

Fires and explosions may be occur in the flammable atmosphere when the potential of charged insulator exceeds certain value. It is important to determinate the safety electrostatic potential of charged insulator for the prevention of electrostatic accidents and evaluation of antistatic procedures.

Many authors have studied the energy of discharge from charged insulator [1-3]. But few of them investigated the safety potential of charged insulator. Y.Tabata (1983) [4] reported that the hydrogen gases with ignition energy of several ten microjoules will be ignited when the potential of charged insulator exceeds 10 kV, and for gases with ignition energy of several hundred microjoules, the potential is 30 kV. Y.Tabata and S.Masuda (1984) [2] presented a equation with a coefficient k (k<1) for calculating the energy of discharge from charged insulator and estimated the minimum ignition potential of charged insulator to cause incendiary discharge. Some safety potential values had been proposed in Japan's book [5], for hydrogen gas with minimum ignition energy of several ten microjoules, the potential should be less than 1 kV, for hydrocarbon gases with minimum ignition energy of several hundred microjoules, the potential should be less than 5 kV. But up to now no detailed experimental results have been reported about the safety potential of no incendiary discharges from charged insulators. Because the measurement of potential is convenient in the worksite, this paper concentrates on the study of safety potential of charged insulator.

Experimental Method

Preliminary results showed that a charge transfer of 0.1 μc in a single brush discharge from charged insulator may ignite flammable gases with minimum ignition energy of 0.2 mJ [1] [6] [7]. This paper experimentally investigated many factors which have an effect upon the charge transfers such as electrostatic potential of the charged insulator, diameter of the grounding electrode, surface size of the charged insulator, distance between the surface of charged insulator and the earthed metal plate, surface resistivity of the insulator and the methodes of charging etc. The potential of charged insulator corresponding to charge transfer of 0.1 μc is regarded as the minimum ignition potential. The safety potential will be determined after making some allowance for a safety margin.

The testing insulator, a square plate with 3 mm thick, is discharged by static eliminator of induction type before charging it in order to charge uniformly. There are four nylon threads tied each corner of the insulator to prevent the fingers of operator from discharging with the charged insulator and contaminating the surface of insulator when moving and lifting the charged insulator. The insulator is placed on a large grounded metal plate and fixed by plastics to avoid moving when rubbing by mohair sheet. The charged insulator plate is lifted to a plastic support and its centre potential is registered by a electrostatic voltmeter. Discharge is produced by approaching the charged insulator plate to an earthed metal electrode which connected to a capacitor so the charge transferred is calculated by the formula Q=CV, C is the electrical capacity of the measuring system and V is the maximum voltage which recorded by a storage oscilloscope. Repeated several hundred experiments. Take the average value of the larger charge transfer from several hundred discharges under the same potential of charged insulator as the maximum charge transfer of the potential. The ambient temperature was 25±2 C and relative humidity 30±5% during all experiments.

Experimental Results

Effect of the electrode diameter Fig.(1) is the experimental results of discharges between polyvinyle chloride (PVC) plate with area of 14.1X 14.1 cm² and electrodes with diameter of 50, 30, 20 and 8 mm. The potential of the PVC plate corresponding charge transfer of 0.1 μc is plotted on the ordinate. The diameter of electrodes is plotted on the abscissa. As shown in Fig.(1), the potential of PVC plate with charge transfer of 0.1 μc get minimum when the electrode diameter is 20 mm. It means under the same conditions the charge transfer is greater than that with other electrodes

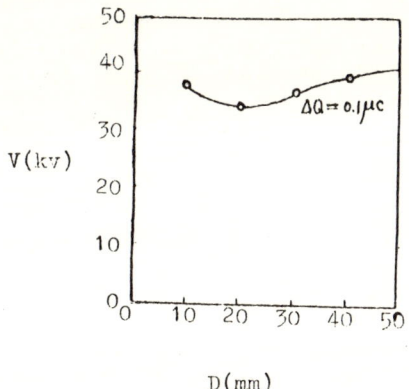

Fig.(1). Potential, V, of the PVC plate corresponding maximum charge transfer of 0.1 μc as a function of the diameter, D, of the electrode. The PVC plate is 14.1×14.1 cm², negatively charged by rubbing.

Effect of the area of charged insulator The experimental results of discharge between charged PVC plates with area of 50, 100, 200, 300 and 400 cm² and the electrode with diameter of 20 mm is shown in Fig.(2). The charge transfers did not reach to 0.1 μc although several hundreds tests have been done with 50 cm² PVC plate (Fig.(2) does not indicate) as shown in Fig.(2). when the area of PVC plate is 200 cm², the potential of PVC plate corresponding charge transfer of 0.1 μc is lower than other plates. It means that under the same conditions the charge transfer will be greater than that of others when PVC plate area is 200 cm².

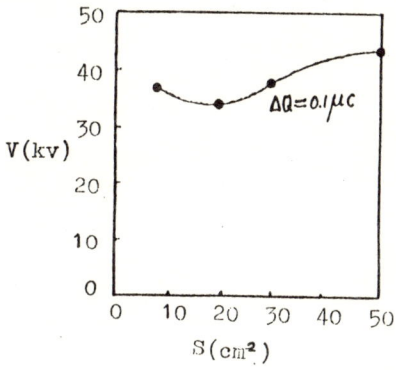

Fig.(2). Potential, V, of the PVC plate corresponding maximum charge transfer of 0.1 μc as a function of the area, S, of charged PVC plate. Electrode diameter is 20 mm, negatively charged by rubbing.

Effect of the distance from charged insulator surface to the earthed metal plate Fig.(3) shows the results of discharges from 20×20 cm² PVC plate behind which there is an earthed metal plate which parallel to the PVC plate. The electrode diameter is 20 mm. From Fig.(3), it can be seen that the larger the distance from the charged surface of PVC plate to the earthed metal plate, the higher the potential of PVC plate corresponding charge transfer of 0.1 μc. The rate of increase of potential tend to zero when the distance greater than 15 cm. This means the influence of the earthed metal plate on the charge transfer can be neglected when the distance is greater than 15 cm. The tests also show that when the distance is less than 4 mm, the charge transfer can not achieve 0.1 μc if it is charged by rubbing.

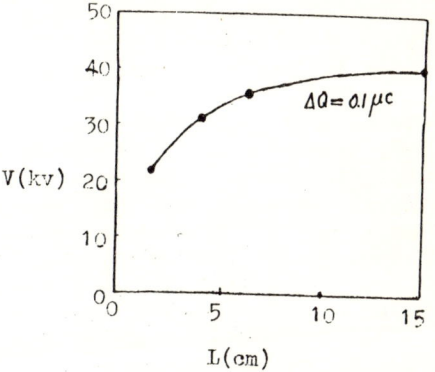

Fig.(3) Potential, V, of the PVC plate corresponding maximum charge transfer of 0.1 μc as a function of the distance, L, from charged surface of PVC plate to earthed metal plate, PVC plate is 20×20 cm², Electrode diameter is 20 mm.

Effect of the surface resistivity of the insulate plate Experimented results indicate that under the same conditions the charge transferred from charged PVC plate with surface resistivity of 10^{16} –10^{17} ohms is slightly greater than that of charged polyethylene plate with surface resistivity 10^{17} –10^{18} ohms.

Effect of charging method The results show that the charge transfer of discharges from insulator charged by rubbing is a little larger than that by corona charging, but the charge transfer from corona charged plate will be many times greater than that by rubbing when there is a grounding metal plate behind the charged surface , especially the distance between them is very short. However in practice electrostatic charge is usually generated by rubbing.

Estimation of minimum ignition potential and the determination of safety potential The results above show that the charge transfer will be greater when the diameter of electrode is 20 mm, area of charged insulator is 200 cm², the plate is negatively charged by rubbing, the relative humidity is less than 30%. The results under these conditions are shown in Fig.(4). Only parts of the results choosed from several hundred tests is marked in the figure. The out envelope curve in Fig.(4) means no matter what size is the electrode, what area is the insulator, how is the charging method and any polarity of charges on the insulator, what relative humidity of the atmosphere is, the charge transfer in single discharge will not greater than that showed by the curve when the distance between charged surface of insulator and earthed metal is greater than 15 cm. This is called the maximum charge transfer curve under the optimum condititions. The potential of charged insulator corresponding charge transfer of 0.1 μc is about 35 kV. That is, the minimum potential to ignite the flammable gases with minimum ignition energy of 0.2 mJ is about 35 kV. It can be considered to have an equivalent electrical energy of 0.2 mJ. We can also see from Fig.(4) that the potential of insulator corresponding the maximum charge transfer of 0.02 μc is about 15 kV. Nevertheless the charge transfer to ignite hydrogen gases of minimum ignition energy of 0.02 mJ is about 0.02 μc [7]. It means the equivalent electrical energy is about 0.02 mJ. Therefore, in the view of equivalent electrical energy, if we take 15 kV as the safety potential of insulator for the flammable gases with minimum ignition energy of 0.2 mJ. The safety margin is 10. So we suggest 15 kV as the safety potential of insulator which can not ignite flammable gases with minimum ignition energy more than 0.2 mJ.

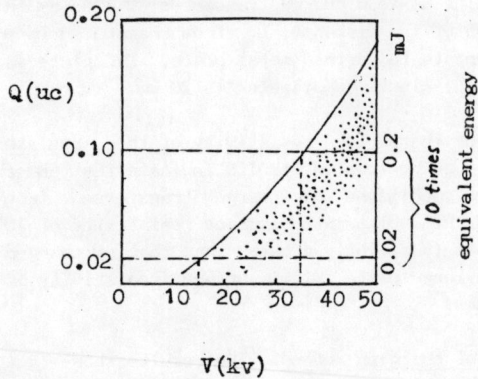

Fig.(4) Charge transfer, Q, as a function of potential, V, of the charged insulator plate.

Conclusion

This paper investigated the safety potential of charged insulator based on the charge transfer of 0.1 μc as the ignition criterion for flammable gases with minimum ignition energy of 0.2 mJ. Many factors which have influences upon the charge transfer have been tested. The relationship between the maximum charge transfer and the potential of charged insulator had been founded. The minimum potential to ignite flammable gases with minimum ignition energy of 0.2 mJ is about 35 kV. In view of possible experimental error and other imponderabilities, some allowance has been made for a safety margin. We recommend that the safety potential of charged insulator is 15 kV for gases and vapors of minimum ignition energy greater than 0.2 mJ.

References

[1] K.G. Lovstrand, J.Electrostatics, vol. 10, p.161-168, 1981.
[2] Y.Tabata; S.Masuda, IEEE Tran IA-20, no.5, p.1206-1211, 1984.
[3] Tang An-zhen; Zhou Xi-zhong, Physics, vol.10, no.9, p.543-549, 1981 (Published in China).
[4] Y. Tabata, High pressure gas, vol.20, no.12, p.16, 1983. (Published in Japan).
[5] Industrial Safety Institute of the Ministry of Labour Japan, Guidance of Electrostatic Safety, 1978, Published in Japan.
[6] N.Gibson; F.C.Lloyd, Br.J.Appl.physi, vol.16, p.1619-1613, 1965.
[7] N.Wilson, J.Electrostatics, 16, p.231-235, 1985.

THE PRINCIPLE AND APPLICATION OF α ION SOURCE STATIC ELIMINATOR

Zhang HaiYing
NORICO Chao Yang, Instrument Factory
China

Abstract

This article introduces the outline of using radioisotopes to eliminate static electricity in general. It stresses the principle that α ion source eliminates static electricity. The radioisotope is chosen in terms of the specific limit current I_{lim} and the radiation dose. The inner radius of the chamber is designed in light of Bragg's curve. The structure of ion-blower eliminatior and its effects on eliminating static charge are discussed on whole. In addition, its applications in defense industry and opitc industry are stated in brief.

Introduction

The technology, making use of radioisotope to eliminte static electricity, has been used for more than 30 years. Since α and β rays have high ability in ionization and the radiation protection is simple, α and β radioisotopes are widely used in static eliminator. As far as using α radioisotopes to make the eliminator is concerned, there generally exist two kinds of eliminator—fixed style and ion blower style. The effective range of elimination with the fixed style is very small and its application is limited. However, α radioisotope is widely used in ion blower style. This paper only introduces the principle and applications of ion blower static eliminator.

Principle

Generation of ion   When α particles undergo collision with atoms and moleculars in air, ion pairs are generated in the air. One α particle with 5 Mev energy can generate 1.5×10^5 ion pairs in the air. These ion pairs are distributed around the track of the α particle inhomogeneously.

The number of ion pairs within each unit length of the track is called specific ionization Sp. The ralationship between Sp and the residual range R was firstly measyred by Bragg's with the experiment. At the beginning of the residual range (at the surface of source), Sp is smaller, with decrease of R, Sp rises gradually. As soon as R is very closed to zero, Sp rises greatly to the maximum value and then drops down at zero.

Principle   If ion pair fill the space around the charged object, the opposite ions in comparision with those on the object must be attracted to the object and neutralized with the charges on the object.

The specific limit current I_{lim} If the energy of a radioactive source is all consumed for generating ion pairs and each of the ion pair is used in eliminating static electricity, the maximum current supplied by the radioactive source is called the maximum current I_{max}.

In fact, since the total energy of α particles can not be used to generate ion pairs in the air, and some ion pairs do not eliminate static charges due to their diffuse migration and recombination. The useful current in eliminating static charge is much less than I_{max}. The maximum current corresponding with each unit activity of radioactive source is called the specific limit current I_{lim} in the light of following definition.

$$I_{lim} = I_{max}/A \quad (A/B_q)$$

bing on of the important index in choosing radioisotope for eliminator, the I_{lim} values of some radioisotopes used in elimintor is listed in Table 1. Where A refer to the activity of radioactive source.

Table 1. The I_{lim} values of some radioisotopes

Isotope	$T_{1/2}$	Ray	E(Mev)	$I_{lim}(\times 10^{-19} A)$
^{90}Sr	28Y	β	0.51 2.2	4.4
$^{204}T_l$	4.1y	β	0.76	1.2
$^{210}P_o$	138.4d	α	5.3	25.1
$^{238}P_u$	87y	α	5.5	26.0
$^{241}A_m$	433y	α	5.45	25.9

Applications

Structure of α ion source
Based on the advantages of high specific limit current, ling half life, very week γ radiation, as given by Table 1, the ^{238}Pu could be selected to satisfy effective elimination.

In consideration of the structure of α ion source, the radius of its inner chamber should be larger than the maximum range of α particle in the air so that the track corresponding to maximum specific ionization ends up in the air.

The effect on eliminating static charge can be evaluated by half time $T_{1/2}$. The relationship between $T_{1/2}$ and the inner radius of the chamber has been pointed out. The smaller the radius, the longer the $T_{1/2}$, which results in worse effect on eliminating electrostatic. With increase of inner radius, $T_{1/2}$ drops down very quickly. The inner radius of α ioniic sourcse has been optimized so as to get the best effect of unit activity of the radioactive source.

Influence of distance R on eliminating static electricity The relationship between $T_{1/2}$ and the distance from the ionic source to the charged object is given in Table 2. Obviously, the shorter the distance, the better the effect. This is because the higher the ionic concentration around the charged object, the better the effect on eliminating static charge. Opositely, the longer the distance, the lower the ionic concentration, the worse the effect.

Application in industry of explosive

In explosive industry there exists serious electrostatic hazards in the fabrication and usage of explosive products. The ion blower eliminator is widely used in the case of eliminating static electricity. It has advantages of wide effective range, obvious effect, simple and convenient operation, and has been used in our military manufactories.

Application in optic industry

In the production of optic instruments, static charges is always generated on glass by rubbing or moving, which causes absorbing dust and decreases the working efficiency. For this reason, the ion blower eliminator has been used in many factories such as the Chao Yang Instrument Factory. It has been proved that the working efficiency is increased while the labour intensity is decreased. The effect on eliminating static charge with our made eliminator is excellent. Actually, its characteristics are superiou to the same kind products made by 3M Company of America in 1986.

Table 2. Relationship between $T_{1/2}$ and R

R(m)	1.5	1.2	1.5	1.8	2.0	3.0	4.0
n	6	8	12	8	6	6	6
$T_{1/2}$(s)	1.6	2.4	2.4	3.2	3.3	4.4	7.8

n—the tested number

Experimental conditions:
P=4.0 kg/cm^2,   V=5 kv,
T=27°C,   R.H.=37%

References

Lu Xi-Ting,
A manual about nuclear parameters, Atomic Energy Publishing House, 1981.

Wang Tong-Sheng,
The basis of nuclear radiation protection, p 53-63, Atomic Energy Publishing House, 1983.

Nuclear Energy Institute, Experiment report about ^{210}Po source used in eliminator, 1983.

Proceedings of International Conference on Modern Electrostatics

Po-210 STATIC ELIMINATOR AND ITS INDUSTRIAL APPLICATION

Cai Shanyu, Zhou Zhenghe, Hao Shiqi, Sun Shuzheng
Isotope Division, Institute of Atomic Energy
P. O. Box 275/66, Beijing, P. R. China

Abstract

The construction, characteristics and features of Po-210 static eliminator made in IAE of China, are simply described in this paper. Applications of Po-210 static eliminator in printing, paper, photographic film, plastic, rubber, light guide fibre, textile, instrument and package, etc; are summarized. Tests indicate that radioactive static eliminator is one of the effective methods for eliminating harmful static electricity in industry. Good results has been obtained in reducing consume, saving power, increasing yields, enhancing quality, eliminating electric shock and ensuring safety in production. Po-210 static eliminator has enormus potential application.

Introduction

Static electricity problems become much more serious along with the development of modern science and technology, because the new types of processes, equipments, materials and techniques are constantly emerged, as well as automation and high speed are gradually relized. In many cases, the accumulated static charge not ouly causes electric shock to personnel, but also effects the productivity and quality. Furthermore, static electricity can result in fire and explosion.

Isotope Division of IAE began to develop the radioactive static eliminator in 1980, in order to solve the harm of static electricity in the processes. The principle of radioactive static eliminator is that the accumulated static charge on the surface of insulator can be neutralized by positive or negative ions which are generated by the rays emmission from the radioactive source in the air. In comparison with the ionization capability of three different kinds of rays, alpha particle ($\sim 4\times 10^4$ ion pairs/cm) is stronger then beta particle (several hundreds ion pairs/cm) and gamma ray (several ion pairs/cm). On the other hand, Po-210 is almost a pure alpha emitter comparing with the other emitter nuclides, such as Am-241, Pu-238 and Cm-244, and its high specific radioactivity. It is suitable to choose Po-210 for this form of eliminator as result of its intense ionizing capability, simple shielding, relative cheap and easier availability.

One of the most important things is to prepare safer radioactive source, so as to utilize radioactive static eliminator in industry. At present, Po-210 static eliminator has been produced by Nuclear products Department 3M Company[1] using microsphere technology. The Amersham International Ltd. has been preparing Po-210 static eliminator for many years by means of powder metallurgy-rolling technique.[2,3] In addition, the enamel method is being used to make Pu-239 and Pm-147 static eliminator in Soviet Union.[4] The Institute of Atomic Energy used spontaneous-nickel plating method to manufacture Po-210 static eliminator. latter, it was replaced by powder metallurgy technique, so that the safety of Po-210(α)source was improved. the quality control and contamination control of Po-210 static eliminator products have reached the national and international standard through the special examination and long-term monitoring.[5,6] Products of Po-210 static eliminator passed the technical identification and started to put into the market in 1983.

In the present years, polonium static eliminator has been widely used in printing, paper, photographic film, plastic, rubber, light guide fibre, textile, instrument and package, etc; We are now looking for the new applications to expend usage areas of Po-210 static eliminator.[7,8]

Construction of Po-210 Static Eliminator

Different models and specifications of Po-210 static eliminator were developed by IAE in China. These Products are available as bar, ring, half-ring, disc, miniature, ionized air gun and ionized air nozzle,etc(see Fig1).

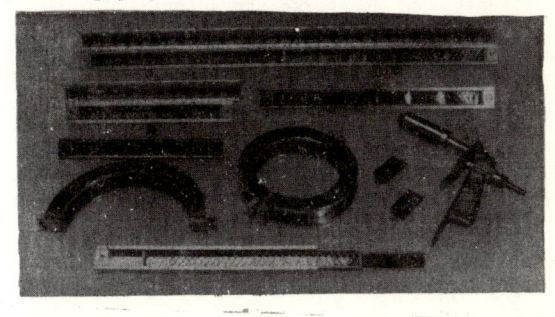

Fig 1. Po-210 static eliminator

The main part of Po-210 static eliminator is Po-210 (α)foil source which construction is shown in Fig 2.

-477-

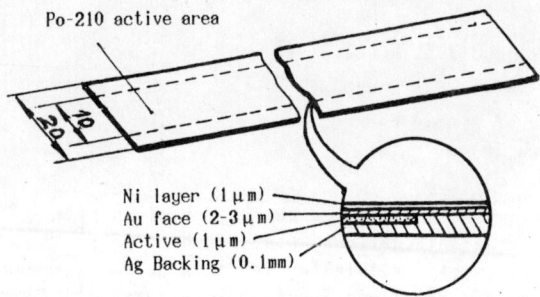

Fig 2. Construction of Po-210(α)foil

The Polonium, as a relatively stable, inert compound, is intimately bonded between the corrosion resistant gold face and the silver backing. Further protection is provided by a nickel coating.

The total thickness of Po-210(α) foil is about 0.1 mm. the activity loading of foil is controlled at $2.96 \times 10^9 Bq(80mCi)/m$ considering the safety in production and application. The maximum ion current (i.e. saturated ion current) providing by Po-210 static eliminator is about $0.1 \mu A/3.7 \times 10^7 Bq(mCi)$.

Characteristics of Po-210 Static Eliminator

The half-life of Po-210 is 138.4 days. The range of 5.305 Mev Po-210 alpha particle is 3.93cm and about 1.5×10^5 ion pairs per alpha particle along a 2-3 cm track.

The working ability of Po-210 static eliminator is considerably related to the number of ion pairs generated by alpha particles from eliminator in the air. In other words, the ionization capability of Po-210 static eliminator depends on the activity of Po-210(α) foil source, efficiency of ionization and fixed position.

The typical discharge characteristic of a single unit(bar) Po-210 static eliminator is illustrated in the following figure.

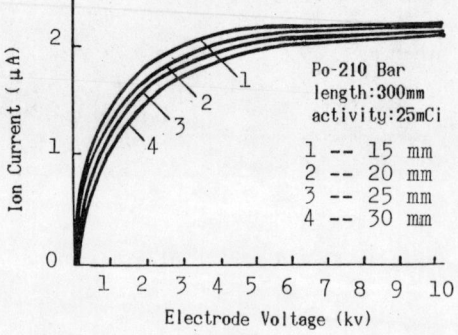

Fig 3. Discharge characteristic of a Po-210 bar alone

Fig 3 shows the ion current of Po-210 static eliminator increases with increasing the voltage between bar and electrode; the ion current gradually reaches stauration when the voltage of electrodes continuously increases; the ion current decreases with increasing the bar-electrode distance when the voltage is constant. So, the large ion current can also be obtained in a weak electric field as long as shortening the distance of electrodes. In fact, the optimum distance from a Po-210 static eliminator to the insulator is about 2-3 cm.

It is very effective using Po-210 static eliminator with induction discharge needles when the ion current generated by Po-210 bar alone is not enough to eliminate the accumulated static charge on the surface of non conductor. Fig 4 shows the discharge characteristic for the combination unit Po-210 bar.

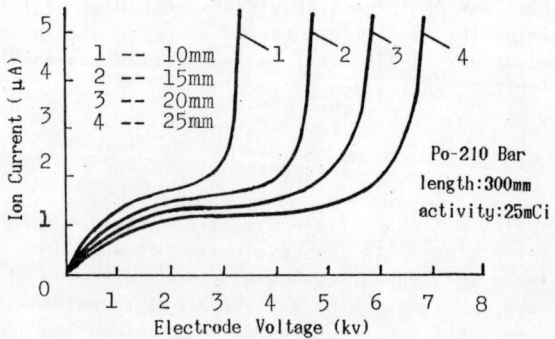

Fig 4. Discharge characteristic of combination static eliminator

It can be seen that the starting voltage of passive bar will drop down and the residual charge will reduce when single unit Po-210 static eliminator and induction discharge needles are used together. So, the combination unit offers a simple and versatile solution, and it is capable of effective neutralization in a wide range of industrial static electricity problems.

Features of Po-210 Static Eliminator

The features of the Po-210 static eliminator are as follows:

(1) It is simple to fix, easy to maintain and convenient to use.

(2) No power supply is required and no electric shock is generated.

(3) It does not produce heat and spark, so, it is adaptable to inflammable and explosive atmospheres.

(4) No matter what the sign of ion is, and no matter how the change of charge density is, the same effect on eliminating the static charge can be obtained.

Table 1. The Status of Industrial Application for Po-210 Static Eliminator

Industry	Processes	Static problems	Solutions	Results
Printing	Paper printing	Adhering to each other or mutual repelling, Putting paper out of order, Disrupting normal production.	Bars Combination	Avoiding adhesion, Putting paper in order Increasing speed, Raising quality of printing.
	Color printing (on plastic film)	Untidy moving, Dust, Printing disorder, Hard to open or seal, Effecting operation.	Bars Combination	Uniform moving, Reducing dust, Increasing yields, Improving quality.
Paper	Cutting Pressing Dring, winding	Electric shock, Paper disorder, Effecting the work of thickness meter.	Bars Combination Ionized air gun	Eliminating electric shock, Reducing losses, Convenient to open paper.
Film	Coating Cutting	Discharge sparking, Static mark(blemish), Attracting dust, Easy to fire.	Bars Ionized air fans	Preventing sparking, Improving the quality, Reducing the waste, Obtaining good benefit.
Plastic	Blowing Plastic film Stretching	Electric shock, Hard to wind up and open. Attracting dust.	Bars Combination	Eliminating electric shock. Reducing dust attraction.
	Paint spraying	Uneven paint coat, Easy to peel off.	Ionized air guns	Removing dust, Paint adhesion.
Rubber	Mangling (tire cord)	Discharge sparking, Easy to fire, Dust, Electric shock.	Combination	Eliminating shock, Reducing dust. Preventing the firing.
Camera	Cleaning and Fixing	Attracting dust, Hard to clean, Effecting quality.	Ionized air guns nozzles	Eliminating static, Removing dust, Raising the quality.
Light guide fibre	Wiredrawing Winding	Hard to gather bundle, Broken end.	Discs	Easy to gather bundle, Smoothing.
Sanitation material	Coating	Discharge Sparking, Easy to fire.	Bars	Preventing the firing Operating safety.
Instrument	Fine balances	static disturbing, Effecting operation.	Miniatures	Eliminating static, Weighing precision.
Textile	Drawing	Flyings, Broken end, Cling, Hard to gather bundle, Dust, Production suspension,	Bars Ionized air nozzles	Reducing flyings, broken end, cling, Reducing consume, Improving quality.
	Heat setting (printing and dyeing)	Hard to fold, Eletric shock, Dust, Effecting operation.	combination	Good folding, Eliminating serious shock, Reducing dust.
Package	Cigarette packing line	Glass paper adhesion, Production suspension.	Bars	Packing in order, Increasing yields.
	Gourment powder packing line	Ununiform moving, Hard to seal bag, Adhering.	Bars Rings	Reducing losses, Normal production.

For the radioactive static eliminator, safety is the prime consideration. The International Organization for Standardization (ISO) specifies that the prototype testing should be carried out before the sealed sources are put into actual use. The substance of testing consists of temperature, external pressure, impact, vibration and puncture. According to the classification of sealed source performance standard, it is C22222 for the radioactive static eliminator. The classification of IAE products is up to or higher then that of ISO standard through identification.

The radioactive source and static eliminator are respectively wiped with cotton wool or tissue, then, the activity removed is measured. Wipe testing shows that the acitvity removed from Po-210(α) sources is less then 185 Bq (0.005 μCi) and the activity removed from Po-210 static eliminator is less then 1.85 Bq (0.05 nCi).

In addition, the exposure rate of Po-210 static eliminator does not exceed 0.75 mrem/hr in 1 cm (this is the maximum exposure rate permitted for non accupationally exposed workers).

The activity will decrease 16% after a year, because the half-life of Po-210 is 138.4 days. At that time, the ion current of Po-210 static eliminator will reduce correspondingly. In general the old unit should be substituted.

Application of Po-210 Static Eliminator

The applications of Po-210 static eliminator in production processes are listed in table 1. Practice indicates that industrial static electricity is very complicated. In some processes, the effect is obvious for eliminating static elec tricity; but, in some processes, the effect is not too satisfied. It is necessary to continue investigation.

In order to get good results, it is important to choose the fixed position. In general, Po-210 static eliminator is fixed in the suitable place that is required for processes. Static measuring is needed to understand whether the fixed position is resonable or not by means of static detecting device. In addition, the Po-210 static eliminator itself must be well earthed.

References

(1) J. P. Ryan, T. H. Lahr; "Microspherical Carrier Carriers for Radioisotopes", Nucl. News, Vol. 8, No. 7, P13 (1965).
(2) Amersham International Ltd.; "Static eliminators", Catalogues RS 16-3 (1981).
(3) R. J. B. Hadden and B. J. Senior; "Static Elimination: A guide to theory and practice", TRC Report No 413 (1979).
(4) К.Д.Писманник и др;"Radioisotopic Neutralization of electrostatic Charges in Industry", Proceedings of the Fouth International Conference, Geneva, Vol. 14, P331(1971).
(5) International Atomic Energy Agency; "Seaaled Radioisotope Sources-Classification", ISO-2919 (1980).
(6) GB-4075" Sealed Radioisotope Sources-Ciassification",(1983).
(7) Sun Shu-Zheng;"The Usage Guide of Radioactive Static Eliminator", Technical Information, (1983).
(8) Cai Shan-Yu; "A New Effective Method of Eliminating Static Electricity—Po-210 Static Eliminator", Communication of Weaving Machine, No 1 (1983).

STUDY ON THE CRITICAL VOLTAGE AND ELECTROSTATIC ELIMINATION IN RELEASE OF HYDROGEN

Liu Jun Tao and Liu Bing
101 Station of Ministry of Astronautics Industry
Beijing, P. R. China

ABSTRACT

Based upon the experimental results of minimum ignition energy, limits of inflammability voltage of spark breakdown, the quenching distance and optimum ignition concentration, both electrostatic spark ignition limit (voltage limit) in hydrogen release below -170 °C and a provisional standard in safety voltage are established. The paper also gives the test results before and after adoption of electrostatic eliminator, introduces the principle and method of its design, installation and restriction of discharge energy from it. The spatial static electric charge in the hydrogen released into atmosphere at a high speed of 100--400 m/s and other charged gases can be safely and effectively eliminated.

Introduction

One of the important measures in engineering treatment of liquid hydrogen (LH_2) is to decrease the pressure of its storage tank and let the evaporated hydrogen escape into air from a vertical tube. The review on 96 hydrogen accidents by American NASA indicated that as hydrogen is released into air, fire accidents of the mixture made up 62% of the total and 17.2% of the fires is caused by static electricity[1]. Since we employed LH_2, many fires, deflagrations and detonations[2] have happened and caused equipment damage, staff workers to be wounded and even led to test failure. Therefore it is necessary to define the electrostatic ignition limit of hydrogen and voltage limit, establish provisional standard of safty voltage and develop safe-spark model self-discharge type eliminator so as to ensure the safety of hydrogen release.

Voltage Limit of Hydrogen

Incendivity of static electricity is basically expressed in terms of its energy. However in fact, although the charge carried by the low-temperature hydrogen surpasses 10^{-8} coulomb, its release is still safe so long as the measured potentials at the prescribed points are lower than the voltage limit of ± 5kv. Therefore, in a sense, measurement of electrostatic paramenters can be attributed to the measurement of electrostatic voltage. The lowest voltage to ignite the gaseous hydrogen (GH_2)-air mixture (i.e. voltage limit) depends upon temperature, pressure and humidity and is determined by the minimum ignition energy, limits of inflammability, breakdown voltage and quenching distance.

1. Minimum ignition energy

Minimum ignition energy is the least energy to ignite nearly stoichiometric combustible gaseous mixture by a capacitive discharge spark of 1 ms duration across a gap of about quenching distance composed of a pair of free electrode tips. Its magnitude is

Table 1 Electric ignition parameters of GH_2-air mixture[4]

15 °C					-170 °C				
volumetric conc. of H_2 in air(V%)	quenching distance (mm)	breakdown voltage (kv)	minimum ignition (mJ)	voltage limit (kv)	volumetric conc. of H_2 in air(V%)	quenching distance (mm)	breakdown voltage (kv)	minimum ignition (mJ)	voltage limit (kv)
15	0.89	4	0.046	5.00**	15	1.4	13	1.6	5.00***
30	0.61	2.8	0.019		30	0.89	8.5	0.145*	
45	0.91	3.4	0.043		45	1.47	1.17	1.48	
60	2.08	5.9	0.030		60	3.86	22.8	10.05	

* An estimated value based on the reults of experiments and calculation.
** From a standard of The Research Institute of Industrial Safety of Japanese Ministry of Labour.
*** From our provisional standard.

(4) In order to avoid frictional sparking and impact sparking and reduce the capacitance of the eliminator to the ground, all construction materials used in the eliminator should be soft non-ferrous metals.

In fact, each point mentioned above is not only related to the construction but also to the installation methods. For example, the smaller the radius soft metal support, the smaller the capacitance of the eliminator. The shorter the distance between the first layed electrodes and the top end of the eliminator, the larger the discharge current, the better the elimination. The sharper the point electrode, the shorter the distance between the point electrode and the charged hydrogen, The smaller the critical breakdown voltage, the smaller the discharge energy for each pulse.

In practical application, if the following requirements could be assured, the static eliminator could be used safely.
(1) Adopting a microampere meter connected in serise with the eliminator and the ground, the discharging patterns are recorded.
(2) If the discharging current is less than 2 µA for each point electrode, the corona is a safe discharge, which is remarkably distinguished with spark or brush discharge by its discharge energy (≤ 0.012 mJ).
(3) If the released hydrogen is charged densely, the discharging curent from each point electrode will be larger than 2 µA. In order to reduce the discharging energy, the charge leakage resistance should be used to assure the safety.

2. The Principle of Static Elimination

When the charged LH_2 approaching the earthed point electrode, the field strength gets stronger near the point electrode which would cause the corona discharge as soon as the field strength becomes larger than the critical ionization field. The ionized pairs are attracted to the LH_2 or point electrodes respectively, and by this way the static charge on the LH_2 could be nutralized with the coming ions.

3. The Installation of the Static Eliminator

Generally speaking, the static eliminator had better be installed at the position 30 mm away from the pipe end.

4. The Elimination Results

Contresting experiments have been conducted for various kinds of charged LH_2. The successful elimination results are listed in Table 5. It can be seen that the potential on the LH_2 could be reduced to that below the safty limits with the corona discharging current of 1--5µA.

Table 5 The elimination results under the normal atomospere temperature

No.	potential (kv) before	after	No.	potential(kv) before	after	corona current(µA)
1	5.7~5.0	-1.1	9	*	-3.5	-1.0
2	0.9~-2.9	-0.9	10	*	-4.0	-1.5
3	5.2~0.7	-2.1~0.6	11	*	-1.2	-1.5
4	17~-8.0	1.5~-3.5	12	*	-2.4	-5
5	25	-2.6	13	*	0	-4
6	10~-5.7	0.6~-3.5	14	*	0.8	2
7	5.4	4				
8	3.3	1.8				

In order to vertify the eliminator is of explosion-proof, special experiments under the critical situations for explosion are conducted as shown in Table 6. It is shown that even under the release pressure ≥ 16 kg/cm^2, the Mach Number $M \geq 0.26$ and without N_2 protection, the ignition caused by static discharge has not been observed.

Table 6 Experimental results under the normal atomsephere temperature the Mach number M= 0.31

No.	time (s)	measured GH_2 pressure (kg/cm^2) direct	indirect	discharging current(µA)	ignition result	Note: with eliminator
1	10	16	16	------	yes	no
2	10	16	16	------	yes	no
3	10	16	16	------	yes	no
4	10	11.5	>16	0	no	yes
5	10	15	16	-4	no	yes
6	10	>16	>16	-8	no	yes
7	40	16	16	-8	no	yes
8	40	>16	16	*	no	yes
9	40	16	15	-8	no	yes
10	40	15	14	*	no	yes

Note: The critical limit for ignition of the GH_2-air mixture is the release pressure of 16 kg/cm^2 and the Mach Number= 0.26, at which the electrostatic discharge can always cause the ignition or explosion.

The total charge transfer of the corona discharge, or the quantity of nutralized charge on LH_2 could be obtained with the following formula:

$$Q = \int_0^t i(t)dt = \frac{1}{n} \sum_{k=1}^{n} I_k \Delta t = n \Delta t \bar{I}$$

where Q is the total charge tansfer,
i(t) -- the current, t-- the time
and $\bar{I}$ -- the average corona current.

proporsional to voltage squared in the capacitive circuit and decreases with increase of temperature as shown in Table 1 and Fig.1, 2.

The smaller the minimum ignition energy, the danger in ignition. From this, the Research Institute of Industrial Safety of Japanese Ministry of Labour establishes voltage limit.

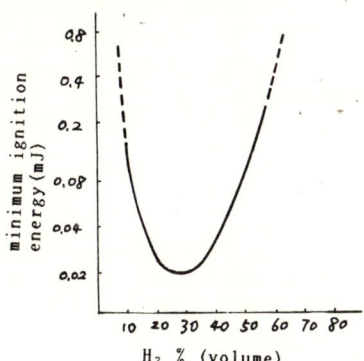

Fig.1 Limits of inflammability of H_2 in air and the minimum ignition energy versus volumetric concentration.

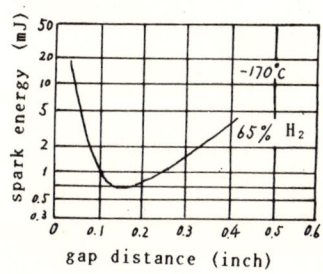

Fig. 2 Spark energy in H_2 versus gap distance

2. Limits of inflammability and ignition concentration

Minimum ignition energy of GH_2-air mixture reaches its optimum when near its stoichiometric concentration (29--30%) and increases with the concentration approaching the limits of inflammability (Fig.1). At the same temperature, the wider the range of inflammability, the more danger of ignition in which it is. Between the lower and upper limits of inflammability, the most ignitable concentration varies with ignition sources (Table 2)[4].

Table 2 Optimum concentration of GH_2-air mixture for various ignition sources

Ignition modes	Flat flame	Incandescence lamp	Electric filament	Friction and collision
Optimum conc.(V%)	30.0	10--15	21--22	13--18

3. Spark breakdown voltage

Spark breakdown voltage is referred to as the lowest voltage which causes dielectric within the electrode gap abrupt destruction (short-circuit) and creates a instant discharge current. Its value depends upon the materials and configuration of electrodes, the product of gaseous pressure and the electrodes gap as shown in Table 1, Table 3 and in Fig.3 and Fig.4.

Table 3 The relation between breakdown voltage and the gap distance

breakdown voltage(kv)	2.8	4.6	8.6	11.1	14.1	17.1	20.0	22.7
gap distance(mm)	0.05	0.10	0.20	0.30	0.40	0.50	0.60	0.70

Generally speaking, under the same situations the smaller the breakdown voltage, the larger the incendiary probability.

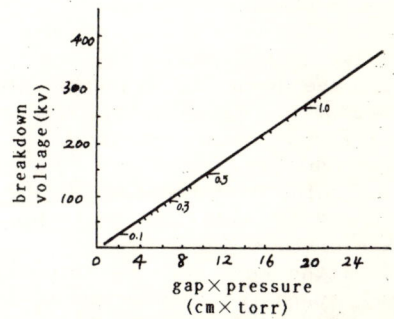

Fig. 3 The dependence of breakdown voltage on the product of gap distance and the pressure.

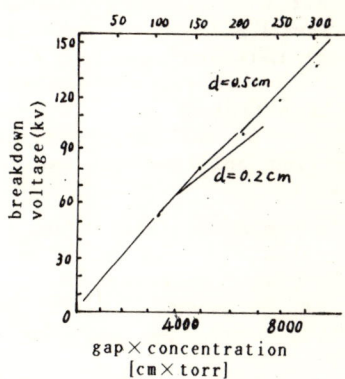

Fig.4 The dependence of breakdown voltage on the product of gap-distance and the concentration

4. Quenching distance

Quenching distance is defined by the critical propagation width of flame between the plate electrodes without thermal exchange with surroundings. It depends on compents of the gas mixture, the relative humidity, the configuration and materials of the electrodes, and decreases with the increase of temperature or the pressure. When the gap-distance is less than the quenching distance, due to the heat loss to the electrodes, the gas mixture could not be ignited even with a larger discharge energy as shown in Table 1 and in Fig.5.

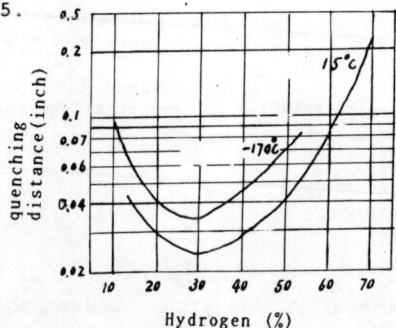

Fig.5 The dependence of quenching distance on the hydrogen concentration under the standard atomsphere.

Under the same temperature, the shorter the quenching distance, the larger the incendiary probability. The quenching distance is mainly controlled by the burning velocity of flame. For example, the quenching distance for LH_2 air mixture is only 0.33 times of that for propane and air mixture.

From the mentioned results (see Table 1) and supposing the critical breakdown voltage is ± 8.5 kv, the limit potential for ignition could be Vs = ±8.5/1.7 =± 5kv, when taking the safty coefficient as 1.7.

In view of the above results we can point out the following conclusion:

1. The limit safty potential for the hydrogen released into air should be less than 2.8 kv because of the effect of the safty coefficient. It is very dangerous when adopting the safty potential to be 5 kv derived only by the minimum ignition energy[7].

It is not sufficient enough to cause the electrostatic ignition when only the static charge is accammlated. If must be under the following situations, the electrostatic discharge could be an ignition suorce.

(1) The static potential is larger than its critical value.
(2) The discharge energy is larger than the minimum ignition energy.
(3) There is exists combustible material in the discharge channel.
(4) The discharge period should not be less than 1ms for the unflowing mixture.

Experiments show that the minimum ignition energy decreases with the increase of the discharge period as shown in Table 4.

Table 4 The relation between the minimum ignition energy and the discharging period

minimum ignition energy (J)	discharging period (sec.)	reference
1.9×10^{-5}	1×10^{-3}	[8]
5.2×10^{-4}	$10^{-4} - 10^{-5}$	[9]
10^{-3}	10^{-6}	[9]

If any one of the above situations could be controlled, the accidents caused by electrostatic discharge could be avoided.

Safely Self-maintain Spark Discharge Electrostatic Eliminator

Three kinds of safely self-maintain spark discharge electrostatic eliminator have been designed which can be safely used to eliminate static charge on the hydrogen released.

1. The construction of the static eliminator

The eliminator consists of earthed point electrodes and soft metal support as shown in Fig.6.

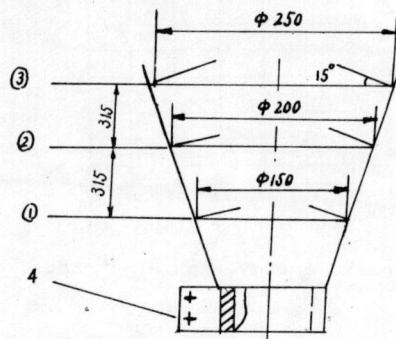

Fig.6 Schematic diagram of the static eliminator welded on the lathe dog 4.
There are 9 point electrodes welded on the support wall(1) (φ =150 mm), 15 point electrodes on wall(2) (φ =200 mm) and 21 on wall(3) (φ =250 mm). All the length of the point electrode is equal to 40 mm.

The reasonable design and installation should be according the following points:

(1) Select proper materials in order to decrease as much as possible the capacitance between the eliminator and the charged gaseous hydrogen.
(2) Select proper installation position to avoid the shielding to the point electrodes by the surrounding earthed conductive materials.
(3) Optimize the electrodes array in order to get the iminmum critical voltage of breakdown.

It is shown by computation that even with the weaker corona discharge (current -- 0.1 μA), the total charge of 8×10^{-5} c on the LH_2 could be eliminated for 10 sec.

Experiments show that with our static eliminator the amount of treated LH_2 per second is larger than that with the similar products made in USA. And the residual potential after elimination is reduced to below the safty limits.

Reference

1. "用氢安全和事故分析"
国防科委情报研所，Sept.,1978
2. "1.28 氢爆轰现象实例分析", 低温工程 No.1,1979
3. Spark Ignition Parameters of Cryogemic Hydrogen in Oxygen and Nitrogen mixtures, Adv in Cryogemic Eng. 10, 265 (1965)
4. "从可燃极限看氢的着火危险性", 低温工程, No.3,1981.
5. Meek J.M. and Graggs, Electrical break down of gases, 1978
6. Hazards of Liquid Hydrogen in Research and Development Facilities, AD-294451 (1962)
7. "静電気安全指針のねらいと企業における活用ガイド
日本 OHM No.3, 1979
8. Electrostatic Hazards Associated with the Transfer and Storage of Liqued Hydrogen,
Adv in Cryogen Eng. 7, 327, 1960
9. Ignition of Gaseous Mixtures by Electric Spark of Various Duration, Chem. Abstract 82, 751282 (1975)
10. 除電器とその使用法, 日本 OHM No.7, 1973

X. Electrostatic Materials

THE RESEARCH, DEVELOPMENT AND APPLICATION OF ELECTRICALLY CONDUCTIVE FIBRE

Shen Meixia, Chen Meihua
China Textile University,
Shanghai

Zhang Shu
Shanghai Municipal
Institute of
Labour Proctection

Jiang Shangxin
Institute of Textile,
Shanghai University of
Engineering Science

Abstract

Anti electron static fibre was manufactured by PAN fibre treated with cuprous solution and subsequently treated with sulfur-containing reducing agent. This polyacrylonitrile copper sulfide compound fibre possess electric resistivity of 10^{-2} $\Omega \cdot cm$ and hence with petroleum resisting, wear resisting, washing resisting and a chemical stability. Then it is good for static charge dissipation work dress, petroleum sampling rope, antistatic filter material and microwave shielding materials respectively.

Polyacrylonitrile cuprous complex fiber obtained by treating PAN fibre with copper salt solution and subsquently reduced with sulfur-containing reduction agent possess electric resistivity of $10^{-2} \Omega \cdot cm$ and hence exhibits good antielectro static property. New chemical bondings were formed in these fibres which were proved to be Cu-CN and Cu-S-Cu as revealed by light electron energy spectrum and the chemical structure is proposed as the following:

$$2\{CH_2-CH-CH_2-CH\}_n + Cu^+ \rightarrow \begin{array}{c} \cdots-CH_2-CH-CH_2-CH-\cdots \\ | \quad\quad | \\ CN \quad CN \\ | \quad\quad | \\ Cu^+ \\ | \quad\quad | \\ CN \quad CN \\ | \quad\quad | \\ \cdots-CH_2-CH-CH_2-CH-\cdots \end{array}$$

This modified PAN fibre possess petroleum resisting, washing resisting and a good chemical stability. When static electricity occurs from friction a slight corona discharge in this fibre will eliminate the static electricity, so the modified PAN fibre is one of the ideal anti-eletrostatic materials.

Fibre Manufacture

32^S PAN fibre is washed with detergent and put into copper sulfate solution, heated, mixed and then reducted with one of the following sulfur containing agent: sodium sulfide, hydrogen sulfide, sodium hydrosulphite, sodium hyposulfite etc. The pH value of reaction bath is kept below 6. When the reaction is complete the electrically conductive PAN fibre is washed with water and dry in air. This modified PAN fibre possess the similar physical parameter as that of the origin and has a good quality of weaving.

Physical Properties and the Test

With the help of electronic microscope the phenomenon of weak corona discharging was observed in the white feather shape part of the electrically conductive fibre which is showed in Fig. 1.

Fig. 1. Corona discharge phenomenon in white shape part of electrically conductive fibre under electronic microscope.

The influence of various conditions on the electrically conductive fibre such as temperature and relative humidity of air, exposed time, friction and the washing etc. were also tested. Results revealed that there are no significant relationship between the electrically conductive property of the fibre and the temperature, relative humidity and the exposed time. Results are showed in Table 1 and Table 2

Table 1. The relationship between resistance of the electrically conductive fibre and the temperature

RH%	62				
Temperature (°C)	0	13	20	32	40
Resistance of electrically conductive fibre (Ω)	6.4×10^2	6.6×10^2	7.8×10^2	8.8×10^2	9.0×10^2

Table 2. The relationship between resistance of the electrically conductive fibre and relative humidity ()T: 20°C)

Temperature (°C)	20			
RH (%)	40	50	60	70
Resistance of electrically conductive fibre(Ω)	1.0×10^3	0.6×10^3	0.8×10^3	0.8×10^3

Table 3. Exposion time and electric resistance relation of antielectrostatic fibre

Exposion time (day)	new prepared	10	20	30
Ω, sample	9.5×10^3	1.0×10^4	1.1×10^4	1.15×10^4
Ω, sample	63			

40	50	60	70	R/R₀
1.20×10^4	1.20×10^4	1.25×10^4	1.35×10^4	1.4
			176(180day)	2.8

Table 4. Variation of resistance of electrically conductive fibre with the wash time.

Wash times	0	10	20	30	40	R/R₀
Resistance Ω	212	248	274	355	370	1.7

Table 3 and Table 4 are the effects of exposion time and washing times on the resistance of antielectrostatic fibres. From the ratio of R/R_0 it may be seen that fibres obtained from the above method show fairly good stabilityes both for washing and long exposion in air.

The Development and Application of the Electrically Conductive Fibre

1. Antielectrostatic clothes.

Antielectrostatic fibre prepared by means of the above mentioned method can be interwoven longitudinally with the distance of about two cm. into terylene rayon and terylene cotton blend clothes or other mixed fabric chemical fibres. As may be seen in Table 1 that in comparing with general working dress the noval working dress made with this blend fabric has good antielectrostatic property and is suitable for the preparation of the up-to-date antielectrostatic clothes.

Table 5. The comparison of antielectrostatic properties between general working dress and those made of polyacrylonitrile cuprous complex fibre

Test no. volt/μcoulomb clothes sample	1	2	3	4	5
1. terylene-cotton khaki coat	-380/0.77	-400/0.81	-380/0.77	-380/0.77	-360/0.73
2. Anti electron static Terylene-cotton khaki	+25/0.051	+25/0.051	+28/0.057	+25/0.051	+22/0.044
3. Antielectro static Terylene-rayon coat	+60/0.12	+64/0.13	+60/0.12	+60/0.12	+62/0.13
4. Terylene-Rayon trousers	-410/0.83	-412/0.83	-410/0.83	-428/0.86	-406/0.82
5. Anti electron static trousers	+80/0.16	+78/0.158	+76/0.153	+85/0.17	+82/0.166
6. Anti electron static Terylene cotton khaki trousers	+20/0.04	+22/0.044	+18/0.036	+18/0.036	+22/0.044

6	7	8	9	10
-380/0.77	-380/0.77	-380/0.77	-385/0.78	-395/0.80
+25/0.051	+25/0.051	+26/0.053	+25/0.051	+25/0.051
+68/0.14	+68/0.14	+62/0.13	+64/0.13	+62/0.13
-412/0.83	-412/0.83	-412/0.83	-420/0.85	-406/0.82
+78/0.158	+79/0.158	+82/0.166	+80/0.16	+80/0.16
+20/0.04	+20/0.04	+22/0.044	+22/0.044	+25/1.051

2. Electrically conductive Antielectrostatic Ropes for petroleum sample cellection.

For the sake of safty, the sampling rope in petroleum production has to be one of the antielectrostatic materials. Polyacrylonitrile cuprous complex fibre was mixed with other fibre to form such a rope. By doing so, sampling process will be eliminated through a weak corona discharge, Experimental results indicated that factors such as relative humidity of air, exposion time, mechanical friction etc. do not show much influence on the resistance of the rope. Exposion tests were under load and immersed in different organic solvents. Table 6 is the results in which the ratio of resistance before and after test was taken as the comparision parameter. It is evident that this kind of rope is good for petroleum sampling and has been widely used in petroleum industry in China.

Table 6. Resistance along with immersion time relation of #1 petroleum sampling rope in different solvent(Length and diameter of rope 100×4mm)

Time (day) \ solvent Ω	Benzene	Toluene	Aviation gasoline	Gasoline	Diesel oil
new prep.	7.8×10^2	5.7×10^2	1.7×10^3	8.7×10^2	2.3×10^3
30	1.3×10^3	1.1×10^3	3.3×10^3	1.7×10^3	5.5×10^3
60	2.3×10^3	1.2×10^3	6.5×10^3	1.8×10^3	7.5×10^3
90	1.4×10^3	1.3×10^3	8.5×10^3	2.5×10^3	1.5×10^4
R/R_o	1.8	2.2	3.9	2.9	6.5

3. Antielectrostatic filtering cloth for dust

Because the powder dust usually have large sulface, their chemical activity is inhanced. When the concentration of dust in the air reaches certain values where the explosion limits are set, a spark can lead to an explosion. In order to eliminate such dust a bag collector, which made of the antielectrostatic cloth inlayed in the material every 2 cms, is often used. It has been proved by practical test that the efficiency of eliminating dust of antielectrostatic cloth is as good as that of ordinary filterring cloth, and has obvious effects in eliminating the powder dust.

4. Shielded material for microwave

If we mixed the electrically conductive fibre, terylene-rayon fibre and polypropylene fibre together in a certain proportion the product shows a markable obvious shielded effect to protect from microwave. As showed in Table 7. all the samples exhibit high shielding abilitives, ie. higher than 50 decibel.

Table 7. Shielding efficency of mixed electrically conductive fibre

Sample No.	1	2	3	4	5	6	7
Shielding ability (dB)	>50	>50	>50	>50	>50	>50	>50
reflective percentage(%)	3	13	13	/	12	/	7
absorbence percentage(%)	97	87	87	/	88	/	93

The Studies of the Properties of PVC Conducting Plastic and Its Application

Wu Xuezheng　Chen Jingliang
Liu Xuezhong　Liu wenbing

Department of Electrical Engineering, Xi'an Jiaotong University

Abstract

The conducting properties of PVC conducting plastic which will be used for semiconducting shielding layers of 10KV PVC cable have been studied by the authors since 1986. According to the relevant standard, the qualified PVC conducting plastic was developed in 1987. The PVC conducting plastic can be used to improve the partial discharge characteristics of cable, it is also functional material for antielectrostatic properties. The conducting and antielectrostatic mechanism of this material is discussed in this paper.

Polyvinyl-Chloride(PVC)—a polar dielectrics, after having its properties improved by mixing with some of additives, can be used as intermediate voltage cable (up to 10KV) insulation material(1). According to IEC-502 specification, it is required that the qualified PVC semiconductive material must be used for 10KV cable shielding to ensure the cable to have excellent partial discharge characteristics(2)(3). For this reason, the thermoplastic PVC semiconductive cable material was developed in 1987 by the authors cooperating in work since 1986. This material may be used as shielding materials for PVC cable conductor and insulation as well as jacket materials for antielectrostatic cable. This composition is made by PVC resin, mixed with some kinds of additives, and has conducting charecteristics, so that it has no physical and chemical interference with PVC insulation material. According to Standard for PVC Semicoductive Cable Material(published by Shanhai Cable Research center), with reference to IEC-502, AEIC-CS 5-82 and Specification for 10 KV Cable Insulation, the Standard for 10 KV cable was made by the authors. The Standard and test results are shown in table 1. To lower the high resistivity of PVC composition ($\rho = 10^{13} \Omega \cdot cm$) to lower order($\rho = 10^0 - 10^4 \Omega \cdot cm$), it is required to improve the conducting property of PVC. Three specimens were developed by the authors by mixing this composition with carbon black.

Table 1 Standards of PVC semiconducting plastic properties and test results
(Test condition, ambient temperature 25℃, relative humidity 55%)

The property parameters of the semiconducting plastic for cable shielding	Equal standards with IEC of PVC	Enterprise standards of PE	Test results of three PVC specimens			Standards of the conducting (1) and anti-electrostatic (2) PVC plastic
			S86-1	S87-1	S87-2	
Volume resistivity ($\Omega \cdot cm$)	$= 10^3 \sim 10^4$	$\leq 1 \times 10^3$	8.8	70.5	3×10^3	(1) $10^0 < \rho < 10^3$ (2) $10^3 < \rho < 10^{10}$
Tensile strength δ_s (N/mm²)	≥ 10	≥ 10	16.3	12.0	13.0	
Elongation at break ε_s (%)	≥ 150	≥ 200	150	206	298	
Cold brittle temperature T_b (℃)	$\leq -15 \sim -20$	≤ -45	-12	-30	-33	
Softening Temperature T_s (℃)	$= 170 \sim 190$	/	206	190	175	
Aging Temperature(℃)×Duration(hs)	110×168	90×168	110×168	110×168	110×168	as that of semi-conducting PVC
Aging coefficient K_1 of δ_s (%)	≥ 100	≥ 80	101	105	110	
Aging coefficient K_2 of ε_s (%)	≥ 70	≥ 80	83	94	99	
Variation of mass $\geq 5\%$(or 2mg/cm²)	5 (2)	/	4.6	1.2(0.7)	2.4(1.5)	
Thermal stability at 200℃ (min)	60	30	60	71	66	

From Table 1 we can see that the resistivity of PVC will increase when T_s decrease and k_a increase. This shows that the conducting property of the specimens depends on the carbon black contents in the composition. Clearly, S 87-2 is a qualified PVC semiconducting plastic. The value of its resistivity ρ is between that of conductors and semiconductors(4).

The authors find that S 87-2 and others are also an antielctrostatic functional materials, according to the defination in (5).

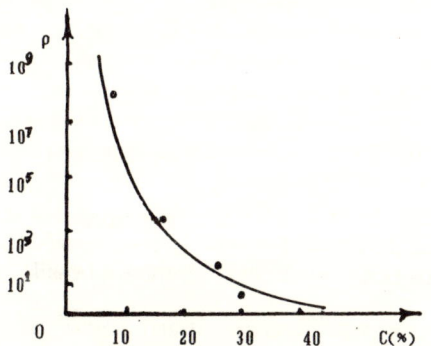

Fig. 1. Variation of resistivity ρ of PVC semiconducting, conducting plastic with carbon black contents.

The resistivity ρ, as in semiconducting materials, will change between $10^{-4} - 10^9 \Omega \cdot cm$ with the carbon black contents in the compositions. The effect of increasing the carbon black contents on resistivity ρ is shown clearly in figure 1. Moreover, the materials are softened with increasing of resistivity ρ(6). The relation between the resistivity ρ of three specimens and barbon black content corresponds with the curve in figure 1 (see black points). The S 86-1(carbon content about 30%) and the s 87-1 (carbon content about 25%), can not be used as a semiconductive material for 10 KV cable. But added to S 87-2(carbon is about 17%), they could be used as PVC conducting plastic. If the carbon blackcontent in the S 87-2 specimen is decreased to half of its original, a new type of antielectrostatic plastic might be developed(see the white point). Because carbon black is a kind of volume active additives, it has good physical and chemical stability compared with surface active additives. The specimen S 87-2, for instance, retmains its stability over a long period of time(after being stored 150 days, the test result shows no change). This material, by the carbon black contents, could be used as some kind of antielectrostatic product to eliminate the electrostatic hazards in the elecrical, electronic, petro-chemical industrial fields etc., and accidents due to the personal electrostatic. For example, it could be used for jackets of mineral cables to prevent accumulation of electrostatic charges. Hence, we could develop colour PVC conducting plastic by mixing with some sorts of metal powder into PVC plastics. The colour PVC composition could not only be made for shielding layer of 10 KV PVC cables but also be developd as one kinds of antielectrostatic products for special use.

The conducting mechanism of PVC conducting plastic is of two kinds of models. Speaking briefly, one of them is the contacting conductive mechanism, and the other is the mechanism of thermal activated electron hopping and tunnel effect.

1. The high conductive mechanism of contacting among conductive particles. Carbon black particles in three-D of PVC compositions are distributed uniformly, therefor, it could be considered that the carbon black particles are distributed continuously in all directions, from the point of sub-micromechanism. The instinctive resistance of carbon black paticles is extremely small so that it could be omited. The resistivity of semiconducting compositions is contributed mainly by contacting resistance among the carbon particles. Therefor, the high conducting mechanism, for the two specimen S 86-1 and S 87-1, could be considered to be formed by a kind of 'United Bridgre' conducting network of this composition, as in conducting rubber(7).

2. The conducting mechanosm of thermal activated electron hopping and tunnel effect.

It is suggested that some particles in the compsition might form a 'Small Section' ; the gaps among these small sections might be ' thick ' or 'thin' (order of μm). The thermo-activated electrons are activated by the thermal motion energy (provided the system temperature $T>0(k)$), consideration of this mechanism suggests that it be thermal activated electron hopping at low field strength (measuring field strength $E<2KV/cm$), or tunnel effect at high field strength (partial high field strength) (4)(8). From above discussions, the authors suggest that the conducting mechacism of these compositions be "United Bridge" conductive network i.e. contacting conduction mainly combined with electron hopping and tunnal effect. However, besides the mechanism of electron conduction, there is ionic conduction. The small amount of particles, even though, could contribute to lowering the insulating resistance, because of the high resistance of the plastic involved a small amount of impure ions, and have little effect on the average volume resistivity, so that it could be omited. In view of that the carbon black particles in plastic are distributed

uniformly, the surface resistivity ρ_s is the same as that of volume resistivity ρ.

Conclusion

Seven kinds of improved PVC insulation materials

have been developed by the authors cooperating in work, and the physical and chemical properties have fulfilled the Specification or Standard for 10 KV PVC cable. The semiconducting plastic, mixed with conducting carbon black, could be used as shielding material for 10 KV PVC cable. It was found that the requirments of electrical properties for PVC semiconducting plasstic were similar to that of conducting plastic. The specimens S 87-1 and S 87-2, for instance, could be developed as a kind of conducting plastic for preventing electrostatic hazards. The conducting mechanism of these compositions suggested to be composed mainly by the contacting conduction of conductive particles combined with the thermal activated electron hopping at low field strength and tunnel effect at high field strength.

References

(1) Wu Xuezhen, The Study of the properties for 10 KV PVC Cable Materials, Electrical Technical. No.6,1987.

(2) F. F. Humbel etc., Polyethlene-Insulated power Cable in Europe, Wire and Wire Products, February 1967.

(3) Wu Xuezheng, Wu jong, The Study of Void Partial Discharge in XLPE, National Dielectrical Materials Symposium in 1983, Published by Xi' an Jiaotong University Press, August.1984.

(4) Chen Jidan, Liu Ziyu, Dielectric Physics, Published by Mechanical Industry Press, Nov.1982.

(5) G.B. (Electrostatic Safety Terminology), Electrostatic, No. 2, 1986.

(6) Suge Gioto, Electrostatic Handbook, Published by Science Press, Apr. 1983.

(7) Лукьянова А. М. и др., Исследование механизма действия графита в электропроводящих резиновых композициях, Каучук и резина, 1983, №7

(8) SHIN' NOSUKE MIYAUCHI and EIKI TOGASHI, The Conduction Mechanism of Polymer-Filler Particles, Jounnal of Applied Polymer Science, Vol.30, No. 8,1985.

USE OF PIEZO-SENSITIVE MATERIAL TO MAKE ANTI-STATIC DETONATORS

Li Xuerong   He Fu   Lu Chenzu
Huainan Mining Engineering college, Anhui, China

Abstract

This paper describes the electro-static hazard of the cap with metal shell, which is greater than that of the cap with paper shell. The principal static hazard is from the discharge of the unsymmetrical sparks between the lead wire and the shell, accompanied by the current of the bridge wire. Some efforts for the solutions of the problem have been done. The antistatic characteristcs of the cap with pressure-sensitive plastics plug as well as its mechanism are mainly introduced.

I. problems

It is well known that static shock may cause electric cap to detonate. It has also been proved both practically and theoretically that it is more unlnerable to static shock for the electric cap with metal shell. The paper shell of the electric cap used in coal mines is now being replaced by copper-plated iron shell. Thus potential static hazard is increasing, which should be dealt with seriously.

The discharge voltage of the igniting sparks between the lead wires and the shell of the metal cap owing to static induction, is much lower than that of paper shell cap, that means it is easier to occur spark discharge for a lower static shock.

Again owing to the discrepancy in process, the distance between the lead wires and the cap shell is not the same. According to $E=U/d$, discharge often occurs on the side of the smaller gap between the lead wire and shell. And because of the larger curvature at the bonding of the bridge wires, discharge often occurs near the fire head as well. Meanwhile, this unsymmetrical one-side discharge will cause bridge wire current. Upon the observation of the bottomless cap with the naked eye, the bridge wire current produced by unsymmetrical discharge may heat the bridge wire glowingly. The current of the glown bridge wire is about 350mA. The unsymmetrical discharge sparks accompanied by bridge wire current may ignite the floated explosives or the fire head, resulting in static hazard. The equivalent circuit of the instantaneous discharge of the single discharge path is shown in Fig. 1.

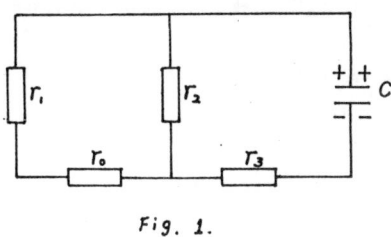

Fig. 1.

II. Measures

1. Cut off the dangerous discharge path i.e, separate the fire head and the lead wires from the metal shell with good insulator to ensure no discharge occuring inside the shell.[1]

2. Build up protecting path, that is, make two optimum symmetrical discharge paths between the plastic plug and the metal shell. For example, drill holes symmetrically at the two opposite sides of the insulating plastic plug, and put in discharge electrodes; reduce the insulating strength of the plastic plug appropriately; introduce optimum symmetrical discharge path into the cap along both sides of the bridge wires through the process of pressured joint, and make use of pressure-sensitive plastics plug, etc.

After treatment by the above mentioned measures discharge is mainly symmetrical and occurs inside the plug only. And a little bridge current exists. the equivalent circuit is shown in Fig. 2. The antistatic ability has met with the international tersting standards of the antistatic caps.[2]

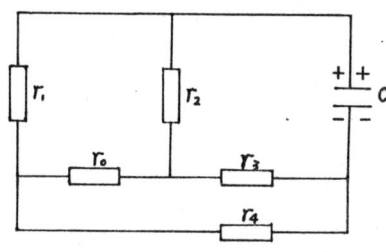

Fig. 2

III. the pressure-sensitive plastic plug

It is worth mentioning here that we have developed a kind of pressure-sensitive plastics which is of non-linear resistance. Under low voltage, it behaves as an insulator. When the voltage is above some critical value, its resistance will reduce abruptly and behaves as a static conductor. If the cap plug is made of such kind of material, and when the static voltage is over the critical voltage, the charges will leak out towards the shell through the lead wires. By doing so, the spark discharge is eliminated. Besides, its leakage resistance of r_a and r_b is nearly equal, the bridge current is almost zero.

A. the composition and V-A characteristics of the plug

The plug is mainly composed of PVC and zinc oxide, added with various additives which has the ability to change the conductivity, non-linearity and breakdown voltage. Through orthogonal test, the plug is moulded under heat and pressure with optimum ingredients, then fitted at the shell, as shown in Fig.3.

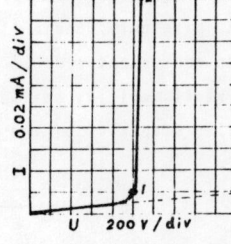

Fig.3 structure of copper cap with pressure-sensitive plug
1. leg wire
2. plug with pressure-sensitive material
3. cap shell
4. bridge wire
5. fire head

The V-A characteristic curves are displayed on a transistor characteristic displayer (QT-14) and photographed. Take sample No.6 as an example as shown in Fig.4. Let's analyze its features in positive direction.

Fig. 4. sample 6 positive direction

It gives its resistance before breakdown

$$R = \frac{U}{I} = \frac{200 \times 10}{0.02 \times 10^{-3}} = 10^8 \Omega > 10^6 \Omega$$

From its V-A characteristics, the non-linear coefficient α can be determined, which agrees with the relationship

$$I = \left(\frac{U}{C}\right)^\alpha, \text{ that is } \alpha = \frac{lgI_2 - lgI_1}{lgV_2 - lgV_1}$$

Assume $I_2 = 10 I_1$, $V_1 = 1000^v$, we obtain $V = 1067^v$

$$\therefore \alpha = \frac{1}{lg1067 - lg1000} = 35.5 > 25$$

This shows that the pressure-sensitive plastics is of little leakage current before breakdown, which presents high resistance, about $10^8 \Omega$, while after breakdown the voltage increases a little, and resistance of the plastics will reduce enormously. The non-linear efficiency $\alpha \doteq 35$ shows that the resistance drops to $1/10^{\alpha-1}$ of the initial value, that is $1/10^{34}$, while the voltage increases ten times, which is equivalent to short circuit, and the current rises sharply, while the voltage remains nearly unchanged. Furthermore, the characteristics of both positive and negative directions are basically symmetrical.

B. stability

The stability here refers to the ability of automatical recovery to its normal state after shock with high voltage impulses. Five, ten and twenty minutes after the first shock to the sample with high voltage, the tests are done orderly. The results show that the pressure-sensitive element possesses the ability of automatical recovery to its initial features after the high voltage is removed, but it takes some time. Further experiments show that additives Cr_2O_3 and MnO_2 are advantageous to the stability and they can shorten the recovery time. Taking the cost and some finite recovery time into account, addition quantities of Cr_2O_3 and MnO_2 may be regulated accordingly.

C. experiment on the antistatic characteristics of the semi-finished cap

To make test on the antistatic ability of the cap, we have prepared a hundred shots of copper-plated caps with pressure-sensitive plugs, which do not contain floated explosives but fire heads (semifinished products). Apply high voltage impulse to the sensitive static spark meter to examine its striking voltage. The testing results are compared with that of the copper-plated cap with common plastic plug. The results show that the bridge current is an important factor of the static hazard. The bridge current can be greatly reduced by short circuit connection of the lead wire. Under such condition, the antistatic ability of the copper-plated cap with pressure-sensitive plug is much more improved than that of the common copper-plated cap. When the series resistance is 5kΩ and the discharge capacitor is 1000pF , no ignition occurs under the shock of 30kv , which meets with the international standards of non-ignition

R = 5kΩ , C =500pF , 25kv.

D. the normal detonating characteristics

By means of the detonator nominated with 950v and 6ms, we detonate ten shots of finished caps in series and parallel connections respectively. The test is successful. It proves that the pressure-sensitive plug does not affect the detonating characteristics.

E. study on the non-linear conductive mechanism of the pressure-sensitive elements[1][2]

We have employed microphotography to study the characteristic mechanism of the pressure-sensitive elements. A group of coloured photoes with microstructure of the cross-sections of the pressur-sensitive plug are obtained (Fig. 5)

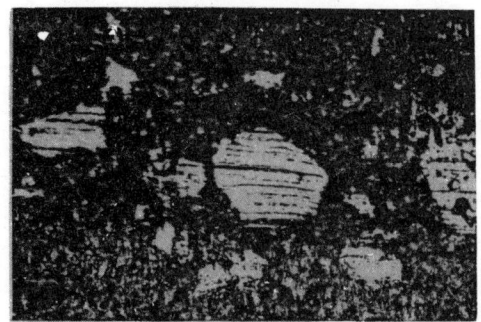

Fig.5 microstructure of the cross-sections of the pressure-sensitive plug

The model of typical structure of the pressure-sensitive element is shown in Fig.6. It can be seen from Fig. 6 that the pressure-sensitive element is of an orgnization with multiple phases. One is conductive material, another is non-conductive. PVC is an adhesive, and serves as the isulation layer. The conductive particles are non-uniformly distributed in the orgnization, most of which are not in contact and separated by non-conductive phases, and form discrete particles. The smaller conductive particles are able to form link structure spontaneously, while the bigger conductive ones may also break up the thicker films of the adhesive. ZnO particles are conductors of n-type with low resistance, which exist in the state of masses. The size of the particles directly affects the magnitude of the limit voltage. The other two kinds of non-conductive materials are oxides of p-type which is sensitive to voltage. As a kind of material of the secondary electron emitting, the oxide will release some electrons when attacked by the first energy electron and electrons are distributed over the surface of the pressure-sensitive substance, ZnO, forming a boundary layer. And electrons are in the chain structure as well. It discharges when subjected to the energy shock supplied by the conductive materials on both sides, which causes to form conductive path between the conductive materials and the ZnO massess with pressure-sensitive feature. This structure is not only similar to "the combined group of Zener diode" , but also of chain structure of Sic pressure-sensitive resistor, so the pressure-sensitive material possesses both advantages. It has been proved from experiments that it is necessary to adjust the ratio between the thickness of the boundary layer and the particle diameter to the optimum value. The adjustment of the ratios between the conductive and the non-linear compositions may obtain satisfactory pressure-sensitive features. Therefore, this new material is a good antistatic material. It can not only be used to make plugs for antistatic caps, but also find practical applications in the fields of industries of electronics, space and mining.

Reference

[1] Liu Bei-te; Huang Jiu-hua Research on Antistatic Caps (III) and the Antistatic caps with Insulated Structure, Modern Chemistry Research Corporation, Lanchow.

[2] Huafong Information, 1981

[3] Resistors, 1984 Technical Standard Press

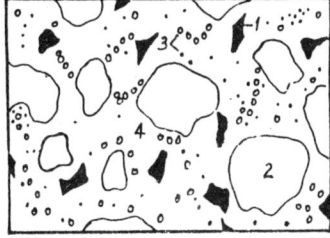

Fig.6 structure model for pressure-sensitive material
1. ---oxide for P-type 3. ---conductive substance
2. ---zinc oxide   4. ---PVC

Research on Antistatic Pressure-sensitive Materials

Lu Chen-zu
Huainan Mining Institute, P.R. China

Abstract

The pressure-sensitive material discussed in this paper is mainly composed of PVC and Z_nO. The discovery of this kind of material discussed is considered a breakthrough. Based on Volt-Ampere characteristics of the material as well as other features found in the experiments, the present paper discusses the conduction mechanism of the material in terms of barrier, energy band and tunnel effect.

Preparation of the Sample

The sample is mainly composed of PVC and Z_nO and some metal oxide additives. Big granular PVC are first put into acetone until they dissolve and expand. Then they are ground into powder in a pot before Z_nO powder and the additives are added. These are mixed in proper portion. Both Z_nO and the additives added are either analytically pure or chemically pure. Blending is done then while grinding. The mixture is then evenly heated at 180-200 °C till it turns into a paste. When cooled, it is compressed so that samples in rod shape are obtained.

V-I Characteristics of the Samples

With a transistor curve tracer (type TQ-14) the typical V-I characteristics is obtained and shown in Fig.1. It clearly shows that when a small voltage is applied to the sample, it behaves as an insulating solid. When the applied voltage is higher than 700 v, the sample turns a low resistance state and becomes a stabilivolt conductor. This shows that the sample is a pressure-sensitive material. Furthermore, the V-I characteristics is basically symmetrical with respect to the origin.

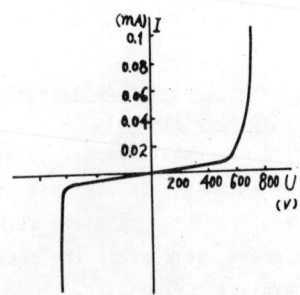

Fig. 1 V-I characteristics of the sample (6 mm in length, mm in dia.)

The V-I properties shown in Fig.1 could be expressed as:

$$I = \left(\frac{U}{C}\right)^{\alpha} \quad (1)$$

where α is the non-linear coefficent and can be defined as:

$$\alpha = \frac{dI}{I} \bigg/ \frac{dU}{U} \quad (2)$$

The bigger the α, the greater the relative current variation caused by the increase in voltage and the better the non-linearity or pressure-sensitivity. Using equation (2), we get α equal to 41 when I= 0.1 mA in the positive direction and $\alpha \rightarrow \infty$ at 0.1 mA in negative direction.

Characteristics of the Samples

1. Experiments show that the breakdown voltage (impulse voltage) for the virgin samples is much higher than that obtained in the successive tests. This is shown in Table 1. Therefore, the samples can only acquire normal (or stable) pressure-sensitive features after being broken down by an appropriately high voltage.

Table 1 Comparison of the breakdown voltage of virgin samples with their stable breakdown voltage

sample	length(mm)	breakdown voltage for virgin samples (kv)	stable breakdown voltage (kv)
1	22.5	21	1.5
2	4	14	1.2
3	6	32	2.5
Remark: Ingredients in different samples are different.			

2. Recovery period effect. Experiments show that a stable sample can not recover its stable breakdown voltage and non-linear properies right after a breakdown test till it is laid aside for an adequate period of time. Table 2 shows this effect.

Table 2 Breakdown voltage after different recovery period

sample	breakdown voltage (kv)	breakdown voltages (kv) after different recovery period (in min.)							
		0	5min	10	20	30	60	120	180	240
1	1.7	1.2	1.5	1.6	1.6	1.7				
2	2.5	1.5	1.6	1.75		1.85	2.0	2.2	2.3	2.5

3. If an impulse voltage over the critical breakdown voltage is applied to the sample, the higher the voltage exerted, the lower wold be the critical breakdown voltage when the sample is tested immediately after the initial over-voltage breakdown. This is shown in Table 3.

Table 3 The critical breakdown voltage after over-voltage breakdown

Sample	Critical breakdown voltage(kv)	Voltage for over-voltage breakdown(kv)	Critical breakdown voltage right after its over-voltage breakdown (kv)
1	1.5	1.5	1.25
	1.5	5.0	0.5
2	1.5	1.5	1.05
	1.5	5.0	0.6

4. Table 4 shows that the critical breakdown voltage of the samples is not susceptable to the humidity of atmosphere.

Table 4 The breakdown voltage at different relative humidities

breakdown voltage(kv) \ Sample	R.H(%)						
	25	30	40	50	60	70	80
1	0.8	0.8	0.8	0.8	0.8	0.79	0.8
2	1	1	1	1	1	1.05	1

Micro-Structure of the Sample

Computer-qualitative analysis of a sample is conducted with a combined device of electron microprobe and scanning electron microscope (Model JCXA-733 from Japan). The micro-structure is shown in Fig. 2. A back electron identification (BEI) of the sample surface is magnified 200 times.

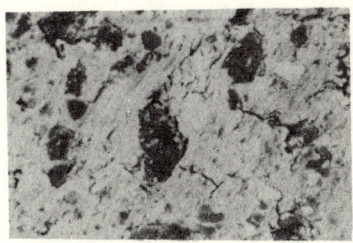

Fig.2 The back electron identification of the sample surface (BEI), 200x

It can be seen from the micrograph that the surface of the sample appears to have three different contrasts, which corresponds to the three phases in its composition. Analysis of each phase by computer-qualitative program indicates that the white phase is rich in Cu, while the light color phase is rich in Zn and the phase in dark color is mainly made up of organic matter(PVC). No phase rich in bismuth can be found, which seems to indicate that most of Bi_2O_3 is dissolved in the ZnO particles. Fig. 2 also shows that ZnO particles are bound by PVC.

Electric Conduction Mechanism of the Pressure-Sensitive Material

ZnO is a semi-conductor of n-type. On the surface there exist acceptor surface states. The amount of electrons filled in the surface states within the ZnO particles depends on the location of

Fermi-Level. The electrons entering the surface states cause the surface of ZnO particles to be charged negatively, while its inside to be charged positively. Thus a negative surface potential forms. This potential causes the energy band of ZnO to bend upward on the surface and results in the so-called Schottsky Barrier shown in Fig.3, where E_F is the Fermi Energy.

When PVC touches ZnO particles, the electrons of PVC (certainly few) will enter the surface states of the ZnO particle, if the Fermi level of PVC is higher than that of the surface state of the ZnO particle (Then the Fermi level becomes the same as that within the ZnO particle.).

If a voltage is applied to the pressure-sensitive material, the energy band in Fig.4 will distort. Suppose negative bias is applied to the right barrier, the left one will be affected by positive bias. Being affected by bias, the depletion layer on the side of negative bias will be thicker while the one at the positive side will be thinner. Hence, the height of the barrier ϕ_R seen from the side of the negative bias becomes much higher than ϕ_0 while that on the side of positive bias is a bit lower than ϕ_0 as shown in Fig.5, where ϕ_B is the height from the Fermi Level to the barrier top at the boundary.

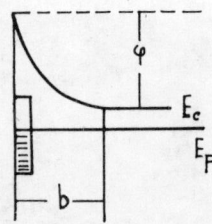

Fig.3 The Schottsky barrier of the isolated ZnO

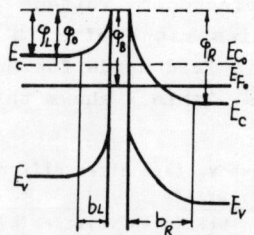

Fig.5 Barrier of ZnO-PVC-ZnO affected by applied voltage

Under this condition, the electrons on the ZnO particle surface will increase, the energy band on the surface will bend much more upward, and the height of the Schottsky barrier will be also raised. However, in the sample, ZnO crystals and the PVC layer form a separated back-to-back double Schottsky barrier, as shown in Fig. 4. The height of the barrier potential at the crystal boundary is ϕ:

$$\phi = e_2 n_s / 2 \epsilon_0 \epsilon_r N_d \quad (3)$$

where: n_s - concentration of the acceptors on the surface; N_d - concentration of the doners within the ZnO particle.

When subjected to a low voltage, the depletion layer on both sides of positive and negative bias is quite wide, and the tunnel effect is little enough to be neglected. The current mainly comes from thermally excited free electrons. Therefore, as the voltage increases, little changes in current occur. The sample is of high resistance with only little electric leakage.

Under high bias the depletion layer on the side of positive bias becomes thin owing to the action of the intense electric field ($E > 10^6$ v/cm). The tunnel effect will occur when b is narrow enough. The tunnel current brings sufficiently great number of electrons through the barrier layer, and causes a rapid increase in current. Under this condition the sample falls to a low-resistance state, the breakdown state. This is called the tunnel breakdown or Zener effect. If the bias

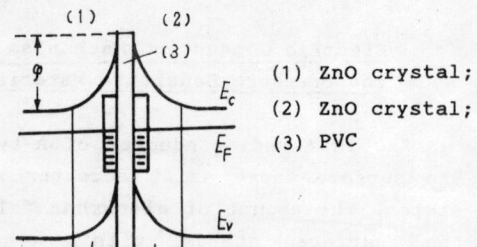

(1) ZnO crystal;
(2) ZnO crystal;
(3) PVC

Fig.4 Barrier of ZnO-PVC-ZnO

electrode changes its position, the thick and thin depletion layers will also change their positions, and the sample again experiences a reverse breakdown. The sample appears clearly to have a symmetrical pressure-sensitivity.

It is theoretically proved that the width of the depletion layer b(i.e. the tunnel length) depends upon both the field intensity in the depletion layer and the charge number No. Within the unit volume. The more intense the electric field the bigger No. and the samller b will be.

According to the above conclusion, it can be seen why the breakdown voltage of the virgin sample is very high and its stable breakdown voltage is lower than the former.

Discussion

1. ZnO is a semi-conductor of n-type, the crystalline surface of which carries surface states so that a surface barrier (or Schottsky barrier) forms. Form this model, it can be assumed that the close combination between sample ZnO particles must also possess pressure-sensitive features when the very thin air layer exists between the particles. However, due to less space charges in the depletion layer, the barrier tunnel is too thick and breakdown voltage is high.

2. Three possible situations may result, i.e. the Fermi Level of the dielectric is either lower, or equal to or higher than that of ZnO crystalline surface when the dielectric exists between ZnO crystalline particles. In the former two cases, neither has the dielectric electrons to fill in the ZnO surface states, nor does the ZnO surface backfills the dielectric with electrons. Hence, it is difficult or even impossible for them to have pressure-sensitivity. The third case is the theoretical foundation of pressure-sensitivity for the above sample. The higher the Fermi level of the dielectric when compared with that of ZnO crystalline surface states, the lower will be the breakdown voltage of the virgin samples and the stable breakdown voltage.

3. The stable breakdown voltage of the sample is also related to the applied voltage exerted on the virgin sample. When the applied voltage is high, the higher the Fermi Level of the dielectric when compared with that of the ZnO surface states, the lower will be the stable breakdown voltage. THerefore, it would be better to treat the virgin samples with the lowest breakdown voltage (critical breakdown voltage) so as not to cause the stable breakdown voltage to be too low.

4. The addition of additives is to change the height and width of the Schottsky barrier so as to change the disruptive voltage of the sample.

The added conductor powder (Cu powder for example) to the composite material functions as some kind of additive on one hand and it helps the material to connect the junctions formed by the ZnO particles and their boundary layers, which behave as Zener diodes on the other. In this way the whole sample becomes a stable conductor combined with many Zener diodes.

Reference

1 Li Biaorong, Inoroganic Dielectric Material, 1986(in chinese)
2 Huang Kun, Physics Foundation of Semi-Conductor,1979 (in chinese)

A STUDY ON USING RADIATIVE EFFECT TO IMPROVE THE ELECTROSTATIC PROPERTIES OF POLYMER SURFACE

Liu Shang-he, Liu Zhi-cheng, Zhai Bing-xun, Wei Guang-hui, Zhang Ling-zhen

Ordnance Engineering College, Shijiazhuang, Hebei, China.

ABSTRACT

When special polyvinyl chloride (R-PVC) and polymethylmethacrylate (PMMA) are irradiated by ion beam, at medium doses of 150 Kev Ar^+ (or $N^+ \cdot O^+$) ions, the surface electrostatic properties of these polymers will be dramatically changed. The triboelectrification charge can be decreased from $1.24nc/cm^2$ to $0.02nc/cm^2$ or less, electrostatic surface potential of electrification is decreased from 800-1000V to 0-50V. The halfvalue time of charge decreases from a few hours to less than 1s. The surface resistivity drops down four orders of magnitude. Under proper conditions, anti-static new materials may be produced.

I. INTRODUCTION

This paper deals with the researching results in the modification of the electrostatic property of the R-PVC and PMMA surface by using radiative effect of ion beam. The advantage of this method is that the modification takes place only in the surface layer, the internal structure is not affected. Therefore its mechanical characteristics remains unchanged.

II. EXPERIMENTAL METHOD AND APPARATUS

The samples used in our investigation are $44 \times 44mm^2$ R-PVC and PMMA pieces. After treated appropriately, the samples were implanted in the high-energy ion implantation apparatus at room temperature with 150 Kev Argon or Nitrogen, Oxygen ions. In order to observe the change of anti-static property with irradiation doses, we selected different dose to bombard different samples respectively.

Resistivity and tribo-electrification of all the samples are measured under parallel conditions so as to contrast the properties of the samples with ions implanted and those of the samples without ions implanted.

The voltage is measured by non-contact-typed electrostatic voltmeter together with its calibrating system. The tri-electrode system and extra high resistance, microcurrent meter have been used to measure the resistivity. The amount of the triboelectrification charge is measured by a Faraday cage and an electrostatic voltmeter of Type Q5-V. By using the Type S-5109 Static Honestmeter, the electrostatic halfvalue time is measured. During experiments, ambient temperature and relative humidity are kept unchanged.

III. RESULTS AND ANALYSES

3.1 The dependence of triboelectrification on ion implantation dose

Fig. 1 and table 1 show the relationship between ion implantation doses and the generated charge and surface voltage for various samples due to friction between every sample surface and waterproof polyester fibre-cotton Khaki cloth at normal pressure of 0.5kg with constant velocity of 0.15m/s through a distance of 0.3m. Figure 1 is for R-PVC Samples. From figure 1 it is found that both the charge q and voltage V decrease linearly with increasing dose of the ion implantation. When the dose increases to $3 \times 10^{15} Ar^+/cm^2$, q and v reach to zero. The implanted samples exhibit remarkable anti-static property.

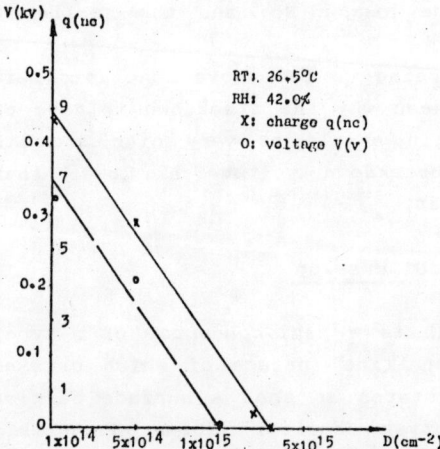

Fig. 1. Charge q and surface voltage V of triboelectrification variation as a function of does for 150Kev Ar^+ beam bombarded R-PVC samples.(sige: $44 \times 44mm^2$)

After friction, the relation between the ion implantation dose and the static parameter of PMMA samples is given in the table 1. It is found that the samples implanted with Ar^+, N^+ or O^+ all have a certain degree of anti-static property. The triboelectrification charge may well decrease to zero, the voltage declines to about 50V. As the dose is larger than $5 \times 10^{15} cm^{-2}$, the anti-static effects is better.

Table 1. Triboelectrification voltage V and charge density q of PMMA samples

Environmental Conditions	RT: 27.0 ℃, RH: 41.5%					
Number	1	2	3	4	5	6
Dose $(\times 10^{14} cm^{-2})$	0	5 Ar^+	10 Ar^+	100 Ar^+	5 O^+	2.5 N^+
V' (V)	0	0	0	0	-10	0
V (V)	+1000	-310	-290	-50	+50	-500
q (nc/cm^2)	0.7	0.023	0.023	0	0.023	0.023

V'----The voltage after elimination

The R-PVC samples unimplanted hold the original static properties — the triboelectrification charge density is about 1.24nc/cm², the voltage ranges from 800 to 1000V.

It is worth to mention that the unimplanted PMMA sample surface is charged positively by friction, while after implantation it is charged negatively under the same experimental conditions. This is because the ion implantation modifies the surface structure and the energy level.

3.2 Dependence of electrostatic charge decay rate on ion implantation dose

The charge decay rate on the charged materials is an important parameter, when evaluating its anti-static properties. Suppose the charge on the sample investigated is Q,

then
$$Q = Q_0 \exp\left(-\frac{t}{\varepsilon_0 \varepsilon_r \rho}\right) \quad (1)$$

Where ε_0 is the vacuum permittivity, ε_r is relative permittivity, ρ is resistivity.

If the charge distribute uniformly on the sample surface, then surface charge density σ can be written as following,

$$\sigma = \sigma_0 \exp\left(-\frac{t}{\varepsilon_0 \varepsilon_r \rho}\right) \quad (2)$$

It is found that the charge density decay exponentially with increasing time t. The time τ_0 is called the static halfvalue time at which the charge Q or charge density σ, voltage V decay to half of the primary value. The electrostatic decay curves of the samples irradiated by ion beam mainly depend on the implanted surface layer. The electrostatic charge decay curves of the samples of R-PVC and PMMA are shown in figure 2 and 3. From figure 2, it is found that the value of τ_0 is few hours or more for the R-PVC samples unimplanted (UNI - 7, UNI - 6), but, for the samples implanted with the dose of $3 \times 10^{15} Ar^+/cm^2$ (I - 5), it is less than 1 s.

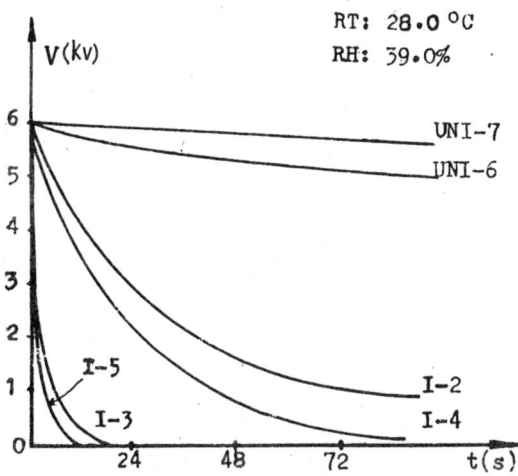

Fig. 2. The chart of electrostatic charge dissipation for R-PVC implanted-samples and unimplanted-samples

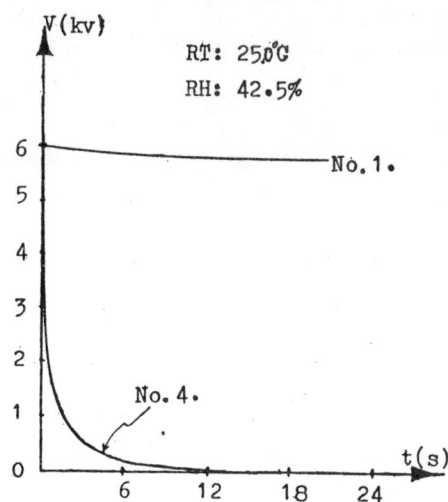

Fig. 3. The electrostatic decay curves for PMMA samples

3.3 Dependence of the surface resistivity ρ_s on ion implantation dose

The relation between the ion implantation dose and surface resistivity ρ_s of the R-PVC and PMMA samples are shown in table 2 and table 3 respectively. From them, it is found that the surface resistivity ρ_s is reduced gradually with increasing dose of the ion implantation. When the dose is increased one order of magnitude, ρ_s drops down about three orders of magnitude. If the radiative dose is $3\times10^{15}Ar^+/cm^2$, the surface resistivity of the R-PVC samples drops down more than four orders of magnitude and reaches $10^{10}\Omega$. These implanted samples have anti-static ability in a certain degree. If the radiative condition is chosen properly, with the increase of ion implantation dose, the surface resistivity of the samples may be reduced further.[1]

Table 2. Surface resistivity ρ_s of R-PVC samples under different doses irradiation

Number	Dose (Ar⁺ Cm⁻²)	500v, ρ_s (Ω) RT: 26.0 ℃, RH: 44.0%	500v, ρ_s (Ω) RT: 22.0 ℃ RH: 58.0%
I-2	1×10^{13}	4.4×10^{14}	1.1×10^{14}
I-3	2×10^{14}	2.8×10^{12}	3.2×10^{11}
I-4	5×10^{14}	3.6×10^{11}	5.6×10^{11}
I-5	3×10^{15}	4.4×10^{10}	1.6×10^{10}
UNI-6	0	4.0×10^{14}	4.4×10^{14}
UNI-7	0	8.0×10^{14}	4.4×10^{14}

Table 3. Surface resistivity ρ_s of PMMA samples under different does irradiation

Experimental Conditions	500V, RT: 26.9 °C, RH: 49.0%					
Number	1	2	3	4	5	6
Dose $\times 10^{14}$(cm^{-2})	0	5 Ar$^+$	10 Ar$^+$	100 Ar$^+$	5 O$^+$	2.5 N$^+$
ρ_s($\times 10^{11}$ Ω)	72	32	0.88	0.028	2.0	2.0

IV. DISCUSSION

With ion beam irradiation on surface of the solid, the depth of the trapped energy level of the electron in the surface layer is changed. The surface energy levels of polymer are essentially dominated by the disposition of the molecular groups in the surface layer. The radiative damage produced by ion implantation is strong enough to break down the bond chain of the polymer molecules, and make the molecular groups be rearranged and recomposed. Furthermore as we known when polymer is implanted, gas molecules like H_2, Cl_2 and so on are selectively released. But the carbon content remains unchanged. In other words, polymer surface layer is changed into a carbon rich material when molecular hydrogen desorption takes place[1,2]. These finally make polymer surface layer permanently modified. New kinds of anti-static materials may be obtained by using ion beam irradiation.

It is worth to mention that the results given in figure 1 and 2 are not essentially consistent with table 2. From figure 1 and 2, it can be seen that the surface implanted has remarkable anti-static ability. But in table 2, it is not so strong. This is just the problem need to study fruther. It is generally considered that the conductance in polymer is divided into ionic type and electronic type. At present, the result measured by using standard method of measuring resistivity is actually due to the jump conductive process of impurity ion in polymer materials, but in the triboelectrification experiment in dry environment only the conduction of electronic jump plays the role[4]. Therefore, the measured resistivity is not enough to be used to evaluate the anti-static materials properties, especially for the ion implantation layer. The complex composition and energy levels or static characteres can not be represented totally with resistivity. Just as a Japanese professor Murasaki advanced that the electrostatic leakage resistance and resistance in the sense of common should be distinguished. In fact, we should not only distinguish the two kinds of resistance from conception, but adopt different measuremental standards.

Reference

1. T. Venkatesan, Nucl. Instr. Meth., B7/8, (1985) 461.
2. T. Venkatesan, J. Appl. Phys., 54(6) (1983) 3150.
3. C.B. Duke, T.J. Fabish, J. Appl. Phys., 49(1978) 315.
4. L. Calcagno and G. Foti, Nucl. Instr. Meth., B19/20, (1987) 895.

XI. Ozone Techniques

Proceedings of International Conference on Modern Electrostatics

QUICK DISINFECTION OF HANDPIECE IN DENTAL USE

Senichi Masuda   Fukui Institute of Technology
                     O.415, 3-2-1, Nishigahara, Kitaku, Tokyo 114, Japan
Endre Kiss   Dunaujvaros College of Miskolc University
                     Dunaujvaros, Hungary
Kengo Ishida   Midori Denshi Co., Ltd.
                     3-10-6, Daikan, Yamato-shi, Kanagawa-ken 242, Japan
Hirokazu Asai   Midori Denshi Co,. Ltd.

Abstract

A high speed zone sterilizing device has been developed, using the ceramic based ozonizer of high frequency surface discharge type. Th advantage of this ozonizer is that it can produce a very high concentration ozone with its very compact size so as to allow its use in a limited space of sterilizers. Oxygen is used to produce the standard ozone concentration of 20,000 ppm. Even at this high concentration a complete sterilization for the reference bacteria - Bacillus Subtiris - in a limited time(ca. 3 - 5 min) is only possible in the case when adequate temperature and humidity are maintained in the ozonied oxygen gas: 50 °C and ca. 80 %. The object of sterilization is cleaned by a wet paper so that the object is wetted to provide such humidity in the sterilizing chamber kept at 50°C. This device produces for the above reference bacteria the standard sterilizing effect of killing one million pieces in 3 - 5 min, with 12 - 14 min in one run including prior and after processings. The conventional method of using an autoclave requires ca. 30 min. The method of using ethylene oxide gas requires long time of ca. 45 min and a more complicated proessing. The prensent high speed sterilizer using high concentration ozone has first been applied in the sterilization of a handpiece in dental use which is used for drilling teeth.

Introduction

The complete sterilization of some of the medical and dental tools are very difficult or almost impossible because of thier complicated mechanical structures. The handpiece being used for drilling tooth in a dental clinik represents one of such tools. As it comprizes a very dilicate high speed air turbine for driving a drill ans as it has plastic sealing materials vulnerable to heating, the use of an autoclave is impossible in most cases. Even if an improved handpiece allowing the use of the autoclave be used, the sterilization with it requires 2 kg/cm (121 °C) to 3 kg/cm (132 °C) pressurized steam with a total of 30 min processing time (heating up: several min; sterilization: 3 - 5 min; cooling: 20 min). The use of ethylene oxide gas requires a preprocessing of evacuating the sterilizing chamber before introducng the gas, and keeping the temperature at 55 - 60 °C and humidity at 50 - 70 %. After sterilizing period the gas must be replaced with clean air, and the air flow must be continued untill the residual toxic ethylene oxide gas is removed from the surface of the handpiece. The toxicity of this gas is the greatest problem. This method requires at least 45 min for one run.

One the other hand the above difficulty in the complete sterilization of the handpiece after its use has become a greatest concern nowadays in view of a possiblity of infection with germs and virures through a blood contaminated handpiece.

One of the authers has developed with his coworkers a very compact high performance ozonizer using a high frequency surfacc dicharge, as illustrated in Fig. 1 (1). A ceramic cylinder (92% alumina) has on its inner surface a number of strip-like corona electrodes (1 mm wide, 30 um thick, 100 mm long, 5mm strip-to-strip;tungsten), and a film-like induction electrode (10 um thick; tungsten). The thickness of the ceramic layer between the corona and induction electrode is 0.5 mm. The induction electrode serves to produce a tangential component of an electric field in the gas at the front of the streamer-like surface discharge during an entire period of its advancement, thereby enhancing its activity. A high frequency high voltage (10 kHz; 5 kV preak) is applied between both electrodes. Then, the intence high frequency surface discharge occuers from the strip-like corona electrode along the inner surface of the cylinder in which oxigen gas is flowing. Ozone is produced by the plasma chemical process in the high frequency surface dicharge, and its concentration easily reac es as high as 20,000 - 30,000 ppm. The use of the surface discharge, easy to cool by the adjacent ceramic surface, enables, in combination with a fairly good heat conductivity of the alumina ceramic , the use of a high frequency without raising temperature up to a limit where ozone is destroyed. This produces a great reduction of the ozonizer size, as the ozone generation rate remains proportional to the frequency so far as the excessive temperature rise be avoided.

It has been known that ozone has a very high sterilizing effect owing its great oxiding ability (next to flourine), and that it leaves no residual toxicity after its decomposition to oxygen.

Stimulated by the above three factors the authors attempted to develop a high speed sterilizing device using our novel ozonizer, primarily for a complete sterilization of the dental handpiece.
The construction of this device, its operating principle, and its effectiveness are described in this paper.

Construction and Operation

Fig. 2 shows the construction of the present sterilizing device. The handpiece 14 after use is cleaned with a wet paper so that it is adequately wetted with water, and placed into the inside of sterilizing chamber 15. The chamber has a double-door 17 for safety not to allow the outleak of high concentration ozone from inside. The gas is continuously suctioned from between the outer and inner partition of the double doors, and it is discharged to an open air through two ozone decomposing catalyzer units 21 and 23 in series, between which an ozone sensor 22 is inserted to monitor the ozone concentration being kept below 0.05 ppm.

The chamber is constantly warmed by a heated clean air to keep its inside temperature at 50 °C. A high grade air filter 8 cleans air, which is heated by a heater 10 regulated by a temperature sensor 11.

The processing starts with oxygen purge to drive air from the chamber to outside. The oxygen from a vessel 1 passes through the ozonizer 4 (see Fig. 1) and heater 10 to enter into the chamber when the ozonizer 4 is operated. The ozonized oxygen passes through the 4-way valve 3 and the ozone decomposing catalyzers 21 and 23 to be discharged to an open air. Heating of the handpiece is made during this purging period which lasts 1 min. The ozone concentraion is 8,000

- 9,000 ppm. The purging is switched to the other pass so that the inner pipes and air turbine part are passed by the ozonized oxygen. The valve 12 is turned so that the ozonized oxygen passes through another heater 13 regulated by a temperature sensor 11 and enters into the inside of the hand -piece from its bottom. The ozonized oxygen leaves from its head into the chamber, and it passes through the valve 3, the catalyzer units 21 and 23 to be discharged into outside.

The next step of the processing is the sterilizing period when the oxygen is circulated in a closed circuit:ozonizer 4 - heater 10 - chamber 15 - ozonizer 4. During this circulation time the ozone concentration rises rapidly to 20,000 ppm, and the humidity of the circulating gas is raised up to a desired level by the evaporation of the water from the surface of the handpiece. The gas and handpiece temperatures are at 50 °C. The outer surface of the handpiece is sterilized. Then, the circulation path is changed by the valve 12 to ozonizer 4 - heater 13 - inside path of handpiece 14 - ozonizer 4. The inside of the air and water tubes as well as the air turbine part are sterilized. This period lasts 3 - 5 min in total.

The final step is the purging of the ozonized oxygen with clean air. The ozonizer is stopped. The clean air flows through a path: air filter 8 - heater 10 - chamber 15 - catalyzers 21 and 23 - open air. The ozonized oxygen is substituted with heated clean air, and the residual ozone layer absorbed on the outer surface of the handpiece is removed by air cleansing. This is followed by the switching to the cleansing of the inner pipes and air turbine part by the clean air. The reason why this air cleansing is needed is explained later. This final process lasts 8 min in total.

As a result the time required for one run amounts to 12 - 14 min in total.

Fig.3. shows another ozone sterilizing device. This device has another components of a ozonized gas service tank and a vacuum pump. By using these components, the sterilization of the device is made more effective as the ozonized gas is able to penetrate sufficiently into the pores and gaps of medical tools or fiber materials placed in sterilizing chamber.

The device is able to generate and compress a large amount of high concentration ozone by both the ozonizer 5 and the compressing pump 6 with the open valve 3 and the closed valve 8 and 9, and hold in the service tank 7 for use. The concentration of the ozonized gas is enriched by continuously recirculating its gas from the service tank 7 into the ozonizsr 5 with closing the volve 3 and opening the valve 8. And decompressing the gas in the sterilizing chamber 11 by the vacuum pump 14 before suppling the ozonized gas is able to make the sterilization high speed and sufficient. Then the temperature and humidity of the ozonized gas in the chamber 11 are necessary to controlled byheater 12, humidifier 10 and inner fan 15 for high speed sterilization. After the sterilization processing, the residual ozonized gas in the chamber 11 could be exhausted by the pump 21 and replaced by the clean air through air filter 17. In the processing, there sidual ozone could be decomposed with catalyzer 19. Room air should be dehumified and absorbed when used as a raw material,by using the dehumidifier 2 and absorber 3 prepared in combinatin with the activated morecular sieves or sirica gel. The air can be replaced by pure oxygen, if used.

In these device the ozone sensors are playing a role of crucial importance as the safety precaution. The high concentration ozone sensor is of an electrolytic membrane type, while the low concentration ozone sensor is a semiconductor sensor detecting the temperature rise produced the decomposition of ozone on the semiconductor element.

Residual Ozone on Handpieace

It is recognized that the residual ozone on the surface of the handpiece after a short time exposure (3-5 min) to the high concentration ozone (20,000 - 30,000 ppm) may produce an irretating stimulation to a mucous membrane of mouth when contacted. Hence, this residual ozone has to be removed somehow, but its quantity is measured at first using a handpiece as shown in Fig. 4.

This handpiece is inserted into a test cell as illustrated in Fig. 5 by 5 cm from its head, and exposed to ozonied gas (room air, dry air, or oxygen) with the ozone concentration in 1,000 - 30,000 pmm at room temperature for 100 seconds. After that the handpiece is rinsed with pottashium iodide solution (40 ml), and the amount of the residual ozone reacted is determined by KI-method.

The results obtained are shown in Fig. 6, where the quantity of the residual ozone on the handpiece surface (ca. 15 cm) is plotted again the product (ozone concentration) X (exposure time) in ppm·seconds. It can be seen that the plots for room air (ca. 60 % relative humidity), dry air (-75 °C dew point), and a comletely dry oxegen from a vessel fall on a common Curve A in the case the handpiece is not wetted. However, when the handpiece is wetted, the amount of the residual ozone becomes almost doubled as shown by Curve B which is for the dry oxygen and wet handpiece.

In addtion, the Curve A indicates the amount of the residual ozone for dry handpiece tends to satureted at about 30 um. This corresponds to a more than 10 molecular thickness of ozone film in its liquid phase. There is no such saturation tendency for the Curve B for a wetted handpiece, and the ozone concentration inside the water film seems to be very high, possibly more than several hundred ppm. This suggests that wetting of the handpiece and increase of the relative humidity of the ozonized gas should be important factors to achieve a satisfactory sterilizing effect.

Removal of Residual Ozone

The amount of residual ozone on the surface of the handpiece is ca. 100 ug (ca. 9 ug/cm 2) or less in most of the cases. But, it gives a strong odor of ozone when approched to the nose, and an irritating stimulation on the sensitive part of the mucous membrane, such as tongue, in mouth. In consideration of this fact and also of a possible harmful effect of the residual ozone, attemts are made to remove it.

The methods of ozone removal tested are:

1) Exposure to a hot air stream (0.7 m/s at handpiece),
2) Immersion in boiling water (95 °C),
3) Immersion in pure water (15 °C),
4) Cleansing by compressed air (50 l/min),
5) Warming up by heated air from a hair drier,
6) Radiation heating of the surface with air flow (100 °C, 9 l/min),using a heating element (not lamp),
7) Radiation heating of the surface with air flow (85 °C, 50 l/min), using a heating element, and
8) Wiping of the surface with a wet paper immersed in:
   8a) pure water,
   8b) 1 % sodium bicarbonate solution, and
   8c) 5 % sodium bicarbonate solution.

Table 1 shows the results obtained with the methods 1) - 10), when the handpiece has been exposed to the ozonized oxygen with a very high concentration ozone in 30,000 ppm for 50 seconds and 100 seconds, and the processing for removal is continued for 5 - 100 seconds. The most effective of these 10 methods is the exposure to a 355 °C hot air stream with 0.7 m/s velocity at the handpiece, which results in zero residual ozone at 30 seconds processing time both for 100 and 50 seconds exposure to the ozone. The handpiece is either wetted before ozone exposure (marked by "wet"), or dried with blowing room air (marked by "dry"). Lowering the air stream temperature produced a reduction of ozone removal with a fluctuating value of the residual ozone after processing.

A much more stable effect is obtained with the method 2) using boiling water immersion and the method 3) using a pure water immersion. Both methods produce the ozone reduction down to 5 - 12 ug (0.3 - 1.0 ug/cm 2).

The method 4) using compressed air blowing also produces the same effect, but the heating by hot air stream (Method 5) and radiation with a heating element (not UR-lamp) produce much less effects.

Tests are made also with a UR-lamp (500 W) and a UV-lamp, both with 30 seconds irradiation time, but no effct is obtained.

Table 2 shows the results obtained by wiping with a wet paper (Method 11). The handpiece is immmersed in pure water, and is wiped with paper before exposure to the ozonized oxygen. After the exposure it is piped 2 - 3 times with a wet paper carefully, and its residual ozone determined with the KI-method.

It can be seen on Table 2 that no difference exists between pure water and sodium bicarbonate solution, both producing the reduction of residual ozone down to about 10 ug (0.6 ug/cm). The following effects are obtained in commom to these wiping methods:
a) A strong odor of ozone from the handpiece is substantially reduced.
b) No irritation is felt when the mucous membrane in mouth is attached with the handpiece.

When wiping is made 5 - 6 times with a wet paper using a sodium bicarbonate solution with a higher sodium concentration (several %), the ozone odor disappears completely.

After evaluating these test results, choice is made of the hot air exposure method in consideration of its simplicity and ease in use. A longer time in combination with a lower temperature provides an equally satisfactory effect, giving our standard processing procedure described in the previous section (8 min; 50 °C).

Sterilizing Effect

A large number of sterilizing tests are made using the reference bacteria (Bacillus Subtiris of Blastomyce) under a wide range of experimental conditions: ozone concentraion, gas temperature, gas humidity, wetted and dried handpiece, treatment time, etc.

The target is placed to meet the sterilization standard that 10 germs of the above reference bateria are completely killed, which must be proven by culture after the treatment.

The tests are made in collaboration with dentists, biologists and medical doctors. The detailed test results are to be reported separately, and the essential points are described here:

1) Without prewetting of the handpiece, its warming and drying during the ozone exposure using a heated ozonized gas, and preferably raising its humidity, a complete sterilization of the handpiece is not possible even with the use of very high concentration ozone in as high as 20,000 - 30,000 ppm.
2) It is possible to achieve a complete sterilization in the case when the requirements given in 1) are met during the processing.

Tests made in a rainy season when ambient temperature and humidity are quite high (30 °C and 80 - 90 % RH) produce the complete sterilization easily, even without pre-wetting of the handpiece and raising gas temperature. However, tests made in a cool and dry season never provide the same effect, unless the requirements of 1) be met.

It is evident that the existence of water film on the surface of the handpiece is very important, as it absorbs a much greater quantity of ozone, as decribed in the previous section. In addtion, the outer shell of the reference bacteria might be swellen and soften by such wetting so as to easily absorb ozone into its inside. As the test sample of the bacteria forms an aggregation layer (see Fig.7), it is not easy for ozone to penetrate into the bottom of such layer to produce the complete sterilizing effect. However, in case the water film exists to absorb the high concentration ozone at its outer surface region, and this ozonized water region with a very high concentration ozone sweep the aggregate layer of the bateria during its drying porcess by heating, all of the bacteria may contact with such ozonized water region to subject its effect.

The true mechanism of the necessity of the water film and gas heating is yet to be investigated, but it is extremely interesting to make a comparative observation of the surviving and killed germs in the electron microscope pictures (see Fig. 8). The killed germ always indicate a spot which seems to be an inner substance spilled out of its membrane through a tiny puncture possibly produced by ozone attack. Where as the surving cell never indicates such spot.

It should be pointed out that the things seem to be much more involved and complicated in the ozone sterilizing process, or in the sterilizing process as a whole, but these are beyond the scope of this paper and are not decribed here.

Anyway, the sterilizng conditions decribed in the preceding section are fixed after evaluation of all the data obtained in these sterilizing tests.

Conclusion

A series of tests are made to develop a high speed sterilization equipments for dental and medical use, using very high concentration ozone produced with a very small sized novel ceramic based ozonizer. The emphasis in the first phase of test series is placed on a complete sterilization of the dental handpiece with a target that 10 cells of Bacillus subtiris of Blastromyce are killed.

The following conclusions are obtained from these tests:

(1) A complete sterilization of the dental handiece is possible by meeting the requirements:
 a) the handpiece is prewetted by wiping wet paper,
 b) the sterilization cell is prewarm by flowing a heated clean air,
 c) the purging ozonized oxygen is preheated,
 d) the sterilizing process is made with a very high concentration ozone (20,000 - 30,000 ppm) for about 3 - 5 minutes.
(2) After sterilization process the residual ozone remaining on the handpiece surface must be removed so as to reduce

its level down to at least 0.5 (ug/cm 2) so as not to cause an irritation when the handpiece is contacted to a sensitive mucous membrane such as tongue in mouth.
(3) The ozone sterilizing equipment reported in this paper uses 20,000 ppm ozone in its sterilizing period by circulating the ozonied oxygen through its ozonizer and sterilizing chamber to save the use of oxygen.
(4) The present equipment uses for removal of the residual ozone from the handpiece the cleaning by heated clean air (50 °C).
(5) The inner passages and air turbine part are also ozone sterilizied in this equipment.
(6) The total sterilizinng time for one run amounts to 12 - 14 minutes, much shorter compared to the autoclave method (30 min) and ethylene oxide method (45 min).

The ozone sterilizing equipment can be used also in many other dental and medical areas, where the object specific design and application mode must be developed indivisually.

It is hoped that the advantage of ozone sterilization - a high speed, simplicity in construction and use, and no harmful residues to be left after use - would serve various needs of sterilizing the tools against various germs and viruses.

Acknowledgements

The authors acknowledge valuable advises and suggestion given by Prof. Nozomu Hoashi of Dental Department of Saitama College of Health, and Mr. Yuzaburo Seike of Midori Anzen Co., Ltd. They also appreciate the collaboration and help given by Dr. Masaru Kasiwagi of Tokyo Sanitary Research Center and Dr. H. Yamaguchi of Showa Medical University.

Reference

(1) S. Masuda, K. Akutu, M. Kuroda, Y. Awatsu and Y. Shibuya: A Ceramic-Based Ozonizer Using High Frequency Discharge, IEEE-IA Transactions, Vol. 24, No. 2 (March/April, 1988), pp.233-231 (1988)

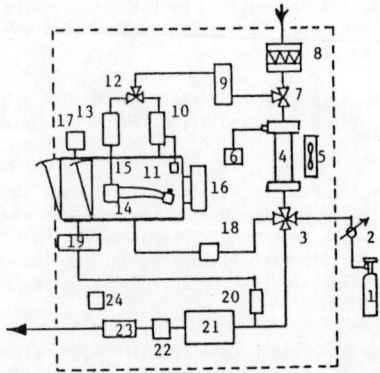

1: oxygen vessel, 2: pressure reguration volve, 3: 4-way valve, 5: cooling fan, 6: pressure guage, 7: 3-way valve, 8: air filter, 9: pump, 10: heater, 11: temperature sensor, 12: 3-way valve, 13: heater, 14: handpiece, 15: sterilizing chamber, 16: fan, 17: interrock for double doors, 18: high concentration ozone sensor, 19: fan, 20: check valve, 21: ozone decomposing catalyzer, 22: low concentration ozone sensor, 23: ozone decomposing catalyzer, 24: low temperature sensor.

Fig. 2 Construction of Ozone Sterilizing Equipment for Handpiece for Dental Use.

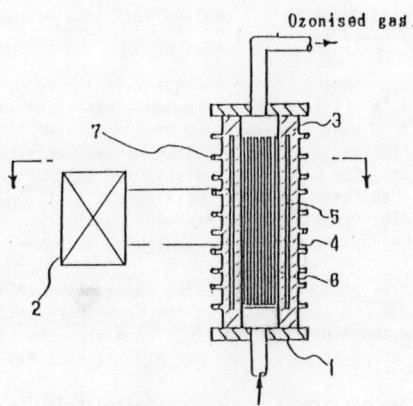

(a) Vertical Cross-Section

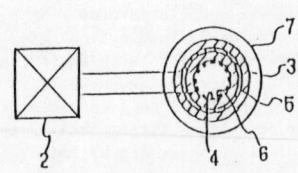

(b) Horizontal Cross-Section

1: end plate, 2: high frequency high voltage power supply, 3: alumina ceramic cylinder, 4: strip-like corona electrodes, 5: film-like induction electrode, 6: inner surface of ceramic cylinder, 7: cooling fin.

Fig. 1 Ceramic Based Ozonizer Using High Frequecy Surface Discharge (Cylinder Type; OC-10).

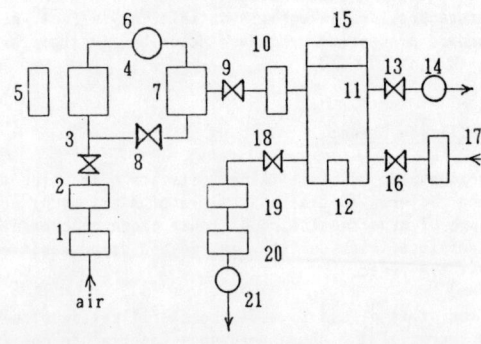

1: absorber, 2: dehumidifier, 3: supply valve, 4: ozonizer, 5: cooling fan, 6: supply and compressing pump, 7: ozonized gas service tank, 8: recirculation valve, 9: supply valve, 10: humidifier, 11: sterilizing chamber, 12: heater, 13: inner fan, 14: vacuum volve, 15: vacuum pump, 16: air valve, 17: air filter, 18: exhaust valve, 19: catalyzer, 20: ozone monitor, 21: exhaust pump.

Fig. 3 Construction of Ozone Sterilizing Equipment for Medical Use.

Fig. 4 Handpiece for Drilling Teeth.

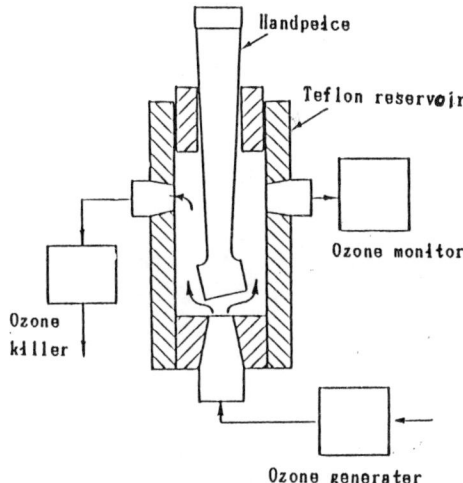

Fig. 5 Experimental Setup for Measuring Residual Ozone on Handpiece.

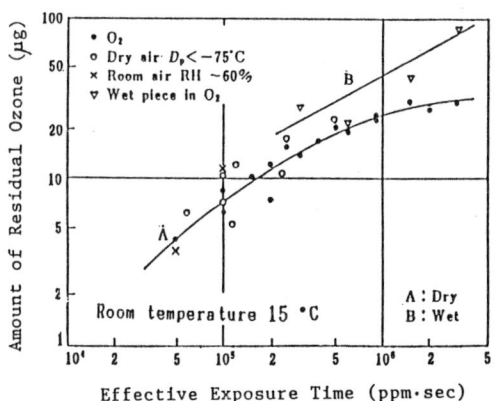

Fig. 6 Amount of Residual Ozone vs. Effective Exposure Time Te (Te= (exposure time) X (zone concentration) in ppm·sec).

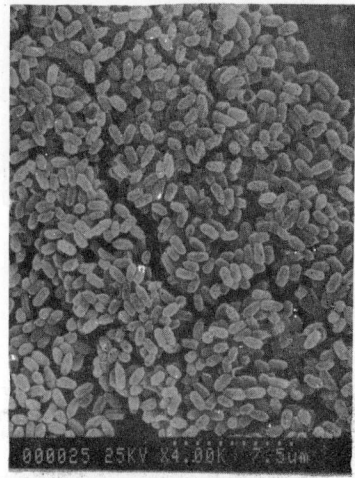

Fig. 7 Electron Microscope Photogragh of Aggregate Deposit of Bacillus Subtilis after Drying.

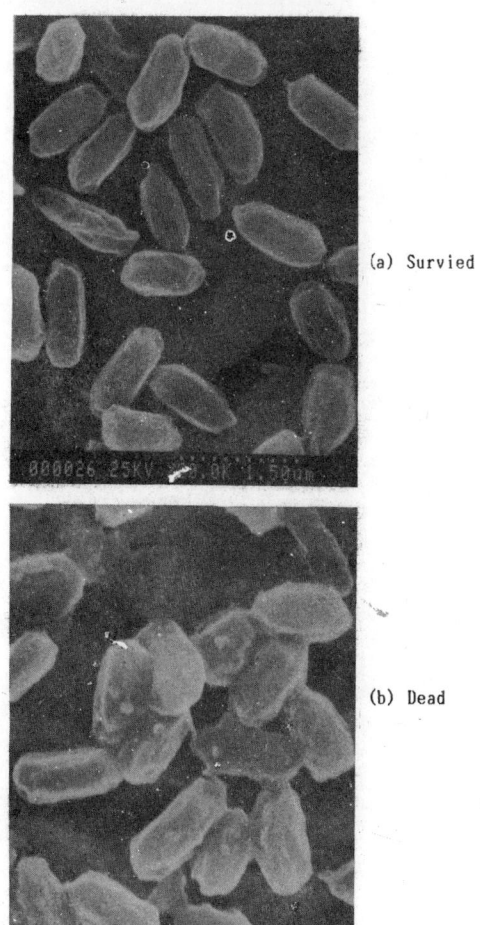

(a) Survied

(b) Dead

Fig. 8 Electron Microscope Photograghs of Survied and Dead Bacillus Subtilis Cells.

Table 1 Methods of Removal of Residual Ozone from Surface of Dental Handpiece

Method No.	Method	Ozone Exposure Time (s)	Removal Processing Time (s)	Residual Ozone (ug)
1	exposure to hot air stream (0.7 m/s at handpiece)	100 (dry)	30 (355°C)	0
		50 (dry)	30 (355°C)	0
		100 (dry)	5 (355°C)	15
		50 (dry)	5 (355°C)	9
		100 (dry)	30 (180°C)	0 - 20
		50 (dry)	30 (180°C)	0 - 19
		100 (wet)	30 (180°C)	0 - 30
		50 (wet)	30 (180°C)	0 - 30
2	immersion in boiling water (95°C)	100 (dry)	5	12
		100 (wet)	5	4
		50 (wet)	5	3.6
3	immersion in pure water (15°C)	100 (wet)	5	9
		50 (wet)	5	6
4	cleansing by compressed air (0.2 kg/cm; 50 l/min)	100 (dry)	30	10
		100 (wet)	30	9
5	warming by heated air from hair drier	100 (dry)	30	18
		50 (dry)	30	30
		100 (wet)	30	26
		50 (wet)	30	33
6	radiation heating of surface plus air flow (100°C; 9 l/min)	100 (dry)	30	34
		50 (wet)	30	15
7	radiation heating of surface plus air flow (85°C; 50 l/min)	100 (dry)	30	20
		100 (wet)	30	17

Ozone Concentration: 30,000 ppm

Table 2 Removal of Residual Ozone by Wiping with a Wet Paper (Method 11)

Method No.	Liquid	Exposure Time(s)	Residual Ozone (ug)
8a	Pure water	100 (wet)	12
		50 (wet)	11
8b	Water Solution of 1 % Sodium Bicarbonate	100 (wet)	8.5
		50 (wet)	8.6
8c	Water Solution of 5 % Sodium Bicarbonate	100 (wet)	11.4
		50 (wet)	14.4

Ozone concentration: 25,000 ppm.

EXPERIMENTAL STUDY OF IMPROVEMENT IN OZONE YIELD
IN A PARALLEL PLATE OZONIZER WITH A ROTATING PLATE ELECTRODE

Y. Nomoto[1], T. Ohkubo[1], T. Adachi[1], J.S. Chang[2], and M. Akazaki[3]

1) Dept. of Elect., Eng. Faculty of Eng., Oita University,
   700, Dannoharu, Oita, Japan 870-11
2) Dept. of Eng. Phys., Faculty of Eng., McMaster University,
   Hamilton, Ontario, Canada L8S 4M1
3) Graduate School of Engineering Science, Kyushu University,
   6-1, Kasuga-Koen, Kasuga, Japan 816

Abstract

The ozone generation efficiency in an industrial ozonizer is very low compared to the theoretical efficiency for instance, approximately 7.5 % of the theoretical value for air. In this paper, in order to improve the ozone generation efficiency, a new type ozonizer with a rotating plate electrode is proposed. As a result, the ozone generation efficiency is improved by the effect of electrode rotation up to 35 % higher than that of the stationary electrode type ozonizer. Improvement in ozone yield is more effective for high power operation in the ozonizer than for low power operation.

Introduction

Ozone has been produced by a ozonizer with the silent discharge and used widely for the treatment of water and exhausted smoke, elimination of the rank odors, color removal and disinfection etc. because of its strong oxidization. However, an ozone generation efficiency which has been obtained using silent discharge method is approximately 200 g/kWh with oxygen and about 90 g/kWh with air. These values are very small compared with the theoretical value of 1200 g/kWh. In order to improve the efficiency, a lot of attempts have been made [1-4]. It seems that the increase of the ozone yield causes the decrease of ozone generation efficiency. That is, high discharging energy is required to produce much ozone, however the produced ozone is destroyed in turn by the heat produced in the ozonizer and by the electron impact.

In this paper, in order to improve the ozone generation efficiency, a new type ozonizer with a rotating plate electrode is proposed. Since plate spacing of usual ozonizer is set to approximately 1 mm, it is considered that produced ozone does not flow effectively. In the new type ozonizer, it is expected that produced ozone is able to flow effectively by the centrifugal force due to the rotation of the electrode and that the silent discharge in the electrode gap is homogenized by the rotation of the electrode.

Experimental Apparatus

Schematic diagram of experimental apparatus for a parallel plate ozonizer with a rotating electrode is shown in Fig.1. The ozonizer has two aluminum plate electrodes of 60 mm diameter. The upper electrode has a hole of 6 mm diameter at the center through which dry air is drawn in as material gas. The lower electrode on which a quartz glass plate of 80 mm diameter and 1 mm thickness is sticked as a dielectric by a plastic glue, can be rotated at a rotational speed from 0 rpm to 3500 rpm using a dc motor with a pulley and a belt. The gap length between two electrodes is adjusted to the range from 0.5 mm to 3 mm by putting Teflon sheets between the electrodes. The electrode system is enclosed in an acrylic vessel. Ozonized gas in the vessel is inhaled into the ozone monitor with ultraviolet ray absorption method by a pump at the rate of 2 l/min. AC voltage of 60 Hz frequency is applied to the upper electrode and the lower electrode is grounded by a slip-ring and a brush. Discharging power is calculated from the area of the Lissjous's figure which is drawn by using the applied voltage and the voltage in the film capacitor of 1 µF which is connected between the lower electrode and the ground.

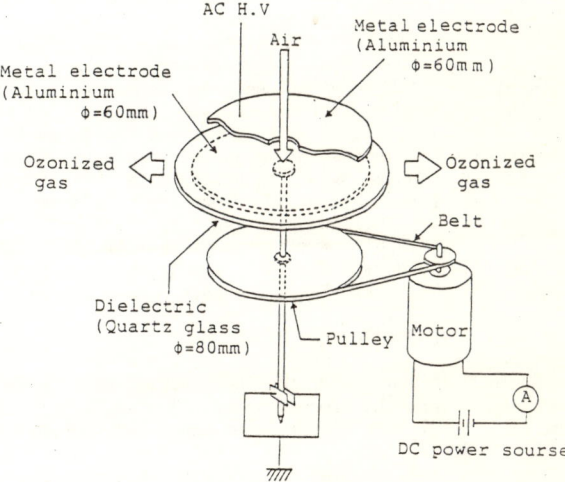

Fig.1 Schematic diagram of experimental apparatus for a ozonizer with a rotating electrode

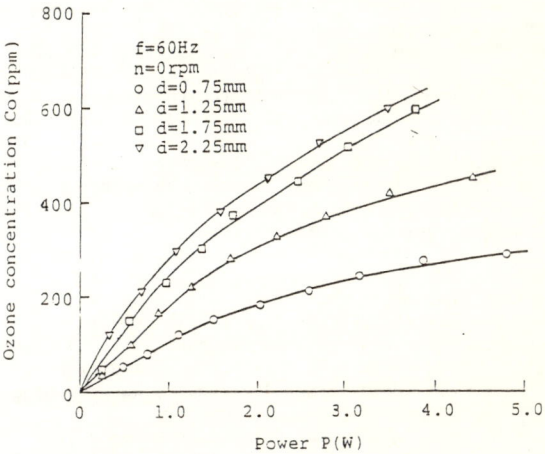

Fig.2 Dependence of ozone concentration on discharging power with d as a variable

-513-

Results and Discussions

Characteristics of the ozonizer with stationary electrode

Relations of ozone concentration Co (ppm) and ozone generation efficiency η (g/kWh) with discharging power in the ozonizer with the stationary electrode, are shown in Fig.2 and in Fig.3 respectively. Discharging power is increased by increasing the applied voltage with constant frequency f=60Hz. As the gap length increases under the condition of constant power, the ozone concentration as well as the ozone generation efficiency increase. Therefore, it is better for ozone production at the constant power to operate the ozonizer with low energy density and large discharge volume than with high energy density and small discharge volume. As discharging power increases at any gap length, the ozone concentration increases and approachs a constant value determined by the gap length, however the ozone genaration efficiency decreases contrary. It is seen from these results that high discharging power is necessary to product much ozone, however it causes a lowering of the ozóne generation efficiency because high discharging power heats the electrodes of the ozonizer and generates NOx.

Figure 4 shows the relations between the ozone generation efficiency and the electric field strength in the ozonizer gap for various gap length. At the low electric field strength, the ozone generation efficiency has a constant value determined by the gap length, independently of the electric field strength. The constant values are 30 g/kWh for d=0.75 mm, 50 g/kWh for d=1.25 mm and 62 g/kWh for d=1.75 mm. In the region of higher electric field strength, the ozone generation efficiency η is in inverse proportion to the electric field strength E as follows:

$$E \cdot \eta = 220 \quad (kV \cdot g/kWh \cdot mm) \quad (1)$$

So, η becomes lower for the higher electric field strength where the ozone concentration is high. The solid line in the figure shows the curve corresponding to equation (1).

Characteristics of the ozonizer with a rotating electrode

Voltage-current characteristics of the ozonizer are shown with gap length as a parameter in Fig.5. As the applied voltage is increased for the constant gap length, the alternating current proportional to the applied voltage flows which is determined by the capacity of the parallel plate electrodes. Gradient of the current with respect to the voltage becomes steeper when the ozonizer discharge begins at the voltage of 4 to 5 kV. The onset voltage of the ozonizer discharge is higher for longer gap length. In this figure, the voltage-current characteristics for the rotational electrode with 1260 rpm are also shown as well as for the stationary one. It is seen from this figure that the voltage-current characteristics for the rotational electrode are almost same as ones for the stationary electrode. It is also confirmed that voltage-charge Lissjous's figure hardly change when the lower electrode is rotated. It can be considered from these results that discharging power in ozonizer discharge is not affected by the rotation of the electrode.

The ozone generation efficiency becomes low when the discharging power or the electric field strength in the ozonizer is high, as mentioned in preceding section. The reasons causing lower ozone generation efficiency could be as follows: first, the tempera-

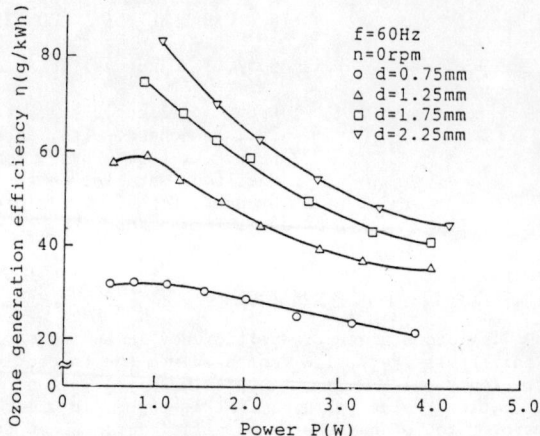

Fig.3 Dependence of ozone generation efficiency on discharging power with d as a variable

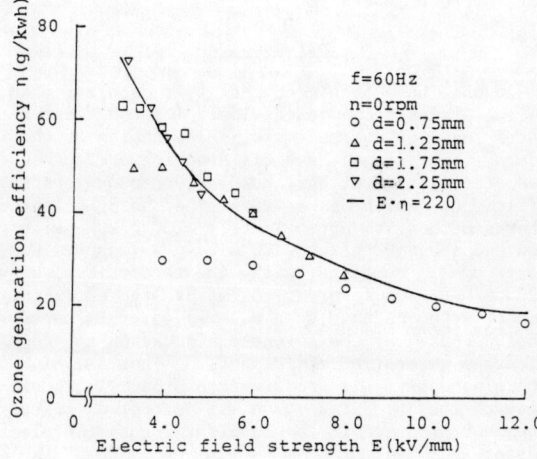

Fig.4 Dependence of ozone generation efficiency on electric field strength with d as a variable

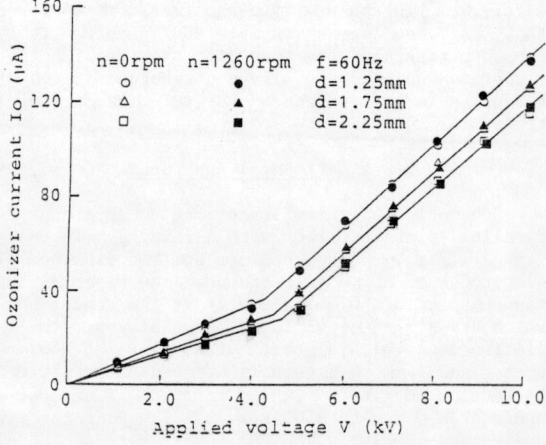

Fig.5 Current-voltage characteristics of the ozonizer with d as a variable

ture in the ozonizer become higher due to increase of the discharging power, so ozone is destroyed into oxygen by heat; secondarily, produced ozone does not flow effectively because of narrow gap length of approximately 1 mm in a usual ozonizer; thirdly ozone generated by the silent discharge is decomposed into oxygen by the following electron impact,

$$O_3 + e \longrightarrow O_2 + O + e. \quad (2)$$

It is considered that the rate of electron impact dissociation of ozone increase if the second microdischarge happens at the same place where the first microdischarge has occurred and produced ozone. So, the electron impact dissociation of ozone is expected to reduce by rotating electrode in the parallel plate ozonizer and homogenizing discharge.

Figure 6 and Fig.7 show the dependence of ozone concentration and ozone generation efficiency on the rotational speed of the lower electrode with gap length as a parameter. The ozone concentration and the ozone generation efficiency have a maximum value at the revolution of about 1000 rpm with increasing revolution of the electrode. Hence, rotating electrode makes the ozone generation efficiency increase by approximately 20 % from the value in stationary electrode.

Effects of the electrode rotation on ozone concentration-voltage characteristics are shown in Fig.8. The rotating electrode with n=1260 rpm has no effects upon the ozone concentration for the applied voltage below 8 kV, however, increases the ozone concentration with the voltage above 8 kV; Increment rate of ozone concentration is approximately 20 % at V=12 kV.

The ozone generation efficiency increases with the rotational speed of the electrode, then the efficiency decreases gradually at higher rotational speed. That is, the rotational speed N_{max} exists where the maximum ozone generation efficiency η_{max} is accomplished. N_{max} depends on a gap length and applied voltage. Relations between N_{max} and gap length and relations between the maximum value η_{max} of ozone generation efficiency corresponding to the N_{max} and the gap length are shown in Fig.9 and in Fig.10 respectively with applied voltage as a parameter. For V=8.0 kV, N_{max} has an almost constant value of 1000 rpm independently of gap length. N_{max} has the maximum value for V=6.0 kV as gap length is increased.

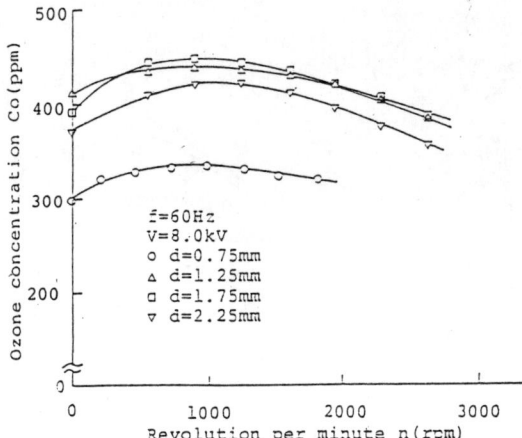

Fig.6 Effects of rotational speed on ozone concentration with d as a variable

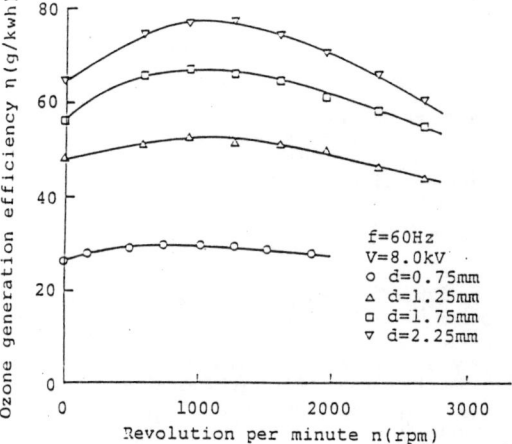

Fig.7 Effects of rotational speed on ozone generation efficiency with d as a variable

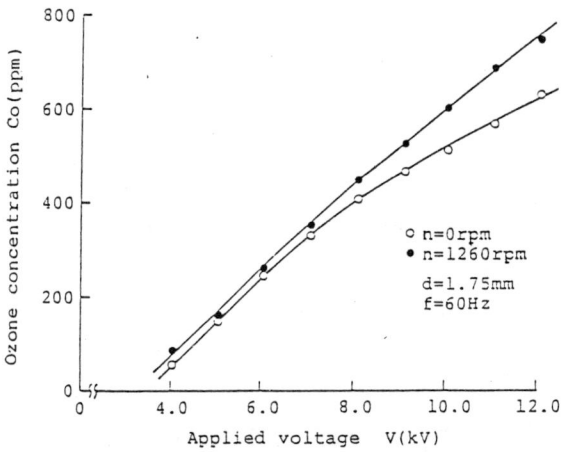

Fig.8 Effects of electrode rotation on ozone concentration-applied voltage characteristics

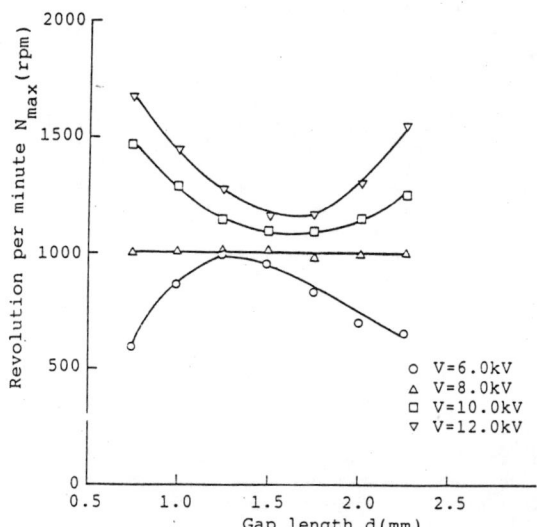

Fig.9 Rotational speed N_{max} for η to take the maximum value

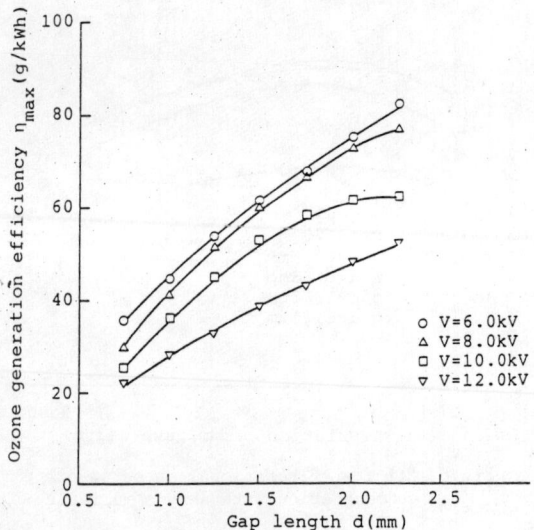

Fig.10 Dependence of η_{max} on d with V as a variable

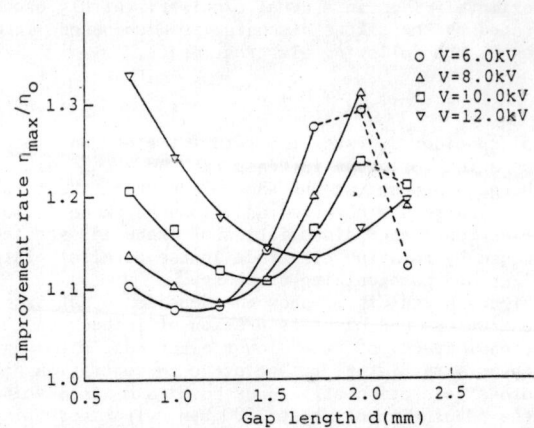

Fig.11 Improvement rate of ozone generation efficiency η_{max}/η_o in the rotating electrode ozonizer

On the other hand, N_{max} has the minimum value for V=10.0 kV and V=12.0 kV. It is considered that these results show the improvement mechanism in the ozone generation efficiency for V=6.0 kV is different from that for higher applied voltage. N_{max} has a larger value for higher voltage at the constant gap length. N_{max} changes in a wide range at d=0.75 mm and d=2.25 mm for the change of the applied voltage from 6 kV to 12kV, however, N_{max} changes in a narrow range of 950rpm to 1150rpm at d=1.5mm. So that, for about 1.5 mm gap length the ozone generation efficiency can be improved for a wide range of the applied voltage by the small change of rotational speed in the electrode. The maximum value η_{max} of ozone generation efficiency becomes low with decreasing gap length and increasing applied voltage. So, the ozone generation efficiency is lower for the higher electric field strength in the gap. That is in agreement with the results obtained in Fig.4.

The rate of improvement in the ozone generation efficiency by the rotational electrode which is defined by the ratio of η_{max} expressed in Fig.11 to η_o for the stationary electrode, is plotted with respect to gap length as shown in Fig.11. The effect of the rotational electrode on the improvement of the efficiency is considerable for high electric field strength in the gap. The maximum improvement rate of 35% is obtained for d=0.75mm and V=12 kV. As the gap length increases under the constant applied voltage, the improvement rate of the efficiency become low, then has the minimum value determined by the applied voltage. For the longer gap length, the ozone generation becomes unstable due to the lowering electric field strength.

Conclusions

A new type ozonizer with a rotating plate electrode is proposed in order to improve the ozone generation efficiency. Effects of the rotational speed on the ozone concentration and the ozone generation efficiency in the ozonizer is experimentally investigated. The results are summarized as follows.

(1) High discharging power is necessary to product much ozone, however it generally causes lowering of the ozone generation efficiency η.
(2) η becomes lower for the higher electric field strength E where the ozone concentration is high. So, η is approximately in inverse proportion to E.
(3) It can be considered that discharging power in ozonizer discharge is not affected by the rotation of the electrode.
(4) The rotational speed N_{max} exists where the maximum ozone generation efficiency is accomplished. N_{max} depends on a gap length and applied voltage.
(5) The effect of the rotational electrode on the improvement of the efficiency is considerable for high electric field strength in the gap.
The maximum improvement of 35 % is obtained for d=0.75 mm and V=12 kV

Acknowledgements

The authors are pleased to acknowledge for useful discussions with Dr. M.Hara of Kyusyu University. The authors also wish to express thanks to Mr. S.Akamine of Oita University for technical help with the experimental programme.

References

[1] S. Masuda, K. Akutsu and M. Kuroda : Proc. 8th JSPC (Sept., 1987)
[2] S. Masuda and S. Koizumi : Proc. IEEE/IAS 1986 Annual Conf.
[3] J. Salge and P. Braumann: Proc. 4th Int. Symp. on Plasma Chemistry, 735 (1979)
[4] Y. Kondo, S. Kajita and S. Ushiroda: J. Japan Research-Group of Elect. Discharge No.114, 12(1987)

INVESTIGATION OF DISCHARGE CURRENT OF SURFACE DISCHARGE TYPE OZONISER

Endre Kiss* and Senichi Masuda**

* Dunaujváros College of Miskolc University, H 2401 Dunaujváros, Hungary
** Fukui Institute of Technology, Gakuen, Fukui, 910 Japan

Abstract

High frequency surface discharge is generated on the surface of a high purity alumina ceramic sheet between a discharge electrode situated on the surface and an induction electrode embedded inside. High concentration of ozone gas is produced in discharge filaments in the presence of oxygen. In order to better understand the ozone forming and destroying processes in these filaments the streamer discharge current and the intensity, as well as the spectral distribution of the light generated by the discharge is investigated at different frequencies and applied voltages. The peak value of the current of an elementary discharge seems to be characteristic for ozone production, and is a linear function of the reciprocal pressure, similarly to the average length of the streamers. The spectral distribution of light is changing with increasing applied voltage.

Introduction

It is well known since decades that ozone is the most reliable oxidizing material, because of its effectiveness, and because it leaves no hazardous materials after the treatment and, after a certain time, decomposes into oxygen which is an ordinary component of the environment. Therefore the demands for ozone generators are steadily increasing nowadays. A great effort is beeing done world-wide to satisfy these demands.. One of the most promising development is the ceramic-made ozoniser device in which high frequency surface discharge is generated on the surface of a high purity alumina ceramic between a discharge electrode situated on the surface and an induction sheet electrode embedded inside. As the ceramic is similar to that used in IC package, the ozoniser is very durable against mechanical, thermal and electric shocks, and it allows the use of a rather high exciting frequency causing a considerably high grade reduction in its dimensions and cost. This type of ozoniser works well, and has been investigated extensively (1, 2, 3), but the streamer discharge of this construction and the light emitted by that have more thoroughly to be investigated. This report describes the work made to fulfill this task.

Materials and Methods

The ceramic-based ozoniser devices have various geometrical shapes and sizes (1,2). The basic design of them is the plate type ozoniser presented in Fig. 1. The most useful one is the cylindrical type (Fig. 2). For small power ozoniser a linear type is used (Fig. 3), because of its small dimension and cheap price. In the present investigation the cylindrical (Fig. 2) and the linear types (Fig. 3) are tested with an active electrode length of 1m and 40mm, respectively.

The discharge current and the light emitted by the streamers are measured in the experimental arrangement illustrated in Fig. 4. The spectral distribution of light is investigated by using a Hitachi 100-20 type monochromator with a spectral bandwidth of 7nm (Fig. 5). The photomultiplier used is a Hamamatsu R212. The

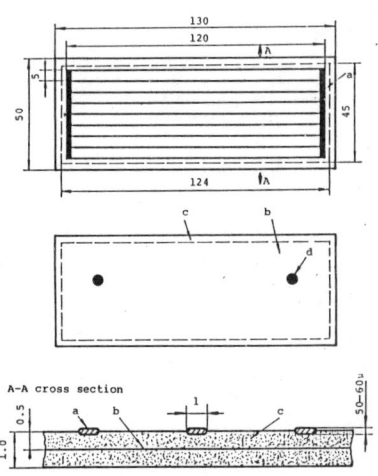

a: discharge electrode; b: induction electrode; c: high purity alumina ceramics; terminal for induction electrode

Fig. 1 Basic Design of Ozonisers Tested

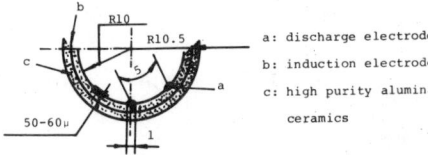

Fig. 2 Cross Sectional View of Cylindrical Ozoniser Tested

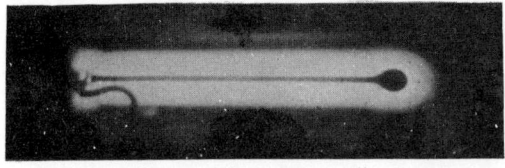

Fig. 3 Linear Type Ozoniser and the Light Emitted by the Discharge

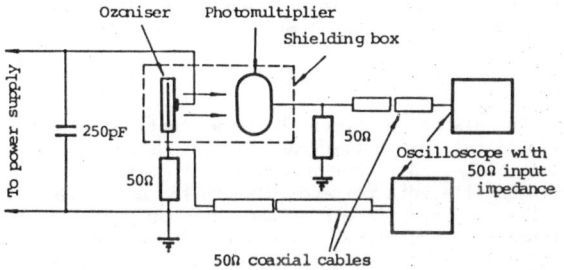

Fig. 4 Experimental Arrangement to Measure Discharge Current and Light

oscilloscopes used are a Tektronix 7912AD Programmable Digitizer with a No. 604 monitor, and a Tektronix 7844 Dual Beam oscilloscope. To excite the ozonisers the applied voltage is changed from 0 to $15kV_{pp}$, the frequency is varied between 50Hz and 10kHz in the cases of both dry and room air, as well as oxygen feed gases at a pressure range from 35 to 450kPa.

To obtain information from the ozone destroying processes, the linear type ozoniser is irradiated by a separate light source emitting intensively the 254nm wavelength. The ozoniser is adjusted such a direction which allows only the reflected light to enter the monocromator of Fig. 5.

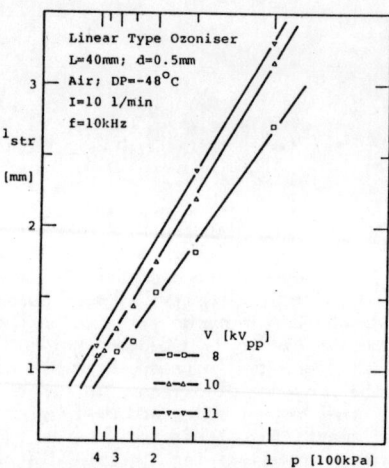

Fig. 7 Average Length of the Luminous Part of the Discharge Filaments vs. Pressure in Dry Air

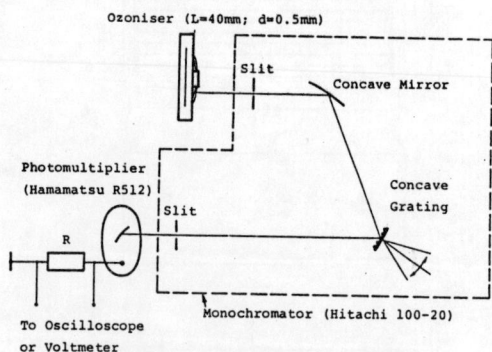

Fig. 5 Experimental Arrangement to Investigate Spectral Distribution of Discharge Light

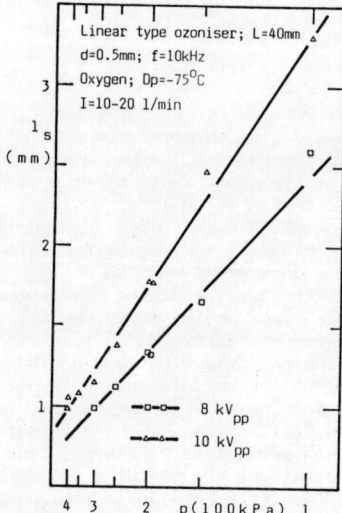

Fig. 8 Average Length of the Luminous Part of the Discharge Filaments vs. Pressure in Oxygen

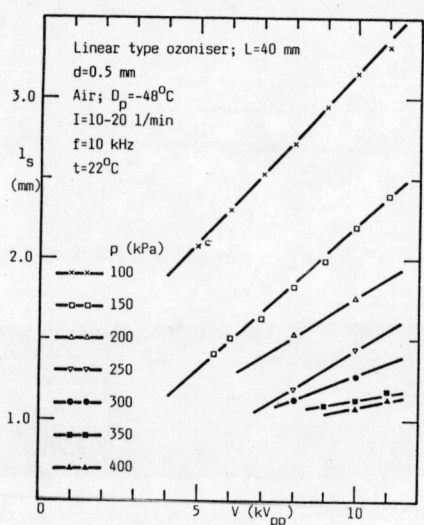

Fig. 6 Average Length of the Luminous Part of the Discharge Filaments vs. Applied Voltage in Dry Air

Results

The discharges appear to be a continuous stripe surrounding the entire edge of the discharge electrodes (Fig. 3), but in reality it consists of many luminous filaments perpendicular to the electrodes.

The average length of the luminous part of the filaments is dependent on the applied voltage, pressure and feed gases. If the voltage is increasing linearly at a constant pressure, the length increases also linearly both in air (Fig. 6) and in oxygen.

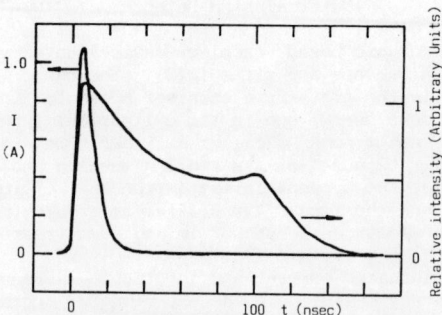

Fig. 9 Time Dependence of Overall Discharge Current and Light of the Linear Type Ozoniser

The average length of the luminous part of the filaments seems to be proportional to the reciprocal pressure (Fig. 7 and 8).

The time dependence of the overall discharge current and the light emitted by the discharge in the case of the linear type ozoniser with an active electrode length (L) of 40mm is presented in Fig. 9. The current is a very short pulse with a fast rise and decay. The light impulse is similar in rising, but has a considerable after glow.

The relative intensity of the light emitted by the entire linear type ozoniser increases with increasing applied voltage (Fig. 10 and 11), but if the cooling conditions are not perfect (Fig. 10 at 100-200kPa) the proportionality becomes wrong.

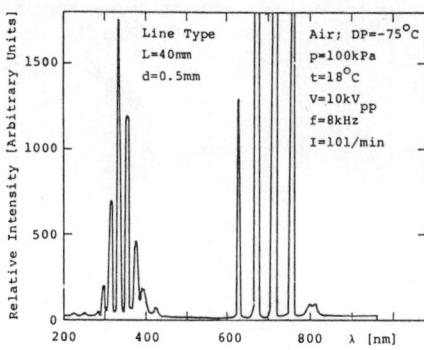

Fig. 12 Spectral Distribution of Light Emitted by the Discharge in Air

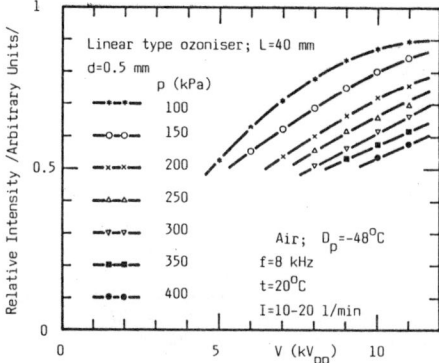

Fig. 10 Relative Intensity of Light Emitted by Linear Type Ozoniser vs. Voltage in Air

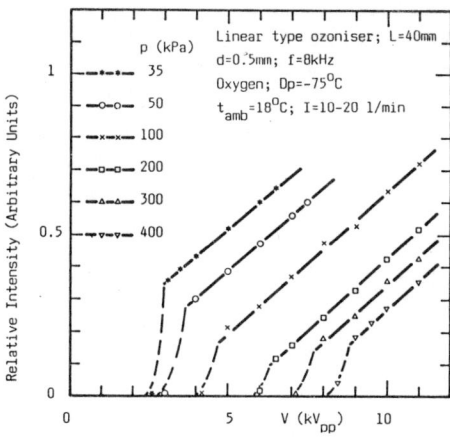

Fig. 11 Relative Intensity of Light Emitted by Linear Type Ozoniser vs. Voltage in Oxygen

The spectral distribution of the light emitted by the discharge is presented in Fig. 12 for dry air. The normalised relative intensities of some characteristic lines are plotted against applied voltage in Fig. 13. It seems to be clear that the spectral distribution is changing by applied voltage.

The light impulses at different characteristic wavelengths are fundamentally similar to that of Fig. 9.

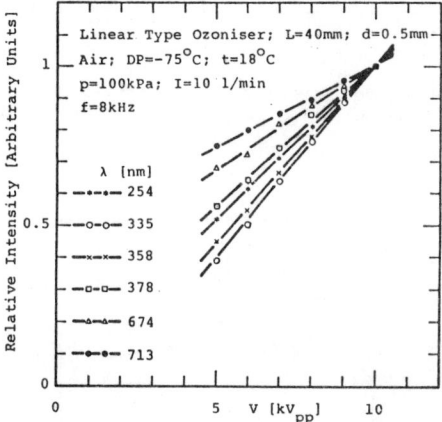

Fig. 13 Relative Intensity of the Characteristic Lines vs. Applied Voltage

The drop in the relative intensity of the light reflected at 254nm from the linear ozoniser (which is proportional to the ozone concentration in the place of reflection) is plotted against applied voltage together with the ozone concentration measured simultaneously at 100 and 300 kPa (Fig. 14). As the concentration of ozone is measured not very close to the electrodes but the light is investigated in the discharge region, it can be said that the ozone destroying processes are taking place far from the discharge region.

The peak value of the overall discharge current of the cylindrical type ozoniser (Fig. 2) depends on the applied voltage (2), and on flow rates (Fig. 15). It is very interesting that the discharge current vs. flow rates and the ozone productivity vs. flow rates (3) curves are rather similar.

The peak value of the overall discharge current pulse seems to be proportional to the reciprocal pressure (Fig. 16 and 17), and a similar relationship can be observed between the RMS of the discharge current and the reciprocal pressure (Fig. 17)

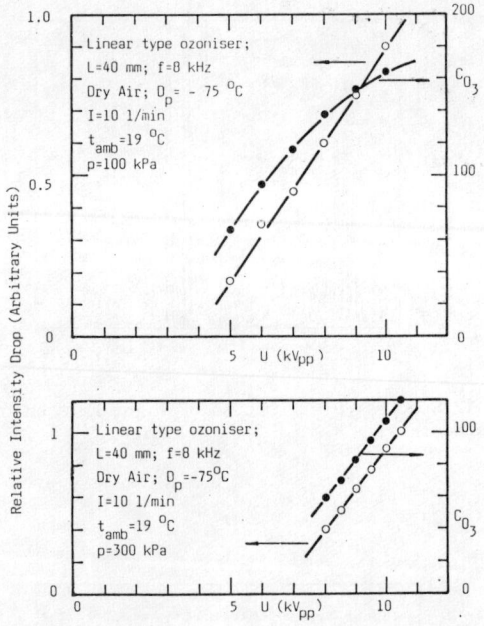

Fig. 14 Relative Intensity Drop in Reflected UV Light, and Ozone Concentration vs. Voltage in Air at 100 and 300kPa

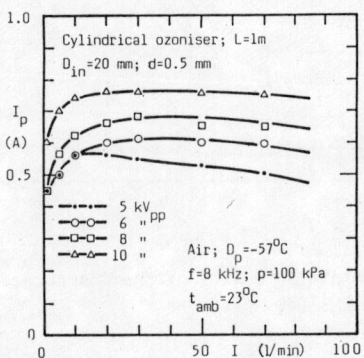

Fig. 15 Peak Value of Overall Discharge Current of Cylindrical Ozoniser vs. Flow Rates

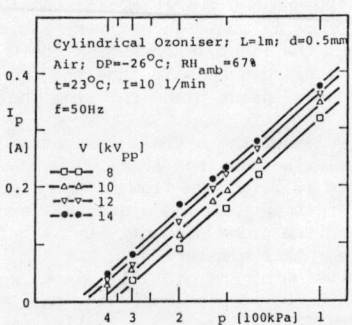

Fig. 16 Peak Value of Overall Discharge Current of Cylindrical Ozoniser vs. Pressure at 50Hz

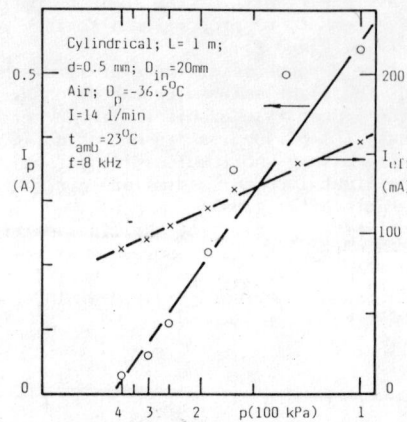

Fig. 17 Peak Value of Overall Discharge Current and RMS of that of Cylindrical Ozoniser vs. Pressure at 8kHz

Conclusions

The average length of the luminous part of the discharge filaments, and the peak value of the overall discharge current, as well as the RMS of that are proportional to the reciprocal pressure.

The relative intensity of the light emitted by the discharges increases with increasing applied voltage.

The spectral intensity of the discharge light changes by applied voltage.

The ozone destroying processes take place mainly out of the discharge region.

The peak value of the overall discharge current depends similarly on the flow rates as the ozone productivity does.

References

S. Masuda, E. Kiss: Proc. of 1983 Annual Meeting of the Institute of Electrostatics Japan (Okt. 22-23, 1983 in Tokyo)

E. Kiss, S. Masuda: Proc. of 1987 Annual Meeting of the Institute of Electrostatics Japan (Okt. 13-15, 1987 in Fukui)

S. Masuda, E. Kiss: Proc. of 3rd Int. Conf. on Electrostatic Precipitation (Okt. 25-29, 1987 in Abano)

DELAYED PAPERS

DEODORIZATION OF GASES BY PULSE CORONA INDUCED PLASMA CHEMICAL PROCESS

Senichi Masuda, Fukui Institute of Technology
3-6-1, Gakuen, Fukui, Japan 910

J.-Z. Wang, Anshan Research Institute of
Applied Electrostatics
Anshan, Liao-Ning, China

Z.-M. Zhou, Department of Chemistry
Beijing Institute of Technology
Beijing, China

Tests were made of deodorization of gases using a novel method based on an intense corona discharge induced by a very fast rising, narrow pulse high voltage. The rise time was about 100 nanoseconds, while the duration time about 300 - 500 nanoseconds. The peak field strength in terms of a corona gap was about 12 kV/cm. A wire-to-cylinder electrode system with 0.5 mm wire and 10 mm cylinder diameters and 200 mm effective length was used.

The odor gas samples were NH_3, HS and methylmercaptan. The results obtained were very satisfactory. For example NH_3 gas with 10000 ppm high concentration could be completely removed with only 0.5 residence time. Thus, this method proved to be a very promising means of odorization.

The principle of this process and its deodorizing mechanism will be discussed, with the detailed descriptions of the experimental data.

EVALUATION AND MONITORING OF AIR IONS IN PRODUCTION ENVIRONMENT

Li Anbo
The Faculty of Public Health
Xian Medical University, Xian, China

ABSTRACT

This articale deals with the methodology of the evaluation of air ions in industrial premises. Air ions critical index is convenient for the evaluation. Air ions concentration in different places such as waterfall, sanatorium, park and workshop were detected. Values of air ions in those places were various. By way of critical index, we found that air ions concentration was too high or too low and the ratio between negative and positive was imbalance in industrial workshops. These phenomena would be looked at closely.

The air ions concentration in production environment is vary from residential area. Whether or not any harmful effect on health of the workers is a new problem and would be looked at closely.

Sourses of Air Ionization in Production Environment

Some physical factors such as electrical discharge; emitting, short wave, ultravialet light, thermal response, high energy radiation, flame fusion, spray, effux, "Lanard effect", high voltage static electricity, high or low frequency electric-magnetic wave, microwave and laser etc. are the sources of air ionization in industrial environment. Air ions can be classified into three groups according to their size and mobilities.

Table 1 Classification of air ions

Classification of air ions	Diameter (μm)	Mobility ($cm^2/v.sec$)
Large (heavy)	0.03-0.1	0.005
Intermediate	0.003-0.01	0.05
Small (light)	0.001-0.003	1.0

In industrial environment, the small air ions are absorbed on the particles of industrial dust, fume, gas to form large air ions. They do people great harm.

Air Ion Conditions in Some Production Environment

Method In 1982-1986, we investigated on air ions in the different industrial workshops. Air ions in a blank control (outdoor), a standard control (have air ionization sources) and an experimental control (no air ionization sources) places were detected in every industrial factory by DLY-1A model air ion counter and mobility discriminator. Monitoring time were twice before, on, and after work hours. Totaly six times a day, and three days a course. Microclimate was recorded at same time.

Results

1. There were significant differences among workshops containing air ionization sources and no sources and outdoor. The air ions concentrations were higher in the former than the latter. ($p < 0.01$).

2. Both negative and positive air ions concentrations were higher during work hours than go off work, especially found in electric welding workshop, textile mile and neutron emission field.

3. All values of positive air ions concentrations were higher than negative air ions except electric welding workshop where, on the contrary, negative air ions was 2.6 times higher than positive. While in supercleaning workshops of color TV kinescope and power capacitor works, both negative and positive air ions were very low and lower than outdoor.

4. The single polar index of air ions in workshops were vary from 1.25 to 3.45, and outdoor 1.17 to 2.73. There were no significant differences among them.
(see Table 2)

Air Ions Conditions in Some Natural Environment in Our Country

The basic air ion concentration of natural environment might be seen as the parameters as a sanitary standard.

The highest negative air ions concentration was found near the waterfall. Air ions concentrations in sanatoriums, mountains and parks were higher than residential areas.
(see Table 3)

Sanitory Evaluation in Industrial Premises of Air Ions

There are no agreemental sanitary standard of air ions in industrial pre-

Table 2 Determination of air ions concentration in different industries

Industry	Workshop	air ions concentration (particles/cm³) negative	positive	q	CI
Textile	spining	62233	61716	0.99	62.68
	weaving	90000	310500	3.45	26.08
	outdoor	216	590	2.73	0.08
Neutron	detecting chamber	40000	50000	1.25	32.00
emission	control chamber	433	625	1.44	0.30
	outdoor	360	519	1.44	0.25
Machining	machine	276	881	3.19	0.09
	grind	376	1044	2.84	0.13
	tools	313	589	1.88	0.17
	outdoor	323	740	2.29	0.14
Steel	electric welding	2373	647	0.27	8.78
structure	outdoor	208	490	2.36	0.09
Enamel	refractory coating	227	778	3.43	0.07
	blank making	347	602	1.64	0.21
	outdoor	331	687	1.96	0.17
Petroleum	stick on a membrane	260	728	2.80	0.09
	air compressor	373	800	2.14	0.17
	outdoor	229	633	2.77	0.08
Televisor	supercleaning	112	272	2.43	0.05
	outdoor	311	374	1.17	0.27
Power	supercleaning	33	39	1.18	0.04
capacitor	outdoor	180	120	0.67	0.12

Table 3 Determination of air ions concentrations in nature

Area place	Air ions concentration (particle/cm³) negative	positive	q	CI
Waterfall	50000	0	—	—
Sanatorium				
Qingdao	2759	5541	2.00	1.38
Lushan	855	665	0.80	1.09
Xiamen	1600	1400	0.88	1.82
Mountain Park				
North Sea	882	1206	1.37	0.64
People	572	615	1.08	0.53
Residential				
Beijing	299	371	1.24	0.24
Shanghai	216	287	1.33	0.16
Xian	202	314	1.55	0.17

mises. Those methods as follows may be used.

1. Degree of air ionization

(1) Total air ions count: It is indicated by negative or positive particles per cm.

(2) Single polar index (q): $q = \frac{n^+}{n^-}$

(3) Mean single polar index (q_z):

$$q_z = \frac{\sum(N^+ + n^+)}{\sum(N^- + n^-)}$$

(4) Relative density (D%): $D\% = \frac{\sum N^+}{\sum n^+}$

Here n^+: positive small air ions;
n^-: negative small air ions;
N^+: positive large air ions;
N^-: negative large air ions.

2. Air ions critical index (CI)

$$CI = \frac{\text{detecting air ions concentration}}{1000} \times \frac{1}{q}$$

Table 4 Air clean and critical index

Degree of air clean	CI
most	>1.0
more	0.7--1.0
intermetant	0.5--0.69
allowable	0.3--0.49
critical	<0.29

Discussion

The characteristics and concentration of air ions in cities and production environments are closely related to the resident health. It is clearly indicated that air ions are definitely capable of working a wide range of physiolgical and biochemical response in human body. Air ions have been proved to interact with cellular enzymes, hormous and other substances related to metabolic and physiological activities.

It was convenient to evaluate the air

ions in natural environment by critical index(CI) as in waterfall, sanatorium etc. places. Their CI were more than 0.5. But in production environment, the critical index were less than 0.3. Some workshops that contained strong air ionization sources such as textile mile, neutron detecting chamber, electric welding had CI large in number. Supercleaning workshops in color TV and power capacitor works had CI 0.05.

The fact that both combined and complex exposure of the human body to the higher and lower air ions concentrations and imbalance of the negative and positive polarity of air ions may occur makes their hygienic evaluation a matter of urgency.

References

1. Masser SH, Negative ions 3rd. Cambriedge University Press 1976,264-415
2. Arthure Stern, Air Pollution 3rd ed. Academic Press,New York, San Francisco, Landon
3. Krueger AP, The biological effects of air ions. Int. J. Biometeoro 1985, 29(3):205-206
4. Hawkinson TE, The industrial hygiene significance of small air ions. Hyg. Assoc.J.1981, 42(10):759-762
5. Kroling P. Nature and artificially produced air ions --- A biologically relevant climate factor? Int. J.Biometeoro 1985,29(3):233-242
6. Kataka S. et al. Effects of air ions on microorganisms and other biological materials, CRC Cri. Rew. Microbiol. 1978, 6(2):109-149
7. Li Anbo et al. Investigation on air ions concentration in Qindao, Lushan mountain and Linton sanatorium. Chinese J. of Physical Therapy. 1985,(2):86
8. Li Anbo et al. A monitoring on negative and positive air ions in several industrial workshops, Labor Hygiene, Bulletin of Railway. 1984,41(3):24

AN INSTRUMENT FOR MEASURING EMULSION STABILITY

J. F. Hughes and T. Hirata

Department of Electrical Engineering, University of Southampton, Southampton, SO9 5NH, UK

ABSTRACT

There are no perfect rapid method to measure a stability of emulsion. It is considered that apply an electric field to an emulsion is one of a rapid method to measure a stability of emulsion. An electric field separates an emulsion and measureing a separation time of emulsion gives a yardstick of stability of emulsion. This method is effective to W/O emulsions.

INTRODUCTION

There are some methods to measur a stability of emulsion, such as a centrifuge and leave alone, however those methods are not perfect or take a long time to measure. It is possible to demulsify an emulsion by using an electric field. It is considered that an electric field demulsification technique is useful to measure a stsbility of emulsion. An emulsion was put into a test cell and an electric field was applied to an emultion, and emultion was separated. It is possible to measure a stability of emulsion from the separation time of an emulsion.

EXPERIMENTAL APPARATUS

Figure 1 shows the outline of experimental apparatus. The experimental apparatus were equipped with four equipments, the test cell, the control box, a power supply, and a pen recorder. The photograph of the test cell and the control box were shown in figure 2. A measureing emulsion was poured into a container. A lower part of a container was negative electrode and an upper part was covered with a positive electrode. A container was inserted to the test cell. The test cell has an electric bulb with a concave lens which gives a parallel beam pass through a container, and the other side of the test cell is installed an optical detector. A container was made of transparent

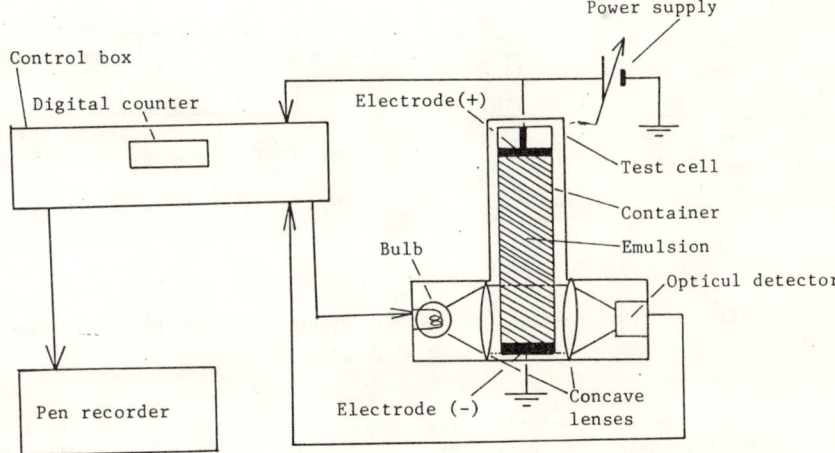

Fig. 1 Schematic diagram of the instrument.

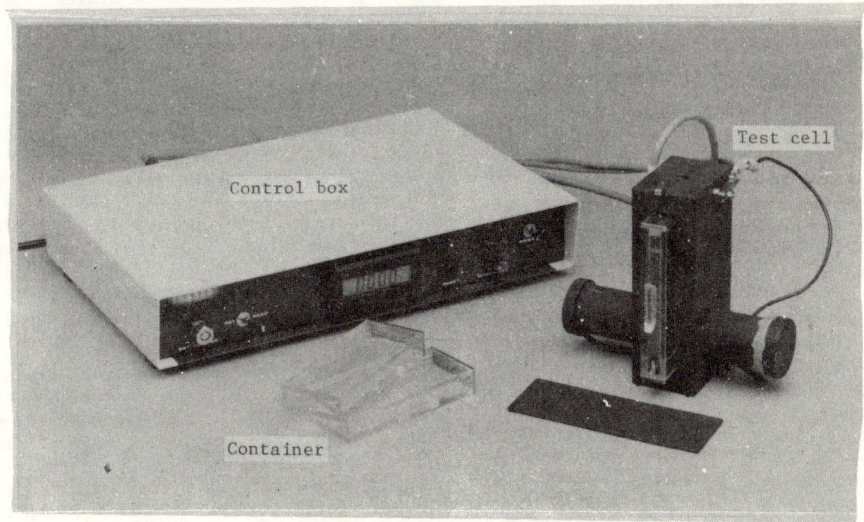

Fig.2 A photograph of the control box and the test cell.

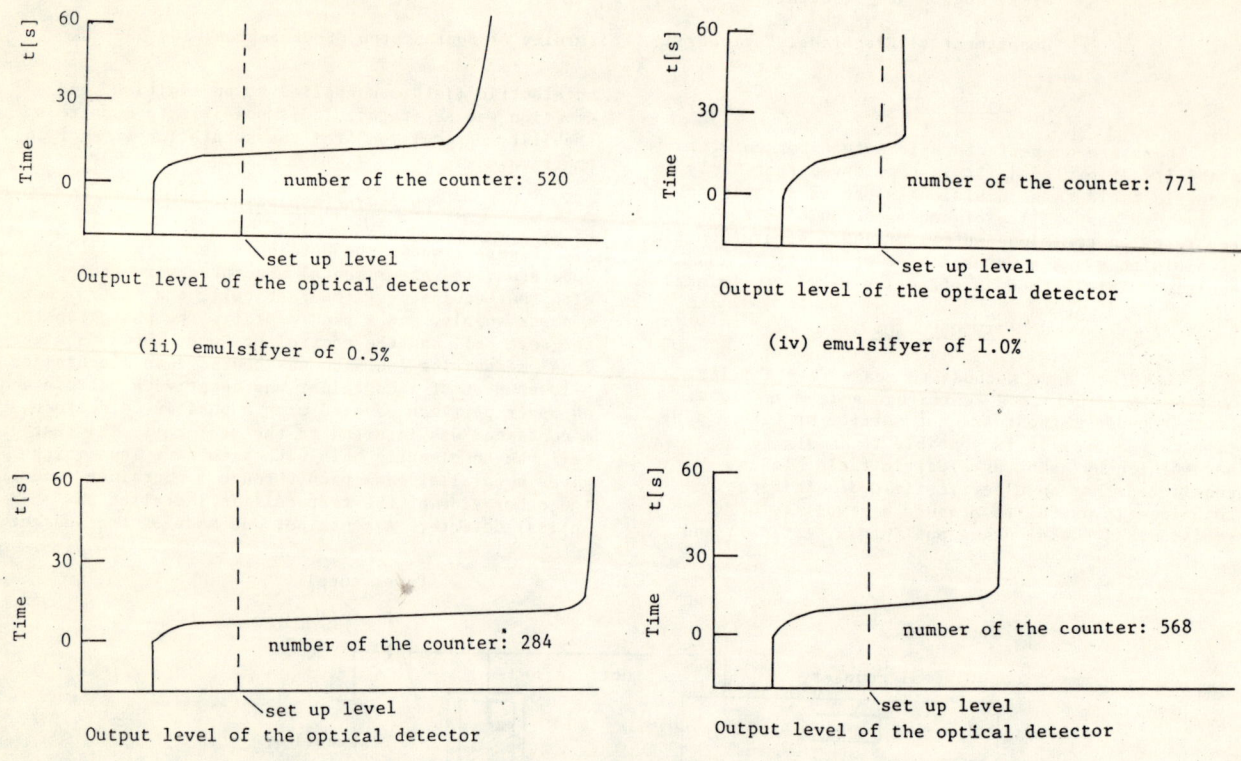

Fig.3 Shift of output level of the optical detector when an applied voltage of 320V.

acrylic. An signal of optical detector which detects an quantity of light past through a container, goes to the control box. A power supply was connected to both the test cell and the control box. A power supply produced an electric field to an emulsion inside a container, and gave a signal to the control box if there was an elecrric field inside a container. The digital counter was equipped in the control box. When an electric field is applied to an emulsion poured in a conteainer, the digital counter begins counting a demulsification time. An every 50 counts gives 1 second. A transparency between an emulsion and oil or water phase is different. If an emulsion start separating, an oil phase appear at an upper side and water phase occurrence at lower part of a container, so that the optical detector detects separation. An output level of the optical detector become higher while detecting part of an emulsion is separating. If an output level of optical detector reaches a set up level, the digital counter stop counting, so that the number of the counter gives a yardstick of separation time of an emulsion. An output level of the optical detector also goes to the pen recorder. The pen recorder writes down an output level of the optical detector which equipped in the test cell. A recorded sheet of the pen recorder shows a shift of an output level of the optical detector which gives a change of a transparency of emulsion. A separation time of an emulsion also realized from a recorded sheet. A set up level of digital counter was set through the control box. A set up level is set at a proper level which should be lower than the level of a separated emulsion. A power of elecric bulb was set through the control box. When an output of power supply is switched on, the digital counter begins counting. The pen recorder writes down an output level of optical detector which shows a transparency of testing emulsion. Then an output level of optical detector gets to the set up level, the digital counter stops counting, and the number of the counter tells a yardstick of separation time of an emulsion. The recorded sheet of pen recorder also gives a separation time of an emulsion. It is possible to determine which kind of emulsion has best stability by compareing both a number of the digital counter and a recorded sheet of each kind of emulsion. However good result are obtained in case of only W/O emulsion.

EXPERIMENTAL RESULTS

Figure 3 and 4 show output levels of the optical detector equipped in the test cell. Those figures were acquired by the pen recorder. Figure 3 was obtained when an applied voltage of electric field was 320 V. Figure 3(i) was in case of an emulsion of which emulsifyer was 0.25 %, (ii) in figure 3 was 0.5 %, (iii) in figure 3 was 0.75 %, and (iv) was 1.0 %. The number of digital counter equipped in the

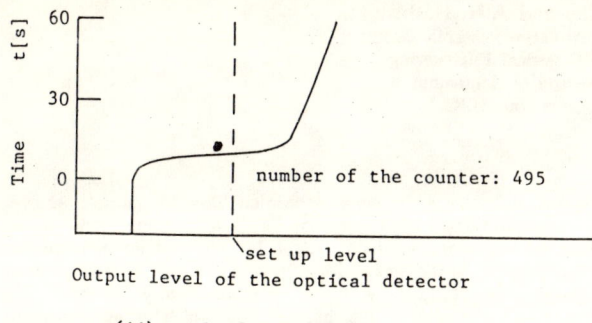

(ii) emulsifyer of 0.5%

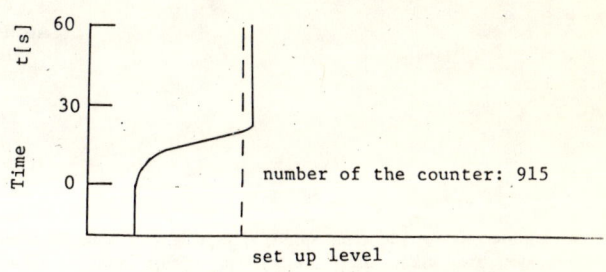

(iv) emulsifyer of 1.0%

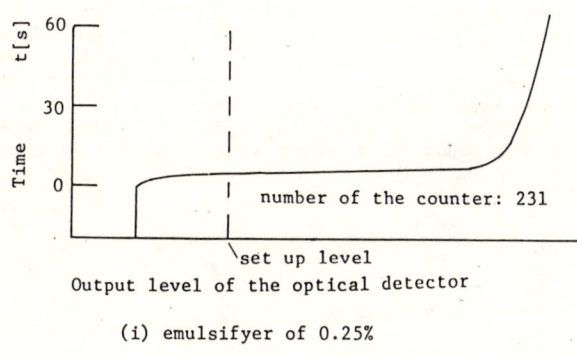

(i) emulsifyer of 0.25%

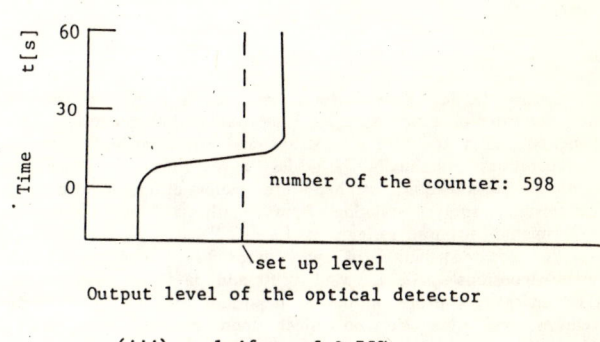

(iii) emulsifyer of 0.75%

Fig.4 Shift of output level of the optical detector when an applied voltage of 380V.

control box was 284, 520, 568, and 771 respectively. From those four figures in figure 3 and the numbers of the digital counter, an emulsion which emulsifyer was 1.0 %, took longest time to separate in four kind of emulsions. It can be concluded that an emulsion which emulsifyer of 1.0 % has best stability of four. Figure 4 was obtained with an applied voltage of electric field was 380 V. Figure 4 (i) was in case of an emulsion which emulsifyer was 0.25 %, (ii) in figure 4 was 0.5 %, (iii) was 0.75 %, and (iv) was 1.0 %. The digital counter of control box indicated 231, 495, 598, and 915 respectively. From figure 4 and numbers of the digital counter, emulsion of which emulsifyer was 0.1 % has best stability in four. It can be noted that in case of both applied voltage of electric field shows same concludion, so that this instrument is able to measure a stability of emulsion.

CONCLUSION

It is possible to measure a stability of emulsion by using an electric field. An electric field demulsify an emulsion and separation time which was mesured by both the digital counter and a pen recorder gives a yardstick of a stability of an emulsion. It is easy to compare between similar kind of emulsions, however this method is only effective to W/O emulsion.

From this experiment, four different concentration of emulsifyer were investigated and emulsifyer of 1.0 % emulsion had best stability.

For further examination, it should be investigate about O/W emulsion.

HEALTH HAZARDS ASSOCIATED WITH ELECTROSTATIC CROP SPRAYING

A.G. Bailey and A.H. Hashish
Applied Electrostatics Research Group
Dept. of Electrical Engineering
The University of Southampton
Southampton, U.K.

Abstract

During electrostatic crop spraying, especially if aqueous-based pesticides are used, drop evaporation occurs and significant reductions in drop size take place in less than one second. Rayleigh break up mechanisms lead to further size reductions. Charged droplets of size well below $5\mu m$ can be generated; they have low settling velocities and are susceptible to wind. Inhalation of these fine droplets may easily result. A computer model of charged droplet deposition within the human respiratory tract shows that charged droplets of diameter $<5\mu m$ can readily be deposited within the lung. A health hazard due to the inadvertant inhalation of charged pesticide spray can arise.

Introduction

During the last decade, there has been a revival of interest in electrostatic crop spraying, motivated by potential cost benefits, health and safety legislation and environmental pollution considerations. Usually pesticides have either a water or oil-based formulation. There are many different forms of electrostatic sprayer ranging from small hand-held units to experimental airborne systems.

The size distribution of spray drops has a dominant effect upon deposition pattern, wind drift and in-flight evaporation. The charge on drops influences deposition pattern, very likely reduces drift, but has no effect upon in-flight evaporation. However, as in-flight evaporation occurs, drops may reach the Rayleigh limit of charge and disrupt, thereby causing significant changes in drop size distributions. Evaporation rates depend upon drop size and are often only significant when aqueous formulations are used.

The droplet size range of interest, in chemical pesticide applications, is from $1\mu m$ to $1000\mu m$ diameter (Law, 1980). The drops produced by typical nozzles, such as the Law nozzle for example, are charged to about 20% of the Rayleigh limit under typical operating conditions (Bailey, 1986) although some devices such as the Electrodyn (Coffee, 1980) may produce oil-based drops charged to 50-70% of the Rayleigh limit.

As the drops used in electrostatic crop spraying are usually well above a median size of about $5\mu m$ they are not considered to be a respirable hazard. However, reductions in drop size due to evaporation invariably occurs. As evaporation proceeds, mass is lost but charge is conserved. When the Rayleigh limit is reached a drop disintegrates. Abbass and Latham (1967) found that about 25% of drop mass, and about 30% of drop charge (Doyle et al., 1964) are lost in the form of one or more charged satellite droplets. Charged drop disintegration has been investigated by many workers subsequently, but there is still doubt as to the number of satellite droplets that are ejected. Following an ejection, the residual drop is stable and evaporation proceeds towards another instability. Charged drops produced during crop spraying may take tens of milliseconds or even seconds to reach their targets, especially during aerial spraying and in windy conditions. Multiple disintegrations will occur with the production of many satellite droplets most of which will be respirable. Since the level of charge they carry is high, it is probable that complete retention in the human respiratory system will occur if they are inhaled. The concentration of toxic material within the drops will be high as the evaporation process leads only to the loss of solvent.

In vivo experiments in humans (Melandri et al., 1977 and 1983) and animals (Ferin et al., 1983) have all shown a significant increase in lung deposition due to particle charge. Theoretical studies by Yu, (1985) and others support this finding and indicate that the main reason for electrostatic enhancement of deposition is the image force which acts on charged particles moving near to the walls of the respiratory airways. A mathematical model of the human lung which accounts for the depositional effect of charge on aerosol particles has been developed by Hashish et al. (1988).

Lung Model

The anatomical model of the respiratory tract represents each airway generation by a cylinder whose length decreases and diameter increases with penetration into the lung. The branching structure and airway dimensions are based upon Weibel's model (1963). The deposition of droplets within the lung depends upon the mechanisms of impaction, sedimentation, diffusion and electrostatic image forces. In addition the respiratory rate, lung tidal volume and any breathing pause between inhalation and exhalation all affect particle deposition (Hashish et al., 1988).

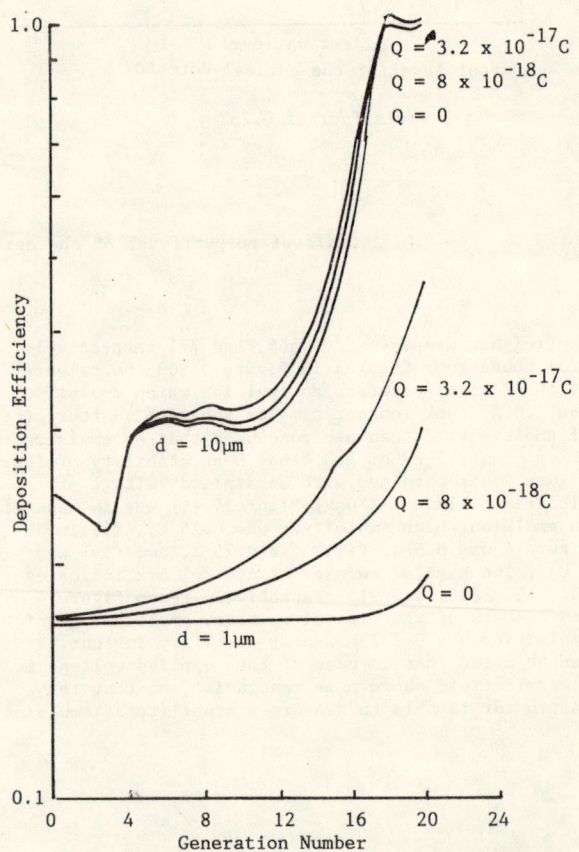

Fig. 1 The effect of aerosol droplet charge on deposition efficiency in the human respiratory tract.

The fractional deposition of particles in a respiratory cycle, is determined by integrating the particle loss over the entire airway length for the residence time involved, and dividing by the number of droplets inhaled, $Q_o c_o \tau$. For the jth airway generation it can be shown that the fraction of droplets deposited is given by:-

$$PD_j = \frac{1}{Q_o c_o \tau} \iint \lambda(x) c(x,t) dx dt \qquad (1)$$

The fractional deposition of droplets in the tracheo-bronchial (TB) region and the alveolar (A) region of the lung, during a respiratory cycle are given by the following expressions:-

$$TB = \sum_{j=0}^{16} PD_j, \qquad A = \sum_{j=17}^{23} PD_j \qquad (2)$$

The lung system has 23 branching sections with the alveolar region being represented by sections 17 to 23.

Using the lung model (Hashish et al., 1988) implemented on a personal computer, inhalation deposition efficiencies were computed for droplet diameters of 1 and 10μm and various charge levels, at a breathing air flow rate of $0.4\ell.s^{-1}$. The results are shown in Fig. 1. For the 1μm diameter droplets a significant increase in deposition for a charge level of only 8×10^{-18}C (50 electronic units) of charge, occurs in all generations of the lung, the greatest effect being in the alveolar region. Conversely charge has negligible effect on droplet deposition for droplets of 10μm diameter. The effect of charge on deposition efficiency increases as droplet charge increases, but decreases as size increases.

Drop Evaporation

Many workers have considered the problem of drop evaporation. Evaporation rate depends upon droplet surface area, vapour pressure and the degree of saturation of the air environment. Other complicating factors such as a reduction in drop temperature and the cooling effect of drop motion through the air may be important. When drop size is significantly greater than the mean free path of air molecules it may be assumed that the rate of reduction of drop surface area is constant. For a spherical drop the rate of change of diameter may be found and it can be shown that the rate of decrease of drop diameter accelerates as size diminishes. When volatile liquids such as water are considered significant evaporation cooling occurs and the rate of change of drop diameter is given by the expression (Hinds, 1982):-

$$\frac{dd}{dt} = -\frac{4D_v M}{R \rho d} \left[\frac{P_\infty}{T} - \frac{P_d}{T_d} \right] \left[\frac{2\lambda + d}{d + 5.33(\lambda^2/d) + 3.42\lambda} \right] \qquad (3)$$

This expression is in cgs units.

Evaluating eqn. (3) and accounting for evaporation cooling is difficult. The steady-state temperature of an evaporating drop is as follows:-

$$T_\infty - T_d = \frac{D_v MH}{RK} \left[\frac{P_d}{T_d} - \frac{P_\infty}{T} \right] \qquad (4)$$

Hinds (1982) has evaluated the above expression for water and presented the information graphically.

Eqn. (3) may be integrated to enable drop lifetimes to be calculated and appropriate corrections can be made for evaporation cooling. This has been done for water at 20°C and RH values of 0% and 50% in Table 1.

Initial diameter of drop	Lifetime (s)	
	RH=0%	RH=50%
0.01	2×10^{-6}	1.6×10^{-6}
0.1	3×10^{-5}	4.7×10^{-5}
1	0.001	1.7×10^{-3}
10	0.08	0.15
40	1.3	2.3

Table 1 Lifetimes of water drops at 20°C

Case History of a Charged, Evaporating Drop

Assume that a 100μm diameter water drop is released into the atmosphere at 20°C and RH=50%. A charge typical of crop sprayer systems is assumed i.e. 13×10^{-13}C (20% Rayleigh limit). Evaporation reduces drop diameter until it reaches the Rayleigh limit d_R of 34.2μm after a time of 0.94s. The drop then ejects about 25% of its mass to produce a satellite droplet of diameter 21.5μm and charged to a level of 4.2×10^{-13}C, which corresponds to 30% of the initial charge. This scenario is of course the simplest as several even smaller satellite droplets may be produced. The initial drop is reduced in diameter from 34.2μm to 31.1μm and retains a charge of 9.9×10^{-13}C. As evaporation proceeds further instabilities and subdivisions occur as shown in Figs. 2a and 2b.

Conclusions

During electrostatic crop spraying operations, although relatively large charged drops may be produced by the spray nozzle, evaporation and drop instabilities lead to the production of copious quantities of very fine, highly charged droplets in less than one second. Such droplets settle very slowly in still air or are easily blown about by wind so that the probability of inhalation by operatives may be high. For charged droplets below a diameter of about 5μm the probability of retention in the lung is very high due to electrostatic image forces causing deposition, especially in the alveolar region. It is feasible to measure the size distribution of the output from any sprayer and, given ambient and spraying conditions, to predict the changes in size distribution as a function of time, that arise due to evaporation and electrostatic instabilities. Using a computer model of the human respiratory system the deposition of inhaled droplets can be quantified.

Nomenclature

A = Alveolar fractional deposition of droplets
c_o = aerosol concentration at airway entrance
c(x,t) = aerosol concentration at x and t
D_v = diffusion coefficient of drop vapour in air
d = droplet diameter
d_R = droplet diameter at Rayleigh limit
H = latent heat of vaporisation
j = airway generation number
K = thermal conductivity of air
M = molecular weight of liquid
PD_j = particle deposition fraction in j^{th} airway
P_d = partial vapour pressure at drop surface
P_∞ = partial vapour pressure away from drop
Q = drop charge
Q_o = initial air flow rate
R = gas constant
T = ambient temperature
T_d = drop temperature (below ambient)
TB = Tracheo-bronchial fractional deposition of droplets
t = time
x = distance into lung
ρ = density of drop liquid
λ = mean free path of air
λ(x) = droplet loss rate
τ = residence time of aerosol in lung

References

Abbass M A and Latham J (1967). The instability of evaporating charged drops. J. Fluid Mech., 30, 663-670.

Bailey A G (1986). The theory and practice of electrostatic spraying. Atomization and spray technology, 2, 95-134.

Coffee R A (1980). Electrodynamic spraying. Brit. Crop Protec. Conf. - Spraying Systems for the 1980's. Walker J O (Ed). 95-107.

Doyle A, Moffett D R and Vonnegut B (1964). Behaviour of evaporating electrically charged droplets. J. Colliod Sci., 19, 136-143.

Ferrin J, Mercer T T and Leach L J (1983). The effect of aerosol charge on the deposition and clearance of TiO_2 particles in rats. Environ. Res., 31, 148-151.

Hashish A, Bailey A G and Williams T J (1988). A mathematical model of the human lung which accounts for the depositional effect of charge on aerosol particles. Aerosol Soc. Second Conf. Bournemouth, 121-126.

Hinds W C (1982). Aerosol Technology. John Wiley & Sons.

Law S E (1980). Droplet charging and electrostatic deposition of pesticide sprays. Brit. Crop Protec. Conf. - Spraying Systems for the 1980's. Walker J O (Ed)., 85-94.

Melandri C., Prodi V, Torroni G, Formignani M, De Zalacomo T, Bompane G F and Maestri G (1977). On the deposition of unipolar charged particles in the human respiratory tract. Inhaled particles IV. Walton W H (Ed). Perg. Press (Oxford). 193-200

Melandri C, Torroni G, Prodi V, De Zalacomo T, Formignani M and Lombardi C C (1983). Deposition of charged particles in the human airways. J. Aerosol Sci. 14, 657-669.

Weibel E R (1963). Morphometry of the human lung. Acad. Press NY, 136-140.

Yu C P (1985). Theories of electrostatic lung deposition of inhaled aerosols. Ann. Occup. Hyg. 29, 219-227.

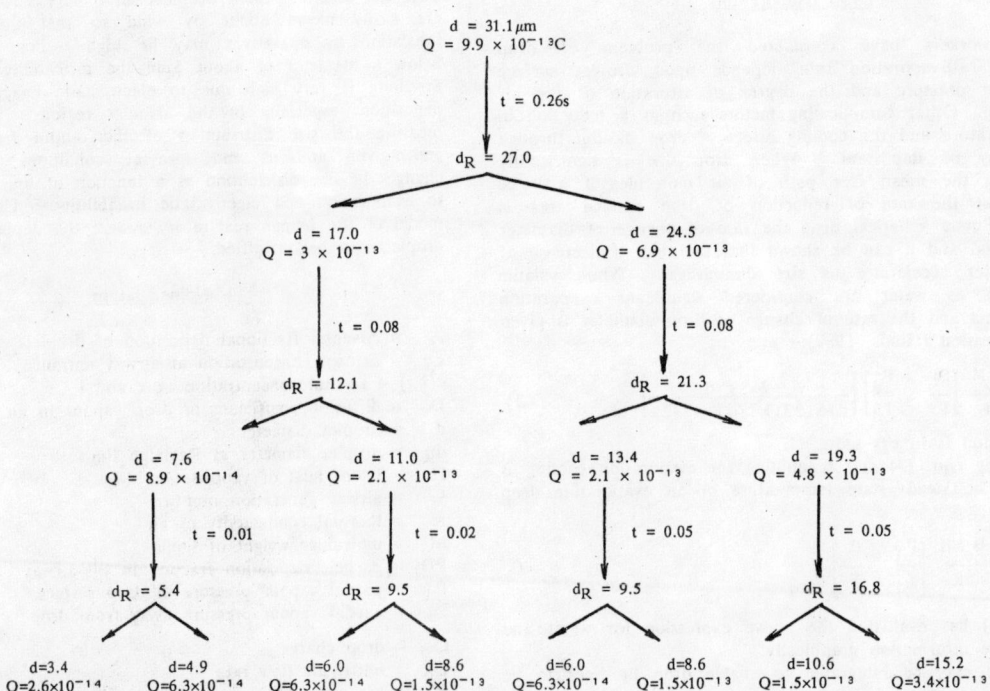

Fig. 2a Case history of residual drop from 100μm mother drop

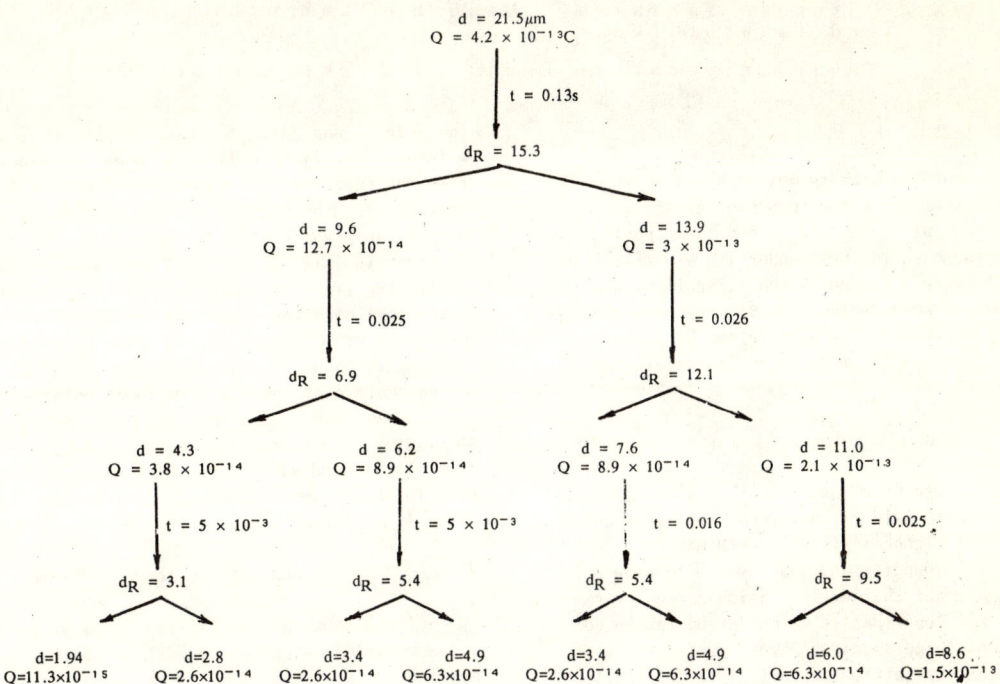

Fig. 2b Case history of satellite droplet from 100μm mother drop

Charge measurement of triboelectric charged toner in electrophotographic developer

Yasusuke Takahashi, Masashi Taniguchi, Takashi Ikeda and Takashi Ito

Tokai University, Dept. of Electro-Photo-Optics

Abstract

New toner charge measuring method by using the laser light scatter and video recording system is described. The free fall locus of charged toner particles introduced into measuring cell under the electric field and gravity were observed by the laser light scattering and video camera system. The electric charge q and radious r of toner could be obtained by applying image analysis system to video locus images displayed on CRT.

Introduction

Electric charge of a toner particle and a charge and size distribution in dry developer are important parameters in electrophotographic performance, relating to both image density and quality. The blow-off method[1] gives a toner charge/mass, which is the average charge of all toner particles contained in the developer. The charge spectrogrphy[2] gives a relative charge distribution of toners, but it is difficult to measure the exact toner charge of each particle.

We have studied the method for the simultaneous measuremet of a toner size and charge in two-and mono-component developer.

Experimental

Fig. 1 shows the schematic model of toner charge measuring system. The most important point of the toner charge measurement is how to separete only the toner from the carrier in two-component developer or from the sleeve in mono-component developer. In this experiment, we designed the new method of toner-carrier separation, which was composed of charge sheet development and deposited toner separation process .

Organic photorecepter was charge in the area of 4×4 cm^2 at the center of photoreceptor and its potential was about ± 500 V. The charge area was developed by magnetic brush method with a test developer. Toner deposited area was exposed to white light and the excess charge which had no contribution to holding of toner particles disappeared.

As shown in Fig. 1, the toner deposited sample is placed above the glass tube (Φ=4.5 cm) at a distance of 2 cm. The toner deposited side is downward and the deposited toner are mechanically separeded from the photoreceptor surface with insulative powder brush. The separated toner particles fall into free space of the glass tube of length 45 cm. The inside air of the glass tube dose not fluctuate, because other sides of measuring system expect top side of the glass tube are seeled against air flow.

Metal electrode section of the system is illustrated in Fig. 2. Metal electrodes, 60×60 mm^2, are seprated by 10 mm. The free falling toner pass through between the parallel electrodes and deposites on the tray of the bottom of electrode cell. As shown in Fig. 1 and 2, He-Ne laser light conducted by optical fiber is introduced between the electrodes from the lower side to the upper at 45° to the horizontal axis. The falling toner pass through the diffused laser light flux. At the video camera position as shown in Fig. 1, we could observed only a front scattering light by toner particles. The falling toner particles were recorded on video tape as a locus of the small light spot by close-up video camera system, which was operated with the fixed focus. A micro-scale positioned at the focal point was recorded on video tape to correct a distance obtained from the display image on CRT, in advance. Laser light was chopped at the frequency of 22.5 Hz, according to obtain the velocity of a toner particle from the CRT display image.

Locus of the toner particle

Fig. 3 shown a free falling locus of toner particles recorded on photographic film in the case of no applied voltage, and Fig. 4 shows the locus in the case of the applied voltage of 100 volts. Since the laser beam was chopped in constant cycle, the light spot locus was recorded as a straght long chain on the film. It is shown in both photographes that the moveing toner has a uniform velocity under the gravity and the electic field plus gravity, respectively. The locus length of black line plus white one is one cycle of chopped light and its time T is 44.4 m sec. Consequently, we can obtain the terminal velocity V by using the toner moving time T and the moved length measured from an enlaged image on photographic paper. It can be possible to apply the same proceddure to still video image on CRT display. On the other hand, The locus length of a moving toner displayed on CRT was obtained by measuring the moving distance from one frame to next frame (frame changing time 1/30 sec).

Theory of measurement of toner charge and radiaus

As shown in Fig. 5, a spherical toner particle, having charge q (c), mass m (kg) and radiaus r (m), falls in air under the gravity and is reached a terminal velocity V . When the moving-down toner particle which has the uniform velocity is passing through between electrodes, the direction of the toner motion is charged by the force of q·E under the uniform field E. The electric field is transverse to the direction of free fall.

The mortion equation of a charged toner for the accelerating forces (electric field and gravity) under the influence of the Stokes drag force are then given by

$$m \frac{dV_y}{dt} = mg - 6\pi\eta r V_y \quad \text{(1)}$$

$$m \frac{dV_x}{dt} = q \cdot E - 6\pi \eta r V_x \quad \text{------(2)}$$

where, g is the acceleration of gravity, η is the shear viscosity of air. On the basis of the results shown in Fig. 1 and Fig. 2, the left side of equation (1) and (2), respectivety, is zero. From equation (1), the particle radius r can be obtained as follows,

$$mg = 6\pi r \eta V_y = 6\pi r \eta V \cos\theta \quad \text{------(3)}$$

$$\therefore r = \sqrt{9/2 \cdot \eta V \cos\theta / \rho g} \quad \text{------(4)}$$

where, ρ is the density of the particle, θ is the angle between the direction of toner motion and that of gravity. From equation of toner charge q of one toner having radius r can be obtained as follows,

$$q \cdot E = 6\pi r \eta V_x = 6\pi r \eta V \sin\theta \quad \text{------(5)}$$

$$\therefore q = \frac{18\pi (\eta V)^{3/2}}{\sqrt{2} \cdot E} \sin\theta \left\{ \frac{\cos\theta}{\rho \cdot g} \right\}^{1/2} \quad \text{------(6)}$$

The V_y is the velocity component of the gravitational direction, the V_x is that of the field direction, respectively.

On the other hand, we can obtain the following equation from Fig. 5,

$$\tan\theta = \frac{q \cdot E}{m \cdot g} \quad \text{------(7)}$$

$$\therefore q = \frac{m \cdot g}{E} \tan\theta$$

$$= \frac{4\pi \rho g}{3 \cdot E} r^3 \tan\theta \quad \text{------(8)}$$

Equation (8) represents that toner charge q is in proportion to r^3 ($\log q = 3\log r + K$).

Results and Discussion

Several kinds of toner and an iron carrier were used. The toners were for Se-drum copy machine and the carrier was an average particle size of 200 mesh. The toner content of the tested developer was about 3 wt. %. The q/m of A-toner measured by Blow-off method was 17.0 μc/g. A and B toner had almost same value of q/m, but the image quality of B-toner's hard copy was better than that of A-toner.

The particle size distributions (R=2r) of both A and B toner are shown in Fig. 7, respectively, which were obtained by photographic camera system. Fig. 6 was plotted on number of 419 toner particles, and Fig. 7 was plotted on number of 282 particles. It is seemed from Fig. 6 and Fig. 7 that A and B toner have almost same size-disstribution and same mean particle-size.

On the other hand, the particle size and its distribution of A and B toner were measured by microscopic method, respectively. The number of measured toner particles was about 200. Fig. 8 shows the particle size distribution of A toner in maximum diameter measured by microscopic method. Fig. 9 shows the particle size distribution of A toner in minimum diameter. The particle size distribution of B toner in maximum and minimum is shown in Fig. 10 and Fig. 11, respectively. It can be seen from Fig. 8 and Fig. 10 that the particle size distribution measured by this laser scattering method is almost the same as that of the toner measured by microscopic method. It is assumed that this coincidence of toner distribution may be caused by rotating fall of particles, which is displaced the center of gravity. We consider that it may be possible by the laser scattering method to measure a resonably exact size of toner particles. Fig. 12 shows the charge distribution of A toner, and Fig. 13 shows that of B toner. It is possible to measure the true charge distribution of an electrophotographic developer, that is, positive, negative and zero charged toner be contained in the developer at the same time. A and B toner had almost same value of q/m, but it is possible to evaluate exactly the difference of developer characteristics by comparison of Fig. 12 and Fig. 13. The B'-toner (q/m= 23.0 μc/g) was the used B-toner after 20,000 copies. Fig. 14 shows the charge distribution of B'-toner, which is more broadly than that of B-toner. The value of peak percent is decreased and the toners having larger charge is increased, compared with the data of B-toner. Fig. 15 shows an approximately linear relationship exists between log q and log r, and the slop of the curve is equal to about 3 and the theoritical equation (8) is satisfied. New toner charge measurement method by the laser scattering video recording system will provide the exact evaluation method for developer propertiy, toner charging propertiy, etc.

Refference

1) D.K.Donald, J.Appl. Phys., 40, 7, 3013 (1967)

2) R.B.Lewis, etal, 4th Int. Conf. Electrophotogr., paper #33, spse, Washington, D.C. (1981)

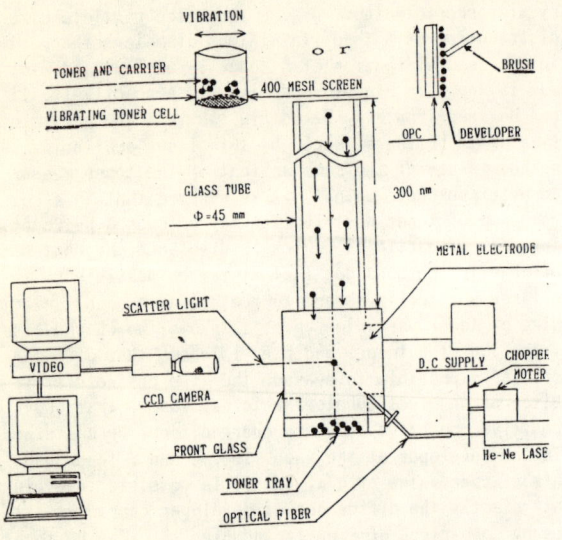

Fig. 1 Schematic model of toner charge measuring system

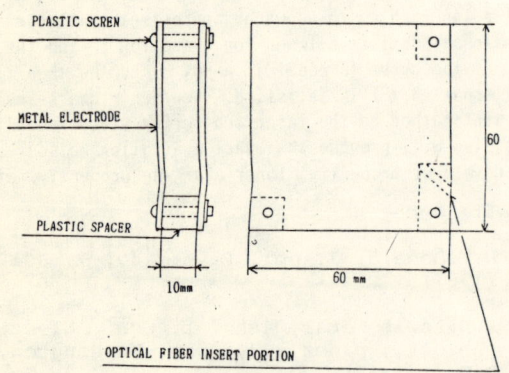

Fig. 2 Metal electrode cell

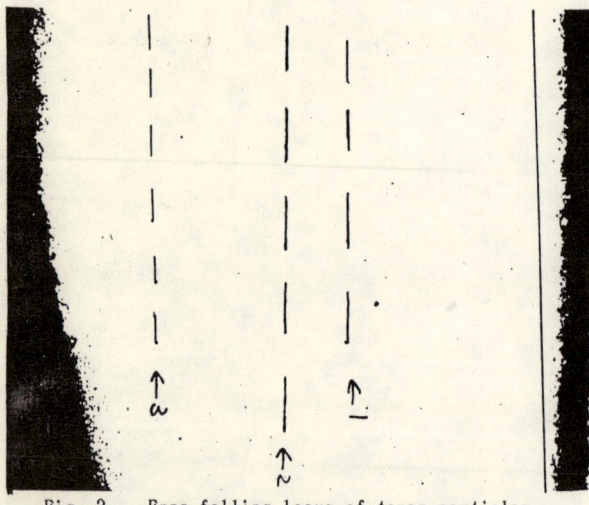

Fig. 3 Free falling locus of toner particles recorded on photographic film in the case of no applied voltage

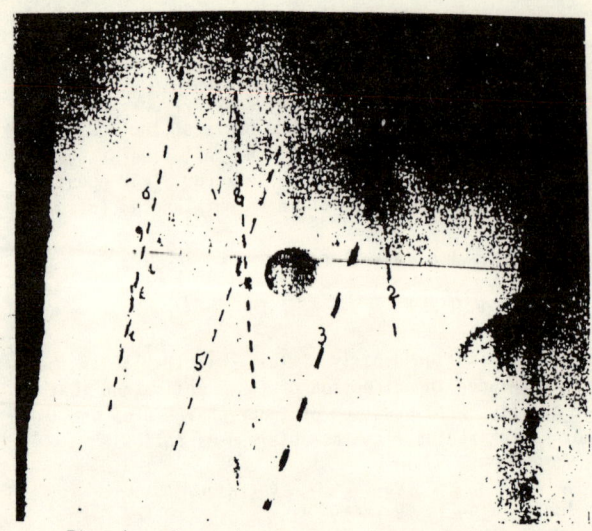

Fig. 4 Locus in the case of the applied voltage of 100 volts

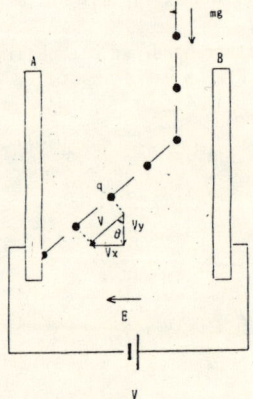

Fig. 5 Motion of a charge toner particle under the gravity and electric field

-536-

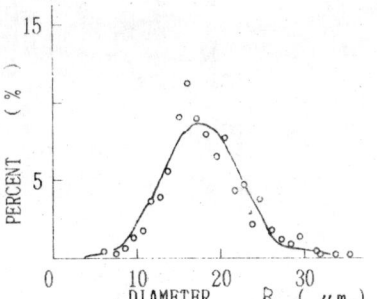
Fig. 6　particle size distribution of A toner

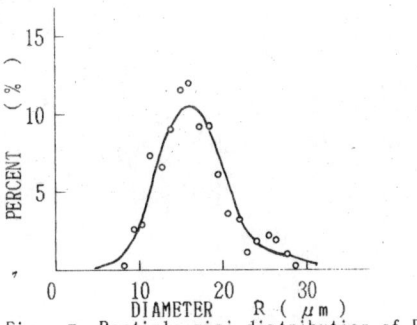
Fig. 7　Particle sizi distribution of B toner

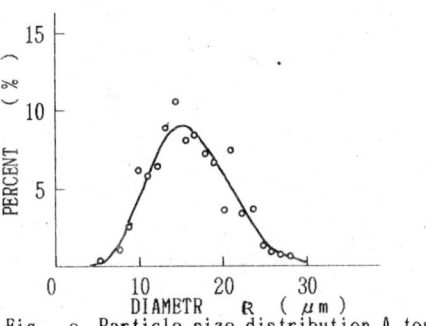
Fig. 8　Particle size distribution A toner measured by microscope (maximum diametr)

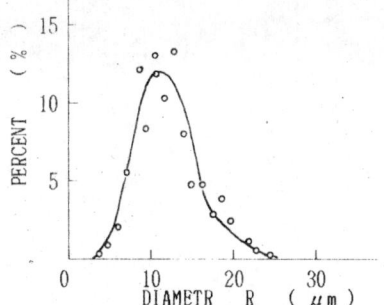
Fig. 9　Particle size distribution of A toner measured by microscope (minimum diameter)

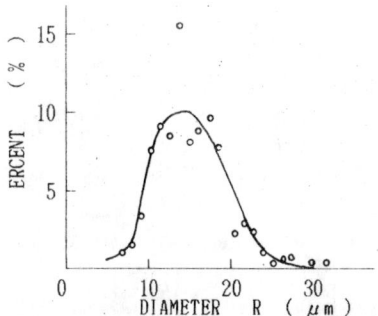
Fig.10　Particle size distribution of B toner measured by microscope (maximum diameter)

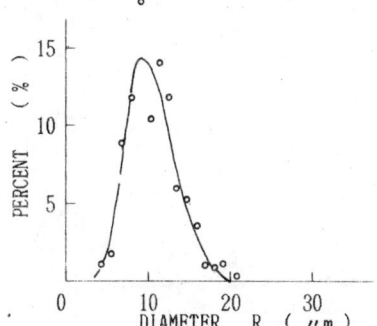
Fig.11　Particle size distribution of B toner measured by microscope (minimum diameter)

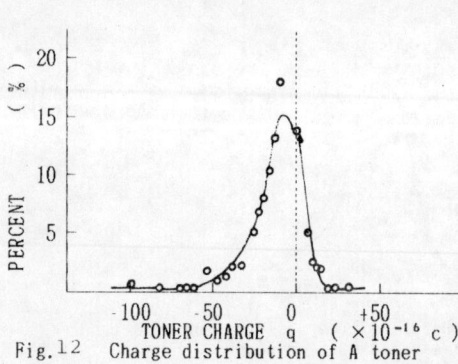
Fig. 12 Charge distribution of A toner

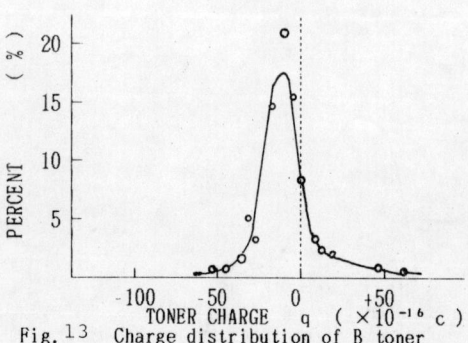
Fig. 13 Charge distribution of B toner

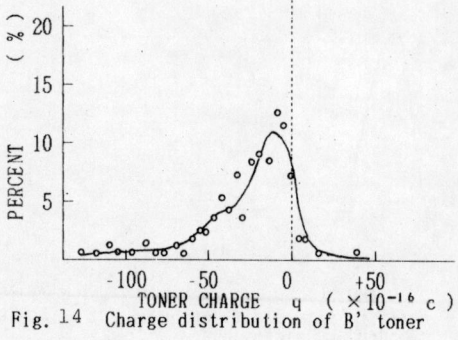
Fig. 14 Charge distribution of B' toner

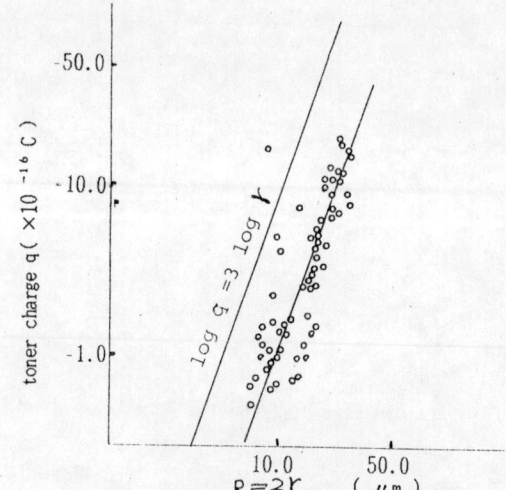
Fig. 15 Relationship between particle size and toner charge of A toner

Proceedings of International Conference on Modern Electrostatics

THE ROLE OF TURBULENCE IN ELECTROSTATIC PRECIPITATORS

by

M. Mitchner, S. A. Self, and K. D. Kihm

High Temperature Gasdynamics Laboratory
Department of Mechanical Engineering
Stanford University
Stanford, California 94305, U.S.A.

ABSTRACT

The present work extends a previous study of the role of turbulence in particle transport under idealized conditions, to examine the effects of nonideal mechanisms that include nonuniform electric fields and diffusivity, nonzero space charge, and the nonuniform particle charge in single-stage precipitators (even with monodisperse particles) that results from a finite charging rate and finite particle diffusivity. The approach to achieving an improved understanding of actual electrostatic precipitators has been to start with a system where only the essential mechanisms are active, and then to add the effects of additional mechanisms one at-a-time, performing experiments and making theoretical calculations corresponding to the conditions of each stage, to the extent possible, to test our understanding. Several important simplifications have been found that will aid in the development of models of electrostatic precipitators. This work has also shown that despite these nonideal effects, precipitator performance exceeding that predicted by the so-called Deutsch model, is still possible.

1.0 INTRODUCTION

Despite their widespread and longstanding use, and not withstanding an increased research effort in recent years, the practical design of electrostatic precipitators remains essentially an art based on a highly oversimplified model (yielding the so-called Deutsch equation) supplemented by extensive empirical data compiled over the years from the performance of existing installations. When a new precipitator is constructed that is to operate under conditions that differ significantly from those in existing installations, the initial performance is often unacceptable, and very expensive procedures are required to rectify the situation. The objective of the work at Stanford has been to develop an improved understanding of the dominant processes that determine precipitator performance, with the goal of transforming precipitator design from an art form into an engineering science.

The work to be described in the present paper is concerned with one of the most important fundamental aspects of the precipitation process, viz. the transport of charged particles by the Coulomb force of the electric field through the turbulent flow of the carrier gas. This work can be summarized with reference to Table 1.

The transport of a dilute concentration of monodisperse particles is governed by the convective-diffusion equation, which describes how the ensemble-averaged particle number density $n(\vec{x},t)$ depends on position and time. This equation represents a statement of conservation of particles. Because the flow in any practical precipitator is necessarily turbulent, the particle number density, as well as other flow quantities, fluctuate in a random fashion on time and distance scales that are too small to be of practical interest. The particle conservation equation for a particular realization of the flow must therefore be ensemble-averaged to obtain an equation that involves just ensemble-averaged quantities. This procedure leads to the main assumption underlying the convective-diffusion equation, that the ensemble-averaged particle flux associated with the turbulent fluctuations can be described by Fick's law, i.e., by the expression $-D\nabla n$, where D is referred to as the particle turbulent diffusivity or diffusion coefficient. (Another assumption, justified for dilute dispersions, is that there exists no agglomeration or de-agglomeration of particles belonging to the size class of interest.)

Shown in Table 1 are the individual terms that constitute the steady-state convective-diffusion equation [1], and their physical interpretation. The term $\vec{u}\cdot\nabla n$ describes the effect of convection of particles by the mean gas velocity $\vec{u}$. The term $\mu\vec{E}\cdot\nabla n$ describes the effect of the convection of particles by the electric field $\vec{E}$, where

$$\mu = \frac{q\,C}{6\pi a \eta_g} \tag{1}$$

denotes the particle mobility and where

$$\vec{w} = \mu\vec{E} \tag{2}$$

is the particle velocity relative to the gas velocity--i.e., the so-called particle electric migration velocity. Here q and a denote the particle charge and radius, respectively, η_g ($\approx 1.8\times10^{-5}$ kg/ms^2, at room temperature) is the gas dynamic viscosity, and C is the Cunningham slip factor [2]. (For the particle sizes of interest in this report, $C \approx 1$.)

Because of Poisson's equation for the electrostatic field, $\nabla\cdot\vec{E} = \rho_c/\varepsilon_g$, the third term reflects the effect of the existence of a space-charge density ρ_c. (Here ε_g is the permittivity of free space.) If the distribution of particle charge is nonuniform, $\nabla\mu$ is nonzero, and the fourth term $n\vec{E}\cdot\nabla\mu$ accounts for this effect. The fifth term $\nabla\cdot(D\nabla n)$ accounts for the effect of either a uniform or nonuniform particle diffusivity.

Our approach to achieving an improved understanding of actual electrostatic precipitators (wherein all of the above mechanisms participate jointly) has been to start with a system where only the essential mechanisms were active, and to then add the effects of additional mechanisms one at-a-time, performing experiments and making theoretical calculations corresponding to the conditions of each step, to the extent possible. These steps are indicated in the remainder of Table 1. (Here the coordinates x and y refer to the streamwise and to the transverse plate-to-plate directions, respectively; the gas speed $u = |\vec{u}|$ is assumed uniform.) In the spirit of this approach, our initial efforts were directed toward studies of particle transport under conditions where both the electric field and the particle diffusivity were uniform.

A low-speed flow facility, shown schematically in Figure 1, was constructed to provide a uniform low-turbulence level ($\leq 0.3\%$) flow at the inlet of a test-section where various electrostatic precipitator configurations could be inserted. A novel method was developed to provide a monodisperse aerosol and to pre-charge the individual particles to the same value. Particle mobilities were determined directly by using a laser anemometer to measure the transverse velocity of single particles as they passed through a known transverse electric field applied between two 0.75-inch-diameter disks, spread 0.25 inches apart, that could be inserted into the laminar flow field just downstream of the nozzle exit, and then removed. To obtain a uniform electric precipitation field, a high voltage was applied between opposed parallel plates located at the test-section. Because the particle concentration was low and because there were no ions in the flow, the charge density was very small and its effects could be neglected. To obtain a uniform particle diffusivity in the core of the flow, grid-generated turbulence (which has been studied extensively) was produced upstream of the test-section. With these idealized conditions realized, it is possible to obtain an analytical solution of the convective-diffusion equation to compare with the experimental data. Because these results provide a base-case, they are summarized briefly in Sec. 2.0.

The grid-generated turbulence employed in our initial experiments resulted in relatively low values of the particle diffusivity ($D \lesssim 1.0$ cm^2/s). Because the turbulence levels in actual precipitators are much larger, we wished to extend our initial work to consider such conditions. For this purpose, we employed several baffle configurations in place of the grids, obtaining values of D up to about 20 cm^2/s at the location of the plates. To understand these data, it was found necessary to account for a nonuniform distribution of diffusivity. These results are summarized in Sec. 3.0.

The wire-plate configuration employed in actual one-stage precipitators results in a nonuniform distribution of $\vec{E}$. By employing relatively large-diameter wires and sufficiently low voltages, we were able to study these effects in the absence of a corona discharge. These results are described in Sec. 4.0.

The presence of a corona-discharge produces two new mechanisms--a nonzero ionic space charge and a nonuniform particle charge distribution. To separate the effects of these mechanisms, we employed uniformly pre-charged particles, charged to a value that was significantly larger than the charge a particle could acquire in the corona current. The results showing primarily the effects of space charge are described in Sec. 5.0. In Sec. 6.0, we present results obtained without pre-charging the particles and with both mechanisms active, as they would be in a single-stage precipitator.

2.0 UNIFORM PARTICLE DIFFUSIVITY AND UNIFORM ELECTRIC FIELD

The earliest theoretical model of precipitator performance was developed by Deutsch [3] in 1922, and this model is still widely used to design and evaluate electrostatic precipitators. In addition to the assumptions of uniform flow speed u, particle diffusivity, and electric field, the Deutsch model assumes that transverse turbulent mixing is sufficiently intense to maintain a uniform particle concentration distribution across the duct width at every axial location. This model leads to the result that the particle concentration decays exponentially in the x-direction according to the relation

$$n/n_0 = \exp(-wx/ud) \tag{3}$$

and that the collection efficiency for a precipitator of length $x = L$ is given by the equation

$$\eta = 1 - \exp(-De) \tag{4}$$

where the "Deutsch number"

$$De = \frac{wL}{ud} \tag{5}$$

may be viewed as a dimensionless length. Here n_0 denotes the particle number density at the precipitator inlet and d is half the plate-to-plate spacing for a wire-plate precipitator. (For a plate-plate collector, d will denote the full plate-to-plate spacing.)

The measured efficiencies of commercial precipitators are commonly found to be less than that predicted by the Deutsch equation, and sometimes considerably less. This experience has led to a widespread belief in the industry, that the Deutsch efficiency represents a theoretical optimal level of performance.

It is known for a uniform laminar flow without mixing that the efficiency for De $\leq$ 1 is given by

$$\eta = (1 - De) \tag{6}$$

and that for De > 1 the efficiency is 100%. This observation suggests for finite, but nonzero, values of the particle diffusivity that the efficiency should fall between the laminar and Deutsch limits. With the assumptions of this section and assuming a plate-plate collector, the appropriate form of the convective-diffusion equation is shown in Table 1. (For precipitator conditions, the axial diffusion term $D(\partial^2 n/\partial x^2)$ may often be neglected.)

Strictly speaking, the particle diffusivity cannot be uniform over the entire flow region since at the duct walls, it must approach the Brownian diffusivity D_B which, for 4-μm diameter particles, has the value 6.0×10^{-7} cm^2/s. The diffusivity must decrease from a core value on the order of 1 cm^2/s to D_B in a region close to the plate that one may refer to as a "diffusivity layer." Near the plate $(\partial n/\partial y)$ is expected to be much larger than $(\partial n/\partial x)$, which suggests that the term $u(\partial n/\partial x)$ may be neglected in comparison with $w(\partial n/\partial y)$. If this condition is satisfied in the diffusivity layer and if this layer is thin, then one may show [4,5] that the boundary conditions for the noncollecting plate and the collecting plate, respectively, are

$$D\frac{\partial n}{\partial y} - wn = 0 \quad \text{(zero flux condition)} \tag{7}$$

$$\frac{\partial n}{\partial y} = 0 \quad \text{(zero slope condition)} \tag{8}$$

Here D denotes the core-value of the diffusivity.

A rigorous analysis of the process of precipitation for flow with arbitrary values of D based on the preceding assumptions is presented in references [4] and [5]. In addition to the dimensionless Deutsch number, a second dimensionless number

$$Pe = wd/D \tag{9}$$

appears, that we refer to as the electric Peclet number. Physically Pe provides a measure of the ratio of transverse particle transport resulting from the electric field to that resulting from diffusion. Thus Pe $\rightarrow \infty$ (D $\rightarrow$ 0) corresponds to laminar flow, while Pe $\rightarrow$ 0 (D $\rightarrow \infty$) corresponds to the Deutsch model.

For a uniform particle number density profile at the entrance, the analysis leads to a normal mode expansion. The profile evolves with axial distance until the dominant mode remains as a self-similar solution. Calculated profiles for the dominant mode for different Peclet numbers are shown in Figure 2. For large Pe (small D) the particles are "heaped up" towards the collecting wall (y = d), while for low Pe (large D) the profile is flatter and tends to a uniform profile in the Deutsch limit Pe $\rightarrow$ 0. The efficiency as a function of De and Pe, obtained by integrating the concentration profiles, is shown in Figure 3.

In the entrance region, where the profile evolves from the uniform profile assumed by Deutsch to that of the dominant mode, the efficiency increases from an initial Deutsch value to the higher value of the dominant mode, as shown in Figure 3. In the limit, as the Peclet number becomes very large, the curves approach the result for laminar flow. It should be noted that typical modern high-efficiency precipitators employ a value of the Deutsch number of about 5 or 6 to achieve efficiencies of ~99.5%. Figure 3 clearly indicates that such efficiencies should be attainable with values of the Deutsch number a factor 3 or 4 smaller if turbulent diffusion could be controlled to give Peclet numbers in the range Pe = 20 to 30.

A comprehensive experimental check, using the apparatus previously described, of the validity of the theoretical predictions of the analysis based on the convective-diffusion equation has been carried out, as reported in references [4], [6], and [7]. The theory was tested by measuring the concentration profiles n(y) (by a non-intrusive light-scattering technique) which constitutes a more rigorous test than just an efficiency measurement. Typical concentration profiles at different stations along the flow direction are shown in Figure 4. The solid curves are calculations for an assumed turbulent diffusivity of D = 1.2 cm^2/s, which value is close to that to be expected for the large grid used here, and corresponds to a Peclet number of Pe = 125. Similar measurements were made over a range of (measured) migration velocities, stations (i.e., x-positions), and diffusivities (different grids) spanning the range of Deutsch parameter 0.05-1.0 and Peclet number 30-350. All these profile measurements can be compared with theory in a normalized plot as shown in Figure 5. The fact that the data fall close to the theoretical curve for a wide range of the parameters is strong evidence of the essential correctness of the theory. The efficiencies corresponding to the three conditions of Figure 4 are shown plotted as points in Figure 3. Since for Pe = 125 the efficiency curve is very close to the laminar limit, we see that the agreement between experiment and theory is also very good when examined in these terms.

For the low levels of grid-generated turbulence used in these experiments, corresponding to De in the range 0.6-1.2 cm^2/s, the Peclet numbers are large (Pe > 30). On the other hand, commercial precipitators almost certainly have much larger values of diffusivity, and lower values of Pe, because they are designed with little regard for flow quality, principally because there has been little recognition of its importance. Even for relatively moderate values of Pe, we see that there exists the potential for significant improvement in precipitator efficiency and size over that predicted by the Deutsch equation.

3.0 NONUNIFORM DIFFUSIVITY AND UNIFORM ELECTRIC FIELD

Because existing commercial electrostatic precipitators almost certainly have much smaller values of Pe than those obtained in the work described in the previous section, experiments were undertaken in which the turbulent diffusivity was much larger than that which could be obtained with grids. It is known for isotropic homogeneous turbulence that

$$D \propto V^* L^* \tag{10}$$

Here V^* and L^* are characteristic values of the turbulent velocity and length scale (of the energy-containing eddies), respectively. Using this relationship as a guide, experiments were undertaken using the three turbulence-generating configurations shown schematically in Figure 6.

Shown in Figure 7 are hot-wire anemometer measurements of the mean streamwise gas velocity and the r.m.s. intensity of the streamwise turbulence fluctuations made at several downstream locations for the case of the 2-cm single baffles. Similar measurements were made for the other two baffle configurations. The flow separation and reattachment caused by the baffles result in the formation of large eddies that interact and then tend to dissipate as the gas proceeds downstream. As a result, asymmetric and unstable three-dimensional flows are observed initially within a certain distance from the baffles. The upstream edge of the plate-plate collector, used to produce a uniform electric field, was therefore placed 40 cm downstream of the last baffle, where the flows exhibited reasonably uniform core conditions but where the turbulence intensity was still quite large.

Particle concentration measurements for the 2-cm single baffles are shown in Figure 8(a) for a migration velocity w = 7.5 cm/s and in Figure 8(b) for w = 20 cm/s.

In each figure, data are shown for measurements made 20 cm and 40 cm from the collector inlet. All measurements shown in these figures were made for an average gas speed of 200 cm/s. Similar concentration profile measurements were made for the other baffle configurations and for other values of the average gas speeds, for a total of six conditions in all.

The dashed curves shown in Figure 8 were calculated using the uniform diffusivity theory dicussed in the previous section. (For these calculations the value used for the particle diffusivity, $D = 17.5$ cm^2/s, was obtained by the method described below. The theoretical profiles are relatively insensitive to variations in the value of the diffusivity within a factor of about two.) We see that the theoretial curves agree reasonably well with the data near the collecting wall, but that there is a marked discrepancy near the noncollecting walls. Similar behavior was exhibited for all six measurement conditions.

The discrepancy at the noncollecting wall appears to stem from the assumption in the theory that the particle diffusivity is uniform, whereas in fact, D must become very small at the noncollecting wall. This too-large value of D causes the predicted turbulent diffusion flux of particles from the core region toward the noncollecting wall to be too large, and therefore the predicted particle concentration in the near-wall region will also be too large. The reason that the uniform diffusivity theory agreed well with the data for the grid-generated turbulence described in the previous section was that the thickness of the region of rapidly varying diffusivity (that we called the diffusivity layer) was small.

In order to seek better agreement with the concentration profile data for the baffle-generated turbulence, the nonuniform diffusivity model shown in Figure 9 was assumed. The value of the particle diffusivity at the wall is taken to be the value actually at the edge of the viscous sublayer, $D_w \simeq 0.019$ cm^2/s, and the boundary conditions at this location can be shown to be the same as those employed for the uniform diffusivity model, viz. zero flux at the noncollecting wall and zero gradient at the collecting wall.

To apply this model, it is necessary to specify the values of two parameters, the diffusivity layer thickness ξ and the core diffusivity D. On the basis of flow measurements, such as those shown in Figure 7, and as a result of some exploratory calculations, it was determined that a reasonable value for the diffusivity layer thickness for all the experimental conditions examined was $\xi = 0.3d$. The core diffusivity was determined for each baffle-configuration, mean gas speed, and axial location, by fitting the calculated concentration profile to the data for the lower value of migration velocity w. Comparison of the data with a calculation performed for the higher value of w (but with the same value of D previously determined) then provided an independent measure of the success of the assumed nonuniform diffusivity model.

The solid curves in Figure 8 show the results of the calculations for one set of conditions, and are representative of the results obtained for all six cases examined. We see that the nonuniform and uniform diffusivity calculations agree approximately with each other and with the data near the collecting wall, but that the nonuniformity model provides much better agreement with the data near the noncollecting wall. Values of the core diffusivity determined by the fitting procedure described above for $w = 7.5$ cm/s are seen to do a good job of predicting the profiles for $w = 20$ cm/s.

By integrating the experimental concentration profiles one can obtain the corresponding experimental efficiencies. These values measured at $x = 40$ cm for two different values of w are shown plotted in Figure 10 vs. the Deutsch number (with x corrected for the upstream fringing electric field) for the three different baffle configurations. Even for these highly turbulent flows the measured efficiencies are seen to be significantly larger than those predicted by the Deutsch model.

It is also of interest to compare the measured efficiencies with the uniform diffusivity model. This comparison is provided by the lines in Figure 10 labeled by different values of the electric Schmidt number Sc. The results presented in terms of the Peclet number in Figure 3 are presented here in terms of an alternative dimensionless parameter

$$Sc \equiv \frac{Pe}{De} = \frac{ud^2}{DL} \tag{11}$$

in order to be able to extract the value of the uniform diffusivity D that best approximates the data. We see that the experimental points fall approximately on lines of constant Sc having values that are roughly a factor of half of those calculated using the core values of diffusivity obtained by fitting the measured concentration profiles with the nonuniform diffusivity model (and hence values of diffusivity that are roughly twice the previous values).

4.0 NONUNIFORM ELECTRIC FIELD AND UNIFORM PARTICLE DIFFUSIVITY

The preceding sections have dealt with cases for which the electric field is uniform, as indicated schematically in Figure 11(a), and can be approximated in the laboratory by applying a voltage between opposed parallel plates. Such a configuration would correspond to a plate-plate collector in a 2-stage precipitator.

For large single-stage electrostatic precipitators, the electric field distribution is more like that shown schematically in Figure 11(b), and is clearly nonuniform. In such precipitators the voltage applied between the wires and plates is sufficiently high to produce a corona current, so that the particles are both charged and transported to the plates in the same device. (The regular field lines depicted in Figure 10(a) would actually correspond only to the case of a positive corona.) To study first just the effects of the nonuniform electric field, the simplified case has been considered where the voltage, in effect, is below the threshold for corona onset and the particles are pre-charged.

For the electric field we may use the known closed-form expression [8] for an infinite array of wires of radius a and wire-to-wire spacing 2h placed along the centerplane between infinite parallel plates separated by a distance 2d, where (a/d) << 1. (Since the plates in our case are idealized as semi-infinite, this expression approximates the electric field in the vicinity of the inlet.) The case being considered here would therefore correspond to a wire-plate collector in a 2-stage precipitator. As an aid to our understanding, it is useful first to consider laminar flow (D = 0) and then to describe the turbulent flow work (D ≠ 0).

LAMINAR FLOW

The particle trajectories for the two types of collector are shown in Figure 12. For the plate-plate collector the trajectories, shown as oblique straight lines in Figure 12(a), result from the superposition of a uniform transverse particle migration velocity w (which is proportional to the electric field vector $\vec{E}$) and a uniform axial particle velocity u in the flow direction. The normalized particle flux profile $\Gamma(y)/\Gamma_0$ equals unity in the region where particles are present, and is zero elsewhere. The collection is completed in a normalized length $\bar{x}_1$ (= x_1/d) = $1/\epsilon$, corresponding to a value of the Deutsch number $(De)_1 \equiv \bar{x}_1 \epsilon \doteq 1$. Here $\epsilon \equiv w/u$ is the ratio of the particle migration velocity to the inlet streamwise velocity.

For the wire-plate collector in the case of laminar flow, the particle continuity equation for the particle number density n(x,y) reduces to the form

$$\frac{\partial n}{\partial x} + F(x,y) \frac{\partial n}{\partial y} = 0$$

where $F(x,y) = \varepsilon_0 G_y/(1 + \varepsilon_0 G_x)$. Here $\varepsilon_0 = w_0/u$ is the ratio of the particle electric migration velocity w_0 to the mean inlet gas speed u. The particle velocity w_0 corresponds to the value of electric field $E_0 = V_0/d$, where V_0 is the applied potential between the wires and the plates. The vector $\vec{G} = (G_x, G_y)$ is the nondimensional wire-plate electric field distribution $\vec{E}/E_0$, which depends on the three geometric parameters h, a, and d. If all the lengths are nondimensionalized by d, then $\vec{G} = \vec{G}(\bar{x},\bar{y},\bar{h},\bar{a})$, where $\bar{x} = x/d$, $\bar{y} = y/d$, $\bar{h} = h/d$, and $\bar{a} = a/d$.

The method of characteristics applied to this equation yields the result that the particle number density $n(x,y)$ is constant along characteristic lines given by the differential equation $(dy/dx) = F(x,y)$. The characteristics are identical with the particle trajectories. Thus along a trajectory, n remains constant with the value n_0 set by the inlet boundary, which we shall take to be a constant, independent of the coordinate y.

As shown in Figure 12(b), the particle trajectories for the wire-plate collector, corresponding to the geometry of Figure 11(b), are no longer straight lines, but exhibit a "wavy" character [9,10]. (The trajectories are shown only for one-half of the space between the plates. Because of symmetry, the trajectories reflected in the plane of the wires are the same.) As the particles approach a wire, they experience a larger electric field and therefore a greater force. The spatial variation in the transverse electric field component E_y and the resulting transverse repulsive force causes the trajectories to be wavy, as shown, with the particles nearest the wires showing the largest effects.

The streamwise component of the electric field E_x (absent for a plate-plate collector) causes the particles, on the average, to be slowed, thereby reducing the average particle streamwise velocity $u(x,y)$ and causing the normalized particle flux profile $\Gamma(y)/\Gamma_0$ at an x-position to exhibit the shape indicated. Since in the region where particles are located, the particle concentration retains its inlet value, the spatial variation of $\Gamma(x,y) = n_0 u(x,y)$ results directly from the variation in the average particle streamwise velocity.

The particle penetration is obtained by performing the integration $P = \int_0^1 (\Gamma(\bar{y})/\Gamma_0)d\bar{y}$. The precipitation efficiency must then be of the form $\eta \equiv 1 - P = \eta(\bar{x},\varepsilon_0,\bar{h},\bar{a})$. Replacing $\bar{x}$ by $De_c = \bar{x}\varepsilon_c$, where $\varepsilon_c = w_c/u$ and where w_c is the migration velocity for some characteristic electric field E_c, we would have for the functional form of the efficiency $\eta = \eta(De_c,\varepsilon_0,\varepsilon_c,\bar{h},\bar{a})$.

The simplest choice for ε_c would appear to be $\varepsilon_c = \varepsilon_0$. Shown in Figure 13(a) are the results of calculation [9,10] for η as a function of $De_0 = \bar{x}\varepsilon_0$, for representative values of ε_0 and $\bar{h}$, and for the case $\bar{a} = 0.01$. As $\bar{h} \to \bar{a}$, corresponding to very close-spaced wires, the efficiency curve tends to that of a plate-plate collector. As the wire spacing h increases, the Deutsch number De_0 for 100% efficiency increases. This behavior occurs because for a fixed $E_0 \equiv (V_0/d)$, the surface-averaged values of the precipitating field $E_y(x,y=d)$ on the collector decreases as the wire spacing increases. The dependence on the parameter $\varepsilon_0 \equiv (w_0/u)$ is weak when the results are plotted in this fashion so that to a good approximation, the number of independent dimensionless variables are reduced to three, and we may write $\eta \equiv \eta(De_0,\bar{h},\bar{a})$.

The foregoing results suggest that instead of the selection $E_c = E_0$, a more appropriate choice would be $E_c = E_{av} \equiv \langle E_y(x,y=d)\rangle$, where E_{av} denotes the surface-averaged precipitating field on the collector. When the results of Figure 13(a) are replotted against the Deutsch number $De_{av} \equiv (w_{av}\bar{x}/u)$, based on $w_{av} = \mu E_{av}$ (here μ is the particle mobility), the efficiency curves for all values of $\bar{h}$ collapse onto the curve for the plate-plate case, as shown in Figure 13(b). Moreover, the solutions for other values of $\bar{a}$ of interest exhibit approximately this same behavior. Thus the number of independent dimensionless variables is reduced from an original four to one, and we may write $\eta = \eta(De_{av})$. We see that with this choice of variables, a very significant simplification is achieved.

TURBULENT FLOW

Theory

To explain the particle number density profiles measured in a plate-plate collector with highly turbulent flows (see Sec. 3.0) required that a nonuniform model be employed only near the noncollecting plate--that the description of particle transport in terms of a uniform diffusivity was quite acceptable near the collecting plate. In a wire-plate collector the noncollecting surface consists of the relatively very small surface of the wires, and additionally, there will be very few particles near the wires (except, possibly, the first one or two wires near the precipitator inlet). Therefore the uniform diffusivity model should provide a good approximation for charged particle transport in a wire-plate collector.

The form of the convective-diffusion equation that applies in this case is shown in Table 1. In addition to the new terms accounting for a nonzero component of the electric field in the x-direction E_x, the term $\partial^2 n/\partial x^2$ must be retained in the diffusion term for at least two reasons. First, it is not evident that this term is small in the vicinity of a wire. Second, the wires exert a repulsive Coulomb force on incoming particles: In order to account for this "upwind" effect, the convective-diffusion equation must be of elliptic type.

This equation has been solved numerically [9,11] in conjunction with the appropriate boundary conditions, for a collector with 20 wires (modeled as strips) and with an effective normalized wire diameter $(2a/d) = 0.027$. Figure 14 shows the particle concentration as a function of $\bar{x}$ in four planes, $\bar{y} = 0$, 0.15, 0.5, and 1.0, for the typical parameter values $\epsilon_0 = 0.1$ and $Pe_0 = 2.5$. It is seen that, as expected, the particle concentration generally decreases along the flow direction ($\bar{x}$) for all $\bar{y}$ planes, and increases towards the collector ($\bar{y} = 1$). However, the concentration profile in the x-direction shows marked periodic behavior near the centerplane (small $\bar{y}$) while near the collector ($\bar{y} = 1$) it decreases monotonically. Along the centerline ($\bar{y} = 0$) the concentration decreases suddenly near the upstream edge of the strip where there is a strong repulsive force on the particles. Downstream of an electrode the concentration increases again due to turbulent diffusion into the region of low field midway between electrodes, until the upstream edge of the next strip is approached. This is the origin of the successively decreasing peaks in $\bar{n}(\bar{x})$. Far downstream of the last electrode the concentration tends to a constant value, independent of $\bar{y}$, since there is negligible electric field and turbulent diffusion causes the profile to be flat.

Shown in Figure 15(a) is the efficiency as a function of the Deutsch number De_0 (based on $E_0 = V_0/d$), for the typical values $h = 1$, $\bar{a} = 0.02$, $\epsilon_0 = (w_0/u) = 0.1$, and for several values of the Peclet number $Pe_0 \equiv (w_0 d/D)$. (The break in the curves at $De_0 = 4$, beyond which the efficiency remains constant, corresponds to the x-coordinate of the last wire.) The general trend of the efficiency curves with De_0 and Pe_0 is similar to that for a plate-plate collector, but the efficiency, for the same values of De and Pe is significantly lower. This behavior reflects the fact that for equal applied voltages, the wire-plate geometry has a significantly lower surface-averaged electric field at the collector surface than the uniform field E_0 of the plate-plate collector.

As for the laminar flow case, one may replot the results with De and Pe based on the surface-averaged field on the collector E_{av}, instead of E_0. When this replotting is done the results, as shown in Figure 15(b), agree closely with those calculated for a plate-plate collector (that has a uniform field). Thus, for calculating the efficiency it is a good approximation to use the simpler uniform electric field theory provided this field is identified with the surface-averaged field $E_{av} = \langle E_y(x,y=d)\rangle$.

Experiment

The plate-plate collector apparatus previously described was converted to a wire-plate collector by installing ten wire (of diameter equal to 0.89 mm) electrodes along the midplane of the plate-plate configuration. Experiments were performed with applied voltages V_0 less than the corona threshold voltage of 14.0 kV, and mono-charged monodisperse pre-charged particles were employed. Particle concentration profiles were measured at the transverse midplane between the eighth and ninth wires, where the mean particle velocity equals the mean gas velocity (since the axial electric field component E_x is ideally equal to zero at this location). Measurements were made both for low turbulence levels, by accepting whatever free-stream turbulence conditions were produced by the wind tunnel itself, and for high turbulence levels by employing the 2-cm baffle upstream-turbulence-generating configuration previously discussed.

Shown in Figure 16 are the particle concentration profiles measured with low and with high turbulence levels, for different values of the applied voltage expressed in terms of the dimensionless parameter $\varepsilon_0 = w_0/u$. Here u is the mean gas speed and $w_0 = \mu V_0/d$ is the particle migration velocity for a reference electric field equal to V_0/d. The quantity d denotes the wire-to-plate distance. The diffusivity values D, shown in the captions, were determined by fitting the experimental efficiencies to those predicted by the uniform diffusivity and uniform electric field theory discussed in Sec. 2.0. The value of the effective uniform field was chosen to equal the surface-averaged electric field of the actual nonuniform field distribution, as supported by theory.

For the low-turbulence case, shown in Figure 16(a), the steep gradients in the particle concentration profiles indicate that the effect of the higher turbulence level existing in the wire wakes is small. The large electric field existing near the wires quickly depletes these regions of particles so that after passing the first one or two upstream wires, most of the particles are located outside the wake region where the free-stream turbulence level is relatively low, and uniform. For the high-turbulence case, shown in Figure 16(b), the baffle-generated turbulence level dominates the wake turbulence so that the diffusivity is again relatively uniform. The larger diffusivity causes the particle concentration profiles to be smoothed out in comparison to the results shown in Figure 16(a).

The efficiencies for the two cases discussed above, and for a second high-turbulence case at a different mean gas velocity (u = 100 cm/s), are shown plotted in Figure 17 as a function of the Deutsch number based on w_{av}, i.e., the particle migration velocity for the surface-averaged electric field. Also shown are the lines of constant electric Schmidt number ($Sc = ud^2/DL$) given by the uniform properties theory for a plate-plate collector discussed in Sec. 2.0. We can see first that the efficiencies, even for the large-turbulence-level cases, significantly exceed the Deutsch limit. Second, all the experimental points corresponding to the same flow conditions fall approximately on a single Schmidt number line. This behavior supports the applicability of the idealized plate-plate collector theory to the description of a wire-plate collector based on the concept of the equivalence of space-averaged fields.

Since the Sc number depends on the single fluid property D, these results can be used to back out the effective values of diffusivity shown in the caption of Figure 16. If we regard the wire-plate collector as two plate-plate collectors, with a common noncollecting plate occupying the symmetry plane passing through the wires, then we can use these values of D to obtain the theoretical particle concentration profiles shown in Figure 16. The theory and data are seen to agree very well, except possibly near the centerline. The cusps in the theory curves are artificial and result from the zero flux boundary condition imposed separately on either side of the centerline.

The theory described at the beginning of this section with a zero-slope boundary condition imposed at the symmetry plane would represent more closely the actual experimental configuration, but such calculations are lengthy and were not performed in this study. The uniform diffusivity plate-plate model with the electric field set equal to the surface-averaged field at the collecting wall is seen to provide a good approximation for predicting not only the global efficiencies for wire-plate collectors, but also the detailed particle concentration profiles.

Before moving to the next topic, it will be useful to describe here an alternative method of presentation of the theoretical efficiencies η (or penetrations P) predicted by the uniform diffusivity and uniform electric field model. The results of this theory, as shown graphically in Fig. 18, are of the form $P \equiv (1 - \eta) = P(De, Sc)$. Instead of expressing precipitator performance in terms of P (or η), one can introduce the concept of the effective electric migration velocity, w_{eff}. This quantity is defined as the value the migration velocity would need to be in order that the predicted (or measured) P (or η) be given by the Deutsch theory, i.e., $P = \exp(-w_{eff} L/ud)$. Equating these two expressions for P, one can write

$$\left(\frac{w_{eff}}{w}\right) = -\frac{\ln P(De, Sc)}{De} \quad (12)$$

This form of presentation therefore removes the exponential-like dependence of P on De. Since $P = n/n_o$, $(-dn)/n = (w_{eff}/ud)dL$: We may therefore interpret the fractional collection per unit length as proportional to w_{eff}.

The relation (12) is shown plotted in Figure 18, together with the experimental points taken from Figure 17. The line Sc = 0, for which $(w_{eff}/w) = 1$, corresponds to the Deutsch model ($D \to \infty$). With increasing Sc (decreasing D), the curves (w_{eff}/w) rise increasingly above $(w_{eff}/w) = 1$ with increasing De (i.e., increasing length or migration velocity).

5.0 EFFECT OF SPACE CHARGE

In the experiments described to this point, conditions were selected so as to prevent the onset of corona discharges within the collector volume. The only source of space charge would have been the pre-charged particles, but the particle number density was so small that the effects of the associated space charge were negligible. However, for experiments that correspond to a single-stage precipitator, with particles being charged in corona discharges and being collected in a single device, the ionic space charge is usually sufficiently large that its effects need to be examined. However, the problem is complicated by the fact that the two effects of space charge and of nonuniform particle charge are simultaneously present, making it difficult to sort out the separate effects of each mechanism acting alone.

To overcome this difficulty, experiments were performed using corona discharges to produce a space charge, but employing particles that were pre-charged to a value larger than could be achieved by the corona discharge acting alone (except for possibly a relatively small fraction of particles). A uniform, spatial distribution of particle charge was therefore maintained and the only new feature was the presence of the space charge mechanism.

The saturated value of charge acquired by a particle in an electric field E_c by ionic bombardment is given by the expression [2]

$$q_{sat} = 12\pi \varepsilon_g \left(\frac{\kappa}{\kappa + 2}\right) E_c a^2 \quad (13)$$

Here $\varepsilon_g = 8.85 \times 10^{-12}$ F/m is the permittivity of free space, a is the particle radius (2 μm for this work), and κ is the relative dielectric constant of the particle (κ = 2.46 for oleic acid, used in this work). To keep the value of q_{sat} low, it was necessary to keep the value of E_c low. The wires employed in the wire-plate precipitator apparatus described in Sec. 4.0 were therefore replaced by wires having a diameter of 0.10 mm. The threshold corona onset voltage for this geometry was measured to be 5.0 kV (in contrast to 14.0 kV for the 0.089-cm diameter wires used in previous experiments).

The amount of pre-charge was set to a value three times the approximate maximum value of q_{sat} calculated by using for the charging field the value V_0/d. For these conditions the particles were collected so rapidly that very few remained in the gas at the measurement location used previously--i.e., between the eighth and ninth wires. For these experiments the measurements of particle concentration were therefore made at the midplane between the second and third wires, corresponding to $x = 12.7$ cm.

The measured particle concentration profiles for both positive and negative coronas are shown in Figure 19. These data are for low turbulence conditions, without upstream turbulence-generating obstacles, and should be compared to the profiles shown in Figure 16(a) that were obtained for similar conditions, but with no space charge effects. We may note first that the profiles for positive and negative coronas are quite similar, with the increased scatter shown for the negative corona attributable to the nonuniform tuft-like discharge structure associated with the negative corona.

Comparing Figure 16(a) and Figure 19, it appears that the presence of space charge does not affect the particle concentration profiles in a major way, although there are significant differences. The absence of particles from the vicinity of the midplane passing through the wires is much more pronounced with space charge present, and the transition to the particle-containing regions much more abrupt.

The behavior in the near-wall region is also different. We may note first that although the profiles shown in Figure 16(a) and Figure 19 were obtained at fixed locations for increasing values of the applied potential, the profiles obtained for a fixed potential and increasing values of x would be very similar. It is somewhat simpler to describe the difference between these figures in terms of the latter picture. At $x = x_0$, $n/n_0 = 1$, and as x increases, in the absence of space charge, the value of n/n_0 initially remains unity within near-wall plateau regions that gradually shrink to zero when the transition transverse concentration gradients move sufficiently close to the collecting walls that they begin to reduce the value of n at the walls. By contrast, in the presence of space charge, the constant n/n_0 plateau region is replaced by a region where n/n_0 decreases in the transverse direction with a small value of $|\partial n/\partial y|$ away from the walls, and at the same time the wall-value of n/n_0 decreases in the streamwise direction.

A numerical analysis of the particle continuity equation for the conditions of these experiments, as given in Table 1, was not undertaken. The effort to develop a numerical method for solving this equation coupled to Poisson's equation would have exceeded the scope of the present study. However, it is relatively simple to show that the decrease of the near-wall value of n with increasing x is consistent with nonzero values for the space charge and for the transverse gradient in n. (It should be pointed out that with corona discharges present, it is possible to generate corona winds that affect the distribution of gas velocity. However, the corona discharge in these experiments were relatively weak [4,7], and no effects attributable to corona winds were evident in the data.)

6.0 EXPERIMENTS FOR SINGLE-STAGE PRECIPITATOR CONDITIONS

For bombardment charging in a corona discharge (which is the dominant charging process for the 4.0-μm diameter particles employed here), the saturation charge is proportional to the charging field, as shown by Eq. (13). The charge is acquired as a function of time according to the relation

$$q(t) = \frac{q_{sat}}{1 + (\tau/t)} \tag{14a}$$

where the characteristic time for charging (i.e., the time for the particle to acquire half its saturated value) is

$$\tau = \frac{4\epsilon_g E_c}{J} \quad (14b)$$

Here J is the ion current density. The values of both E_c and J are large near the wire and are much smaller near the collector plates. These nonuniformities in E_c and J result in a nonuniform distribution of particle charge and, hence, of particle mobility, which is accounted for by the term $n\vec{E} \cdot \nabla\mu$ in the equation governing particle transport shown in Table 1.

Particles entering near the centerline will tend to quickly get charged to the saturation level, whereas particles entering near the plates will tend to be charged more slowly and to a lower level. The charging process is clearly quite complicated, and no attempt has been made in the present work to undertake any modeling studies. However, exploratory experiments were performed for both positive and negative corona discharge, and for both low and high levels of turbulence. The measurements point out some of the dominant physical features resulting from charging in a precipitator environment and shed new light on the importance of gaining a more complete understanding of the role of charging on precipitator performance.

The major new feature of the experiments to be described in this section was the use of a radioactive source of Kr 85 to neutralize any charge the particles may have acquired prior to their passage into the precipitator. Thus the entire charging occurred as the particle passed through regions of space occupied by corona discharge. The precipitator consisted of ten 0.1-mm diameter wires, and measurements were made in the midplane between the eighth and ninth wires.

The particle concentration profiles measured with a positive corona for increasing values of the current per unit length (i/ℓ) are shown in Figure 20(a). These profiles are seen to exhibit extremely sharp and well-defined peaks on either side of the centerline, with maximum values of n/n_0 as large as 1.8--i.e., a higher density than the inlet value. These peaks, which were not observed in any measurements with pre-charged particles, including those with space charge present, must occur as a result of a nonuniform particle charge (and, hence, mobility) distribution.

In the presence of the strong and highly nonuniform electric field and ion current near the wire electrodes, particles closest to the wire are most highly charged, thereby leading to a rapidly decreasing distribution in particle mobility towards the grounded walls. The more highly charged particles are forced to move rapidly away from the wires by the stronger electric field. Particles further away from the wires, which have a lower mobility and are exposed to a relatively lower electric field, are not forced to move as quickly. As a result, the particle number density forms a peak at a certain location where particles are piling up. The distance of the peak formation from the centerline increases with increasing linear current density.

The heights of the peaks decrease and the widths increase with increasing discharge current, and the peaks eventually disappear when the linear current density is very large. This transition occurs when most particles have been collected (owing to the high current density) and insufficient numbers of particles remain for the piling-up process to take place. Since the data are steady and highly reproducible, it can be inferred that the core flow must have remained non-turbulent in the positive corona discharge. This conclusion is consistent with the laser Doppler measurements of turbulence intensity presented in Figure 21(a), which show that the positive corona does not increase the turbulence level (except near the walls, where an instability exists [4,7]).

Measurements of the concentration profiles for the negative corona discharge are presented in Figure 20(b). The inlet conditions are identical to those for the case of the positive corona results shown in Figure 20(a). The concentration peaks, which were observed with the positive corona, have now been broadened by the particle mixing that results from turbulence induced by the corona tufts, as shown in Figure 21(b), and vestiges of the peaks are barely discernible. It is not possible to obtain good reproducibility as demonstrated by a comparison of the two sets of measurements for i/ℓ = 0.04 mA/m.

Another factor which may also contribute to the loss of well-defined concentration peaks is the nonuniform particle charging that occurs as a result of the spatially inhomogeneous structure of the negative corona discharge tufts: Particles, which are exposed to the locally intensified ion current from a corona tuft, will attain a higher charge, compared with particles which are traveling in the regions between the tufts. It should also be noted that the method of measurement (counting laser-scattering events) used in these experiments is actually proportional to the mean particle flux--i.e., the product of the mean velocity and the particle number density. Hence the ordinates in Figure 20 are more precisely equal to $(un)_{corona\ on}/(un)_{corona\ off}$, which equals n/n_0 only if the mean velocity remains unchanged with the wires energized or unenergized. Since the tufts are observed to change positions on a time scale on the order of a few seconds, quasi-steady corona winds could be set up by the tufts that would cause this condition to be violated.

Measurements of the turbulence induced by negative corona discharge tufts are plotted in Figure 21(b). As expected [4], the negative corona increases the flow and the particle turbulence level significantly beyond that measured without the corona discharge. The measurements are not well reproducible and the velocity fluctuation level increases as the grounded collector walls are approached. Since the characteristic time scale for the velocity fluctuations is very much smaller than that for movement of the discharge tufts (on the order of seconds), it appears that the increased turbulence in the core results from flow instabilities associated with the interaction of the main gas flow and corona-winds induced by the discharge tufts. The increased fluctuation levels near the walls could result from the impingement on the walls of jets of the corona-wind induced by the tufts.

To examine the effect of highly turbulent flow entering the precipitator, two opposed 2-cm-wide baffles [see Figure 6(b)] located approximately 40 cm upstream of the precipitator were employed to generate the turbulence. Particle concentration profiles measured between the eighth and ninth wires (as before) are shown in Figure 22 for several different values of the linear current density i/ℓ. The results for the positive corona in Figure 22(a) show, first, that the turbulence eliminates the concentration peaks observed for low-turbulence flow, as presented in Figure 20(a). We may note also the similarity between these profiles and those obtained for the same turbulence-generating conditions, but with pre-charged particles, as shown in Figure 16(b). Since highly charged particles retain their charge even when moved to the near-wall region, this similarity may be interpreted as showing that the mixing action of the turbulence tends to homogenize the spatial distribution of particle charge, thereby producing approximately the uniform charge distribution of the pre-charged particles.

Concentration profiles for the negative corona discharge are shown in Figure 22(b). Except for the change in polarity, the conditions for Figures 22(b) and 22(a) are identical. Unlike the results for the corresponding negative corona low-turbulence flow presented in Figure 20(b), the measurements are now quite reproducible and show little scatter in the data. The particle concentration profiles now for positive and negative corona are in fact very similar, although there is a marked difference in the nature of these two kinds of corona discharge. This behavior occurs because the turbulence induced by the negative corona discharge is dominated by the baffle-generated turbulence. As shown in Figure 21(b) the induced turbulence intensity was about 5% over much of the core, whereas the baffle-generated turbulence level at the precipitator inlet was measured to be about 12%.

The precipitation efficiencies determined from the foregoing concentration profiles are given in Table 2. The fact that the applied voltage for a given linear current density is larger for positive corona than for negative corona probably is the main reason for the generally higher efficiencies obtained with positive corona. Another factor may be that with positive corona, particles are almost always passing through regions of ion current, whereas with negative corona, particles may spend an appreciable fraction of their transit times in regions between discharge tufts where they would temporarily cease being charged. (On the other hand, while within a discharge tuft the current density would be larger and therefore the charging rate would also be larger.)

Perhaps the most striking feature of these data is that for otherwise identical conditions, the efficiencies are higher for high turbulence conditions than for low turbulence. This aspect of the data is completely reversed from the large body of theory and experiments obtained with pre-charged particles that shows high turbulence levels to be detrimental to precipitator performance. There is not a contradiction here, because the latter experiments were focused on the question of particle transport, and all the particles were specifically prepared so as to carry the same value of charge. In fact, the mixing property of turbulence, as referred to earlier, may be advantageous during the charging process that occurs primarily within some inlet region of a precipitator. The extent of this charging region can be estimated using Eq. (14). Thus, for example, for $i/\ell = 0.04$ mA/m, the particle must travel about 3.4 wire spacings to reach 80% of the saturation charge; for $i/\ell = 0.6$ mA/m, the particle must travel about half a wire-wire spacing to become 80% charged.

To describe the charging process that takes place in the entrance region of a precipitator, we may employ a highly simplified model that is based on the Deutsch model, but accounts for increasing particle charge by assuming that the electric migration velocity w is an increasing function of the axial distance x. The model does not account for variations of w in the transverse direction y, and assumes that field charging occurs under conditions of uniform electric field and current density. The assumption of an approximately flat particle number density profile near the precipitator inlet is not unreasonable.

Since a particle located at position x has been charging for a time $t = x/u$, we may write in accord with Eq. (14), that

$$w(x) = \frac{w}{1 + \tau/t} = w\left[\frac{x}{x + u\tau}\right] \qquad (15)$$

Here w is the migration velocity of a particle [see Eqs. (1) and (2)] carrying the saturated charge q_{sat} given by Eq. (13).

On the basis of the Deutsch number model expressed in differential form, we can write

$$\frac{dn}{n} = -\frac{w(x)dx}{ud} = -\frac{w}{ud}\frac{x\,dx}{x + u\tau} \qquad (16)$$

Integrating both sides of this equation from $x = 0$ to $x = x$, we have

$$\ln\left[\frac{n(x)}{n_0}\right] = -\left(\frac{w}{ud}\right)\left[x - u\tau\cdot\ln\left(\frac{x + u\tau}{u\tau}\right)\right] \equiv -\frac{w_{eff}\,x}{ud}$$

Thus the finite charging rate effect can be expressed in terms of an effective migration velocity given by the relation

$$\frac{w_{eff}}{w} = 1 - \frac{\ln[1 + (x/u\tau)]}{(x/u\tau)} \qquad (17)$$

The finite charging rate effect introduces a new dimensionless parameter $x' = x/u\tau$ which represents the axial distance into the precipitator in units of the characteristic charging distance $u\tau$ (i.e., the distance the particle travels at the gas velocity in the time required for the particle to acquire half its saturation charge).

The relationship (17), expressing the effect of finite charging rate is shown plotted in Figure 23. The quantity (w_{eff}/w) starts at zero when $(x/u\tau) = 0$, increases rapidly, and then slowly approaches the asymptotic value of unity. It would be useful to compare this behavior with experimentally derived values, $(w_{eff}/w)_{exp}$. We were able to follow this approach for the experiments using pre-charged particles because the value of the electric migration velocity w was measured directly. In the present experiments the particles have different migration velocities, and we have no experimental measurements for w. However, we can proceed in an approximate manner using theory as a guide.

If we take for the charge q in Eq. (1) the value of the saturation charge q_{sat} given by Eq. (13), we can write (in SI units)

$$w_{th} \simeq \left\{ \frac{12\pi \epsilon_g [\kappa/(\kappa+2)]}{6\pi n_g} \right\} a\, E_c E_p \simeq 1.13\times 10^{-12}\, E_p E_c \text{ m/s} \tag{18}$$

The subscript (th) is used to indicate that this value of the migration velocity is based on theoretical considerations. The quantity E_p denotes the precipitation electric field, and is interpreted as the surface-averaged field at the collecting wall. It is known that this field in the presence of space charge (as described by Poisson's equation) is significantly larger than the field that exists in the absence of space charge (as described by Laplace's equation). The value of the charging field E_c will also be affected by space charge. The detailed numerical calculation of these fields for the conditions of the experiments reported here was beyond the scope of present work. However, we can use the reference field $E_0 = V_0/d$ as a rough estimate of E_p and E_c, to obtain the value (in SI units)

$$w_{th} \simeq 1.13 \times 10^{-12}\, E_0^2 \tag{19}$$

Similarly, the characteristic charging time can be estimated (in SI units) as shown below:

$$\tau = \frac{4\epsilon_0 E_c}{J} \approx 4\epsilon_0 E_c \frac{4h}{i/\ell} = 16 \times 8.85 \times 10^{-12} \times .0254 \times \frac{E_c}{i/\ell} = 3.60 \times 10^{-12} \times \frac{E_0}{i/\ell} \tag{20}$$

(Here the wire-to-wire spacing 2h is taken equal to the plate-to-plate spacing 2d.) The values calculated in this manner for (w_{eff}/w_{th}) and for $(x/u\tau)$ (with x = .406 m) are shown in Table 3 and are plotted in Figure 23.

For lower values of the charging parameter $(x/u\tau)$, the data appear to conform approximately in shape to the theory curve,[1] but for larger values of $(x/u\tau)$, the "experimental" (w_{eff}/w_{th}) values increase more rapidly than theory and, in fact, exceed the value one, which is the asymptotic maximum value of the finite charging rate theory. Although the number of data points for large $(x/u\tau)$ are fewer than desirable, there is a clear indication that the behavior in this range represents the effect of finite diffusivity in particle transport, which we know leads to values of (w_{eff}/w) greater than one.

To obtain a "theory" that combines the effects of both finite charging rate and finite diffusivity, we may employ the ad hoc procedure of simply multiplying the values of (w_{eff}/w) based on finite diffusivity by the finite charging rate factor given in Eq. (17). Although not rigorous, this procedure should give a result which is approximately correct. Since the results of the finite diffusivity theory are expressed as a function of the Deutsch number De (and the dimensionless number Sc = ud^2/xD, where D is the particle diffusivity--as in Figure 18), the finite charging rate effect needs to be expressed in terms of De instead of $(x/u\tau)$. For the conditions of these experiments, the transformation between $(x/u\tau)$ and De is defined by the values for these quantities shown in Table 3. The theory for the combined

---

[1] Given the major simplifications necessary to adjust the data so as to permit a comparison with the theory, good agreement with respect to magnitude would not be expected, but a comparison with general trends is justified.

effects of finite charging rate and finite diffusivity, obtained in this manner, are shown by the solid curves in Figure 24. Also shown, for comparison, are the corresponding (dotted) curves for finite diffusivity acting alone, and the experimental points for (w_{eff}/w_{th}) replotted as a function of De.

The similarity in the shapes of the theory and data curves supports the assertion previously made that data for low values of De reflect primarily the effects of finite charging, and for higher values of De one sees that the effects of finite diffusivity dominate. It is of interest to note from the theory curves that a full recovery from the poor performance near the entrance of the precipitator owing to the finite charging rate is not made, even for relatively large values of De (because of the slow variation in the log function that governs the approach of the particle charge to its saturation value).

For both positive and negative corona, and irrespective of turbulence level, (w_{eff}/w_{th}) starts from zero at about the corona onset voltage V_{crit} corresponding to De $\approx$ 0.37. Then (w_{eff}/w_{th}) increases quite rapidly with increasing $V_0 > V_{crit}$. The fact that (w_{eff}/w_{th}) increases more rapidly for negative than positive corona (irrespective of the turbulence level) may be due, at least in part, to the fact that the current density is higher for negative corona at the same V_0.

It is also quite evident (except at the highest values of De) that (w_{eff}/w_{th}) is larger for high turbulence than for low turbulence. This behavior supports the hypothesis that turbulence, by its enhanced mixing action, promotes more effective charging for both polarities. It should also be noted that in the range where De is not too large, the beneficial effect of low turbulence on particle transport (see Figure 18) is relatively small. This beneficial effect becomes relatively stronger for larger value of De.

Although the number of experimental points is few, nevertheless there is reasonably clear evidence for both positive and negative corona that at the largest De values obtained experimentally, there is a cross-over in the values of (w_{eff}/w_{th}) for the low and high turbulence cases. Clearly, at the largest De values, the beneficial effect of low turbulence on transport more than compensates for the detrimental effect of low turbulence on particle charging. Moreover, the low turbulence curves are increasing with a significantly greater slope than the high-turbulence curves, as one would expect from the effect of turbulence on particle transport (see Figure 18).

In the experiments, the maximum value of De was set by the onset of sparking when $V_0 \sim 2 V_{crit}$ (corresponding to $De_{max} \sim 4 De_{min}$), so that unfortunately the data are not as complete as desirable to establish firmly that for larger De values the beneficial effect of low turbulence on particle transport increasingly overrides the beneficial effect of high turbulence on particle charging. This situation stems from the fact that only a relatively short precipitator, L = 40.6 cm (8 wires), was used. To obtain data at larger (and more realistic) values of De measurements should be made for larger L. However, under these conditions, the efficiency is very high and it is not easy to obtain accurate measurements of penetration (and w_{eff}) by measuring the concentration (or flux) profile by the laser-scattering technique.

In conclusion, it should be noted that the beneficial effect of high turbulence on particle charging is mainly operative near the precipitator entrance, whereas the beneficial effect of low turbulence on particle transport is operative throughout the precipitator. Thus, an effective strategy may well be to exploit both beneficial effects simultaneously, by deliberately designing for a reasonably high turbulence level near the entrance in a duct which is otherwise designed for low turbulence. The foregoing analysis points out the need for a more systematic study of the effect of turbulence on particle charging in single-stage precipitators.

ACKNOWLEDGMENTS

This work was supported by the National Science Foundation under grants No. CPE-7926290 and No. CBT-8217719, and in part by the Electric Power Research Institute under contract No. RP-533-1. We wish to acknowledge also the valuable laboratory support provided by Rodman Leach.

The work described in this paper was not funded by the U.S. Environmental Protection Agency and therefore the contents do not necessarily reflect the views of the Agency and no official endorsement should be inferred.

REFERENCES

1. Friedlander, S.K. Smoke, Dust and Haze, John Wiley & Sons, New York, 1977. pp. 162-164.

2. White, H.J. Industrial Electrostatic Precipitators, Addison-Wesley, New York 1963.

3. Deutsch, W. Bewegung und Ladung der Electricitatstrager in Zylinder Kondensator. Ann. Physik 68: 335-344, 1922.

4. Leonard, G.L. Effect of turbulence on electrostatic precipitator performance. HTGL Topical Report 196, Stanford University, April 1982.

5. Leonard, G.L., Mitchner, M., and Self, S.A. Particle transport in electrostatic precipitators. Atmos. Envir. 14(11): 1289-1299, 1980.

6. Leonard, G.L., Mitchner, M., and Self, S.A. Experimental study of the effect of turbulent diffusion on precipitator efficiency. J. Aerosol Science 13: 271-284, 1982.

7. Leonard, G.L., Mitchner, M., and Self, S.A. An experimental study of the electrohydrodynamics flow in electrostatic precipitators. J. Fluid Mech. 127: 123-140, 1983.

8. Spiegel, M.R. Complex Variables (Schaum Series). McGraw-Hill, New York, 1974. pp. 252-253.

9. Kihm, K.D. Effects of Nonuniformities on Particle Transport in Electrostatic Precipitators. Ph.D. Thesis, Stanford University, March 1987.

10. Kihm, K.D., Mitchner, M., and Self, S.A. Comparison of wire-plate and plate-plate electrostatic precipitators in laminar flow. Jour. Electrostatics 17: 193-208, 1985.

11. Kihm, K.D., Mitchner, M., and Self, S.A. Comparison of wire-plate and plate-plate electrostatic precipitators in turbulent flow. Jour. Electrostatics 19: 21-32, 1987.

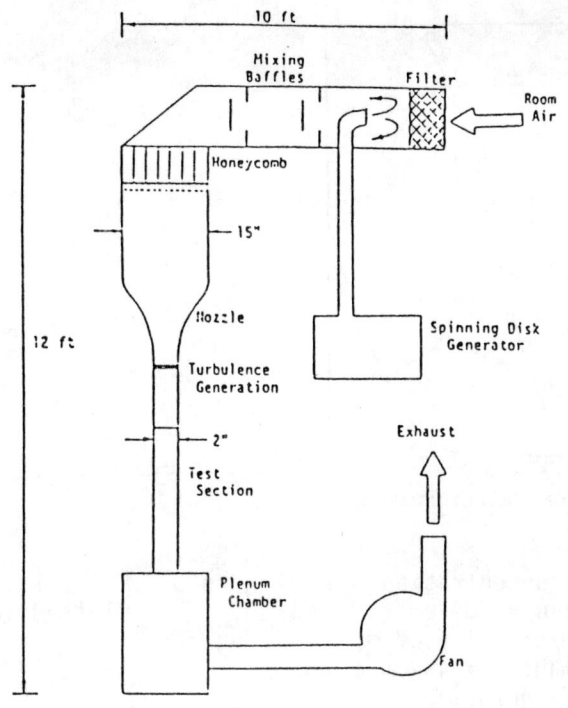

Figure 1. A schematic drawing of the experimental facility.

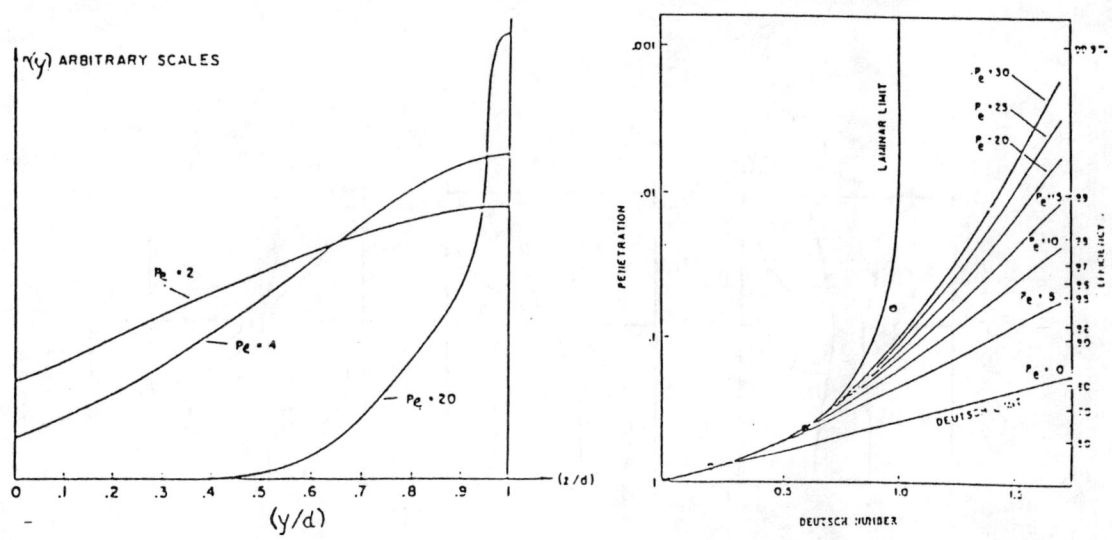

Figure 2. Concentration profiles for various values of Pe.

Figure 3. Collector efficiency as a function of the Deutsch number for different values of the Peclet number.

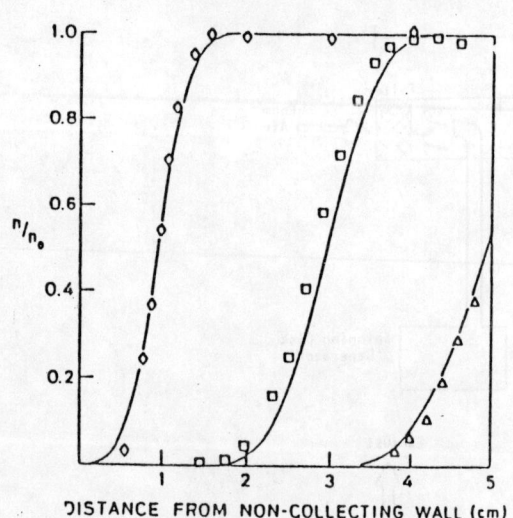
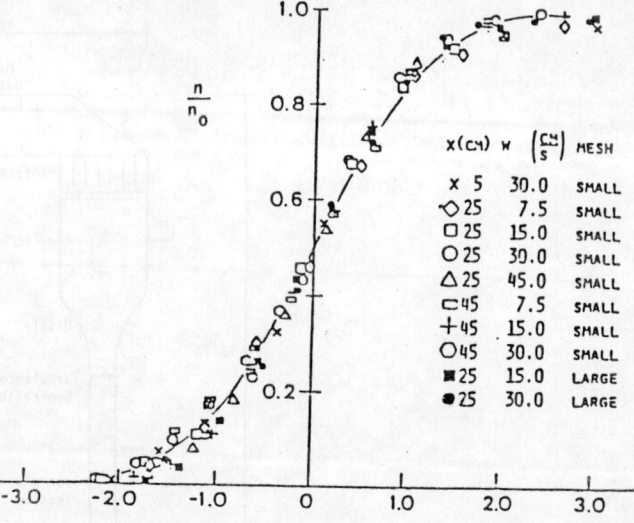

Figure 4. Particle concentration profiles: u = 300 cm/s, w = 30 cm/s; x = 5 cm ◇; x = 25 cm □; x = 45 cm △. Large mesh grid, $D = 1.2$ cm^2/s.

Figure 5. Particle concentration profiles plotted in self-similar form.

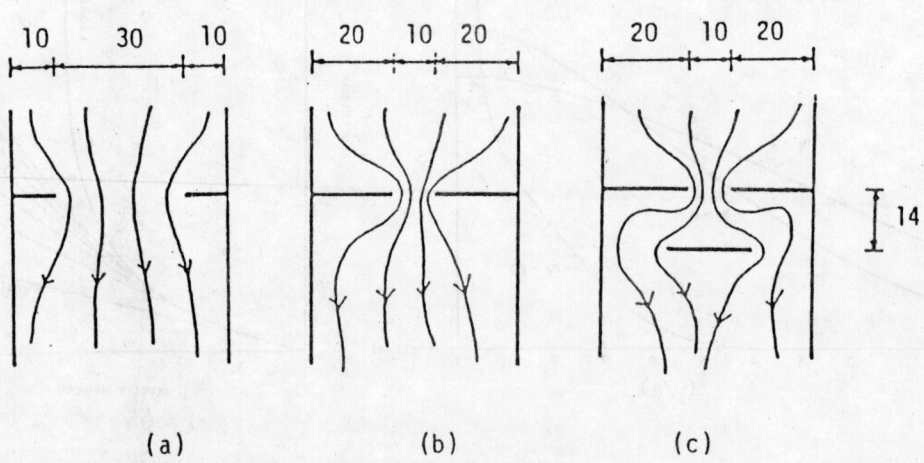

Figure 6. The turbulence-producing baffle configurations: (a) 1-cm baffles; (b) 2-cm baffles; and (c) 2-cm double baffles.

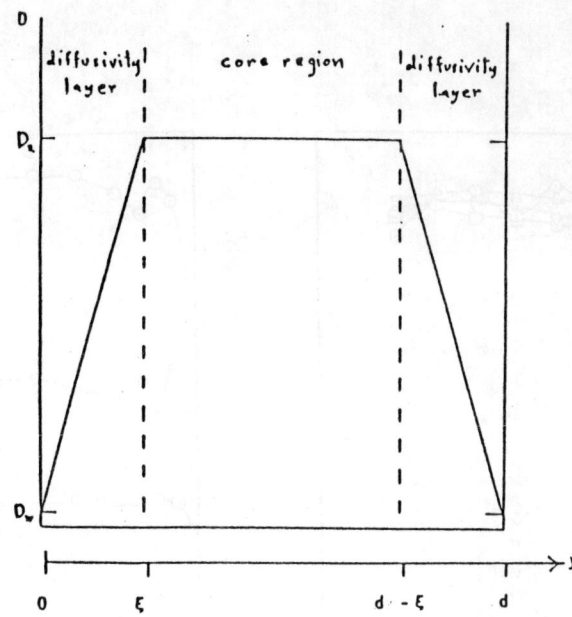

Figure 9. A schematic distribution of the nonuniform particle diffusivity in the plate-plate collector geometry.

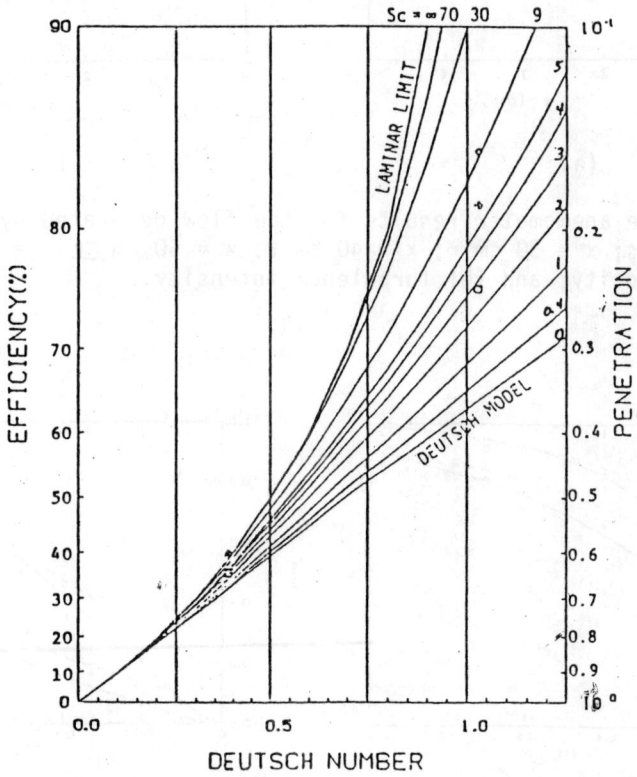

Figure 10. The plate-plate collector efficiency as a function of the Deutsch number for three different incoming turbulence conditions generated by three different baffle configurations: x = 40 cm from the inlet; 1-cm baffles ◊ (Sc = 14.8); 2-cm double baffles ✠ (Sc_c = 11.1); 2-cm baffles ☐ (Sc = 6.35) (Schmidt numbers listed here are obtained from the estimated values of the core diffusivity by use of the nonuniform diffusivity model calculations.)

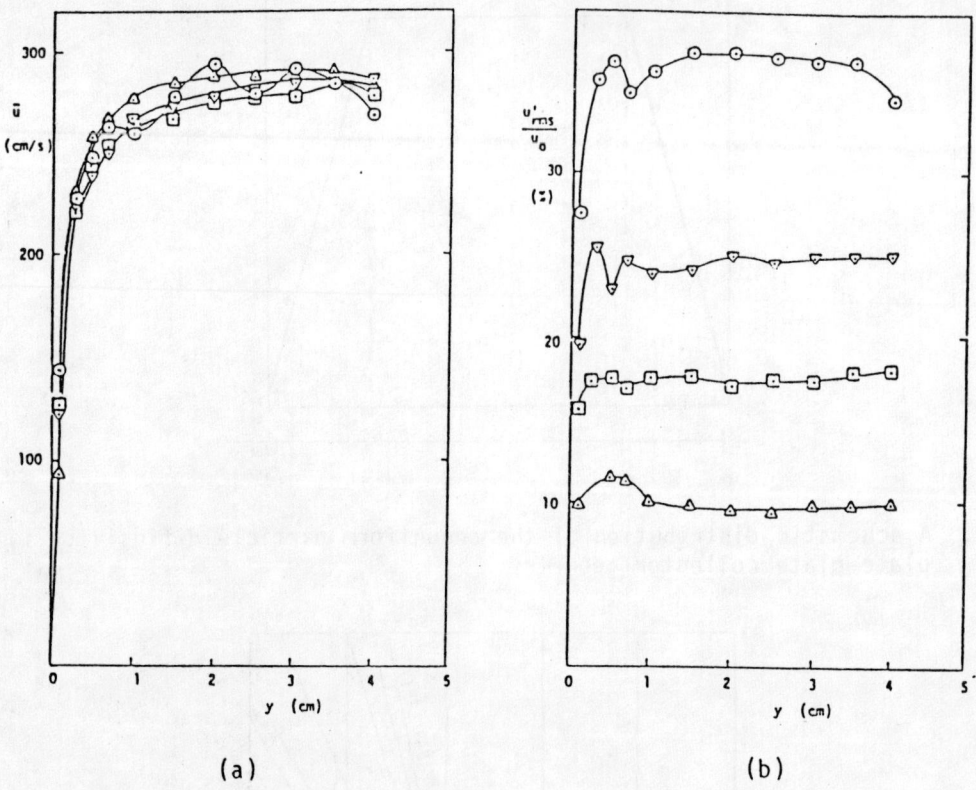

Figure 7. Hot-wire anemometer results for the flow generated by 2-cm baffles: u = 280 cm/s; x = 30 cm ⊙; x = 40 cm ▽; x = 50 cm ☐; x = 90 cm ▲; (a) mean gas velocity; and (b) turbulence intensity.

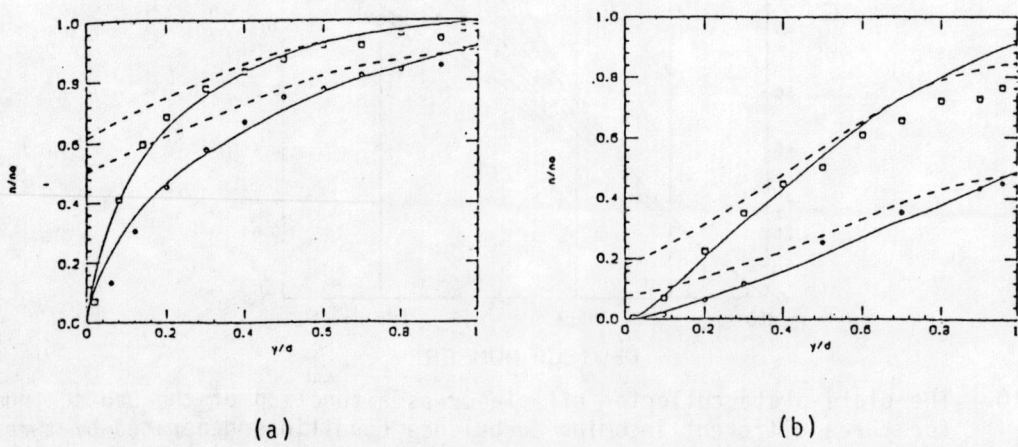

Figure 8. Particle concentration profiles for 2-cm baffles: u = 200 cm/s; D = 17.5 cm^2/s, ξ/d = 0.3; x = 20 cm ☐; x = 40 cm ◇; (a) w = 7.5 cm/s; and (b) w = 20 cm/s. Theory: uniform diffusivity -----; nonuniform diffusivity ———.

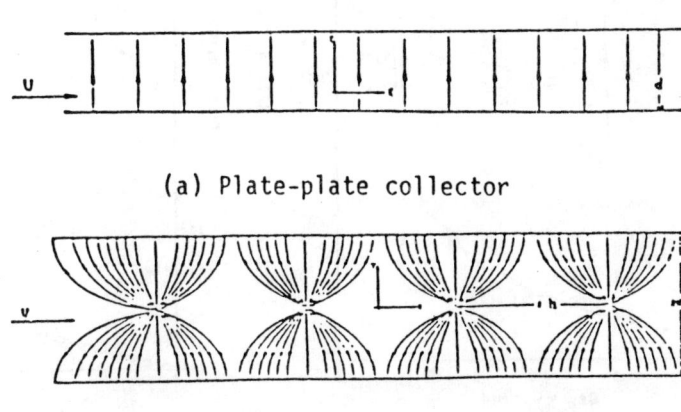

Figure 11. Comparison of electric field lines for plate-plate and wire-plate laminar-flow collectors (h = d).

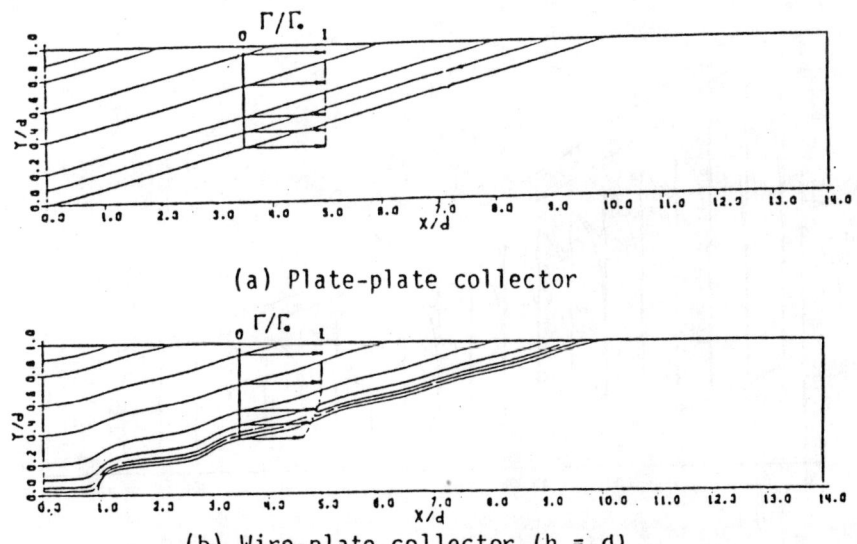

Figure 12. Comparison of particle trajectories and fluxes for plate-plate and wire-plate laminar flow collectors.

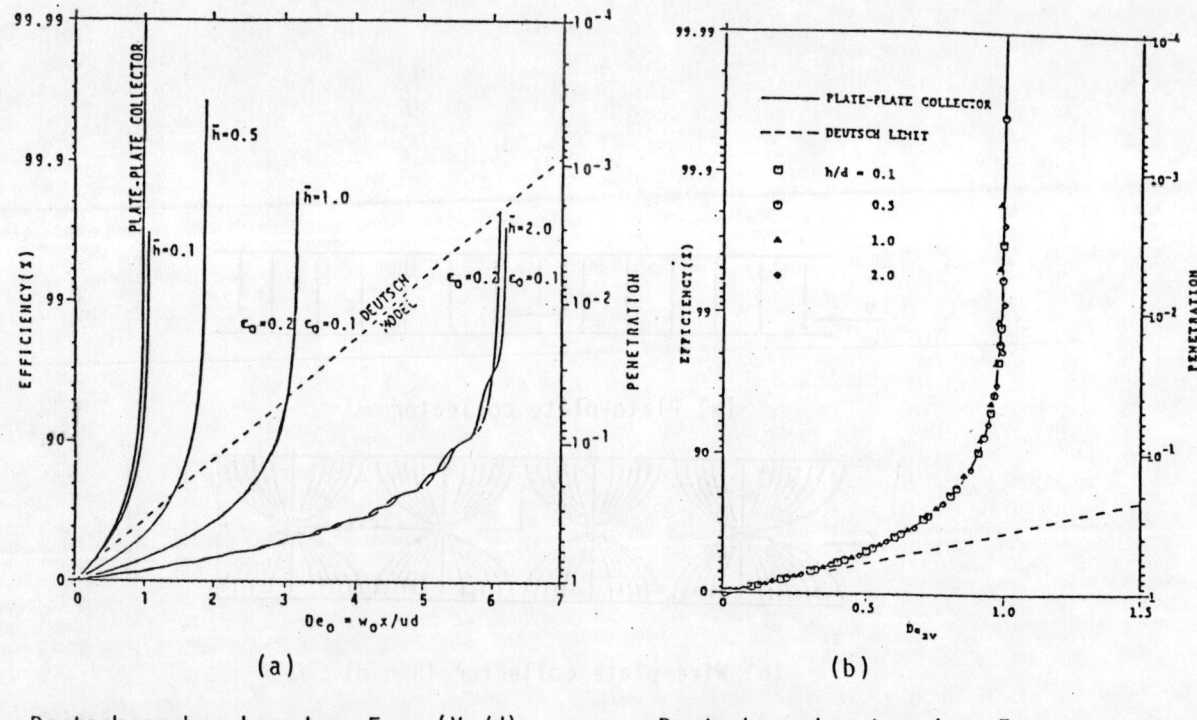

Figure 13. Dependence of collection efficiency on Deutsch number for laminar flow wire-plate precipitators ($\bar{a} = 0.01$). Also shown are the plate-plate and Deutsch model (very large transverse turbulent diffusivity) cases.

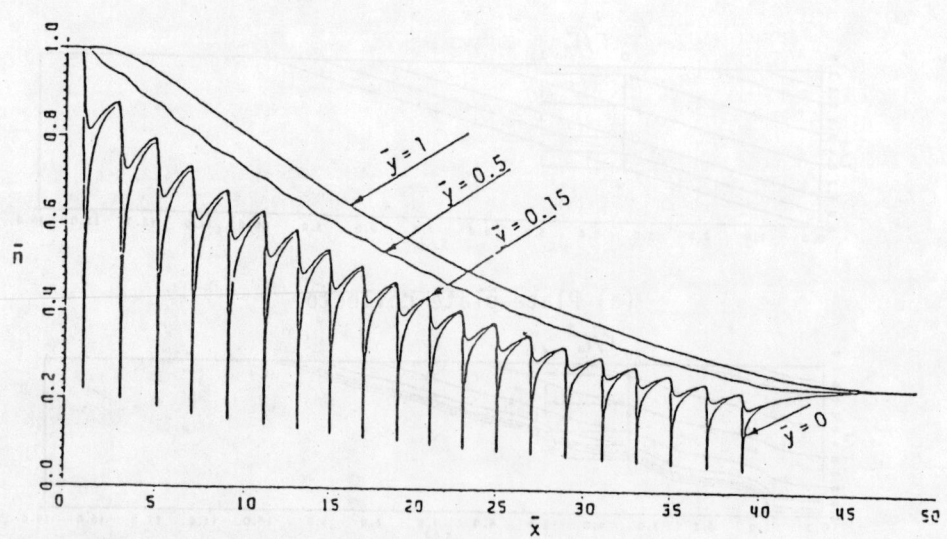

Figure 14. Distribution of the particle concentration profiles for the strip-plate collector as a function of x/d at $y/d = 0$, 0.15, 0.5, and 1.0, $b/d = 0.02$, $\varepsilon_0 = (w_0/u) = 0.1$ and $Pe_0 = 2.5$.

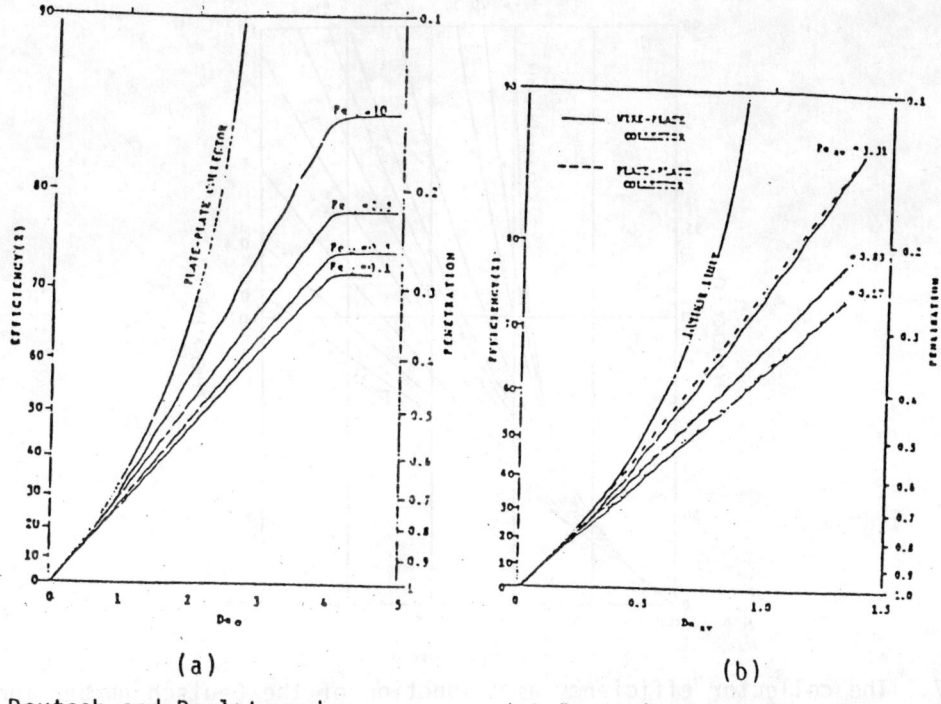

(a) Deutsch and Peclet numbers based on E_0

(b) Deutsch and Peclet numbers based on E_{av}

Figure 15. Dependence of collection efficiency on Deutsch number and Peclet number for turbulent flow wire-plate precipitators ($\bar{a} \simeq 0.02$), assuming a uniform value of diffusivity.

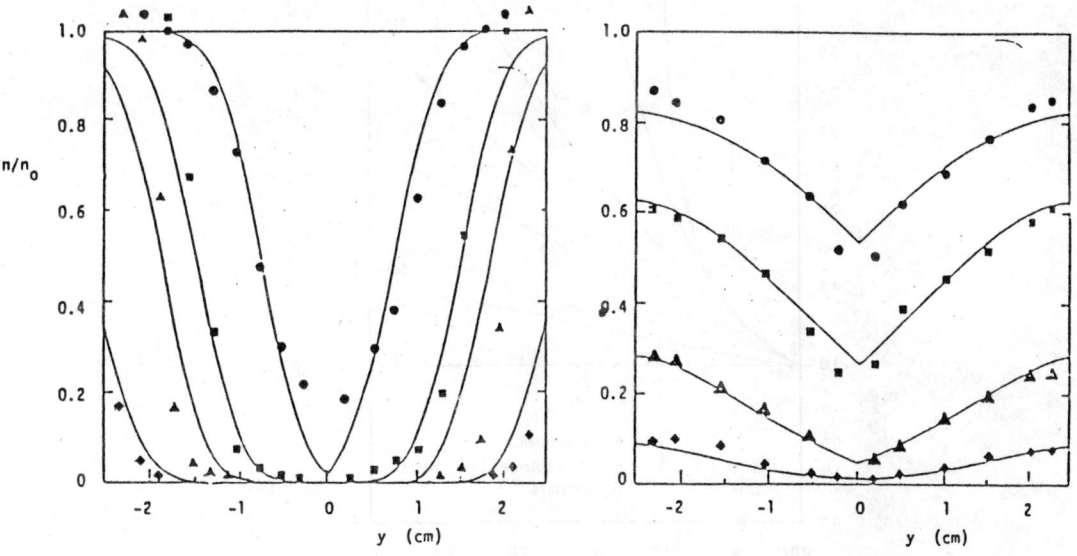

(a) Low inlet turbulence:

$D = 0.4$ cm²/s; $\varepsilon_0 = 0.05$ ●; $\varepsilon_0 = 0.10$ ■; $\varepsilon_0 = 0.125$ ▲; and $\varepsilon_0 = 0.175$ ◆

(b) High inlet turbulence generated by the 2-cm baffle configuration:

$D = 10.0$ cm²/s; $\varepsilon_0 = 0.05$ ●; $\varepsilon_0 = 0.10$ ■; $\varepsilon_0 = 0.20$ ▲; and $\varepsilon_0 = 0.30$ ◆

Figure 16. Particle concentration profiles in the wire-plate collector: $u = 200$ cm/s, $x = 43.2$ cm. Theory ———.

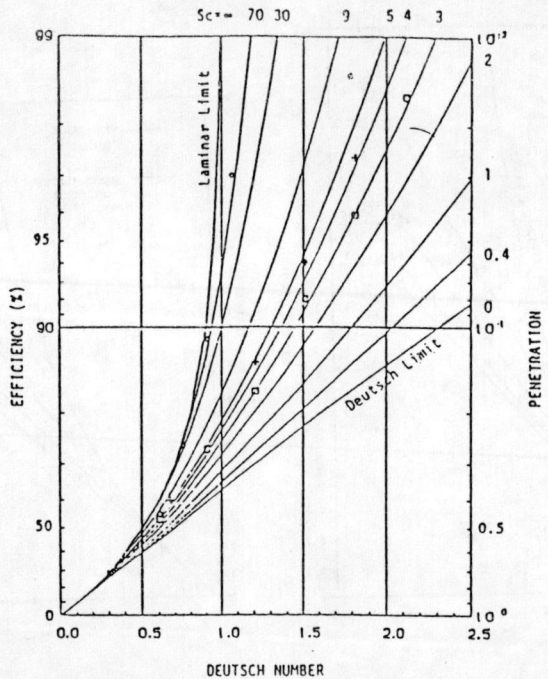

Figure 17. The collector efficiency as a function of the Deutsch number for different values of the Schmidt number: low turbulent flow (u = 200 cm/s) ◇; highly turbulent flow (u = 100 cm/s) ⊕; and highly turbulent flow (u = 200 cm/s) ☐. The flow turbulence is generated by 2-cm baffles.

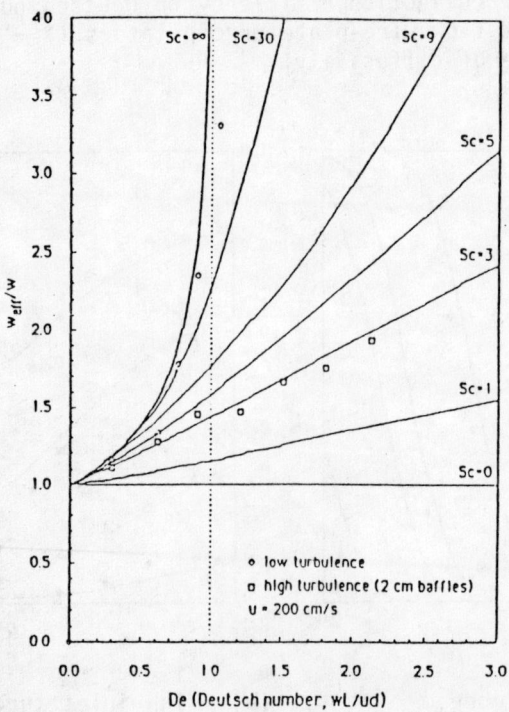

Figure 18. Precipitator performance expressed in terms of the effective migration velocity w_{eff}, given by the theoretical model that assumes uniform diffusivity and uniform electric field. Here w denotes the actual particle migration velocity, De is the Deutsch number, and Sc is the electric Schmidt number. The points represent experimentally determined values of the ordinate and abscissa, as discussed in this section.

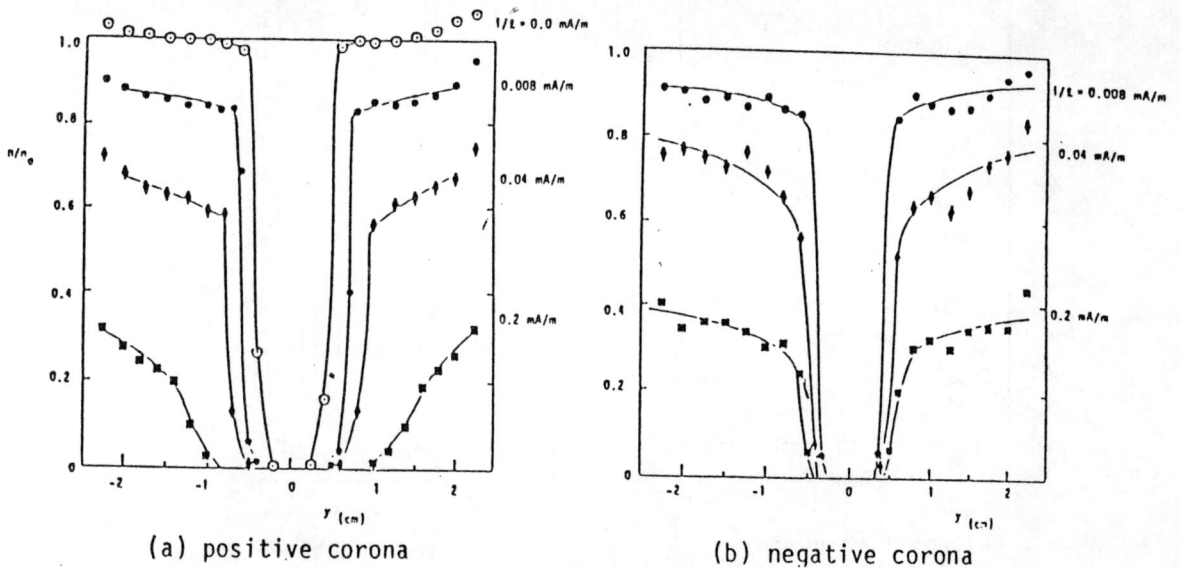

(a) positive corona
(b) negative corona

Figure 19. Particle concentration profiles in the wire-plate precipitator with precharged particles: $u = 200$ cm/s, $x = 12.7$ cm, wire diameter = 0.1 mm; $i/\ell = 0$ ◉; $i/\ell = 0.008$ mA/m ●; $i/\ell = 0.04$ mA/m ◆, and $i/\ell = 0.2$ mA/m ■.

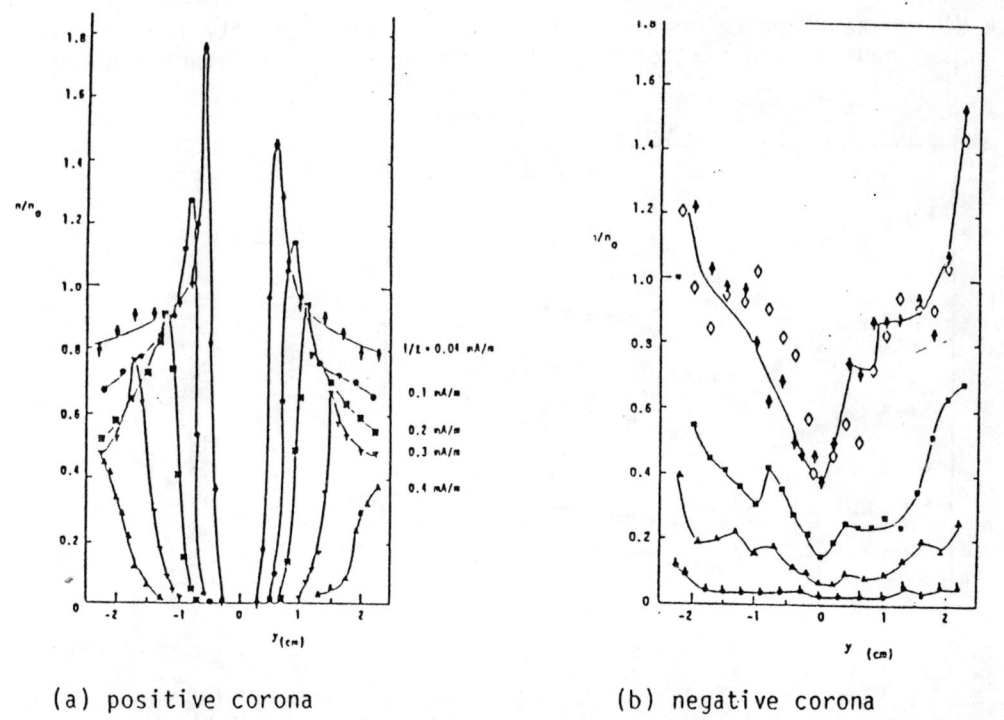

(a) positive corona
(b) negative corona

Figure 20. Particle concentration profiles in the wire-plate precipitator with initially neutralized particles: $u = 200$ cm/s, $x = 43.2$ cm, wire diameter = 0.1 mm; $i/\ell = 0.04$ mA/m ◆, ◇; $i/\ell = 0.1$ mA/m ●; $i/\ell = 0.2$ mA/m ■; $i/\ell = 0.3$ mA/m ▼; $i/\ell = 0.4$ mA/m ▲; and $i/\ell = 0.6$ mA/m ♦.

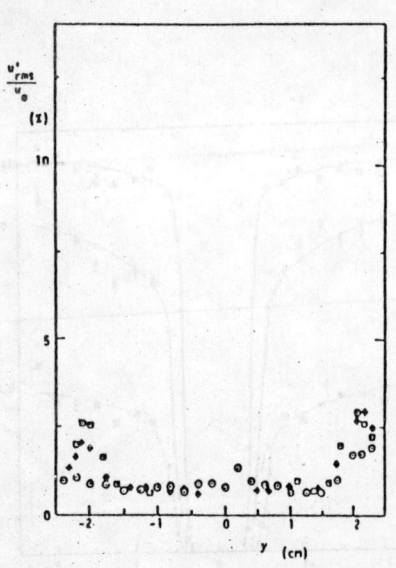
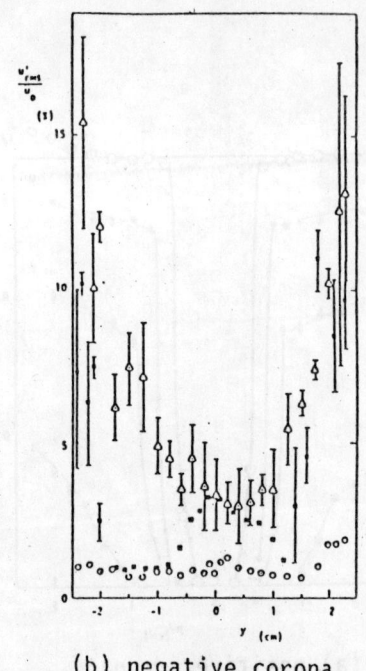

(a) positive corona　　　　　　　　(b) negative corona

$i/\ell = 0$ ◐; $i/\ell = 0.04$ mA/m ◆;　　$i/\ell = 0$ ◯; $i/\ell = 0.04$ mA/m ■; $i/\ell = 0.2$
$i/\ell = 0.1$ mA/m ⊡.　　　　　　　△, ◆. (Extended bars represent the data
　　　　　　　　　　　　　　　　　　fluctuations of repeated measurements.)

Figure 21. Laser Doppler measurements of turbulence intensity for initially
neutralized particles in the wire-plate precipitator: u = 200 cm/s.

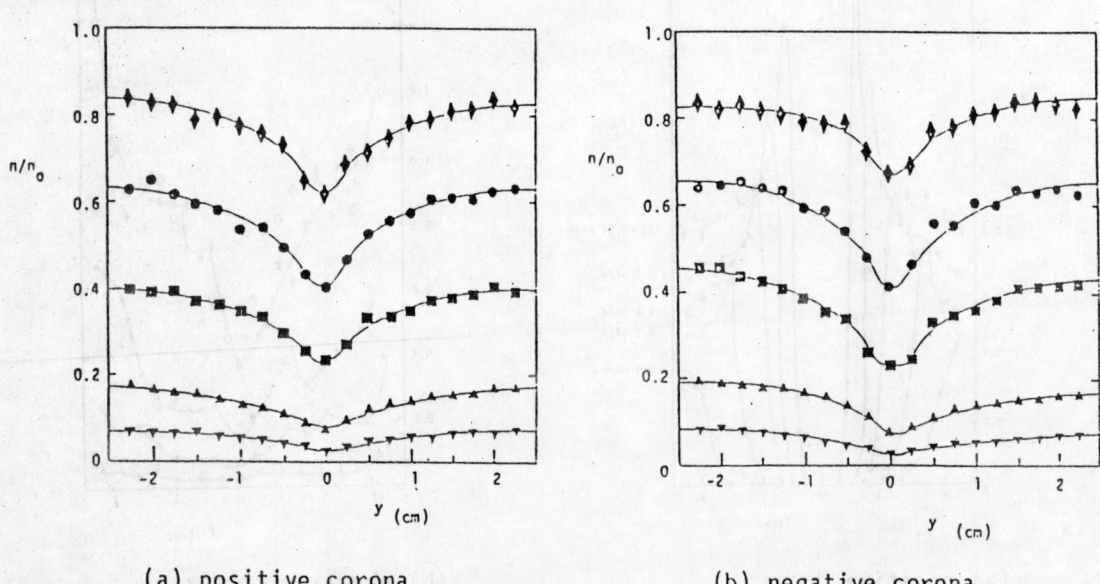

(a) positive corona　　　　　　　　(b) negative corona

Figure 22. Particle concentration profiles in the wire-plate precipitator with
turbulent flow generated by the 2-cm baffles: u = 200 cm/s, x = 43.2 cm,
wire diameter = 0.1 mm; $i/\ell = 0.04$ mA/m ◆; $i/\ell = 0.1$ mA/m ●; $i/\ell = 0.2$
mA/m ■; $i/\ell = 0.4$ mA/m ▲; $i/\ell = 0.6$ mA/m ▼.

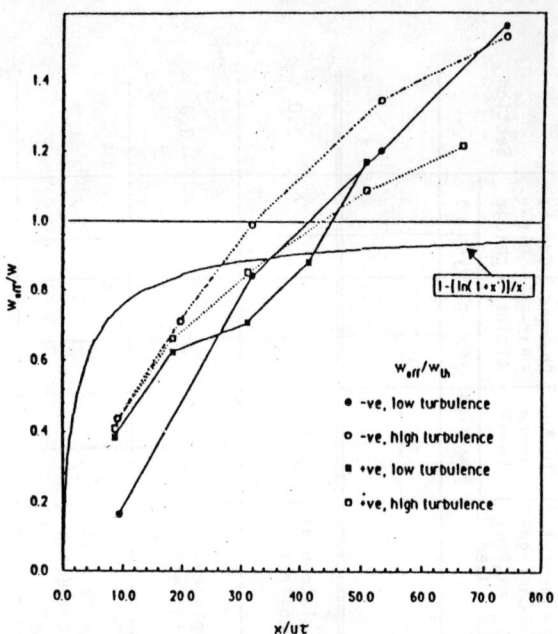

Figure 23. Reduction in precipitator efficiency, expressed in terms of w_{eff}/w, resulting from the finite particle charging rate. Theory is based on an extension of the Deutsch model. Experimental points are the values of w_{eff}/w_{th} determined from measurements with 0.1-mm wires (made between the 8th and 9th wires) for increasing values of corona current.

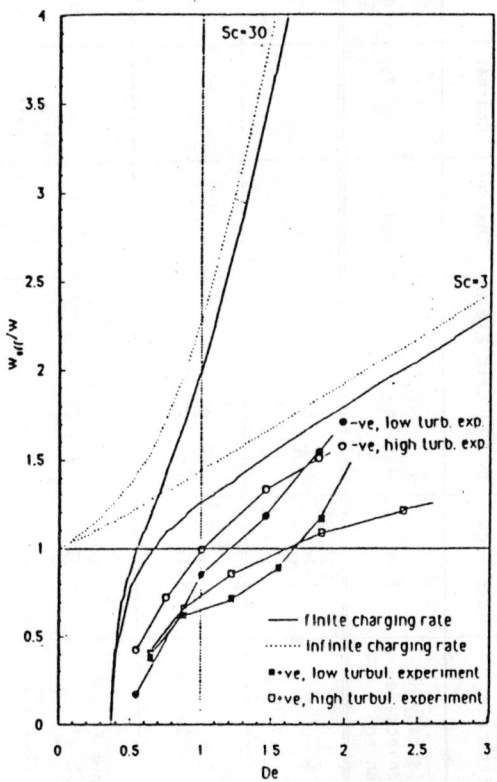

Figure 24. Precipitator performance showing the combined effects of both finite charging rate and finite diffusivity for the conditions of the experiments described in this section (x constant, and increasing corona current).

TABLE 1. TERM-BY-TERM ANALYSIS OF MECHANISMS GOVERNING PARTICLE TRANSPORT IN ELECTROSTATIC PRECIPITATORS

Main Focus of Interest	Mechanism					Experimental Conditions							Section
	Particle Convection by Mean Gas Flow	Particle Convection by Electric Field	Effect of Space Charge	Effect of Non-uniform Particle Charge	Turbulent Particle Diffusion	Experimental Configuration Used	Exper. Turbulence Level		Pre-Charged Particles		Corona Discharge		
							low	high	yes	no	yes	no	
general steady-state convective-diffusion eqn.	$\vec{u}\cdot\nabla n$	$+\mu\vec{E}\cdot\nabla n$	$+n\mu\nabla\cdot\vec{E}$	$+n\vec{E}\cdot\nabla\mu$	$=\nabla\cdot(D\nabla n)$								
uniform E uniform D	$u\frac{\partial n}{\partial x}$	$+\mu E_y\frac{\partial n}{\partial y}$			$=D\frac{\partial^2 n}{\partial y^2}$	plate-plate collector	✓			✓		✓	2.0
uniform E nonuniform D	$u\frac{\partial n}{\partial x}$	$+\mu E_y\frac{\partial n}{\partial y}$			$=D\frac{\partial^2 n}{\partial y^2}+\frac{\partial D}{\partial y}\frac{\partial n}{\partial y}$	plate-plate collector	✓			✓		✓	3.0
nonuniform E uniform D	$u\frac{\partial n}{\partial x}$	$+\mu E_y\frac{\partial n}{\partial y}+\mu E_x\frac{\partial n}{\partial x}$			$=D(\frac{\partial^2 n}{\partial y^2}+\frac{\partial^2 n}{\partial x^2})$	wire-plate collector	✓			✓		✓	4.0
space-charge $\neq 0$ uniform particle charge	$u\frac{\partial n}{\partial x}$	$+\mu E_y\frac{\partial n}{\partial y}+\mu E_x\frac{\partial n}{\partial x}$	$+n\mu\frac{\rho_c}{\epsilon_v}$		$=D\frac{\partial^2 n}{\partial y^2}$	wire-plate charger/collector	✓			✓	✓		5.0
space-charge $\neq 0$ nonuniform particle charge	$u\frac{\partial n}{\partial x}$	$+\mu E_y\frac{\partial n}{\partial y}+\mu E_x\frac{\partial n}{\partial x}$	$+n\mu\frac{\rho_c}{\epsilon_v}$	$+n\vec{E}\cdot\nabla\mu$	$=D\frac{\partial^2 n}{\partial y^2}$	wire-plate charger/collector	✓			✓	✓		6.0

TABLE 2. EFFICIENCIES (%) DETERMINED FOR SINGLE-STAGE WIRE-PLATE EXPERIMENTS

i/ℓ (mA/m)	Positive Corona			Negative Corona		
	V_0 (kV)	low turb.	high turb.	V_0 (kV)	low turb.	high turb.
0.04	6.57	21.9	23.1	6.11	8.7	20.7
0.1	7.71	41.9	43.7	7.14	-	41.4
0.2	9.10	58.0	64.8	8.22	57.0	62.5
0.3	10.24	74.4	-	-	-	-
0.4	11.18	88.3	86.5	9.89	81.9	85.4
0.6	12.78	-	94.6	11.15	94.2	93.9

Wire diameter, 2a = 0.1 mm; precipitator length, L ≈ 40.6 cm; precipitator breadth (height), b = 25.4 cm; plate-plate spacing, 2d = 5 cm; wire-to-wire spacing, 2h = 5.08 cm; gas velocity, u = 200 cm/s.

TABLE 3. CALCULATED VALUES OF w_{th}, De $(=w_{th}x/ud)$, AND $x/u\tau$, BASED ON MEASURED ELECTRICAL CONDITIONS AND EFFICIENCIES (TABLE 2)
The values of w_{eff} are obtained using Eq. (12)

Electrical Conditions		Charging Time	Charging Parameter	Migr. Velocity	Deutsch Number	w_{eff}, cm/s		w_{eff}/w_{th}	
i/ℓ, mA/m	V_0, kV	τ, ms	$x/u\tau$	w_{th}, cm/s	De	lo trb	hi trb	lo trb	hi trb
0.00	+5.00	∞	0.00	4.5	0.37	--	--	--	--
0.04	+6.57	23.63	8.60	7.8	0.63	3.0	3.2	0.38	0.41
0.10	+7.71	11.09	18.32	10.7	0.87	6.7	7.1	0.62	0.66
0.20	+9.10	6.55	31.04	15.0	1.22	10.7	12.8	0.71	0.85
0.30	+10.24	4.91	41.38	19.0	1.54	16.8	--	0.89	--
0.40	+11.18	4.02	50.53	22.6	1.84	26.4	24.6	1.17	1.09
0.60	+12.78	3.06	66.31	29.5	2.40	--	35.9	--	1.22
0.04	-6.11	21.98	9.25	6.7	0.55	1.1	2.9	0.16	0.43
0.10	-7.14	10.27	19.78	9.2	0.75	--	6.6	--	0.72
0.20	-8.22	5.91	34.37	12.2	0.99	10.4	12.1	0.85	0.99
0.30	--	--	--	--	--	--	--	--	--
0.40	-9.89	3.56	57.13	17.7	1.44	21.0	23.7	1.19	1.34
0.60	-11.15	2.67	76.01	22.5	1.83	35.0	34.4	1.56	1.53